Jim Hoffman

Conversions *(continued)*

1 gram/cm^3	= 62.4 . . . lb$_m$/ft^3
	= 1000 kg/m^3
	= 1 Mg/m^3
1 inch	= 0.0254 m
1 joule	= 0.947 . . . $\times 10^{-3}$ Btu
	= 0.239 . . . cal, gram
	= 6.24 . . . $\times 10^{18}$ eV
	= 0.737 . . . ft·lb$_f$
	= 1 watt·sec
1 joule/meter^2	= 8.80 . . . $\times 10^{-5}$ Btu/ft^2
1 [joule/(m^2·s)]/[°C/m]	= 1.92 . . . $\times 10^{-3}$ [Btu/(ft^2·s)]/[°F/in.]
1 kilogram	= 2.20 . . . lb$_m$
1 megagram/meter^3	= 1 g/cm^3
	= 10^6 g/m^3
	= 1000 kg/m^3
1 meter	= 10^{10} Å
	= 10^9 nm
	= 3.28 . . . ft
	= 39.37 in.
1 micrometer	= 10^{-6} m
1 nanometer	= 10^{-9} m
1 newton	= 0.224 . . . lb$_f$
1 ohm·inch	= 0.0254 . . . Ω·m
1 ohm·meter	= 39.37 Ω·in.
1 pascal	= 0.145 . . . $\times 10^{-3}$ lb$_f$/in.^{2}
1 poise	= 0.1 Pa·s
1 pound (force)	= 4.44 . . . newtons
1 pound (mass)	= 0.453 . . . kg
1 pound/foot^3	= 16.0 . . . kg/m^3
1 pound/inch^2	= 6.89 . . . $\times 10^{-3}$ MPa
1 watt	= 1 J/s
1 (watt/m^2)/(°C/m)	= 1.92 . . . $\times 10^{-3}$ [Btu/(ft^2·s)]/[°F/in.]

SI prefixes

giga	G	10^9
mega	M	10^6
kilo	k	10^3
milli	m	10^{-3}
micro	μ	10^{-6}
nano	n	10^{-9}
pico	p	10^{-12}

MATERIALS FOR ENGINEERING: CONCEPTS AND APPLICATIONS

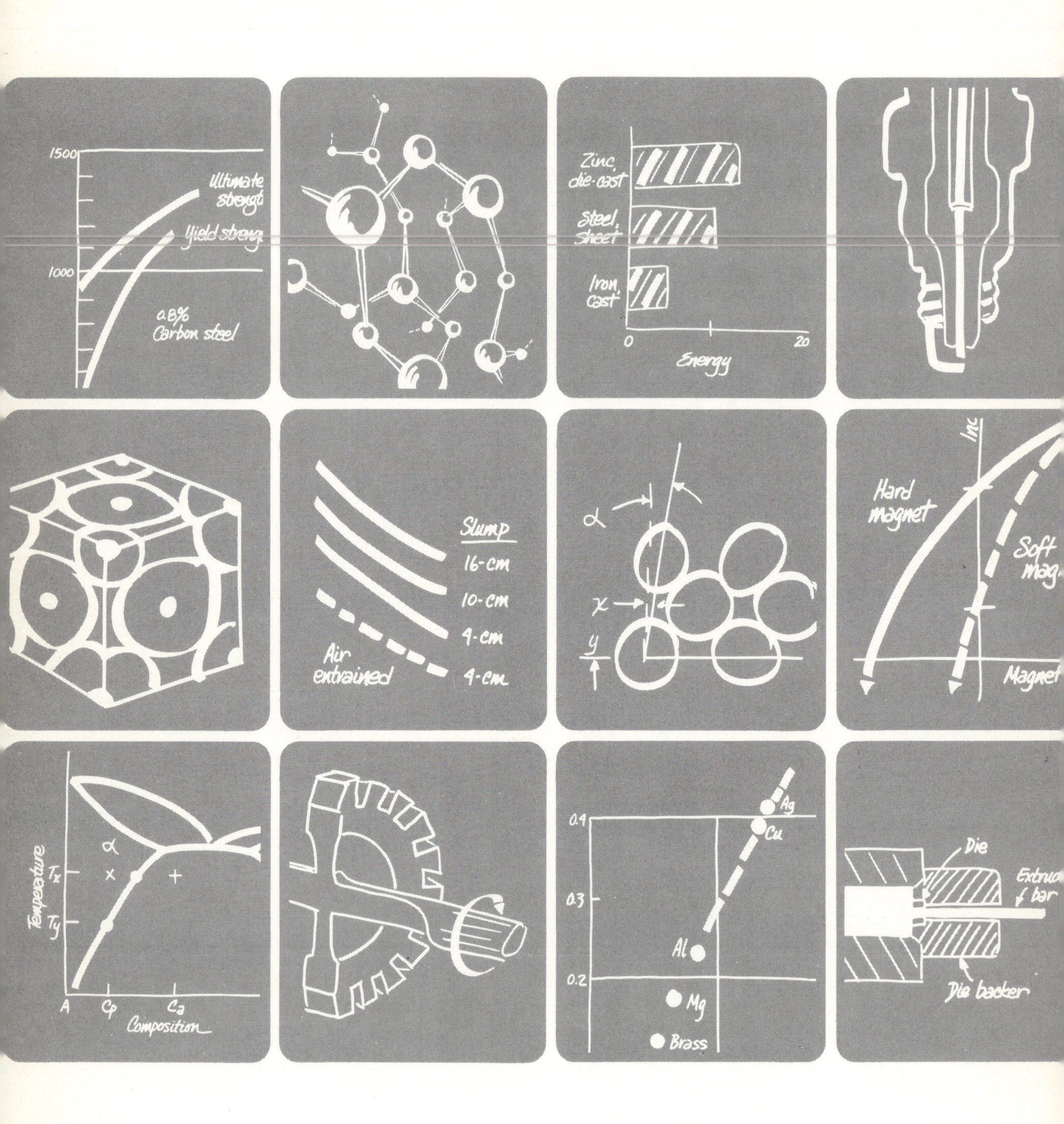
1500
1000
Ultimate
0.8%
Carbon steel
Zinc,
die-cast
Steel,
sheet
Iron,
cast
0
20
Energy
Slump
16-cm
10-cm
4-cm
Air
entrained
4-cm
α
x
y
Hard
magnet
Soft
Temperature
α
T_x
T_y
A
C_p
C_a
Composition
0.4
0.3
0.2
Ag
Cu
Al
Mg
Brass
Die
Die backer

MATERIALS FOR ENGINEERING: CONCEPTS AND APPLICATIONS

THE UNIVERSITY OF MICHIGAN, ANN ARBOR, MICHIGAN Lawrence H. Van Vlack

Addison-Wesley Publishing Company
READING, MASSACHUSETTS
MENLO PARK, CALIFORNIA
LONDON • AMSTERDAM
DON MILLS, ONTARIO • SYDNEY

This book is in the
ADDISON-WESLEY SERIES IN
METALLURGY AND MATERIALS

Consulting Editor
Morris Cohen

Sponsoring Editor: Thomas Robbins
Production Editor: Marion E. Howe
Designer: Margaret Ong Tsao
Illustrator: Oxford Illustrators, Ltd.
Susanah H. Michener
Dick Morton
Cover designer: Richard Hannus
Cover illustrations, Frontispiece, Chapter opening illustrations: Gail Ellison for Chartex/Boston

Library of Congress Cataloging in Publication Data

Van Vlack, Lawrence H.
Materials for engineering.

Includes index.
1. Materials. I. Title.
TA403.V357 620.1'1 81-2194
ISBN 0-201-08065-6 AACR2

ABCDEFGHIJ-DO-898765432

To Fran,
for tolerance, day after day;
and to our Mothers—
RLVV and LTR

PREFACE

There are multiple dimensions to all technical subjects. This is particularly true in the field of Materials Science and Engineering where the range of academic sophistication extends from descriptive to highly quantitative; where the focus can be on the science side, in the engineering arena, or at any location along the intervening spectrum; and where the attention may be specifically in one technical field, such as structures or electronic devices, or broadly across all technology.

Like the textbook *Elements of Materials Science and Engineering,* this textbook follows College Chemistry and General Physics. It is recommended that it precede courses in engineering design. Thus, it is aimed for the 3rd-to-5th semester student.

In contrast to the above text, the focus of this book is located further toward the engineering region of the science-engineering spectrum. Thus, there is more emphasis on applications. The science, or the "why" of behavior, is used where necessary but does not form the outline of the text. This has brought about significant differences in organization. Metals are presented first because their property–structure–performance relationships are more easily presented than for polymers and ceramics, which come later. The introductory concepts of processing are presented as additional units for each of the above material groups. Separate units are included for some of the widely encountered materials—Steel and Iron (6), Wood (10), Glass (13), and Construction Materials (15). Their use is optional with the instructor. Units have been included for specific types of applications; for example, where strength, toughness, etc., are critical (Unit 14), when hostile environments must be met (Unit 16), or where electrical properties are prime (Units 17, 18, 19). Here again the instructor has the option of including or skipping the units according to local needs.

As just indicated, this text has been written for more flexible use than was *Elements of Materials Science and Engineering*. There has long been an expressed desire for a modular approach to the teaching of Materials Science and Engineering, so that the instructor could tailor a course to the local constraints that exist in time, audience, etc. The national effort to this end, EMMSE (with which I have been associated), is rapidly accumulating modules for this purpose. However, it is apparent that those modules are most effective for second-level and advanced-level courses. In fact, with only a few exceptions, it is now assumed that those modules will not be used in introductory materials courses. This textbook attempts to fill that gap. It is written so that the instructor may choose from 20 units those that fulfill the local specifications for an introductory materials course. This has meant that the units in this text must be more self-standing than chapters in a normal text. At the same time, however, the units are coordinated to avoid redundancy and to provide support where necessary.

The final unit (20) in this book bears special comment. Over the past decade there has been an emerging realization the engineer must view Materials more broadly than solely their science and technology. The last unit gives the instructor an opportunity to "open the door" so that the student may have a glimpse at the broad interface between Materials and Society. Admittedly, it is a preview only, but one that many feel should be introduced as early as possible.

Several features have been included in this textbook to facilitate the learning process. Each unit has its own introduction, outline, and stated objectives for the student. The principal prerequisites are cited. Each unit is followed by a Review and Study section that includes a Summary, a glossary of new Technical Terms, a set of Checks for student use, and Study Problems for home assignment.

The engineer must solve problems. Therefore, it is natural that some technical concepts are more easily illustrated by a calculation than by verbage in a paragraph. To this end, the units include Illustrative Problems where additional information may be presented. The students should include these I.P.'s in their study and not consider them solely as examples to mimic when they are given home assignments. Problem-solving abilities are achieved progressively. The first encounter is to observe someone else's solution. The I.P.'s can serve this initial stage. In a second stage, there is a place for "equation-plugging" or for the mimicking of existing solutions. Study Problems of this nature are found at the end of each unit. Those Study Problems that require more analysis and synthesis by the student are identified with a vertical rule. Of course the engineering student should reach this stage in the process of mastering a subject.

SI units are the primary dimensional mode in this text. However, English Units are used parenthetically where such dimensions persist in practice, since the engineer will continue to profit by being able to communicate with those individuals who have a wealth of experience in that mode.

Students at The University of Michigan have made a major contribution to this text, because the classroom has been its development laboratory. They deserve a special thanks from me, and also, I am certain, from the future students who will be using this text. Several of the units have benefitted through extensive suggestions and criticisms of experts in related materials fields. These are: Unit 10, "Wood," that was critiqued by Dr. G. Marra, on leave from Washington State to the U.S. Forest Products Laboratory; Unit 13, "Glass and Vitreous Products," Dr. H. R. Swift of LOF Glass; Unit 15, "Concrete and Materials of Construction," Prof. E. Tons, P.E., The University of Michigan; and Unit 20, Section 4, "Failure Analysis," Prof. W. Larson, Iowa State University. Each is to be sincerely thanked. In addition, I wish to acknowledge the input of my Ann Arbor colleagues, the encouragement of Prof. M. Cohen, plus the help beyond the call of duty by Ardis Vukas, Marion Howe, Dick Morton, and Tom Robbins.

Ann Arbor, Michigan
October 1981

V^2

CONTENTS

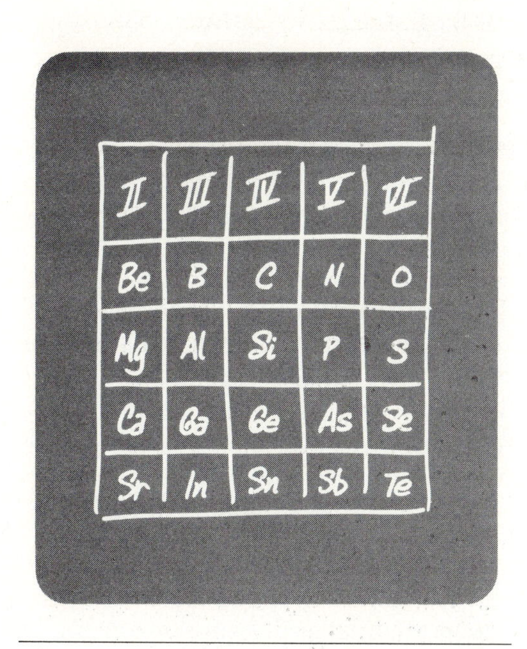

UNIT ONE

INTRODUCTION

All engineers use materials. It is possible to obtain a better understanding of the properties and behavior of materials in engineering design if we analyze their internal structure. That will be the approach in this text, rather than to cite the various properties for the thousands of available materials. Materials are conveniently divided into metals, polymers, and ceramics. We will also group materials according to design function—strength, service stability, or electrical characteristics.

Contents

Prerequisite for this unit None.

From Unit 1, the engineer should

1) Recognize that *all* technical products and systems contain materials—from calculators to steam power plants to highways.
2) Appreciate that for a product or system to be functional, materials must be specified, selected, produced, tested, assembled, and monitored in use.
3) Become aware that materials have internal structures that involve atoms, crystals, molecules, etc., because these structures will affect properties.
4) Be able to place materials into metal, polymer (plastics), and ceramic categories.

1-1 MATERIALS AND ENGINEERING

Many prospective engineering students have asked, "What do engineers do?" The simplest, most general answer could be

> Engineers adapt materials and/or energy for society's needs.

More specifically, engineers design products and systems, make them, and monitor their usage. Every product is made with materials; and energy is involved in production, in use, and even in communication. This association with design, production, and usage is the reason that engineering students take a course in materials as part of their undergraduate curricula.

The various engineering disciplines focus on their own subset of products and systems. As examples, the M.E. student gives attention to automobiles, power generating equipment, and other mechanical products, while the naval architect considers ships for marine transport, and the E.E. will construct circuits ranging from megawatt distribution systems to nanowatt computer signals (Fig. 1-1.1). The petroleum refinery, designed by the Ch.E.; the bridge girder, by the C.E.; and the landing craft, by the Aerospace Engineer are further examples. All of these products must be made out of materials. The materials that are specified must have the correct properties—first for production, and subsequently for service. They must not fail in use, producing a liability. Costs of the processed materials must be acceptable, and nowadays the choices of materials must be compatible with the environmental standards—from raw material sources,

Figure 1-1.1 Engineering products (high-density semiconductor chip). Every branch of engineering adapts materials and/or energy into useful products or purposes. All engineering products—from a 64,000-bit, RAM chip (left) to superhighway bridges and nuclear reactors—are made of materials with specified properties, shapes, and functions. (Courtesy of IBM Corp.)

Figure 1-1.2 Materials in design (extruded aluminum face bars for automobile bumpers). Weight is saved by replacing steel with aluminum; however, the redesign is not one of simple substitution. Since the properties are different for the two metals, adjustments must be made in both design and accompanying manufacturing procedures. (Courtesy Kaiser Aluminum and Sales, Inc.)

through manufacturing steps and product usage, to eventual discard. This means that those newer technical fields, such as "environmental engineering" and "public-policy engineering" must also give consideration to materials.

The age of technology is also an age of materials barriers. Nowhere is this more dramatically revealed than in energy conversion. The development of the solar cell requires that the engineer perfect the process for semiconductor purification and crystal growth still further than the achievements to date. Coal-to-watts conversion becomes more efficient at higher steam temperatures. Thus, we should not be surprised that as soon as materials were developed for 400°C (750°F) the design engineers then asked for 450°C (840°F) materials. Those are now available. The next step and the following step will always call for more and more stringent demands. We will never overcome the final materials barrier. However, design engineers and the materials engineers are always working in that direction.

Materials substitution is another arena of engineering activity. Aluminum substitution for steel is not automatic in order to develop lighter weight, more fuel-efficient cars (Fig. 1-1.2). Aluminum and steel are not processed identically, so there must be changes in capital equipment. They have different rigidities, which must be considered in the technicalities of design. The substitution of stainless steel for regular steel could appear to be rational for longer life of bridges or ocean-going boats until the engineer compares cost, fabricability, and the sources of our raw materials with their international complications. The role of materials in modern society is manifold.

Some engineers will not encounter materials in their first assignment after graduating, and probably a few of those engineers will not advance to more responsible positions where they consider materials; however, sooner or later the majority of engineers will be involved in materials selection or performance, either directly or with colleagues in their technical organizations. Materials (and energy) are seldom avoided by today's engineer.

1-2 MATERIALS FOR THE ENGINEER

Every known type of material is available to the engineer. Admittedly, some are not used widely because of availability, initial properties, cost, or service performance. Others, like iron and steel, paper, concrete, water, and vinyl plastics, find extensive uses. Of course, some of these categories can be subdivided. For example, it has been suggested that there are as many as 2000 types of steel, 5000 types of plastics, and 10,000 kinds of glass. Whether or not these numbers are realistic, the engineer has a large number of choices for the product being designed.

To the above numbers, we must add the capability of modifying properties of materials, as in the case of gears (Fig. 1-2.1). They must be weak and soft to be machinable during manufacture, but strong and wear-resistant for service in the power train of the heavy construction equipment. Thus, the task of knowing the properties and behavior of all types of available materials becomes enormous. In addition, several hundred new varieties of materials appear on the market each month. This means that individual engineers cannot hope to be familiar with the properties and performance of all types of materials in their numerous forms. However, the engineer can learn principles for guidance in the selection and application of materials.

The purpose of this text is to familiarize technical students with those factors that affect properties and to give the student an idea of the service potential of various materials. When equipped with this knowledge, the technical person should be able (1) to select more readily and intelligently those materials that are best for the purpose, and (2) to avoid service conditions that might be damaging to materials.

The *properties* of a material originate from the internal structure of that material. This is analogous to saying that the operation of a D/A converter depends on the component parts and the internal circuits within that set (Fig. 1-2.2). The internal structures of materials involve atoms, and the way atoms are associated with their neighbors into crystals, molecules, and microstructures. In the following chapters we shall devote much attention to these structures, because a technical person must understand them in order to produce and use materials, just as an electronic technician must understand radio circuits if that person is going to design, assemble, or repair a radio. Before we talk about structures, however, let us look briefly at the variety of materials available. This will enable us to discuss the relationship of properties to structures more intelligently.

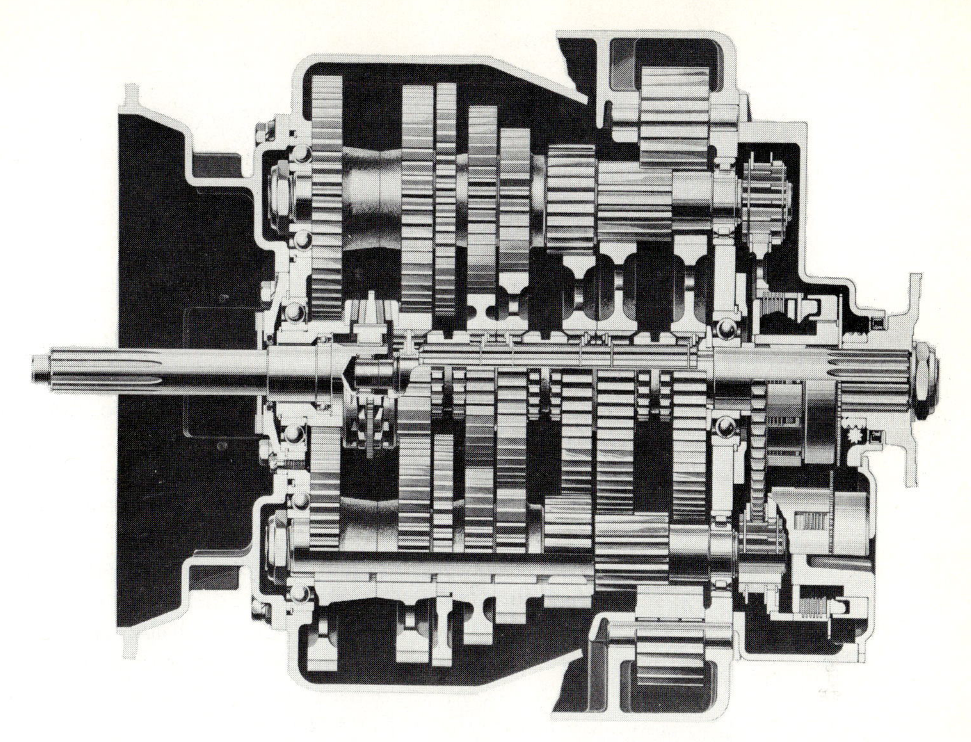

Figure 1-2.1 Power transmission equipment, like all engineering products, requires the use of materials. For this product, the metal must be strong, tough, and wear-resistant for dependable service. During production, however, the steel had to be weak and soft so that the gears could be forged and machined. In order to change the properties after processing, the engineer had to change the internal structure of the steel. (Courtesy of Spicer Transmission Division of Dana Corporation.)

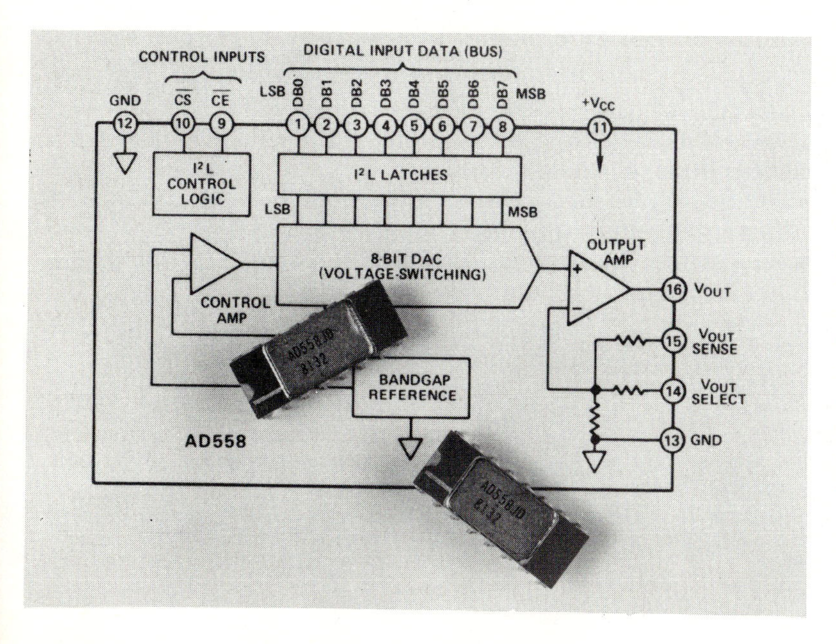

Figure 1-2.2 Digital/analog converter. The performance of this microprocessor device depends upon its internal circuits. Likewise, the performance of a material depends on the structure of its internal components. We shall see that these internal arrangements involve electrons, atoms, crystals, and microstructures. The engineer can select and modify these internal structures, just as a circuit designer can select and modify internal circuits. (Courtesy of Analog Devices.)

Illustrative problem 1-2.1* Take a cord from some appliance, such as a toaster or coffee maker. List the chief materials used and the probable reason for their selection.

SOLUTION
a) Copper wire: Copper offers low resistance, so that wire with a small cross section can be used. The small diameter of the wire strands lends flexibility.
b) Brass prongs: These prongs must have greater dimensions so that they will be rigid. With these larger dimensions, brass may be used, even though it is more resistive than copper. Brass is also stronger and cheaper than copper.
c) Rubber insulation: Rubber is a nonconductor that has the added advantage of flexibility. Also it may be readily applied as a coating to the wire during the initial stages of production.
d) Bakelite receptacle housing: Bakelite is a good insulator, offers stability of temperature, and is rigid.
e) Steel spring: Steel can be thin, flexible, and not subject to permanent bending.

ADDED INFORMATION At this stage we can only note some of the characteristics of materials. Later we shall examine *why* copper has a lower resistivity, and *why* a bakelite-type plastic does not soften when it gets hot while other types of plastics do. ◀

1-3 TYPES OF MATERIALS

With so many types of materials available to the engineer, it is desirable to group them into categories for a more systematic presentation. Various groupings are possible, and none is ideal. For our purposes, we shall consider metals (Units 2-6), polymers (plastics) (Units 7-9), and ceramics (Units 11-13) as the main categories (Fig. 1-3.1). We use this division because a number of the common processing procedures are different for metals, polymers, and ceramics, particularly in their initial steps that involve raw material processing and the primary forming operations.†

Materials for some applications do not fall neatly into the metal-polymer-ceramic trio. Therefore, we will consider these along application lines. This will include the "strong, tough, and hard materials" (Unit 14), those "materials for hostile environments" (Unit 16), metallic conductors

* Illustrative problems will be given throughout the text, usually followed by both a solution and comments. The purpose is to give the reader additional points and information. Thus they should be studied along with the text itself. Additional problems appear at the end of each unit for purposes of individual study.

† We shall find that the final processing steps are less distinct; e.g., powder processing is used for both ceramics and metals, blow molding for glass and plastic, and sheet forming for polymers and metals.

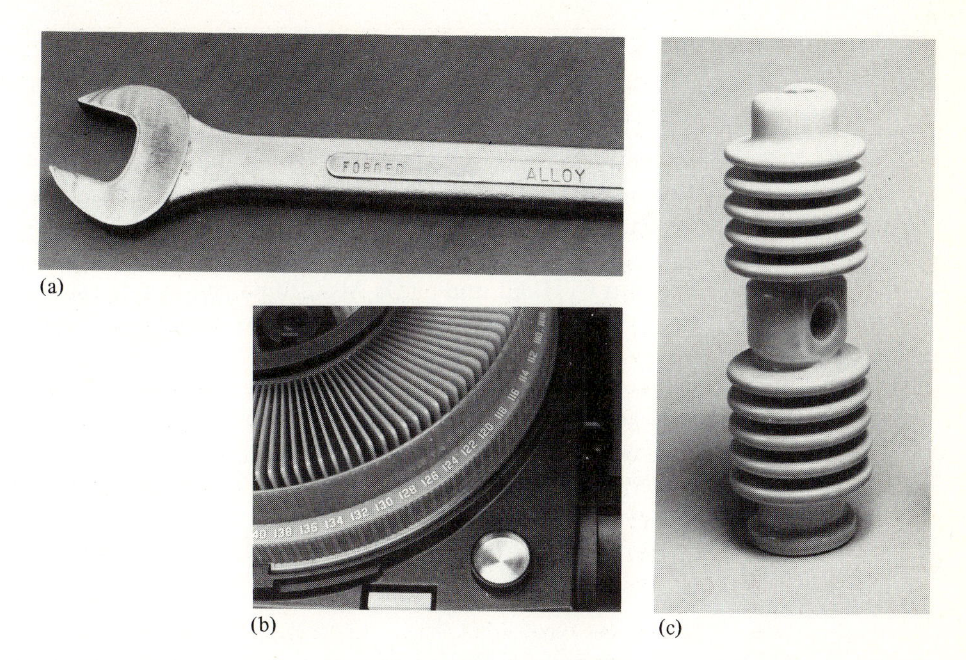

Figure 1-3.1 Familiar examples of (a) metals (alloy wrench), (b) polymers (slide reel), and (c) ceramics (electrical insulator). Metallic elements readily release valence electrons for conduction, or to make positive ions. Polymers, commonly called plastics, contain nonmetallic elements that share valence electrons. Ceramics, which are compounds of metallic and nonmetallic elements, are commonly harder and more inert to severe service conditions than are elemental materials. (Insulator courtesy of The Edward Orton Jr. Ceramic Foundation.)

(Unit 17), semiconductors (Unit 18), and dielectric and magnetic materials (Unit 19). These groupings also include materials from each of the three previous categories—metals, polymers, and ceramics.

Metals

We recognize that a metal has high thermal conductivity because it quickly conducts heat to or from one's hand. We also know that a metal has high electrical conductivity, and that it is opaque. Usually it may be polished to a high luster. All the above traits may be attributed to the fact that electrons have greater freedom of movement in metals than in polymers or ceramics. Often, but not always, metals are heavy and deformable; however, these two characteristics are not definitive.

Chemically, metals fall on the left and lower portions of the periodic table as conventionally presented (Fig. 1-3.2).

Polymers (often called plastics)

Polymers are characterized by their low density and their use as insulators, both thermal and electrical. Those materials that we call polymers contain predominant numbers of covalent bonds. These are bonds where adjacent atoms share pairs of electrons and are typical of the elements in the upper-right corner of the periodic table (Fig. 1–3.2). Most polymers (plastics) contain C, H, Cl and/or O. Also, some contain nitrogen, sulfur, and fluorine.

Those materials that contain only the above nonmetallic elements commonly form large molecules, called *macromolecules*. We shall see in Units 7 through 9 that these large molecules contain many repeating units called *mers*, from which we get the term, *polymers*. Further, these molecules commonly have structures that can be deformed; hence the term *plastic*. By learning more of the architecture of these molecules, the engineer has been able to produce plastics that have the required properties for an ever-expanding set of technical applications.

Ceramics

The most simple definition of a ceramic is that it is a material that is a compound of a metal and a nonmetal. Examples include spark plug insulators (Al_2O_3), temperature-resistant brick (MgO), SiO_2-based glass, nuclear fuels (UO_2), concrete (and even rocks themselves).

Each of these materials is relatively hard and brittle. Indeed, hardness and brittleness are general attributes of ceramics, along with the fact that they tend to be more resistant than either metals or polymers to high temperatures and to severe environments. The basis for these characteristics is again the electronic behavior of the constituent atoms. Consistent with their natural tendencies, the metallic elements release their outermost electrons and give them to the nonmetallic atoms, which retain them. The result is that these electrons are immobilized, so that the typical ceramic material is a good insulator, both electrically and thermally.

Equally important, the positive metallic ions (atoms that have lost electrons) and the negative nonmetallic ions (atoms that have gained electrons) develop strong attractions for each other. Each *cation* (positive) surrounds itself with *anions* (negative). Considerable energy (and therefore considerable force) is usually required to separate them. It is not surprising that ceramic materials tend to be hard (mechanically resistant), refractory (thermally resistant), and inert (chemically resistant).

Illustrative problem 1-3.1 Polyvinyl chloride is a common polymer. Its structural unit (called a mer) has a composition of $(C_2H_3Cl)_n$. What is the weight percent (w/o) carbon? (Use atomic mass units—amu.)

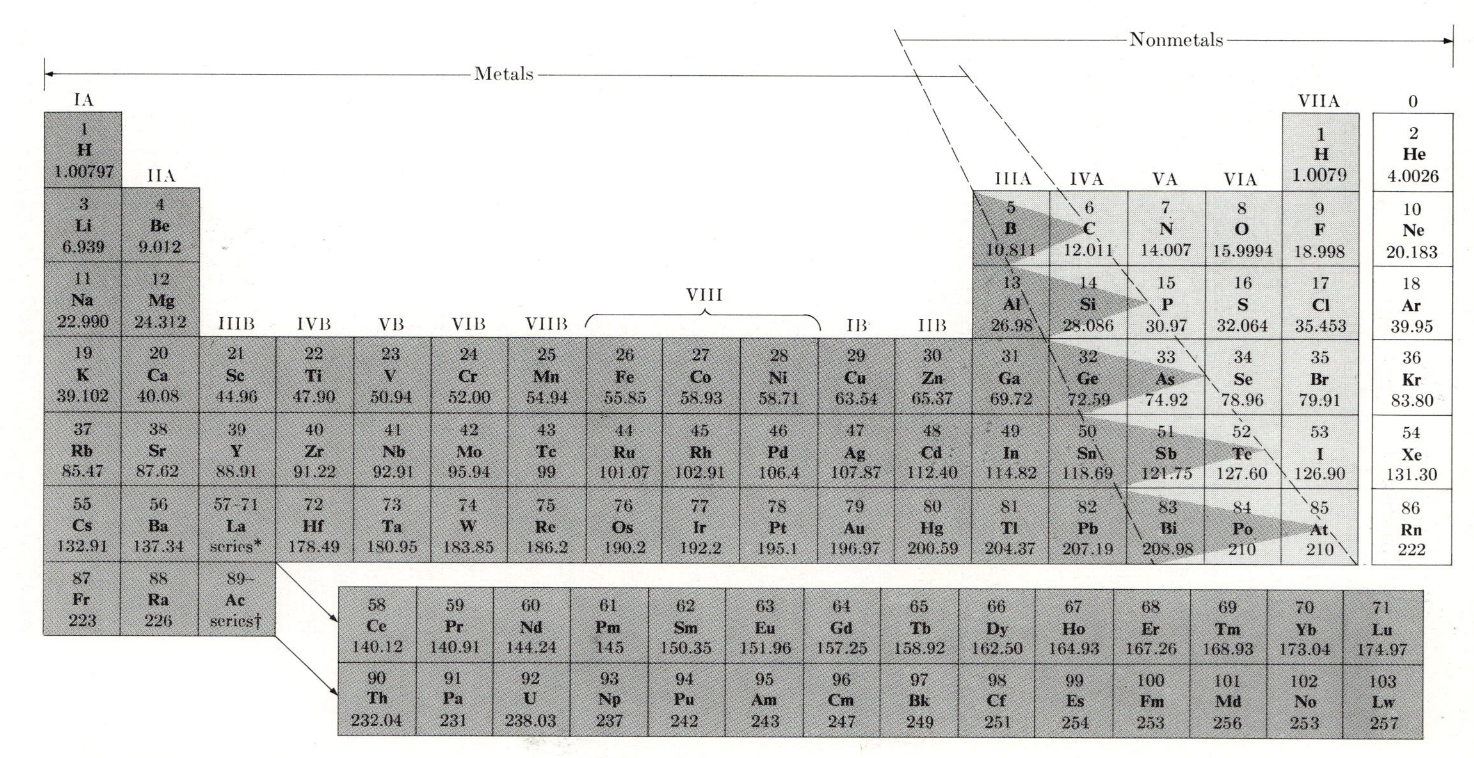

Figure 1-3.2 Periodic table of elements. Included are the atomic numbers and the atomic weights (in amu). There are 0.602×10^{24} atomic mass units per gram, and 0.602×10^{24} atoms per atomic weight (in grams). The darker-shaded elements are metals, which readily release their valence electrons. The nonmetals readily accept or share electrons.

SOLUTION Basis: 1 mer = 2C + 3H + 1Cl.

Carbon:	2(12 amu) = 24 amu;	w/o C = 38.
Hydrogen:	3(1 amu) = 3 amu;	w/o H = 5.
Chlorine:	= 35.5 amu;	w/o Cl = 57.
	Total 62.5 amu	100

ADDED INFORMATION Polyvinyl chloride contains 50 atomic percent (a/o) hydrogen and only 17 a/o chlorine. Since the hydrogen atoms are very much lighter than the others and the chlorine atom is heavier, the weight percent of hydrogen is less than the 50 a/o and the weight percent of chlorine is greater than the 17 a/o.

By convention, compositional percentages in condensed phases (liquids and solids) are expressed in weight percent, w/o, *unless* stated otherwise. ◀

Illustrative problem 1-3.2 Magnesium with a density of 1.74 g/cm³ oxidizes to MgO with a density of 3.58 g/cm³. Determine the percent volume change with the reaction

$$Mg + \frac{1}{2} O_2 \rightarrow MgO. \tag{1-3.1}$$

SOLUTION Basis: 24.3 g Mg + 16 g O → (24.3 + 16) g MgO.

Volume Mg = 24.3 g/(1.74 g/cm³) = 13.97 cm³ Mg
Volume MgO = 40.3 g/(3.58 g/cm³) = 11.26 cm³ MgO
ΔV = (11.26 cm³ − 13.97 cm³/(13.97 cm³) = −0.19 (or −19 v/o).

ADDED INFORMATION Less volume is required for MgO than for Mg, even though oxygen has been added!! This is because the Mg^{2+}—O^{2-} bond is so strong that it pulls the ions very close together (0.21 nm, center-to-center). In metallic magnesium, where the bonds are weaker, the two Mg atoms are separated by ~0.32 nm (center-to-center).

The stronger bonds influence properties such as the melting temperatures, T_m. The metal, Mg, melts at 650°C; the oxide, MgO, melts at 2800°C. Likewise, elastic properties, hardness, etc., are also affected by the strength of the bond. ◀

Review and Study

SUMMARY

This unit outlines the role of materials to the engineering profession.

1. Engineers adapt materials and energy to society's needs. Since all disciplines of engineering are involved with the design of technical products, their produc-

tion, and/or their use in technical systems, all engineering disciplines encounter materials.

2. There are innumerable types of materials; therefore, it is impossible for the engineer to be familiar with each and all. However, the engineer can be acquainted with the principles that govern the properties and performance of materials in engineering applications. Those principles are based on the internal structures of materials—atom, crystalline, and microstructures. In this text, we shall relate those concepts to applications.
3. For many purposes we can divide materials into the categories of metals, polymers (plastics), or ceramics. These groupings will be useful to us in the early chapters of this text when we consider structure and properties. Later, it will be convenient to examine materials on the basis of their properties and behavior, for example, strength and toughness, service stability to the surrounding environment, and electrical and magnetic characteristics.
4. In general, metals reside in the left side or in the lower part of the periodic table as it is conventionally presented. Polymers contain nonmetallic elements located in the upper-right corner of the table. Ceramics are compounds of metals and nonmetals. Of course, some materials fall in border areas and are not as definitively categorized. Even so, we will be able to relate the structures and properties of these materials to metals, polymers, and ceramics.

TECHNICAL TERMS

It is not necessary to memorize definitions verbatim. However, engineering students must become familiar with the meaning of technical terms in context (a) so that he or she does not have to look up terms each time one is mentioned in the text, in lectures, or in study problems, and (b) so that communication is possible in future engineering jobs with supervisors and colleagues.

Selected terms are defined. Also check their use in context in the proceding sections.

Ceramics Materials consisting of compounds of metallic and nonmetallic elements.

Engineering The adaptation of materials and/or energy for society's needs. (Other definitions are also possible.)

Materials (engineering) Substances for use in technical products. (These include metals, ceramics, polymers, semiconductors, glasses, and natural substances like wood and stone; but generally exclude food, clothing, drugs, and related substances.)

Metals Materials characterized by their high electrical and thermal conductivities because their electrons are mobile. Metallic elements are located in the left and lower portions of periodic tables as conventionally presented.

Periodic table Systematic presentation of the elements by atomic number in periods (horizontal) and groups (vertical). Fig. 1–3.2.

Percents (convention) Unless stated otherwise, compositions of solids (and liquids) are expressed in weight percent, w/o. Also by convention, compositions of gases are expressed by volume percent, v/o (or mole percent, m/o). Other abbreviations used in this text include a/o for atom percent, l/o for linear percent, etc.

Plastics *See* polymers.

Polymers Large molecules of nonmetallic elements composed of many repetitive units (mers). Commonly called plastics.

STUDY PROBLEMS

1-2.1. (a) Dismantle a cheap pen. List the materials that are used. (b) For each material, cite service conditions to which attention must be given in the material selection.

1-2.2. (a) Go to your car (or one owned by a friend). List all the materials you can identify. (Do not list names of parts.) (b) Compare your list with lists by your classmates. (c) Discuss the attributes that were important for their selection. (d) Suggest plausible substitutes for each. Why was the substitute *not* used?

1-3.1. Cellulose, a constituent of wood, has the composition $C_6H_{10}O_5$. Express its carbon content in both (a) a/o, and (b) w/o. (c) Repeat for hydrogen and oxygen.

Answer: (a) 29 a/o (b) 44 w/o

1-3.2. Examine Appendix A. (a) Make generalizations about the comparative properties of metals, polymers, and ceramics. (b) From your own experience, cite exceptions to the generalizations you just made.

1-3.3. Take your list from S.P. 1-2.2 above and categorize the materials as (a) metals, (b) polymers (plastics), or (c) ceramics.

UNIT TWO

METALLIC MATERIALS

Metals contain atoms in arrangements we call crystals. By knowing these arrangements, we can determine density and related properties. These crystalline arrangements are destroyed above the melting temperature. Also, metals such as iron change their arrangements as solids without melting, with corresponding changes in properties. It is necessary to understand these changes in order to know how properties can be controlled in engineering materials. Solid metals contain many grains of individual crystals.

Contents

Prerequisite General chemistry.

From Unit 2, the engineer should be able

1) To calculate the lattice constants and packing factors for cubic metals from their atomic radii and crystal structure.
2) To use atomic radii, lattice constants, packing factors, and densities in solving crystal structure problems.

• Topics or sections marked with this symbol contain subject matter that is not a primary prerequisite for later (unmarked) sections.

3) To calculate atom percent from weight percent and vice versa, and to calculate the size of the interstitial space among atoms.
4) To determine the volume and density changes that occur in a polymorphic reaction.
5) To estimate the boundary area in a solid from a 2-D section through the solid.
6) To understand technical terms pertaining to simple metal crystals.

2-1 SINGLE-COMPONENT METALS

We shall look first at the materials that have predominantly one kind of atom. These are the commercially pure metals such as aluminum that is used in kitchen utensils, copper that is found in electrical wire, 24-carat gold, and iron that is made into car fenders. Since each of these metals contains only minor amounts of other components, the properties of these metals arise from the principal component only. We shall not have to consider composition as a variable for the time being. Further, every atom has only one kind of neighbor—an atom identical to itself.

Atoms

When all the atoms are the same, we readily calculate the number of atoms that are present per cubic centimeter (or any other volume). To do this (Illustrative Problem 2–1.1), we must recall that the atomic weight of an atom, in amu's (Fig. 1–3.2), is equal to the mass in grams of 6.02×10^{23} atoms. This number is called *Avogadro's number*, N, and will be used frequently in the discussions and calculations of this text. As examples, from Fig. 1–3.2 we observe that it takes 63.54 g of copper to provide 6.02×10^{23} atoms of copper; and 26.98 g of aluminum to make 6.02×10^{23} atoms of aluminum; and 55.85 g of iron to make 6.02×10^{23} atoms of iron. Thus, since 1 cm^3 of copper has a mass of 8.92 g, we can calculate that each cm^3 has (8.92 g/cm^3)/(63.54 g/6.02×10^{23} atoms), or 0.85×10^{23} atoms of copper.

Atomic radii

The center-to-center distances between atoms can be measured by x-ray diffraction. We shall not determine those distances in this text, but the procedure is well established, so that we have access to data such as found in Table 2–1.1. Although the analogy has shortcomings, it is convenient for us to envisage a *hard-ball model*, in which the atoms in simple metals are

Table 2–1.1 Center-to-center (interatomic) distances and atomic radii of common metals (20°C)

Metal	Structure	Distance, nm**	Radius, nm
Ag	fcc	0.2888	0.1444
Al	fcc	0.2862	0.1431
Au	fcc	0.2882	0.1441
Be	hcp	0.228*	0.114*
Cd	hcp	0.296*	0.158*
Co	hcp	0.250*	0.125*
Cr	bcc	0.2498	0.1249
Cu	fcc	0.2556	0.1278
Fe (α)	bcc	0.24824	0.12412
Fe (γ)	fcc	0.2540	0.1270
K	bcc	0.4624	0.2312
Li	bcc	0.3038	0.1519
Mg	hcp	0.322*	0.161*
Mo	bcc	0.2725	0.1362
Na	bcc	0.3714	0.1857
Ni	fcc	0.2491	0.1246
Pb	fcc	0.3499	0.1750
Pt	fcc	0.2775	0.1386
Ti (α)	hcp	0.293*	0.146*
Ti (β)	bcc	0.285	0.1425
V	bcc	0.2632	0.1316
W	bcc	0.2734	0.1367
Zn	hcp	0.278*	0.139*
Zr	hcp	0.324*	0.162*

* The average dimension is given for hcp, since the radius varies somewhat with direction; that is, the hard-ball atom is not quite spherical, but ellipsoidal.

** A nanometer, nm, is 10^{-9} m (= 3.937×10^{-8} in.).

spherical. Thus, since the interatomic distance in copper is 0.2556 nm, the radius of a copper atom is simply half that dimension, or 0.1278 nm. (One nanometer, nm, equals 10^{-9} m. Although we shall use the simpler nanometer units whenever possible, at times it will be necessary to revert to other units; for example, $R_{Cu} = 0.1278 \times 10^{-6}$ mm.) The volume of a sphere with the above radius is 0.9×10^{-20} mm^3. Realizing that space can not be fully packed with spheres, we can quickly *estimate* that there may be slightly more than 10^{-20} mm^3 per copper atom, or slightly fewer than 10^{20} copper atoms per mm^3. In the preceding paragraph, we calculated 0.85×10^{23} per cm^3 (or 0.85×10^{20}/mm^3) as the more precise number.

Illustrative problem 2-1.1* An aluminum wire is 30.48 cm long and 2.58 mm in diameter. How many atoms are present?

SOLUTION See Appendix B for elemental data.

$$\text{No. of Al atoms} = \frac{(2.70\ \text{g/cm}^3)[\pi(0.129\ \text{cm})^2(30.48\ \text{cm})]}{(26.98\ \text{g}/6.02 \times 10^{23}\ \text{atoms})}$$
$$= 0.96 \times 10^{23}\ \text{atoms.}$$

ADDED INFORMATION There are 0.6×10^{23} Al atoms/cm^3 compared to 0.85×10^{23} Cu atoms/cm^3, indicating that $R_{\text{Al}} > R_{\text{Cu}}$. This is verified by data in Table 2-1.1. ◄

Illustrative problem 2-1.2 Assume spherical copper atoms with a hard-ball radius of 0.1278 nm. We know from the previous discussion that there are 0.85×10^{23} atoms of copper per cm^3. What is the atomic *packing factor* (PF) for copper; i.e., what fraction of the volume is occupied by the ball-like atoms?

SOLUTION

$$\left(\frac{\text{Volume of atoms}}{\text{Total volume}}\right) = \frac{(0.85 \times 10^{23}\ \text{atoms})(4\pi/3)(0.1278 \times 10^{-7}\ \text{cm})^3/\text{atom}}{1\ \text{cm}^3}$$
$$= 0.74\ \text{cm}^3/\text{cm}^3;$$
$$\text{PF} = 0.74.$$

ADDED INFORMATION The value of 0.74 is the highest atomic packing factor possible for spheres of only one size. (See Section 2-2.)

Our hard-ball model of atoms simply notes that we cannot force two copper atoms with their charged nuclei and accompanying electrons closer together than 0.2556 nm. However, an *uncharged* particle, such as a neutron, can move through this spherical volume, indicating that open space remains within the atom. ◄

2-2 CUBIC METALS

Bcc metals

The iron atoms in a car fender are most stable when each individual iron atom has eight neighbors. If each of the eight atoms surrounding the initial atom also has eight neighbors, and likewise for those neighbor's neighbors, a *long-range order* is established throughout the metal. This is called a *crystal lattice* and is shown in Fig. 2-2.1 for the first few atoms. In this case,

* Not uncommonly, a calculation is a more direct way to introduce a new concept than is textual verbage. In this text, the illustrative problems are often used in this role (in addition to serving as examples for study assignments). Typically, illustrative problems also permit the inclusion of additional information. Therefore, the student is advised to include these I.P.'s as an integral part of any reading assignment.

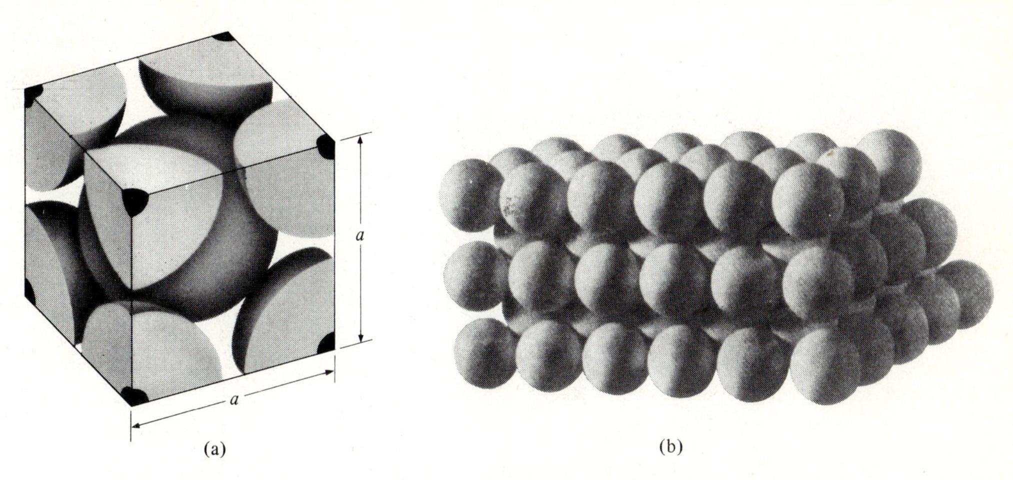

Figure 2-2.1 Body-centered cubic (metal). (a) Unit cell. (b) Hard-ball model showing the long-range order of ~100 atoms. This lattice can be repeated indefinitely. Each atom is coordinated with eight neighbors (CN = 8). The corner atoms and the body-centered atoms are identically located. (G. R. Fitterer. Modified and reproduced by permission from Bruce Rogers, *The Nature of Metals,* 2d ed., Metals Park, Ohio: American Society for Metals.)

we observe that the eight neighbors are arranged as cube corners around the center atom. The lattice is called *body-centered cubic* (bcc). The metals that possess the bcc lattice include

Ba	Fe <912°C	Na	Ta	W
Ca >448°C	K	Nb	Th >1345°C	Zr >862°C.
Cr	Li	Rb	Ti > 882°C	
Cs	Mo	Sr >557°C	V	

(Some metals, e.g., calcium, have this bcc structure only at high temperatures; the lowest-temperature form (<912°C) of iron is bcc.)

Note several things about the crystal structure of Fig. 2–2.1: (a) the cube, called a *unit cell,* could have been centered on *any* of the atoms and the structure would still be bcc; (b) eight corner atoms of the unit cell are all equal distance from the center atom; and (c) each body-centered atom lies at the midpoint of the four body diagonals of the cube. This structure permits us to establish several simple relationships in the first two illustrative problems at the end of this section.

Fcc metals

Atoms of copper in an electrical wire have a *coordination number* of 12, that is, each atom has 12 immediate neighbors. This pattern of 12 neighbors is more than twice as common among metals as the pattern of eight neighbors

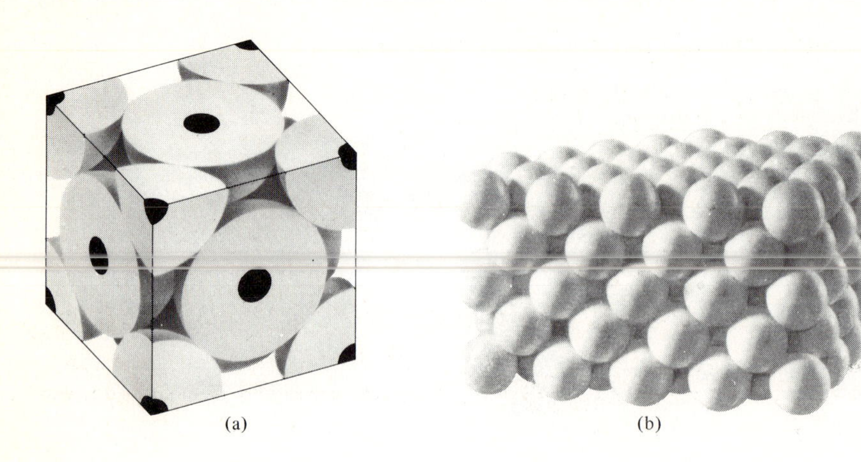

Figure 2–2.2 Face-centered cubic (metal). (a) Unit cell. (b) Hard-ball model showing CN = 12. The face-centered position can be the location of the unit cell corner if desired; the lattice would be the identical in every respect. (G. R. Fitterer. Reproduced by permission from Bruce Rogers, *The Nature of Metals*, 2d ed., Metals Park, Ohio: American Society for Metals.)

(CN = 8). However, there are two crystal lattices with a coordination number (CN) of 12. Initial attention will be given to the one with a cubic lattice.

Figure 2–2.2 shows the *face-centered cubic* (fcc) unit cell. Focus your attention on the atom at the center of the top face. The four top corner atoms are all contacting neighbors. Also, when viewed in 3-dimensions, there are four contacting neighbors halfway down the unit cell, centered on its midplane. A final group of four atoms would make contact with the atom of our focus if the structure were extended upward. The metals that possess this fcc lattice include

Ag	Co >417°C	Ir	Pt
Al	Cu	Ni	Rh
Au	Fe > 912°C	Pb	Sr < 557°C
Ca <448°C	(& <1400°C)	Pd	Th <1345°C.

Observe from Fig. 2–2.2 that (a) there are four metal atoms per unit cell—$\frac{1}{8}$ atom at each of eight corners, plus $\frac{1}{2}$ atom with each of six faces; (b) the unit cell could be placed to have its corner among *any* of the atoms, and (c) each face-centered atom lies at the midpoints of the two face diagonals of the cube. Again, as with bcc, we can make several basic calculations (Illustrative problems 2–2.3 and 2–2.4).

Illustrative problem 2–2.1 Calculate the dimensions of the edge of the unit cell of bcc titanium. (This edge, a, of Fig. 2–2.1, is called the *lattice constant.*)

SOLUTION The interatomic distance of titanium (bcc) given in Table 2–1.1 is 0.285 nm. That is one half of the body diagonal.

From the Pythagorean theorem of geometry,

$$a^2 + a^2 + a^2 = (\text{body diagonal})^2$$
$$3a^2 = [2(0.285 \text{ nm})]^2$$
$$a = 0.329 \text{ nm}.$$

ADDED INFORMATION For a body-centered cubic *metal*,

$$a\sqrt{3} = 4R. \blacktriangleleft \tag{2–2.1}$$

Illustrative problem 2–2.2 Determine the packing factor, PF, of bcc iron.

SOLUTION There are two atoms per unit cell of a bcc metal (Fig. 2–2.1a), because only an octant of each of the eight corner atoms resides in the unit cell.

$$\text{Center atom} + \tfrac{1}{8}\,(8 \text{ corner atoms}) = 2 \text{ atoms/u.c.}$$
$$\text{Volume of 2 atoms} = 2(4\pi/3)R^3.$$

From Eq. (2–2.1),

$$\text{Volume of unit cell} = a^3 = (4R/\sqrt{3})^3.$$

$$\text{Therefore, PF} = \frac{2(4\pi/3)R^3}{(4R/\sqrt{3})^3} = \frac{\pi\sqrt{3}}{8} = 0.68.$$

ADDED INFORMATION Note that we do not need to know the radius when calculating the packing factor of a *metal*, since all of the atoms are identical. Thus, this packing factor applies to all *bcc metals*. ◀

Illustrative problem 2–2.3 The edge of an fcc unit cell of calcium is 0.5569 nm. Calculate the atomic radius and the interatomic distances.

SOLUTION From the Pythagorean theorem of geometry,

$$a^2 + a^2 = [\text{face diagonal}]^2$$
$$2a^2 = (4R)^2,$$

or

$$a\sqrt{2} = 4R. \tag{2–2.2}$$

$$R = (0.5569 \text{ nm}/4)\sqrt{2} = 0.1969 \text{ nm}.$$

$$\text{Distance} = 2(0.1969 \text{ nm}) = 0.3938 \text{ nm}.$$

ADDED INFORMATION Observe that these dimensions can be calculated to four significant figures. This is not the case in very many engineering calculations. Be cautious in later problems so that you do not give answers with more digits than are warranted from the original data. ◀

Illustrative problem 2–2.4 The lattice constant, a, for copper may be determined from the data of Table 2–1.1 as being 0.3615 nm. Calculate the density of copper to verify this 0.3615 nm value.

SOLUTION As discussed above, there are four atoms per fcc unit cell. Based on its atomic mass (App. B), each copper atom has a mass of 63.54 amu, or $(63.54/0.602 \times 10^{24})$g.

$$\rho = \frac{m}{V} = \frac{4(63.54 \text{ amu}/0.602 \times 10^{24} \text{ amu per g})}{(0.3615 \times 10^{-9} \text{ m})^3}$$
$$= 8.94 \times 10^6 \text{ g/m}^3, \qquad (\text{or } 8.94 \text{ g/cm}^3).$$

ADDED INFORMATION The above calculation, which utilized x-ray diffraction data and the fcc structure, almost duplicates the density value in the appendix (8.92 g/cm³), which arose from mass and volume determinations. This should lend credibility to our picture of the fcc structures since the data are from independent sources. ◄

2-3 NONCUBIC METALS

Some important metals have noncubic structures. One such structure is shown in Fig. 2-3.1, which has the shape of a hexagonal prism. A quick check shows that each atom has twelve neighbors. Specifically, the center top atom is surrounded by six contacting neighbors. There are also three that make contact below; likewise, an upward continuation of the structure brings three atoms in contact above our reference atom. This structure is *hexagonal close-packed* (hcp) because it has the highest atomic packing factor of any hexagonal metal. That packing factor is 0.74 and matches the packing factor of fcc metals, which also have CN = 12. (Some people call the fcc lattice a *cubic close-packed* (ccp) lattice. That alternate name is apt.)

There are nearly equal numbers of hcp and fcc metals. However, many hcp metals (like Ce and Pr) are not well-known. The more familiar hcp

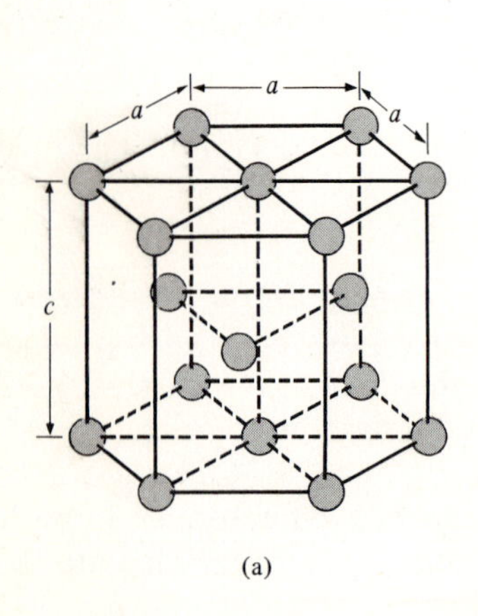

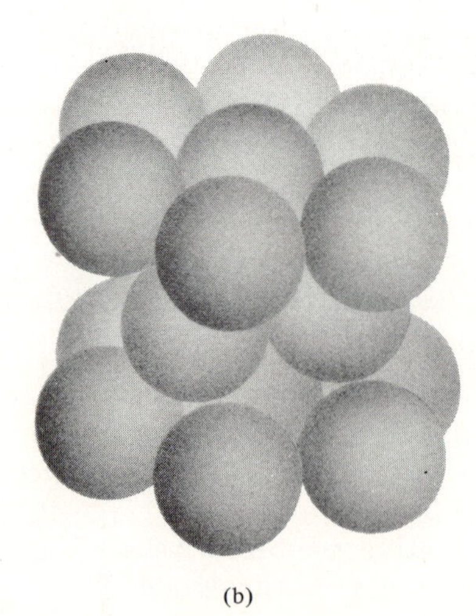

Figure 2-3.1 Hexagonal close-packed (metal). (a) Unit cell. (b) Hard-ball model showing CN = 12. As with fcc, the packing factor is 0.74. (Typically the atoms of hcp metals are not quite spherical. See the added information with I. P. 2-3.1.)

metals include

Be	Co <417°C	Mg	Zn
Cd	Hf	Ti <882°C	Zr <862°C.

If the preceding lists of bcc, fcc, and hcp metals were complete, we would find that nearly 90 percent of the metals are included. This is fortunate since these lattices are relatively simple and will serve the majority of our needs in this text. There is no predominant pattern among the remaining ten percent. We will simply note here that ordinary tin, a quenched form of iron, and uranium are among this minority. Their structures will be presented later within other contexts.

Illustrative problem 2–3.1 Titanium, which is hcp at normal temperatures, has an average radius of approximately 0.146 nm and a packing factor, PF, of 0.74. Calculate the density ρ of titanium.

SOLUTION Basis: 1 atom.

$$\text{PF} = 0.74 = \frac{(4\pi/3)(0.146 \times 10^{-9}\ \text{m})^3}{x\ \text{m}^3},$$

where x is the total "space" per atom;

$$x = 17.6 \times 10^{-30}\ \text{m}^3/\text{atom},$$

$$\rho = \frac{47.9\ \text{g}/0.6 \times 10^{24}\ \text{atoms}}{17.6 \times 10^{-30}\ \text{m}^3/\text{atom}} = 4.5 \times 10^6\ \text{g/m}^3, \qquad (\text{or } 4.5\ \text{g/cm}^3).$$

ADDED INFORMATION Atoms of hexagonal metals are usually not quite spherical. For example, titanium is slightly "squashed" in one direction, with the largest radius equal to 0.147 nm and the shortest radius equal to 0.145 nm. Except for magnesium and titanium, which are slightly compressed, most other hexagonal metals are distorted by being slightly elongated in the hexagonal prism direction. ◄

Illustrative problem 2–3.2 Tin is noncubic with a unit cell that is 0.582 nm × 0.582 nm × 0.318 nm. How many tin atoms are there per unit cell?

SOLUTION From App. B, $\rho = 7.3\ \text{g/cm}^3$ (or $7.3\ \text{Mg/m}^3$)

and

atomic mass = 118.69 amu.

$$\text{Mass per unit cell} = (7.3 \times 10^6\ \text{g/m}^3)(0.582 \times 10^{-9}\ \text{m})^2(0.318 \times 10^{-9}\ \text{m})$$
$$= 7.86 \times 10^{-22}\ \text{g} = \frac{n(118.69\ \text{amu})}{(0.602 \times 10^{24}\ \text{amu/g})}$$
$$n = 4\ \text{atoms/u.c.}$$

ADDED INFORMATION This unit cell is *tetragonal*. Its base is square; however, the third dimension differs from that of the base. ◄

2-4 HEATING AND MELTING OF METALS

Higher temperatures bring about increased thermal agitation within any material. The atoms vibrate more vigorously. This produces an expansion of the crystal lattice and allows us to determine a *thermal expansion coefficient*, that is, the fractional change in length per degree—cm/cm/°C, or simply, $°C^{-1}$. The thermal expansion coefficient may be either on a linear basis, α_L; or it may be expressed on a volume basis, α_V, for example, $cm^3/cm^3/°C$. Illustrative problem 2-4.1 shows how the two are related in a cubic material.

Melting temperature

With increased temperatures and more and more thermal agitation, the point is eventually reached where the atoms no longer can keep their optimum number of neighbors, and each atom moves among other atoms. That temperature is the *melting temperature*, T_m. The melting point is a sharp, precise temperature. In fact, we use the melting temperatures of selected materials to calibrate thermometers [Hg, −38.862°C; H_2O, 0.000°C; Sn, 231.968°C; Pb, 327.502°C; Ag, 961.93°C; Cu, 1084.5°C; etc.].

Liquid metals

Above the melting temperature, the metal is no longer crystalline. It has fluidity since the atoms are not locked into positions with a fixed number of neighbors. Therefore, the metal flows and has a number of other changes in properties that we associate with noncrystalline materials. For example, all of the bcc, fcc, and hcp metals expand significantly on melting (Fig. 2-4.1). This is because the moving atom has fewer contacting neighbors and slightly greater interatomic distances than found in Figs. 2-2.1, 2-2.2, and 2-3.1.*

An examination of Fig. 2-4.1 also shows that the liquid has a greater thermal expansion coefficient than does the solid, that is, the slope of V vs. T is greater. We can attribute this to the fact that in a liquid each atom moves more or less independently. Within a crystal, the thermal vibrations of many atoms are coordinated, and less expansion is required to accomodate the increased thermal energy.

* Materials such as ice and silicon have low coordination numbers as crystals because they have directional bonds (Unit 7). Therefore, their structures collapse into a smaller volume when they melt. This unusual behavior is not found among the typical metals that we are considering in this unit.

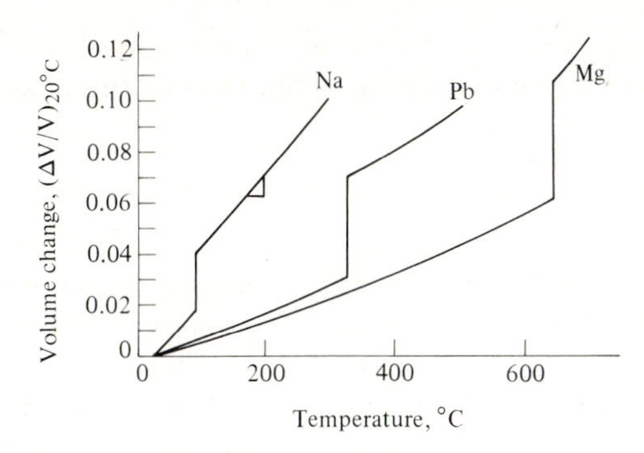

Figure 2-4.1 Volume changes with temperature (sodium—bcc, lead—fcc, magnesium—hcp). Metals with these structures expand upon melting. The expansion coefficient, dV/dT, is greater for the liquids than for the solids.

• Heat capacity

Energy is required to raise the temperature of every material. We call this energy the *heat capacity*. Since all thermal agitation is vibrational in the simple solids, we find the amounts of energy required per atom per degree is comparable. Thus, the required energy per °C per mole (0.6×10^{24} atoms) of aluminum is nearly the same as the energy required per mole of lead. This is ~25 J/°C.* Of course, a mole of aluminum has a mass of only 27 g, while a mole of lead has a mass of 207 g; therefore the J/g·°C will differ.

• Heat of fusion

Extra energy is required to melt a material since the regular crystal structure must be destroyed. This extra energy is called the *heat of fusion*. We know that refractory metals, that is, those with high melting temperatures, have stronger interatomic bonds than low-melting metals, simply because it takes more thermal agitation to destroy the crystal structure. Therefore, we should expect that refractory metals will have higher heats of fusion than lower melting metals. Table 2–4.1 reveals that this is generally true.

Illustrative problem 2–4.1 The linear thermal expansion coefficient, α_L, is *isotropic*, that is, it is the same in all directions. Show the relationship between the value of α_L and α_V, the volume thermal expansion coefficient.

SOLUTION Consider a 1° temperature rise.

At T: $V = L^3$.

At $T + 1°C$: $V + \Delta V = (L + \Delta L)^3$.

Dividing: $1 + \Delta V/V = (1 + \Delta L/L)^3$,

or $1 + \alpha_V = (1 + \alpha_L)^3 = 1 + 3\,\alpha_L + \cdots;$

and $$\alpha_V = {\sim}3\,\alpha_L. \tag{2-4.1}$$

• Sections marked with this symbol contain subject matter that is not a primary prerequisite for later (unmarked) sections.

* This generalization does not hold at cryogenic temperatures where quantum effects are predominant. Also, the value of ~25 J/mole·°C increases slightly at elevated temperatures, because the electrons pick up additional energy.

Table 2-4.1
Heats of fusion of metals

Metal	Melting temperature, °C (°F)	Heat of fusion, joule/mole*
Tungsten, W	3410 (6170)	32,000
Molybdenum, Mo	2610 (4730)	28,000
Chromium, Cr	1875 (3407)	21,000
Titanium, Ti	1668 (3035)	21,000
Iron, Fe	1537 (2798)	15,300
Nickel, Ni	1453 (2647)	17,900
Copper, Cu	1083 (1981)	13,500
Aluminum, Al	660 (1220)	10,500
Magnesium, Mg	650 (1202)	9,000
Zinc, Zn	419 (787)	6,600
Lead, Pb	327 (621)	5,400
Mercury, Hg	−38.9 (−38)	2,340

* joule/(6×10^{23} atoms). It requires 4.18 joules to heat 1 g of water 1°C. To obtain cal/mole, divide joule/mole by 4.18.

For cubic crystals (and isotropic solids) the volume expansion coefficient is three times the linear value.

ADDED INFORMATION In a noncubic material, the volume expansion coefficient approximates the sum of the three orthogonal linear coefficients,

$$\alpha_V \simeq \alpha_x + \alpha_y + \alpha_z. \tag{2-4.2}$$

This, of course, reduces to Eq. (2-4.1) when $\alpha_x = \alpha_y = \alpha_z$. ◄

• **Illustrative problem 2-4.2** (a) Compare the heat capacity, c, of copper and of silver on a gram basis. (b) On a volume basis.

SOLUTION (a) Since metals possess a heat capacity of ~25 J/mole·°C, that is, (25 J/°C)/(0.6×10^{24} atoms)

$$c_{Cu} = (\sim 25\ \text{J/°C})/63.54\ \text{g} = \sim 0.39\ \text{J/g·°C};$$

and

$$c_{Ag} = (\sim 25\ \text{J/°C})/107.87\ \text{g} = \sim 0.23\ \text{J/g·°C}.$$

b) With density data from App. B,

$$c_{Cu} = (\sim 0.39\ \text{J/g·°C})(8.9\ \text{g/cm}^3) = \sim 3.5\ \text{J/cm}^3\text{·°C};$$
$$c_{Ag} = (\sim 0.23\ \text{J/g·°C})(10.5\ \text{g/cm}^3) = \sim 2.4\ \text{J/cm}^3\text{·°C}.$$

ADDED INFORMATION Experimental values are 0.386 J/g·°C and 0.234 J/g·°C, respectively, for copper and silver at 20°C. ◄

2–5 IMPURE METALS

In some metals, impurities are undesirable; in other metals, they are intentionally added as alloying elements. Furthermore, even commercially pure metals have 0.01 to 1 percent of other kinds of atoms present. In this section we shall pay particular attention to *where* the impurity atoms are located, and what percentage may be present. In Unit 5 we shall look at *alloys* in which impurities are added intentionally.

Solid solutions

Our concept of solutions from high school chemistry involves liquids. For example, salt dissolves in water and loses its identity. Water absorbs the sodium and chlorine ions of the salt within the structure of the liquid. A similar situation can exist in solids. We shall observe in Unit 5 that *brass* is fcc copper in which some (≈35%) of the copper atoms have been replaced by zinc atoms. As shown in Fig. 2–5.1, the fcc structure accommodates the zinc atoms because they are approximately the same size as the copper atoms (R_{Zn} = 0.139 nm versus R_{Cu} = 0.1278 nm), and they have similar tendencies to release electrons. We call this structure a *substitutional solid solution*. As a rule of thumb, the ability of two elements to coexist in a substitutional solid solution is very limited if their radii differ more than 12–15 percent in size.

A structure of one element may also dissolve atoms of a second element to form an *interstitial solid solution*. If the impurity atom is very small compared to the host atoms, it can reside in the vacant spaces (or interstices) among the large atoms. Carbon atoms dissolve in fcc iron in this manner at elevated temperatures (Fig. 2–5.2). However, the carbon atoms are large enough so that there is some crowding. As a result, it is impossible to dissolve carbon atoms into more than 5–10 percent of the interstices at any

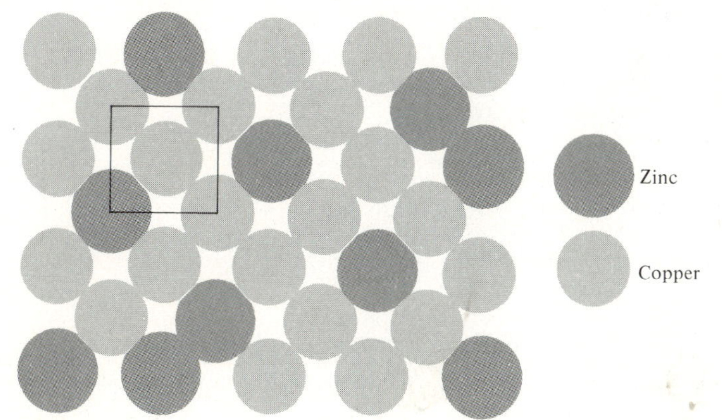

Figure 2–5.1 Substitution solid solution (zinc in copper). The fcc lattice of copper accommodates up to ~35% zinc atoms as substitutes for copper atoms.

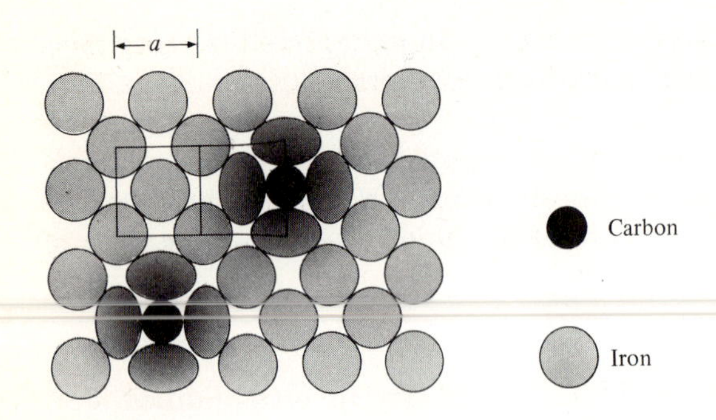

Figure 2-5.2 Interstitial solid solution (carbon in fcc iron). The fcc lattice of iron accommodates a limited number (<10 a/o) of the smaller carbon atoms in the interstices among the iron atoms.

one time. The interstices in bcc iron, although they are more numerous than they are in fcc iron, are smaller; thus an interstitial solid solution of iron and carbon at normal temperatures is extremely limited. (See Section 6-2.) In Unit 6 we shall see that heat treatment of steels involves the dissolving of carbon in fcc iron at high temperatures, followed by various cooling processes in which the carbon atoms come out of solid solution as the iron changes to a body-centered crystal structure.

We can express the composition of solid solutions (or any material) in terms of weight percent (w/o) or atom percent (a/o). By convention, weight percent is used for solids and liquids, *unless* we indicate otherwise. In contrast, convention uses atom percent (or mole percent) for gases unless indicated otherwise.*

Illustrative problem 2-5.1 Iron may contain nitrogen as an interstitial impurity. Analysis of one sample shows 0.05 w/o nitrogen; what is the atom percent nitrogen?

SOLUTION Basis: 100,000 amu = 99,950 amu Fe + 50 amu N. Using the atomic weights of Fig. 1-3.2, we have

$$\frac{99{,}950 \text{ amu}}{55.85 \text{ amu/Fe atom}} = 1790 \text{ Fe atoms;}$$

$$\frac{50 \text{ amu}}{14.01 \text{ amu/N atom}} = 3.5 \text{ N atoms.}$$

Therefore

$$\text{a/o N} = (3.5/1793)100 = 0.2 \text{ a/o.}$$

ADDED INFORMATION Only boron, carbon, nitrogen, and sometimes oxygen are small enough to dissolve interstitially into iron. Other common alloying

---

* In this text we shall use the abbreviations of w/o for weight percent, a/o for atom percent, m/o—molecular percent, l/o—linear percent, v/o—volume percent, etc.

elements, such as nickel, manganese, and chromium, replace the iron atoms in the bcc or fcc structure to form substitutional solid solutions. ◀

Illustrative problem 2–5.2 Assume no crowding in fcc iron at high temperature. What is the radius of the largest atom which can be introduced interstitially? (The radius of iron at 1000°C is 0.129 nm; a = dimension of unit cell.)

SOLUTION Refer to Fig. 2–5.2.

$$a = 4(0.129\ \text{nm})/\sqrt{2} = 0.365\ \text{nm},$$
$$r_{\text{hole}} = [0.365\ \text{nm} - 2(0.129\ \text{nm})]/2 = 0.053\ \text{nm}.$$

ADDED INFORMATION A commonly accepted value for the radius of a carbon atom *in iron* is 0.07 nm. Thus there must be some local strain of the iron lattice around each carbon atom (Fig. 2–5.2).

The 0.258 nm between the centers of the iron atoms is larger than given in Table 2–1.1 because (1) there is thermal expansion, and (2) there is an increase in radius when there are 12 neighbors (fcc) instead of 8 neighbors (bcc). ◀

2–6 PHASES

We have described several basic types of metal crystals—bcc, fcc, and hcp. There are other less simple types. Also, a liquid metal has a structure dissimilar to these crystalline solids. You can cite examples of materials that have still other structures and, therefore, do not mutually dissolve with those structures we have described to date. Included would be water, glass, and air. Liquid metal cannot be dissolved into water. Likewise, there is a sharp distinction between glass, which is an immobile supercooled liquid, and other liquids, such as oil. Air represents a special case, because it is devoid of any structure larger than the molecular species, for example, an O_2 molecule of air does not coordinate with N_2 molecules. Therefore, the "structure" of a gas, such as air, is random compared to crystalline solids.

We use the term *phase* to connote the distinct structures that have just been described. There is only one gas phase of a material. It is capable of receiving any and all atoms or molecules that have the necessary thermal energy to vaporize.

Several insoluble liquid phases may coexist, for example, a mixture* of oil and water, of mercury and glass, or of liquid solder and a rosin flux. Often, however, liquid phases have the ability to accommodate large amounts of solute within their structures, because these structures do not have specific 3-D coordinational requirements with neighboring atoms or

* We distinguish between a mixture and a solution. A mixture has physically separable phases.

molecules. Thus, we have solutions of brine containing salt in water, or syrup composed of sugar in water, each of which is a single-phase liquid with two components.

Solid phases

There are many solid phases, that is, many different solid structures. We have already encountered some of these—bcc iron, fcc iron, hcp titanium, fcc silver, and brass (fcc copper with zinc in solid solution), to name a few. The number of crystalline phases certainly runs into the thousands and are seemingly limitless. There are many, because when atoms are arranged in highly packed structures and have specific electronic orbitals, there are only limited possibilities for solid solutions.

Phase boundaries

The spatial transition from ice to water (two phases of H_2O)is abrupt. In ice, the H_2O molecules coordinate as a crystal. A nanometer away, across the phase boundary and in the liquid, the structure lacks the precise coordination, permits flow, and has other attributes of a fluid. Likewise at 20°C, a copper-silver alloy has a sharp transition at the boundary from nearly pure copper (fcc with a = 0.3615 nm) to silver (fcc with a = 0.4084 nm). A *phase boundary* defines a *discontinuity in structure and/or composition* within a material.

Polymorphs

All engineering materials can exist as more than one phase, e.g., ice, water, and steam are the solid, liquid, and gas phases of H_2O. Many engineering materials can also have multiple solid phases; for example, solid carbon, either as graphite or as diamond. These multiple phases are called *polymorphs* (or *allotropes* by some metallurgists).

Iron is polymorphic. When iron solidifies, it forms a bcc structure. Each atom has eight neighbors until the temperature drops to 1400°C (2550°F); then its structure changes to fcc, has a CN = 12, and has a packing factor of 0.74—greater than the 0.68 for bcc. Iron happens to be unique in that it transforms back to bcc at 912°C (1673°F). No other material is known to revert to its initial high-temperature polymorph as it cools to lower temperatures.* These transformations produce volume changes (Fig. 2–6.1).

* This peculiarity of iron is very fortunate, since almost all heat-treatment procedures for irons and steels are based on this fcc → bcc transformation near the lower end of the "red-hot" temperature range. Because of this structural change, we can make steels sufficiently hard to machine other steels. This "bootstrap operation" is not possible with other metals. Our whole technological history and development would be completely different were it not for this freak characteristic of iron.

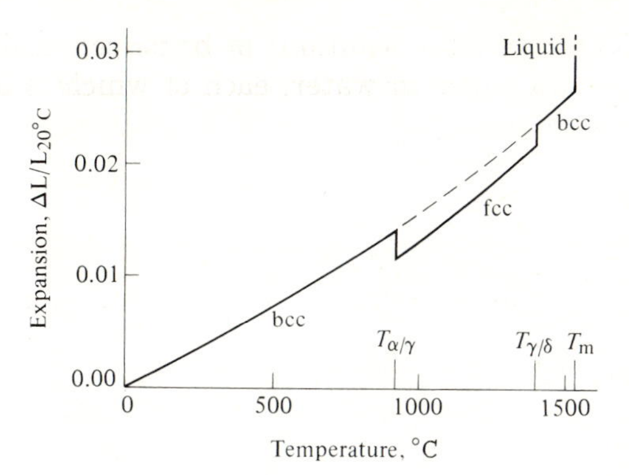

Figure 2-6.1 Thermal expansion of iron. Since iron changes from bcc to fcc between 912°C and 1400°C, there is a contraction in that temperature range. The fcc polymorph has a higher packing factor than the bcc polymorph, 0.74 versus 0.68. Its atoms also have larger radii. (See I.P. 2-6.1.)

We can identify other polymorphic metals from Sections 2-2 and 2-3, where we saw that calcium is fcc below 448°C and bcc above that temperature. Likewise, cobalt, strontium, thorium, titanium, and zirconium have two polymorphs. Generally, the less densely packed bcc form is the high-temperature form, because the atoms tend to pack more closely (fcc or hcp) when the thermal agitation decreases. Iron's 912°C transformation to a low-temperature bcc form is unique.

Illustrative problem 2-6.1 During heating, pure iron changes from bcc to fcc at 912°C (1673°F). Will the iron expand or contract when it makes that change? (At 912°C, the radii are 0.1258 nm and 0.1291 nm, respectively, for bcc and fcc iron.)

SOLUTION Basis: 4 atoms = 2 bcc unit cells = 1 fcc unit cell.

$$V_{\text{bcc}} = 2\, a_{\text{bcc}}^3 = 2(4R/\sqrt{3})^3 = 24.6\, R_{\text{bcc}}^3.$$

$$V_{\text{fcc}} = a_{\text{fcc}}^3 = (4R/\sqrt{2})^3 = 22.6\, R_{\text{fcc}}^3.$$

$$V_{\text{fcc}} = \frac{22.6}{24.6}\left(\frac{0.1291}{0.1258}\right)^3 V_{\text{bcc}} = 0.993\, V_{\text{bcc}}.$$

Thus, 0.7 v/o contraction (or −0.24 l/o, Fig. 2-6.1).

ADDED INFORMATION The contraction occurs because an fcc structure has a packing factor of 0.74 versus 0.68 for bcc. However, note that partial compensation arises because in fcc, with CN = 12, atoms have larger radii than in bcc with CN = 8 (Table 2-1.1).

This abrupt volume change can lead to quench-cracking in some steels. ◀

2-7 MICRO-STRUCTURES IN SINGLE-PHASE METALS

The several arrangements of metal atoms that were described in the previous sections led us to three crystal structures—bcc, fcc, and hcp—that are commonly encountered among pure metals and their solid solutions. This does not say, however, that the aluminum wire of I.P. 2-1.1 contains only one crystal. Rather, there are many crystals in any commercial aluminum wire—all with the same fcc structure. The individual crystals within a wire are called *grains*. Typically, the grains in metals are sufficiently small so that we must use a microscope to see them (Fig. 2-7.1). These granular crystals do not have the regular faceted surfaces that we associate with gem crystals, such as ruby and diamond, simply because each crystal grows until it encounters the adjacent growing grain.

In order to see the structure of Fig. 2-7.1, called the *microstructure* of this metal, it was necessary to cut and polish the metal to have a mirror-like flat surface. The surface was etched with a weak acid. The acid corroded the boundaries between adjacent grains. The corroded *grain boundary* does not give specular reflection as does the grain itself (Fig. 2-7.2); therefore, we see a dark line at the grain boundary when we view the polished and etched metal through a microscope.

Grain boundaries

If we were to view adjacent grains and the grain boundary at an atomic level, we would observe that the latter is a zone of mismatch (Fig. 2-7.3). The packing factor is lower along the grain boundary; in fact, this provides a

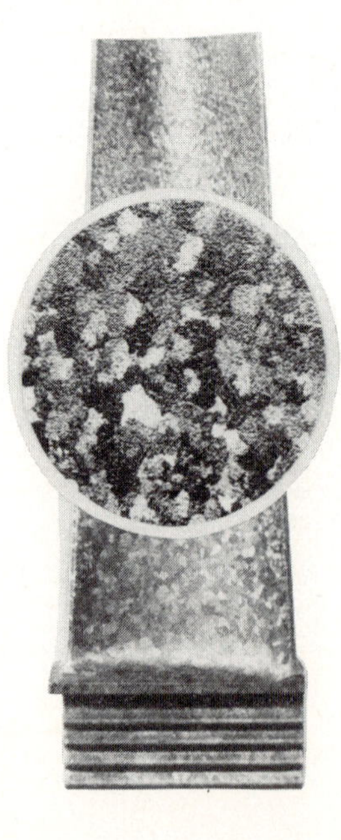

Figure 2-7.1 Grains in metal (turbine blade). The microstructure of this alloy contains many small grains, each an individual crystal. Low magnification reveals these grains more readily. In many materials, higher magnifications are required. (Courtesy of Pratt and Whitney Aircraft Group, United Technologies.)

qualitative explanation of the preferential etching described in Fig. 2-7.2. The grain boundary also provides a path of easier atom movements, and a location for initiating chemical reactions. Finally, the crystal directions change at the grain boundaries. All of these factors become important when

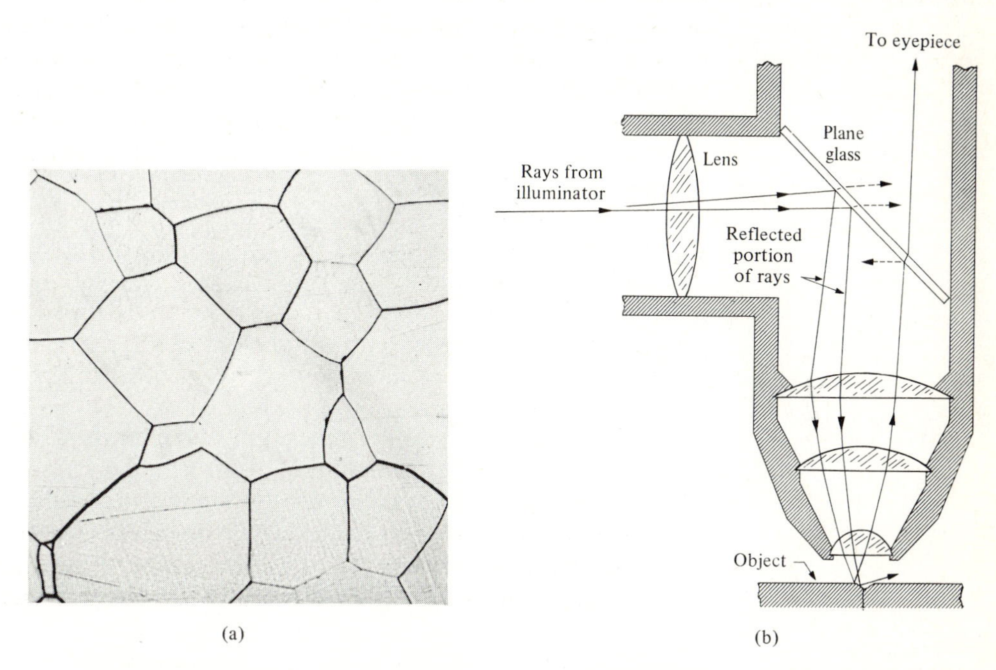

Figure 2-7.2 Photomicrograph of metal. (a) Molybdenum (×250), polished and etched. (Courtesy O. K. Riegger) (b) The grain boundaries of (a) are dark because, after being etched, they do not reflect light through the microscope. (Bruce Rogers, *The Nature of Metals,* Metals Park, Ohio: American Society for Metals.)

Figure 2-7.3 Grain boundaries (schematic). The atoms along the boundary are in a mismatch zone. This irregularity affects the properties and behavior of the metal. (Clyde Mason, *Introductory Physical Metallurgy,* Metals Park, Ohio: American Society for Metals.)

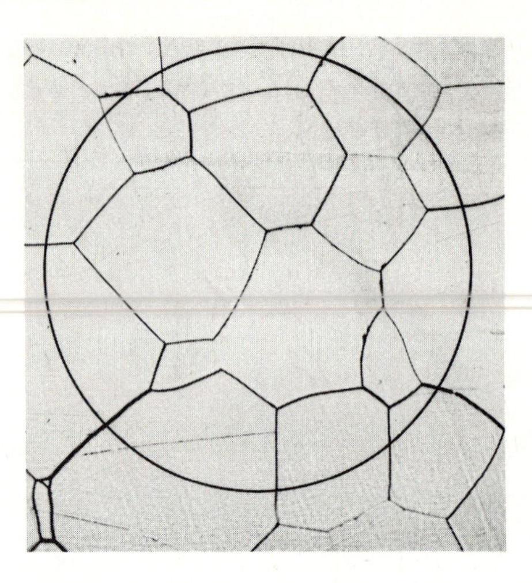

Figure 2-7.4 Calculation of the grain boundary area (×250). The randomly placed circle interrupts 11 grain boundaries in a distance of 0.64 mm (when the ×250 magnification is considered). We could have sampled P_L by other methods. (See I.P. 2-7.1.)

we relate properties to microstructure in later units. Therefore, it is useful to be able to measure the amount of grain-boundary area within material.

Fine-grained metals have more *grain-boundary area* per unit volume than do coarse-grained metals. In fact, if one piece of metal has grains that are twice the dimensions of the grains in another metal, it will possess half of the grain-boundary surface per mm^3. Thus, atomic movements, reactions, etc., are going to be affected. Fortunately, it is very easy to estimate the amount of grain-boundary surface. We can do this by simply placing a line of known length randomly across a microstructure; then count the number of intersection points that line has with the boundaries. The number of such points per unit length of the line is P_L. The boundary surface area, S_V, per unit volume, is numerically equal to twice P_L:

$$S_V = 2P_L. \qquad (2\text{-}7.1)^*$$

The units for the two sides of the equations are mm^2/mm^3 and mm^{-1}, respectively (or other comparable dimensions, $in.^2/in.^3$, m^2/m^3, etc.).

From the several relationships described in the previous paragraph, we may express *grain size* as $1/P_L$, which is the mean intercept chord between grain boundaries.

Illustrative problem 2-7.1 (a) Estimate the grain-boundary area per mm^3 in the molybdenum of Fig. 2-7.2. (b) Check your answer with a second sampling from the figure.

* We shall not derive this equation; however, it is interesting to know that this useful relationship was discovered independently at least half a dozen times—in this country, in Russia, by metallurgists, by mathematicians, by botanists, etc.—all of whom had reason to want to know the amount of boundary area, but none of whom had technical contact with the others.

SOLUTION (a) Place a circle at random on the microstructure (Fig. 2–7.4). The circle (51 mm) intercepts 11 boundaries. Since the magnification is ×250,

$$P_L = 11/(51\pi \text{ mm}/250) = 17/\text{mm},$$
$$S_V = 2(17/\text{mm}) = 34 \text{ mm}^2/\text{mm}^3.$$

In inches,

$$S_V = 2P_L = 2(11)/(2\pi \text{ in.}/250) = 880 \text{ in.}^2/\text{in.}^3$$

b) Use the perimeter as your sample:

$$S_V = 2P_L = 2(16)/(4 \times 57 \text{ mm}/250) = 35 \text{ mm}^2/\text{mm}^3.$$

b′) Use two diagonals as your sample:

$$S_V = 2(15)/(2 \times 81 \text{ mm}/250) = 46 \text{ mm}^2/\text{mm}^3.$$

ADDED INFORMATION These three answers are samplings; therefore, they should not be expected to be identical. In fact, the 34 mm²/mm³ and 35 mm²/mm³ answers are coincidentally close. The same three procedures on a second photomicrograph of the same material would give somewhat different samples and results, especially since we see a range of grain sizes with the field of Fig. 2–7.4. However, that does not detract from the procedure because we shall be concerned with order of magnitude differences in boundary area/volume in later units. ◂

Figure 2–7.5 Study problem 2–7.1 (iron). (a) Photomicrograph of polished and etched cross section (×500). Each grain is a single crystal. (Courtesy of U.S. Steel Corp.) (b) Grain boundaries (sketched). The boundary is the surface of mismatch between grains.

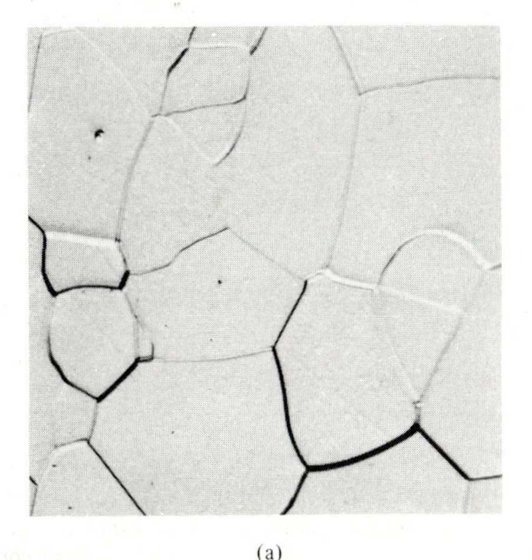

(a)

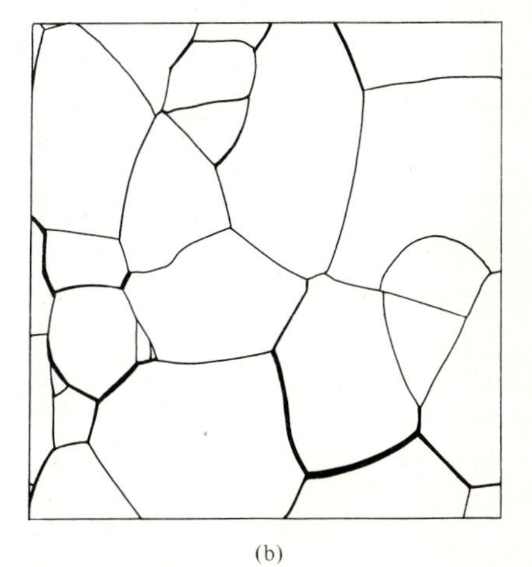

(b)

Review and Study

SUMMARY

This unit has presented the following information about metals:

1. Atoms are spaced at very specific distances in metals. We consider half of this distance to be the atomic radius.
2. About 30% of the metals possess a body-centered cubic structure of atoms. Another 30% possess a face-centered cubic structure. The atoms in a bcc metal are coordinated with eight neighboring atoms and have a packing factor of 68%. The coordination number in an fcc metal is 12; thus, the packing factor is higher—74%.
3. Still another 25%–30% of the metals are hexagonal close-packed. Since atoms in hcp metals—like fcc metal—are coordinated with 12 neighbors, their packing factor is also 74%. A few metals possess other noncubic crystal lattices.
4. Increased thermal energy expands the crystal lattice, eventually destroying it at the melting temperature. • The heat of fusion, which is the energy required to "shake the crystal lattice apart," is greatest for those metals with a high melting temperature because those metals have the strongest bonds.
5. Foreign atoms may enter a crystal (a) by substitution if they are similar in size and electronically to the original atoms, or (b) interstitially if they are sufficiently small. These solid solutions will play an important role in our understanding of alloys.
6. A phase refers to a structurally distinct part of a material. There is an abrupt discontinuity in structure and/or composition at the boundary between two phases. An individual metal will have only one gas phase and one liquid phase, but it may have more than one solid phase; for example, carbon as graphite or diamond, or hcp and bcc titanium. We call these polymorphs. Although their composition is the same, their properties differ.
7. Commercial metals contain a large number of grains, each of which is an individual crystal. The boundary between adjacent grains influences the properties of the metal in numerous ways. Therefore, we must pay attention to grain size, and to the amount of grain-boundary area per unit volume.

TECHNICAL TERMS

Read again the comments with the Technical Terms in Unit 1.

Alloy A metal containing two or more elements.

Atomic radius (*R*) In metals, half of the interatomic distance (center-to-center).

Avogadro's Number (*N*) Number of amu's per gram (0.602×10^{24}). Number of molecules per mole.

Body-centered cubic (bcc) A cubic lattice with the center position fully equivalent to each of the eight corners.

Brass An alloy of copper and zinc.

Component (phases) The basic chemical substances required to create a chemical solution or mixture.

Coordination number (CN) The number of closest atomic or ionic neighbors.

Crystal lattice The spatial arrangement of equivalent sites within a crystal.

Face-centered cubic (fcc) A unit cell with face positions fully equivalent to each of the eight corners.

Grain (metals and ceramics) Individual crystal of a microstructure.

Grain boundary The surface of crystal mismatch between adjacent grains.

Grain-boundary area (S_V) Area of grain boundary surface per unit volume; for example, mm^2/mm^3, or $in.^2/in.^3$.

Grain size ($1/P_L$) Mean intercept chord length of grains.

Hard-ball model The assumption that atoms are hard spheres. This is a very useful concept. However, it is not universally applicable, e.g., for neutron radiation exposure.

• **Heat capacity (c)** Energy per unit temperature change, dH/dT, (per unit mass, or unit volume).

• **Heat of fusion (ΔH_f)** Energy required to melt a solid (per gram, or per mole).

Hexagonal close-packed (hcp) The structure of a hexagonal metal with atoms located at the corners *and* in offset positions at midheight. (See Fig. 2–3.1.)

Interatomic distance Center-to-center distance of adjacent atoms.

Interstice Unoccupied space between atoms or ions.

Interstitial solid solution Crystals that contain a second component in their interstices. The basic structure is unaltered.

Lattice The space arrangement of equivalent sites in a crystal.

Lattice constant (a) Dimensions of the unit cell.

Long-range order The repetitive pattern over many atomic distances.

Microstructure Structure of grains and phases. Generally requires magnification for observation.

Noncubic lattices Crystals with two or more unequal lattice constants and/or non-90° axial angles.

Packing factor (PF) Occupied volume per unit of total volume.

Phase (material) A structurally distinct part of a materials system.

Phase boundary Compositional and/or structural discontinuity between two phases.

Polymorphism (allotropism) Multiple solid phases for a single composition.

Single-phase materials Materials containing only one atomic arrangement of atoms.

Solid solution A crystalline phase with a second component.

Substitutional solid solutions Crystals with a second component substituted for the atoms of the basic structure.

Tetragonal (crystal) Two of three lattice constants equal; all three axes at 90°.

Thermal expansion coefficient (α) (Change in dimensions)/(change in temperature).

Unit cell A small repetitive volume that contains the complete lattice pattern of the crystal.

CHECKS*

2A Avodagro's number is the number of ___***___ per atomic weight, or the number of ___***___ per gram.

2B A density of 1 Mg/m^3 is equal to a density of ___***___ g/cm^3.

2C The closest approach of the atomic nuclei in metallic lead is 0.35×10^{-6} mm. The radius of a lead atom is ___***___ nm.

2D With only one kind of metallic atom present, the coordination numbers for fcc and bcc metals are ___***___ and ___***___, respectively.

2E With only one kind of metallic atom present, there are four metal atoms per unit cell of a ___***___-centered unit cell.

2F With only one kind of metallic atom present, the atomic packing factor in fcc metals is ___***___ than in bcc metals.

2G With only one kind of metallic atom present, the atomic packing factor is ___***___ in hcp metals and the coordination number is ___***___.

• 2H Among all of the metals we have encountered in this unit, those with higher ___***___ also have higher ___***___.

2I In cubic materials, the ___***___ is one third the ___***___ thermal expansion coefficient.

2J Most materials ___***___ on melting; ___***___ is a notable exception.

2K When comparing solids and liquids, the ___***___ typically have greater volume expansion coefficients.

• 2L The value of 25 J/°C per ___***___ is an approximate value of the ___***___ of simple solids such as metals. (It does not hold for polymers.)

• 2M To a first approximation, the heat of fusion of metals is a function of ___***___ and therefore of the ___***___ temperature.

2N Solid solutions fall into two categories, ___***___ and ___***___.

2O Since a carbon atom is light, its ___***___ percent in steel will be greater than its ___***___ percent in a steel.

2P A steel contains 0.5 percent carbon. By this we mean ___***___ percent.

2Q Copper and nickel both prefer CN = 12; another similarity that facilitates complete solid solubility is their ___***___.

2R Any two phases differ in ___***___ and/or ___***___.

2S ___***___ means "multiple structure of the same composition." Examples include ___***___ and ___***___, or ___***___ and ___***___.

2T Atomic radii are larger when the ___***___ is higher.

---

* Since some concepts are more readily introduced through calculations, these checks assume the reader has studied the illustrative problems as well as the textual material.

2U Each __***__ of a microstructure is a separate crystal.

2V If the grain-boundary area per mm^3 of a microstructure is decreased by a factor of two, the average "diameter" changes by a factor of __***__ and the number of grains per mm^2 of polished surface changes by a factor of __***__.

STUDY PROBLEMS*

2-1.1. How long should a 2.3-mm copper wire be to contain 10^{24} atoms? (By convention, wire dimensions are given as diameters.)

Answer: 2.8 m†

2-1.2. One-tenth gram of nickel is plated onto a surface (18 cm × 2.7 cm). (a) How thick is the plated nickel? (b) How many atoms per mm^2? (See Appendix B for elemental data.)

* | 2-1.3. The density of iron is 7.87 g/cm³. (a) How many atoms are there per cm³? (b) What is the packing factor?

Answer: (a) 0.8483×10^{23} Fe/cm³ (b) 0.68

| 2-1.4. Solid argon has the same packing factor as aluminum (0.74.). Calculate its density from its atomic radius (Appendix B).

2-2.1. Calculate the dimensions of the edge of the unit cell of silver.

Answer: 0.4084 nm

2-2.2. The atomic radius of cesium (bcc) is 0.267 nm. Calculate its density.

2-2.3. Calculate the density of one of the fcc metals in Table 2-1.1.

Answer: See Appendix.

2-2.4. Calculate the density of one of the bcc metals in Table 2-1.1.

| 2-2.5. From the data in Table 2-1.1, determine the number of gold atoms per mm³.

Answer: 0.591×10^{20}/mm³

| 2-2.6. Below −249°C (<24 K), neon is solid with a density of 1.45 g/cm³. Its lattice constant (cubic) is 0.452 nm and its atomic mass is 20.18 amu. From these data (and without knowing its lattice), (a) what is the mass per unit cell? (b) How many atoms per unit cell?

2-3.1. Zinc is hcp with the density shown in App. B. Verify that 0.139 nm is the average atomic radius, assuming spheres.

---

* Many study problems, S.P., are introductory in that they either mimic illustrative problems closely, or come directly from equations. Those preceded by a vertical rule require more analysis and synthesis by the student. The student should be able to solve these. Data available in the accompanying tables, or in the appendices, generally will *not* be repeated in the problem statement.

† Answers should *not* contain excessive digits. For example, an answer for S.P. 2-1.1 of 284.8002142 cm (which could be read from your calculator) has more figures than are warranted and would therefore be erroneous. Many engineers follow this practice: (a) Carry all figures in your calculator during the initial steps. (b) After the final calculation, round off to the *number* of significant figures, *plus one*, in the least accurate (noninteger) data. The extra digit lets one see which way the last significant figure "tilts."

2-3.2. Zirconium is hcp with an average atomic radius of ~0.16 nm. Calculate the density from the data in Fig. 1-3.2.

2-3.3. Uranium is noncubic but with axes at right angles. Dimensions are 0.2854 nm × 0.5869 nm × 0.4956 nm. How many atoms are there per unit cell?

Answer: 4

2-3.4. The height of the zinc unit cell is 0.494 nm. What are the dimensions of the edges of the base?

2-3.5. Hcp titanium can be represented by Fig. 2-3.1. The ratio of its height to the base edge is 1.59; the unit cell volume is 0.106 nm^3. (a) What are the values of the height, c, and the base edge, a? (b) What is the radius of the titanium atom in the a-direction?

Answer: (a) 0.469 nm, 0.295 nm (b) 0.1475 nm

2-4.1. (a) From Fig. 2-4.1, calculate the linear expansion coefficient of magnesium between the 20° and the melting temperature. (b) Compare this with data in Appendix A, and explain any difference.

Answer: (a) 33×10^{-6}/°C

• 2-4.2. Estimate the approximate heat of fusion of gold.

• 2-4.3. Estimate the approximate heat capacity of gold. (a) In J/g·°C; (b) J/cm^3·°C.

Answer: (a) ~0.13 J/g·°C (b) ~2.5 J/cm^3·°C

2-5.1. A steel contains 0.04% carbon and 0.26% molybdenum (balance iron). Which alloying element is present in greater numbers?

Answer (partial): C = 0.19 a/o

2-5.2. What are the atom percentages of the alloying elements in an 18-8 stainless steel (18% Cr–8% Ni-balance iron).

2-5.3. A sterling silver (92.5 Ag–7.5 Cu) is an fcc substitutional solid solution when cooled rapidly. What is the unit cell dimension?

Answer: 0.403 nm

2-5.4. The largest symmetrical interstice in bcc iron is located among four atoms. (Two at adjacent cell corners; one at the cell center; and one at the center of the adjoining unit cell.) What is the radius of the largest atom that can be interstitially without crowding? (a = 0.2866 nm)

2-5.5. Using the answer of S.P. 2-5.3, calculate the density of sterling silver.

Answer: 10.4 g/cm^3

2-6.1. Calcium changes from fcc to bcc as it is heated past 448°C. Will there be an expansion or contraction? How much? (With 12 neighbors, the radii are about 2% larger than with 8 neighbors.)

Answer: +2.6 v/o

2-6.2. Refer to S.P. 2-6.1. Assume the atoms do not change size. How much volume change?

2–7.1. Refer to Fig. 2–7.5. What is the grain boundary area per mm^3? The magnification is ×500.

Answer: 75 mm^2/mm^3

2–7.2. Refer to Figs. 6–3.3(a) and (b). The whiter matrix and the grayer areas are different phases, that is, they have different crystal structures. There is a *phase boundary* between them. (a) Sample the P_L's for this boundary by placing six vertical lines across each figure. (b) Sample the P_L's for this boundary by placing three horizontal lines across each figure. (c) Explain the differences, if any, that you obtain. (The percent gray phase is the same in each figure.)

CHECK OUTS

2A	atoms atomic mass units (amu)
2B	one
2C	0.175
2D	12 8
2E	face
2F	greater
2G	0.74 12
2H	melting temperature heat of fusion (coordination numbers) (packing factors)
2I	linear thermal expansion coefficient volume
2J	expand water
2K	liquids
2L	atomic weight heat capacity
2M	bond strength melting temperature
2N	substitutional interstitial
2O	atom weight
2P	weight
2Q	radii
2R	structure composition
2S	polymorph graphite, diamond bcc Fe, fcc Fe
2T	coordination number (packing factor)
2U	grain
2V	2 4

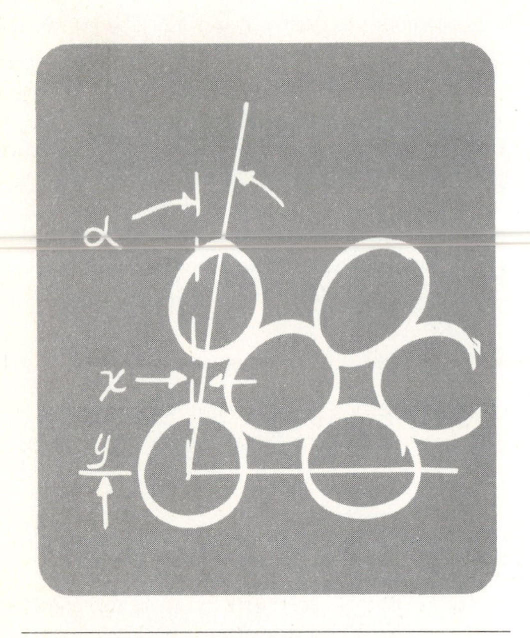

UNIT THREE

METAL DEFORMATION

Metals under load are stressed; therefore, they are deformed. The resulting strain may be elastic (recoverable) and/or plastic (permanent). The engineer must have quantitative figures of stresses for failure (strength), strain prior to fracture (ductility), and the interrelationships of the two. In metals, the mechanism of deformation is related to the crystal structure.

Contents

Prerequisites Mechanics in general physics; Section 2-2.

From Unit 3, the engineer should

1) Understand the various measures of strength, elasticity, and ductility with which the engineer works.
2) Be able to calculate stresses, strains, elastic moduli, and to differentiate between the nominal and true bases.
3) Know the factors that affect elastic and plastic deformation.
4) Be familiar with the mechanism of slip.
5) Understand the meaning of the technical terms associated with deformation.

3-1 STRESS AND STRAIN

We are familiar with many engineering products that are designed to carry a load. Examples include not only bridges and airplane wings but also metals for car bumpers and for gas pipelines, which must be given specifications for strength. Likewise, the design engineer must anticipate the forces on a spring clip behind a calculator key, and on a valve in a nuclear reactor.

Stress

In order to define the effect of loads upon materials, we speak of *stress*, s, which is force, F, per unit area, A:

$$s = F/A. \tag{3-1.1}$$

This permits us to use some of the same considerations, whether we are working with a wire for a steel-belted tire or a massive underframe for a railroad diesel locomotive.

Strain

The material responds to a stress by undergoing strain. Most simply stated, a *tensile* stress (+) stretches the interatomic distances to lengthen the metal; and a *compression* stress (−) squeezes the atoms closer together to shorten the interatomic distances in the metal. The *strain*, e, is the unit deformation:

$$e = \Delta L/L. \tag{3-1.2}$$

The units for stress are newtons per square meter, N/m^2, which are called *pascals*, Pa. Since a pascal is small in comparison to stresses typically encountered, we commonly use megapascals, MPa. (The corresponding English units for stress are $lb_f/in.^2$, or psi. To convert, 1000 psi = 6.894 MPa, or 1 MPa = 145 psi.) Strain is *dimensionless*.

Young's elastic modulus

The first strain that we observe in most metals is *elastic*. That is, the strain is *recoverable* when the load or force is removed. According to Hooke's law of physics, the amount of elastic strain is proportional to the applied stress (Fig. 3-1.1). The proportionality constant is called *Young's modulus*, E, and is one of three *moduli of elasticity* that engineers use:

$$E = s/e. \tag{3-1.3}$$

Since strain is dimensionless, this elastic modulus has the same units as stress, MPa (or psi).

Table 3-1.1 lists the Young's moduli for several common cubic metals. Others are found in Appendix A. This property, together with the shape and dimensions, accounts for the deflection of a beam or of other structural

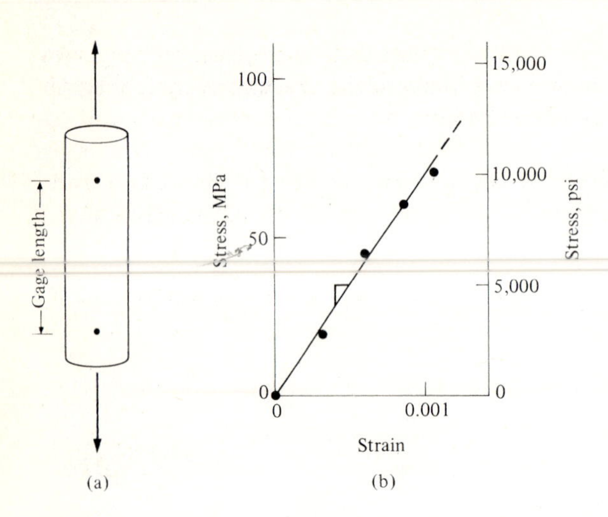

Figure 3-1.1 Elastic deformation (aluminum). The test specimen was strained in the testing machine; the required force was read and the stress calculated (Eq. 3-1.1). Elastic strain is proportional to the stress, so the slope of the curve is Young's modulus (Eq. 3-1.3).

Table 3-1.1
Young's moduli of selected cubic metals

	Young's modulus	
Metal	**MPa**	**(10^6 psi)**
Tungsten	345,000	(50)
Iron	205,000	(30)
Copper	110,000	(16)
Aluminum	70,000	(10)
Lead	14,000	(2)

members. When the Young's modulus is high, it takes considerable stress for a given amount of deflection; conversely stated, an aluminum beam with $E = 70{,}000$ MPa (or 10×10^6 psi) is only one third as rigid as a steel beam of the same dimensions, since E for the latter is 205,000 MPa (or 30×10^6 psi).

Illustrative problem 3-1.1 What is the tensile stress on a wire with a diameter of 1.0 mm (or 0.039 in.) that supports 15 kg (or 33 lb)?

SOLUTION

$$s = \frac{\text{force}}{\text{area}} = \frac{+15 \text{ kg } (9.8 \text{ m/s}^2)}{\pi(0.001 \text{ m})^2/4} = \frac{+147 \text{ newtons}}{0.785 \times 10^{-6} \text{ m}^2}$$
$$= +190 \times 10^6 \text{ Pa} = +190 \text{ MPa};$$

or,

$$s = \frac{+33 \text{ lbs}}{(\pi/4)(0.039 \text{ in.})^2} = +27{,}000 \text{ psi}.$$

ADDED INFORMATION Note that our original data are given to only two significant figures. Therefore, one should avoid the practice of including all of the numbers in the display of the calculator. ◄

Illustrative problem 3-1.2 Suppose that the wire in I.P. 3-1.1 is aluminum with a length of 10.0 m (or 32.8 ft). (a) How much will it stretch? (b) Repeat for an iron wire.

SOLUTION From the elastic modulus data of Table 3-1.1, the answer of I.P. 3-1.1, and Eq. (3-1.3),

a) $e = \frac{s}{E} = \frac{+190 \text{ MPa}}{70{,}000 \text{ MPa}} = +0.0027;$

$$\Delta L = (+0.0027)(10.0 \text{ m}) = +0.027 \text{ m} \qquad \text{(or } +27 \text{ mm).}$$

In English units, $e = +27{,}000 \text{ psi}/10^7 \text{ psi} = +0.0027;$

$$\Delta L = (+0.0027)(32.8 \text{ ft}) = 0.089 \text{ ft} \qquad \text{(or 1.1 in.).}$$

b) In iron,

$$\Delta L = (+190 \text{ MPa}/205{,}000 \text{ MPa})(10.0 \text{ m}) = +0.009 \text{ m} \qquad \text{(or 9 mm).}$$

ADDED INFORMATION Note that the dimensional units on each side of the equation must cancel each other. ◄

3-2 THE STRESS-STRAIN CURVE

The curve of Fig. 3-1.1 was obtained by straining a test specimen and reading the required force from the testing machine. The strain had to be calculated from the extension over a measured *gage length*, since the strain is $\Delta L/L$. The stress had to be converted from the force by considering the cross-sectional area.

As long as atoms are not separated from their immediate neighbors, both the resisting force and the strain drop back to zero if the testing machine is unloaded. However, if a metal is stretched too far, it becomes permanently deformed. This is shown with the test data of Fig. 3-2.1. The metal underwent deformation in excess of that arising from Young's modulus. This *plastic deformation* is not recoverable when the load is removed. The unloading curve (dotted line) shows a permanent set. This is the type of deformation that occurs when a wire is bent too far. In each case, this plastic deformation occurs only if some of the atoms are permanently displaced from several or all of their former neighbors.

Yield strength

We use the term *yield strength*, S_y, to indicate the stress level that initiates the yield, or plastic deformation. The yield strength is an important prop-

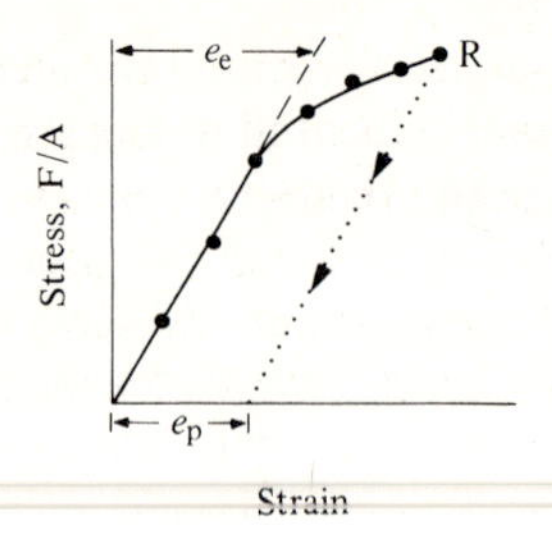

Figure 3-2.1 Plastic deformation. Initial deformation in most metals is by elastic strain, which is proportional to stress. At higher stresses, beyond the *proportional limit*, there is additional strain. On unloading, the elastic strain, e_e, is recovered according to Hooke's law (dotted curve). The remainder of the strain is permanent, e_p.

erty of a material, because design engineers commonly must design their products to preclude permanent deformation.

If a test specimen is stressed beyond the yield strength, we observe that the strain exceeds that predicted solely by Young's modulus (dashed line of Fig. 3-2.1). However, equally important is the fact that the metal gains strength as it is plastically deformed. Specifically with plastic deformation, the stress is *higher* than the yield strength. This characteristic of many metals is particularly important to the engineer. First, it means that a part may be designed to "fail-safe." As an example, a rung on a ladder can bend without actually breaking if the yield strength is exceeded. This may prevent a bad fall. Secondly, since the metal gets stronger as it deforms, it provides a means of strain-hardening metals to higher performance levels. This will be discussed in the next unit.

Some metals, notably steels, have a sharp *yield point*, YP, where plastic deformation starts (Fig. 3-2.2b). The yield strength for these metals is easily

Figure 3-2.2 Stress-strain curves. (a) Curve showing yield strength, S_y, ultimate tensile strength, S_u, and breaking strength, S_f. The stresses for each of these is calculated on the basis of the original area. (b) Yield point, YP. Plastic deformation is initiated with an obvious strain, called yield. This is typical of plain-carbon steels. (c) Offset yield strength. When the start of plastic deformation is gradual, an arbitrary offset (for example, 0.2%) is used to define the yield strength.

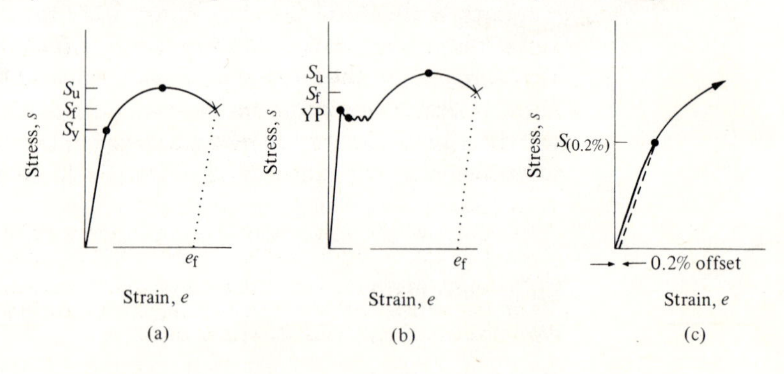

identified. The *proportional limit*, that is, the limit on the stress-strain curve for *only* elastic strain, is less readily located in the stress-strain, *s-e*, curve of Fig. 3–2.2(a), because the plastic strain starts very gradually and sometimes at very low levels. The curve of Fig. 3–2.2(a) is typical of copper, aluminum, nickel and most of the common nonferrous metals.*

The engineer overcomes the uncertainty in locating the yield strength of these metals in a very pragmatic way. The engineer concludes that the product can tolerate a minor amount of permanent strain, say 0.2%, and specifies a tolerable offset. The yield strength is now defined as the intersection of the offset line and the *s-e* curve (Fig. 3–2.2c). It is a reproducible stress that will be read the same by the aluminum producer and the aluminum user.†

Ultimate tensile strength

Even though the engineer generally designs on the basis of yield strength, it is commonly desirable to know the maximum load a metal will support without complete failure.

Beyond the yield strength (Fig. 3–2.2), the *s-e* curve rises to a maximum and then drops off. It ends when the specimen breaks. This curve was plotted by dividing the load by the *original* area, A_0. The *maximum* stress calculated in this manner is called the *ultimate tensile strength*, *UTS*, or more simply, S_u. The *breaking strength* is the strength for fracture, S_f. Note that the *original* area is used in the calculations for this curve.

It is the ultimate strength rather than the breaking strength that is critical in most designs that permit permanent deformation. A load that will permit the stress to rise to the maximum will spontaneously continue to deform the metal until fracture. Only by reducing the load after the maximum is exceeded can the fracture be avoided.

True stress-strain

Since the ductile metal is plastically deformed (Fig. 3–2.3a) the breaking strength would be higher if we used the *true* area in our stress calculations for a *true stress*, σ, rather than the *nominal* (original) area. Likewise, the *true strain*, ϵ, at the point of fracture is greater than the nominal strain. This leads to the *true stress-strain curve* of Fig. 3–2.3. (See I.P. 3–2.2.)

The design engineer almost always uses the nominal *s-e* curve, rather than the true σ-ϵ curve. This is partially because design calculations for

* As an extreme, the first irreversible strain occurs in zinc so early in the loading process that it is difficult to determine Young's modulus; but that is an exception.

† The design engineer has the option of specifying some other offset value, for example, 0.1% if the design warrants. However, excessively restrictive limits are seldom satisfactory, because they tend to add rejections, as well as costs.

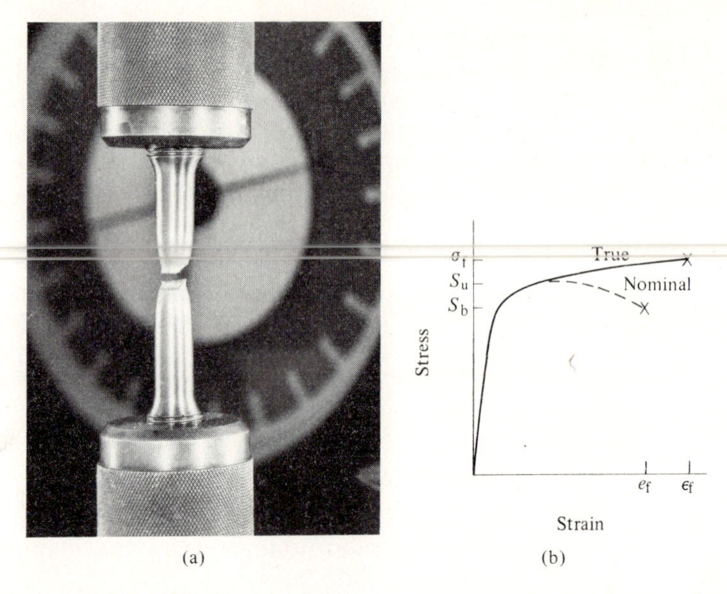

Figure 3-2.3 Necking and fracture. (a) A ductile material decreases in cross-sectional area before it fractures. (Courtesy of U.S. Steel Corp.) (b) True stress-strain curve (solid curve). The true stress, σ, is based on the actual area, rather than the nominal (original) area. Therefore the true fracture stress, σ_f, exceeds the nominal breaking strength, S_b. Also, since the strain is localized, the true strain at the point of fracture, ϵ_f, exceeds the nominal strain for fracture, e_f. (See I.P. 3-2.3.)

engineering products are based on the original dimensions. More importantly, it would be impractical to reduce the load as the material plastically deforms during the last moments before complete failure.*

Although products are never designed on the basis of the true fracture strength, σ_f, (Fig. 3-2.3), the knowledge of this stress is useful in the design of materials. We shall observe in the next unit that it is possible to make materials in which the ultimate strength, S_u, is raised toward that limit.

Ductility

This is a measure of the total *plastic* strain that accompanies fracture. This strain may be expressed as *elongation*,

$$\text{El.} = (L_f - L_o)/L_o. \tag{3-2.1}$$

The subscripts are for original and fractured lengths; elongation is commonly expressed as percent.

---

* In a facetious but realistic vein, the design engineer will not volunteer to crawl out on the bridge to measure the diameter of the tie bar just before it breaks and the bridge falls into the river so that the *true* stress and strain are known. Sometimes it does not pay to know the truth!

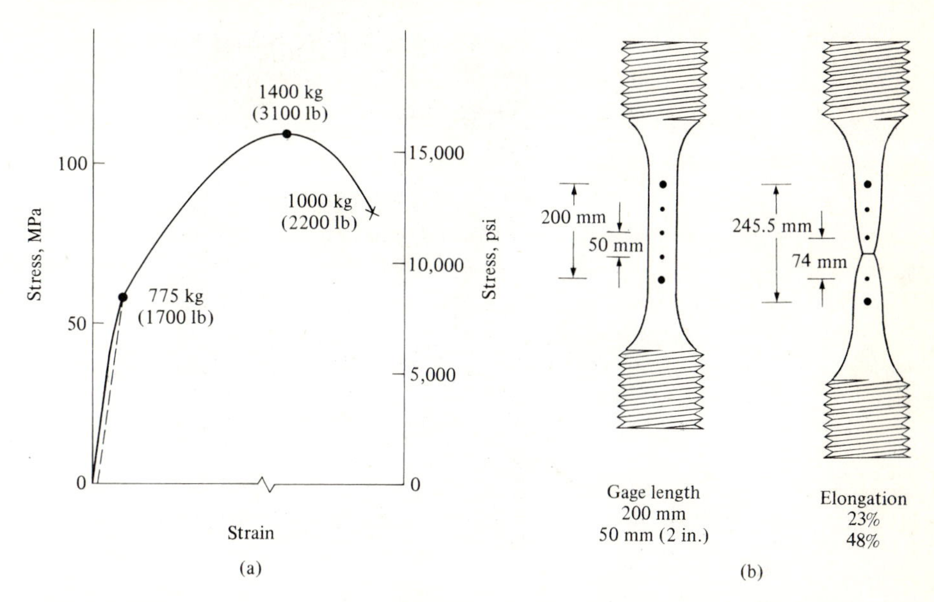

Figure 3–2.4 (a) Loads are from I.P. 3–2.1. (Dia. = 12.8 mm, or 0.505 in.) (b) Elongation versus gage length. (Data in (b) are from I.P. 3–2.2.)

From Fig. 3–2.4, observe that since the ductility is commonly localized in the necked-down region, the amount of elongation is a function of the *gage length*. Whenever reporting elongation, one must be specific about the gage length.

Ductility may also be measured by the amount of *reduction in area* (R of A) at the point of fracture.

$$\text{R of A} = (A_o - A_f)/A_o. \tag{3–2.2}$$

The subscripts follow those of Eq. (3–2.1). The calculation of the % R of A is somewhat less accurate than that of % elongation; however, it does not require a gage length because the test specimen had a specific initial diameter. Also, the R of A can be used to determine the strain at any particular point along the test specimen. Therefore, it is a useful measure of ductility.

One cannot establish an exact correlation between elongation and reduction in area, since ductility may be highly localized. Elongation is a measure of plastic "stretching," whereas reduction of area is a measure of plastic "contraction." Of course, a highly ductile material has high values of each, and a nonductile material has near-zero values of each.

Illustrative problem 3–2.1 A test bar of aluminum has a diameter of 12.8 mm (0.505 in.). Yielding causes a 0.2% offset when the load is 7600 N (775 kg, or 1700 lb_f); the maximum load during testing is 1400 kg (3100 lb_f); and the load is 1000 kg (2200 lb_f) when the bar breaks. (a) What is the yield strength? (b) The ultimate strength? (c) The breaking strength?

SOLUTION Initial cross-sectional area $= \pi(0.0128\ \text{m})^2/4$
$= 130 \times 10^{-6}\ \text{m}^2$ (or 0.200 in.^{2}).

$S_y = 7600\ \text{N}/(130 \times 10^{-6}\ \text{m}^2) = 59\ \text{MPa}$ (or 8500 psi).
$S_u = 3100\ \text{lb}_f/(0.200\ \text{in}^2) = 15{,}500\ \text{psi}$ (or 110 MPa).
$S_b = (1000\ \text{kg})(9.8\ \text{m/s}^2)/(130 \times 10^{-6}\ \text{m}^2) = 76\ \text{MPa}$ (or 11,000 psi).

See Fig. 3–2.4.

ADDED INFORMATION After necking starts, we do not know the true stress, that is, the force per actual area, because we do not know (from this problem statement) what the true area was. ◂

Illustrative problem 3–2.2 Before testing the aluminum test bar of the previous problem, an engineering technician placed five gage marks at 50-mm (2-in.) centers along the bar. (Cf. Fig. 3–2.4b.) After the bar was broken, the four spacings are 54 mm (2.13 in.), 74 mm (2.91 in.), 64 mm (2.52 in.), and 53.5 mm (2.11 in.), respectively, or 245.5 mm for all four. (Fracture occurred in the second spacing.) Also the diameter of the bar at the fracture was 4.3 mm (0.169 in.), as compared to the initial diameter of 0.505 in. (12.8 mm).
a) What is the 50-mm (2-in.) elongation? 200-mm (8-in.) elongation?
b) What is the reduction of area? The true stress, σ_f, at fracture?

SOLUTION

a) $\text{El.}_{50\text{-mm}} = 24\ \text{mm}/50\ \text{mm} = 48\%.$
$\text{El.}_{200\text{-mm}} = (245.5\ \text{mm} - 200\ \text{mm})/200\ \text{mm} = 23\%.$

b) $$\text{R of A} = \frac{[\pi(12.8\ \text{mm})^2 - \pi(4.3\ \text{mm})^2]/4}{\pi(12.8\ \text{mm})^2/4} = 89\%.$$

$$\text{True breaking stress} = \frac{2200\ \text{lb}_f}{(\pi/4)(0.169\ \text{in.})^2} = 98{,}000\ \text{psi;}$$

$$\text{or, in SI, } \sigma_f = \frac{(1000)(9.8)\text{N}}{(\pi/4)(0.0043\ \text{m})^2} = 675\ \text{MPa}.$$

ADDED INFORMATION The figure for 50-mm (2-inch) elongation is always greater than the figure for 200-mm (8-inch) elongation. The actual stress in the neck is always higher than the nominal stress calculated as the breaking strength in I.P. 3–2.1. ◂

• **Illustrative problem 3–2.3** (a) Derive a relationship for the *true strain*, ϵ, of a material. (b) The 12.8 mm-diameter test bar of I.P. 3–2.2 had a fracture diameter of 4.3 mm (0.168 in.). What was the true strain at fracture, ϵ_f?

SOLUTION (a) If we define true strain ϵ as

$$\epsilon \equiv \int_{l_o}^{l} dl/l = \ln(l/l_o),$$

and assume constant volume, $Al = A_o l_o$; then

$$\epsilon = \ln (A_o/A).$$

b) $\epsilon_f = \ln (A_o/A_f) = \ln (d_o^2/d_f^2)$
$= \ln (12.8 \text{ mm}/4.3 \text{ mm})^2 = 2.18$ (or 218%).

ADDED INFORMATION The value of ϵ_f exceeds either measure of ductility in I.P. 3–2.2, because ϵ_f is based solely on the diameter undergoing greatest deformation. ◄

3–3 HARDNESS

The hardness of a material is important to the engineer in several obvious ways; in addition, it is easily measured and may often be related to the mechanical strength of a material.

Various indexes of hardness may be reported. Most commonly, an engineer uses either a *Brinell hardness number*, BHN, or a *Rockwell hardness number*, R. Both of these are determined by forcing a hard indenter into the surface of the material to be measured (Fig. 3–3.1). The Brinell indenter is a steel ball 10 mm (0.39 in.) in diameter, usually pressed under a load of 3000 kg. (A 500-kg load is used for soft materials.) The diameter of the indentation is related to the hardness;* a hard material naturally exhibits a smaller indentation than a soft one. With the Rockwell system, the indenter is much smaller, and the index of hardness is related to the depth of penetration. Several scales are available, depending on the load applied and the exact shape of the indenter. The most common scale for reporting the hardnesses of steels is the Rockwell-C scale, R_C. It will be widely used in this book. Table 3–3.1 gives approximate relationships between various hardness scales.

Figure 3–3.2 shows several correlations between hardness and strength. It is apparent that the same relationship does not hold for all materials; however, for a given type of ductile material, the strength can be estimated from the hardness data within 5% to 10%. Since the hardness test may be conducted without machining a test bar, the hardness test is widely used as an indicator of strength. In fact, the strength of certain ductile products such

---

* The BHN is the load F in kg divided by the *surface area* of the indentation, where d is the indentation diameter (mm) and D is the ball diameter:

$$\text{BHN} = \frac{F}{(\pi D/2)(D - \sqrt{D^2 - d^2})}. \tag{3-3.1}$$

The BHN is an index rather than a stress. (The load is not in newtons.)

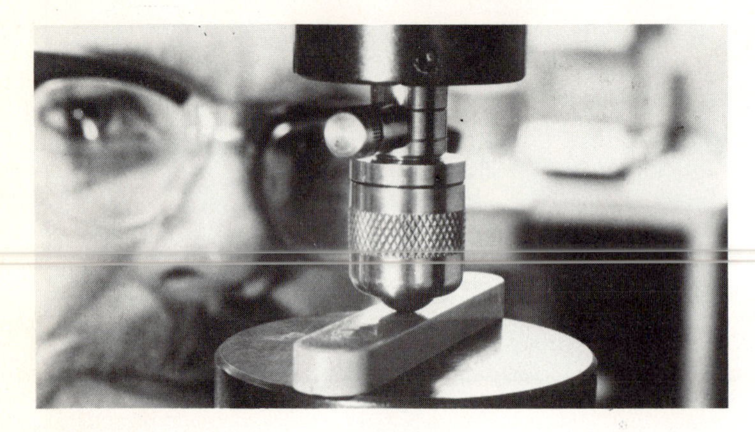

Figure 3-3.1 Hardness testing. The hardness is determined by the size of the indentation. (Courtesy of Measurement Systems Division, ACCO Industries.)

Table 3-3.1
Hardness conversion scales†

BHN	Rockwell hardness R_C^1	R_B^2	R_F^3	KHN[4]
	80			
	70			972
614	60			732
484	50			542
372	40			402
283	30			311
230	20	98		251
185	10	90		201
150		80		164
125		70	97	139
107		60	91	120
100		55	88	112
83‡		50	85	107
75‡		41	80	98
63‡		23	70	82
59‡		14	65	75
50‡			55	65

† These conversions vary from material to material. For more accurate comparisons for specific materials, consult the *Metals Handbook* (published by ASM).

‡ Load = 500 kg; higher BHN values use a 3000-kg load.

[1] Indenter: diamond "brale"; load, 150 kg.

[2] Indenter: ball, $\frac{1}{16}$ in. (1.6-mm) diameter; load, 100 kg.

[3] Indenter: ball, $\frac{1}{16}$ in. (1.6-mm) diameter; load, 60 kg.

[4] Knoop hardness number. This hardness measurement is made by indenting small areas of the material under a microscope.

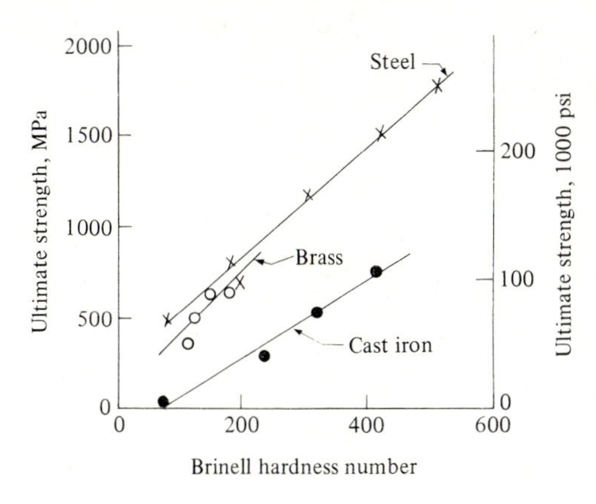

Figure 3-3.2 Ultimate strength versus hardness (BHN). A rough correlation exists between the tensile strength of *metals* and their hardness. This correlation is less specific for nonmetallic materials.

as steel beams can be estimated without preparing a sample, as would be necessary for tensile testing.

Brittle materials, however, provide erratic data with respect to hardness and strength because cracks may form that serve as stress-raisers for propagation of fractures. Therefore, in brittle materials, simple correlations do not occur between hardness and strength.

Illustrative problem 3-3.1 Basing your solution on the data for steel in Fig. 3-3.2, establish an empirical formula for the ultimate strength, S_u, of *steel* based on its Brinell hardness.

SOLUTION The slope of the curve for steel approximates 3.5 MPa (or 500 psi) per 1 BHN:

$$S_u \text{ in MPa} \cong 3.5 \times \text{BHN}; \qquad (3\text{-}3.2a)$$
$$S_u \text{ in psi} \cong 500 \times \text{BHN}. \qquad (3\text{-}3.2b)$$

ADDED INFORMATION This formula does not necessarily hold for other materials. (See Fig. 3-3.2.) ◀

3-4 ELASTIC BEHAVIOR OF METALS

The modulus of elasticity that was presented in the first section of this chapter is *Young's modulus, E*. We only considered that a stress along the axis of a test specimen will strain the material in proportion to the amount of stress (Eq. 3-1.3). In this section we shall examine Young's modulus more closely for the effect of several variables. In addition, attention will be given to the *shear modulus, G*. The reader is referred to other texts for information on the third elastic modulus, the *bulk modulus, K*, which arises from hydrostatic compression.*

* The bulk modulus, the reciprocal of *compressibility*, β, is the ratio of hydrostatic pressure to $\Delta V/V$. [See Van Vlack, *Materials Science for Engineers*, or *Elements of Materials Science and Engineering*, 4th ed., Reading, Mass.: Addison-Wesley.]

Table 3-4.1
Young's moduli vs. melting temperatures of cubic metals

Metal	E, MPa	T_m, °C (°F)
W	345,000	3410 (6170)
Fe	205,000	1538 (2800)
Cu	110,000	1084 (1984)
Al	70,000	660 (1220)
Pb	14,000	327 (621)
Na	<7,000	98 (208)

Table 3-4.1 lists the elastic moduli for several common metals in relation to their melting temperatures. An examination quickly reveals that there is a close relationship between the two. This should be expected, since a more strongly bonded material requires not only a higher stress to produce strain but also a higher temperature to produce the thermal agitation that destroys the crystal lattice, that is, to cause melting. [The correlation between Young's modulus E and melting temperature T_m, shown in Table 3-4.1, is much better for metals than for other materials, in which a variety of bonds may be present. Therefore one should hesitate to predict the modulus elasticity for polymers and ceramics on the basis of their melting points alone.]

Compressive moduli

Values of compressive Young's moduli, where both the stress and the strain have negative signs, are essentially identical to values obtained in tension *for metals*. That is, the stress to pull atoms apart is nearly identical to the stress required to squeeze them together for the same strain. We would have to impose major stresses, $s > E/100$, and major strains, $e > 0.01$, in order to find a difference. In metals, this would first bring about plastic deformation and/or fracture in almost every case. Thus in calculations for metals, we do not need to concern ourselves with different modulus values in compression and tension.

Effect of temperature

As long as the structure does not change, the elastic modulus decreases at higher temperatures (Fig. 3-4.1). The discontinuity in the curve for iron is due to the change of iron from bcc to fcc at 912°C (1673°F). Since the atoms of fcc iron have twelve neighbors rather than eight, as in bcc iron, the fcc

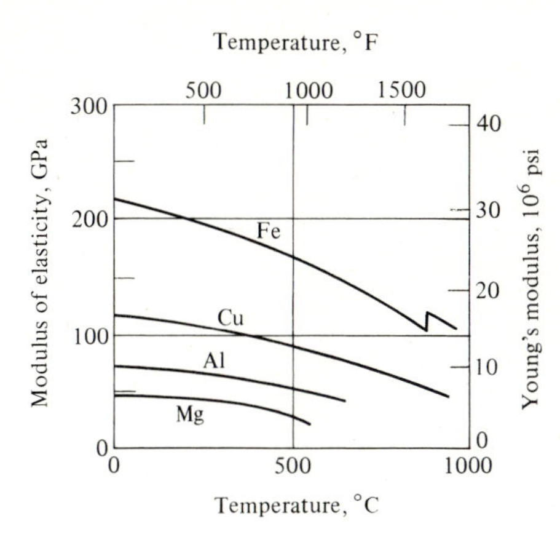

Figure 3-4.1 Elastic modulus (Young's) versus temperature. In general, the elastic modulus of metals decreases at elevated temperatures. Exceptions occur only when there is a change in structure. (A. G. Guy, *Elements of Physical Metallurgy*, Reading, Mass.: Addison-Wesley)

structure is more rigid (has a higher modulus) and responds less to applied stresses. The normal decrease in moduli with increased temperatures may be explained in part by the thermal expansion, which allows more strain for a given stress (and thus a lower modulus).

Poisson's ratio

A stretching produces an accompanying elastic contraction in the two lateral directions (and a compression in one direction produces an accompanying elastic expansion in the other two directions, Fig. 3–4.2). If an elastic strain involved absolutely no change in volume, the two lateral strains would be equal to one half the tensile strain (but of the opposite sign). However, when we stretch a material in one direction, the cross-sectional area is not reduced quite as much proportionately. Therefore the ratio of strains, $-e_x/e_z$, is always less than 0.5. This ratio is called *Poisson's ratio*, ν, and e_z is the strain in the direction of applied load. The two lateral strains, e_x and e_y, are generally equal if the stress is applied in only one direction,

$$\nu = -e_x/e_z. \tag{3–4.1}$$

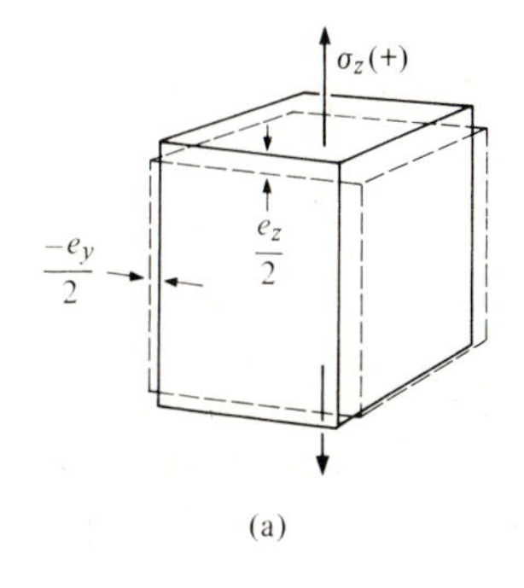

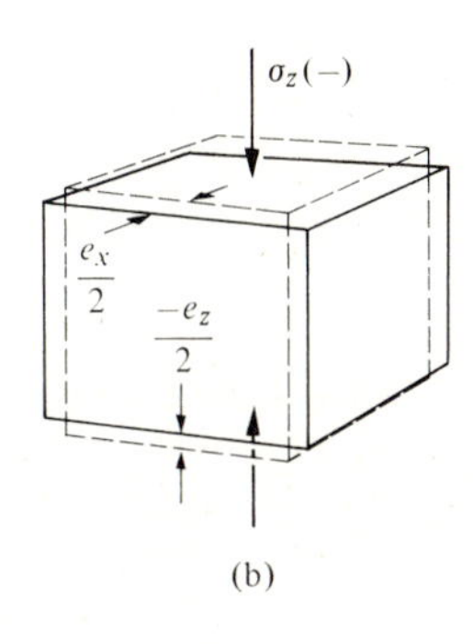

Figure 3-4.2 Poisson's ratio. This is the ratio between lateral strain and axial strain (Eq. 3-4.1). With no volume change, $\nu = 0.5$. Normally the volume changes slightly, so $\nu < 0.5$.

Table 3-4.2
Poisson's ratios of cubic metals

Metal	Poisson's ratio
Tungsten	0.28
Iron and Steel	0.29
Copper	0.35
Brass	0.37
Aluminum	0.34
Lead	0.45

Under tension the signs are

$$\nu = -(-)/(+);$$

and with compression, where e_z is negative,

$$\nu = -(+)/(-).$$

Therefore Poisson's ratio is always positive. Typical values are shown in Table 3-4.2.

Shear modulus

To date we have considered only stresses that are axial, i.e., the forces are perpendicular to the area on which they act. In many engineering applications there are shear forces in which the forces have components that are parallel to the area of interest. As shown in Fig. 3-4.3(b), a *shear stress*, τ, produces a strain angle, α. The *shear strain*, γ, is defined in terms of that angle,

$$\gamma = \tan \alpha, \tag{3-4.2}$$

and the amount of elastic strain is initially proportional to the amount of elastic shear stress,

$$G = \tau/\gamma. \tag{3-4.3}$$

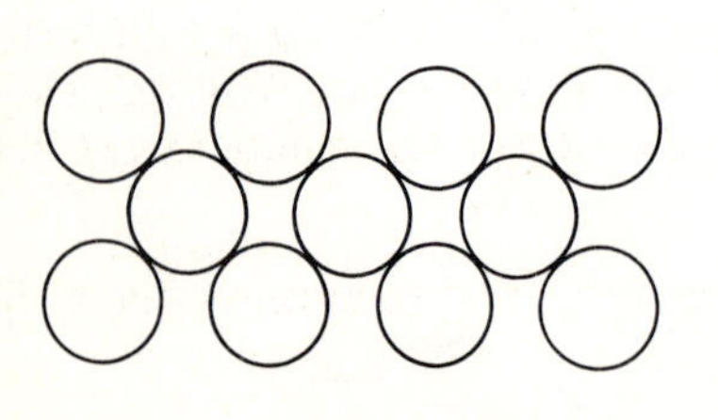

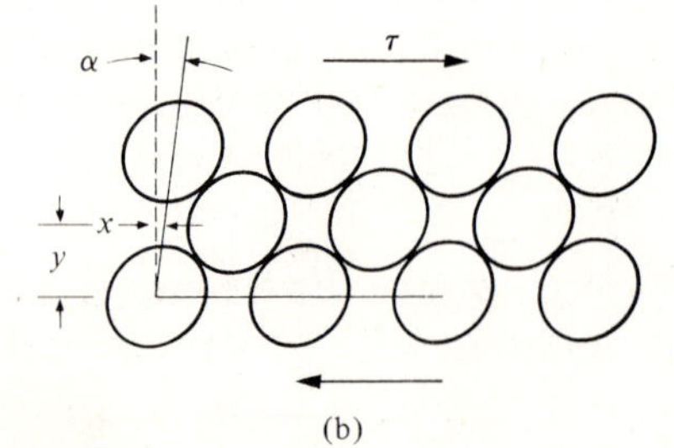

Figure 3-4.3 Elastic shear strain. (a) No strain. (b) Shear strain. The strain is elastic as long as the atoms maintain their original neighbors. (See Eqs. 3-4.2 and 3-4.3.)

This shear modulus, G, is different from Young's modulus, E; however, the two are related as follows:

$$E = 2G(1 + \nu). \qquad (3\text{–}4.4)$$

Since Poisson's ratio, ν, is normally between 0.25 and 0.5 (Table 3–4.2), the value of G is between 0.4 and 0.33 of E.

Illustrative problem 3–4.1 A precisely machined steel rod has a specified diameter of 18.6 mm (0.732 in.). It is to be elastically loaded longitudinally with a force of 670,000 N (150,000 lb_f). What percent will its diameter d change? Its area A?

SOLUTION

$$A_o = (\pi/4)(0.0186\text{ m})^2 = 272 \times 10^{-6}\text{ m}^2 \qquad (\text{or } 272\text{ mm}^2).$$

$$e_z = s/E = (F/A)/E$$

$$= \frac{(670{,}000\text{ N})/(272 \times 10^{-6}\text{ m}^2)}{205{,}000 \times 10^6\text{ Pa}} = 0.012 \qquad (\text{or } 1.2\%).$$

$$e_x = -(0.29)e_z = -0.0035 \qquad (\text{or } -0.35\%).$$

$$A_e = (\pi/4)(0.0186\text{ m} \times 0.9965)^2 = 270 \times 10^{-6}\text{ m}^2 \qquad (\text{or } 270\text{ mm}^2).$$

$$\Delta A = (270\text{ mm}^2 - 272\text{ mm}^2)/(272\text{ mm}^2) = -0.007 \qquad (\text{or } -0.7\%).$$

Alternatively,

$$1 + \Delta A = (1 - 0.0035)^2$$

$$\Delta A = -0.007 \qquad (\text{or } -0.7\%).$$

ADDED INFORMATION This change in area is smaller than the indicated accuracy of the load; therefore a correction does not need to be made to determine the actual stress. However, if there were a reason for specifying a $\pm 0.5\%$ dimension, the engineer should point out that the area contraction exceeds that limit.

The 1.2% increase in length exceeds the 0.7% decrease in area; therefore, there is a slight increase in volume. ◀

Illustrative problem 3–4.2 The elastic modulus of copper drops from 110,000 MPa (16×10^6 psi) at 20°C (68°F) to 107,000 MPa (15.5×10^6 psi) at 50°C (122°F). What is the change in the total length of a 1575-mm (62-in.) bar if the stress is held constant at 165 MPa while the temperature rises those 30°C?

SOLUTION Assume the load is removed at 20°C, and then reapplied at 50°. Elastic strain at 20°C (to be recovered):

$$\Delta L/L = -165\text{ MPa}/110{,}000\text{ MPa} = -0.00150.$$

(Zero load at 20°C to zero load at 50°C) + elastic strain at 50°C:

$$\Delta L/L = (17 \times 10^{-6}/°C)(+30°C) + (165 \text{ MPa}/107{,}000 \text{ MPa})$$
$$= 0.00051 + 0.00154 = 0.00205.$$

Therefore,

$$\Delta L = (0.00205 - 0.00150 \text{ mm/mm})(1575 \text{ mm}) = 0.87 \text{ mm}.$$

ADDED INFORMATION Designs for many high-temperature applications—for example, turbine blades (Fig. 2-7.1)—must take into account both thermal expansion and the change in Young's modulus. ◄

3-5 PLASTIC DEFORMATION OF DUCTILE METALS

A crystal may be viewed as a collection of parallel planes containing layers of atoms (Fig. 3-5.1). Plastic deformation commonly occurs by the *slip* of one plane over another. The plane does not slip all at once, but with the localized motion shown schematically in Fig. 3-5.2, where the edge rows of four planes are shown responding to shear stresses. Note that in the shear process only a portion of the atoms are displaced at any one time, and that in effect extra atoms are squeezed into the upper rows. The crystal is not perfect in this region. We call this type of imperfection an *edge dislocation.*

Slip

Deformation is very nonuniform within a polycrystalline solid for several reasons: (1) When tensile (or compressive) stresses are applied, they must be resolved as shear stresses onto the slip planes (Fig. 3-5.3). The greatest shear stress develops at 45° to the direction of tension; no shear stress is

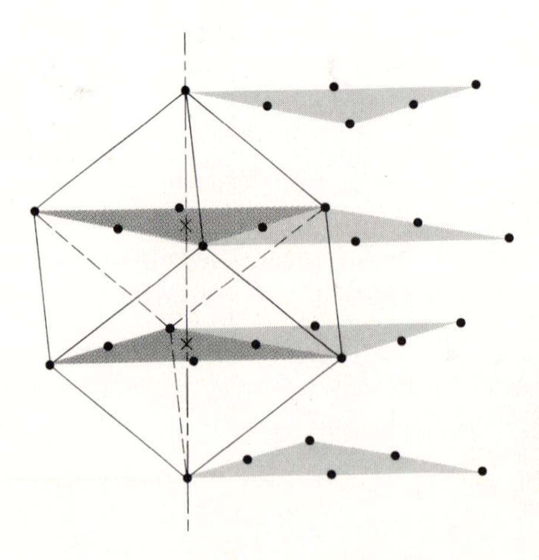

Figure 3-5.1 Crystal planes. The atoms have a regular arrangement on each plane. The planes are stacked and may slide over one another by shear stresses.

developed on planes parallel or perpendicular to the direction of tension. Thus slip cannot occur on all of the crystal planes within a metal grain. (2) Some grains are oriented more favorably for deformation than others. (3) Plastic deformation results from dislocation movements and occurs with lower applied stress on crystal planes that are farthest apart and in direc-

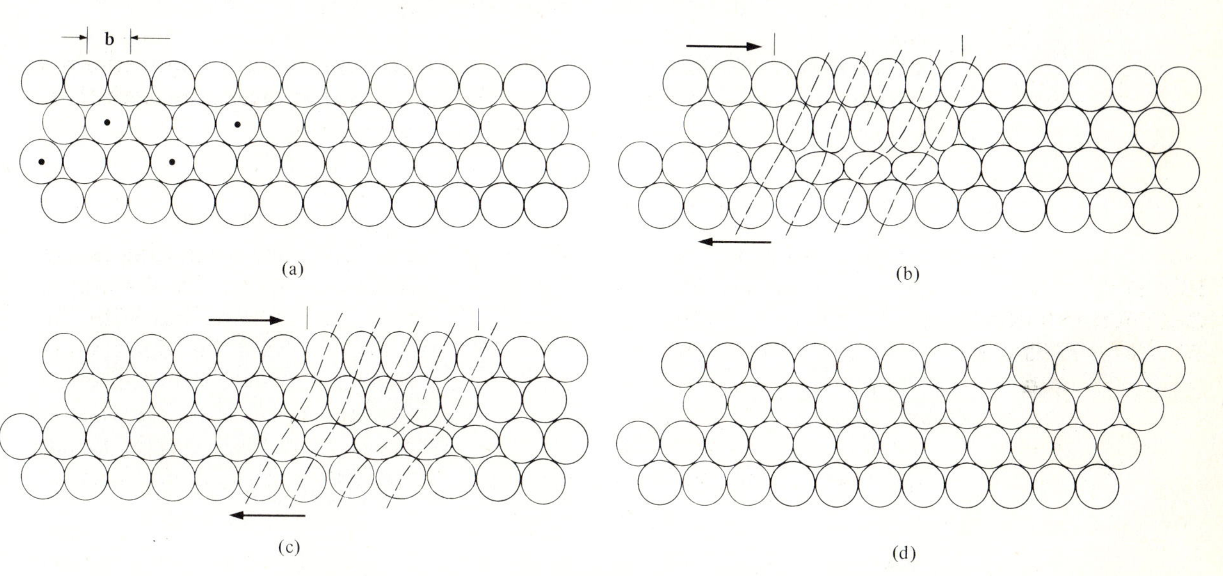

Figure 3-5.2 Plastic deformation by shear. The edges of four planes are shown. Slip of this type does not occur simultaneously along the total slip plane. Rather, the shear force produces an imperfection, called a *dislocation*, which moves along the slip plane.

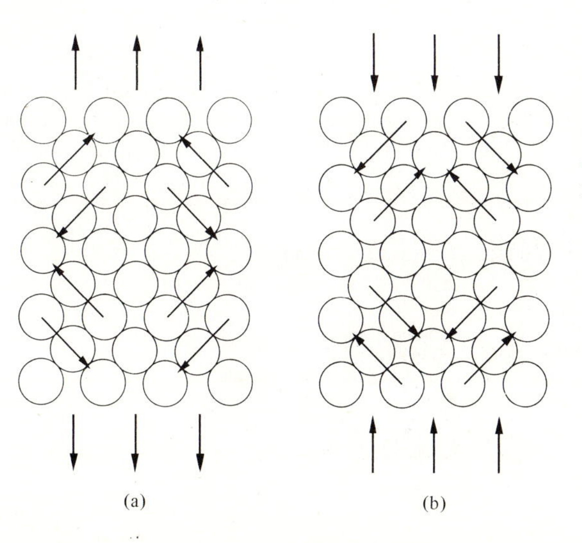

Figure 3-5.3 Resolved shear stresses. A tensile stress (a), or a compressive stress (b), also provides shear stresses that are at a maximum at 45°.

tions in which the displacement step, $\boldsymbol{b}$, is shortest. In other words, slip occurs preferentially on the most densely packed planes (most atoms per unit area), and in the most densely packed directions of those planes (most atoms per unit length).

Illustrative problem 3-5.1 Refer to Fig. 3-5.4, which shows three different sets of planes commonly referred to in an fcc crystal. Assume that the lattice is that of copper, with $a = 0.3615$ nm. What are the perpendicular distances, p, between the sets of planes labeled (010), (110), and $(\bar{1}11)$?

SOLUTION

$$p_{010} = a/2 = 0.1807 \text{ nm},$$
$$p_{110} = a\sqrt{2}/4 = 0.1278 \text{ nm},$$
$$p_{\bar{1}11} = a\sqrt{3}/3 = 0.2087 \text{ nm}.$$

ADDED INFORMATION Note that the (010) planes cut the y-axis only. The (110) planes cut the x-and y-axes at equal distances from the origin but are parallel to the z-axis. The (111) planes cut all three axes at equal distances from the origin. We call these notations for planes *Miller indices*.

In fcc metals, slip occurs almost exclusively on (111)-type planes, which have greater interplanar spacings than the (010) and (110)-type planes. There are *four* (111)-type planes with various orientations around the origin.

On a (111)-type plane, stress for slip is lowest in the direction of the *face diagonal* of the cube. In those directions, the atoms touch (cf. Fig. 2-2.2) and therefore have the shortest displacement step $\boldsymbol{b}$ for dislocations (cf. Fig. 3-5.2). Each of the four (111)-type planes has *three* of these easy directions (each with a plus and a minus sense). Consequently an fcc crystal has 4×3, or twelve, possible slip combinations. This is one reason why fcc metals are among the most ductile of materials. ◄

Figure 3-5.4 Planes (low-index) in an fcc metal. (a) (010), (b) (110), (c) $(\bar{1}11)$.

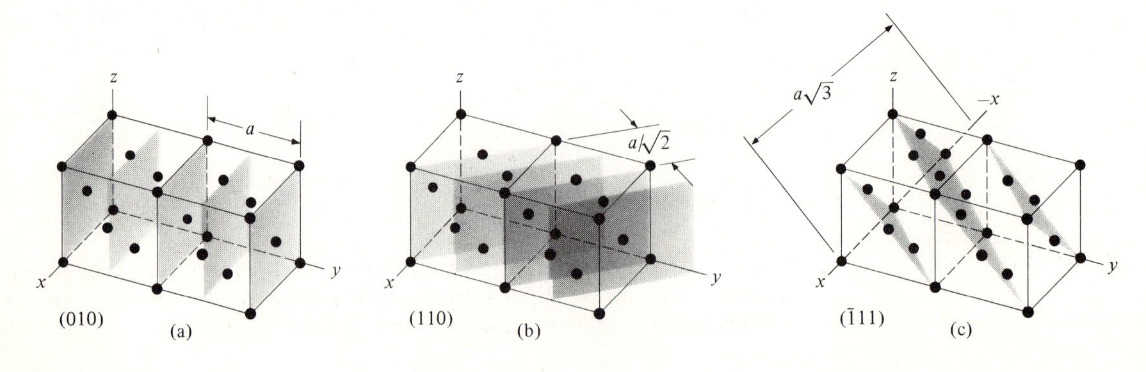

Illustrative problem 3-5.2 What is the copper atom packing, i.e., atoms/cm^2, on the (111)-type planes shown in Fig. 3-5.4?

SOLUTION NO. 1 There are 4 atoms per unit cell, which is a $(3.615 \times 10^{-8}\text{cm})^3$ cube. Therefore, one calculates 0.85×10^{23} atoms/cm^3 to match the figure we obtained in Section 2-1.

From I.P. 3-5.1, the interplanar spacing is 0.2087 nm. Thus there are 0.48×10^8 planes per centimeter.

$$\text{Atoms/cm}^2 = \frac{0.85 \times 10^{23}\ \text{atoms/cm}^3}{0.48 \times 10^8/\text{cm}} = 1.8 \times 10^{15}\ \text{atoms/cm}^2.$$

SOLUTION NO. 2 The $(\bar{1}11)$ plane cuts each unit cell as an equilateral triangle which is $a\sqrt{2}$ (or 0.51 nm) on each side. If one draws a sketch, it can become apparent that this triangle contains $(\frac{3}{6} + \frac{3}{2})$, or two atoms:

$$\begin{aligned}\text{Atoms/cm}^2 &= \frac{2\ \text{atoms}}{\frac{1}{2}[0.51\ \text{nm}][0.866 \times 0.51\ \text{nm}]} \\ &= 18\ \text{atoms/nm}^2 \qquad (\text{or } 1.8 \times 10^{15}\ \text{atoms/cm}^2).\end{aligned}$$

ADDED INFORMATION The (010) and (110) planes of copper have 1.5×10^{15} atoms/cm^2 and 1.1×10^{15} atoms/cm^2, respectively. Slip occurs most readily on those planes with the highest atom packing, as well as in directions with closest atom contact. ◀

Illustrative problem 3-5.3 A force of 100 newtons is directed along the x-axis of Fig. 3-5.4(b). What is the resolved force in the direction of the body diagonal?

SOLUTION

$$\frac{\text{Diagonal force}}{\text{Force along axis}} = \text{cos of angle between directions}; \qquad (3\text{-}5.1)$$

$$\text{Diagonal force} = (100\ \text{N})(a/a\sqrt{3}) = 58\ \text{N}.$$

ADDED INFORMATION If we were to calculate a resolved *stress*, we would have to take into account not only the resolved force but also the resolved area. (See Schmid's law in other textbooks on materials science.) ◀

Review and Study

SUMMARY

1. All solid materials deform when they are loaded. In order to make direct comparisons, we convert the load to a stress—the force per unit area; we convert deformation to strain—the deformation per unit dimension.

2. Stress and strain are proportional as long as the deformation does not relocate atoms with respect to their neighbors. This strain is elastic because it is recoverable when the stress is removed. The proportionality constant is called the modulus of elasticity. Young's modulus is (axial stress)/(axial strain); the shear modulus is (shear stress)/(shear strain); the bulk modulus, which is not discussed in this text, is (hydrostatic pressure)/(volume compression). Elastic moduli are greatest for strongly bonded materials, for example, those with a high melting temperature. The elastic moduli of all metals decrease as the temperature is increased.

3. The yield strength is the stress limit prior to plastic strain, that is, prior to permanent deformation in which some of the atoms slip to positions with new neighbors. This is the design strength required for many engineering applications, since plastic strain is not recoverable after the stress is removed.

4. Ductility is the total plastic strain that is accumulated prior to fracture. This may be measured as the unit stretching, called elongation, or as the cross-sectional contraction, called the reduction in area. Both are dimensionless and usually expressed in percent.

5. When stressed beyond the yield strength, the strength of a ductile metal increases. Elastic strain continues to increase with these higher stresses; however, it is usually masked by the greater plastic strain. Reference is therefore (incorrectly) made to the elastic limit. In reality, it is the limit of elastic strain alone. This proportional limit is very obvious in many steels in which there is a yield point that defines the yield strength. For other metals, it is necessary to define the yield strength by an arbitrary amount of plastic offset, usually 0.2 percent.

6. The stress that is calculated on the basis of load per original area rises to a maximum, called the ultimate strength. This limit is important to the engineer, since additional strain leads to necking and breaking. The breaking strength has little meaning to the engineer because it is calculated on the basis of the original area. The true stress (and the true strain) is calculated on the basis of the true area. The true stress is not used in design, but is significant (a) for understanding the potentialities of a metal, and (b) for controlling metal-forming processes.

7. Hardness is an index of the resistance that a material has to penetration and to scratching. It is related to the ultimate strength, since both involve plastic deformation. Although less accurate than a tension test, it is commonly simpler to determine; therefore, it is a useful index of strength for ductile materials.

8. Plastic deformation occurs in metals as slip along crystal planes. The total plane does not move simultaneously; rather slip is by dislocation movements that displace only part of the atoms from their equilibrium distance at any one time. Slip occurs most readily along planes with the greatest number of atoms (planes are farthest apart), and in directions with the greatest atom density (shortest displacement).

TECHNICAL TERMS

Dislocation A defect that arises because part of the crystal is displaced (dislocated) with respect to the balance of the crystal. Plastic deformation occurs by dislocation movements.

Ductility Total permanent strain prior to fracture—measured as elongation or as reduction of area.

Elastic deformation Strain that is recoverable when the load is removed.

Elongation (El.) Total plastic strain before fracture, measured as the percent of axial strain. (A gage length must be stated.)

Gage length Initial dimensions for determining strain. (See Fig. 3-2.4)

Hardness Resistance to penetration (or to scratching if the indentor moves). Three common procedures are used (Table 3-3.1).

Modulus of elasticity (E) The ratio of stress to elastic strain. Young's modulus = (axial stress)/(axial strain); and shear modulus = (shear stress)/(shear strain).

Plastic deformation Permanent strain. It is not recoverable, since some atoms have moved to new neighbors.

Poisson's ratio (ν) Ratio (negative) of lateral strain to axial strain.

Proportional limit The limit of the proportional range of the stress-strain curve. (Elastic strain increases beyond this "limit" but is masked by plastic strain.)

Reduction of area (R of A) Total plastic strain before fracture, measured as the percent decrease in cross-sectional area.

Slip Plastic deformation by shear along a crystal plane. (Slip occurs by dislocation movements.)

Strain (e) Unit deformation. Elastic strain is recoverable; plastic strain is permanent. Axial strain is $\Delta L/L$; shear strain is tan α, the displacment angle.

Strength (S) Stress to produce failure. Yield strength, S_y, is the stress to initiate the first plastic deformation; ultimate tensile strength, S_u, is the maximum stress on the basis of the original area.

Stress (s) Force per unit area. Nominal stress is based on the original design area; true stress is based on the actual area.

Yield point (YP) The point at the proportional limit where there is sudden plastic deformation (observed principally in low-carbon steels.)

CHECKS

3A Stress is __***__ per unit __***__, while strain is __***__ per unit __***__, and a pascal is a __***__ per m^2.

3B Stress is the product of __***__ and Young's modulus.

3C The elastic modulus of a rubber band is __***__er than that of a copper wire of the same dimensions, because it takes smaller __***__ to produce 1% strain.

3D The SI units of stress are __***__; of force, __***__; of strain, __***__; of Young's modulus, __***__.

3E The ___***___ strength is the stress required to initiate plastic deformation. Above this stress, the strain is both ___***___ and plastic.

3F The ___***___ strength is the stress calculated as maximum load per unit of ___***___ area.

3G In a ductile material the breaking strength, S_b, is less than the ultimate strength, S_u, because ___***___ . . . .

3H A specified offset strain is commonly used to determine the ___***___ strength when plastic strain appears gradually over a range of stresses.

3I Two measures of ductility are ___***___ and ___***___. These are strains that precede ___***___.

3J True stress in tension is ___* er___ than the nominal stress; true strain in tension is ___* er___ than the nominal strain.

3K Hardness is commonly expressed as either a ___***___ hardness number or a ___***___ hardness. The first is based on the ___***___ of the hardness indentation; the second upon the ___***___ of penetration.

3L Since the structures of nickel and lead are both fcc, we may relate their elastic moduli to their melting temperatures (1455°C and 327°C, respectively). On that basis, ___***___ will deform more for a given stress in its elastic range.

3M Since Poisson's ratio is less than 0.5, the volume of a bronze block ___* creases___ when loaded in axial compression.

3N The shear modulus, G, is ___***___% of Young's modulus, E, for a material that shows negligible volume change on compression.

3O Refer to I.P. 3–5.1 and Fig. 3–5.4. An fcc crystal of metal with a = 0.40 nm has ___***___ atoms per a^2 on the (010) plane, and ___***___ atoms per a^2 on the (110) plane.

3P A ___***___ is a linear defect in a crystal that accompanies the displacement of part of the crystal with respect to the remainder of the crystal.

3Q The permanent displacement of one plane of atoms over another is called ___***___.

3R Slip occurs most readily on the ___***___ densely packed planes, and in directions with the ___***___ repeat distances.

STUDY PROBLEMS

3–1.1. A copper bar has only elastic deformation if it is stressed less than 95 MPa (<13,800 psi). (a) What square cross-sectional area is the minimum possible to support a load of 1340 kg (2950 lb) without exceeding the above stress? (b) What is the strain with this load?

Answer: (a) 11.8 mm (or 0.46 in.) (b) 0.00086

3–1.2. Assume the bar in the previous problem is 6.1 m (20 ft) long and made of aluminum. How much would the length change when the load is applied?

* 3–1.3. A ship lowers a 1-mm (0.04 in.) aluminum wire over its side. How much wire (length) must hang down before the stress of 95 MPa (13,800 psi) is developed at the upper end?

Answer: 5700 m (or 18,700 ft)

* 3–1.4. A copper wire is stressed to 1.5 MPa (220 psi) when it is stretched between two buildings. The air temperature drops 10°C. What is the new stress?

3–2.1. The loads for initial plastic deformation of a 1.0-mm (0.039-in.) copper wire and a 0.69-mm (0.027-in.) steel wire are 18 kg (40 lb) and 14 kg (31 lb), respectively. Which wire has the lower yield strength?

Answer: Copper, 225 MPa (or 33,000 psi)

3–2.2. The maximum possible loads carried by the two wires (copper and iron) in the previous problem are 30 kg (66 lb) and 25 kg (55 lb), respectively. Will the ultimate strength of the copper be greater or less than the yield strength of the steel?

3–2.3. The diameter at the fracture point of the 1.0-mm copper wire in the previous two problems is 0.66 mm (0.026 in.). The load at fracture was 22 kg (49 lb). (a) What was the breaking strength, S_f? The true stress at the point of fracture? (b) What was the reduction in area?

Answer: (a) 275 MPa and 630 MPa (or 41,000 psi and 92,000 psi)
(b) 56% Reduction of Area

3–2.4. Change the data of Fig. 3–2.4(b) from 245.5 mm and 74 mm to 218 mm and 63 mm, respectively. The diameter at the break is 9.9 mm. Calculate *three* ductility values.

3–2.5. The offset (0.2%) yield strength of an aluminum alloy rod is 104 MPa (15,100 psi). What is the total strain when the 0.2% permanent strain is attained?

Answer: 0.0035

3–2.6. The yield strength of a copper alloy is 222 MPa (32,200 psi). When it is stressed to 238 MPa (34,500 psi), the total strain is 1.2%. What fraction of that strain is elastic?

3–3.1. The BHN of a steel is 305. Estimate its ultimate strength (a) in psi. (b) in MPa.

Answer: (a) ~150,000 psi

3–4.1. An aluminum rod is stressed 210 MPa (30,000 psi) in tension. Its initial diameter was 24.02 mm (0.946 in.). What is the change in diameter as a result of the load?

Answer: −0.025 mm (−0.001 in.)

3–4.2. What is the change in cross-sectional area if a hardened steel wire is stressed to its yield strength of 1400 MPa (203,000 psi)? (Use appendix data for E.)

---

* See footnote with S.P. 2–1.3.

3-4.3. Provide a "ball-park" figure for Young's elastic modulus (a) of chromium; (b) of silver.

Answer: (a) Estimate 230,000 MPa (vs. 250,000 MPa by experiment)

3-5.1. The perpendicular distance (Fig. 3-5.4) between the planes labeled $(\bar{1}11)$ in fcc aluminum is 0.2338 nm. What is the lattice constant?

Answer: 0.4049 nm

3-5.2. Nickel is fcc (a = 0.35 nm) and contains 0.9×10^{23} atoms/cm^3. Use two separate methods to calculate the number of atoms per nm^2 on the (010) plane.

3-5.3. A force of 24 newtons is required in the direction of the cube edge of a crystal. What force must be applied in a face-diagonal direction to realize the 24 newtons?

Answer: 34 N

3-5.4. The lattice constant for bcc iron is 0.287 nm. How many atoms are there per nm^2 (a) on the (110) plane? (b) on the (010) plane?

3-5.5. Iron is bcc (a = 0.287 nm). How far apart are (a) the closest (010) planes? (b) The closest (110) planes? [*Hint:* Make a sketch for bcc comparable to the fcc in Fig. 3-5.4.]

Answer: (a) 0.144 nm (b) 0.203 nm

CHECK OUTS

3A force, area; deformation, length; newton

3B strain

3C lower; stresses

3D N/m^2 (pascals); newtons; dimensionless; N/m^2 (pascals)

3E yield; elastic

3F ultimate; original

3G the area decreases before breaking (necking occurs)

3H yield

3I elongation, reduction of area; fracture

3J greater; greater

3K Brinell; Rockwell; diameter; depth

3L lead

3M decreases

3N 33

3O 2; 1.4

3P dislocation

3Q slip

3R most; shortest

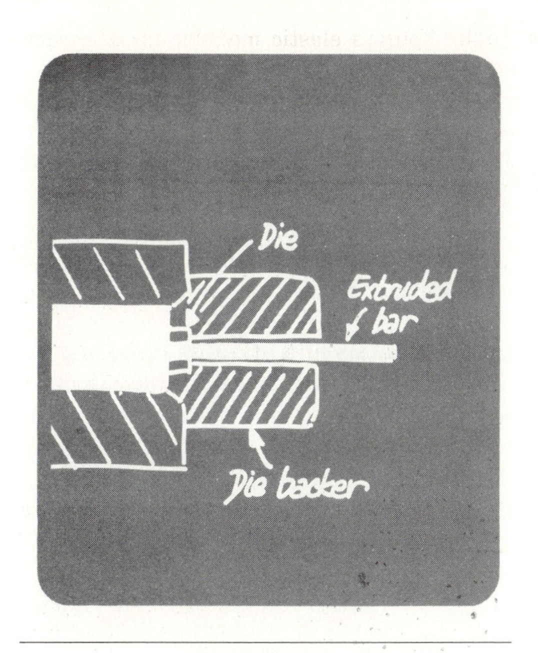

UNIT FOUR

MAKING AND SHAPING OF METALS

Metals must be extracted from ores, purified, and then shaped into the desired product. Refining is primarily a chemical process. Forming is principally a mechanical process using plastic deformation. This deformation strain-hardens a metal and makes it less ductile; however, the metal may be softened by annealing. Thus, it is possible to control properties by cold-work/anneal operations. Annealing following cold work produces a new generation of grains within the metal. At high temperatures, recrystallization occurs almost simultaneously so that there is no strain hardening.

Contents

Prerequisite Units 2 and 3.

From Unit 4, the engineer should

• 1) Recognize the basic steps required to extract and refine metals, and be able to handle the accompanying calculations.

2) Understand the distinction between hot and cold working in terms of (a) relative temperatures, (b) effects on strength and ductility, and (c) their general use in production.

3) Identify the principal factors that affect recrystallization.

4) Know how to predict properties from cold-working data, and set up cold-work/anneal processes to meet specifications.
5) Be able to participate in conversations containing the technical terms at the end of the Unit.

4-1 METAL PRODUCTION

A major operation in the processing of metals and alloys is removing them from the original ores, and refining them into the compositions required by the user. The engineering and scientific accomplishments required in order to achieve this are considerable. We cannot even begin to outline them all here. However, we should cite the principles that underlie the production of various metals because this information often relates to the selection and use of metals.

• Extraction from ores

The ores of metals—except for gold, and a small fraction of our copper and silver supply—are not metallic phases. Most commonly they are oxides. However, they may be sulfides. Copper, silver, lead, and zinc sulfides are significant sources of those metals. In either case, the metal must be extracted from its ore. Even when the ore is a sulfide (MS, where M is the metal ion and S is the sulfide ion), it is common to first oxidize the ore to MO, the comparable oxide. Thus, the principal extraction step is one of *reduction*, in which the oxygen is removed, and the positive metal ions, M^{2+}, are changed to the metal atoms, M:

$$M^{2+} + 2e^{-} \rightarrow M. \qquad (4\text{-}1.1)$$

The ease with which reduction occurs varies from metal to metal. Lead oxide reduces readily to metallic lead, whereas it requires an input of considerable energy to effect the reduction of alumina (Al_2O_3). A measure of the stability of the oxides, or of the amount of energy required for extraction, is available from the *free energy of formation* of the various oxides (Table 4-1.1). These are the reaction energies that are *released* as the metal is burned to form the oxide (therefore, minus quantities). The same amount of energy is *required* to separate the oxygen from the metal (therefore, positive quantities).

The metallurgist has several choices of ways to force oxide reduction. As observed in Table 4-1.1, the energy required for reduction decreases as the melting temperature of the metal is approached. In fact, for silver the sign of the energy requirement reverses, and reduction of the oxide to the metal

Table 4-1.1
Energy* of formation of oxides

	Energy released on formation	Energy required to reduce oxide to metal			
	kJ/mole oxide†	kJ/mole oxide		kJ/g metal	
Oxide	25°C	25°C	T_m‡	25°C	T_m‡
Al_2O_3	−1580	+1580	+1380	+29	+26
MgO	− 570	+ 570	+ 500	+23	+21
Fe_2O_3	− 740	+ 740	+ 365	+ 6.6	+ 3.3
FeO	− 240	+ 240	+ 150	+ 4.3	+ 2.7
SnO	− 260	+ 260	+ 240	+ 2.2	+ 2.0
CuO	− 135	+ 135	+ 35	+ 2.1	+ 0.6
PbO	− 190	+ 190	+ 160	+ 0.9	+ 0.8
Ag_2O	− 10	+ 10	− 45	+ 0.05	− 0.21
CO_2	− 394	+ 394	—	+32.8	—
CO_2(327°C, 620°F)	− 395	+ 395	—	+32.9	—
CO_2(1000°C, 1832°F)	− 396	+ 396	—	+33.0	—

* Free energy. ($\Delta G_f^0 \equiv \Delta H_f^0 - T\Delta S_f^0$; however, it will not be necessary for us to use this equation.)

† 1 joule = 0.239 calories. Calories are the metric units of thermal energy still commonly used by chemists. A mole is 6×10^{23} atoms, or 6×10^{23} formula units of oxide.

‡ At melting temperature of metal.

proceeds naturally at T_m. However, this is not a very efficient way to remove the oxygen, because the consumption of energy by the furnace and the losses to the surrounding atmosphere would run high, even if elaborate designs were used for the equipment.

A more practical method of reduction involves the use of some *reducing agent*, such as carbon. In effect, carbon, which is readily available, attracts to itself the oxygen that is in the oxide. Using PbO as an example,

$$C + 2PbO \rightarrow 2Pb + CO_{2(gas)}. \tag{4-1.2}$$

Observe from Table 4-1.1 that the energy released by forming CO_2 is more than is required to separate oxygen from the lead. [These figures are −395 kJ/mole for the CO_2 versus (2)(+160 kJ/mole) for the two PbO's at the melting temperature of lead, 327°C.] Furthermore, when the resulting CO_2 gas is removed, there is no tendency for the reaction to reverse.

Reduction by carbon may also occur after the initial formation of CO:

$$3CO_{(gas)} + Fe_2O_3 \xrightarrow[\sim 2900°F]{\sim 1600°C} 2Fe_{(liquid)} + 3CO_{2(gas)}. \tag{4-1.3}$$

Although there are a number of variants, these types of reactions are the ones most widely encountered in metal production. Note that elevated temperatures favor this reaction, since the energy required for the reduction of

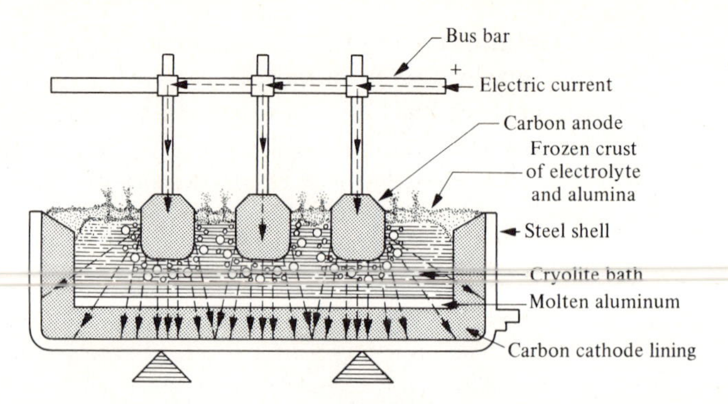

Figure 4-1.1 Electric cell for producing aluminum. Since the energy required to remove aluminum from its raw material (Al_2O_3) is great, a high-temperature electroplating process is used. Molten aluminum collects at the cathode, which is at the base of the cell, and CO gas is evolved at the anode (Eqs. 4-1.4 and 4-1.5). The critical feature of Hall's (and Heroult's) revolutionary process was the use of fluxes to make a water-free electrolyte. Water cannot be present or else hydrogen gas (H_2) will "plate" at the electrode ahead of the aluminum. (Courtesy of Aluminum Company of America.)

ore decreases markedly, while the energy released by the formation of CO_2 increases slightly.

The reduction reaction just described is not always feasible. Sometimes the oxidation of carbon does not supply enough energy. Sometimes the necessary temperature is excessively high. Sometimes the molten metal reacts with the container. Electrical energy offers an alternative, provided that the oxide can be dissolved in a flux. For example, although aluminum-bearing ores had been known for years, the commercial production of aluminum had to await the development of a suitable flux to dissolve the $Al_2^{3+}O_3^{2-}$ for *electrolytic extraction* (Fig. 4-1.1). Here the reaction at 980°C (1800°F) is the same as in Eq. (4-1.1), except that the metal, M, is trivalent aluminum. Carbon and an electric current are used to release the electrons from the oxygen ions and transfer them to the aluminum:

$$3O^{2-} + 3C \rightarrow CO_{(gas)} + 6e^- \qquad (4\text{-}1.4)$$

$$6e^- + 2Al^{3+} \rightarrow 6Al_{(liquid)}. \qquad (4\text{-}1.5)$$

A few metals, such as zinc, lend themselves to *vaporization* as a means of extraction. Vaporization also may be aided by carbon as a reducing agent because carbon removes the oxygen in the form of CO or CO_2 (I.P. 4-1.1).

• Refining

Most metals, even when they are reduced to remove the oxygen, are not pure enough for commercial use; they must be refined. This is illustrated by an aluminum ore that contains some iron. Since iron reduces more readily than aluminum, any iron in the ore will also appear in the metal. Likewise, a certain fraction of the silicon or phosphorus in an iron ore appears in extracted metallic iron, because these elements are partially reduced along with the iron. As a rule, these impurities are not desired; they must therefore be removed.

Again, there are a variety of metallurgical processes that are available. We shall describe only one, the *basic oxygen process*, which is the result of modern technology (Fig. 4-1.2). Let us assume that we have molten iron containing two weight percent (2 w/o) silicon and 3.5 w/o carbon. This composition is suitable, provided that we want to use it for certain cast irons (Section 6-6); but we must lower the silicon and carbon contents if we are to use the iron in low-alloy steels (Table 6-1.1). We invariably use oxygen to remove these two impurities selectively. Note, however, that we cannot have just a straight reversal of the initial reduction process we described earlier, or the iron would be oxidized along with the silicon and carbon. In simplest terms we use a CaO-rich slag because it has a great affinity for SiO_2. Thus, as the dissolved silicon and oxygen react, the SiO_2 that forms is deactivated by the CaO in the slag. The oxidation of the carbon produces CO and CO_2, which leave the furnace in the form of gases. Although some of the remaining iron will be oxidized and dissolved in the slag, the iron losses can be minimized by appropriate control of temperature and time.

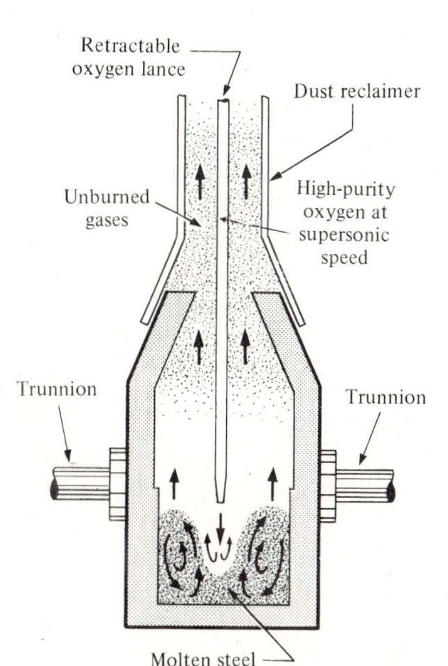

Figure 4-1.2 "Basic oxygen" refining furnace (iron). Oxygen that is blown into the molten metal oxidizes the dissolved carbon (to CO) and dissolved silicon (to SiO_2). The CO leaves the vessel as a gas and completes its combusion to CO_2 in the air. The SiO_2 is dissolved into the basic, CaO-rich slag. (A. G. Guy, *Physical Metallurgy for Engineers,* Reading, Mass.: Addison-Wesley.)

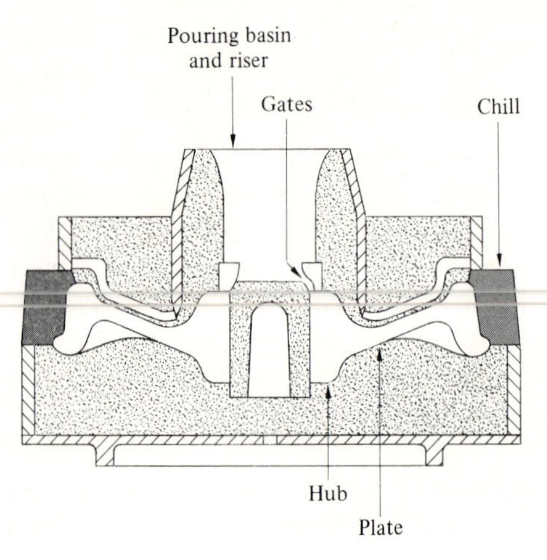

Figure 4-1.3 Mold (made of sand) for casting metal (railroad car wheel). The pouring basin serves as a riser, or reservoir, to supply metal to compensate for the shrinkage during solidification. The chills direct the solidification from the rim inward. (R.A. Flinn, *Fundamentals of Metal Casting*, Reading, Mass.: Addison-Wesley.)

• Casting

The next step in metal production is to *cast* the metal as an ingot, or to mold it directly into the desired shape. An *ingot* is simply a large solidified mass of metal that can subsequently be mechanically deformed by rolling or forging.

One of the prime technical considerations in making a casting is that the volume changes when the metal changes from liquid to solid (Fig. 2–4.1). With rare exceptions—for example, "type metal" used in printing—shrinkage accompanies solidification. Therefore a reservoir, usually called a *riser*, must be available that will feed additional molten metal into the casting while freezing (that is, solidification) progresses (Fig. 4–1.3). Also care must be taken that constrictions in the feed channels (*gates*) do not freeze or solidify completely and choke off the inflow of molten metal from the riser to larger sections that are not frozen. The metallurgist can partially compensate for this problem by using *chills*, or *heat sinks*, adjacent to the thicker sections of the casting, thereby increasing the freezing rate in those zones. A casting system, if it is to be satisfactory, must be designed by someone with considerable technical knowledge and accumulated experience about the behavior of molten and solid metals.

Illustrative problem 4–1.1 Zinc oxide (ZnO) and carbon are mixed and placed in a retort, where they are heated so that a reaction similar to Eq. (4–1.2) can occur. Both the zinc and the CO that result are vapors initially; however, the metallic zinc condenses in the cooler part of the retort. (a) How much carbon is required per 100 kg of zinc oxide? (b) How much zinc is reduced?

SOLUTION

$$ZnO + C \rightarrow CO + Zn. \qquad (4\text{–}1.6)$$

Using atomic masses:

$$(65.37 + 16) + 12 \rightarrow (12 + 16) + 65.37.$$

a) $\dfrac{C}{100 \text{ kg ZnO}} = \dfrac{12 \text{ amu}}{81.37 \text{ amu}}$, $C = 14.75$ kg.

b) $\dfrac{Zn}{100 \text{ kg ZnO}} = \dfrac{65.37 \text{ amu}}{81.37 \text{ amu}}$, $Zn = 80.3$ kg.

ADDED INFORMATION If the reaction temperature can be kept low, some of the CO will react with more ZnO to give CO_2 and additional zinc. The lower the temperature, however, the slower the reaction. If the temperature is raised, zinc vapors may escape. The metallurgist must optimize and control the temperature for maximum production efficiency. ◄

Illustrative problem 4-1.2 According to Eq. (4-1.5), six electrons are required for every two aluminum atoms. How many coulombs are required to extract 27 g of aluminum?

SOLUTION The atomic mass of aluminum is 27 amu (Fig. 1-3.2); thus 27 g have 6×10^{23} atoms.

$$\text{Electrons required} = (6 \times 10^{23} \text{ atoms})(3 \text{ electrons/atom}) = 18 \times 10^{23} \text{ electrons}.$$

$$\text{Coulombs required} = (18 \times 10^{23} \text{ electrons})(1.6 \times 10^{-19} \text{ coul/electron}) = 2.88 \times 10^{5} \text{ coul/27 g Al}.$$

ADDED INFORMATION Since a coulomb is 1 amp · sec, a current of one ampere would be required for 288,000 sec (80 hr), or 80 amp for 1 hr (3600 sec).

The term *faraday* ($\mathfrak{F}$) is used for 96,000 coulombs, the charge for 6×10^{23} electrons. The above answer is equal to three faradays, since each aluminum atom requires three electrons. ◄

4-2 SHAPING METALS BY COLD DEFORMATION

Although deformation is to be avoided in most service conditions, it is one of the principal mechanisms of metal processing. An automobile fender, which is sheet-thin, was initially a large ingot that was first rolled to a slab, then to a plate, and finally to the required thinness. A No. 14 electrical wire was initially cast as a wire-bar with a 20-cm (8-in.) cross section. The copper was then rolled and drawn through sequential dies to the specified gage.

In effect, each of the shaping steps just described involves "controlled failure," since the metal is permanently deformed in the process. *Rolling*, *forging*, *extrusion*, *drawing*, and *spinning* (Fig. 4-2.1) are among the more common reshaping processes that transform bulk metal into the geometry desired for the engineering product. Since these shaping processes require the mechanical deformation of metal, engineers need to use the information of the previous unit in their designs.

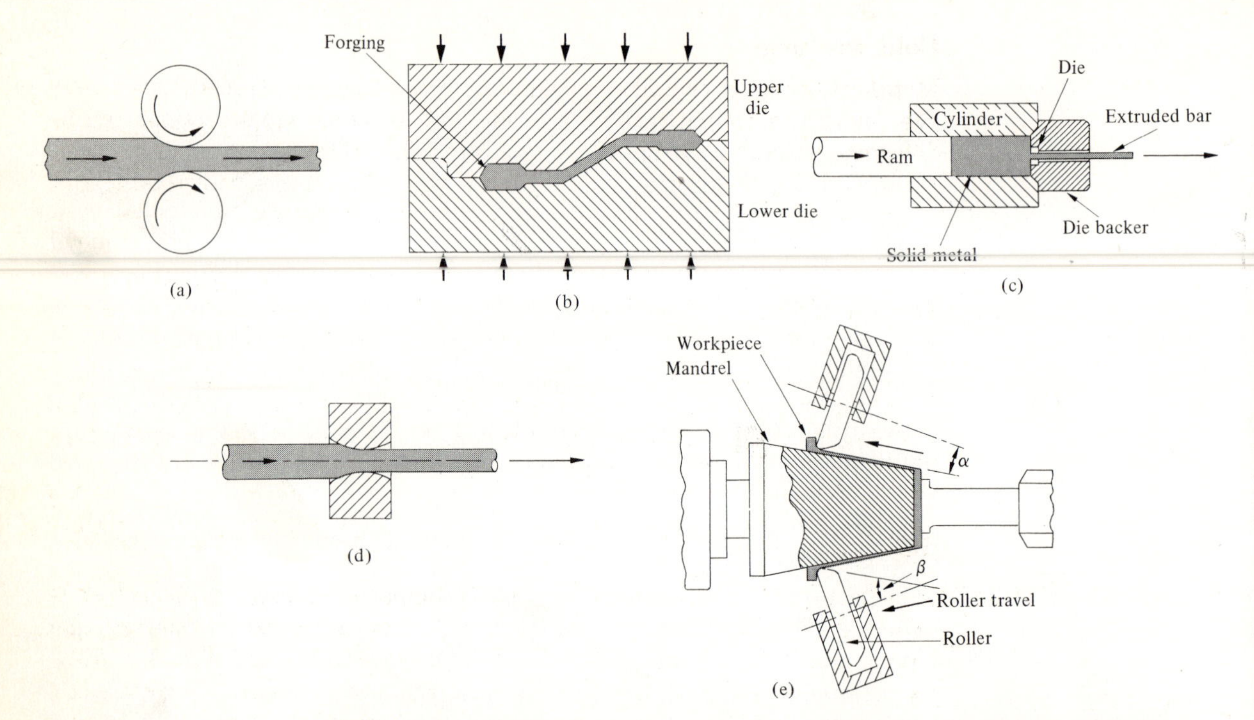

Figure 4-2.1 Common shaping processes. (a) Rolling. (b) Forging (adapted from ASM *Metals Handbook,* Vol. 5). (c) Extrusion (A. Guy, *Physical Metallurgy for Engineers*). (d) Wire drawing. (e) Spinning (adapted from ASM *Metals Handbook,* Vol. 4). Often two or more of these processes are used. Hot working—i.e., shaping above the recrystallization temperature—is necessary for shaping products with large cross sections, or for shaping without the introduction of strain hardening.

Figure 4-2.2 Strain hardening. Cold work (a) increases the hardness, (b) increases the strength, and (c) decreases the ductility.

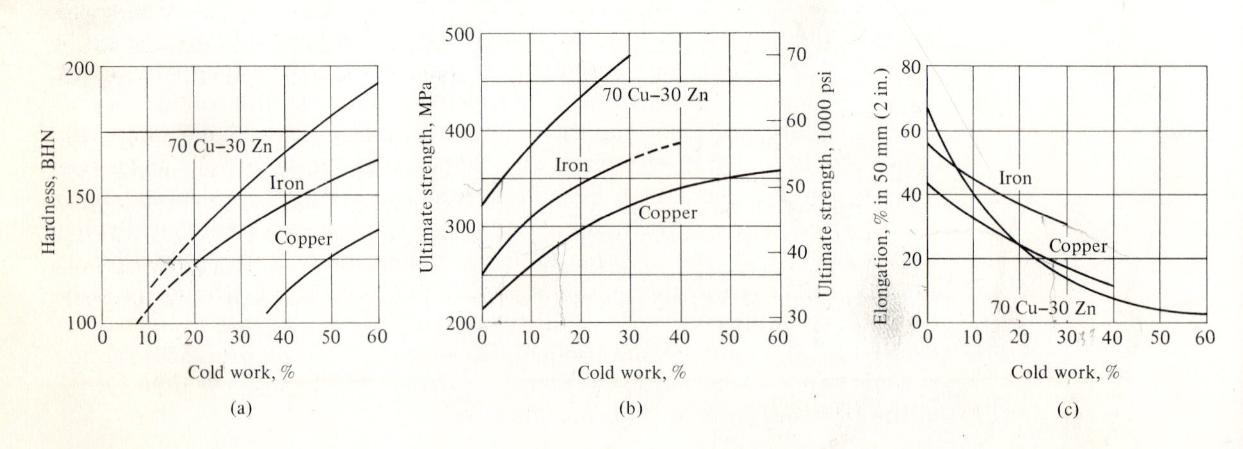

Cold working

Metals become softer and more deformable at elevated temperatures; however, with the correct techniques, metals with small cross sections can be deformed without heating. When this is possible, the production operation is simplified. Since the craftsman "works" the metal into the desired shape, this process of forming at ambient temperatures is commonly called *cold working*.

Cold working hardens the metal. We have already observed this in our discussion of the stress-strain curve (Fig. 3–2.2). As the yield strength is exceeded, more and more stress is required to continue plastic strain, which in turn means that the metal will have a greater hardness. The reader may have had occasion to realize this hardening effect directly while bending a wire to break it. The first bend occurs readily; however, subsequent bends at the same location require progressively more force.

Strain hardening

The phenomenon of hardening that accompanies plastic deformation is called *strain hardening*. The engineer makes use of this as a strengthening mechanism. For example, if a rod is drawn through a die (Fig. 4–2.1d) it becomes stronger as its diameter is reduced (and the length is increased). The amount of this strengthening is a function of the amount of strain. Hence, we commonly use the nondimensional change in cross-sectional area as a measure of strain during cold working, e_{cw}:

$$e_{cw} = \Delta A/A_o. \tag{4–2.1}$$

This calculation is the same as for the reduction of area (Eq. 3–2.2), except that earlier calculation was for the strain at fracture. In cold working, the process must be stopped short of fracture.

Figure 4–2.2 shows the amount of strain hardening that accompanies the cold working of iron, copper, and brass. These data were obtained by drawing these metals through a die (Fig. 4–2.1d); however, the same results could have been obtained by cold rolling (Fig. 4–2.1a). Of course, the $\Delta A/A_o$ calculations are made from different cross-sectional geometries (Study Problem 4–2.1).

Strain hardening is accompanied by a decrease in ductility (Fig. 4–2.2c).* This loss of ductility is to be expected, because part of the total plastic deformation that is inherent in a material has been expended in the cold-working process. Thus, if we place gage marks on a test specimen of copper that has been cold rolled, there will be less calculated elongation (Eq. 3–2.1) than there would be on a specimen of copper that had not been previously cold worked.

* Again this may be familiar to you, if you have broken a wire by repeated bending. The wire, which was originally ductile, finally breaks with a brittle fracture.

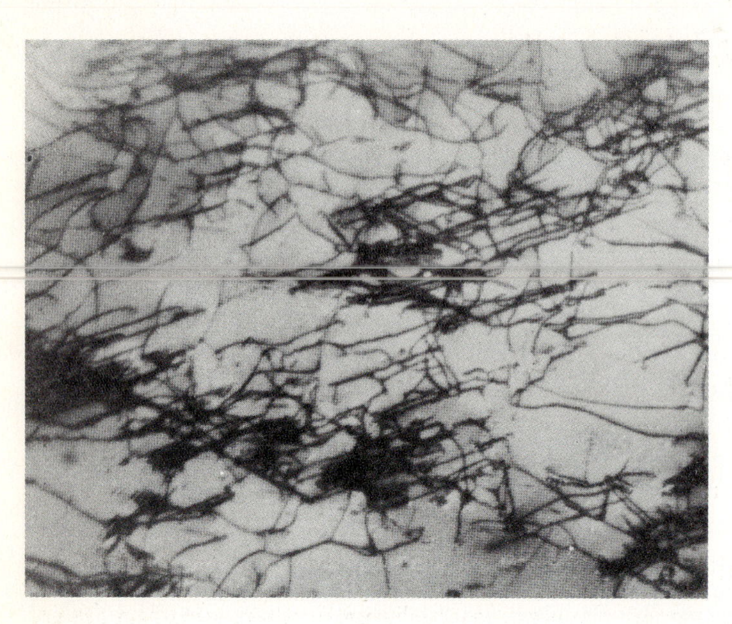

Figure 4-2.3 Dislocations in plastically deformed metal (stainless steel, × 30,000). Additional deformation introduces more dislocations. Also, additional dislocations interfere with further deformation. *Strain hardening* is the result. (M.J. Whelan, *Proc. Roy. Soc.*, Series A Volume 249.)

As a general rule, a metal that has been strengthened will also be low in ductility. Fortunately this is not an absolute rule, and considerable metallurgical effort is directed toward obtaining strong metals that are not brittle (Unit 14). However, it does mean that the engineer who is designing a product must commonly make decisions on whether strength is preferred over ductility or vice versa. A material that is strong may break with no advance warning if it is nonductile, often with dire consequences.

Strain hardening occurs because dislocations are introduced during plastic deformation. Although dislocations are required for slip that produces plastic deformation (Fig. 3-5.2), they become entangled when there are large numbers of them (Fig. 4-2.3). Thus, as their numbers and lengths increase, it takes more and more stress to cause further slip (Fig. 4-2.4).

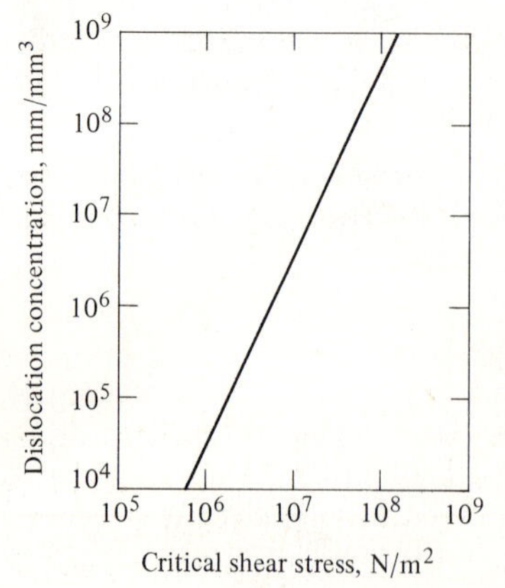

Figure 4-2.4 Shear stress versus dislocation density (copper). Slip occurs through dislocation *movements*. Therefore a high shear stress is required for deformation when (1) *no* dislocations are present, or (2) the numbers of dislocations are such that they become *entangled*. In the latter cases strain hardening (cold work) results. [Adapted from H. Wiedersich, "Hardening Mechanisms and Theory of Deformation," *J. Metals*.]

Illustrative problem 4-2.1 Two 20-m (65-ft) aluminum rods are each 14.0 mm (0.551 in.) in diameter. One bar is drawn through a 12.7-mm (0.50-in.) die. (a) What are the new dimensions of that bar? (b) Assume identical test samples are made from each bar (deformed and undeformed) and marked with 50-mm gage lengths. Which one, if either, of the bars will have the greater ductility? The higher yield strength?

SOLUTION (a) Since the volume does not change,

$$
\begin{aligned}
A_o L_0 &= (AL)_{\text{drawn}} \\
L_{\text{drawn}} &= L_o\,(A_o/A_{\text{drawn}}) \\
&= (20\text{ m})(14.0^2\pi/4)/(12.7^2\pi/4) = 24.3\text{ m}.
\end{aligned}
$$

b) The *undeformed* bar will be *more* ductile. The deformed bar already "used" some of the ductility before making the gage marks for elongation. The *deformed* bar will have the *higher* yield strength. ◂

Illustrative problem 4-2.2 A copper wire must have an ultimate strength of at least 280 MPa (40,000 psi), a ductility of at least 20% elongation (50-mm), and be 2.0 mm (0.08 in.) in diameter. Propose the necessary processing by cold working.

SOLUTION From Fig. 4-2.2, we observe the following requirements.

For $S_u > 280$ MPa: cold-work > 16%.
For elongation > 20% (50-mm): cold-work < 26%.
Use ~21% cold-work to obtain final diameter $D_f = 2.0$ mm.

$$
0.21 = \frac{(\pi/4)(D_o)^2 - (\pi/4)(2.0\text{ mm})^2}{(\pi/4)(D_o)^2};
$$

$$
D_o = 2.25\text{ mm} \qquad \text{(or 0.09 in.).}
$$

(Select a 2.25-mm wire, then cold-work it 21% to 2.0 mm.)

ADDED INFORMATION We determined a specification "window" for an allowable working range from Fig. 4-2.2. In the absence of other requirements, we choose a percent cold-work near the center. This permits some latitude in processing. In I.P. 5-2.2, there will be an economic incentive to work close to one side of the "window."

The initial wire with a diameter of 2.25 mm must be annealed so that it contains no strain hardening. We shall discuss this in Section 4-3. ◂

4-3 ANNEALING BY RECRYSTALLIZATION

Strain hardening can be removed by annealing. *Annealing* is a semitechnical term that broadly means "heating a material so it will fracture less readily." The internal structure of different materials responds to heating differently. Therefore, all annealing does not reduce the tendency to frac-

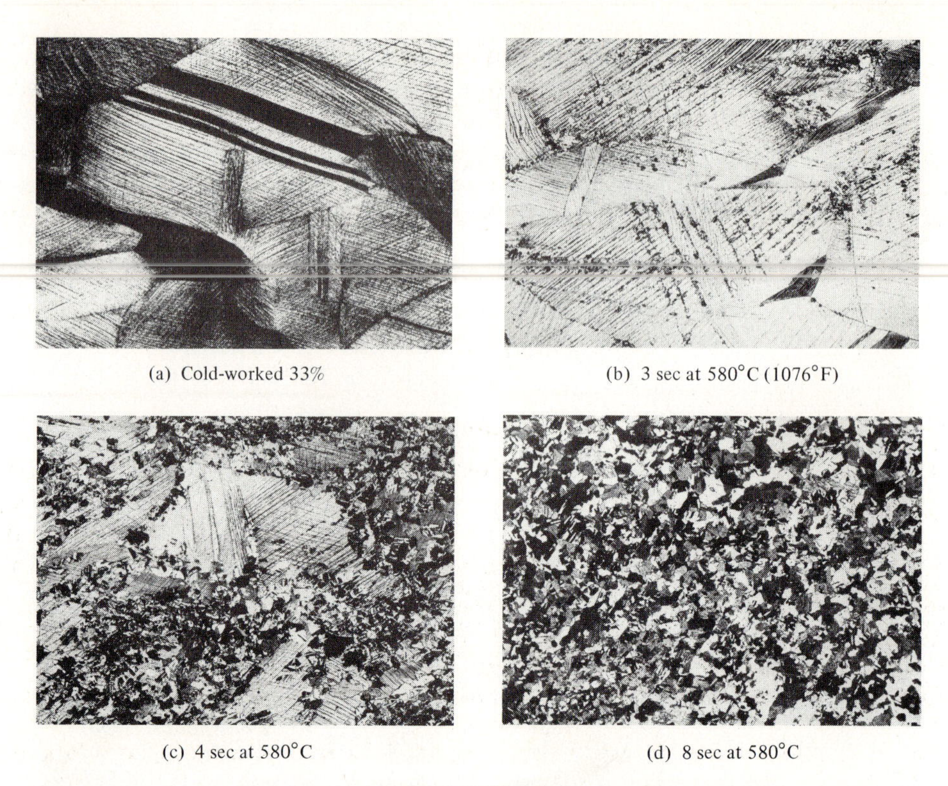

(a) Cold-worked 33%

(b) 3 sec at 580°C (1076°F)

(c) 4 sec at 580°C

(d) 8 sec at 580°C

Figure 4-3.1 Recrystallization (brass, × 40). The deformed brass of (a) is gradually replaced by new, strain-free grains that have few dislocations. This softens the metal (cf. Fig. 4-3.4). (J. E. Burke, General Electric Co.)

ture in the same manner. When annealing follows strain hardening, the most obvious change is an increase in ductility. For example, take a wire coat hanger in which the steel wire had been heavily cold-worked for high strength (and, therefore, requires less material). Such a wire (cold-worked 75%) may have a strength close to 500 MPa (>70,000 psi), but a reduction of area of only 10%. Therefore, it may fail by fracturing. By annealing that wire, the ductility can be increased to the original value of more than 50% R of A; therefore, the wire fractures less readily because it deforms plastically. (Of course, the ductile annealed wire is weaker.)

Recrystallization

Annealing a cold-drawn wire produces a change in properties because the internal structure of the wire is changed. When annealing follows cold working, the metal is *recrystallized*. That is, new crystals form within the wire. Figure 4-3.1 shows this for a cold-worked brass. The brass of part (a) of the figure had been deformed 33%, but not reheated. The figure shows the slip lines within the grains, where crystal planes have undergone shear (Fig. 3-5.2). The brass of Fig. 4-3.1(b) has been reheated 3 sec at 580°C (~1075°F). New grains have started to form along the slip lines. Approximately 50% of the brass has been recrystallized in Fig. 4-3.1(c). It had been reheated slightly longer (4 sec) at 580°C. Recrystallization is complete after 8 sec at this temperature (Fig. 4-3.1d). The hardness dropped significantly with this recrystallization sequence, all due to the change in internal structure that occurred with the new grain growth. The time requirement for annealing is longer at lower temperatures.

Cold-work/anneal cycles

We use cold work advantageously to increase the strength of a metal; however, a problem is encountered in using cold work as a shaping process. Consider drawing a 2.5-mm (0.1 in.) iron wire into a finer wire (~0.5 mm) for a tire cord. This would involve 96% cold work!! Theoretically, the wire would be strong, but possess nil ductility. In practice, it would be very difficult to cold-work the wire that much. Being nonductile, it would fracture in the process of drawing it through the die. Even if we manage to get it through the die, we would have to expect short die life because the hardened wire would wear and enlarge the die perimeter.

To circumvent these problems, we use *cold-work-and-anneal cycles*. Starting with the above 2.5-mm wire, the wire is drawn through a series of dies to reduce it to an intermediate diameter, say 1.8 mm. From Eq. (4-2.1), there has been 48% cold work. By extrapolating the data in Fig. 4-2.2, we could expect a strength of ~400 MPa (~60,000 psi) and an elongation of ~20% (50-mm). Now by annealing the wire, we remove this strain hardening (and are back to 0% cold work) so that we have iron with 250 MPa (36,000 psi) and 55% elongation. This 1.8-mm wire can now be drawn again. Cold drawing to a 1.4-mm diameter would introduce another 40% cold work. A following anneal would again establish the strength at 250 MPa and the ductility at 55% (50-mm). Such cold-work/anneal cycles can be repeated as often as necessary to gain the final dimensions. As shown in I.P. 4-3.1, it is commonly desirable to end the cycles with the correct amount of cold work to meet specifications for strength and ductility.

Recrystallization temperature range

LAB

The new crystals that formed in the brass of Fig. 4-3.1 were the result of atom movements. The former plastically deformed grains were no longer nearly perfect crystals but contained many imperfections, particularly dislocations (Fig. 4-2.3). The new crystals grew by the atoms taking on new neighbors in a nearly perfect fcc (for brass) pattern.

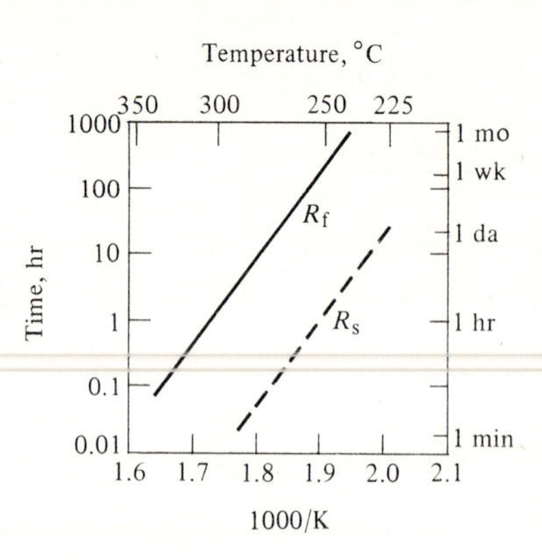

Figure 4-3.2 Recrystallization time versus temperature (aluminum, 75% cold-worked). Dashed line: start of recrystallization, R_s. Solid line: recrystallization finished, R_f. The straight-line relationship between log time and reciprocal temperature ($1/T$) indicates that the atom movements govern the recrystallization. (Adapted from *Aluminum*, Vol. 1, Metals Park, Ohio: American Society for Metals.)

The rate at which atoms move is highly sensitive to temperature. This is shown in Fig. 4-3.2 for aluminum that had been cold-worked 75%. For this aluminum, recrystallization started, R_s, in a few minutes at 275°C (525°F), but required 24 hours at 225°C (440°F). Figure 4-3.3 shows the change in yield strength with recrystallization. Samples that had been previously cold worked 75% were held for 1 hour at each testing temperature (circles of Fig. 4-3.3) and then cooled to room temperature for tensile testing. The most notable change occurred between 250°C and 300°C, which is 60% of the absolute melting temperature (660°C + 273°C = 933 K). Had we allowed a year or more for recrystallization, the most noticeable change would have occurred near 200°C. We thus speak of a *recrystallization temperature range*. The upper end is hot enough for convenient annealing times (~1 hr); the lower end is a temperature at which the strain-hardened metal will retain its desired properties for the service life and service temperature of the product for which it is designed. (1 yr?, 10 yrs? It depends on what you, the engineer, are designing.) Thus, it is common to define the recrystallization temperature as being between ~0.4 and ~0.6 of the absolute melting temperature, T_m.

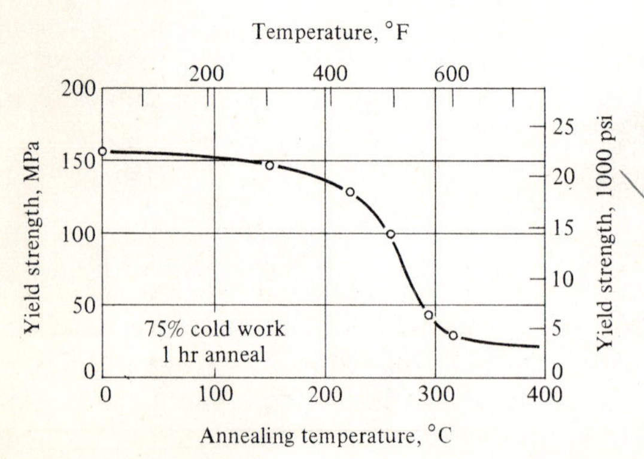

Figure 4-3.3 Yield strength versus recrystallization (aluminum). Initially cold-worked 75%, the metal was reheated to the indicated temperatures for one hour. This was enough time to recrystallize the metal at 300°C and above. The yield strength decreases and the strain hardening disappears with the development of the new grains. Cf. Fig. 4-3.1. (Adapted from *Aluminum*, Vol. 1, Metals Park, Ohio: American Society for Metals.)

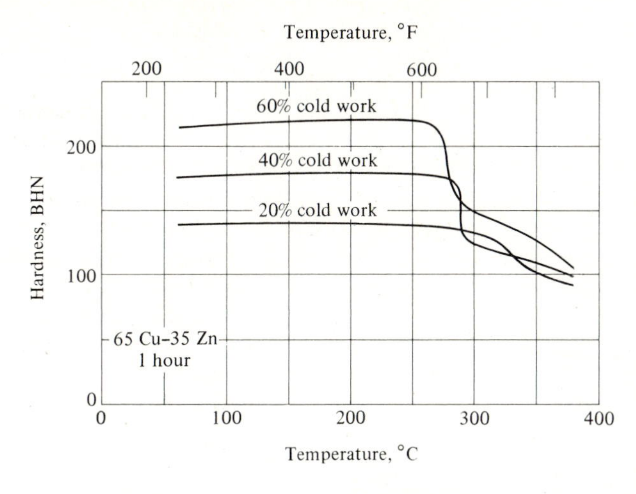

Figure 4–3.4 Softening during annealing (65Cu–35Zn brass). The more severely cold-worked brass (60%) softens and recrystallizes at lower temperatures than a brass with 20% cold work, ~275°C versus ~325°C (ASM data).

The recrystallization temperature also depends upon the amount of plastic strain in the metal. Figure 4–3.4 shows the recrystallization softening for a 65 Cu–35Zn brass that had e_{cw} = 20%, 40%, and 60%. Recrystallization occurs at lower temperatures when the metal had more plastic deformation—~325°C, ~290°C, and ~275°C, respectively. This is because the crystal distortion during plastic deformation introduces strain energy. That is, some of the atoms are closer together; some are farther apart than equilibrium distances (Fig. 4–3.5). Either variant requires extra energy. As a result, less thermal energy and, therefore, lower temperatures are required to start and complete recrystallization.

Figure 4–3.5 Strain energy in dislocations. Plastic deformation introduces dislocations (Fig. 3–5.2). (a) Atoms do not have the equilibrium interatomic distances in the vicinity of the dislocation. Whether they are compressed (above) or under tension (below), they possess extra energy. With this strain energy, less thermal energy is required for recrystallization. (b) Just for fun! (L. S. Darken in the style of L. Matteoli.)

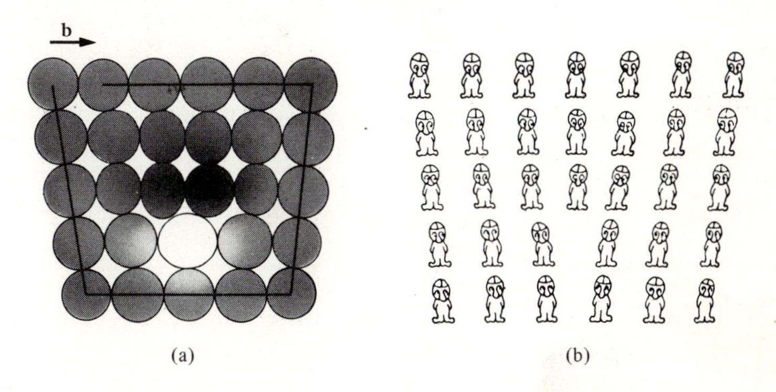

Finally, the recrystallization temperature is lower for pure metals than for alloys with considerable solid solution, for example, pure copper vs. a 65Cu–35Zn brass.

Illustrative problem 4–3.1 A sheet of copper is to be 2.0 mm (0.08 in.) thick and have a tensile strength at least 275 MPa (40,000 psi), and a ductility of 25% elongation (50-mm). The plate stock from which this sheet can be rolled is 5 mm (0.197 in.) thick. For the rolling mill to make a satisfactory product without cracking the metal, the metal to be rolled must have a ductility of at least 15% elongation (50-mm). Specify a processing procedure.

PROCEDURE To process satisfactorily according to Fig. 4–2.2(c):

$$\text{El.(50-mm)} \geq 15\% \qquad e_{cw} \leq 33\%.$$

To meet the product specifications (Fig. 4–2.2):

$$\left.\begin{array}{ll} S_u \geq 275 \text{ MPa} & e_{cw} \geq 14\% \\ \text{El.(50-mm)} \geq 25\% & e_{cw} \leq 19\% \end{array}\right\} \text{ use } 17\%.$$

Determine t_x:

$$e_{cw} = 0.17 = \Delta A/A_o = (t_x w - t_f w)/t_x w$$
$$0.17 = (t_x - 2.0 \text{ mm})/t_x$$
$$t_x = 2.4 \text{ mm} \qquad \text{(or 0.096 in.).}$$

The sheet must be annealed when it is 2.4 mm thick. However, it cannot be cold worked directly from 5.0 mm to 2.4 mm, because that would involve 52% cold work and could cause cracking during rolling. Therefore:

Cold work	5.0 to ~3.5 mm;	$e_{cw} = \sim 30\%$.
Anneal		
Cold work	~3.5 to 2.4 mm;	$e_{cw} = \sim 31\%$.
Anneal		
Cold work	2.4 to 2.0;	$e_{cw} = 17\%$. ◄

4–4 SHAPING METALS BY HOT DEFORMATION

The cold forming of metals is a very useful shaping process since preheating is not necessary, and since strain hardening can enhance strengths. There are, however, situations where it is preferable to hot-form metals. For example, it would require prohibitively massive rolling mills to reshape a 10-ton ingot into a bridge beam, if the beam simultaneously strain-hardened to several hundred MPa (up to 50,000 psi). Likewise, sufficient ductility would not be available in a cold-worked metal to shape a 310,000-kg (680,000-lb) rotor for an electric generator (Fig. 4–4.1).

Figure 4–4.1 Hot-worked metal (rotor forging for a steam turbine of an electric power generator). The initial ingot, which weighed more than 300,000 kg (~680,000 lbs), had to be hot-forged into the approximate shape of the rotor before the preliminary machining. Not only is the metal softer and more ductile at hot-working temperatures, the deformed metal does not strain harden. (Courtesy of S. Yukawa, General Electric Co.)

In general, cold forming is restricted to metals of relatively small cross-sectional dimensions, for example, plate, sheet, and wire, or to products that are only locally deformed, for example, the cold heading of bolts. The more massive shaping operations must be performed at elevated temperatures where the metal is softer and more ductile. Equally important, the metal does not strain harden at the elevated temperatures but remains soft and ductile.

Hot working

We saw in Fig. 4–3.2 that recrystallization started, R_s, more rapidly and was completed sooner as the temperature was raised. A mathematical relationship can be established:

$$\ln t = A + B/T. \qquad (4\text{–}4.1)$$

In this equation t is time in hours, and T is absolute temperature, K. The constants A and B are *unique for each metal or alloy* and are also affected by the amount of cold work. They must be calculated from experimental data. The calculated values of A and B for the R_s curve of aluminum in Fig. 4-3.2 are -57 and 30,000 K, respectively. These numbers were obtained from the following data, and from $T_m = 660°C$ (or 933 K):

200°C (473 K), which is 0.51 T_m, ~600 hr;
250°C (523 K), which is 0.56 T_m, ~ 1.5 hr;
300°C (573 K), which is 0.61 T_m, ~ 0.01 hr.

Immediately, it is apparent that an extrapolation to still higher temperature indicates that there will be a completion of recrystallization in sufficiently short time so that hardening is lost during the deformation process:

300°C, which is 0.61 T_m, ~35 sec;
350°C, which is 0.67 T_m, ~ 0.5 sec;
400°C, which is 0.72 T_m, ~ 0.01 sec.

If rolled at 450°C, which is 77% of the absolute melting temperature (933 K), recrystallization starts in milliseconds (and is completed before the aluminum leaves the rolls). (See I.P. 4-4.1 and Fig. 4-2.1a.) This permits large reductions with each rolling pass; production mills can be lighter; less energy is required; and the more ductile metal is less likely to crack during deformation than is cold-worked metal.

Temperatures for cold working some metals are higher than for hot working other metals. For example, 150°C is more than 0.7 T_m for lead, since its melting temperature, T_m, is 327°C (600 K). In contrast, 150°C is less than 0.35 T_m for copper with a melting temperature of 1084°C (1357 K). At 150°C lead is hot worked, and copper is cold worked.

Illustrative problem 4-4.1 (a) Use the values of

250°C ~200 hr,
350°C ~ 0.02 hr,

to calculate the A and B constants of Eq. (4-4.1) for the final recrystallization, R_f, of the aluminum in Fig. 4-3.2. (b) How long will it take to complete recrystallization at 450°C?

SOLUTION

a) $\ln 200 = A + B/523\ K$,
$\ln 0.02 = A + B/623\ K$.

Solving simultaneously,

$$A = -52; \qquad B = 30{,}000\ K.$$

b) $\ln t = -52 + 30{,}000\text{ K}/723\text{ K} = -10.5,$
$t = 2.7 \times 10^{-5}\text{ hr}$ (or 0.1 sec).

ADDITIONAL INFORMATION The values of B for the two curves of Fig. 4–3.2 are both 30,000 K because the curves are parallel and have the same slope. This value will be different for other metals (as will the intercept value, A). The constants A and B must be calculated from experimental data. ◀

• 4–5 SHAPING BY METAL REMOVAL

Machining is by far the most widely used metal-removal process. It may be performed on a lathe, in a milling machine, by drilling, or by several other variations. It is a combination of cutting and mechanical deformation, as shown in Fig. 4–5.1, in which various terms associated with cutting tools are listed, and in Fig. 4–5.2, which illustrates three types of chip formation and distorted steel.

In practice, the process engineer prefers a chip that breaks as it is cut (Fig. 4–5.2a). This means that it is often desirable to specify some cold work prior to machining, to reduce the ductility of the metal. It also means that

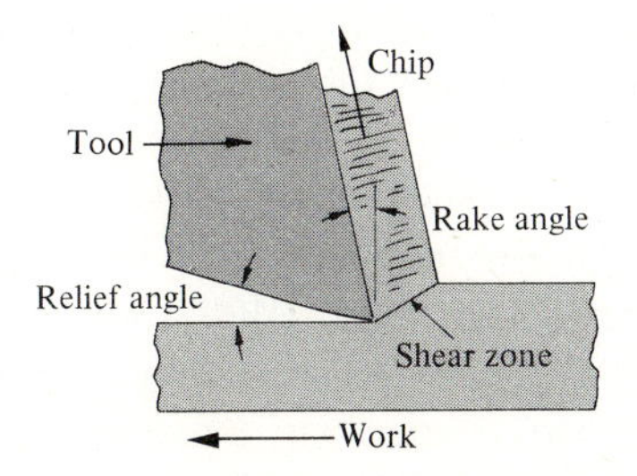

Figure 4–5.1 Machining. This metal-removal process, which involves both cutting and deformation, depends on the properties of the metal and the geometry of the cutting tool.

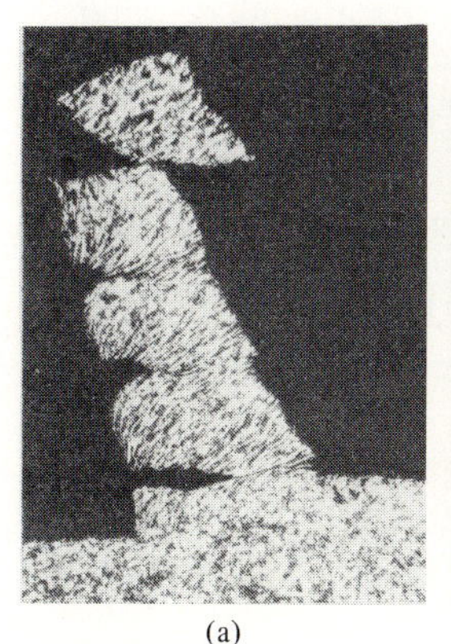
(a)

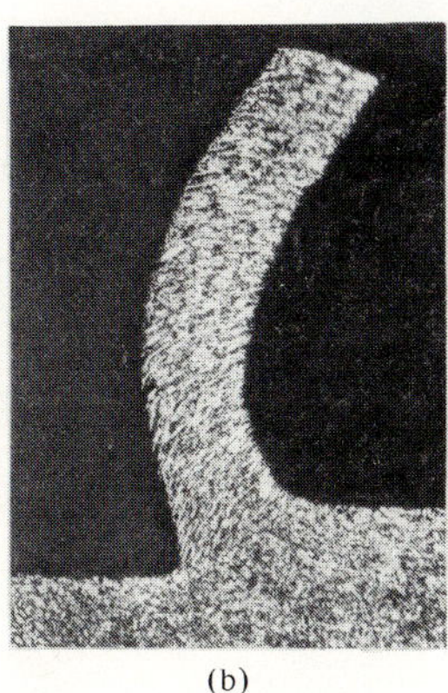
(b)

(c)

Figure 4–5.2 Chip formation. (a) Discontinuous. (b) Continuous. (c) Tool-edge buildup. (Cincinnati-Milacron, Inc.)

certain impurity phases—for example, MnS—may be specified as additions to the steel to aid chip formation, producing a "free-machining" steel. Chips rather than long turnings not only facilitate the handling of the removed metal but also provide less friction and therefore less heating at the tool tip. This is important because excessive heating of the tool tip decreases its hardness, which permits faster dulling and more friction and heat, leading to rapid failure.

• 4-6 POWDER FABRICATION

Powders of a material may be compacted into a product and then bonded by *sintering* (Fig. 4-6.1). This procedure has been used throughout history for producing ceramic products from clays and/or other solid particles (Section 12-4). In recent years, it has been used more and more in metallurgy, so that now *P/M* (*powdered metal*) products constitute a significant fraction of design components. The metal powders are produced by atomizing liquid metal in a blast of inert gas while the metal solidifies.

In principle, P/M products are made by compacting metal powders into a mold. Those metals that are ductile deform at points of contact; however, it is usually impossible to bring about full density—that is, remove all porosity—by this procedure. Compaction is followed by sintering, which (1) binds the powder particles together, and (2) densifies the *compact*, thus reducing the porosity. Here the engineer has some choice. In designing products such as filters (for example, in the gas feed-line of a car), or "oilless" bearings (which actually maintain a lifetime supply of oil in the pores), the engineer may specify porosity. Alternatively, near-zero porosity may be

Figure 4-6.1 Powdered-metal gear. Processing steps include (1) compaction into a mold, (2) solid sintering in an oxygen-free atmosphere.

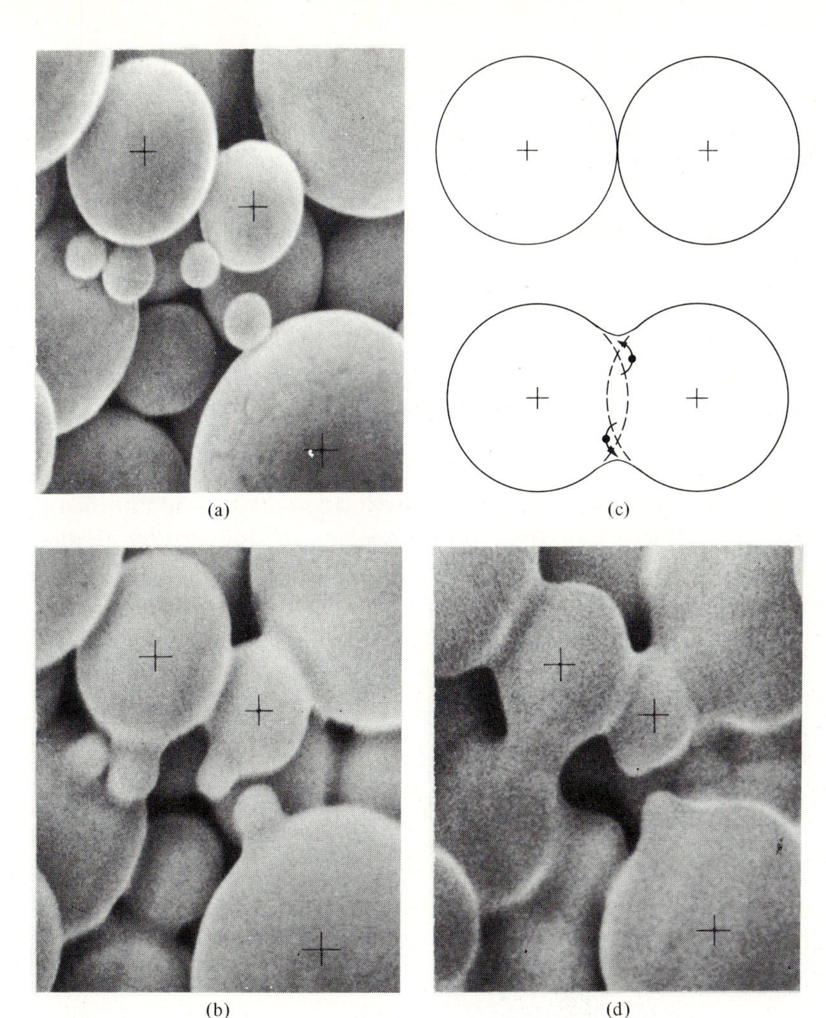

Figure 4-6.2 The sintering process (nickel powder). The initial points of contact between metal particles in (a) became areas of contact in (c) and (d) after the powder had been heated at 1100°C–1200°C for approximately an hour. In order to attain this solid-to-solid bond, atoms must diffuse as sketched in (b), with the resulting shrinkage of dimensions. (Approximately × 1500) (R. M. Fulrath, Inorganic Materials Research Division of the Lawrence Radiation Laboratory, University of California, Berkeley.)

specified if strength is important (such a zero-porosity material requires higher-temperature heating for densification).

The sintering process requires extensive diffusion of atoms, as illustrated in Fig. 4–6.2, which shows nickel powder in the sintering process. Initially (Fig. 4–6.2a), the particles were simply in contact (in this case not compacted together). The points of contact become areas of contact (Figs. 4–6.2c and d). This requires a diffusion of nickel, as sketched in Fig. 4–6.2(b). We shall not examine the details of this diffusion, except to note that (1) diffusion, and therefore sintering, occurs more readily at high temperatures, and (2) shrinkage accompanies sintering by these mechanisms. The shrinkage becomes evident when one observes the distance between the centers (+) of the nickel particles in Fig. 4–6.2.

The process of sintering metals must be performed in an atmosphere that is devoid of oxygen. Any oxygen present would react with the metal and preclude metal-to-metal bonding of adjacent particles, as developed in Fig. 4-6.2(c). Generally a hydrogen atmosphere is used (with appropriate precautions). The temperature of sintering is normally close to—but short of—the melting point of the metal. This accelerates the required diffusion.

A development of major importance involves the use of P/M *preforms* for hot pressing. A compact is made from powdered metal, followed by a heating step that initiates sintering. The hot preform is then pressed in a closed-die mold to full density. There are two advantages of this process over cast or wrought metals. First, nearly 100% of the metal goes into the product. Second, the composition can be more uniform throughout the part than is possible with metal that originates in a casting.

Illustrative problem 4-6.1 The powdered metal gear (brass) of Fig. 4-6.1 must be 41.0 mm (1.614 in.) in diameter with near-zero porosity as a final sintered product. What diameter of die is required if the compacted porosity is 20 v/o?

SOLUTION Basis: A cube, 1 mm × 1 mm × 1 mm, containing 20 v/o porosity. Actual metal volume = 0.80 mm³. A cube with 0.80 mm³ and zero porosity = (0.928 mm)³. The compact diameter, d, is proportional:

$$\frac{1.00 \text{ mm}}{0.928 \text{ mm}} = \frac{d \text{ mm}}{41.0 \text{ mm}} = \frac{d \text{ in.}}{1.614 \text{ in.}},$$

$$d = 44.2 \text{ mm} \qquad \text{(or 1.74 in.)}.$$

ADDED INFORMATION If specifications had called for 5 v/o porosity in the final product, the 0.80 mm³ of actual metal volume would have occupied [0.80 mm³/0.95], or 0.842 mm³. This is (0.944 mm)³. Following the above proportions, the initial dimension d would have to be 43.4 mm (or 1.71 in.). ◀

Review and Study

SUMMARY

• 1. Most metals are extracted from their ores and refined as liquids at elevated temperatures. Some are cast in molds with the shape of the product. Others are solidified as large ingots and mechanically deformed to their required shapes.

2. Cold-worked metals are deformed at ambient temperatures. Thus, they become strain-hardened; they also lose part of their ductility. Annealing will soften a cold-worked metal and regain the original ductility because the metal recrystallizes.

3. The recrystallization temperature varies from 0.3 to 0.6 of the absolute melting temperature depending on (a) the available time, (b) the amount of cold work,

and (c) the purity of the metal. Recrystallization is faster at higher temperatures, with more cold work, and in pure metals.

4. Above approximately 0.7 T_m, recrystallization occurs during deformation. Strain hardening is not observed. We call this hot working. Hot working must be used for large ingots or forgings; cold working is more adaptable for the final shaping process of sheet, wire, and other small products.

• 5. Machining is a combination of deformation and cutting for shaping by metal removal. Powder-metal processes require the compaction and sintering of metal particles. The latter step requires atomic diffusion at elevated processing temperatures. Shrinkage occurs.

TECHNICAL TERMS

Annealing Heating and cooling to reduce fracturing.

• **Casting (metal)** The process of pouring a liquid metal into a mold, or the object produced by this process.

Cold work ($\%e_{cw}$) Plastic strain calculated from the change in cross-sectional area.

Cold working Deformation below the recrystallization temperature.

Cold-work/anneal cycle Repeated cold-work-and-anneal sequence to process a metal without fracturing and with smaller equipment.

Drawing Mechanical forming by tension, for example, by pulling a wire through a die, or by stretching sheet metal between dies. Usually drawing is a cold-forming process.

Extrusion Shaping a metal by pushing it through an open-end die.

Forging Shaping by compression between opposing dies.

Hot working Deformation above the recrystallization temperature, so that recrystallization occurs nearly simultaneously.

• **Ingot** A large casting that is subsequently to be rolled or forged.

• **Machining** Metal-removal process by cutting.

• **Powder metallurgy** Powder, which was made by the atomization of liquid metals, is compacted in a die and bonded by sintering.

Recrystallization The formation of new, annealed grains from a previously strain-hardened solid.

Recrystallization temperature, (T_R) Temperature at which recrystallization is spontaneous—0.3 T_m to 0.6 T_m depending on available time, strain hardening, purity, etc.

• **Reduction** Removal of oxygen from an oxide; the lowering of the valence level of an element.

• **Refining** Purification of a previously reduced metal.

Rolling Mechanical working between two cylindrical rolls.

Spinning Mechanical working of sheet metal on a rotating mandrel.

Strain hardening Increased hardness (and strength) as a result of plastic deformation.

CHECKS

• 4A Most metals are combined with _***_ or sulfur in their original ores. These are commonly reduced by _***_ at elevated temperatures.

• 4B The process of _***_ is one in which a molten metal is purified.

• 4C Since most molten metals _***_ during solidification, it is commonly necessary to use risers in a casting mold.

4D The _***_ and _***_ increase during cold working.

4E Strain hardening introduces a decrease in _***_.

4F The numbers of _***_ increase with plastic deformation. As a result, they cannot move as readily and more _***_ is required for slip.

4G In general, _***_ means heating a material so it will fracture less readily.

4H The process of annealing can eliminate _***_ in cold-worked metals.

4I Annealing increases the _***_ of a previously cold-worked metal.

4J Within a previously cold-worked metal, annealing causes _***_.

4K Recrystallization occurs within a metal by the growth of new _***_, which generally start along _***_ lines.

4L The recrystallization temperature depends on the length of _***_ available, the amount of _***_, and it is lower for _***_ metals.

4M The hot-working temperature for _***_ is higher than for aluminum and lower than for _***_.

4N The time for recrystallization follows a $\ln t$ vs. _***_ relationship, where temperature is expressed on the _***_ scale.

• 4O In machining metals, _***_ are preferred since the cut metal is easier to handle, and there is less _***_ at the tool tip.

• 4P In powder fabrication, the powders are generally _***_ and then bonded by _***_.

• 4Q _***_ and _***_ accompany the sintering of powder at high temperatures.

4R P/M preforms followed by hot pressing have two advantages over cast metal parts. These are _***_ and _***_.

STUDY PROBLEMS

• 4–1.1 Carbon monoxide (CO) reacts with iron ore (Fe_2O_3) to produce metallic iron (and CO_2). (a) How much carbon is required to form the CO necessary for reduction of 1000 kg of Fe_2O_3? (b) For the production of 1000 kg of Fe?

Answer: (a) 225 kg (b) 322 kg

• 4–1.2. At their melting temperatures, (a) how much energy must be supplied to separate 1 cm^3 of iron from Fe_2O_3? (b) 1 cm^3 of aluminum from Al_2O_3?

Answer: (a) 26 kJ/cm^3 [*Note:* This energy does not include the energy required for heating, or the energy losses that do not go to reduction.]

• 4–1.3. A wire must carry a load of 50 kg (110 lb) without permanent deformation. Assume the yield strength of iron is 50% greater than of aluminum. (a) What is the ratio of weight requirements for iron and for aluminum? (b) What is the ratio of energy requirements to reduce the metal (25°C)?

Answer: (a) 2-to-1 (b) 4-to-9

• 4–1.4. Joe is a Civil War buff and casts miniature artillery models out of lead. A mold for a cannon ball has a cavity diameter that is 4.9 mm. Assume the solidification is uniform from the surface inward. (a) What will be the final diameter of the ball? (b) What is the diameter of the central void, when solidification is complete? (Refer to Fig. 2–4.1)

4–2.1. (a) A 23-mm aluminum rod is to be cold worked 31%. What is the final diameter? (b) Repeat for the final thickness of a 2.3 mm sheet of aluminum.

Answer: (a) 19.1 mm (b) 1.6 mm

4–2.2. The final diameter of a brass wire is to be 1.3 mm after being cold worked 27%. What should the initial diameter be?

4–2.3. An iron rod is to have an ultimate strength greater than 350 MPa and an elongation of at least 30% (50-mm). (a) How much cold work is required? (b) Repeat for a 70–30 brass wire. (c) How about a copper wire?

Answer: (a) Between 23 and 30% (b) $4\% < e_{cw} < 15\%$

4–2.4. A rod may be either copper or a 70–30 brass but must be cold worked to have a ductility with El. (50-mm) = 12%. The load to be supported without breaking is 100 kg (980 N, or 220 lbs). What is the ratio of required copper for the two choices.

4–2.5. A metal must have a ductility > 40% El. (50-mm), and a S_u/ρ ratio of >40 N·m/g, that is, >40 MPa/(g/cm³). Select the lightest metal from Fig. 4–2.2 that will support a load of 35 kg (343 N or 77 lb).

Answer: (70–30 brass) $d = 1.08$ mm $M = 7.8$ g/m
(iron) $d = 1.15$ mm $M = 8.2$ g/m

4–3.1. (a) What diameter should an annealed iron wire have so that it may be cold drawn to a 2.1-mm diameter and have a hardness equivalent to 125 BHN? (b) Why will Rockwell hardness values (Table 3–3.1) be used for wire, rather than Brinell hardness values?

Answer: 2.36 mm

4–3.2. Can a 1.3-cm iron sheet be cold rolled to 1.0 cm and retain a hardness of less than 125 BHN?

4–3.3. Based on 0.4 T_m to 0.6 T_m, what is the recrystallization temperature range (a) of copper? (b) of lead?

Answer: (a) 270 to 540°C

4–3.4. A 65–35 brass wire must have a diameter of 0.88 mm and an ultimate strength of 424 MPa. How can this be prepared from an available 1.0 mm wire? (Extrapolate the data in Fig. 4–2.2 as is necessary.)

4-4.1. How long will it take the aluminum of Fig. 4-3.2 to start to recrystallize at 225°C?

Answer: 26 hrs (This may be read graphically, or solved from the data following Eq. (4-4.1).)

4-4.2. At what temperature will the aluminum of Fig. 4-3.2 start to recrystallize in 20 minutes? Solve (a) graphically, and (b) by calculation.

4-4.3. A cold-worked alloy is 50% recrystallized in 30 minutes at 130°C and in 24 hours at 90°C. (a) How long will it take at 110°C? (b) At what temperature will it be 50% recrystallized in one minute?

Answer: (a) 190 minutes

4-4.4. The progress of recrystallization is not linear; however, the 50% point is approximately the logarithmic mean of the start and finish. Determine A and B of Eq. (4-4.1) for $R_{50\%}$ for the metal in Fig. 4-3.2.

• 4-6.1. A P/M gear was made in a 2.2-cm diameter mold. When sintered to full density (0% porosity), the diameter is 1.94 cm. What was the initial porosity?

Answer: 31%

• 4-6.2. A second P/M gear was pressed in the same mold as the gear in S.P. 4-6.1. The sintered product has a diameter of 1.98 cm. What porosity remains?

CHECK OUTS

4A oxygen; carbon

4B refining

4C shrink

4D strength; hardness

4E ductility

4F dislocations; stress (force)

4G annealing

4H strain hardening

4I ductility

4J recrystallization (softening)

4K grains; slip

4L time; strain hardening (cold work); pure

4M copper; iron (tungsten)

4N $1/T$; kelvin (absolute)

4O chips; wear (heating)(friction)

4P compacted; sintering

4Q diffusion; shrinkage (densification)

4R more uniform; denser (maximum use of material)

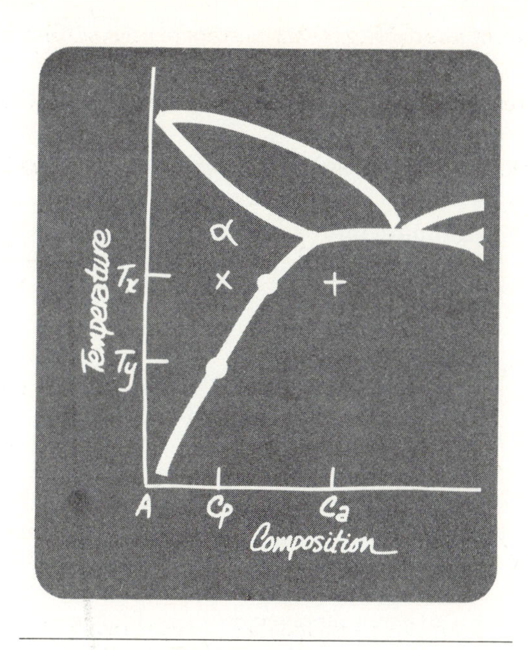

UNIT FIVE

ALLOYS OF METALS

By combining two or more metals into an alloy, the engineer gains various advantages. The number of available materials increases factorially, each with its specific set of properties. Thus, the number of design options increases. Not uncommonly, an alloy is cheaper than a pure metal, and may also have more desirable characteristics. By giving attention to the solubility limits of one metal in another, we shall be able to relate structure to composition and temperature, thus laying the basis for heat treatments to control properties.

Contents

Prerequisites Unit 2, and Sections 4–2 and 4–3.

From Unit 5, the engineer should

1) Be able to use phase diagrams, both qualitatively and quantitatively, since this is a principal tool available (a) to select and tailor alloys, and (b) to anticipate their behavior in service.

2) Understand the mechanism of solution hardening and precipitation hardening, as well as strain hardening (Unit 4), because these additional alternatives for strengthening permit greater engineering design options.

• 3) Become familiar with various commercial alloys such as brass, the bronzes, sterling silver, the light metals, and the cupro-nickels.

4) Solve problems and understand the technical terms at the end of the Unit.

5-1 ALLOYS

Commercially pure metals find many valuable uses, e.g., copper for conductors, aluminum for foil, and chromium for plated surfaces. *Alloys*, metals with two or more components, are even more valuable, if for no other reason than that their numbers far exceed the few dozen pure metals. In addition, the combination of two or more elements into an alloy always alters the properties of the individual metals. Appropriate combinations provide materials with enhanced properties for many applications—properties that would not be available from the pure metals.

The several elemental components that contribute to an alloy may be combined in two distinct ways: (a) as a solid solution, or (b) as a metallic compound. In a *solid solution*, the minor atom either substitutes for one of the major atoms, or it fits interstitially among the major atoms (Section 2-5). The structure of the original phase is retained. An *intermetallic compound* has a crystal structure unlike that of either component metal. It is a different phase. In a pure metal, each atom has its own kind of neighbors. In a compound, each atom has an energy preference for atoms unlike itself. Therefore, in the resulting structure, each atom of a compound surrounds itself with as many unlike neighbors as are available (Fig. 5-1.1b). This preference for unlike neighbors introduces a number of consequences that affect properties. For example, metallic compounds (and the ceramic compounds of Unit 11) are harder and more rigid than pure metals or solid solutions. We shall return to these properties in later sections of this unit, where we shall examine the conditions under which they form.

Illustrative problem 5-1.1 Refer to Fig. 2-5.1. Consider that the solid solution contains silver and gold, respectively, rather than copper and zinc. What is the weight percent of each?

SOLUTION As sketched this alloy contains 10 gold atoms and 22 silver atoms:

$$
\begin{array}{lrlr}
10\ \text{Au}\ (197.0\ \text{amu/Au}) & = 1970\ \text{amu} & = 0.45 & (\text{or } 45\ \text{w/o}) \\
\underline{22}\ \text{Ag}\ (107.87\ \text{amu/Ag}) & = \underline{2373}\ \text{amu} & = 0.55 & (\text{or } 55\ \text{w/o}). \\
32 \qquad\qquad \text{Total} & = 4343 & &
\end{array}
$$

ADDED INFORMATION The atom fraction of gold is 10/32 = 0.31, or 31 a/o. Since the atomic mass of gold is greater than that of silver, the w/o Au will be greater than a/o Au. Convention uses *weight percent*, w/o, for liquids and solids *unless* we specifically state otherwise. Thus, a 70% Cu–30% Zn, or more simply a 70Cu–30Zn, brass means 70 w/o copper and 30 w/o zinc. Had we meant atom percent, we would have used 70a/oCu–30a/oZn, or a statement such as 70/30 atom ratio of copper and zinc. ◄

Illustrative problem 5-1.2 The intermetallic compound that is found in Al–Cu alloys was analyzed to contain 46% Al and 54% Cu. What is the Cu/Al atom ratio?

SOLUTION Basis: 10,000 amu = 5400 amu Cu + 4600 amu Al. Using the atomic masses,

$$
\begin{array}{lr}
5400\ \text{amu}/(63.5\ \text{amu/Cu}) & = \ 85\ \text{Cu atoms} \\
4600\ \text{amu}/(27\ \text{amu/Al}) & = 170\ \text{Al atoms}
\end{array}
$$

The ratio is 1-to-2, Cu-to-Al; therefore, $CuAl_2$.

Figure 5-1.1 Alloys (copper-base). (a) Solid solution (25 a /o Au). The gold atoms substitute for copper atoms in the fcc lattice. At elevated temperatures, each lattice site has a 25% probability of containing a gold atom (and 75% for a copper atom). (b) Compound (CuZn). With equal numbers of copper and zinc atoms and below 460°C (980°F), each copper atom is surrounded by eight zinc atoms, *and* each zinc atom is surrounded by eight copper atoms. As a compound of copper and zinc, this alloy is more stable with *unlike* atomic neighbors.

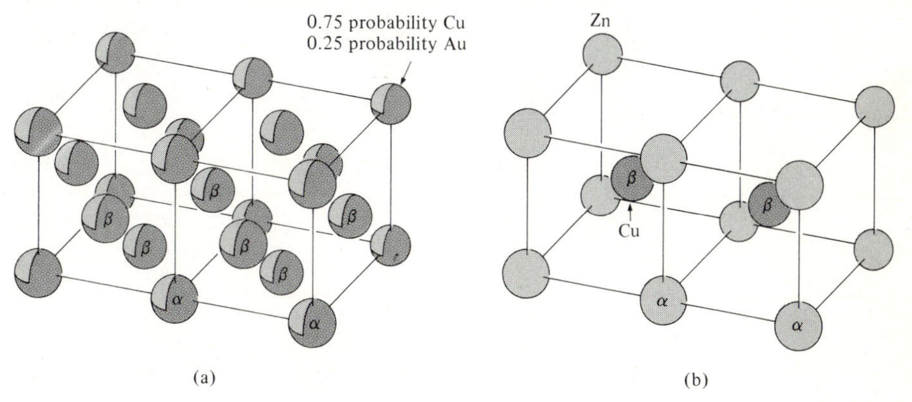

ADDED INFORMATION Commonly, an intermetallic compound varies somewhat from stoichiometry. In the above case, the percent aluminum can range from 45.5 to 47.5 w/o (or 66 to 68 a/o Al). ◂

5-2 SINGLE-PHASE ALLOYS

Ordinary *brass* is an alloy of copper and zinc that has only one phase. That phase is basically a copper phase but may contain up to 35%–40% zinc. The zinc atoms substitute for copper atoms without altering the original fcc structure that has a coordination number of 12 (Section 2-2). The 35%–40% maximum (depending upon the temperature) is the *solubility limit.* Additional zinc beyond that limit cannot be dissolved in the fcc structure but forms a second phase that is not fcc.* The vast majority of commercial brasses have less than 35% zinc. As a result, they possess only one phase, called *α-brass.*

Alloys of copper and nickel are completely soluble across the full composition range. Both copper and nickel are fcc; furthermore, their atomic radii match within a few percent—$R_{Cu} = 0.1278$ nm, and $R_{Ni} = 0.1246$ nm. Thus, an unlimited number of copper atoms may be replaced by nickel atoms and vice versa in *cupronickel.*

Since we can have solid solutions with equal numbers of the two elements (50a/oCu–50a/oNi), we can estimate the lattice constant and the density of such an alloy. This will be done in I.P. 5-2.1. The plot of these variables is given in Fig. 5-2.1. Note that we have used atom fraction rather than the more conventional weight fraction for the abscissa. (See the information added to I.P. 5-1.1.)

Solution hardening

We can also plot other properties for single-phase alloys as a function of composition. This is done in Fig. 5-2.2 for several of the more common mechanical properties. The copper–nickel curve spans the whole composition range from 100% copper to 100% nickel, because the two metals are completely soluble. The Cu–Zn curve extends only to 40% zinc, since that is the solubility limit for zinc in the fcc copper structure, as discussed earlier in this section.

Most notable among the data of Fig. 5-2.2 are the extra strengths (and hardness) that are obtained by solid solution. An alloy of 50Cu–50Ni is stronger than would have been anticipated by straight interpolation. Additions of nickel provide *solution hardening* to copper, *and* additions of copper provide *solution hardening* to nickel. The results are not linear.

---

* The new phase, called *β-brass,* is an intermetallic compound with approximately equal numbers of zinc and copper atoms. At low temperatures (<460°C), each copper atom is surrounded by eight zinc atoms and each zinc atom is surrounded by eight copper atoms (Fig. 5-1.1b).

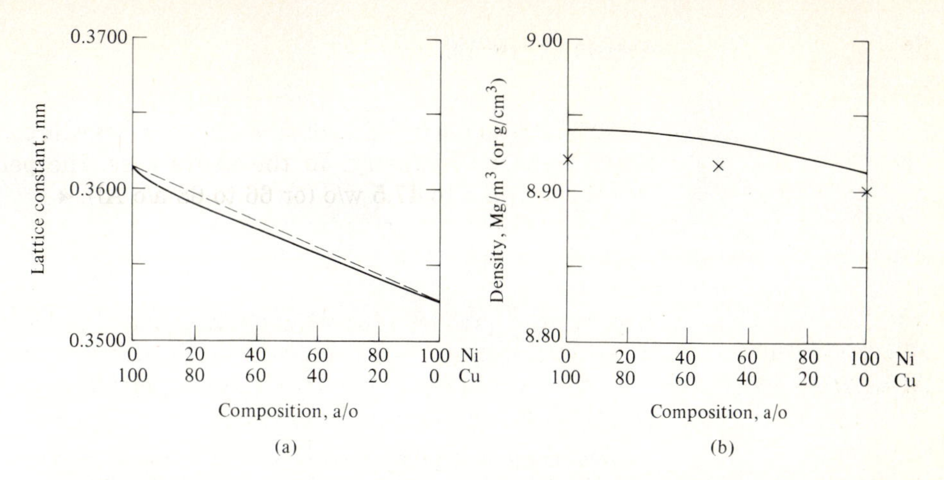

Figure 5-2.1 Solid solutions (Cu-Ni). (a) Lattice constant [dashed line, ideal; solid line, actual]. (b) Density [solid line, calculated from part (a); ×, from buoyancy measurements (± 0.02 g/cm^3 accuracy)]. Note: Since we are using atomic percent, a/o, we must be specific. If % were used, we would assume conventional weight percent (Fig. 5-2.2).

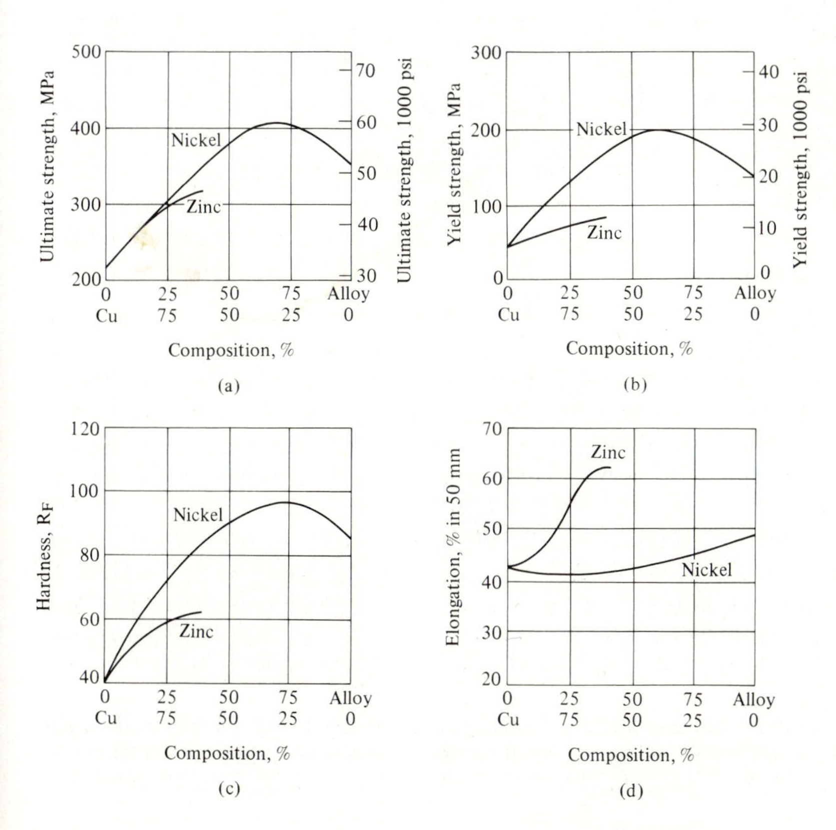

Figure 5-2.2 Solution hardening (annealed copper alloys). Impurity atoms such as Zn, Ni, or Sn serve as anchor points for dislocations. Therefore a greater shear stress is required.

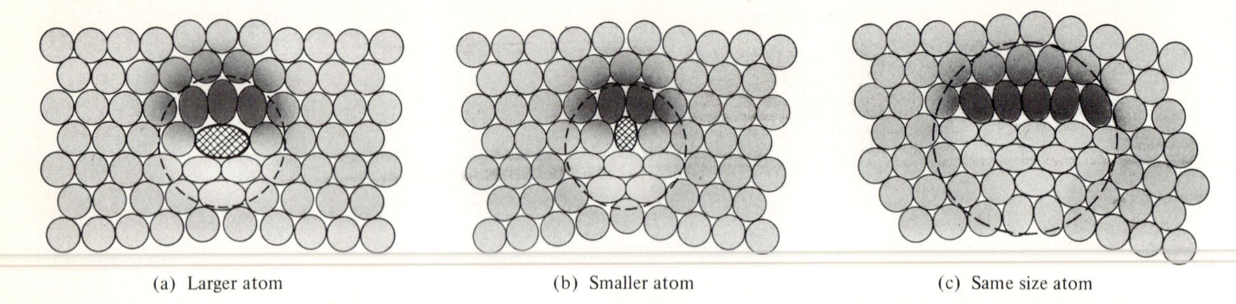

Figure 5-2.3 Impurity atoms and dislocations. An odd-sized atom, larger *or* smaller, decreases the strain around a dislocation. As a result, an increased shear stress is required to move the dislocation beyond the impurity atom.

Solution hardening is widely used by engineers. A brass is not only harder and stronger than is pure copper, but it is also cheaper, since zinc costs less than copper. Furthermore, solution hardening is not removed by annealing. True, a cold-worked brass will recrystallize (Fig. 4-3.1); however, the annealed brass is stronger than annealed copper (Fig. 5-2.2). This is true at elevated temperatures as well as at ambient temperatures.

The solution hardening and strengthening are accomplished because there is a strained region around each dislocation (Fig. 4-3.5). Recall from Section 4-3 that extra energy is required for these strained regions. *Either* a larger *or* a smaller atom reduces that strained region (Fig. 5-2.3). Thus, if a dislocation encounters an impurity atom as it moves during plastic deformation, the strained volume is reduced. The dislocation resists movement beyond the impurity atom because the strained region would have to be enlarged again. Consequently a larger stress is required for continued deformation.

Melting range

A pure metal melts at a unique temperature, for example, T_m = 1084.5°C (1984°F) for copper and 1455°C (2650°F) for nickel. An alloy of copper and nickel melts over a temperature range. To illustrate, on heating, a 30Cu-70Ni alloy starts to melt at 1345°C (2450°F), but does not become fully liquid until 1380°C (2515°F). Between those temperatures, it is a mixture of liquid and solid. The range of melting temperatures for Cu-Ni alloys is shown in Fig. 5-2.4. The lower curve of Fig. 5-2.4 is called the *solidus*, because all compositions are fully solid *below* that curve. The upper curve of Fig. 5-2.4 is called the *liquidus*, because all compositions are fully liquid *above* the temperatures of that curve. The liquidus is the solubility limit for nickel in the liquid.

Figure 5-2.4 is the *phase diagram* for Cu-Ni alloys. Only one solid phase exists in this alloy *system*. It is an fcc solid solution and is labeled α for convenience. Likewise, only one liquid phase exists. We usually do not label

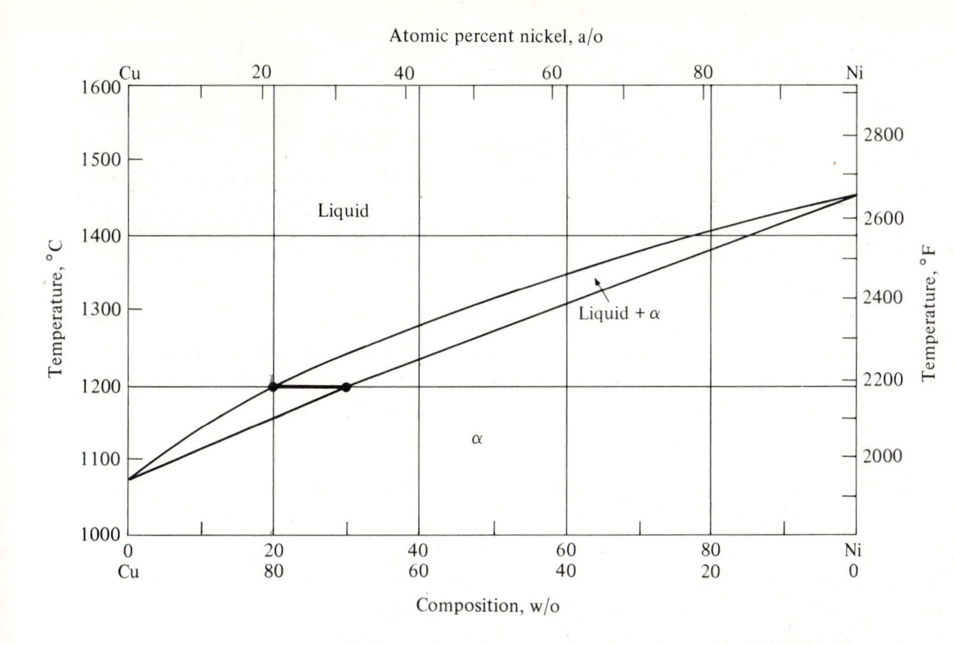

Figure 5-2.4 The Cu-Ni system. The 1200°C tie line is shown between 80 Cu-20 Ni and 70 Cu-30 Ni. (Adapted from ASM *Handbook of Metals*, Metals Park, Ohio: American Society for Metals.)

the liquid phase with a Greek letter, because we seldom encounter more than one within a materials system. (Most systems have more than one solid phase. The Cu-Ni system is an exception.)

Solid-liquid mixtures

Two phases (α + liq.) coexist between the solidus and liquidus of the Cu-Ni system. If the alloy is in equilibrium, the two phases *must* be at the same temperature. We can connect the compositions of the two phases with a *tie line*. One such tie line is shown in Fig. 5-2.4 as a heavy line at 1200°C. The left end of the tie line meets the liquidus at the composition of the liquid. The right end meets the solidus at the composition of the solid.

The relative amounts of the liquid and solid vary with composition and temperature. For example, at 1200°C (2190°F) pure copper is all liquid (Fig. 5-2.5), as is any alloy containing up to 20% Ni (>80% Cu). In contrast, pure nickel is an fcc solid (α) at 1200°C, as is any alloy containing up to 70% Cu (>30% Ni). Between these two limits, there is a mixture of liquid and solid. A 1200°C traverse across the two-phase region changes the alloy from all liquid at 20% Ni to all solid at 30% nickel. Halfway along the tie line between these two limits, at 75Cu-25Ni, there are equal amounts of the two phases, α + liq. By like token, 100 grams of a 73Cu-27Ni alloy contains 30 g of liquid and 70 g of α when in equilibrium at 1200°C.

Since no more than 20% Ni can be present in a Cu-Ni liquid at 1200°C (2190°F) according to Fig. 5-2.5, the liquids in *both* the 75Cu-25Ni and the 73Cu-27Ni alloys have a composition of 80Cu-20Ni. Any extra nickel must reside in the solid, α, phase. Likewise, 70% Cu is the maximum in α at

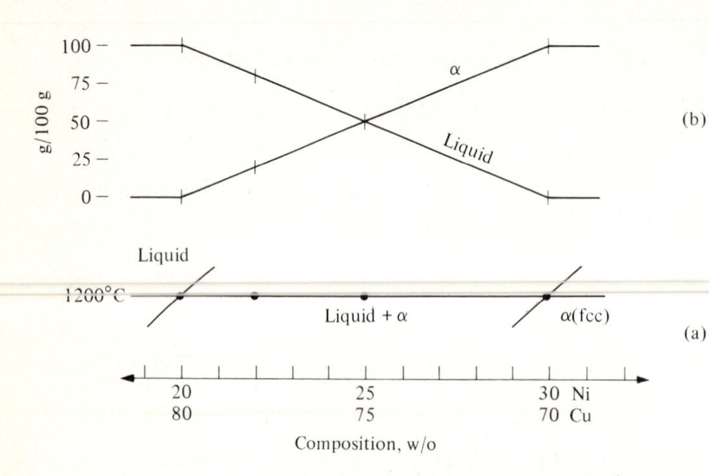

Figure 5-2.5 1200°C isotherm of Cu–Ni system. (a) Pure copper is liquid at 1200°C, as is any alloy containing up to 20% Ni. (Cf. Fig. 5-2.4.) (b) The amount of liquid decreases to zero at 30% Ni, and the amount of α increases from zero to 100 g α per 100 g alloy over the same composition range. The liquid cannot contain more than 20% Ni at 1200°C; the solid must contain at least 30% Ni.

1200°C. Therefore, the α in *both* the 75Cu–25Ni and the 73C–27Ni alloys has a composition of 70Cu–30Ni.

Illustrative problem 5-2.1 (a) Estimate the lattice constant for a solid-solution alloy with a 50/50 ratio of Cu-to-Ni atoms. (b) Based on the above answer, calculate the density of the alloy.

SOLUTION (a) The solid solution has an fcc structure as do copper (R = 0.1278 nm) and nickel (R = 0.1246 nm). Therefore, we are able to interpolate, using an average radius.

$$\bar{R} = 0.5R_{\text{Cu}} + 0.5R_{\text{Ni}} = 0.5(0.1278 \text{ nm}) + 0.5(0.1246 \text{ nm})$$
$$= 0.1262 \text{ nm}.$$

Since the structure is fcc,

$$\bar{a}\sqrt{2} = 4\bar{R},$$

or

$$\bar{a} = 4(0.1262 \text{ nm})/\sqrt{2} = 0.35695 \text{ nm}.$$

b) Basis: 25 unit cells = 100 atoms = 50 Cu + 50 Ni.

$$\bar{\rho} = \frac{[50(63.54 \text{ amu}) + 50(58.71 \text{ amu})]/(0.602 \times 10^{24} \text{ amu/g})}{25[0.35695 \times 10^{-9} \text{ m})^3]}$$
$$= 8.93 \times 10^6 \text{ g/m}^3 \qquad \text{(or } 8.93 \text{ g/cm}^3\text{)}.$$

ADDED INFORMATION The experimental density for this alloy is 8.915 g/cm³. (Cf. Fig. 5-2.1.) ◄

Illustrative problem 5-2.2 Specifications for a product that you are to manufacture call for a Cu-Ni alloy with a ultimate strength of at least 370 MPa (54,000 psi) and a yield strength of at least 140 MPa (20,000 psi). The company you own buys six tons of alloy per year for this product. What kind of alloy would you purchase?

SOLUTION From Fig. 5-2.2, we determine the alloy as follows.

For $S_u > 370$ MPa: 45% Ni $\longleftrightarrow$ 93% Ni.
For $S_y > 140$ MPa: 30% Ni $\longleftrightarrow$ 100% Ni.

Use 45% nickel (55% copper); this alloy meets both specifications and is cheaper than an alloy with a higher percentage of nickel.

ADDED INFORMATION If you are uncertain about the relative cost of copper and nickel, think of the change you have in your pocket. ◀

Illustrative problem 5-2.3 Consider 50 kg of an 20Cu-80Ni alloy.
a) The liquidus temperature is _***_°C.
b) The solidus temperature is _***_°C.
c) At 1400°C (2550°F), the liquid can contain up to _***_% nickel.
d) The maximum copper in α at 1400°C is _***_%.
e) At 1400°C, the major phase in this 20-80 alloy is _***_.
f) The ratio of liquid to solid at 1400°C is _***_ to _***_.

ANSWERS Supply answers for parts (a) through (f) and check yourself at the bottom of page 100. ◀

5-3 SOLUBILITY LIMITS

In the previous section, we read that brass can contain up to 35%–40% zinc, depending upon the temperature. Solid solution of zinc in fcc copper does not occur beyond those values. Thus, there are *solubility limits* within solid solutions just as there are limits of solubility in liquid solutions. For example, the solubility limits for copper in silver, and for silver in copper are

20 °C	<0.1% Cu in Ag-rich solid*	<0.02% Ag in Cu-rich solid†
200	~0.3	~0.1
400	1	0.5
600	3	2
779.4	8.8	8
779.4	28.1% Cu in Ag-Cu liquid	71.9% Ag in Ag-Cu liquid
800	36	75
900	62	92

Not surprisingly, the liquid alloy dissolves more solute than do the corresponding solids, simply because the liquid structure readily adapts to foreign atoms. The crystal structure is locally strained by the introduction of foreign atoms that differ in radius by 11–13% (0.1278 nm versus 0.1444 nm for copper and silver).

The solubility limits are presented on a single graph in Fig. 5-3.1. As with Figs. 5-2.2 and 5-2.5, the abscissa indicates that the weight percent-

* Pure silver is fcc with $a = 0.4084$ nm.

† Pure copper is fcc with $a = 0.3615$ nm.

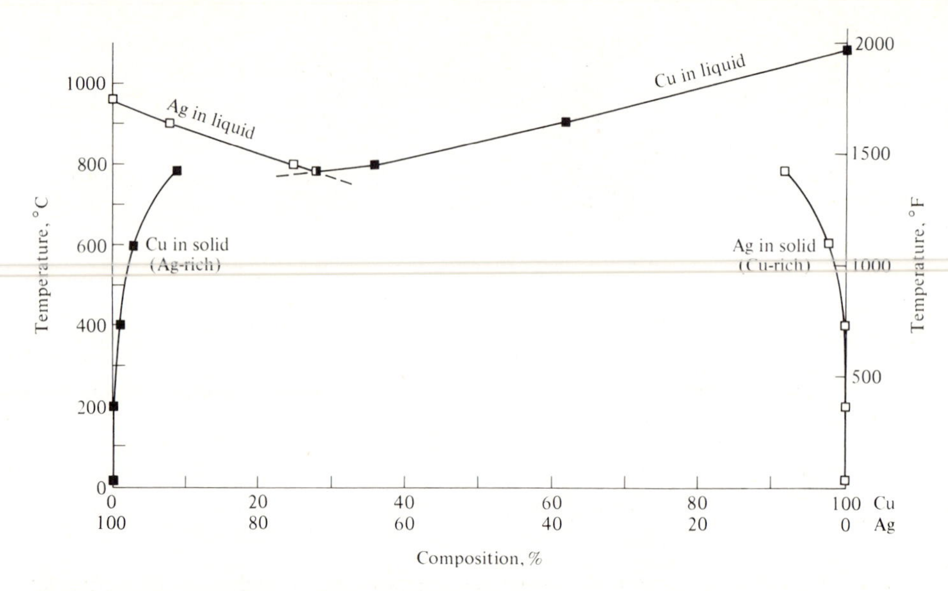

Figure 5-3.1 Solubility limits (Ag–Cu alloys) (a) Copper solubility limit, ■. (b) Silver solubility limit, □.

ages for both components. Increased silver decreases the percentage of copper since, of course, the sum of the two equals 100%.

Eutectic

The melting temperatures for silver and copper are 962°C (1764°F) and 1084°C (1984°F), respectively; however, alloys of the two have reduced liquidus temperatures (Fig. 5-3.1). The lowest melting temperature is 779.4°C (1435°F) for a 71.9Ag–28.1Cu alloy. This low melting temperature is called the *eutectic temperature*, T_e. The 71.9–28.1 composition is the *eutectic composition*, C_e. The two liquid solubility curves cross at this temperature and composition. The eutectic is always a low point on the liquidus.

Not uncommonly we make use of alloys of eutectic composition because of their lower melting temperatures. The eutectic alloy of lead and tin melts at only 183°C (361°F) to provide a *solder* for joining metals. Minimum heating is required in the factory processes and in on-the-job fabrication performed by plumbers and electrical technicians. Cast iron contains enough carbon to lower the liquidus of iron from 1538°C (2800°F) to approximately 1200°C (2200°F). The resulting alloy is made and cast with appreciably less fuel than is a low-carbon steel and also more readily fills complex molds, for example, an engine block for a car (Fig. 6-6.1).

Answers (I.P. 5-2.3) Refer to Fig. 5-2.4.
(a) 1410°C (b) 1380°C (c) 78% Ni (d) ~14% Cu (e) Liquid (f) ~3/1.

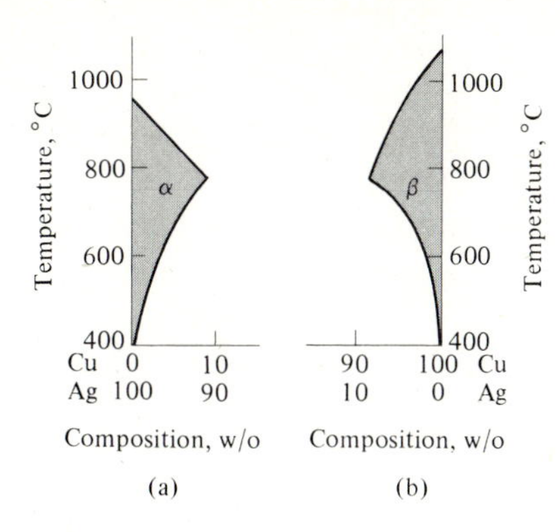

Figure 5-3.2 Solid solubility. (a) Copper in silver. (b) Silver in copper.

Solid solubility limits above T_e

The solid solubility curves in Fig. 5-3.1 are incomplete. Above the eutectic temperature, T_e, the limit of copper content in the Ag-rich phase, α, decreases from 8.8% to zero at T_m. A similar decrease occurs for silver in the Cu-rich phase, β (Fig. 5-3.2). Below T_e, copper is present as the copper-rich β phase when it exceeds the solubility limit for α. Above T_e, the excess copper is present as a liquid. This is shown in Fig. 5-3.3, which is the completed Ag–Cu phase diagram. Observe that excess silver (beyond the solubility limit for the Cu-rich solid solution, β) is present (1) as α below the eutectic temperature, and (2) as liquid above T_e.

Illustrative problem 5-3.1 A syrup may contain only 67% sugar (33% water) at 20°C (68°F), but 83% sugar at 100°C (212°F). One kilogram of sugar and 0.25 kg water are mixed and boiled until all the sugar is dissolved. During cooling, the solubility limit is exceeded so that (with time) excess sugar separates from the syrup. If equilibrium is attained, what is the weight ratio of syrup to sugar at 20°C?

SOLUTION Basis: 1.00 kg sugar + 0.25 kg H_2O = 1.25 kg.

$$\begin{aligned}\text{At } 20°\text{C, amount of syrup} &= 0.25 \text{ kg } H_2O/(0.33 \text{ kg } H_2O/\text{kg syrup}) \\ &= 0.757 \text{ kg syrup} \\ &= 0.493 \text{ kg excess sugar.}\end{aligned}$$

ADDED INFORMATION All the sugar is dissolved at the higher temperature, to give a single phase of syrup. When the mixture cools, the *solubility limit* for an 80 sugar–20 water composition is reached at 87°C (189°F). Typically, however, some *supercooling* is encountered before the excess phase (in this case, sugar) starts to separate. In fact, in this case, supercooling can proceed

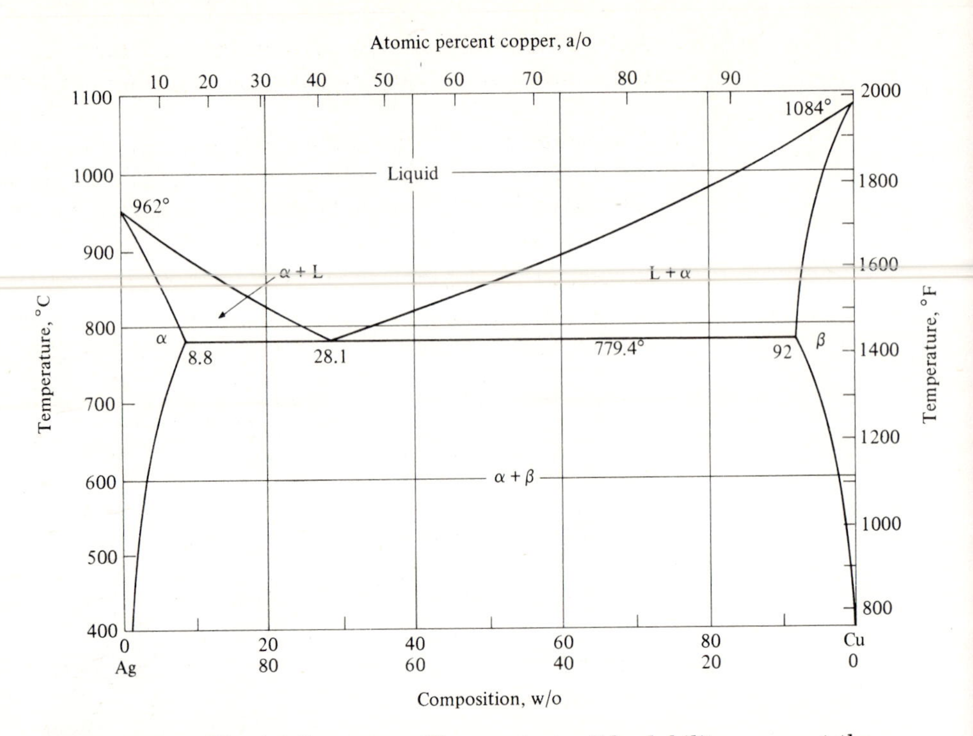

Figure 5-3.3 The Ag–Cu system. The greatest solid solubility occurs at the eutectic temperature. (Adapted from ASM *Handbook of Metals*, Metals Park, Ohio: American Society for Metals.)

to room temperature so that the start of separation may be delayed considerably. Similar supersaturation commonly occurs in metals and other materials. ◀

Illustrative problem 5-3.2 Sixty grams of a 60Ag–40Cu alloy are heated to 900°C (1650°F). How many grams of copper must be added to saturate the liquid?

SOLUTION From Fig. 5-3.3, the copper solubility limit in the liquid alloy is 62%. We will add no silver; therefore, a copper balance in grams gives

$$0.40(60) + x = 0.62(60 + x)$$
$$0.38x = 37.2 - 24$$
$$x = 34.74 \text{ g copper added.}$$

Checking with the silver,

$$0.60(60 \text{ g}) = 0.38(34.74 \text{ g} + 60 \text{ g}).$$

ADDED INFORMATION A *material balance* checks that the total amount of one component is equal to the sums of the amounts in each phase. ◂

• **Illustrative problem 5-3.3** Sterling silver contains 92.5Ag–7.5Cu. At 600°C (1110°F) the solubility limits are 3% Cu in the Ag-rich phase (A), and 2% silver in the Cu-rich phase (C). Consider 50 troy ounces (1555 g) of sterling silver. What fraction of the copper is in the Ag-rich phase?

SOLUTION Basis: 1555 g = 1438 g Ag + 117 g Cu; and A g of the Ag-rich phase + C g of the Cu-rich phase. Since 2% is the maximum Ag in C, and all but 3% of A is silver,

$$
\begin{aligned}
1438 \text{ g Ag} &= 0.97\,A + 0.02\,C \\
&= 0.97\,A + 0.02\,(1555 - A) \\
0.95\,A &= 1438 - 31 \text{ g} = 1407 \text{ g.}
\end{aligned}
$$

$$A = 1481 \text{ g,} \qquad \text{(or 47.6 oz).}$$

Therefore, $C = 1555 \text{ g} - 1481 \text{ g} = 74 \text{ g,}$ (or 2.4 oz).

$$\text{Copper in A} = \frac{(1481 \text{ g})(0.03)}{117 \text{ g}} = 0.38.$$

ADDED INFORMATION Observe that we can speak of various fractions (or percentages):

a) Compositions of phases; for example, 92.5% Ag in sterling silver.
b) Fraction of phases; for example, 74 g/1555 g (or ~5%) being the Cu-rich phase, C.
c) Distribution of copper among phases; for example, 38% in A (and 62% in C). ◂

5-4 USE OF PHASE DIAGRAMS

There are three distinct ways one can use a phase diagram when equilibrium exists. First, the phase diagram indicates *what phase(s)* will be present for any alloy composition and at any selected temperature. Second, the phase diagram shows the *composition of each phase* of the alloy at the selected temperature. Finally, one may calculate the relative *amounts of each phase* for the desired temperature and composition.

(1) What phase(s)?

We simply read the *phase field* from the diagram. Some fields contain only one phase. If solubility limits are exceeded, there are two phases. These 2-phase fields always lie between single-phase fields. We illustrate this by the 200°C isotherm of the phase diagram for the Pb–Sn system (Fig. 5-4.1).

Moving from pure lead to 17% Sn (83% Pb), the alloy is a lead-rich (fcc) phase that we shall call α. Beyond the solubility limit of 17% Sn in α, the 2-phase field contains (α + L) until the amount of lead decreases to 43% (57% Sn). Between 57% Sn and 76% Sn (24% Pb), only liquid is present. Again a 2-phase field exists between 76% Sn and 97% Sn (3% Pb). The two phases are (L + β)—the latter, β, being a tin-rich phase with the same structure as pure metallic tin.

(2) Phase compositions?

By *phase composition* we mean the chemical analysis of the phase expressed *in terms of the two components*. Consider the plumber's solder that contains 85 Pb–15 Sn, which is a single-phase liquid at 325°C. The composition of that phase is 85% Pb–15% Sn. At 250°C, there are two phases (α + L) in equilibrium. According to Fig. 5-4.1, the solid phase, α, can contain no more than 12% Sn, its solubility limit. Hence, the α is 88Pb–12Sn. Conversely, the liquid can contain no more than 64% Pb; thus, it has a composition of 64Pb–36Sn. In a 2-phase field, the compositions of the two phases are de-

Figure 5-4.1 The Pb–Sn system. The eutectic composition (~60 Sn–40 Pb) is used for low-melting solders. The tie line in the 2-phase field (250°C) connects the composition of α (88Pb–12Sn) and the composition of the liquid (64Pb–36Sn).

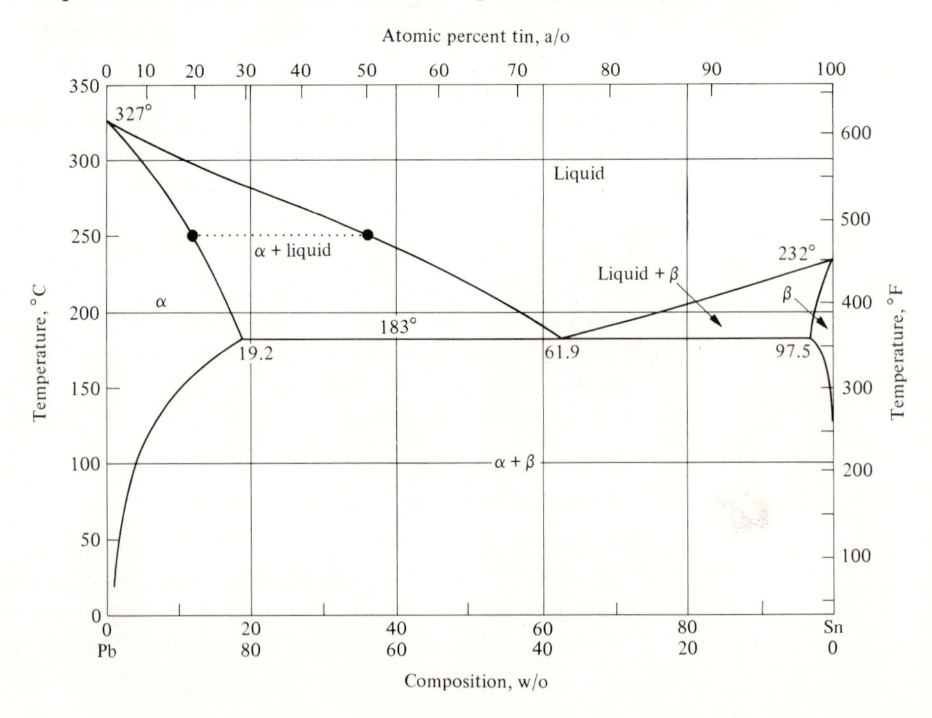

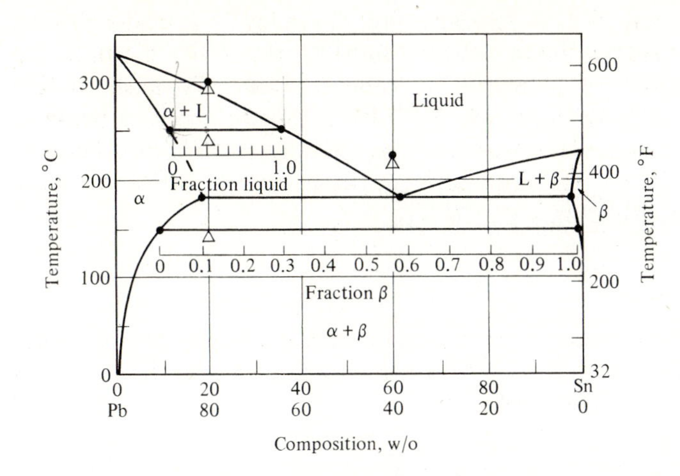

Figure 5-4.2 Quantities of phases (Pb-Sn alloys). An 80Pb-20Sn alloy contains α and β at 150°C. By interpolation, the fraction of β is 0.11. This same alloy contains 0.33 liquid at 250°C (and 0.67 α). See text.

termined by the *intersections of the tie line with the two adjacent solubility curves* (Fig. 5-4.1) for the temperature of concern. (In a 1-phase field, both components are soluble; therefore, we do not use a tie line.)

(3) Amounts of phases?

By the *amount of each phase* we mean the mass, the fraction, or the percentage of each phase. With two phases ($\alpha + \beta$), this involves grams of α and grams of β, or the weight fraction of α and the fraction (or percentage) of β.*

To determine the fraction of each phase of a 2-phase alloy, let us select the desired temperature and progressively change the alloy's composition. For example, at 250°C and with a 90Pb-10Sn alloy, only α exists (Fig. 5-4.2). With added tin, we have only α (100% α) until the tin composition of the alloy is 12%; beyond that, additional tin produces some liquid. There is a mixture of α and liquid between 12% Sn and 36% Sn. With still larger additions of tin, the composition of the alloy is located in the single-phase region of the liquid (Fig. 5-4.2). Along the tie line between 12% Sn (all α) and 36% Sn (all liquid), there is a progressive decrease in solid α from 100% to 0% (and a progressive increase in liquid from 0% to 100%). An alloy of 80Pb-20Sn is one third of the way across the 2-phase area at 250°C; therefore, it contains one-third liquid (and two-thirds α). To determine the *amount* of either phase in 2-phase area of a phase diagram at a given temperature, simply *interpolate along the tie line between the 0% side and the 100% side of the **2-phase** area.*

---

* Observe that to say a phase, α, contains 88% Pb and 12% Sn is different from that of saying that the solder contains $\frac{2}{3}\alpha$ and $\frac{1}{3}$ liquid (or 67% and 33% liquid). The initial statement gives the composition of a phase, α, in terms of the components; the latter statement shows how the alloy is divided between the two phases.

Finally, we can also give the composition of the alloy, e.g., the plumber's solder with 85% Pb and 15% Sn. As with the phase composition, we express the alloy's composition in terms of the chemical components.

There are several procedures for interpolation. The choice is yours. The most obvious is to use a transparent millimeter rule. Alternatively, the abscissa provides a usable scale, e.g., with the above 80Pb–20Sn solder at 250°C:

$$\text{mass fraction liquid} = \frac{20\%\text{Pb} - 12\%\text{Pb}}{36\%\text{Pb} - 12\%\text{Pb}} = \frac{1}{3},$$

or

$$\text{mass fraction liquid} = \frac{80\%\text{Sn} - 88\%\text{Sn}}{64\%\text{Sn} - 88\%\text{Sn}} = \frac{1}{3}.$$

The fraction α would be determined from

$$\text{mass fraction } \alpha = \frac{36\%\text{Sn} - 20\%\text{Sn}}{36\%\text{Sn} - 12\%\text{Sn}} = \frac{2}{3}.$$

This calculation is commonly called the *lever rule.**

• **Fraction charts**

Sometimes it is advantageous to plot the fraction (or percentage) of a phase that is present as a function either of the composition or of the temperature. In Fig. 5–4.3, we have plotted the percentage of α, in Pb–Sn alloys at 150°C, as a function of alloy composition. In Fig. 5–4.4, the percentage of α in a 60Pb–40Sn alloy is plotted as a function of temperature. In either case, the complement percentage indicates the amount of the second phase in the alloy.

Intermetallic compounds

In introductory chemistry we learned that a compound is a phase of two (or more) components in a definite atom ratio. Thus, H_2O possesses a 2:1 ratio of hydrogen and water, and vinyl chloride is C_2H_3Cl.

Compounds may also exist between two (or more) metals. Some of these contain a nearly fixed ratio of atoms, for example, $CuAl_2$, Fe_3C, and Cu_3Sn.† More commonly, an intermetallic compound is *nonstoichiometric* and pos-

---

* Although *interpolation is recommended,* because it is not necessary to recall formal equations, one may establish a formula for the lever rule.

$$\%X = \frac{C_y - C}{C_y - C_x}100, \tag{5–4.1}$$

where C_x is the composition of phase X, C_y is the composition of the second phase, Y, and C is the composition of the alloy. The composition may be expressed in terms of either component, that is, *either* Pb *or* Sn.

† Even these compounds may be slightly nonstoichiometric; for example, the ratio of Al-to-Cu is not precisely 1-to-2 in $CuAl_2$. In contrast, pure water is *stoichiometric* with a 2-to-1 ratio within the detection of any available measuring technique.

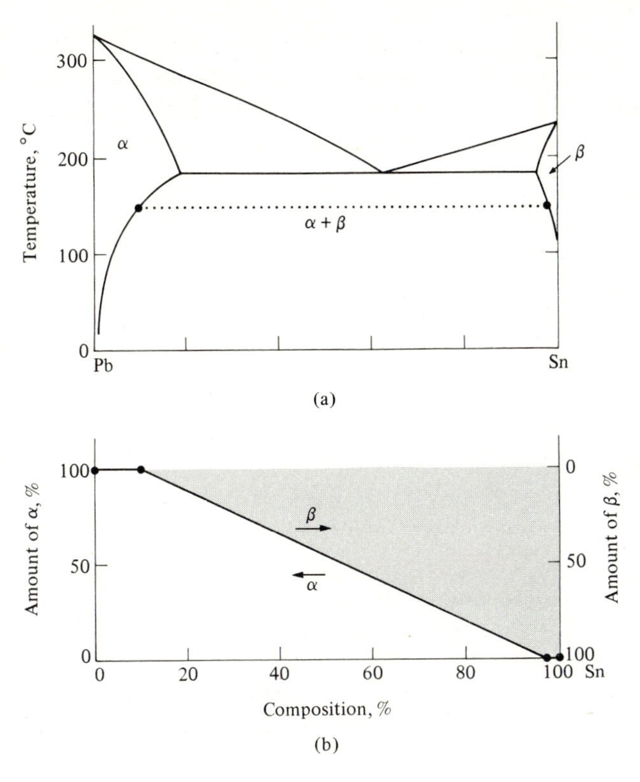

Figure 5-4.3 Amount of α versus compositions (Pb–Sn at 150°C). Linear interpolation may be used along the tie line in the 2-phase field. The amount of β is complementary (100 − % α, right ordinate).

Figure 5-4.4 Amount of α versus temperature (60Pb–40Sn). Interpolation should *not* be made vertically but *must* be performed horizontally along the tie lines for each temperature. (After all, to be in equilibrium, the two phases must be at the same temperature.) The amount of the second phase is complementary (upper abscissa).

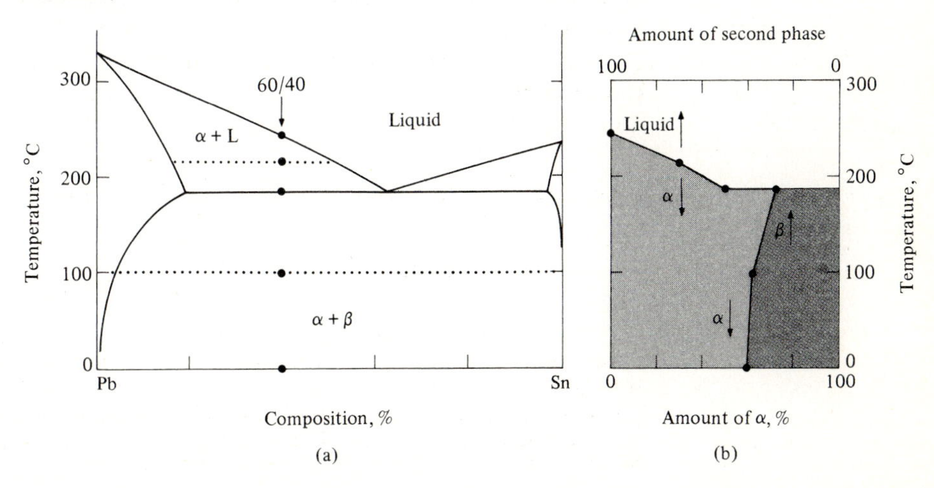

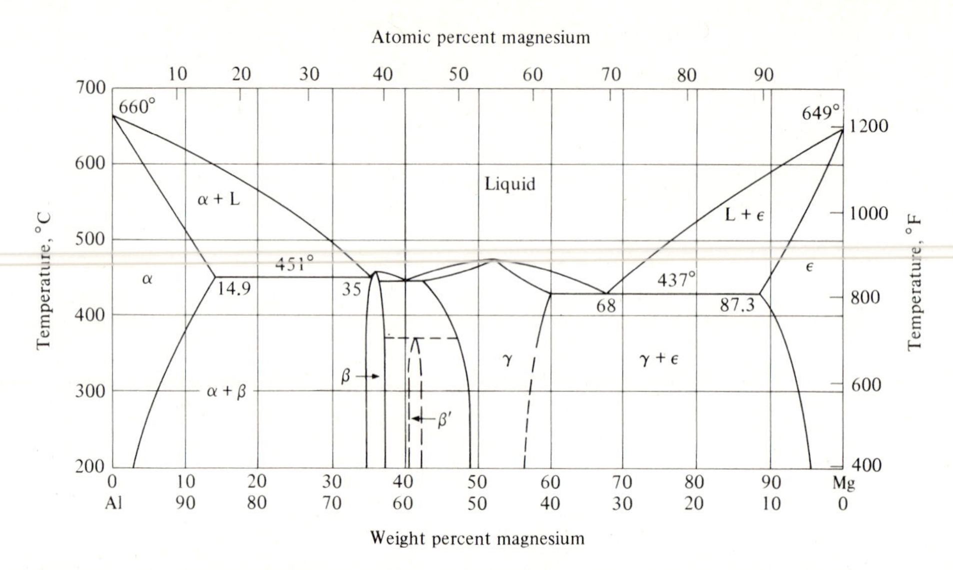

Figure 5-4.5 Al–Mg diagram. (The dashed lines may be subject to a slight revision.) (Adapted from *Metals Handbook*, Metals Park, Ohio: American Society for Metals.)

sesses a compositional range. Thus, the phase called β in the Al–Mg system (Fig. 5-4.5) is Al_5Mg_3; however, its composition ranges from 37 to almost 40 *atom* percent* magnesium (versus the nominal 37.5 a/o Mg).

Intermetallic compounds, like their ceramic cousins, are hard and brittle. We shall examine the reason in Unit 11. In the meantime, we will note that these nonductile phases commonly add strength and decrease the ductility of metals.

Illustrative problem 5-4.1 A solder containing 60Pb–40Sn is melted (a) at 300°C (572°F) and cooled (b) to 200°C (392°F); (c) to 50°C (122°F). Cite the equilibrium phase(s) and their compositions at each of the three temperatures.

SOLUTION Refer to Fig. 5-4.1 and the tie lines of Fig. 5-4.6.

a) 300°C:	Only liquid	60Pb–40Sn
b) 200°C:	α	83Pb–17Sn
	+	
	liquid	43Pb–57Sn
c) 50°C:	α	98Pb–2Sn
	+	
	β	~100Sn

---

* Although compositions are generally presented as weight percent (lower abscissa), the atom percent is also shown in most of the phase diagrams of this text (upper abscissa).

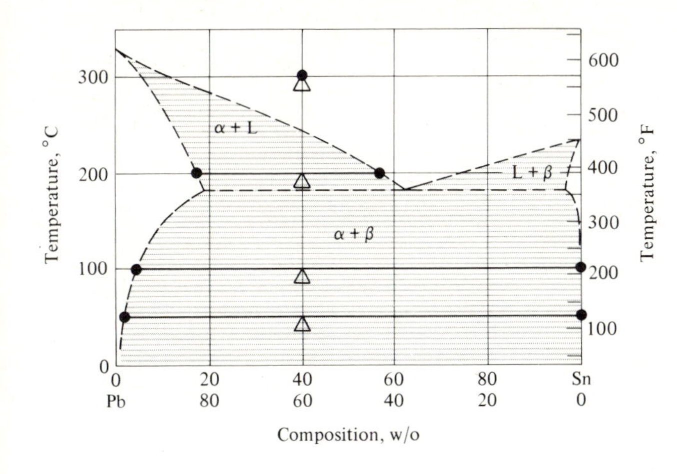

Figure 5-4.6 Temperatures and compositions for I.P. 5-4.1. (Cf. Fig. 5-4.1.)

ADDED INFORMATION The curves at the lower edges of the liquid field in Fig. 5-4.1 are also solubility-limit curves. The one to the left is the solubility-limit curve for lead in the liquid phase. The curve to the right is the solubility-limit curve for tin in the liquid phase.

A common—but not universal—procedure is to label the solid phases as α, β, γ, etc., across the phase diagram from left to right. Thus in the Pb-Sn system (Fig. 5-4.1), α is a lead-rich phase; in the Cu-Zn system (Fig. 5-4.7), α is a brass, a copper-rich phase. This procedure seldom, if ever, causes confusion arising from the multiplicity of α phases. ◄

Illustrative problem 5-4.2 Refer to Fig. 5-4.7 for the Cu-Zn phase diagram. Note that a 70Cu-30Zn brass is single-phase, α, at all temperatures below 920°C. Since this is a copper-rich Cu-Zn solid solution that is contiguous to pure copper, we know that α is fcc.

Cite the phase(s) and their compositions for a 26Cu-74Zn alloy at (a) 600°C (1112°F); (b) at 525°C (977°F).

SOLUTION

a) Only δ, therefore 26Cu-74Zn.
b) γ, which is 30Cu-70Zn;
   and ϵ, which is 21.5Cu-78.5Zn.

ADDED INFORMATION Don't panic because of the apparent complexity of phase diagrams, such as the one for the Cu-Zn system. This problem is included to indicate that even in the more formidable looking areas, one can cite the phase(s) and compositions by focusing on the small area only.

Fortunately the commonly used alloys are found in simpler-looking areas of phase diagrams. Brass is used extensively; the 26Cu-74Zn alloy has extremely rare applications. ◄

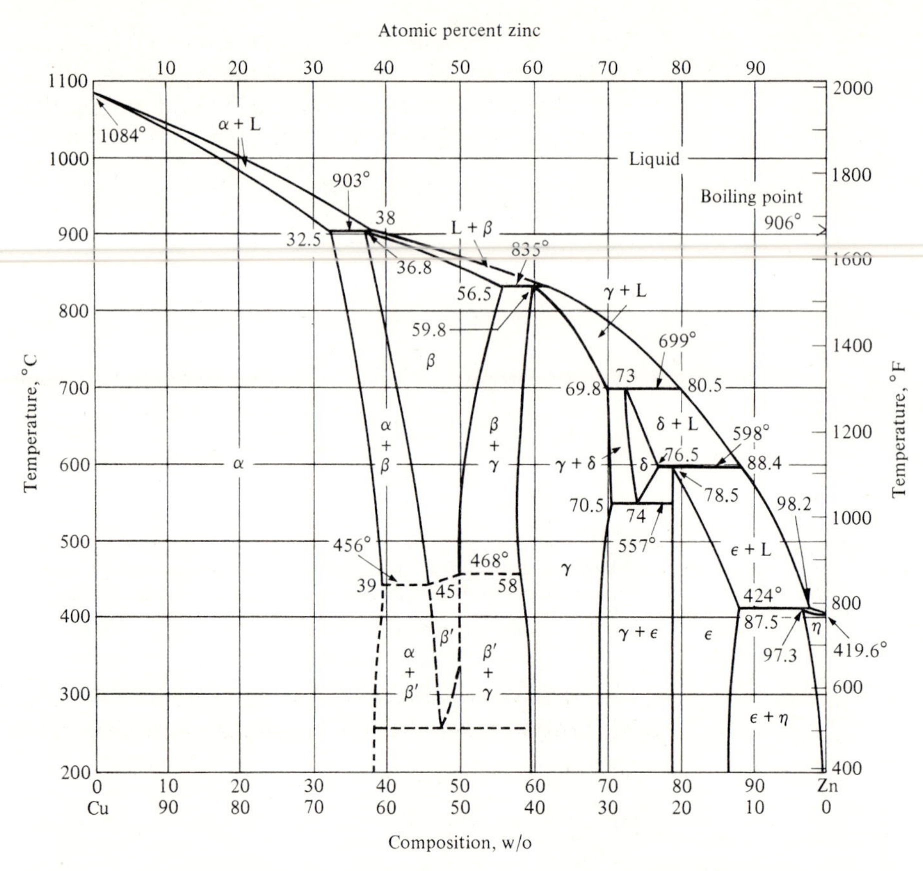

Figure 5-4.7 Cu-Zn diagram. The data for the Cu-Zn alloys in Fig. 5-2.2 were for alloys within the α region of this system. (Cf. Cu-Ni alloys in Figs. 5-2.2 and 5-2.4.) (Adapted from *Metals Handbook*, Metals Park, Ohio: American Society for Metals.)

Illustrative problem 5-4.3 Five kilograms (11 pounds) of a 90Mg-10Al alloy (Fig. 5-4.5) are melted and cooled slowly to 495°C (920°F). By making a material balance on the magnesium, determine the amount of ϵ that is present.

SOLUTION Let E = kilograms ϵ, and (5 − E) = kg liquid.

$$(0.90)(5\text{ kg}) = (0.92)(E) + (0.76)(5 - E),$$
$$(0.90)(5\text{ kg}) - (0.76)(5) = 0.92E - 0.76E,$$
$$E/5\text{ kg} = (0.90 - 0.76)/(0.92 - 0.76)$$
$$= \text{fraction } \epsilon,$$
$$E = 4.38\text{ kg } \epsilon \text{ (and 0.62 kg L).} ◀$$

Illustrative problem 5-4.4 An alloy of 95Al-5Si is to be used as an 8-kg (17.6-lb) casting in an automobile engine. (a) What phases are present at each 100°C interval between 700°C and ambient? (b) What are the phase compositions? (c) How much of each phase at each temperature?

SOLUTION Use Fig. 5-4.8. At 600°C and by interpolation along the tie lines, the amount of liquid equals (4mm/8mm)8000g = 4000g, or using the abscissa [(5−1)/(9−1)]8000g = 4000 g liquid. Use the same pattern at other temperatures.

	Phase(s)	Composition(s)	Amount of each phase*
700°C	Only liquid	95Al-5Si	8000 g L
600°C	L	91Al-9Si	4000 g L
	α	99Al-1Si	4000 g α
500°C	α	99Al-1Si	7676 g α
	β	<0.17Al->99.83Si	324 g β
≤400°C	α	~100Al	7600 g α
	β	~100Si	400 g β ◂

Illustrative problem 5-4.5 Determine the amount of α and β in a 60Pb-40Sn alloy at 100°C (assuming equilibrium).

SOLUTION Refer to Fig. 5-4.6, or 5-4.1.

$$\frac{\beta}{\alpha+\beta} = \frac{0.04 - 0.40}{0.04 - 1.00} = 0.375\beta.$$

$$\frac{\alpha}{\alpha+\beta} = \frac{1.00 - 0.40}{1.00 - 0.04} = 0.625\alpha.$$

ADDED INFORMATION The above calculation was based on the tin contents. We could have made the same calculation based on lead contents.

$$\frac{\beta}{\alpha+\beta} = \frac{0.96 - 0.60}{0.96 - 0} = 0.375\beta.$$

The phrase "assuming equilibrium" will be implied in the remainder of this unit, since a phase diagram is, *per se*, an equilibrium diagram. In Section 5-6 we shall discuss nonequilibrium conditions in metals. ◂

Illustrative problem 5-4.6 Five kilograms (11 lbs) of a 20Ag-80Cu alloy are completely melted and then cooled. How many kilograms of β are there at each 100°C interval from 1000°C (1832°F) down to 400°C (752°F)?

* Sample calculation above

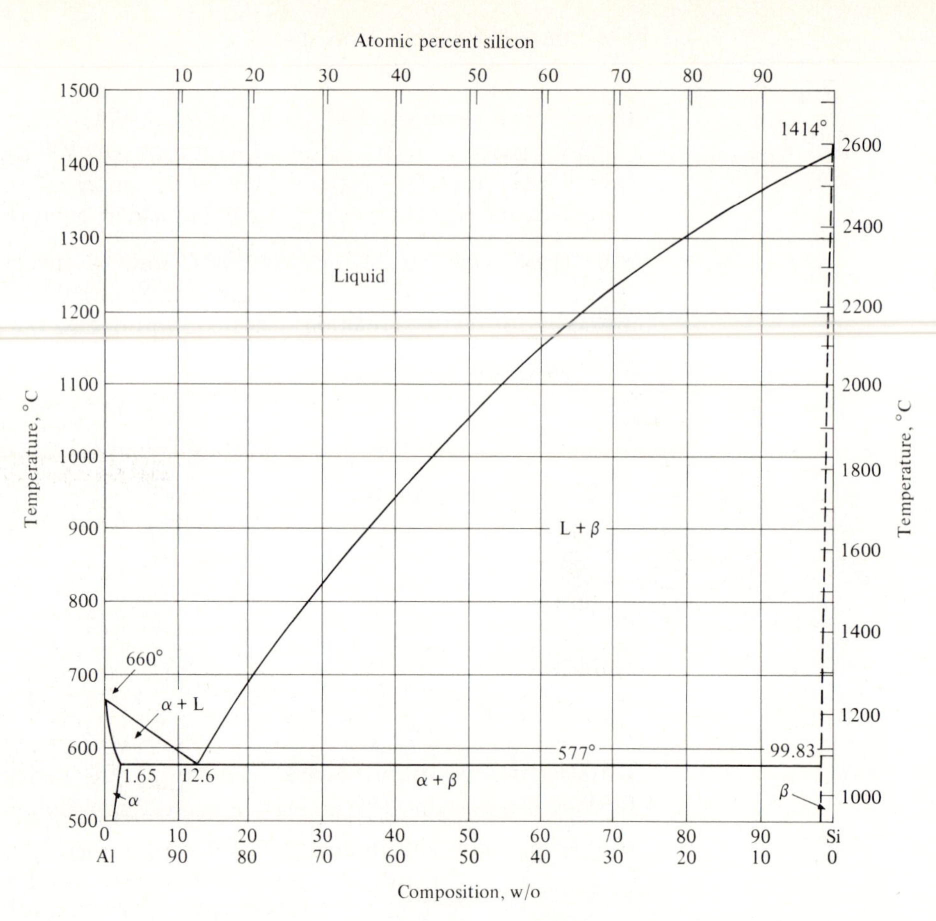

Figure 5-4.8 Al-Si diagram. (Adapted from *Metals Handbook*, Metals Park, Ohio: American Society for Metals.)

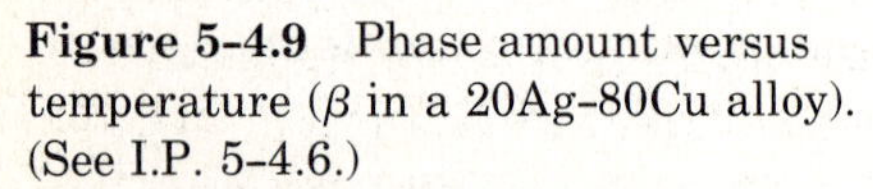

Figure 5-4.9 Phase amount versus temperature (β in a 20Ag-80Cu alloy). (See I.P. 5-4.6.)

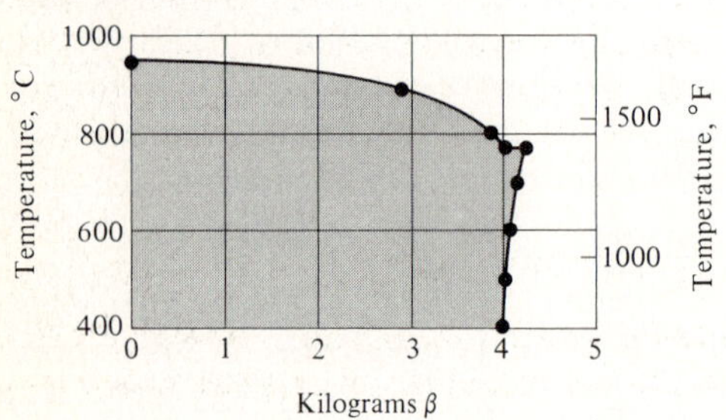

Figure 5-4.10 Phase fractions versus temperature (50Pb-50Sn). (See S.P. 5-4.14.)

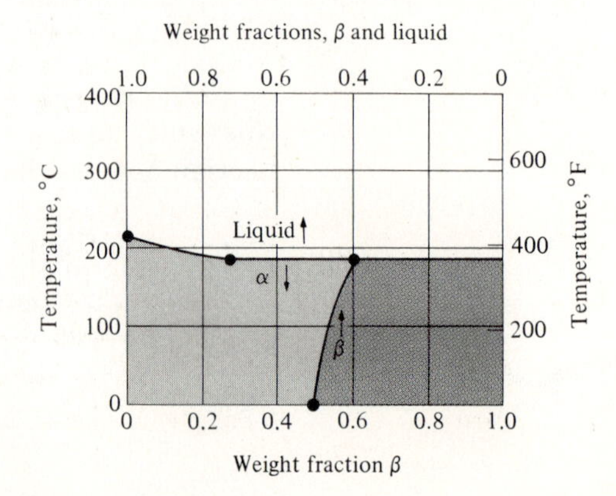

SOLUTION Using the silver content, calculate as follows from Fig. 5-3.3.

1000°C: All liquid $= 0$ kg β
900°C: $5[\beta/(\beta + L)] = (5\text{ kg})(0.37 - 0.20)/(0.37 - 0.07) = 2.83$ kg β
800°C: $5[\beta/(\beta + L)] = (5\text{ kg})(0.66 - 0.20)/(0.66 - 0.08) = 3.96$ kg β
700°C: $5[\beta/(\beta + \alpha)] = (5\text{ kg})(0.94 - 0.20)/0.94 - 0.05) = 4.16$ kg β
600°C: $5[\beta/(\beta + \alpha)] = (5\text{ kg})(0.965 - 0.20)/(0.965 - 0.02) = 4.05$ kg β
500°C: $5[\beta/(\beta + \alpha)] = (5\text{ kg})(0.98 - 0.20)/(0.98 - 0.01) = 4.02$ kg β
400°C: $5[\beta/(\beta + \alpha)] = (5\text{ kg})(0.99 - 0.20)/(0.99 - \text{nil}) = 3.99$ kg β

ADDED INFORMATION We would have obtained the same answers had we used the copper content.

A plot of the weight fraction β versus temperature is shown in Fig. 5-4.9. Note the discontinuity at the eutectic temperature (780°C or 1436°F). ◂

• 5-5 COMMERCIAL ALLOYS

We now have a basis for introducing many commercial alloys that are available to the engineer. These will involve (a) *copper-base alloys*, which include the brasses and bronzes; (b) the *light metals*, predominantly aluminum and magnesium; (c) *titanium alloys*; and (d) the "*white*" *metals*. We shall defer steels until Unit 6, since the iron-carbon phase diagram has some special features that need to be considered first.

Copper-base alloys

The historical importance of copper-base alloys becomes apparent to us when we read about the Bronze Age, in which Cu–Sn alloys significantly affected the advance of civilization. These alloys were primarily copper with 5 to 10 percent tin. Several factors influenced their early development and use: (a) they are readily reduced from their ores; (b) their melting temperatures are easily attained; (c) they are single-phase alloys in the hot-working range (Fig. 5-5.1); and (d) they corrode slowly. The combination of the above factors permitted the early artisan to develop bronze into a widely used material. Nowadays, copper-base alloys have two main types of applications: (1) as corrosion-resistant metals, and (2) as electrically and thermally conducting metals.

Brasses are alloys of copper and zinc. The *α-brasses* contain less than 35% Zn so that they remain single-phase at all temperatures below their solidus (Fig. 5-4.7). The α phase is fcc and may be strain-hardened as shown in Fig. 5-2.2. A *β-brass* has more zinc (~50 a/o) and is therefore cheaper. Its structure is bcc above ~460°C (~860°F). Below that temperature, however, it becomes ordered so that each zinc atom is surrounded by eight copper atoms, and each copper atom is surrounded by eight zinc atoms (Fig.

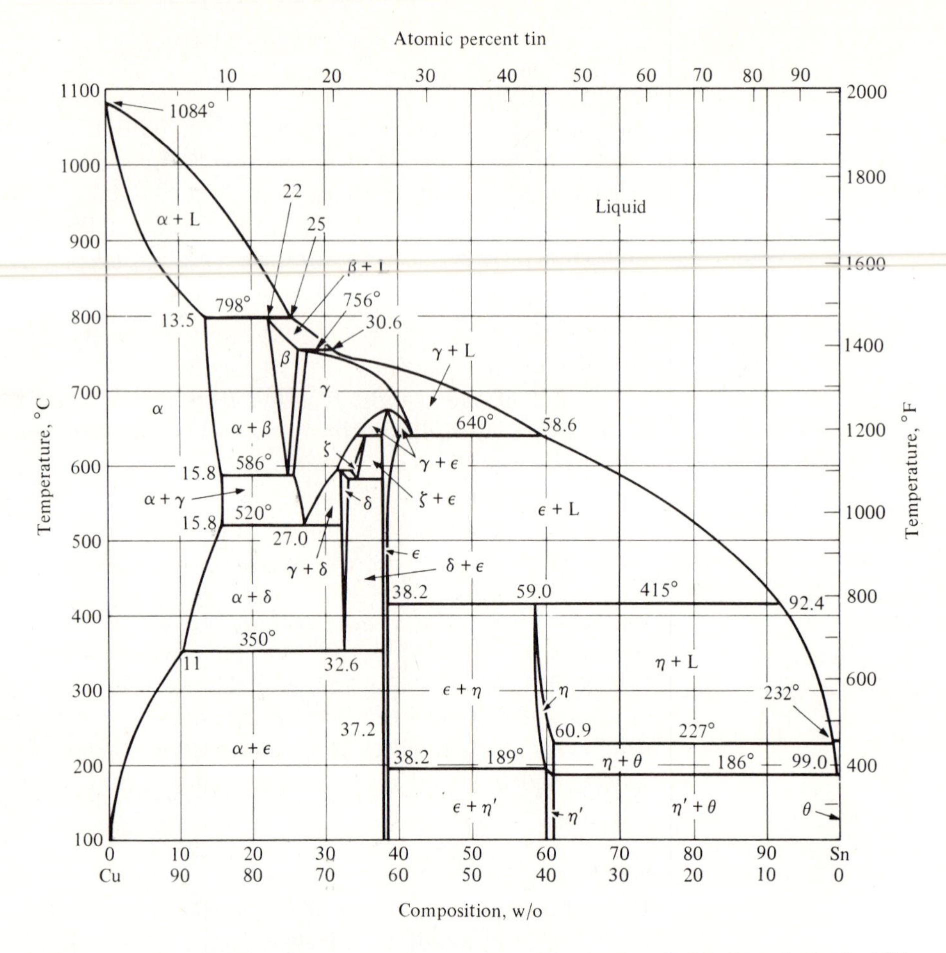

Figure 5-5.1 Cu–Sn diagram. (Adapted from *Metals Handbook*, Metals Park, Ohio: American Society for Metals.)

5-1.1b).* As a result, β-brass is hard and nonductile. Therefore, it cannot be cold-worked. It must either be cast to shape, or hot-worked above 460°C.

There are a variety of bronzes. They are assumed to be Cu–Sn bronzes, unless we specifically state otherwise. Other bronzes include Al-bronzes, Be-bronzes, Mn-bronzes, and Si-bronzes. As with α in Cu–Al alloys (Fig. 5-5.2), each of the bronzes is an fcc copper phase with 5 to 10 (or more) percent solute. Of course, a 95Cu–5Al bronze has more than 10 a/o Al, and a 95Cu–5Sn bronze has less than 3 a/o Sn, simply because of the difference in atomic masses.

---

* The reader is asked to explain why the lower-temperature structure just described is *not* bcc also.

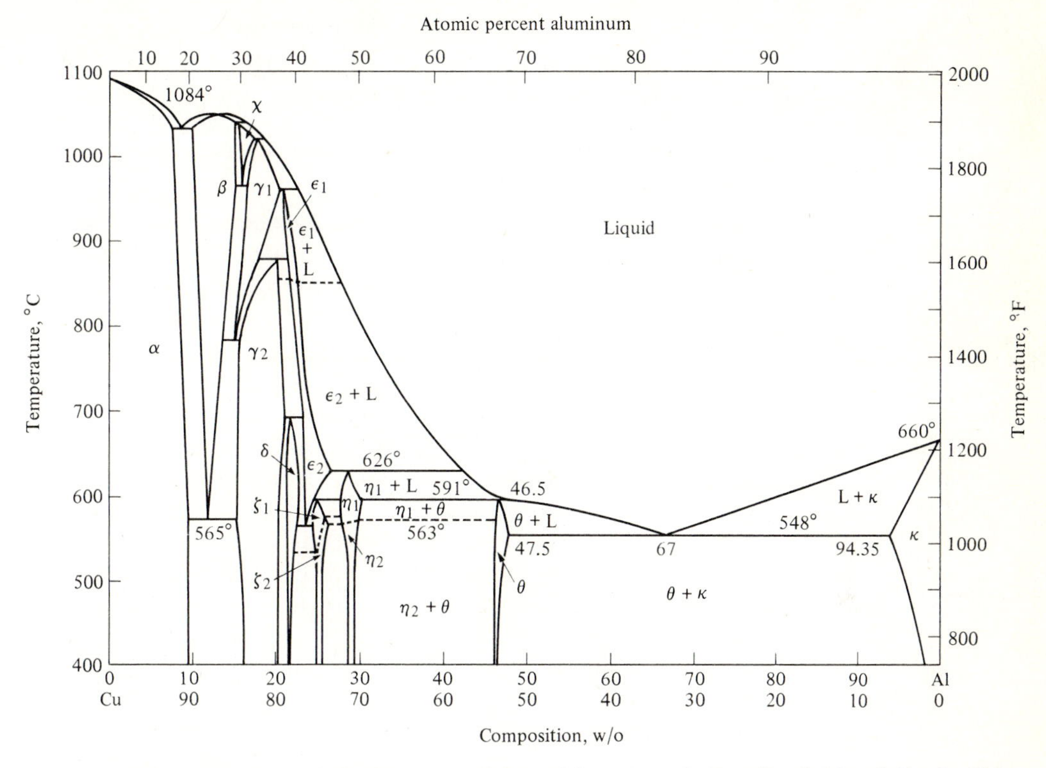

Figure 5-5.2 Al–Cu diagram. (Adapted from *Metals Handbook*, Metals Park, Ohio: American Society for Metals.)

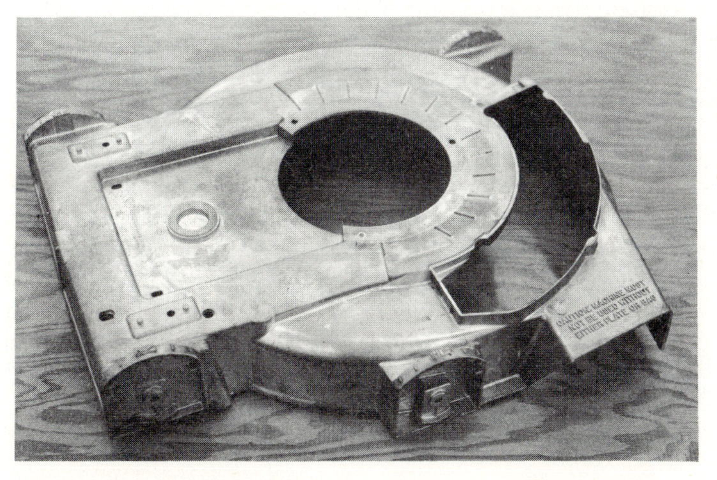

Figure 5-5.3 Magnesium casting (lawnmower housing). Although they are not easily rolled or forged, magnesium alloys are widely used for intricate light-weight castings. (Courtesy of J. D. Hanawalt and Dow Chemical Co.)

The mechanical properties of a variety of brasses and bronzes are presented in Tables 5-5.1 and 5-5.2, respectively. (The precipitation-hardening treatment of Be-bronze will be discussed in Section 5-6.)

Light metals

The light metals are primarily magnesium alloys and aluminum alloys. Magnesium is very attractive for certain applications because of its low density (ρ = 1.74 g/cm^3). The hexagonal structure of magnesium, however, precludes extensive plastic deformation because it lacks the greater symmetry that is present in cubic metals. Therefore magnesium is most commonly used as a cast alloy (Fig. 5-5.3).

Aluminum, somewhat denser (ρ = 2.7 g/cm^3), but still significantly lighter than steels or copper-base alloys, is cubic in structure and thus very amenable to deformation processing. Aluminum sheet and foil illustrate the extensive deformation that is possible. Both aluminum and magnesium are relative latecomers into the materials repertoire of the engineer because they strongly resist separation from their oxide raw materials. Before their use became commercially feasible, it was necessary to develop special technologies. Their affinity for oxygen is reflected in the oxidation and corrosion tendencies of magnesium and aluminum alloys. Fortunately aluminum oxide (Al_2O_3) produces a protective coating on aluminum. A similar coating over the surface of magnesium is not as protective. In salt water, even the protective coating on aluminum may become unstable. The engineer obviously must take these facts into account where materials are specified for designs.

Aluminum can be strengthened by cold work but is subject to annealing at relatively low temperatures because of its low melting temperature (Section 4-3). In general, solid solubility between the light metals and other common metals is limited. Therefore solution hardening is not used extensively. In the next section we shall encounter another strengthening process (precipitation hardening) that is particularly adaptable to the light metals. Tables 5-5.3 and 5-5.4 list various light metal alloys and their properties, and Table 5-5.5 gives the alloy designations.

Titanium alloys

A metal that has come to the forefront relatively recently is titanium. Titanium is interesting to the engineer because it is lighter than steel (4.5 g/cm^3 versus 7.8 g/cm^3), is more rigid than aluminum, and has a high resistance to corrosion (under oxidizing conditions). However, titanium is relatively expensive because it is difficult to remove all the oxygen from its ore.

Table 5-5.1
Properties of several commercial brasses*

Alloy	Composition, % Cu	Zn	Sn	Pb	Mn	Ultimate strength, MPa (lb/in²)	Yield strength, MPa (lb/in²)	Elongation 50-mm (2-in.), %	Typical applications
Wrought alloys (annealed)									
Gilding metal	95	5				205 (30,000)	70 (10,000)	40	Coins, jewelry base for gold plate
Commercial bronze	90	10				230 (33,000)	70 (10,000)	45	Grillwork, costume jewelry
Red brass	85	15				250 (36,000)	70 (10,000)	48	Weatherstrip, plumbing lines
Low brass	80	20				260 (38,000)	80 (12,000)	52	Musical instruments, pump lines
Cartridge brass	70	30				275 (40,000)	75 (11,000)	65	Radiator cores, lamp fixtures
Yellow brass	65	35				295 (43,000)	110 (16,000)	55	Reflectors, fasteners, springs
Muntz metal	60	40				375 (54,000)	145 (21,000)	40	Large nuts and bolts, valve stems
Low-leaded brass	64.5	35		0.5		295 (43,000)	110 (16,000)	55	Ammunition primers, plumbing
Medium-leaded brass	64	35		1.0		295 (43,000)	110 (16,000)	52	Hardware, gears, screws
Free-cutting brass	62	35		3.0		295 (43,000)	117 (17,000)	51	Automatic screw machine stock
Admiralty metal	71	28	1			365 (53,000)	150 (22,000)	65	Condenser tubes, heat exchangers
Naval brass	60	39	1			400 (58,000)	185 (27,000)	45	Marine hardware, valve stems
Manganese bronze	58.5	39.2	1	1 Fe	0.3	450 (65,000)	205 (30,000)	33	Pump rods, shafting rods
Cast alloys									
Cast red brass	85	5	5	5		235 (34,000)	117 (17,000)	25	Pipe fittings, small gears
Cast yellow brass	60	38	1	1		275 (40,000)	96 (14,000)	25	Hardware fittings, ornamental castings
Cast manganese bronze	58	39.7	1 Al	1 Fe	0.3	480 (70,000)	195 (28,000)	30	Propeller hubs and blades

* Adapted from *Physical Metallurgy for Engineers*, by A. G. Guy, Reading, Mass.: Addison-Wesley. The properties will vary somewhat from the properties of laboratory-prepared metals (Fig. 5-2.2).

Table 5-5.2
Properties of several bronzes*

Alloy	Condition	Composition					Ultimate strength, MPa (lb/in^2)	Yield strength, MPa (lb/in^2)	Elongation 50-mm (2-in.), %	Typical applications
		Cu	Sn	Zn	Pb					
Tin bronzes										
5% phosphor bronze	Wrought, annealed	94.8	5			0.2 P	325 (47,000)	130 (19,000)	64	Diaphragms, springs, switch parts
10% phosphor bronze	Wrought, annealed	89.8	10			0.2 P	455 (66,000)	—	68	Bridge bearing plates, special springs
Leaded tin bronze	Sand cast	88	6	4.5	1.5		260 (38,000)	103 (15,000)	35	Valves, gears, bearings
Gun metal	Sand cast	88	10	2			275 (40,000)	103 (15,000)	30	Fittings, bolts, pump parts
Aluminum bronzes										
5% aluminum bronze	Wrought, annealed	95				5 Al	415 (60,000)	175 (25,000)	66	Corrosion-resistant tubing
Silicon bronze										
Silicon bronze, type A	Wrought, annealed	95		1	1 Mn	3 Si	390 (56,000)	145 (21,000)	63	Chemical, equipment, hot water tanks
Nickel bronzes										
30% cupro-nickel	Wrought, annealed	70				30 Ni	415 (60,000)	175 (25,000)	45	Condenser, distiller tubes
18% nickel silver	Wrought, annealed	65		17		18 Ni	390 (56,000)	175 (25,000)	42	Table flatware, zippers
Nickel silver	Sand cast	64	4	8	4	20 Ni	275 (40,000)	175 (25,000)	15	Marine castings, valves
Beryllium bronzes (precipitation hardening)†										
Beryllium copper	Wrought, annealed	98			0.3 Co	1.7 Be	480 (70,000)	205 (30,000)	50	Springs, nonsparking tools
Same	Wrought, hardened						1200 (175,000)	900 (130,000)	5	

* Adapted from *Physical Metallurgy for Engineers*, by A. G. Guy, Reading, Mass.: Addison-Wesley.
† Precipitation hardening is presented in Section 5-6.

Table 5–5.3
Composition and properties of typical magnesium alloys*

Use	ASTM designation†	Composition, %				Condition	Ultimate strength, MPa (lb/in^2)	Yield strength, MPa (lb/in^2)	Elongation 50-mm (2-in.), %
		Al	Zn	Mn	Si				
Sand and permanent mold casting	AZ92	9.0	2.0	0.1	0.2	As-cast	165 (24,000)	95 (14,000)	2
						Solution treated	270 (39,000)	95 (14,000)	10
						Aged‡	270 (39,000)	145 (21,000)	3
Die casting	AZ91	9.0	0.7	0.2	0.2	As-cast	225 (33,000)	145 (21,000)	3
Sheet	AZ31X	3.0	1.0	0.2	0.2	Annealed	240 (35,000)	140 (20,000)	15
						Hard	275 (40,000)	215 (31,000)	8
Structural shapes	AZ80X	8.5	0.5	0.2	0.2	Extruded	330 (48,000)	220 (32,000)	12
						Extruded and aged‡	360 (52,000)	255 (37,000)	8
	ZK60A		6.0		0.6	Extruded	340 (49,000)	260 (38,000)	12
					Zr	Extruded and aged‡	350 (51,000)	290 (42,000)	10

* Adapted from *Physical Metallurgy for Engineers*, by A. G. Guy, Reading, Mass.: Addison-Wesley.
† American Society for Testing and Materials.
‡ Age hardening is presented in Section 5–6.

Table 5-5.4
Some characteristics of several types of aluminum alloys*

Alloy designation	Principal alloying elements	Hardening process†	Range of properties (soft to hard conditions): Ultimate strength, MPa (lb/in^2)	Yield strength, MPa lb/in^2)	Elongation 50-mm (2-in.), %	Endurance limit (5×10^8 cycles), MPa (lb/in^2)	Typical applications
Wrought alloys							
1100	Commercial purity	Cold working	90–165 (13,000–24,000)	35–150 (5,000–22,000)	45–15	35–62 (5,000–9,000)	Cooking utensils
5052	2.5% Mg	Cold working	195–290 (28,000–42,000)	90–255 (13,000–37,000)	30–8	110–140 (16,000–20,000)	Bus and truck bodies
Alclad 2024	4.5% Cu, 1.5% Mg (with protective sheet of pure aluminum)	Precipitation‡	185–470 (27,000–68,000)	75–325 (11,000–47,000)	22–19	90–140 (13,000–20,000)	Aircraft
6061	1.5% Mg_2Si	Precipitation‡	125–310 (18,000–45,000)	55–275 (8,000–40,000)	30–17	62–95 (9,000–14,000)	General structural
7075	5.6% Zn, 2.5% Mg, 1.6% Cu	Precipitation‡	225–570 (33,000–83,000)	105–500 (15,000–73,000)	16–11	160 (23,000)	Aircraft
Cast alloys							
195	4.5% Cu	Precipitation‡	220–280 (32,000–41,000)	110–220 (16,000–32,000)	8.5–2	50–55 (7,000–8,000)	Sand castings
319	3.5% Cu, 6.3% Si	Precipitation‡	185–250 (27,000–36,000)	125–180 (18,000–26,000)	2–1.5	70–75 (10,000–11,000)	Sand castings
356	7% Si, 0.3% Mg	Precipitation‡	260 (38,000)	185 (27,000)	5	90 (13,000)	Permanent mold castings

† In addition to alloy hardening.
* Adapted from *Physical Metallurgy for Engineers*, by A. G. Guy, Reading, Mass.: Addison-Wesley.
‡ Precipitation hardening is presented in Section 5-6.

Table 5-5.5
Aluminum alloy identification codes

Composition designations		
1xyy	Unalloyed aluminum (>99% Al)	
2xxx	Al + Cu as principal alloying element	
3xxx	Al + Mn as principal alloying element	
4xxx	Al + Si as principal alloying element	
5xxx	Al + Mg as principal alloying element	
6xxx	Al + Mg + Si as principal alloying elements	
7xxx	Al + Zn as principal alloying element	
8xxx	Al + other elements	
	yy = points of purity. Thus 1060 = 99.60% Al; 1090 = 99.90% Al; etc.	
Temper designations (suffixes to composition designations)		
—F	As fabricated	
—O	As annealed	
—H	Strain-hardened by a cold-working process	
	—H1X	Hardened only, with X representing fraction hardness (8 = fully hard)
	—H2X	Hardened and partially annealed
	—H3X	Hardened and stabilized
—T	Heat treated*	
	—T2	Annealed (cast alloys)
	—T3	Solution heat-treated and cold-worked
	—T4	Solution heat-treated and aged naturally
	—T5	Artificially aged only
	—T6	Solution heat-treated and artificially aged
	—T7	Solution heat-treated and stabilized
	—T8	Solution heat-treated, cold-worked, and aged
	—T9	Solution heat-treated, aged, and cold-worked
	—T10	Same as —T5, followed by cold working

* Aging is presented in Section 5-6.

Pure titanium transforms from hcp (α) to bcc (β) at 880°C (1620°F):

$$\alpha\text{-Ti} \underset{\text{cooling}}{\overset{\text{heating}}{\rightleftarrows}}_{880^\circ\text{C}} \beta\text{-Ti}. \tag{5-5.1}$$

This transformation in titanium alloys may be compared to the bcc $\rightleftarrows$ fcc transformation in iron (I.P. 2-6.1).

Titanium alloys fall into two categories: α alloys and α–β alloys. The α alloys include (1) commercially pure titanium, which transforms readily to the hexagonal phase on cooling; and (2) a few aluminum- and tin-containing alloys, for example, 92Ti–5Al–2.5Sn, that favor α formation. Most alloying elements favor the β, or bcc, phase. Therefore we find widespread use of α–β

alloys, that is, alloys that retain some β at room temperature. The α–β alloys have yield strengths about twice as high as those of the α alloys—1170 MPa (170,000 psi) versus 550 MPa (80,000 psi) for commercially pure titanium. The α alloys are more desirable for situations that require corrosion resistance.

White metals

The metals of this category not only have a white "color" but also have relatively low melting temperatures. Lead–tin *solders* (Fig. 5–4.1) exemplify much of this group and are specially selected with regard to their low eutectic temperature. *Die-casting alloys* are also selected on the basis of melting temperatures. Here zinc-base alloys are common because they have greater strength than alloys with lead or tin bases. Since die-casting alloys do not have to be plastically deformed, the hexagonal structure of zinc is not a major handicap. (Aluminum die-cast alloys are also common.)

5–6 PRECIPITATION HARDENING (AGE HARDENING)

Several of the tables in the previous section listed property data for aged, or precipitation-hardened, alloys. This hardening mechanism is an important one for many alloys and can be understood on the basis of phase diagrams.

Solution treatment

Consider an alloy of 95.5Al–4.5Cu, which is located near the right side of Fig. 5–5.2. According to that phase diagram, this alloy can be a single-phase solid solution, κ, between approximately 500°C and 570°C (930°F and 1060°F). Below the solubility curve, two phases coexist—called κ and θ. The former, κ, is a ductile, Al-rich, fcc solid solution. The latter, θ, is the hard, intermetallic compound, $CuAl_2$.

Now let us consider some heat-treating procedures. First, we shall *solution-treat* this 95.5Al–4.5Cu alloy, that is, we will heat it into the single-phase temperature range (~540°C) where all of the copper will dissolve into the κ, Al-rich phase (X of Fig. 5–6.1b). An *anneal* with slow cooling after the solution treatment at 540°C (XD) produces a material that is soft, weak, and not especially ductile. In fact, it is less ductile than pure aluminum, and no stronger. This metal, which is generally unsatisfactory for engineering purposes, contains large areas of κ with θ along grain boundaries. The latter is the path of crack propagation when fracture occurs.

Fast cooling (XA in Fig. 5–6.1) provides a stronger alloy that is very ductile (40%–50% elongation in 50-mm). This material is attractive to the

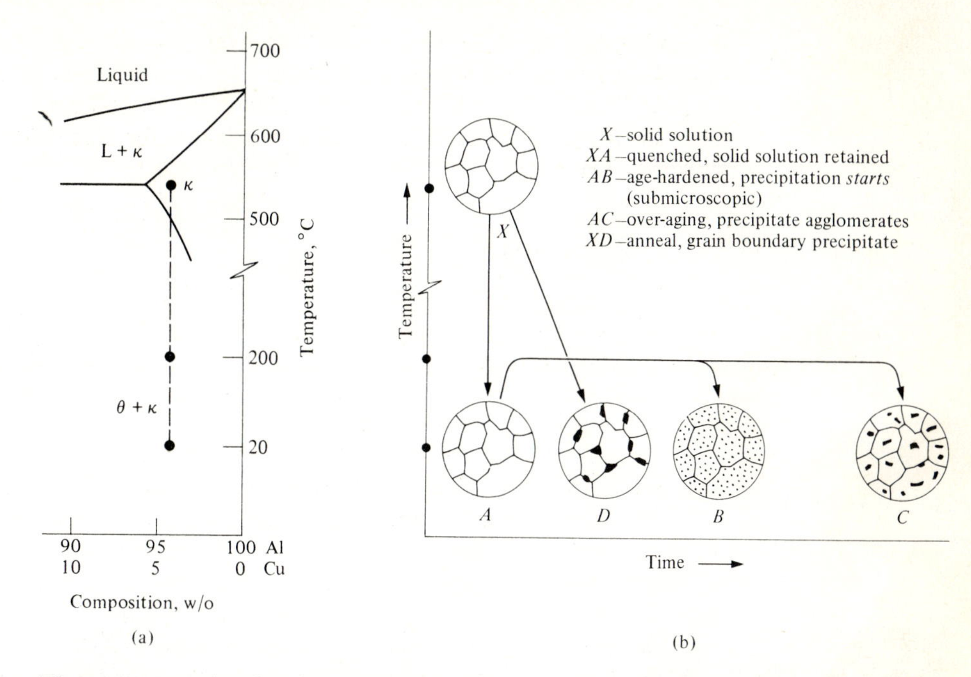

Figure 5-6.1 Age-hardening process (95.5Al–4.5Cu). Maximum hardness occurs when the copper begins to cluster prior to precipitation as $CuAl_2$; that is, θ (cf. Fig. 5-5.2).

engineer because it has the advantage of both strength and ductility. A micro-examination reveals only one phase, κ; x-ray diffraction reveals that all the copper is in solid solution, which indicates that equilibrium of Fig. 5-5.2 has not been reached.

Precipitation

Fast cooling, followed by *reheating to an intermediate temperature, XAB*, just about doubles the strength of the alloy, that is, compared with an alloy processed with the *XA* treatment (Table 5-6.1). This whets the interest of the engineer still further.* In explaining this, the materials scientist concludes that during the *AB* reheating, the supercooled solid solutions *start* to undergo a phase separation, or a *precipitation* of the $CuAl_2$, θ. With the available time and temperature, it is not possible for the copper atoms to diffuse into the grain boundaries and combine there with aluminum to give $CuAl_2$. Rather at innumerable locations throughout the grains, there is a *clustering* of the copper atoms, still in solid solution and not yet separated

* The time requirement for strengthening varies with the temperature (Fig. 5-6.2). Given time, many of these alloys will harden at ambient temperatures, hence the term *age hardening* is appropriate.

Table 5-6.1
Properties of an age-hardenable alloy (95.5 Al-4.5 Cu)

Treatment (See Fig. 5-6.1)		Ultimate strength, MPa (psi)	Yield strength, MPa (psi)	Ductility, % in 50 mm
XA	Solution-treated (540°C) and quenched (20°C)	240(35,000)	105(15,000)	40
XAB	Age-hardened (200°C, 1 hr)	415(60,000)	310(45,000)	20
XAC	Over-aged	~170(~25,000)	~70(~10,000)	~20
XD	Annealed (540°C)	170(25,000)	70(10,000)	15

from the κ into the θ phase. At this stage, the κ lattice is greatly distorted around the copper-enriched clusters, so that dislocation movements, and therefore slip, are severely restricted. The increased strength is shown in Table 5-6.1 for a 95.5Al-4.5Cu alloy. (Admittedly there is a decrease in ductility compared with an alloy treated by *XA*, but improved from one treated by *XD* in Fig. 5-6.1, particularly when the strength level is considered.)

Over-aging

Extended reheating, *XAC*, of the alloy in Fig. 5-6.1 softens this alloy and also reveals the appearance of distinct $CuAl_2$, or θ, particles that continue to grow with time, coalescing into fewer but larger particles. This is referred to as *over-aging*. Figure 5-6.2 shows the aging and over-aging process for one

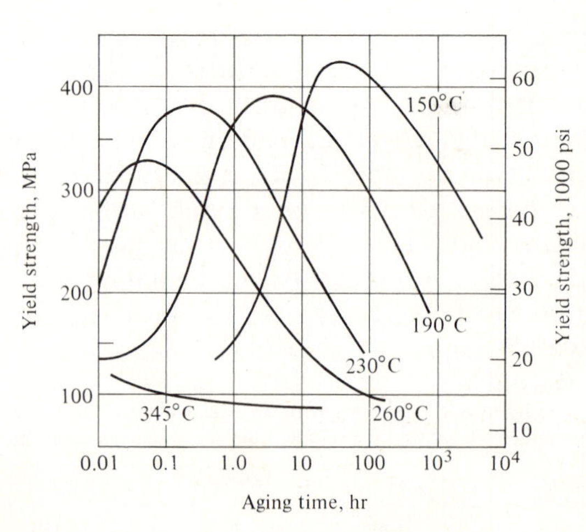

Figure 5-6.2 Over-aging (2014-T4 aluminum). Softening occurs as the precipitated particles grow. This proceeds more rapidly at elevated temperatures. (Adapted from *Aluminum*, Metals Park, Ohio: American Society for Metals.)

aluminum alloy (2014). Note that both aging and over-aging occur in less time at higher temperatures. Also note that the maximum strength is greatest with low aging temperatures. Here, however, the engineer must choose between a slightly greater strength of the product and higher processing costs for the extended heat treatments.

• Combined hardening

Table 5-6.2 summarizes an additional alternative available to the metallurgist for developing strength in certain alloys. The alloy, 98Cu-2Be, is age-hardenable because all the beryllium can be dissolved in the fcc structure at 870°C (1600°F). That solubility drops to near zero, however, at room temperature. The data in Table 5-6.2 show the effect of solution treatment, of age hardening, and also of combinations of age hardening and cold work. Thus when strain hardening follows age hardening, the combined effect is to give an ultimate strength of about six times that of the annealed metal (1380 MPa versus 240 MPa).

Two problems are encountered, however. First, a high-powered production mill is required to cold-work a precipitation-hardened metal that already has 1200 MPa (175,000 psi) ultimate strength. Second, the previously age-hardened alloy has lost considerable ductility, so that there can be cracking during cold working. An alternate sequence—cold working of a solution-treated alloy—is possible and reduces the severity of the above two problems. Much less energy and power is required to cold-work the solution-treated alloy (500 MPa and still ductile). Admittedly the strength finally realized is only 1340 MPa (195,000 psi) rather than 1380 MPa, because there was a slight loss of strain hardening during aging. Fortunately for this purpose, the atom rearrangements during aging are less extensive than during recrystallization. Therefore aging occurs slightly ahead of annealing. Close production control is required.

Table 5-6.2
Ultimate strengths of a strain- and age-hardened alloy (98Cu-2Be)

	MPa	(psi)
Annealed (870°C)	240	(35,000)
Solution-treated (870°C) and rapidly cooled	500	(72,000)
Age-hardened only	1200	(175,000)
Cold-worked only (37%)	740	(107,000)
Age-hardened, then cold-worked (cracked)	1380	(200,000)
Cold-worked, then age-hardened	1340	(195,000)

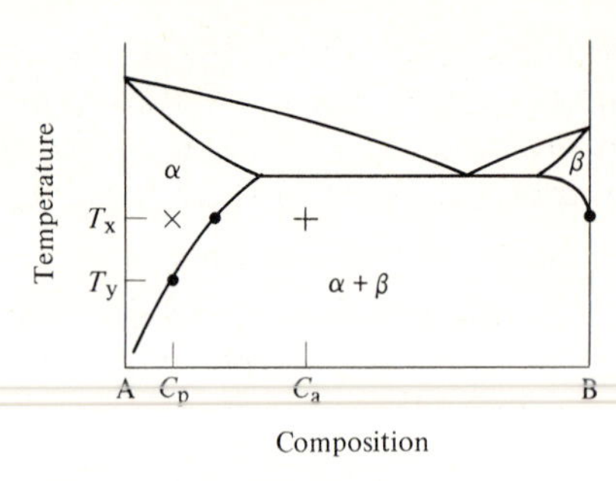

Figure 5-6.3 Precipitation hardening (age hardening) and the solubility limit. If the composition C_p is solution treated at T_x, a single phase results. When this is cooled below the solubility limit at T_y, the precipitate starts to form. Since only part of an alloy, C_a, forms α at T_x, only that part can be precipitation-hardened. The remaining part was pure B at the solution-treatment temperature.

Alloy choices

The prime requirement for precipitation hardening is a decreasing solubility curve for a second alloying component within the principal metal as shown in Fig. 5-6.3. In addition to the Al (+ Cu) and the Be-bronze—that is, Cu (+ Be)—alloys described above, this is a common hardening mechanism for magnesium alloys and certain of the stainless steels. In principle, lead–tin alloys could be hardened in this manner; however, because of their low melting temperatures, aging and over-aging occur too rapidly at room temperature and the envisioned strengths are not retained.

Precipitation-hardened alloys (other than the above-mentioned Pb–Sn alloys) do not over-age at room temperature. However, they all have upper limits of service temperature; for example, aluminum alloys cannot be used in supersonic aircraft, in which aerodynamic heating becomes significant.

Illustrative problem 5-6.1 A 90Al–10Mg alloy is precipitation-hardenable. Indicate plausible changes in temperature, composition, and structure that will produce precipitation hardening.

SOLUTION Refer to Fig. 5-4.5.
Solution treatment (~440°C, or 825°F):

Only one phase, α, 90Al–10Mg.

Quench into water (~20°C):

Still only one phase, α, 90Al–10Mg.

Age at an intermediate temperature (~100°C):

Mg atoms cluster, but remain as part of the α phase.

ADDED INFORMATION Over-aging would produce a matrix of α (low Mg content) containing a dispersion of the β phase, which is approximately Al_5Mg_3 (37 a/o Mg). This latter phase is hard and brittle, as are most intermetallic compounds. ◀

Illustrative problem 5-6.2 An alloy of 98Cu–2Be is solution-treated and aged. Assume that the clusters of the precipitating phase contain an average of 100 beryllium atoms each. (a) Estimate the number of these embryonic particles per mm^3. (b) What is the approximate distance between particles?

SOLUTION First we must estimate how many Be atoms per mm^3. Since the alloy is predominantly copper, we can estimate that it has a density only slightly less than pure copper, say 8.8 g/cm^3, or 0.0088 g/mm^3.

a) Grams $Be/mm^3 = (0.02)(0.0088\ g/mm^3) = 0.176 \times 10^{-3} g$;
Be $atoms/mm^3 = (0.176 \times 10^{-3} g)/(9.01 g/0.6 \times 10^{24}\ atoms)$
$= \sim 12 \times 10^{18}\ Be/mm^3$.
$Clusters/mm^3 = \sim 0.12 \times 10^{18}/mm^3$.

b) $mm^3/cluster = [clusters/mm^3]^{-1} = \sim 8 \times 10^{-18}\ mm^3$.
Assume a cubic array:
$(\sim 8 \times 10^{-18}\ mm^3) = (\sim 2 \times 10^{-6}\ mm)^3$.
Average distance $= \sim 2$ nm.

ADDED INFORMATION This is a typical size of the particles that are effective in age hardening. Since each unit cell is $(0.36\ nm)^3$, or $\sim 5 \times 10^{-20}\ mm^3$, there is one cluster for every 175 unit cells; and the space between particles is 2 nm/0.36 nm, which is 5 or 6 unit cell distances. Part of this distance is distorted by the cluster; therefore dislocation movements are restricted. ◄

Review and Study

SUMMARY

1. Alloys contain two or more metals that combine (a) as a solid solution, or (b) as compounds. In general, the solid solutions are stronger than pure metals and retain usable ductility if they possess a cubic crystal structure. Compounds of two metals are typically hard and nonductile.
2. Alloys may be single-phase, or they may contain a mixture of phases. For any selected temperature and alloy composition, a phase diagram shows us (a) what phase(s), (b) the composition of the phase(s), and (c) the relative amounts of each phase. With two phases, their compositions are dictated by the solubility limits; the amounts of the two phases may be determined by interpolation along the tie line between the solubility limits.
3. Unlike a pure metal, which has a unique melting temperature, an alloy typically melts over a temperature range. On cooling, freezing starts at the liquidus temperature and is completed at the solidus temperature. The liquidus curves are the solubility limits for the components in the liquid.

4. Not uncommonly, alloys melt at lower temperatures than either component metal alone. The lowest melting temperature between two solids is called the eutectic temperature. Our solders, and similar low-melting alloys, utilize these eutectic compositions and temperatures.
5. • Commercial alloys cited in this chapter include (a) copper-base alloys—the brasses (Cu–Zn alloys) and the bronzes (Cu–X alloys), (b) light metals—magnesium alloys and aluminum alloys, (c) titanium alloys, and (d) the white metals. Steel and cast irons will be considered in Unit 6. We will encounter superalloys and refractory metals in Unit 16. These categories are for convenience only; there are alloys in addition to those cited above; and the alloys that were considered could have been placed into different groupings.
6. Alloys, like the metals of Unit 4, may be strengthened by cold work. Solution hardening is possible with alloys of similar metals that possess a significant solid solubility limit, for example, the brasses.
7. Precipitation hardening, also called age hardening, is a very important strengthening process when it is applicable. The alloy is first solution-treated and then cooled rapidly enough for supersaturation. This is followed by the controlled precipitation of the minor second phase. Maximum hardness occurs in the very early stages of separation, in which there are innumerable emerging particles. Over-aging, which arises from a coalescence of the particles (fewer, but larger), produces a softening. For an alloy to be precipitation-hardened, it must be possible to supersaturate one of the phases. The solid solubility curve of the phase diagram must decrease at lower temperatures.

TECHNICAL TERMS

Age hardening *See* precipitation hardening.

Alloy A metal containing two or more elements.

Brass An alloy of copper and zinc.

Bronze An alloy of copper and tin (unless otherwise specified; e.g., an aluminum bronze is an alloy of copper and aluminum).

Clustering The initial stages of precipitation before actual phase separation.

Component (phases) The basic chemical substances required to create a chemical mixture or solution.

Compound A phase composed of two or more elements in a given ratio.

Cupronickel An alloy of copper and nickel.

Die-casting alloys Alloys with melting points sufficiently low so they can be cast into reusable dies—commonly zinc-based or aluminum-based alloys.

Dislocation *See* Technical Terms of Unit 3.

Equilibrium diagram *See* phase diagram.

Eutectic composition (C_e) The composition of the liquid solution with the minimum melting temperature (at the intersection of two liquid solubility curves).

Eutectic temperature (T_e) Melting temperature of an alloy with a eutectic composition. Temperature at the intersection of two liquid solubility curves.

• **Fraction chart** Plot of phase quantities. The sum of all phases must equal 100%.

Isothermal Constant temperature.

Lever rule Calculation method for determining phases. The overall composition is at the fulcrum of the lever. (This is a formalized equation for linear interpolation.)

• **Light metals** Alloys of aluminum and of magnesium.

Liquidus The locus of temperatures above which only liquid is stable. The liquid solubility limit.

Melting range (solidification range) Temperature range between the solidus and the liquidus.

Over-aging Continued aging until softening occurs.

Phase A physically homogeneous part of a materials system.

Phase amount Fraction (or percentage) of phases in a mixture.

Phase composition Analysis of a phase in terms of its components.

Phase diagram Graph of phase stability areas with composition and environment (usually temperature) as coordinates.

Precipitates Dispersion of particles that separated from a supersaturated solution (gas, liquid, or solid).

Precipitation hardening Hardening by the formation of clusters prior to precipitation (also called age hardening).

Solder Metals that melt below 425°C (~800°F) that are used for joining. Commonly, Pb–Sn alloys, but may also be other materials, even glass.

Solid solution A crystalline phase with a second component.

Solidus The locus of temperatures below which only solids are stable.

Solubility limit Maximum solute addition without supersaturation.

Solution hardening Increased strength arising from the creation of solid solutions (or from pinning of dislocations by solute atoms).

Solution treatment Heating to induce solid solution.

Sterling silver An alloy of 92.5 Ag and 7.5 Cu. (This corresponds to the maximum usable solubility of copper in solid silver.)

Supersaturation Excess solute beyond the solubility limit; most commonly achieved by supercooling.

Tie line For a particular temperature, an isothermal line connecting the compositions of the two phases in a two-phase area.

• **White metals** Nonferrous metals other than copper-base alloys. Usually low-melting alloys containing Bi, Cd, Pb, Sn, and/or Zn.

INDEX OF PHASE DIAGRAMS OF METALLIC ALLOYS

CHECKS

5A Alloys are metals with two or more __***__.

5B If an atom of a second component substitutes for an atom in the original structure, we call the alloy a __***__. As discussed in Unit 2, if the atoms of the second component are small, they may locate in the __***__ of the original structure.

5C __***__ is an alloy of copper and zinc. Ordinary bronze is an alloy of copper and __***__. • Other bronzes include alloys of copper and __***__, and of copper and __***__. Sterling silver is an alloy of silver and __***__.

5D A __***__ of two metals has a crystal structure that is unlike the structure of either metal alone.

5E In a compound, there is a preference for __***__ neighbors around each atom.

5F The __***__ of zinc in the fcc structure of copper is between 35 and 40% zinc. We label this solution (with less than 35% Zn) __***__-brass.

5G Copper and nickel are completely __***__ as solids. One factor permitting this is the similarity in their __***__.

5H Because of __***__, brass is stronger than copper.

5I Annealing removes __***__ hardening but does not remove __***__ hardening.

5J Solution hardening occurs because impurity atoms anchor __***__ and hinder their movement.

5K The __***__ is the line(s) on a phase diagram __***__ which there is only liquid; the __***__ is the line(s) on a phase diagram __***__ which there is only solid.

5L Between the __***__ and the __***__, there is a mixture of liquid and solid.

5M A __***__ on a phase diagram connects the compositions of two phases that are at equilibrium.

5N A __***__ alloy has a lower melting liquid than alloys of adjacent compositions.

5O Two __***__ cross at the eutectic temperature and the eutectic composition.

5P During cooling the eutectic reaction involves

___***___ ⟶ 2 solids.

5Q The three types of information that can be obtained from a phase diagram include ___***___, ___***___, and ___***___.

5R The phase compositions in a 2-phase area are determined by the intersection of the ___***___ and the ___***___.

5S The amounts of the phases in a 2-phase area are determined by ___***___ along the tie line between the two solubility limits.

5T $CuAl_2$ is an ___***___ compound of copper and aluminum.

• 5U Aluminum and magnesium fall in the general category of ___***___ metals.

• 5V Aluminum is protected from corrosion by a thin coating of ___***___.

• 5W Although ___***___ alloys have hexagonal structures, they can be used for die casting. Hexagonal metals are not readily ___***___.

5X Age hardening is another name for ___***___ hardening because this hardness develops with time.

5Y In order to have age hardening, the metal must first be ___***___.

5Z Aging occurs in ___***___ time at higher temperatures.

5AA Over-aging softens the metal because the precipitates ___***___ into fewer, but larger, particles.

• 5AB In combining age hardening and strain hardening, less chance for cracking occurs if ___***___ is performed first. Greater hardness is realized if ___***___ is performed first.

STUDY PROBLEMS

5-1.1. What is the atom percent zinc in a 70-30 brass?

Answer: 29 a/o Zn

5-1.2. One of the phases of Cu-Sn alloys contains 38.2% Sn. What is the copper-tin atom ratio?

5-1.3. A compound has a cubic structure with nickel atoms at cube corners and aluminum atoms at face centers. (a) What is the compound? (b) What is the composition of the compound?

Answer: (a) $NiAl_3$ (b) 42Ni-58Al

5-1.4. Estimate the lattice constant, a, for the unit cell of Fig. 5-1.1(a). (Assume a mean radius based on atom fraction.)

5-2.1. Estimate the density of the β-brass in Fig. 5-1.1(b).

Answer: 7.32 Mg/m^3 (or 7.32 g/cm^3)

5-2.2. Estimate the density of the Cu-Au phase in Fig. 5-1.1(a).

5-2.3. What is the atomic packing factor of the β-brass shown in Fig. 5-1.1(b)?

Answer: 68.4% (*Not* identical with I.P. 2-2.2. Why?)

5-2.4. Calculate the atomic packing factor for the $NiAl_3$ compound described in S.P. 5-1.3. [Note: Estimate a from the face diagonal.]

5-2.5. Motorboat production requires seat braces. Iron is excluded because it rusts. Select the most appropriate alloy from Fig. 5-2.2. for producing 10,000 braces. The requirements include an ultimate strength of at least 310 MPa (45,000 psi); a ductility of at least 45% El. (in 50 mm); and low cost. [Note: Zinc is less expensive than copper.]

Answer: 65-35 brass

5-2.6. Copper has an ultimate strength of 220 MPa (Fig. 5-2.2). (a) How much additional strength does each atom percent of zinc add (initially)? (b) Each a/o of nickel?

5-2.7. (a) For a 70-30 Cu-Ni alloy, the liquidus is _***_ °C; (b) the solidus is _***_ °C. (c) At 1400°C (2550°F) the liquidus of Cu-Ni alloys contains _***_% copper; (d) the solidus contains _***_% nickel.

Answer: (b) 1200°C (2190°F) (d) 85% Ni

5-2.8. (a) At 1300°C (2370°F), a cupronickel alloy can contain up to _***_% copper as a solid, and up to _***_% nickel as liquid. (b) The alloy that contains equal amounts of liquid and solid α at 1300°C must contain _***_% Ni and _***_% Cu. (c) At 1300°C the ratio of solid-to-liquid in a 50Cu-50Ni alloy is _***_ -to- _***_.

5-2.9. Twenty-three grams of a 70Cu-30Ni alloy are melted by heating the metal to 1400°C (2550°F). How many grams of nickel must be added to make an alloy that has a liquidus of 1400°C?

Answer: 50.2 g

5-3.1. (a) The liquidus temperature for a 60Ag-40Cu alloy is _***_ °C. (b) A _***_ Ag- _***_ Cu alloy also has this same liquidus temperature.

Answer: (b) 79Ag-21Cu

5-3.2. (a) The solidus temperature for a 60Ag-40Cu alloy is _***_ °C. (b) All alloys between _***_% Cu and _***_% Cu have this same solidus temperature.

5-3.3. Twenty-six grams of sterling silver are melted with 376 g of pure copper. (a) The liquidus temperature becomes _***_ °C. (b) The solidus temperature becomes _***_ °C. (c) What is the new composition?

Answer: (c) 6Ag-94Cu [Note: A transparent millimeter will be useful in this and many of the subsequent problems.]

5-3.4. How much of a 30Ag-70Cu alloy should be melted with 8.5 troy oz. of sterling silver to produce a eutectic alloy? (Your answer may be in ounces, which are still widely used in the commodity markets of precious metals.)

• 5-3.5. Consider 100 g of a eutectic alloy of Ag and Cu. At 600°C, what fraction of the copper is in the Ag-rich phase?

Answer: 8 w/o

• 5-3.6. Repeat S.P. 5-3.5 for 779°C.

5-4.1. What phase(s) are present at each 50°C interval during the heating of an 80Pb-20Sn solder?

Answer: 50°C, 100°C, 150°C—α + β; 200°C, 250°C—α + L; 300°C, 350°C—liquid only.

5-4.2. What phases are present at each 100°C interval during the cooling of a 20Ag-80Cu alloy?

5-4.3. What are the compositions of α in the previous problem?

Answer: 600°C—96.5Ag-3.5Cu, etc.

5-4.4. Cite the phases and their compositions for a 50Pb-50Sn alloy (a) at 184°C (363°F); (b) at 182°C (360°F).

5-4.5. (a) An alloy of 5Al-95Si is cooled slowly from 800°C (1472°F). What phases are present at each 100°C interval if equilibrium is maintained? (b) What are the composition(s) of the phase(s)?

Answer: 700°C—Liq (79Al-21Si) and β(~0.15% Al), etc.

5-4.6. Fifty grams of a 26Cu-74Zn alloy were melted in a large, pure copper mold and held at 800°C (1472°F) until the liquid became saturated with copper. (a) What is the composition of the liquid? (b) What solid phase will form in contact with the liquid?

5-4.7. One hundred grams of a Pb-Sn eutectic liquid are solidified to α + β at 183°C. How many grams of each are present at 182°C?

Answer: 45.5 g α; 54.5 g β

5-4.8. Determine the amount of α and β in 73 kg of a 62Cu-38Zn alloy (a) at 875°C (1607°F); (b) at 700°C (1292°F); (c) at 450°C (842°F).

5-4.9. Lead may dissolve 5 w/o tin in solid solution at 115°C and 18 w/o at 180°C. Scrap (100 kg) containing 95Pb-5Sn is to be melted. How much more tin could one add and still not exceed the solid solubility limit in α after the material has been frozen and cooled to 180°C?

Answer: 15.9 kg

5-4.10. Scrap electrical solder (40Pb-60Sn) is to be mixed with pure lead to make plumber's solder (82Pb-18Sn). How many kilograms of each are required to produce 50 kg of product?

5-4.11. Scrap electrical wire contains a mixture of copper wire (70 w/o), and of aluminum wire 30 w/o. It is available at half of the cost of the two metals separately. How much scrap can be used per 1000 kg of a 96Al-4Cu product?

Answer: 57 kg scrap

5-4.12. (a) Eighty grams of a 50-50 Cu-Ag alloy are first melted and then cooled to 780°C. How many grams are there of eutectic liquid and of β? (b) The eutectic liquid of part (a) is drained off and solidified. What phases are present in the resulting solid? (c) How many grams of each phase are present in this solidified eutectic?

5-4.13. Eighty grams of a eutectic Ag-Cu alloy are solidified. How many grams of the two phases, α and β, are formed?

Answer: 61.4 g α (&18.6 g β)

• 5-4.14. For a 50Pb-50Sn solder, make a fraction chart of α, β, and liquid versus temperature.

Answer: See Fig. 5-4.10.

• 5-5.1. What fraction of the atoms are (a) tin in a 95-5 tin bronze? (b) beryllium in a 98-2 Be-bronze?

Answer: (a) 2.7 a/o

• 5-5.2. Estimate the density of a eutectic Pb-Sn solder at 20°C. [Note: α and β are nearly pure Pb and Sn at 20°C.]

• 5-5.3. Assume no change in atomic radii. What is the volume change as β-Ti goes to α-Ti during cooling?

Answer: −8 v/o

• 5-5.4. Repeat S.P. 5-5.3, but adjust your calculations for a 2% increase in atomic radius when it changes from CN = 8 to CN = 12.

• 5-5.5. Yellow brass costs 90% of that of red brass. How much (%) can be saved by using a yellow brass rod instead of a red brass rod to support a load of 1000 kg (2200 lbs) without breaking? [Assume the same density, and use the data of Table 5-5.1.]

Answer: 24%

• 5-5.6. In order to have minimum weight for a given beam deflection of a cantilevered beam, the value of $\rho/\sqrt{E}$ must be minimum. How do the commercial nonferrous alloys of Appendix A compare with other materials? [ρ and E are density and Young's modulus, respectively.]

5-6.1. A 90Mg-10Al alloy is being considered for age-hardening purposes. Comment on its possibilities.

Answer: Cf. I.P. 5-6.1.

5-6.2. An alloy of 90Al-10Cu is heated to 540°C (1000°F), then given the treatment *XAB* of Fig. 5-6.1. What fraction of the alloy will age harden?

5-6.3. Which of the following alloys cannot undergo age hardening?
(a) 95Cu-5Al (b) 95Al-5Cu (c) 95Cu-5Ni (d) 95Ni-5Cu (e) 95Cu-5Zn
(f) 95Zn-5Cu (g) 95Mg-5Al (h) 95Al-5Mg (i) 95Cu-5Ag (j) 95Ag-5Cu

Answer: Check the phase diagrams for decreasing solubilities.

5-6.4. While they are being stored at an aircraft manufacturing plant, 500 kg of aluminum alloy rivets were exposed to temperatures in excess of 35°C long enough to start them to harden. They must, however, be soft when they are driven into the rivet holes. Can they be salvaged for use? Explain how. (Be specific for 95Al–5Cu rivets.)

5-6.5. Sterling silver is age-hardenable. (a) Indicate plausible temperature, composition, and structural changes that produce the precipitation hardening. (b) The hardness increase is slight when compared to that of a 95Al–5Cu alloy. Suggest why.

5-6.6. It is proposed to use 5% tin to make lead age-hardenable. Comment.

5-6.7. Examine the metallic phase diagrams in this book (listed after the technical terms of this Unit). (a) Indicate compositions that may be considered as candidates for precipitation hardening. (b) Why is a 41Al–59Mg alloy not a probable candidate?

CHECK OUTS

5A components
5B solid solution
interstices
5C brass
tin
Al, Mn (Be, Si)
copper
5D compound
5E unlike
5F solubility limit
α
5G soluble
sizes (electronic structure)
5H solution hardening
5I strain
solution
5J dislocations
5K liquidus, above
solidus, below
5L liquidus
solidus
5M tie line
5N eutectic
5O solubility curves
5P liquid
5Q what phase(s)
phase composition(s)
amount of phase(s)
5R tie line
solubility curves
5S interpolation
5T intermetallic
5U light
5V Al_2O_3
5W zinc (magnesium)
deformed
5X precipitation
5Y supersaturation (solution treated)
5Z shorter
5AA coalesce
5AB strain hardening
age hardening

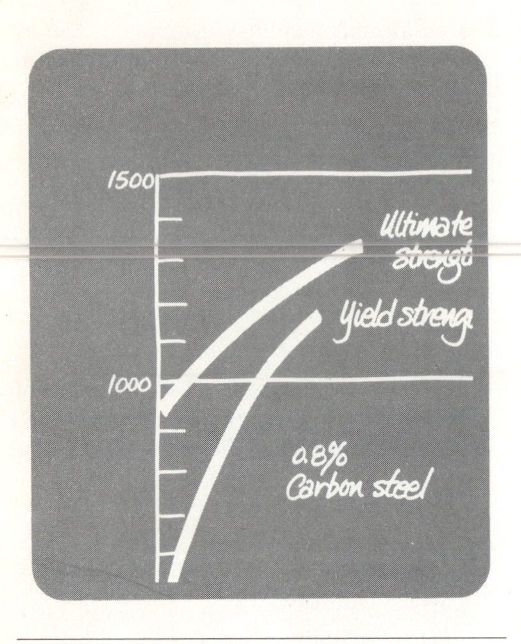

UNIT SIX

STEEL AND IRON

Steels are alloys that contain iron as the principal metal. The carbon content markedly affects the microstructure and therefore properties. In steel, the carbon content is generally below 1 w/o; in cast irons, there is 2%–3.5% carbon. Steel is a very versatile material. It can be deformed extensively to make thin sheet and fine wires; it can be made tough enough for gears; it can be made hard enough to cut other steels, a situation essentially impossible with any other material; it can be made into magnets, either hard or soft; it can be made corrosion resistant; etc. These changes are controlled by composition and by heat treatments; therefore, we will use steel and iron as examples of common materials that can be engineered.

Contents

Prerequisites Units 2 and 5; Section 4–3

From Unit 6, the engineer should

1) Know the details of the unshaded area of Fig. 6-2.2. This will (a) remove the necessity of numerous references to that figure while reading other parts of the text, (b) permit an understanding of the relationships between the microstructures and properties of steels, and (c) serve as a basis for anticipating the heat treatments of low-alloy and medium-alloy steels.
2) Be able to relate heat-treating procedures to microstructural development in steels and cast irons, and therefore be capable of predicting the effect of microstructures upon properties.
3) Have gained a familiarity with the technical terms that pertain to iron and steels.
4) Analyze and solve the study problems at the end of the Unit.

6-1 TYPES OF STEELS

Steels are alloys that contain iron as the principal component. It is convenient to divide steels into several categories: (a) plain-carbon steels, (b) low-alloy steels, and (c) high-alloy, or specialty, steels. Plain-carbon steels are the oldest historically and are somewhat less complex than the others because they are basically two-component alloys—iron and carbon.

Plain-carbon steels

Years ago, the American Iron and Steel Institute, AISI (a group of principal steel manufacturers), and the Society of Automotive Engineers, SAE (engineers of a major steel-consuming industry), established 10xx series as the identification code for *plain-carbon steels*. The final two numbers in the code indicate the amount of carbon in the steel. Thus, a 1050 steel is a plain-carbon steel with 0.50 w/o of carbon. Likewise, a 1080 steel has 0.80% C (or 80 "points" of carbon). This long-used identification code emphasizes the importance of the carbon level in the plain-carbon steels.

Low-alloy steels

Low-alloy steels contain alloying elements in addition to iron and carbon. Commonly used alloying elements include manganese (Mn), nickel (Ni), chromium (Cr), molybdenum (Mo), and silicon (Si), plus others. Typically,

Table 6-1.1
Nomenclature for AISI and SAE steels

AISI or SAE number	Composition
10xx	Plain-carbon steels*
11xx	Plain-carbon (resulfurized for machinability)
15xx	Manganese (1.0–2.0%)
40xx	Molybdenum (0.20–0.30%)
41xx	Chromium (0.40–1.20%), molybdenum (0.08–0.25%)
43xx	Nickel (1.65–2.00%), chromium (0.40–0.90%), molybdenum (0.20–0.30%)
44xx	Molybdenum (0.5%)
46xx	Nickel (1.40–2.00%), molybdenum (0.15–0.30%)
48xx	Nickel (3.25–3.75%), molybdenum (0.20–0.30%)
51xx	Chromium (0.70–1.20%)
61xx	Chromium (0.70–1.10%), vanadium (0.10%)
81xx	Nickel (0.20–0.40%), chromium (0.30–0.55%), molybdenum (0.08–0.15%)
86xx	Nickel (0.30–0.70%), chromium (0.40–0.85%), molybdenum (0.08–0.25%)
87xx	Nickel (0.40–0.70%), chromium (0.40–0.60%), molybdenum (0.20–0.30%)
92xx	Silicon (1.80–2.20%)

xx: carbon content, 0.xx%.

* All plain-carbon steels contain 0.40% to 0.90% manganese, and residual amounts (<0.05 w/o) of other elements.

the amounts of these alloying elements total less than 5%. The formalized identification codes for many of the *low-alloy steels* are shown in Table 6-1.1. These AISI–SAE types specifically include those steels that must undergo *heat treatment* during their manufacture (Section 6-5). Not included are the newer high-strength, low-alloy (HSLA) steels that are available to design engineers working on structures or on automobiles. The first two AISI–SAE digits are simply code numbers for alloy content and do not need to be memorized. As with plain-carbon steels, the last two digits specify the carbon content, for example, 0.40% C in a 4140 steel. This again emphasizes the importance of the carbon level in steels.

Specialty steels

Steels are also designed with alloy contents that are higher than the few percent shown in Table 6-1.1. For example, *stainless steels* always contain at least 12% Cr, and commonly more. (The familiar 18-8 steel has 18% Cr plus 8% Ni.) *Tool steels* contain significant amounts of Cr, W, V, and Mo, all of which form hard, stable carbides that enhance their wear resistance during service. *High-temperature steels* are still another grouping of specialty steels that usually contain high alloy contents. These must resist severe thermal environments (Unit 16), so they require major amounts of elements for solid-solution strengthening and for oxidation retardation.

Irons

Of course, iron is a metallic element and is the major component of steels. In a technological sense, the term, *iron*, can refer (a) to commercially pure sheet iron (C $< 0.1\%$, Fe $> 99.5\%$) such as used for enameling stock for the liners of kitchen ovens, or (b) to cast irons that contain enough carbon (and silicon) to approach the low-melting eutectic (~1200°C, or ~2200°F). These cast alloys will be the subject of Section 6–6.

6–2 Fe-C SYSTEM

We need to give special attention to the Fe–C phase diagram, since it serves as the basis for our knowledge about the internal structure of steels. First, however, let us review several facts about iron itself.

Ferrite (α)

At normal temperatures, iron is bcc (Section 2–2). Because of this, it can dissolve very little carbon interstitially, but it *can* dissolve significant amounts of chromium, silicon, tungsten, and molybdenum by substitution. Two terms are used synonymously to label this phase—α and *ferrite.*

Austenite (γ)

At elevated temperatures iron is fcc. It can dissolve more carbon interstitially than bcc iron can (Fig. 6–2.1). It can also dissolve by substitution large amounts of nickel and manganese—elements that prefer the fcc structures. Here, also, two labels are used interchangeably—γ and *austenite.*

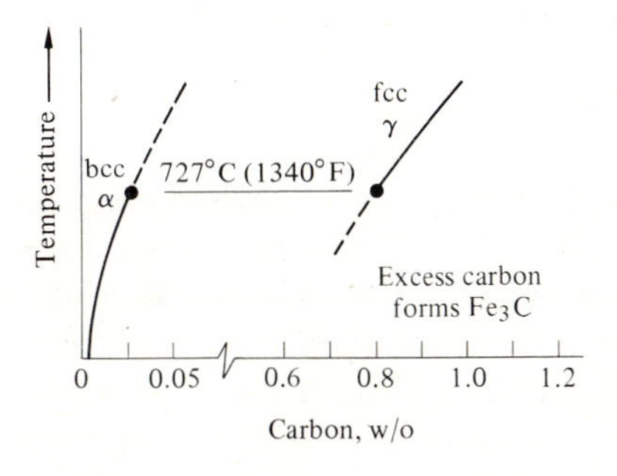

Figure 6–2.1 Solubilities of carbon in ferrite (bcc) and austenite (fcc). Although the packing factor is higher in austenite, the interstices that are present in austenite (γ) are larger (and fewer) than ferrite (α).

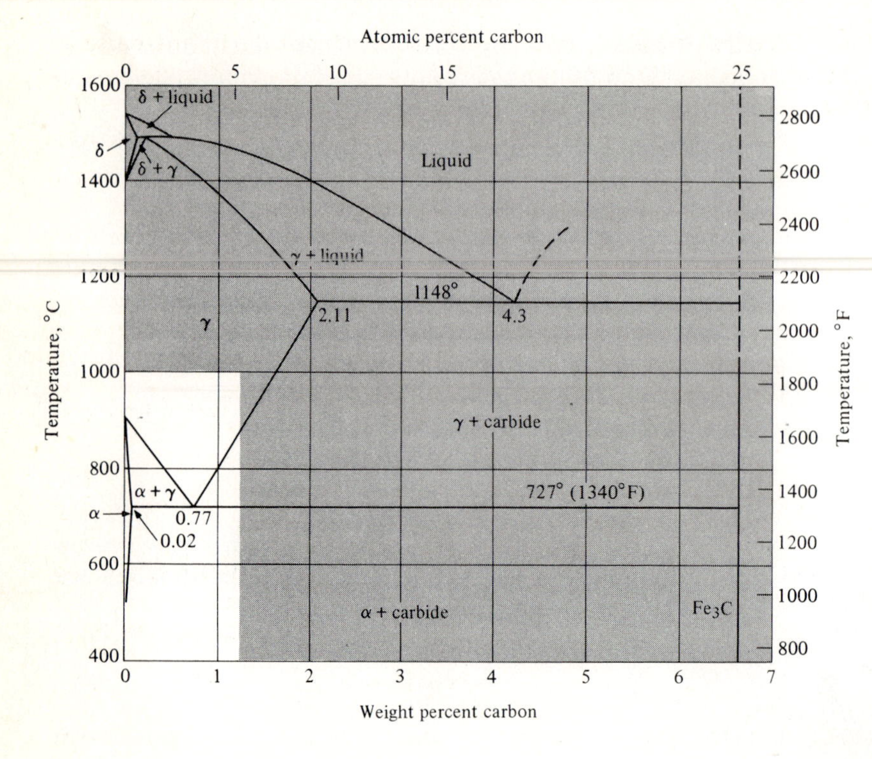

Figure 6-2.2 Fe-Fe_3C phase diagram. The left lower corner receives prime attention in heat treating steels (Fig. 6-2.3). (In calculating, 0.77% is commonly rounded to 0.8%.)

Carbide ($\overline{C}$)

Iron and carbon form the compound Fe_3C (= 93.3 Fe plus 6.7 C) sometimes called *cementite*. This compound is hard and brittle and will play an important role in our examination of steels. In steel, however, there are usually other alloying elements, such as Cr, Mo, and Mn, which also enter into the carbide phase. Therefore the carbide is seldom pure Fe_3C, or cementite. As a result, we shall simply use the term *carbide*, abbreviated by a capital C with an overbar, $\overline{C}$.

The Fe-Fe_3C phase diagram*

There is a eutectic at 1148°C (2100°F) and 95.7Fe–4.3C (Fig. 6-2.2). Since austenite, which is the stable iron-rich phase, can dissolve 2.1 w/o C at that temperature, the eutectic reaction during cooling is

$$\text{Liq (4.3C)} \xrightarrow[2100°\text{F}]{1148°\text{C}} \gamma \text{ (2.1C)} + \text{carbide (6.7C)}. \qquad (6\text{–}2.1)$$

* Our attention will be focused on iron that is below 1350°C (2460°F). At higher temperatures, fcc iron reverts to bcc iron to give what we call δ-ferrite. However, we shall not be involved with δ-ferrite in this text.

During heating, pure bcc iron transforms from ferrite to austenite at 912°C (1673°F). However, this temperature is decreased when carbon is present so that we may find γ (or austenite) stable as low as 727°C (1340°F). See Fig. 6–2.2. This temperature occurs where the solubility limits for carbon in γ and iron in γ cross; specifically, 99.2Fe–0.8C. A steel of this composition dissociates during cooling by the reaction

$$\gamma(\sim 0.8\%\ \text{C}) \xrightarrow[1340^\circ\text{F}]{727^\circ\text{C}} \alpha\ (0.02\%\ \text{C}) + \text{carbide}\ (6.7\%\ \text{C}). \tag{6–2.2}$$

Eutectoid reactions

All *eutectic* reactions involve a liquid and two solid phases. Thus Eq. (6–2.1) may be stated generally as

$$L_2 \underset{\text{heating}}{\overset{\text{cooling}}{\rightleftarrows}} S_1 + S_3, \tag{6–2.3}$$

where the composition of the liquid is between the compositions of the two solid phases (hence the choice of subscripts).

The reaction of Eq. (6–2.2) is very much like the eutectic reaction, with one major exception: All phases are solid.

$$S_2 \underset{\text{heating}}{\overset{\text{cooling}}{\rightleftarrows}} S_1 + S_3, \tag{6–2.4}$$

We call this a *eutectoid* reaction (literally, eutectic-like). It has the eutectic-appearing V-notch in a phase diagram; also the composition of the single higher-temperature phase is between the compositions of the two lower-temperature phases. The reaction of Eq. (6–2.2) is an important reaction during the heat treatment of steels. Therefore, it is recommended that the reader *become fully familiar* with the details of the unshaded part of Fig. 6–2.2. This region is drawn to a larger scale in Fig. 6–2.3.

Illustrative problem 6–2.1 Determine the amount of γ and α in a 3-kg (6.6 lb), 1040 steel casting as it is cooled to (a) 850°C (1560°F), (b) 728°C (1342°F), (c) 726°C (1338°F). [Assume 0.8% C in eutectoid.]

SOLUTION

850°C:	All γ	3.0 kg γ	0 kg α
728°C:	γ(0.8% C) + α(0.02% C)	1.46 kg γ	1.54 kg α
726°C:	α(0.02% C) + $\overline{\text{C}}$(6.7% C)	0 kg γ	2.83 kg α (and 0.17 kg $\overline{\text{C}}$).

ADDED INFORMATION The 1.46 kg of γ at 728°C dissociates to 1.46 kg of pearlite below the eutectoid temperature; this pearlite contains 1.29 kg α and 0.17 kg carbide. The steel thus contains 1.54 kg of *proeutectoid ferrite,*

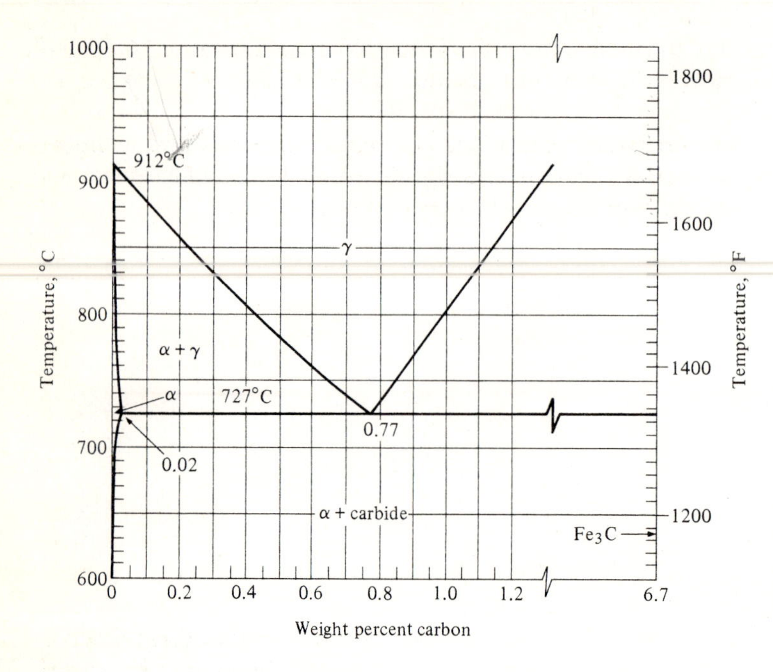

Figure 6-2.3 The eutectoid region of the Fe-Fe_3C phase diagram. (Cf. Fig. 6-2.2.) Steels with ~0.8% C are commonly called *eutectoid* steels. *Hypereutectoid* steels are above that value; *hypoeutectoid* steels, below.

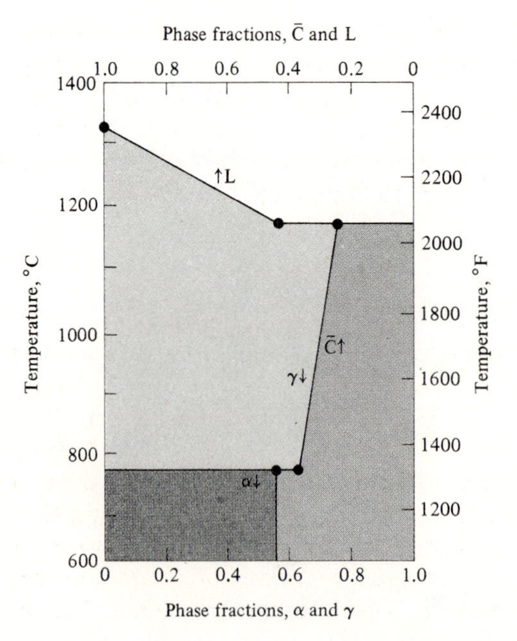

Figure 6-2.4 Phase fractions versus temperature (97Fe-3C). Phases: α (bcc), γ (fcc), read down; $\overline{C}$ (carbide), L (liquid), read up. See I.P. 6-2.2.

that is, ferrite that formed before the eutectoid reaction, and 1.29 kg of *eutectoid ferrite*, which formed from the eutectoid reaction. ◂

Illustrative problem 6-2.2 Calculate the phase fractions and compositions of ferrite, austenite, carbide, and liquid in an Fe-C alloy of 3.0 w/o carbon (97 w/o Fe) as they vary with temperature.

SOLUTION

Above 1300°C	All liquid	
At 1148 (+) °C;	0.41 liquid	(4.3%C)
	0.59 γ	(2.1% C)
At 1148 (−) °C;	0.80 γ	(2.1% C)
	0.20 $\overline{C}$	(6.7% C)
At 727 (+) °C;	0.63 γ	(0.8% C)
	0.37 $\overline{C}$	(6.7% C)
At 727 (−) °C;	0.45 $\overline{C}$	(6.7% C)
	0.55 α	(0.02% C)

See Fig. 6-2.4.

When cooling continues to room temperature under equilibrium conditions, even the minor one-fiftieth of 1% of carbon will leave this ferrite. ◂

6-3 MICROSTRUCTURES OF ANNEALED STEELS

A mixture of ferrite and carbide forms in steels that are equilibrated below the eutectoid temperature in accordance with the Fe-Fe_3C phase diagram. By one procedure, the two phases of this mixture can form simultaneously during the cooling of austenite to ambient temperatures. A mixture of $[\alpha + \overline{C}]$ can also be formed by reheating quenched steel. The microstructures that develop from these two procedures are different. As a result, the properties are different.

Pearlite

The Fe-Fe_3C eutectoid reaction in a plain-carbon steel is

$$\gamma\ (\sim 0.8\%\ C) \underset{\text{heating}}{\overset{\text{cooling}}{\rightleftarrows}} \alpha\ (0.02\%\ C) + \text{carbide}\ (6.7\%C). \qquad (6\text{-}2.2)$$

During cooling, ferrite and carbide form simultaneously from the preexisting austenite. Since the $[\alpha + \overline{C}]$ form concurrently, they are intimately mixed in a characteristic microstructure that we call *pearlite* (Fig. 6-3.1).

Pearlite, P, is a *specific mixture of ferrite and carbide;* the two phases are *lamellar*, that is, layer-like; and they must originate from *austenite of*

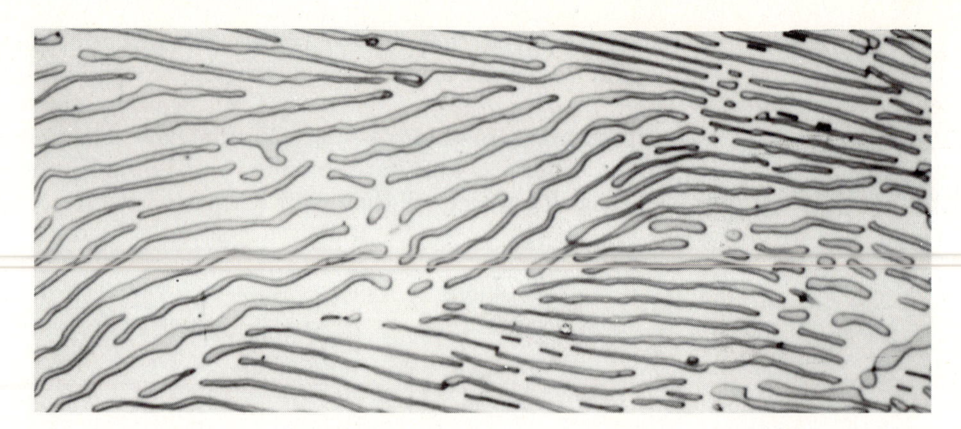

Figure 6-3.1 Pearlite (1080 steel). The lamellar mixtures of ferrite (lighter) and carbide (darker) form from austenite of eutectoid composition (×2500). (Courtesy of U.S. Steel Corp.)

Figure 6-3.2 Ferrite plus pearlite in Fe-C alloys (cf. Fig. 6-2.2) (×500). (a) A 0.20% C alloy contained about 75% α + 25% γ immediately prior to transformation. Since that austenite produced pearlite ($\gamma \rightarrow \alpha + \overline{C}$), the microstructure is 75% proeutectoid ferrite plus 25% pearlite. (b) A 0.50% C alloy had more austenite (γ) than proeutectoid ferrite (α) before transformation. Therefore the major part of it is now pearlite. (c) A 1.2% C alloy contained carbide prior to transformation. The proeutectoid carbide formed at the grain boundaries of the former austenite grains. (Courtesy of U.S. Steel Corp.)

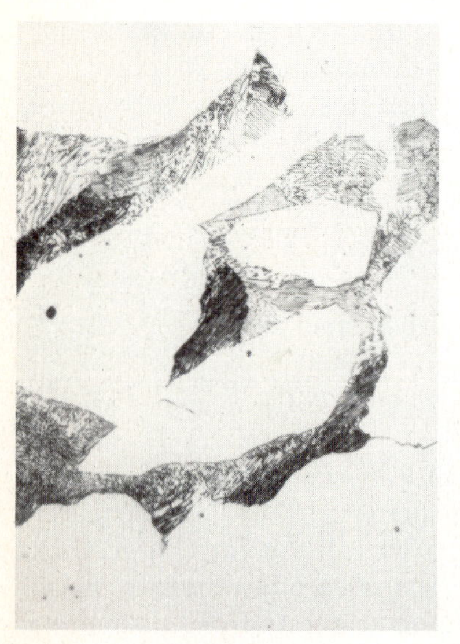

(a) 0.20% C

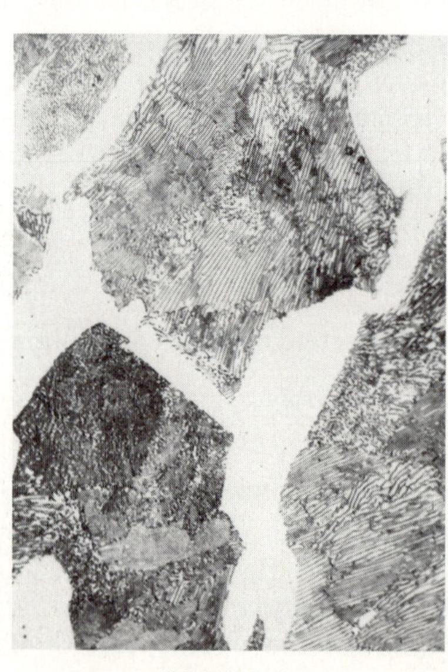

(b) 0.50% C

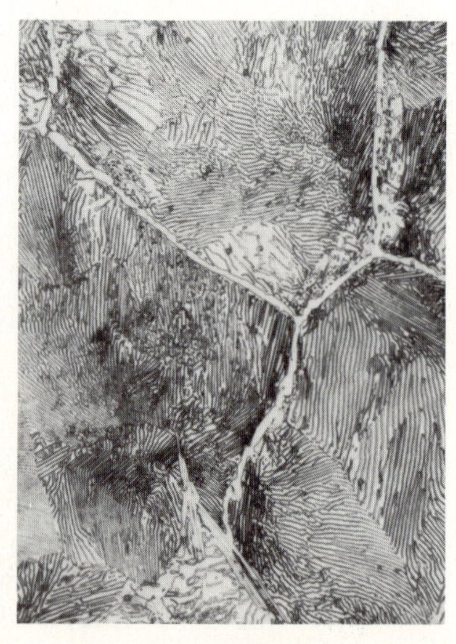

(c) 1.2% C

eutectoid composition. These facts are important, since $[\alpha + \overline{\mathrm{C}}]$ may also form from other than a eutectoid origin. With other origins, the microstructure is not lamellar and therefore has a significantly different set of properties.

Since pearlite comes from austenite of eutectoid composition, the amount of pearlite in a steel *must correspond to the amount of austenite* entering the eutectoid reaction. Figure 6–3.2(a), for example, shows that an 0.20%-carbon steel is approximately three-fourths ferrite (white) and one-fourth pearlite (gray with lamallae of $[\alpha + \overline{\mathrm{C}}]$ barely resolvable at $\times 500$. It had 75% ferrite (proeutectoid) and 25% γ just above 727°C. An 0.50%-carbon steel (Fig. 6–3.2b) has more pearlite than proeutectoid ferrite because it had more than 60% γ just prior to the eutectoid reaction (Fig. 6–2.3).

Spheroidite

The microstructure of pearlite in Fig. 6–3.3(a) may be contrasted with the microstructure in Fig. 6–3.3(b). We call the second microstructure *spheroidite*, S. The two photomicrographs are from the same steel; they have the same composition; and they both show a mixture of ferrite (matrix) and carbide (gray). Whereas pearlite has lamallae of carbides, the spheroidite (Fig. 6–3.3b) has *sphere-like carbides in a matrix of ferrite*. In the next sections we will see that the two microstructures provide their steels with different properties.

The procedure for making spheroidite differs from that for making pearlite. Recall that pearlite forms by cooling austenite of eutectoid composition

Figure 6–3.3 Microstructures of eutectoid steels (0.80% C). Both samples are from the same steel but were heat-treated differently. (a) Pearlite formed by transforming austenite (γ) of eutectoid composition ($\times 2500$). (b) Spheroidite formed by tempering at 700° C ($\times 1000$). (Courtesy of U.S. Steel Corp.)

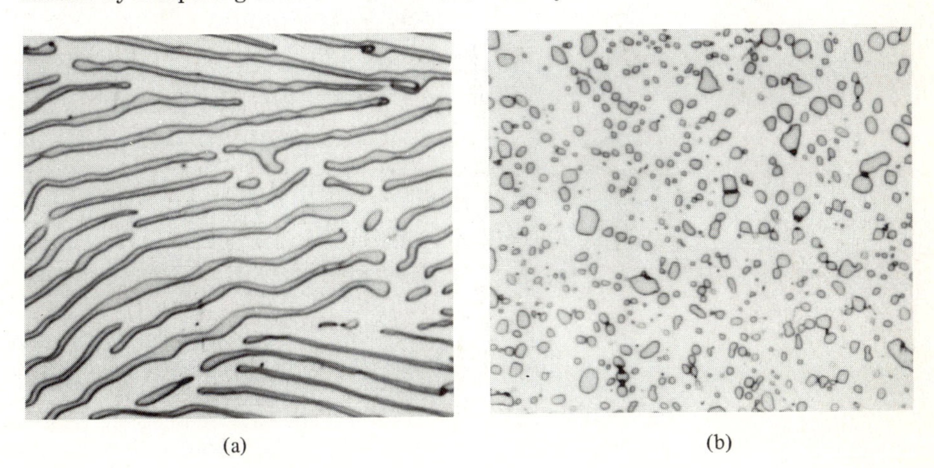

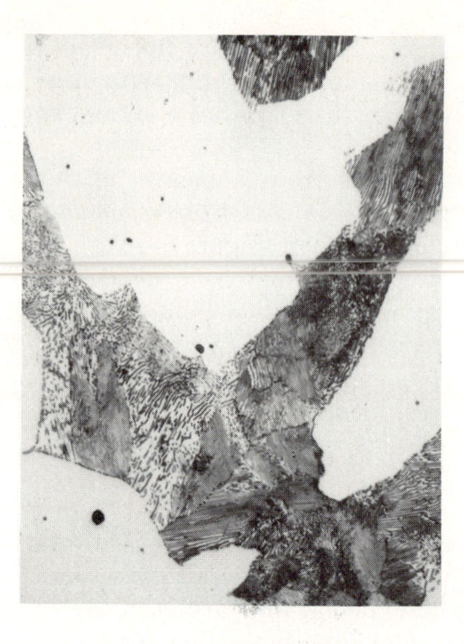

Figure 6-3.4 Microstructure of Fe-C alloy (×500). See I.P. 6-3.2 (cf. Fig. 6-3.2). (Courtesy of U.S. Steel Corp.)

so that the austenite decomposes simultaneously to ferrite and carbide. In order to form spheroidite, it is necessary to heat pearlite for extended periods of time just under the eutectoid temperature (~16 hrs at 700°C). The layers of carbide "ball-up" into a spheroidal shape much as a mist forms droplets of water, but with a longer time requirement. The temperature cannot be raised above the eutectoid, or else the austenite will form again, which in turn would transform on cooling to the pearlite layers.*

Illustrative problem 6-3.1 Determine the amount of pearlite and the amount of proeutectoid ferrite in 800 g (1.75 lbs) of an AISI-SAE 1020 steel.

SOLUTION In a steel with 0.20 C at 728°C,

$$\text{amount of } \alpha = (800\text{ g})(\sim 0.75) = 600\text{ g};$$
$$\text{amount of } \gamma = (800\text{ g})(\sim 0.25) = 200\text{ g}.$$

Below 727°C, these 200 g of γ change to pearlite:

$$\text{amount of } \textit{pearlite} = 200\text{ g}.$$

The 600 g of α formed before the eutectoid reaction; therefore,

$$\text{amount of } \textit{proeutectoid} \text{ ferrite} = 600\text{ g}.$$

ADDED INFORMATION The 200 g of pearlite will contain ~176 g of *eutectoid* ferrite, for a total of ~776 g of ferrite. ◀

---

* Spheroidite can also be obtained by severely quenching steel so that pearlite does not form, then reheating (~1 hr at 700°C) will form spheroidite (Section 6-5).

Illustrative problem 6–3.2 Figure 6–3.4 is a photomicrograph of a steel containing only iron and carbon. It was cooled slowly through the eutectoid temperature and thus contains only ferrite and pearlite. Estimate the carbon content, X. (The densities of α and pearlite are 7.87 and 7.82 g/cm^3, respectively.)

SOLUTION By inspection, there is approximately 45 v/o of the figure that is pearlite and 55 v/o proeutectoid ferrite.

$$\begin{aligned} \text{Basis: 1 cm}^3 \text{ of metal} &= 0.55 \text{ cm}^3\ \alpha + 0.45 \text{ cm}^3 \text{ pearlite} \\ &= 4.33 \text{ g } \alpha + 3.52 \text{ g pearlite} \\ &= 7.85 \text{ g alloy.} \\ \text{Total carbon} &= \text{carbon in } \alpha + \text{carbon in pearlite;} \\ (\text{X w/o})(7.85 \text{ g}) &= (0.02 \text{ w/o})(4.33 \text{ g}) + (0.8 \text{ w/o})(3.52 \text{ g}), \\ \text{X} &= 0.37 \text{ w/o C.} \end{aligned}$$

ADDED INFORMATION There are various methods for estimating the volume percent (v/o). Probably the easiest is to take a piece of semitransparent graph paper, lay it at random over the photomicrograph, and count the grid intersections that lie in each constituent. ◀

6–4 PROPERTIES OF ANNEALED 10xx STEELS

The properties of many ferrous alloys are given in handbooks such as Volume 1 of the *Metals Handbook*, a compendium published by the American Society for Metals. We shall not try to duplicate those data here. It will be useful, however, to examine some of the factors that affect properties in 2-phase alloys. Most obvious among these are (1) the amount of each phase, (2) the size of the minor phase, and (3) the shape or distribution of the minor phase.

Properties versus amounts of phases

We can calculate the density of a 2-phase mixture if we know the volume percent and the density of each phase. In fact, we did this in I.P. 6–3.2, when we determined that 1 cm^3 of that steel had a mass of 7.85 g. As a formula, this calculation is

$$\rho_m = f_1\rho_1 + f_2\rho_2, \tag{6–4.1}$$

where ρ_m is the density of the material and ρ_1 and ρ_2 are the densities of the two phases. The terms f_1 and f_2 represent volume fractions of the two phases. We call Eq. (6–4.1) a *mixture rule,* which in this case is relatively simple.

When other properties are involved, mixture rules are less simple. Rather than making calculations, it is easier for us just to observe the

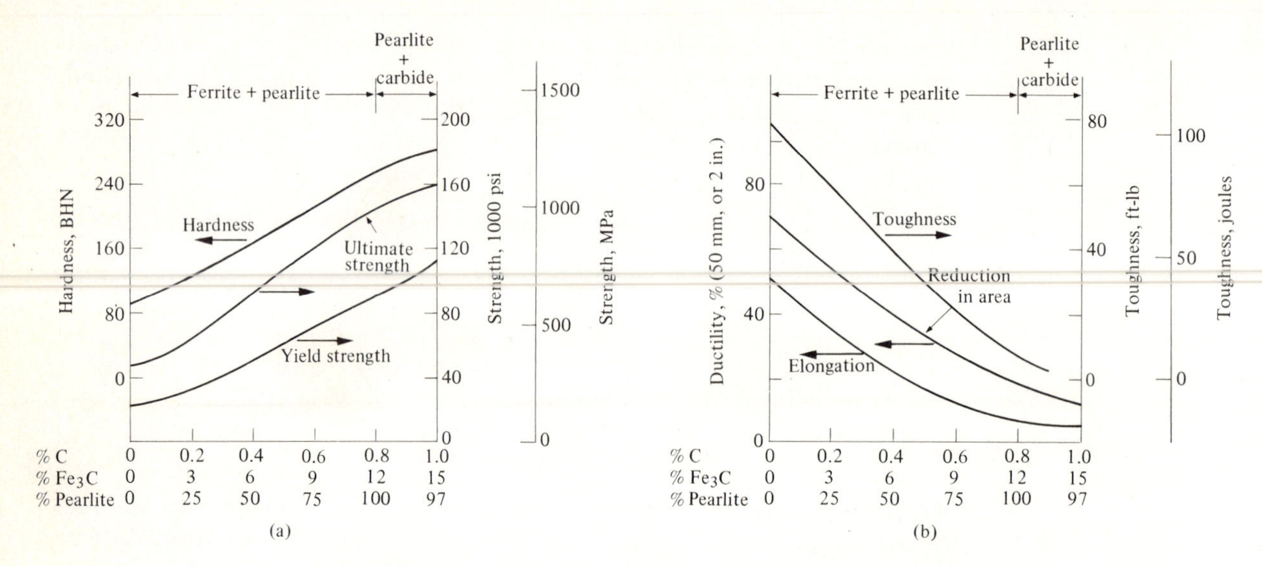

Figure 6-4.1 Properties versus amounts of carbide (and pearlite) in annealed steels. (a) Hardness (read left) and strengths (read right). Refer to Fig. 6-3.2; additional carbide within the pearlite reinforces the weaker ferrite. (b) Ductilities (read left) and toughness (read right). Those steels with less carbide are both more ductile and tougher.

relationship of properties versus amounts of phases. Figure 6-4.1 shows (1) the hardness, (2) the strength, (3) the ductility, and (4) the toughness of annealed Fe-Fe_3C alloys. Carbide is present in pearlite. In view of our description in Section 6-2 of the carbide phase as being hard and brittle, the data of Fig. 6-4.1 are just what we would expect. We should note, however, that 100% Fe_3C would give us a very weak material (~35 MPa, or ~5000 psi) because of the cracking tendency of the brittle carbide. In fact, white cast iron, which has upward of 40% carbide, has very restricted use as a load-bearing material because of its extreme brittleness and lack of strength. (It has important uses as a wear-resistant material.)

Properties versus phase size

When steels are cooled slowly, the austenite transforms more slowly to $[\alpha + \bar{C}]$ and forms much coarser pearlite than normal. Figure 6-4.2 compares the hardness for Fe-C alloys with coarse pearlite versus fine pearlite. For a given composition, fine pearlite has more (though admittedly thinner) layers of carbide and of ferrite than coarse pearlite does. Slip occurs less readily in the finer pearlite, with a resulting increase in hardness.

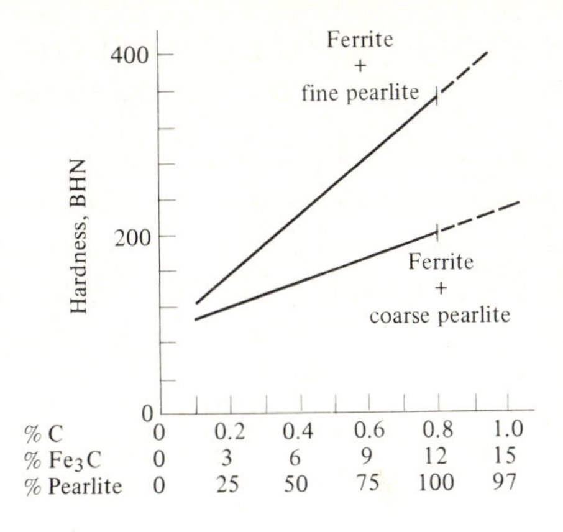

Figure 6-4.2 Effect of microstructural dimensions on hardnesses of annealed steels. The pearlite with thinner lamallae, which is harder, was formed by faster cooling.

Properties versus phase shape

We saw in Section 6-3 that pearlite is lamellar, and that the nonlamellar mixtures of $[\alpha + \overline{C}]$ have markedly different microstructures (Fig. 6-3.3). They also have different properties.

Figure 6-4.3 compares the hardness and toughness properties. The difference is significant. The hardness of [ferrite + pearlite] is higher than that of spheroidite because the two-dimensional carbide lamellae provide more reinforcement to the soft ferrite matrix than do the globular carbide particles in spheroidite. In contrast, the toughness of spheroidite is greater than

Figure 6-4.3 Hardness and toughness versus carbide shape (annealed plain-carbon steels). The carbides in the pearlite are lamellar; those in the spheroidite are sphere-like.

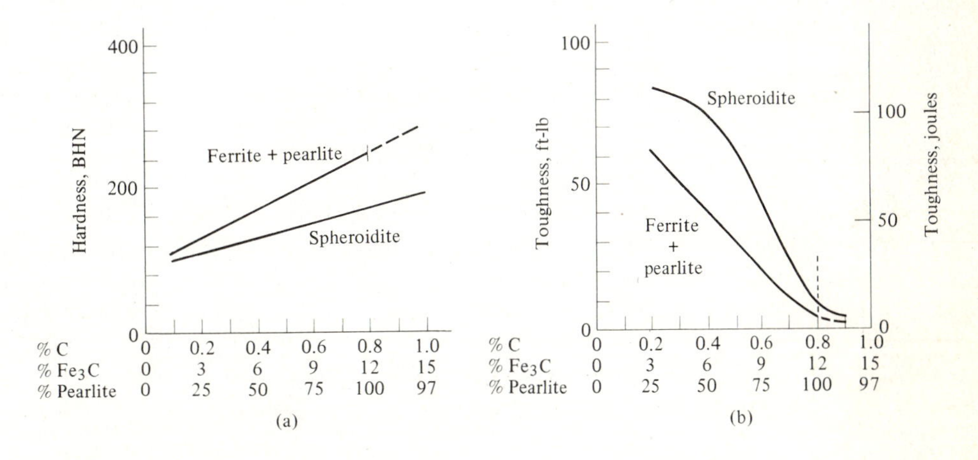

pearlite-containing metal. A progressing crack can follow carbide layers through pearlite, but in spheroidite it must traverse the tougher ferrite.

Illustrative problem 6–4.1 Refer to Fig. 6–6.2(b), which contains ferrite and graphite (dark spheres). Estimate the density of this alloy (97Fe–3C) if the densities of the two phases are 7.8 g/cm³ and 2.0 g/cm³, respectively. Use the grid method to obtain approximately 12 v/o graphite (see information with I.P. 6–3.2).

SOLUTION

$$\rho = (0.88)(7.8 \text{ g/cm}^3) + (0.12)(2.0 \text{ g/cm}^3)$$
$$= 7.1 \text{ g/cm}^3$$

ADDED INFORMATION Equation (6–4.1) is independent of shape and size of the phases when we use it to determine density or specific heat. This particular mixture rule is usually not applicable when we are calculating other properties.

The carbon in this alloy is not in the form of Fe_3C, as in steels; rather it is present in the form of graphite. This difference occurs in cast irons, particularly if there is a high silicon content. We shall see in Section 6–6 that different cast irons have significantly different properties, depending on the shape and amount of this graphite. ◀

• 6–5 QUENCHED AND TEMPERED STEELS

Strengthening of steels

Steels are very versatile; they, like most other metals, can be strengthened by *strain hardening* (Fig. 6–5.1). They can also be *solution-hardened* (Section 5–2). For example, an annealed 18-8 stainless steel, which contains 18% Cr and 8% Ni in solid solution with fcc iron, has a hardness of 200 BHN, in contrast with a hardness of less than 100 BHN for commercially pure iron. Steels can also be hardened by controlling the *microstructure* as shown in Fig. 6–4.2 for coarse and fine pearlite. Although *precipitation hardening* is not commonly used in steels, it provides a useful strengthening procedure with certain stainless steels.*

Still another process is available for strengthening steels that is not available for the majority of other metals. That is by *quenching* and *tempering*. The artisans of history discovered this process, although they did not understand its basis. We now know that the process occurs because iron is that "freak" metal that changes from fcc to a body-centered structure below the eutectoid temperature. With that understanding, the materials engineers are now able to control the hardening process very precisely, to

---

* Also limited amounts of age hardening occur in sheet steels such as those used in food cans, because small amounts of nitrides and carbides start to precipitate.

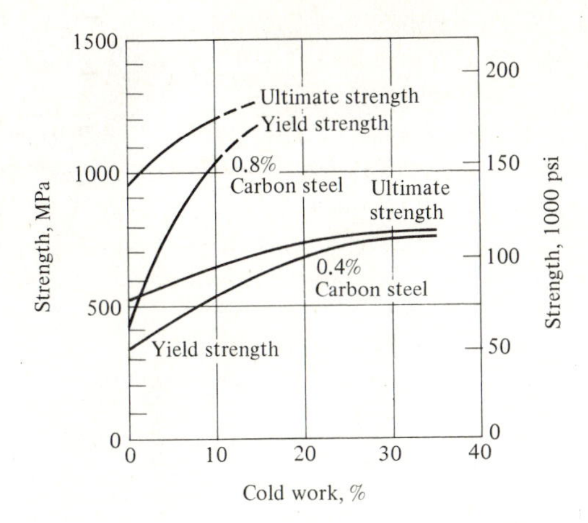

Figure 6-5.1 Cold work versus strength of plain-carbon steels.

predict hardnesses and strengths in their products, and to develop new alloy steels with improved characteristics.

Martensite

When fcc iron or steel is cooled from the austenitic temperature range, carbon and the other alloying elements should relocate from a homogeneous γ solid solution into two phases, $[\alpha + \overline{C}]$. For this to occur, almost all the carbon must diffuse to form carbide. Likewise the carbide formers, Cr, Mo, V, etc., should concentrate in the carbide; whereas nickel and silicon must diffuse into the ferrite. These reactions require time; therefore, with quenching, the $\gamma \rightarrow [\alpha + \overline{C}]$ reaction does not always have time to occur.

If the transformation to $[\alpha + \overline{C}]$ fails to occur before the temperature drops below approximately 400°C (750°F), the opportunity for $[\alpha + \overline{C}]$ to form is forfeited. Below this temperature, diffusion rates become excessively slow. A driving force, however, develops to transform the fcc structure to a body-centered crystal structure. At low enough temperatures this driving force becomes sufficiently demanding to force a transformation by *shear*. The resulting body-centered structure is not cubic but *tetragonal,* a structure in which the unit cell has only two of the three orthogonal axes equal. The resulting body-centered unit cell traps all the carbon and other alloying elements, giving a highly supersaturated composition. This body-centered tetragonal, non-equilibrium phase is called *martensite*.

Martensite is extremely hard for two reasons. (1) Because it is noncubic, martensite does not have as many slip combinations as a bcc structure. (2) The carbon that is entrapped interstitially deters slip. All in all, martensite is the hardest iron-rich phase the engineer has available for widespread use. In Fig. 6-5.2(a), the hardness of martensite is compared with that of ferrite plus carbide. Figure 6-5.2(b) shows the microstructure of martensite at high magnification.

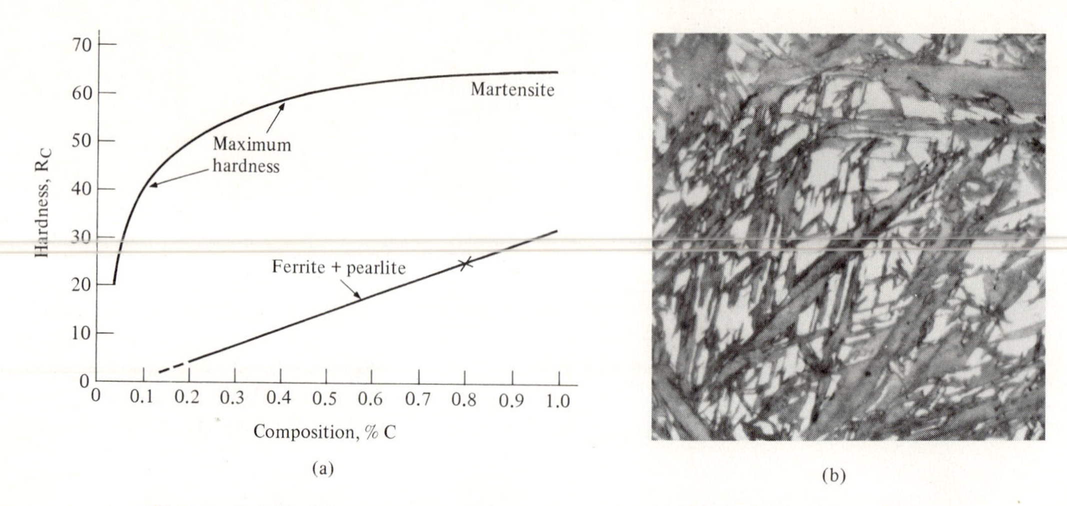

Figure 6-5.2 Martensite in steel. Rapid quenching of austenite can form martensite rather than $[\alpha + \overline{C}]$. (a) The martensite is much harder than the [ferrite + pearlite] that forms in annealed steels. (b) Photomicrograph of martensite (0.8 w/o C). The individual grains of martensite (gray) are platelike, extremely hard and brittle, and have the same composition as the original austenite (white) (×1000). (Courtesy of U.S. Steel Corp.)

Martensite is also relatively brittle, so much so that as-quenched martensite has almost negligible use. Fortunately, *tempering* (to be discussed later) changes the martensite and toughens a quenched steel considerably. If done correctly, tempering does not markedly reduce the hardness excessively. Therefore a steel that is both quenched and tempered is ideal. It is both strong *and* tough. It is this unusual combination of properties that has made steel the prime metal in technology and industry over the past century.

Hardenability

The trick required to form martensite is to cool the steel rapidly enough so that it *cannot* form $[\alpha + \overline{C}]$. The most obvious procedure is to quench the steel in water or an appropriate oil. If you are making a razor blade, for example, you can quench it from 800°C (1470°F) to 20°C in a fraction of a second without ferrite or carbide being formed. A large shaft, however, is much more massive than a razor blade; many more joules (or calories, or Btu's) must be removed through the surface. Even with water quenching, it may be impossible to avoid forming some $[\alpha + \overline{C}]$ in the center of the shaft, where cooling is slowest.

Figure 6-5.3(a) shows a hardness traverse on a 50-mm (2-in.) bar of AISI 1040 steel quenched in water. The surface, which was cooled most

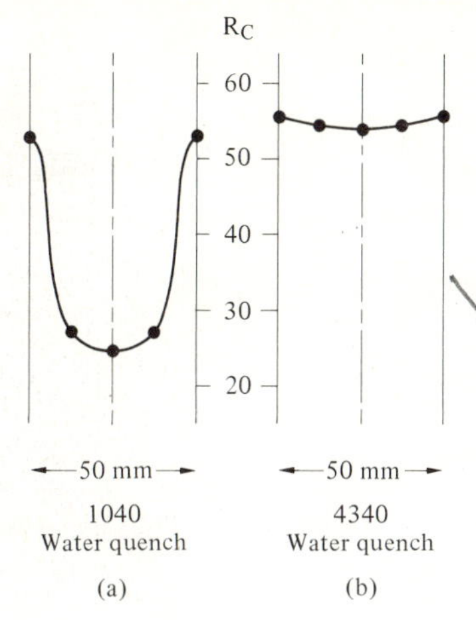

Figure 6-5.3 Hardness profiles. (a) AISI 1040 steel. (b) AISI 4340 steel. Both steels were cooled at the same rate and have the same carbon content. However, the transformation $[\gamma \to \alpha + \bar{C}]$ is slower in the low-alloy steel (4340) than in the plain-carbon steel (1040); therefore 4340 steel forms more martensite. It thus has greater *hardenability*.

rapidly (~300°C/sec), contains a large percentage of martensite and is therefore very hard. The center of the bar cooled more slowly (~25°C/sec) and therefore formed considerable $[\alpha + \bar{C}]$ which, consistent with Fig. 6-5.2(a), is much softer than the martensite. Different steels require quenches of differing severities to form martensite in place of $[\alpha + \bar{C}]$. Since as a general rule any alloying element slows down the rate of the reaction, $\gamma \to [\alpha + \bar{C}]$, alloy steels form the harder martensite more readily.* We speak of this as *hardenability;* it is demonstrated in Fig. 6-5.3(b), which shows the hardness profile of an AISI 4340 steel that received the same quenching and had the same dimensions as the 1040 steel in part (a) of Fig. 6-5.3. Because the $[\gamma \to \alpha + \bar{C}]$ transformation was retarded due to the alloying elements present, more martensite formed and greater hardness was achieved at the center of the alloy steel bar than in the plain-carbon steel bar.

Hardenability curves

Some years ago, in an effort to predict hardness values, a metallurgical engineer by the name of Dr. Walter Jominy devised an end-quench test that is now used as an automobile and steel industry standard. A steel is austenitized, that is, heated to the γ-region to form 100% austenite, then directionally quenched, as shown in Fig. 6-5.4(a). The cooling rates indicated at the various locations behind the quenched end are nearly identical for all types of plain-carbon and low-alloy steels. Therefore when we plot hardness according to the distance behind the quenched end (Fig. 6-5.5—lower abscissa), we are also plotting hardness versus cooling rate (Fig. 6-5.5—upper abscissa). These are called *hardenability curves* and are very

* In alloy steels, the large metal atoms, as well as the carbon atoms, must be segregated into either the ferrite or the carbide. In plain-carbon steels, only the small carbon atoms must be concentrated (into the carbide).

ALLOYING GIVES MORE TIME TO FORM MARTENSITE.

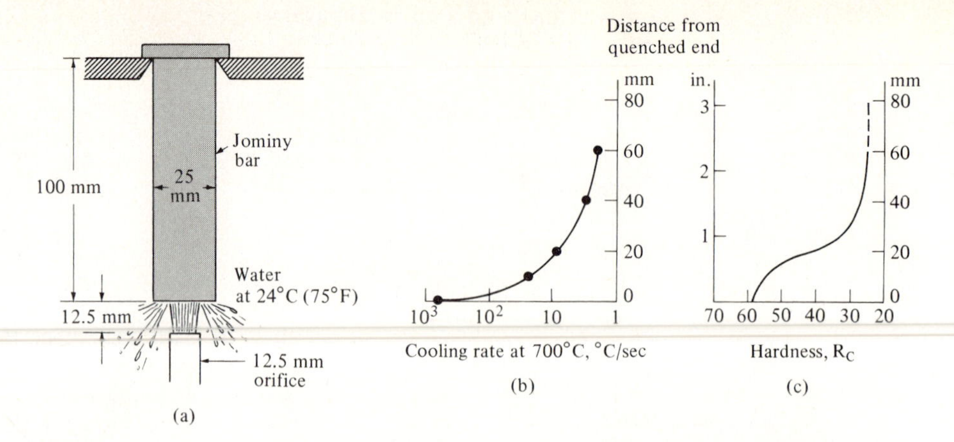

Figure 6-5.4 End-quench (Jominy) test. (a) The "Jominy bar" is heated to form austenite and is dropped into the support over the water stream. (b) The temperature at the quenched end drops through the eutectoid temperature (~700°C) at the rate of 600°C/sec. The cooling rate is slower away from the quenched end. (c) The hardness is greatest where the cooling rate is fastest, because the $\gamma \rightarrow [\alpha + \overline{C}]$ reaction is avoided so martensite is formed.

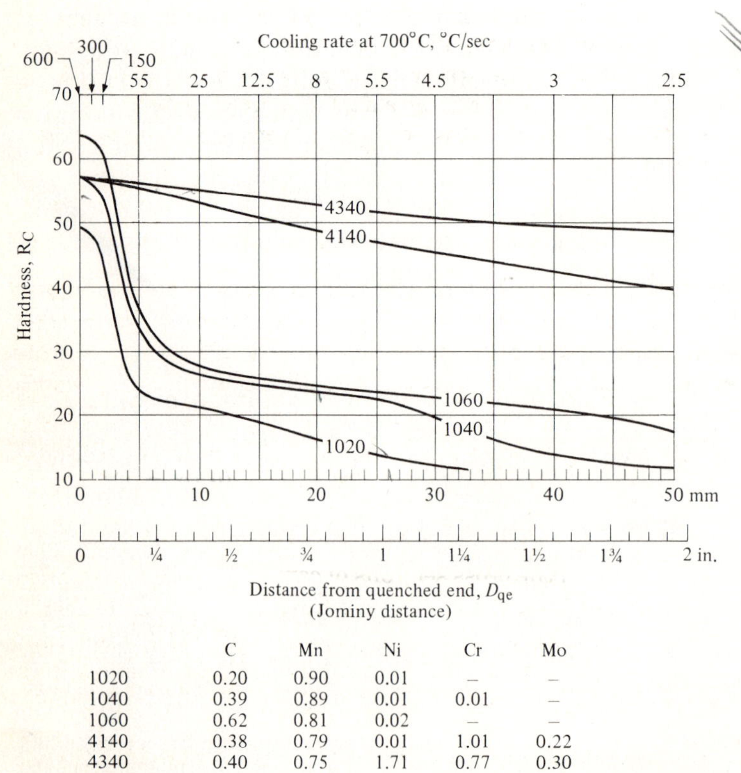

	C	Mn	Ni	Cr	Mo
1020	0.20	0.90	0.01	–	–
1040	0.39	0.89	0.01	0.01	–
1060	0.62	0.81	0.02	–	–
4140	0.38	0.79	0.01	1.01	0.22
4340	0.40	0.75	1.71	0.77	0.30

Figure 6-5.5 Hardenability curves for five steels with the indicated compositions. The steels were end-quenched as shown in Fig. 6-5.4(a). In commercial practice, the hardenability curve of each type of steel varies because of small variations in composition. As a result, hardenability tests are commonly made for each heat of steel that is produced for quench-and-temper applications. (Adapted from U.S. Steel data.)

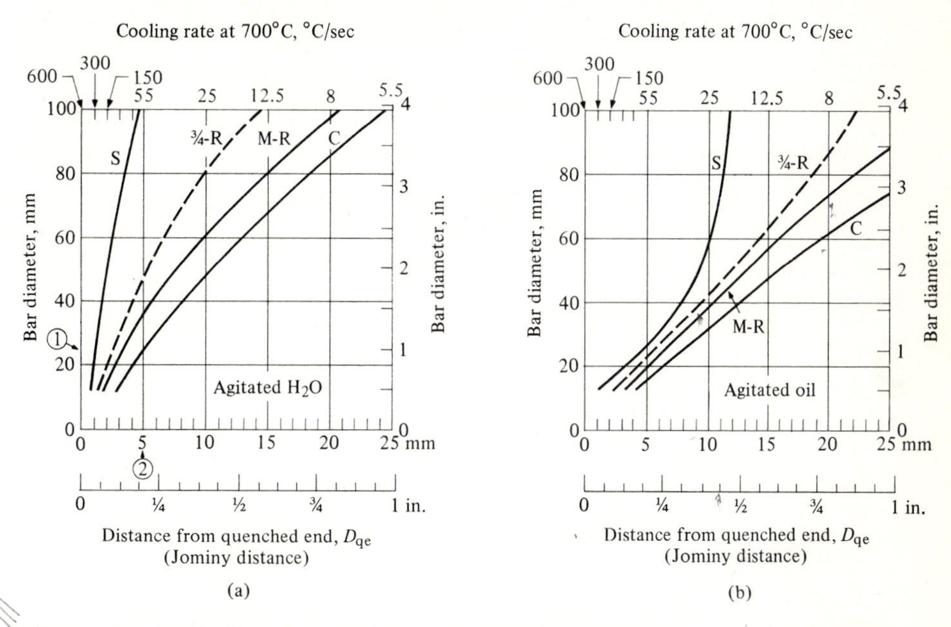

Figure 6–5.6 Cooling rates in long round steel bars quenched in (a) agitated water and (b) agitated oil. Top abscissa, cooling rates at 700°C; bottom abscissa, equivalent positions on an end-quench test bar. (C, center; M-R, mid-radius; S, surface; dashed line, approximate curve for $\frac{3}{4}$-radius). The high heat of vaporization of water produces a severe quench in that quenching medium. (See text regarding locations ① and ②.)

useful to the engineer, because if he or she knows the hardness in a gear tooth, for example, when using one steel, the engineer can readily predict the hardness at a comparable location in a substitute steel that is given the same heat treatment (and therefore has the same cooling rate). For example, any point in a 1060 gear that has a hardness of 30 R_C would have a hardness of 54 R_C in a gear made of 4140 steel (Fig. 6–5.5). The latter has more martensite because of its slower transformation to $[\alpha + \overline{C}]$ with a given cooling rate.

An engineer will find that hardenability curves are also useful in predicting hardness traverses in the cross sections of common steel shapes, such as round bars. Figure 6–5.6 compares cooling rates for bars of different diameters—at their surface S, mid-radius M-R, and center C—when they are quenched in (a) agitated* water and (b) agitated* oil. We observe from the upper abscissa of Fig. 6–5.6(a) that the center of a 25 mm-diameter bar①

* Agitation is vital; otherwise hot liquid and even vapor that is next to the steel will reduce the quenching rate.

cools 55°C/sec when quenched in water. This is the same cooling rate found 5 mm behind the quenched end of a Jominy end-quench bar ②. With this cooling rate for any given steel, these two locations develop identical microstructures and consequently have the same hardness value. Based on Fig. 6–5.5, this hardness value would be 34 R_C in a 1040 steel, 56 R_C in a 4140 steel, etc. Specific examples of calculations will be given at the end of this section.

Tempered martensite

As noted before, martensite is much too brittle to be used just as it is. Therefore we *temper* it, that is, we reheat it to an intermediate temperature. Martensite does not appear in the phase diagram, because it is metastable, that is, it has more energy than $[\alpha + \overline{C}]$. It is also supersaturated with carbon. Given an opportunity at somewhat elevated temperatures, martensite, M, transforms to $[\alpha + \overline{C}]$:

$$\underset{\text{martensite}}{M} \xrightarrow{\text{tempering}} \underset{\text{tempered martensite}}{[\alpha + \overline{C}]} \tag{6–5.1}$$

The tetragonal body-centered martensite changes to bcc ferrite without the excess carbon. The carbon collects in extremely small carbide particles.

Tempered martensite is a 2-phase microstructure of $[\alpha + \overline{C}]$, but one that differs in properties from pearlite and spheroidite because of the size, shape, and distribution of the phases. In tempered martensite, unlike the situation in pearlite, the carbide that forms is not lamellar, but particulate; it is made up of many extremely fine particles that are uniformly distributed on a submicroscopic scale, and that greatly interfere with dislocation movements. Thus tempered martensite remains nearly as hard as martensite just after quenching. In addition any crack that might be initiated must progress through a ductile ferrite matrix as it advances through the steel. This means that tempered martensite is tougher than martensite or pearlite.

At higher temperatures the carbide particles grow (and decrease in number) just as the precipitates do during over-aging of precipitation-hardened alloys (Fig. 6–5.7), with the result that the hardness also decreases. Figure 6–5.8 relates the hardness to both time and temperature. This explains the softening that occurs in many tool steels as they cut at high speeds. A few minutes at 400°C–500°C could decrease the hardness far below the values of 65 R_c available in martensite. Time requirements are increased considerably if the Cr, Mn, Mo, and/or V contents of the carbide are high, because their large atoms must diffuse, as well as the smaller carbon atoms. That is why high-speed tool steels usually contain these alloying elements.

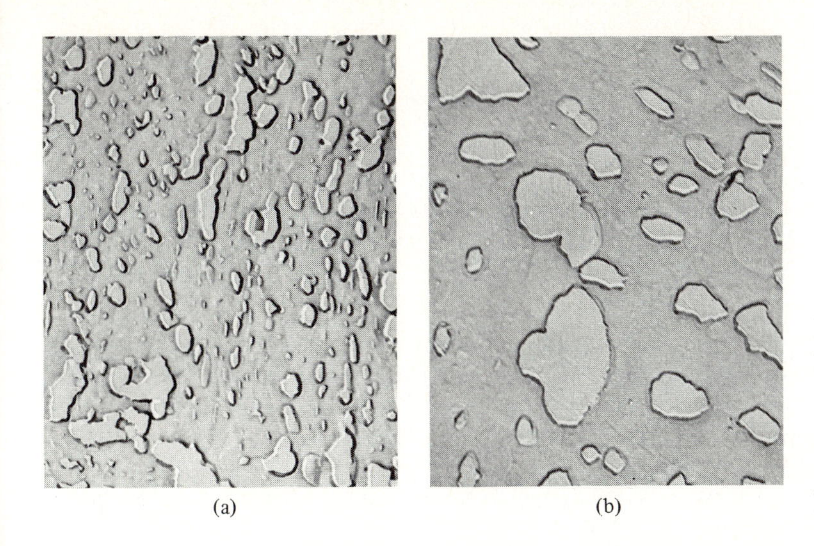

Figure 6-5.7 Growth of carbide particles in tempered martensite (×11,000). The two compositions are the same. (a) One hour at 590°C(1100°F). (b) Twelve hours at 675°C(1250°F). (Courtesy of General Motors Research Laboratories.)

Although there is a slight variation from steel to steel depending on the composition, martensite does not soften below 150°C (300°F) in any reasonable length of time. Therefore, for all intents and purposes, martensite remains indefinitely at ambient temperatures (20°C). Likewise an extrapolation of the data in Fig. 6-5.8 suggests correctly that over-tempering does not occur in reasonable times without reheating. Atoms rearrange themselves too slowly in steel at room temperature to break up the original structures and form new phases. (See I.P. 6-5.4.)

Illustrative problem 6-5.1 Martensite with 3 a/o carbon (~0.66 w/o C) has a body-centered tetragonal unit cell (u.c.) which is 0.283 nm × 0.283 nm × 0.291 nm at room temperature. What is its density ρ?

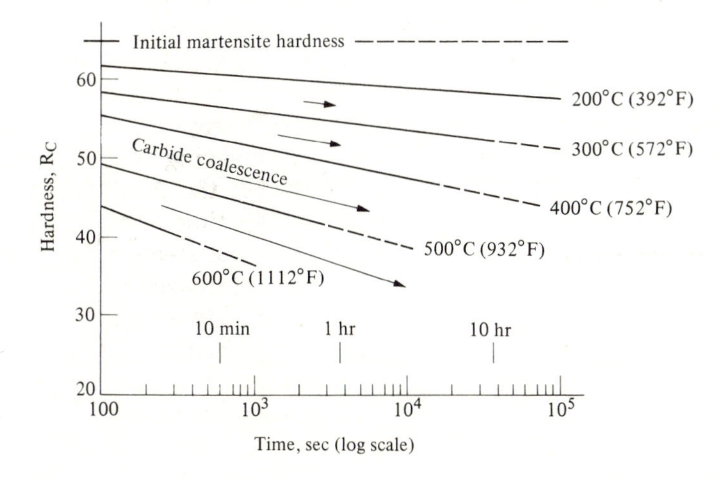

Figure 6-5.8 Hardness of tempered martensite versus tempering time (1080 steel quenched to maximum hardness—65 R_C). Softening occurs because the carbide particles coalesce and grow, giving greater interparticle distances of softer ferrite (Fig. 6-5.7).

SOLUTION

Basis: 200 atoms = 194 Fe atoms + 6 carbon atoms
= 97 unit cells of 2 Fe atoms each, plus 6 interstitial carbon atoms.

$$\rho = \frac{\text{g/97 u.c.}}{\text{vol./97 u.c.}} = \frac{[194(55.85 \text{ amu}) + 6(12 \text{ amu})]/(0.602 \times 10^{24} \text{ amu/g})}{97(0.283 \text{ nm})^2(0.291 \text{ nm})}$$

$$= 8.0 \times 10^6 \text{ g/m}^3. \qquad (\text{or } 8.0 \text{ g/cm}^3).$$

ADDED INFORMATION This density represents an expansion in volume from the original austenite, which is slightly denser (8.15 g/cm^3) at room temperature. Therefore the martensite that forms places the remaining austenite under pressure and, in fact, may force the last of the austenite to be retained at room temperature without change. ◄

Illustrative problem 6–5.2. Figure 6–5.3(a) provides a hardness traverse for a 50-mm-diameter bar of AISI 1040 steel. Based on the hardenability data of Fig. 6–5.5, sketch a traverse for an AISI 1020 steel bar of the same dimensions and quenched in the same manner.

SOLUTION

	AISI 1040, R_C	Cooling rate, °C/sec	Equivalent end-quench position, mm	AISI 1020, R_C
Surface	53	~150	2	42
Mid-radius	28	35	7.5	22
Center	26	22	11	21

ADDED INFORMATION The quenching method affects the cooling rate, and therefore the hardness of the steel. A water quench is more severe than an oil quench because steam forms, requiring a very high heat of vaporization. If the quenching bath is not agitated, the steam can produce insulated spots that slow the heat removal and produce softer spots in the steel. ◄

Illustrative problem 6–5.3 Requirements for a 60-mm-diameter bar (2⅜-in.) call for a surface hardness of more than 50 R_C and a center hardness of less than 32 R_C. Alternatives include the five steels of Fig. 6–5.5 and two methods of quench: agitated oil and agitated water. Make your recommendations.

SOLUTION

	Equivalent distance from quenched end, mm	
For 60-mm bar (Fig. 6-5.6)	Oil quench (o.q.)	Water quench (w.q.)
Surface	10	2.5
Mid-radius	16	10
Center	19	13

From Fig. 6-5.5:
4140 and 4340 have center hardnesses too high, with either an oil quench or a water quench.
1020 has a surface too soft with either quench.
1040 and 1060 have surfaces too soft with an oil quench.

1040 w.q. S: 50 R_C C: 25 R_C
1060 w.q. S: ~55 R_C C: 27 R_C

ADDED INFORMATION The 1060 w.q. is the better choice, since a slight variation in the carbon content of the 1040 steel, say to 0.38% C, could cause the surface hardness to fall below 50 R_C.

Engineers often specify high surface hardness for resistance to wear, and require the center to remain soft for greater toughness. ◂

Illustrative problem 6-5.4 Assume that an extrapolation of the curves in Fig. 6-5.8 is valid. How long will it take at 200°C to soften that 1080 steel to 53 R_C?

SOLUTION We could set up an equation,

$$R_C = A - B \log t;$$

however that would be no more accurate than a graphical extrapolation, which gives 10^8 sec:

$$10^8 \text{ sec} = >3 \text{ yrs.}$$

ADDED INFORMATION Actually, such an extrapolation may underestimate the time. In any event, tempering occurs slowly at this temperature.

A calculation is no more accurate than a graphical solution, since the equation would have to take data from the graph initially. ◂

• 6-6 CAST IRONS

High-carbon iron alloys are the least expensive of all metallic materials. In addition, they have some other advantages that make them attractive for many bulky engineering applications, such as the automobile engine block in Fig. 6-6.1. For example, they readily fill intricate molds; they absorb vibrational energy; and their high elastic modulus makes them more rigid than castings of other common metals of the same dimensions.

Gray cast irons

The most common cast iron is *gray iron*, so-called because its fracture is gray rather than metallic white. An examination reveals that this cast iron, which possesses ~3.25% C (and ~2% Si), contains carbon as graphite and not as Fe_3C, which is found in steels. The fracture path through the cast iron follows the graphite, which is present as flakes (Fig. 6-6.2a). Thus, we see gray graphite on the fractured surface.

We have a right to ask why graphite is present here when it does not appear on the Fe–Fe_3C phase diagram (Fig. 6-2.2). The metallurgists point out that Fe_3C is not completely stable. It forms and exists almost indefinitely in steel. However, with higher temperatures, with plenty of carbon present, and with silicon also present, this carbide gradually decomposes to form the more stable graphite. Above the eutectoid temperature, the decomposition reaction is

$$Fe_3C \rightarrow \gamma + \text{graphite}; \tag{6-6.1}$$

below the eutectoid temperature, the reaction is

$$Fe_3C \rightarrow \alpha + \text{graphite}. \tag{6-6.2}$$

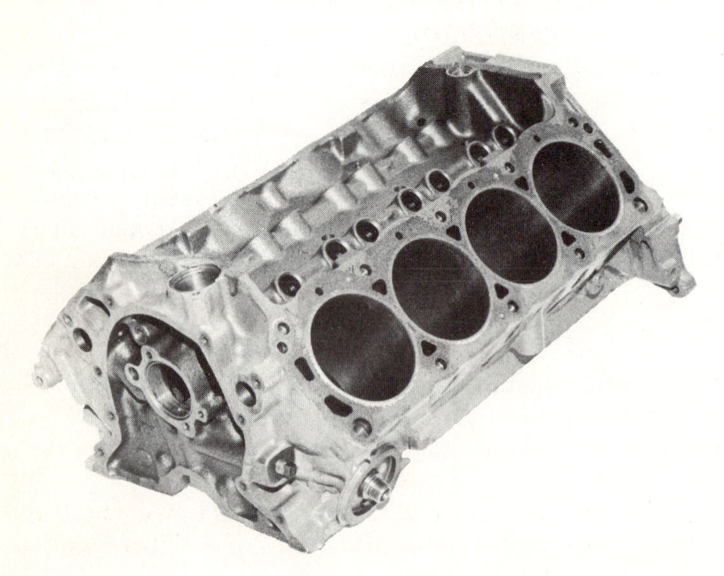

Figure 6-6.1 Cast-iron engine block (V-8). Not only is cast iron one of the most available and least expensive alloys, it has the advantage of castability into intricate designs, thus requiring a minimum of machining before assembling the final engine. (Courtesy of the Ford Motor Company.)

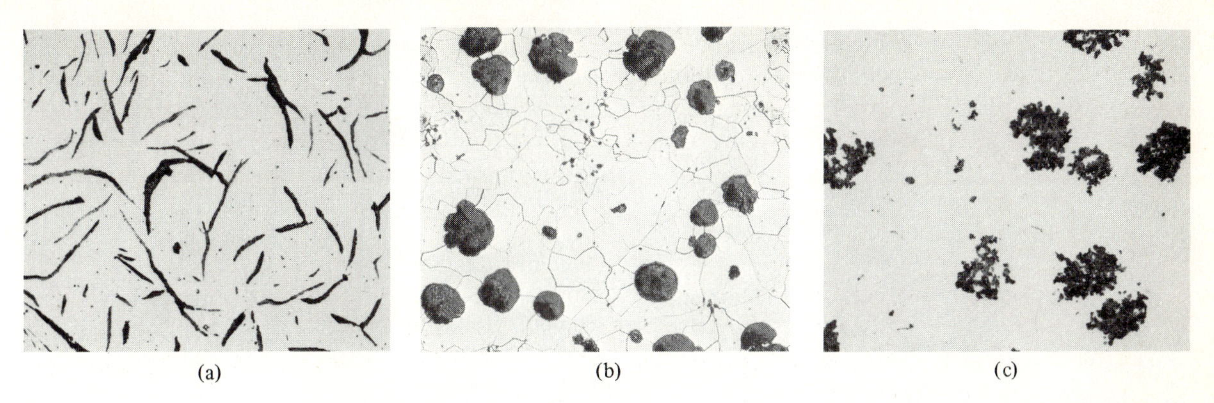

Figure 6–6.2 Microstructures of cast irons (×100). In each photomicrograph, the matrix is ferrite and the dark phase is graphite. (a) Ferritic gray iron. We see the cross sections of graphite flakes. (b) Ferritic nodular (ductile) iron. (c) Ferritic malleable iron. This cast iron solidified as white cast iron with carbon present as Fe_3C rather than as graphite. The metal was heat treated to graphitize the carbon (Eq. 6–6.2). (American Society for Metals.)

The composition previously given for gray cast iron (~3.25% C and 2% Si) accelerates the rate of *graphitization* sufficiently so that it starts simultaneously with solidification. The result is the microstructure of Fig. 6–6.2(a) in which the graphite is present as flakes (in three dimensions).

There are several variants of gray cast iron. The first is *ferritic gray iron.* Assume slow cooling so that both Reactions (6–6.1) and (6–6.2) occur. The final product contains only (α + graphite) and is considered to be "fully gray iron." Such as cast iron is very machinable and has superior damping capacities for applications such as machine bases (Table 6–6.1).

Pearlitic gray iron is solidified sufficiently slowly to complete Reaction (6–6.1) before the eutectoid temperature is reached. The austenite is now eutectoid in composition. This austenite is not significantly different from the γ of steels. Therefore, it changes to pearlite. These pearlitic gray irons (pearlite + graphite) are produced in kiloton amounts in engine blocks and similar castings (Table 6–6.1). It is not ductile, but it is stronger than fully gray iron. It is rigid, machinable, rust-resistant (because of the silicon), and serves the purpose under all normal operational conditions.

Martensitic gray iron initially parallels pearlitic gray iron. However, quenching from (γ + graphite) gives (M + graphite), rather than (P + graphite). It is a very hard cast iron that is used as wearing surfaces.

Nodular cast iron

With special control, for example, through the addition of small amounts of magnesium, or cerium, the graphite that separates during solidification can

Table 6-6.1
Properties of unalloyed cast irons*

Cast iron	Type	Minimum mechanical properties S_u, MPa (psi)	S_y, MPa (psi)	Elong. (50 mm)	Typical uses
Gray (3.2 C–2 Si)	Pearlitic	275 (40,000)	240 (35,000)	<1%	Engine blocks
	Martensitic	550 (80,000)	550 (80,000)	nil	Wearing surfaces
	Ferritic†	172 (25,000)	138 (20,000)	≮1%	Pipe, machine bases
Nodular‡ (3.5 C–2.5 Si)	Ferritic	413 (60,000)	275 (40,000)	18	Pipe
	Pearlitic	550 (80,000)	380 (55,000)	6	Crankshafts
	Tempered martensite	825 (120,000)	620 (90,000)	2	Special machine parts
Malleable (2.2 C–1 Si)	Ferritic	365 (53,000)	240 (35,000)	18	Hardware
	Pearlitic	450 (65,000)	310 (45,000)	10	Railroad equipment
	Tempered martensite	700 (100,000)	550 (80,000)	2	Railroad equipment
White (3.5 C–0.5 Si)	As cast (pearlitic)	275 (40,000)	275 (40,000)	nil	Wear-resistant products

* Adapted from Flinn and Trojan, *Engineering Materials and Their Applications*, and from American Society of Metals.

† Ferritic gray cast iron generally has ~3.5% C and ~2.5% Si.

‡ Also called *ductile* cast iron.

be made to form nodules, or spheres (Fig. 6-6.2b), rather than the flakes of Fig. 6-6.2(a). The resulting metal, called *nodular iron* (or *ductile iron*), is more ductile than the gray cast iron simply because the nodular graphite provides less opportunity for stress concentrations. These premium cast irons can be *ferritic* or *pearlitic* (Table 6-6.1). They are not used in a martensitic form, but they can be treated to give *tempered martensite*.

White and malleable cast iron

Finally, those cast irons that contain <2% Si form graphite so slowly that they solidify as $[\gamma + \overline{C}]$. It is thus possible to have *white cast iron* (no

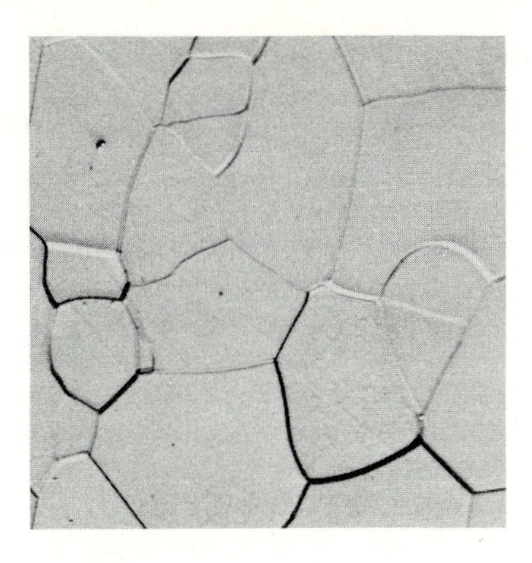

Figure 6-6.3 Ferrite in low-carbon steels (×500). It is more ductile than ferritic cast irons (Table 6-6.2); therefore, it can be rolled into a sheet iron product. (Courtesy of U.S. Steel Corp.)

Table 6-6.2
Ductility of ferritic alloys

Iron type	Ductility % elong. (50-mm)	See Figure
Ferritic steel	40	6-6.3
Ferritic nodular iron	15 to 20	6-6.2(b)
Ferritic malleable iron	15±	6-6.2(c)
Ferritic gray iron	1	6-6.2(a)

graphite), but a cast iron that on extended heating will react according to Eqs. (6-6.1) and/or (6-6.2). The graphite that forms in the solid casting, rather than during solidification, has the microstructure that is shown in Fig. 6-6.2(c). This form of cast iron was the first to exhibit usable ductility (or malleability); hence the name *malleable iron*. Malleable iron may be ferritic, or pearlitic, or may contain tempered martensite (Table 6-6.1).

It is worth pausing to compare properties that arise from the microstructures shown in this section. If each type of cast iron is heat treated to the ferritic form, their ductilities may be compared to that of ferrite alone (Fig. 6-6.3). As shown in Table 6-6.2, the % El. is affected, not only with the presence of graphite, but also by the shape of the graphite.

Illustrative problem 6-6.1 A malleable cast iron containing 3.0 w/o C was initially solidified as [γ + carbide]; it was then reheated (malleablized) to dissociate the carbide. What is the maximum v/o graphite that may form (a) at 800°C (1470°F)? (b) At 680°C (1250°F)? For both α and γ, $\rho \approx 7.6$ g/cm³ at these temperatures. The density of graphite is approximately 2.0 g/cm³.

SOLUTION Basis: 100 g. Refer to Fig. 6–2.2, but with graphite.

a) At 800°C:

$$\text{g } \gamma = \left(\frac{100 - 3.0}{100 - 1.0}\right) 100 \text{ g} = 98 \text{ g } \gamma,$$

$$\text{cm}^3\ \gamma = 98 \text{ g}/(7.6 \text{ g/cm}^3) = 12.9 \text{ cm}^3.$$

$$\text{g graphite} = \left(\frac{3.0 - 1.0}{100 - 1.0}\right) 100 \text{ g} = 2.0 \text{ g graphite.}$$

$$\text{cm}^3 \text{ graphite} = 2.0 \text{ g}/(2.0 \text{ g/cm}^3) = 1.0 \text{ cm}^3.$$

$$\text{v/o graphite} = 100(1.0 \text{ cm}^3)/(13.9 \text{ cm}^3) = 7.2 \text{ v/o graphite.}$$

b) At 680°C:

$$\text{cm}^3\ \alpha = \left(\frac{100 - 3.0}{100 - 0.02}\right) \frac{(100 \text{ g})}{(7.6 \text{ g/cm}^3)} = 12.8 \text{ cm}^3.$$

$$\text{cm}^3 \text{ graphite} = \left(\frac{3.0 - 0.02}{100 - 0.02}\right) \frac{(100 \text{ g})}{(2 \text{ g/cm}^3)} = 1.5 \text{ cm}^3.$$

$$\text{v/o graphite} = 100(1.5 \text{ cm}^3)/(14.3 \text{ cm}^3) = 10.5 \text{ v/o graphite.}$$

ADDED INFORMATION By selecting appropriate processing treatments, one can obtain ferritic malleable irons, pearlitic malleable irons, pearlitic nodular irons, ferritic gray irons, etc. Each has its own set of properties (Table 6–6.1).

Figure 6–2.2 is based on Fe_3C rather than on graphite. There should be a slight modification for graphite. The eutectoid is shifted to 0.68% C and 738°C (1360°F), and the eutectic to 1154°C (2110°F). (See Fig. 13–1.2, Van Vlack, *Elements of Materials Science and Engineering*, 4th ed., Reading, Mass.: Addison-Wesley.). With this change, the answer of part (a) becomes 7.4 v/o graphite. Part (b) is not changed. ◀

• 6–7 PHASE DIAGRAMS OF LOW-ALLOY AND MEDIUM-ALLOY STEELS

The Fe-Fe_3C phase diagram of Fig. 6–2.3 involves iron and carbon alone and does not include effects of other alloying elements. Thus, while we know that the carbon content of an AISI-SAE 5270 steel is 0.70%, one should question using Fig. 6–2.3 to predict phase relationships since a 52xx steel contains approximately 1.5% chromium (1.3% to 1.5%). We can, however, use Fig. 6–2.3 as a base for predicting the eutectoid region of low-alloy and medium-alloy steels (<5% alloying elements).

Eutectoid shifts

The eutectoid temperature of the Fe–Fe_3C phase diagram is 727°C (1340°F); the eutectoid composition is 0.77 w/o C, or approximately 0.8% (Fig. 6–2.3).

The chromium (~1.5%), which is in the 5270 steel of the earlier paragraph, dissolves in both the iron-rich phases—austenite and ferrite. However, chromium, which is bcc itself, favors the bcc ferrite over the fcc austen-

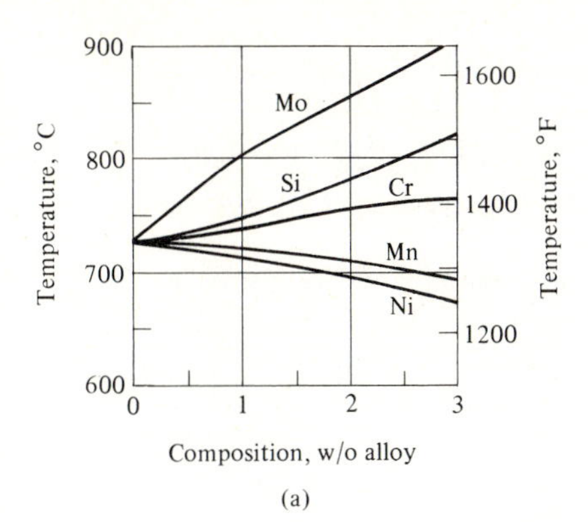

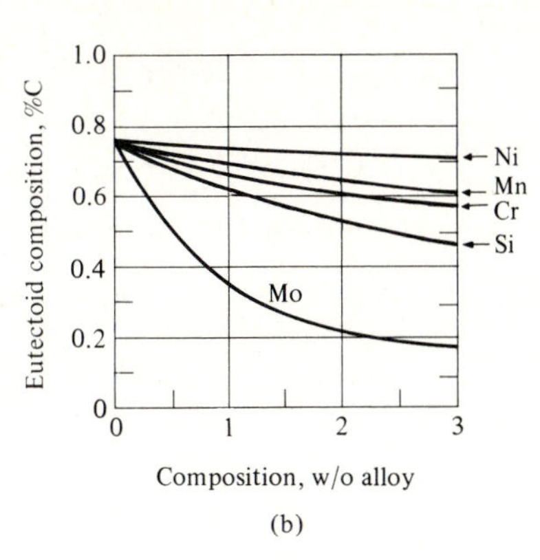

Figure 6-7.1 Eutectoids in Fe-X-C alloys. (X is the third component—Mo, Si, Cr, Mn, or Ni.) Effect of alloy additions on (a) the temperature of the eutectoid reaction and (b) the carbon content of the eutectoid. (Adapted from ASM data.)

ite. Furthermore, chromium forms very stable carbides. Thus, we find that the austenite temperature range of 52xx steels is reduced, and the ferrite-plus-carbide range is expanded from that of Fig. 6-2.2. Overall, the *eutectoid temperature is raised* by chromium. The temperature increase depends on the amount of chromium added. As shown in Fig. 6-7.1(a), 3% Cr raises the eutectoid temperature from 727°C to 765°C (1410°F). The eutectoid temperature of an AISI-SAE 52xx steel with 1.5% Cr is 745°C (1370°F).

The eutectoid composition is also altered in alloy steels. All alloy elements *reduce the carbon content* of the eutectoid (Fig. 6-7.1b). Our AISI--SAE 52xx steels, with ~1.5% Cr, have eutectoid compositions of ~0.65% C. Thus, the carbon content of a 5270 steel is above the eutectoid composition (0.70% vs ~0.65% C).

An examination of the data in Fig. 6-7.1 reveals several predictable facts. (1) Since its own structure is fcc, nickel is an *austenite-former,* thus lowering the eutectoid temperature (and expanding the stable range of the fcc austenite). Manganese also favors fcc over bcc and, therefore, also decreases the eutectoid temperature. (2) In contrast, the structures of pure Cr and Mo are bcc; therefore, they are *ferrite-formers* and raise the eutectoid temperature, expanding the stable range of the bcc ferrite. (3) Molybdenum forms very strong bonds with carbon; nickel almost ignores carbon. We see this effect in Fig. 6-7.1(b) where 1% Mo decreases the eutectoid composition to 0.35% C; that is, at the eutectoid temperature of 800°C (1470°F), the carbon has a solubility limit in γ of 0.35%. Any additional carbon combines with molybdenum as a carbide. In contrast, Fig. 6-7.1 shows that nickel has minimal effect on the carbon solubility in γ at the eutectoid temperature.*

Illustrative problem 6-7.1 (a) Sketch the eutectoid region for 44xx steels. (b) What minimum temperature must a 4465 steel be raised to have 100% γ?

* The slight, barely measurable reduction that occurs (Fig. 6-7.1b) may be attributed to the lowered eutectoid temperature. At 727°C and 1% Ni, the carbon solubility in austenite actually increases slightly from 0.77% to 0.80%.

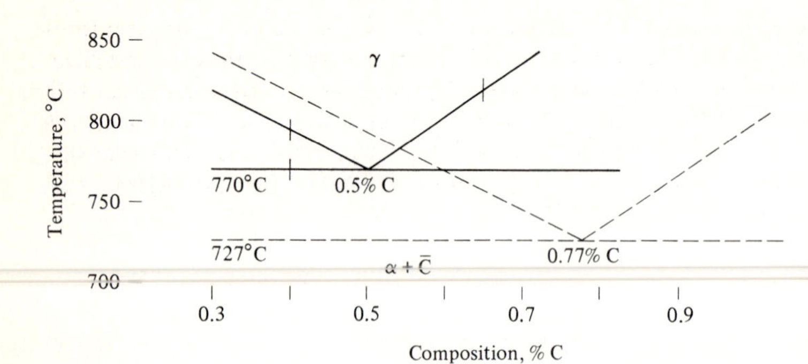

Figure 6-7.2 Eutectoid shift (44xx steels).

----From Fe-Fe_3C (Fig. 6-2.3).
—— Relocated for 0.5% Mo (Fig. 6-7.1).

An addition of 0.5% Mo shifts the eutectoid to ~0.5% carbon at 770°C. The solubility-limit curves remain nearly parallel if the alloy additions are low. (Refer to I.P's. 6-7.1 and 6-7.2.)

(You may assume that the slopes of the solubility curves do not change with low-alloy additions.)

SOLUTION From Table 6-1.1, Mo = ~0.5%. Thus, using data from Fig. 6-7.1:

a) T_e = ~770°C, and C_e = ~0.5% C (see Fig. 6-7.2).
b) From Fig. 6-7.2:
T_γ = 820+°C ◂

Illustrative problem 6-7.2 (a) What is the proeutectoid phase in an AISI-SAE 4440 steel? (b) How much pearlite will form on slow cooling? (c) To what temperature must this steel be raised to have 100% γ?

SOLUTION Use the sketch of Fig. 6-7.2.

a) Just above the eutectoid temperature, 770°C, a 99.1Fe-0.5Mo-0.4C steel will have (α + γ); therefore, ferrite precedes the formation of pearlite.
b) Fraction α = 0.10/0.5 = 0.2;
fraction γ = 0.4/0.5 = 0.8 = fraction pearlite.
c) From Fig. 6-7.2, T_γ = 795°C. ◂

Review and Study

SUMMARY

1) Since iron is polymorphic, the Fe–Fe_3C phase diagram includes a region of austenite, γ. It is fcc and can dissolve the small carbon atoms to the extent of 0.8 w/o at 727°C, and to more than 2 w/o (9 a/o) at the eutectic temperature (1148° C). The bcc phase is called ferrite, α. At 727°C, it can dissolve a maximum of only one carbon atom per 1100 iron atoms, or 0.02 w/o C. Therefore, excess carbon appears as a carbide (Fe_3C, or $\overline{C}$). These facts are fundamental to the understanding and engineering use of steels.

2) Plain-carbon steels (10xx) are iron + carbon alloys. Their carbon contents (0.xx%) permit us to relate their compositions to the equilibrium phases. Most steels contain less than 1% carbon. Low-alloy and medium-alloy steels contain limited amounts (<5%) of other alloying elements along with the iron and carbon. The phase diagram may be modified accordingly. Specialty steels, which are more complex than we will consider, include stainless steels, tool steels, high-temperature steels, etc.

3) The eutectoid reaction of Fe–Fe_3C alloys is

$$\gamma\ (\sim 0.8\%)\ C\ \underset{\text{heating}}{\overset{\text{cooling}}{\rightleftarrows}}\ \ 727^\circ\ (1340^\circ F)\ \ \alpha\ (0.02\% C) + \text{carbide}\ (6.7\% C) \qquad (6\text{–}7.2)$$

Austenite decomposes in this manner when it is cooled slowly enough to permit the carbon to diffuse to the carbide (25 a/o C) and away from the ferrite (0.1 a/o, or 0.02 w/o C). The resulting product is pearlite, P, a microstructure with alternating layers of ferrite and carbide. The amount of pearlite formed under these conditions is the same as the amount of austenite immediately above the eutectoid temperature. Annealed steels with less than eutectoid carbon possess pearlite and proeutectoid ferrite; steels with more than eutectoid carbon possess pearlite and proeutectoid carbide.

4) Raising the carbon content (thus increasing the amounts of carbide and pearlite) in annealed steels increases the strengths and hardness. This change occurs because the carbide particles reinforce the ferrite. However, they also decrease the ductility and toughness by providing a brittle path for fracture. A steel is stronger with thinner pearlite lamallae but is tougher with spheroidized carbides.

5) The $\gamma \rightarrow [\alpha + \overline{C}]$ reaction is avoided by quenching, because there is not time for the diffusion that otherwise is necessary in austenite decomposition. This permits, $\gamma \rightarrow$ martensite. Martensite, M, is a very hard and brittle phase of iron and carbon. Tempering permits the reaction to proceed, $M \rightarrow \alpha + \overline{C}$. The microstructure is not pearlite but a dispersion of small carbide particles in a tough ferrite matrix. With control, so the carbide particles remain very small, the steel has desirable combinations of high strength and good toughness. Steels are quenched and tempered to attain such properties.

6) The $\gamma \rightarrow [\alpha + \overline{C}]$ reaction is slower in alloy steels than in plain-carbon steels, since the alloying atoms must relocate as well as the carbon atoms. Therefore, low-alloy steels produce martensite with less severe quenches than do plain-carbon steels. This gives them greater hardenability—for a given cooling rate, they are harder. We can estimate the quenched hardness of a steel by using a Jominy end-quench test that provides a hardenability curve.

7) Cast irons provide an inexpensive class of alloys for cast shapes. Since they contain more carbon than steels, they melt at lower temperatures and generally contain graphite. The form of the graphite can be varied with corresponding effects on the properties. These variables, plus heat-treating variations permit the production of ferritic, pearlitic, and martensitic cast irons for a wide selection of applications.

TECHNICAL TERMS

AISI-SAE steels Standardized identification code for plain-carbon and low-alloy steels, which is based on composition. Last two numbers indicate the carbon content. (*See* Table 6-1.1.)

Annealing (steels) For a *full* anneal, austenite is formed, then the steel is cooled slowly enough to form pearlite. (Annealing to remove strain hardening is called *process* anneal. The steel is heated just below the eutectoid temperature.)

Austenite (γ) Face-centered cubic iron, or iron alloy based on this structure.

Austenite decomposition Eutectoid reaction that changes austenite to (α + carbide).

Austenization Heat treatment to dissolve carbon into fcc iron, thereby forming austenite.

Carbide ($\bar{C}$) Compound of metal and carbon. Unless specifically stated, it refers to iron-base carbides.

Cast iron Fe–C alloy, sufficiently rich in carbon to produce a eutectic liquid during solidification. In practice, this generally means more carbon than can be dissolved in austenite (>2%).

- **Cast iron (gray)** Cast iron with graphite flakes and therefore a gray fracture surface.
- **Cast iron (malleable)** Cast iron that undergoes graphitization after solidification. Graphite is present as "clusters."
- **Cast iron (nodular)** Cast iron in which graphite spherulites form during solidification. Sometimes called ductile iron because of its properties.
- **Cast iron (white)** Cast iron with Fe_3C rather than graphite.
- **End-quench test** Standardized test, by quenching from one end only, for determining hardenability.

Eutectoid reaction $S_2 \underset{\text{heating}}{\overset{\text{cooling}}{\rightleftharpoons}} S_1 + S_3$.

- **Eutectoid shift** The change in temperature and the carbon content of the eutectoid reaction arising from alloying element additions.

Ferrite (α) Body-centered cubic iron, or an iron alloy based on this structure.

- **Graphitization** Decomposition of iron carbide into iron and carbon (Eqs. 6-6.1 and 6-6.2).
- **Hardenability** The ability to develop maximum hardness by avoiding the ($\gamma \rightarrow \alpha$ + carbide) reaction.
- **Hardenability curve** Hardness profile of end-quench test bar.

Hardness Resistance to penetration.

- **Hardness traverse** Profile of hardness values.

Irons Includes cast irons. Also commercially pure iron. Unlike cast irons, the latter are highly ductile and may be rolled into thin sheets.

- **Jominy bar** *See* end-quench test.
- **Martensite** Metastable body-centered phase of iron supersaturated with carbon; produced from austenite by a shear transformation during quenching.

Mixture rule Simplified equations to relate properties to phase mixtures.

Pearlite (P) A microstructure (ferrite plus lamellar carbide) of eutectoid composition.

Proeutectoid (steel) A phase that forms (on cooling) before the eutectoid austenite decomposes. With less than eutectoid carbon, this phase is ferrite; with more, the proeutectoid phase is carbide.

Quench Cooling accelerated by immersion in agitated water or oil.

Spheroidite Microstructure of coarse spherical carbides in a ferrite matrix.

Steel Iron-base alloys, commonly containing carbon. In practice, the carbon may all be dissolved by heat treatment; hence, <2.0 w/o C.

Steel, low-alloy Steel containing up to 5% alloying elements other than carbon. Phase equilibria are related to the Fe–C diagram.

Steel, plain-carbon Basically Fe–C alloys with minimal alloy content.

Steel, specialty High-alloy steels ($>5\%$ alloying elements). These include stainless steels, tool steels, etc.

• **Tempered martensite** A microstructure of ferrite and dispersed carbides obtained by heating martensite.

• **Tempering** A toughening process in which martensite is heated to initiate a ferrite-plus-carbide microstructure.

CHECKS

6A __***__ steels are basically two-component alloys of iron and carbon.

6B Low-alloy steels contain as much as __***__ percent of metallic elements other than iron.

6C When using the AISI–SAE identification code for plain-carbon and low-alloy steels, the carbon content is indicated by __***__.

6D Pure iron is polymorphic changing from __***__ to __***__ at ~900°C.

6E __***__ is the bcc phase of iron alloys; it is also called __***__.

6F __***__ is the fcc phase of iron alloys; it is also called __***__.

6G Iron and carbon form a compound with a __***__-to-one ratio of iron to carbon atoms.

6H The solubility of carbon in fcc austenite is __***__ than in bcc ferrite.

6I A eutectoid reaction involves __***__ liquid phases and __***__ solid phases.

6J The eutectoid involving iron and iron carbide has a temperature, T_e, of __***__°C and a composition, C_e, of __***__% C.

6K Austenite "of eutectoid composition" in a plain-carbon steel contains __***__% carbon, which decomposes on slow cooling to α (__***__% C) and carbide (__***__% C).

6L __***__ ferrite forms above (before) the eutectoid temperature is reached; __***__ ferrite forms during the eutectoid reaction.

6M The microstructure of $(\alpha + \overline{C})$ that comes from austenite decomposition is called __***__.

6N The amount of pearlite is the same as the amount of ___***___ just above the eutectoid temperature.

6O When austenite decomposes on cooling, the phases involved are ___***___ ⟶ ___***___.

6P The microstructure called ___***___ contains a dispersion of relatively coarse carbide particles in a ferrite matrix; this is unlike pearlite in which the carbides are ___***___.

6Q Spheroidite has ___***___ strength than pearlite, and ___***___ toughness.

6R An annealed 1050 steel has ___***___ strength than a 1025 steel, and ___***___ ductility.

• 6S In order to form ___***___ in steel, the steel must be quenched rapidly enough to avoid the normal $\gamma \rightarrow \alpha + \overline{C}$ reaction.

• 6T During the $\gamma \rightarrow \alpha + \overline{C}$ reaction, ___***___ must occur to segregate carbon and the ___***___-formers from the elements that prefer ferrite.

• 6U The crystal structure of martensite is ___***___-centered ___***___ with a carbon content the same as that in the previous ___***___ phase.

• 6V Martensite is unstable, hence it will change to ___***___ if given an opportunity. That opportunity comes most readily if the ___***___ is increased.

• 6W A steel with a high ___***___ forms martensite with slower cooling rates than a steel with a low ___***___.

• 6X The quenching rate in agitated water is ___***___ than in agitated oil.

• 6Y Tempered martensite has a microstructure containing the phases ___***___ and ___***___.

• 6Z Tempered martensite can have more ___***___ and ___***___ than pearlite; it has ___***___ toughness than pearlite.

• 6AA Tempered martensite softens as the carbide particles ___***___. At low temperatures, this takes ___***___ time than at high temperatures.

• 6AB Graphite is a phase in all cast irons except ___***___ cast iron.

• 6AC The graphite in ___***___ cast iron is in flake form; it has a spherical structure in ___***___ cast iron.

• 6AD In comparing the ferritic types of gray cast iron, nodular cast iron, and malleable cast iron, the ___***___ cast iron is least ductile and the ___***___ cast iron is most ductile.

• 6AE The graphitization of ___***___ cast iron occurs entirely within the solid metal.

• 6AF The presence of ___***___ as an alloying element in cast iron reduces the time required for graphitization.

• 6AG The graphitization of pearlitic malleable iron occurs primarily above the ___***___ temperature.

• 6AH Since chromium and molybdenum are ___***___-formers, they ___***___ the eutectoid temperature, T_e.

• 6AI ___***___ and ___***___ additions to steel lower the eutectoid compositions, C_e, below the value of 0.77% for iron-carbon alloys.

STUDY PROBLEMS

6–1.1. Three steels have the following compositions (in %): (a) C–0.41, Mn–0.69, Cr–0.54, Mo–0.28, Ni–0.61; (b) C–0.39, Mn–1.25, Ni–0.04, Cr–0.02, S–0.015; (c) C–0.79, Mn–0.68, Cu–0.01, Ni–0.03. What are their AISI-SAE numbers?

Answer: (a) 8740 (c) 1080

6–1.2. Which will have the greater density, ferrite or γ? Why?

6–2.1. Calculate (a) the lattice constant, and (b) the density of ferrite (100% Fe).

Answer: (a) 0.2866 nm (b) 7.88 g/cm^3

6–2.2. Calculate (a) the lattice constant, and (b) the density of austenite at 920°C. [$R_{\gamma Fe}$ = ~0.129 nm.]

6–2.3. Calculate the fraction of atoms that are carbon in the Fe–C eutectic.

Answer: 17 a/o

6–2.4. A maximum of 2.1% carbon can dissolve interstitially into austenite (fcc). How many unit cells are there per carbon atom?

6–2.5. A 1030 steel is to be austenized, that is, heated until it is completely austenite. What is the lowest possible temperature for 100% γ?

Answer: 830°C

6–2.6. Determine the fraction of γ and $\overline{C}$ in a 99Fe–1C steel at (a) 850°C (1560°F), (b) 728°C (1342°F), and (c) 726°C (1338°F).

6–2.7. A 1060 steel is equilibrated at 730°C (1345°F). (a) What is the ratio of α to γ? (b) With further cooling, the ferrite does not change; however, the γ gives an intimate mixture of [α + $\overline{C}$]. What is the ratio of α to $\overline{C}$ in that mixture? (c) Of α to $\overline{C}$ in the total steel?

Answer: (a) 23 to 77 (b) 12 to 88 (c) 9 to 91

6–2.8. After slow cooling from 950°C, a steel (with iron and carbon only) was examined through a microscope. Observations showed 35% of the steel was coarse ferrite. The other 65% was an intimate mixture of ferrite and carbide. (a) Does the steel have more or less than eutectoid carbon—is it hypereutectoid or hypoeutectoid? (b) Explain the origin of the structure seen through the microscope. (c) Assign an AISI-SAE number to this steel.

6–3.1. Based on Eq. (6–2.2), what fraction of pearlite is carbide?

Answer: 0.12

6–3.2. What fraction of an annealed 1025 steel can be pearlite?

6–3.3. At what temperature will a 1020 steel contain (a) $\frac{1}{3}\gamma + \frac{2}{3}\alpha$? (b) $\frac{2}{3}\gamma + \frac{1}{3}\alpha$?

Answer: (a) 765°C (1400°F)

6–3.4. Will there be more proeutectoid ferrite or more eutectoid ferrite in a slowly cooled 1040 steel?

6–3.5. An AISI-SAE 10xx steel contains 37% proeutectoid ferrite. What is the total percentage of ferrite?

Answer: AISI-SAE 1050 93% α

6-3.6. A plain-carbon steel contains 2 w/o proeutectoid carbide. Assign an AISI-SAE number.

6-4.1. How much graphite (v/o) should a cast iron have in order to have a density of 7.4 g/cm³? [Cf. I.P. 6-4.1.]

6-4.2. If pearlite has a density of 7.82 g/cm³, and ferrite, 7.87 g/cm³ as shown in I.P. 6-3.2, what is the density of the carbide?

• 6-5.1. A certain steel contains 0.5 w/o carbon. How many carbon atoms are there per 100 unit cells of body-centered tetragonal martensite?

Answer: 4.7 carbon atoms

• 6-5.2. Refer to I.P. 6-5.1. How many carbon atoms are there per (nm)³ in that martensite?

• 6-5.3. Compare the volume of the unit cell of martensite (I.P. 6-5.1) with the volume of a unit cell of ferrite in tempered martensite.

Answer: $V_M = 0.99\ V_\alpha$

• 6-5.4. In martensite, $R_{Fe} = \sim 0.124$ nm. What are the dimensions of the gaps between the iron atoms in the three coordinate directions? [See I.P. 6-5.1 for the unit-cell dimensions of martensite.]

• 6-5.5. A bar of 1060 steel has a center hardness of 25 R_C and a surface hardness of 40 R_C. How fast were the (a) center and (b) surface cooled?

Answer: (a) ~10°C/sec

• 6-5.6. The surface and center of an AISI 1040 steel rod are cooled at a rate of 100°C/sec and 10°C/sec, respectively. What hardnesses result?

• 6-5.7. A gear of 1060 steel had a quenched hardness at its mid-radius of 25 R_C. (a) What hardness would develop if the steel were 4140 and quenched in the same manner? (b) 4340?

Answer: (a) 50 R_C

• 6-5.8. (a) The center of a water-quenched, 60-mm steel bar cools at the same rate as the center of an oil-quenched, __***__-mm steel bar. (b) The center of a water-quenched, 60-mm steel bar has the same hardness as the center of an oil-quenched, __***__-mm steel bar of the same composition.

• 6-5.9. The mid-radius of an oil-quenched, 40-mm steel bar cools at the same rate (a) as the surface of an oil-quenched __***__-mm steel bar; (b) as the center of a water-quenched __***__-mm steel bar.

Answer: (a) 65-mm

• 6-5.10. Plot the hardness traverse for a bar made of 1060 steel, 75-mm (3-in.) in diameter, and quenched in water.

• 6-5.11. Repeat S.P. 6-5.10, but for an oil-quenched bar.

Answer: Surface: 27 R_C, etc.

• 6–5.12. Exact identification was lost concerning a round bar of steel (15-m long × 64-mm diameter). It is known only that it may be either AISI 1040 or AISI 4340. (a) It is proposed that the bar be heated, quenched in oil, and identified by a hardness traverse. Indicate below the hardness values you would expect.

	1040	4340
Surface	*** R_C	*** R_C
16 mm below surface	*** R_C	*** R_C
32 mm below surface	*** R_C	*** R_C

(b) A furnace is not available to heat the 15-m bar. However, since the bar is 10 cm longer than required, it is proposed that 10 cm be cut off for the above identification test. Comment. (c) Suggest and discuss the merits of alternative possibilities for identification.

• 6–5.13. Two round bars, 50-mm and 75-mm in diameter and made of the same steel, have the following hardness traverses:

Diameter	50-mm (2-in.)	75-mm (3-in.)
Quench	Water	Oil
Surface hardness	59 R_C	55 R_C
Mid-rad. hardness	57 R_C	41 R_C
Center hardness	55 R_C	35 R_C

Plot a hardenability curve for this steel. [Note: Your answer should consist of one curve *only*!]

• 6–5.14. Surface hardnesses of 6 bars of the same steel are as follows:

25-mm dia. w.q. 63 R_C	25-mm dia. o.q. 55 R_C
50-mm dia. w.q. 62 R_C	50-mm dia. o.q. 34 R_C
100-mm dia. w.q. 56 R_C	100-mm dia. o.q. 31 R_C

Plot a hardenability curve for this steel. [Note: Your answer should consist of one curve *only*!]

• 6–5.15. (a) Estimate the time necessary at 315°C to temper an AISI 1080 steel to 50 R_C. (b) Estimate the temperature necessary to soften a quenched 1080 wire to 50 R_C in 30 seconds.

Answer: (a) 20 hrs.

• 6–6.1. With graphite, rather than Fe_3C, the eutectoid composition is 0.7% carbon, and the eutectoid temperature is 738°C (2360°F). How much graphite can form in a 96.7Fe–3.3C cast iron (a) at 740°C? (b) At 700°C?

Answer: (a) 2.6 w/o

• 6–6.2. Refer to S.P. 6–6.1 above. What is the volume fraction graphite at 700°C?

• 6–6.3. A white cast iron (97.5Fe–2.5C) slowly changes to (α + graphite). Estimate the volume change. [ρ_α = 7.87 g/cm³; $\rho_{\overline{C}}$ = 7.5 g/cm³; and ρ_{gr} = ~2 g/cm³

Answer: +5 v/o (or +1.7 l/o)

- 6-7.1. What are the eutectoid temperature and eutectoid composition for a 9240 steel?

  Answer: T_e = 780°C

- 6-7.2. A 9240 steel must be fully austenized. To what minimum temperature must it be heated?
- 6-7.3. How much pearlite can form in a 9240 steel?

  Answer: 75%

CHECK OUTS

6A	plain-carbon
6B	five
6C	the last two digits
6D	bcc, fcc (α, γ) (ferrite, austenite)
6E	ferrite, α
6F	austenite, γ
6G	three
6H	greater
6I	no three
6J	727 ~0.8 (0.77)
6K	~0.8 0.02 6.7
6L	proeutectoid eutectoid
6M	pearlite
6N	austenite (γ)
6O	$\gamma \longrightarrow (\alpha + \overline{C})$
6P	spheroidite lamellar
6Q	less more
6R	more less
6S	martensite
6T	diffusion carbide
6U	body tetragonal austenite
6V	$(\alpha + \overline{C})$ temperature
6W	hardenability hardenability
6X	faster
6Y	ferrite carbide
6Z	strength, hardness more
6AA	coalesce more
6AB	white
6AC	gray nodular (ductile)
6AD	gray nodular
6AE	white (malleable)
6AF	silicon
6AG	738 (eutectoid)
6AH	ferrite raise
6AI	Ni, Mn (any alloying element)

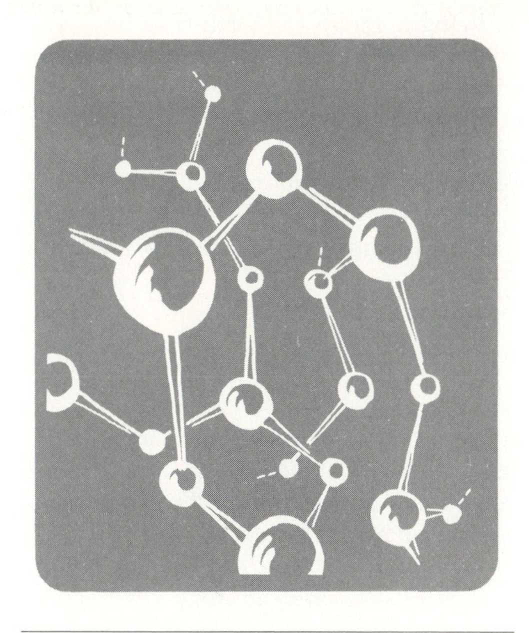

UNIT SEVEN

STRUCTURE OF POLYMERS

Polymers, commonly called plastics, are nonmetallic materials that have received tremendous technical developments in recent years. They are molecular and commonly noncrystalline; therefore, the properties and behavior in service differ significantly from the metals of the previous units. This unit examines their structure; the next unit, their deformation; and Unit 9, their processing.

Contents

Prerequisites General chemistry (organic chemistry is not required); Sections 1–3, 2–2, 2–3

From Unit 7, the engineer should

1) Know the simpler molecules from general chemistry (Fig. 7–1.2) and the vinyl-type molecules of Table 7–1.3, plus similar substitional possibilities for rubbers.
2) Be able to calculate mer mass, average molecular sizes, degree of polymerization, and the mean end-to-end length of linear polymers.
3) Understand the basis and significance of the glass-transition temperature.
4) Be able to use the terms at the end of the Unit in technical conversations, and to solve the study problems.

7-1 GIANT MOLECULES

We shall consider a *molecule* as being a group of atoms that have strong bonds among themselves, but relatively weak bonds to adjacent molecules. Some familiar small molecules include water (H_2O), carbon dioxide (CO_2), methane (CH_4), ethyl alcohol (C_2H_5OH). Plastics are made up of large molecules that may contain thousands to millions of atoms. Sometimes these are referred to as *macro*molecules, or *polymers,* a more widely used term.

A *micro*molecule, such as ethyl alcohol, illustrates the nature of a molecule:

$$\begin{array}{c} \mathrm{H}\;\;\mathrm{H}\;\;\;\;\; \\ \mathrm{H}:\ddot{\underset{..}{\mathrm{C}}}:\ddot{\underset{..}{\mathrm{C}}}:\ddot{\underset{..}{\mathrm{O}}}:\mathrm{H} \\ \mathrm{H}\;\;\mathrm{H}\;\;\;\;\; \end{array} \tag{7-1.1}$$

There are strong bonds between the carbon and hydrogen atoms (also C:C, C:O, and O:H) because two atoms *share* a pair of electrons. (Recall from Fig. 1-3.2 that this is characteristic of nonmetallic elements.) We commonly call this sharing a *covalent bond* and use the presentation:

$$\begin{array}{c} \mathrm{H}\;\;\mathrm{H}\;\;\;\; \\ \mathrm{HC{-}C{-}OH} \\ \mathrm{H}\;\;\mathrm{H}\;\;\;\; \end{array} \tag{7-1.2}$$

A carbon atom may have a maximum of four pairs of shared electrons, or four bonds. Table 7-1.1 shows the maximum numbers of shared electrons for other nonmetallic elements.

Each atom of the above ethyl alcohol molecule is satisfied with its maximum number of bonds. Therefore none of the atoms in the molecule develops strong attractions to adjacent molecules. Weak attractions do exist between molecules, however; ethyl alcohol and water are liquids at room temperature because there is enough attraction at 20°C (68°F) to condense the molecules into close proximity. We know that the attractions are weak because adjacent molecules can flow past each other.

Table 7-1.1
Shared bonds for nonmetallic elements

Element	Shared bonds
H	1
F	1
Cl	1
O	2
S	2
N	3
C	4

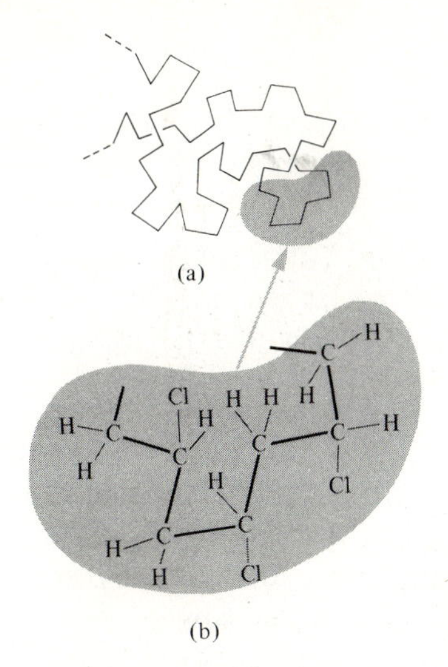

Figure 7-1.1 Linear macromolecules (polyvinyl chloride in liquid solution). Thermal agitation twists the backbone of carbon atoms into a kinked conformation.

A macromolecule commonly has a long sequence of strong covalent bonds in which adjacent atoms share electrons (Fig. 7-1.1). As we encounter longer and longer molecules, however, liquids of these materials become more and more viscous because of entanglements. In fact, even though some polymers retain the structure of liquids, that is, they do not crystallize, they become so rigid as a result of their large molecules that we categorize them among the solids.

Molecular sizes

Based on the data in Fig. 1-3.2, we know that water has a molecular mass of 18 amu, or 18 g per 0.6×10^{24} molecules. The molecular mass of ethyl alcohol, C_2H_5OH, is

$$2(12) + 5(1) + 16 + 1 = 46 \text{ g}/0.6 \times 10^{24} \text{ molecules.} \qquad (7\text{-}1.3)$$

We usually refer simply to this as 46 g/*mole.*

The molecular masses of polymers are invariably more than a thousand grams per mole and may run into the millions. Consider Fig. 7-1.1, and assume a chain of 700 carbon atoms (1050 hydrogen atoms and 350 chlorine atoms). The molecular mass is nearly 22,000 grams per mole. The 700 carbon count is not a limiting number.

Mers

If we look at Fig. 7-1.1, we observe a unit of C_2H_3Cl that repeats itself along the chain. We call this unit a *mer*, and it is the basic unit of structure in *polymers* (literally, *many units*). If we describe the mer, we can describe the whole molecule. Thus

$$\left[\begin{array}{cc} H & H \\ -C- & C- \\ H & Cl \end{array}\right]_n \qquad (7\text{-}1.4)$$

describes the molecule of Fig. 7-1.1, and the mass of each mer is 2(12) + 3(1) + 35.5, or 62.5 amu. If we let n equal 350, the molecular mass is 350 × 62.5 amu or nearly 22,000 g/mole. The polymer with 350 mers linked together has a *degree of polymerization* of 350.

Molecular dimensions

Adjacent carbon atoms have a center-to-center interatomic distance of 0.154 nm. Table 7-1.2 lists comparable distances, called *bond lengths,* for other atom pairs. Double bonds involve the covalent sharing of two pairs of electrons. Carbon-to-carbon bond lengths are of special interest to us because they enable us to estimate the length of a molecule. A first approximation indicates that a molecule with 700 carbon atoms would be 700 × 0.154 nm, or about 108 nm long. Actually the bonds are not 180° across a carbon atom; they have an average angle of 109°. Furthermore, any single bond can rotate in three dimensions, so we see considerable twisting and kinking along the length of the chain (Fig. 7-1.1). This will be important to us because, with kinking (see Fig. 7-1.1), the average end-to-end distance $\overline{L}$ of such molecules in noncrystalline polymers becomes

$$\overline{L} = l\sqrt{m}. \qquad (7\text{-}1.5)$$

Here l is the interatomic distance (0.154 nm for C–C) and m is the number of bonds. Thus instead of 108 nm, the value $\overline{L}$ is (0.154 nm) $\sqrt{700}$ or slightly more than 4 nm. The reason this is so important is that many large molecules may be stretched out from this kinked *conformation* to give high

Table 7-1.2
Bond lengths between nonmetallic atom pairs

Bond	Length, nm	Bond	Length, nm
C–C	0.154	O–H	0.10
C=C	0.13	O–O	0.15
C≡C	0.12	O–Si	0.18
C–H	0.11		
C–N	0.15	N–H	0.10
C–O	0.14	N–O	0.12
C=O	0.12		
C–F	0.14		
C–Cl	0.18	H–H	0.074

strains without noticeably changing the interatomic distances. Rubbers possess this characteristic and may develop high strains at low stresses.

Illustrative problem 7–1.1 Determine the molecular weight of phenol and formaldehyde, two small molecules that are combined to make bakelite.
a) Phenol:

```
        OH
        |
       C
    HC/   \CH
     |  ○  |            (7–1.6)
    HC\   /CH
       C
       H
```

b) Formaldehyde:

```
 H
  C=O                   (7–1.7)
 H
```

SOLUTION

a) $[(6 \times 12) + 16 + (6 \times 1)] = 94$ g/mole
b) $[12 + 16 + (2 \times 1)] = 30$ g/mole

ADDED INFORMATION We will not need to know very many molecular compounds. Figure 7–1.2 lists those that will enter our discussions without specific definition. You are probably already familiar with many of them; if not, you will find it helpful to become aware of them. ◂

```
  H        H H         H H        H H       H H
 HCH      HC–CH       HC–C–OH     C=C       C=C
  H        H H         H H        H H       H |
                                              Cl
 (a)       (b)          (c)        (d)       (e)

      H           OH
      C           C                       H
  HC/   \CH   HC/   \CH                  HCH
   |  ○  |     |  ○  |                 H  |  H H
  HC\   /CH   HC\   /CH     H          C=C–C=C
      C           C          C=O       H      H
      H           H         H
     (f)         (g)        (h)           (i)
```

Figure 7–1.2 Common micromolecules.
(a) Methane, CH_4.
(b) Ethane, C_2H_6.
(c) Ethyl alcohol, C_2H_5OH.
(d) Ethylene, C_2H_4.
(e) Vinyl chloride, C_2H_3Cl.
(f) Benzene, C_6H_6.
(g) Phenol, C_6H_5OH.
(h) Formaldehyde, CH_2O.
(i) Isoprene, C_5H_8.

Illustrative problem 7-1.2 A polyvinyl chloride polymer (Fig. 7-1.1) has an average molecular weight of 72,130 amu (that is, 72,130 g/mole). What is the degree of polymerization, n?

SOLUTION

$$\begin{aligned} C_2H_3Cl \text{ mer weight} &= [2(12) + 3(1) + 35.5] \\ &= 62.5 \text{ amu} = 62.5 \text{ g/mer wt.} \end{aligned}$$

Therefore

$$\begin{aligned} n &= 72{,}130 \text{ amu}/62.5 \text{ amu} \\ &= 1154 \text{ mers/polymer.} \end{aligned}$$

ADDED INFORMATION In a polymer, the molecules are not all the same size. Therefore it is necessary for us to speak of the *average molecular size,* $\overline{M}$. A larger average molecular size generally means a higher melting point, more viscosity, and a more stable material. ◀

Illustrative problem 7-1.3 A plastic contains polystyrene with styrene mers as follows:

$$\left[\begin{array}{cc} H & H \\ -C- & -C- \\ H & | \\ & C \\ & HC \quad CH \\ & HC \quad CH \\ & C \\ & H \end{array}\right]_n \qquad (7\text{-}1.8)$$

The degree of polymerization, n, is 525. How many molecules are there per gram of polystyrene?

SOLUTION Each molecule of polystyrene has

$$\begin{aligned} &\{525[(8 \text{ C/molecule})(12 \text{ amu/C}) + (8 \text{ H/molecule})(1 \text{ amu/H})]\} \\ &\quad = 54{,}600 \text{ amu/molecule.} \end{aligned}$$

Since there are 0.6×10^{24} amu per gram,

$$\begin{aligned} \text{molecules/g} &= \frac{0.6 \times 10^{24} \text{ amu/g}}{5.46 \times 10^4 \text{ amu/molecule}} \\ &= 1.1 \times 10^{19} \text{ molecules/g.} \end{aligned}$$

ADDED INFORMATION Polyvinyl compounds, which include styrene, have the composition

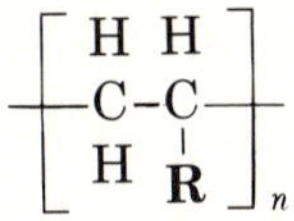

$$\left[\begin{array}{cc} H & H \\ -C- & C- \\ H & R \end{array}\right]_n \tag{7-1.9}$$

where **R** is one of a number of possible *radicals*. Table 7-1.3 lists some commonly encountered ones. ◀

Table 7-1.3

Vinyl compounds $\left(\begin{array}{cc} H & H \\ C= & C \\ H & R \end{array}\right)$

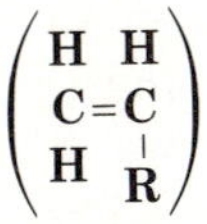

Compound	*R*
Ethylene	–H
Vinyl alcohol	–OH
Vinyl chloride	–Cl
Propylene	$-CH_3$
Vinyl acetate	$-OCOCH_3$
Acrylonitrile	–C≡N
Styrene	 –⟨benzene ring⟩*

* The symbol –⟨benzene ring⟩ is used to denote the benzene ring, which in its more conventional form is

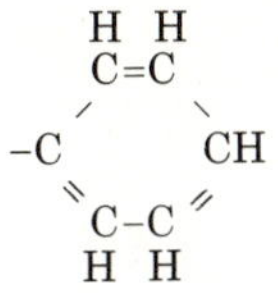

```
       H  H
       C=C
      /    \
  -C          CH    as a radical.
     \\     //
       C-C
       H  H
```

7-2 MOLECULAR CONFIGURATIONS

Just as iron alloys may be either bcc or fcc and retain the same composition, many macromolecules have more than one structure. Consider *natural rubber*, which has a composition [called *isoprene*, $(C_5H_8)_n$] that may be sketched as follows:

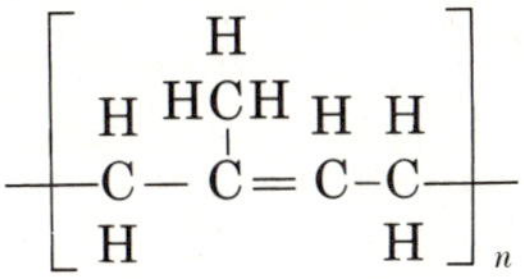

$$\left[\begin{array}{cccc} & H & & \\ H & HCH & H & H \\ -C- & C= & C- & C- \\ H & & & H \end{array}\right]_n \tag{7-2.1}$$

This molecule is highly kinked and exhibits the large strains typical of rubber (Section 8-2). An *isomer* of C_5H_8 is

$$\left[\begin{array}{cccc} & H & & \\ H & HCH & & H \\ -C- & C= & C- & C- \\ H & & H & H \end{array}\right]_n \tag{7-2.2}$$

The composition is the same, $(C_5H_8)_n$, but the structure is slightly different. Because the structure is different, its properties are different. As a result, it is not even called isoprene, but *gutta percha*. It does not have the characteristics of rubber, but is more like the polyvinyl chloride tile we use in floortile. In polymers we shall encounter many different isomers because there are many different ways in which a large number of atoms can be combined.

Stereoisomers

The simplest polymer of everyday usage is polyethylene

$$\left[\begin{array}{cc} H & H \\ -C- & C- \\ H & H \end{array}\right]_n \tag{7-2.3}$$

The mer is considered to be C_2H_4 because this polymer arises from additions of ethylene molecules:

$$\begin{array}{c} \quad H\,H \quad H\,H \quad H\,H \\ \cdots + C{=}C + C{=}C + C{=}C + \cdots \\ \quad H\,H \quad H\,H \quad H\,H \end{array}$$

$$\longrightarrow \cdots \begin{array}{c} H\;H\;H\;H\;H\;H \\ -C-C-C-C-C-C- \\ H\;H\;H\;H\;H\;H \end{array} \cdots \tag{7-2.4}$$

The individual C_2H_4 mer is symmetric with no "front" or "back." In contrast, the majority of vinyl compounds are not symmetrical. Consider propylene:

$$\begin{array}{cc} H & H \\ C= & C \\ H & HCH \\ & H \end{array} \tag{7-2.5}$$

The left and right double-bonded carbons are not identical. One has two hydrogen neighbors; the other has a hydrogen and a CH_3 side group as neighbors. When these are polymerized into chains, there are variations that can occur as shown in Fig. 7-2.1. These variations are called *stereoisomers*. In a later section we shall see that *configuration* differences such as these have an effect on the crystallizability of a polymer and there-

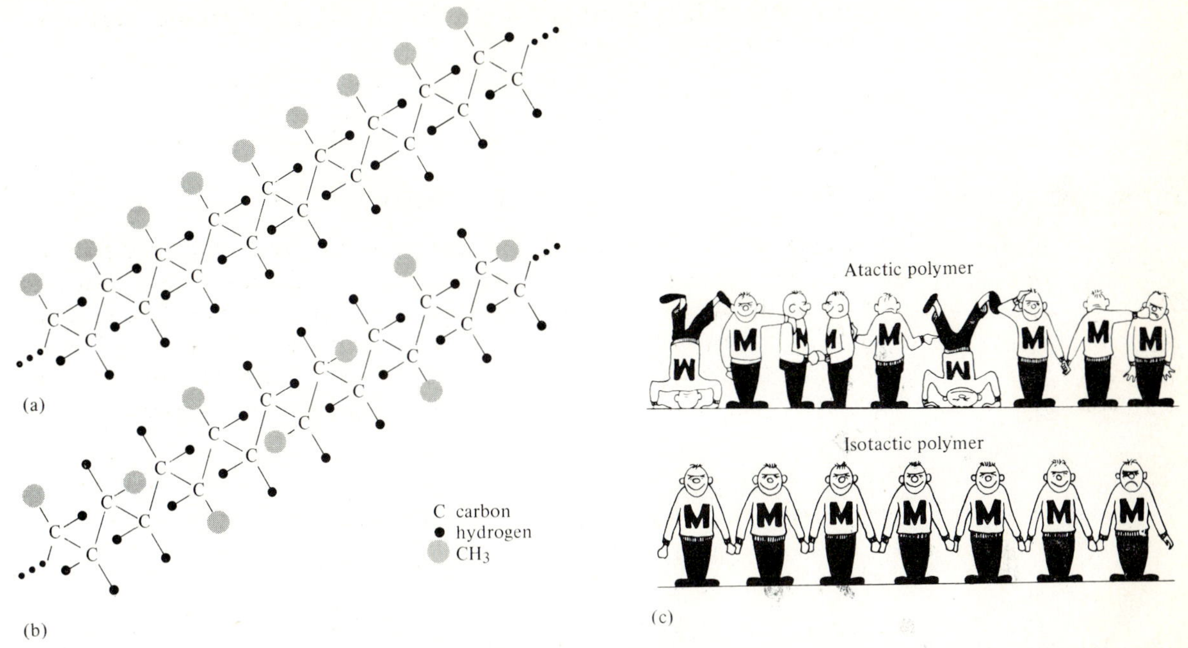

Figure 7-2.1 Stereoisomerism (polypropylene). (a) Isotactic. (b) Atactic. The more regular configurations crystallize more readily. (c) For fun! **M** stands for mer. (Courtesy of General Electric Co.)

fore on the properties. The *isotactic* stereoisomer (Fig. 7-2.1a) crystallizes more readily, is somewhat denser, and is appreciably stronger than the *atactic* isomer (Fig. 7-2.1b).

Branching

Ideally, polymer molecules such as we have studied to date are *linear*,that is, they are two-ended chains. There are cases, however, in which a polymer chain branches. We can indicate this schematically as in Fig. 7-2.2. Although a branch is unusual, once formed it is stable because each carbon atom has its complement of four bonds and each hydrogen atom has one bond. The significance of branching lies in the entanglements that can interfere with plastic deformation. Think of a pile of tree branches compared with a bundle of sticks; it is more difficult to remove the branches than to move the individual sticks.

Cross-linking

Some linear molecules, by virtue of their structure, can be tied together. Consider the molecule of Fig. 7-2.3(a) and its polymerized combination in

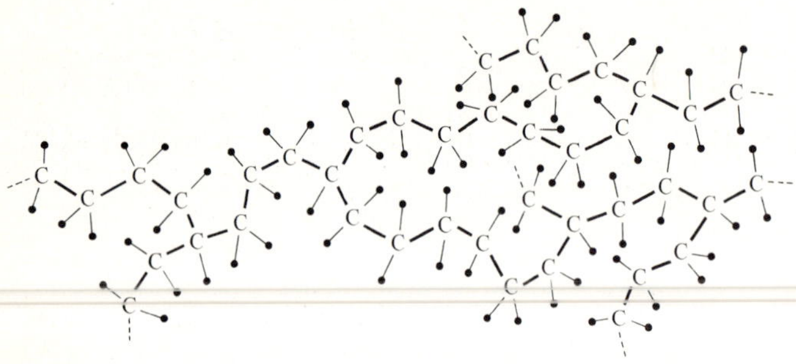

Figure 7-2.2 Branching of polyethylene (with dots representing hydrogen). Branched molecules do not crystallize readily.

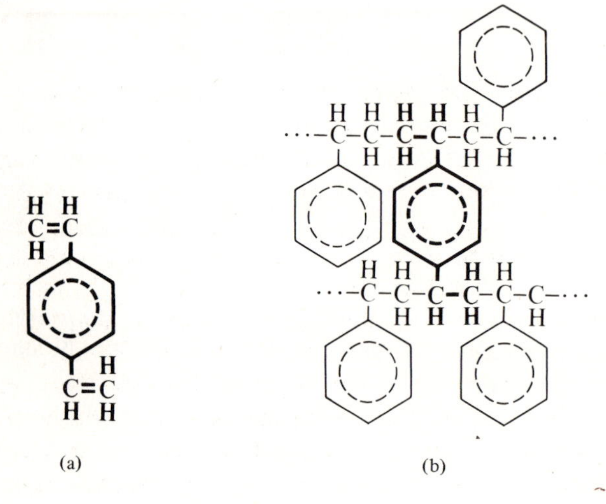

Figure 7-2.3 Cross-linking. (a) Divinyl benzene. (Compare with styrene.) (b) Two chains cross-linked with divinyl benzene. (The symbol ⬡ is used for the benzene ring. Hydrogen atoms are implied at each corner *not* bonded to other atoms.)

Fig. 7-2.3(b). Intentional additives (divinyl benzene in this case, but we do not have to remember the name) tie together two chains of polystyrene. This causes restrictions with respect to plastic deformation.

The *vulcanization* of rubber is a result of cross-linking by sulfur, as shown in Fig. 7-2.4. The effect is pronounced. Without sulfur, rubber is a soft, even sticky material that flows by viscous deformation when it is near room temperature. It could not be used in automobile tires because the service temperature would make it possible for molecules to slide by their neighbors, particularly at the pressures encountered. However, cross-linking by sulfur at about 5% of the possible sites gives the rubber mechanical stability under the above conditions, but still enables it to retain the flexibility that is obviously required. Hard rubber has a much larger percentage of sulfur and appreciably more cross links. You can appreciate the effect of the addition of greater amounts of sulfur on the properties of rubber when you examine a hard rubber product such as a pocket comb.

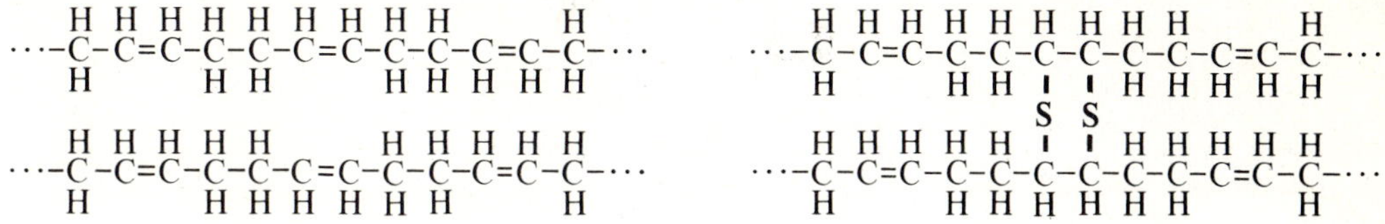

```
    H H H H H H H H H       H
···-C-C=C-C-C-C=C-C-C-C=C-C-···
    H     H H     H H H H H

    H H H H H       H H H H H
···-C-C=C-C-C-C=C-C-C-C=C-C-···
    H     H H H H H H       H

             (a)

    H H H H H H H H H       H
···-C-C=C-C-C-C-C-C-C-C=C-C-···
    H     H H | | H H H H H
              S S
    H H H H H | | H H H H H
···-C-C=C-C-C-C-C-C-C-C=C-C-···
    H     H H H H H H       H

             (b)
```

Figure 7-2.4 Vulcanization (butadiene). Sulfur atoms serve as anchor points between adjacent molecules of rubber. A possible mechanism is a pair of sulfur atoms that react with the double bonds in a pair of adjacent molecules.

Network polymers

The simpler polymers are the linear ones just described. These are *bifunctional,* that is, each mer can connect with two—and only two—other mers. Other types of mers can connect with several adjacent molecules. We saw one such mer in Fig. 7-2.3, in which the divinyl benzene bonds to four styrene mers.

A widely used plastic that has polyfunctional units is phenol formaldehyde (Fig. 7-2.5). The early trade name, *bakelite,* is one name for this combination. The phenol molecules, C_6H_5OH, provide the polymer with *trifunctional* units, and CH_2 groups from formaldehyde, CH_2O, serve as bridges between adjacent phenol rings (Fig. 7-2.5b). The sketches in Fig. 7-2.5 are intended to reveal that a three-dimensional network structure is formed. We shall see in Unit 8 that the thermal behavior of network polymers is markedly different from that of linear polymers. In fact, as you may already have guessed, network polymers are more rigid at high temperatures than linear polymers are.

Illustrative problem 7-2.1 A polyisoprene rubber,

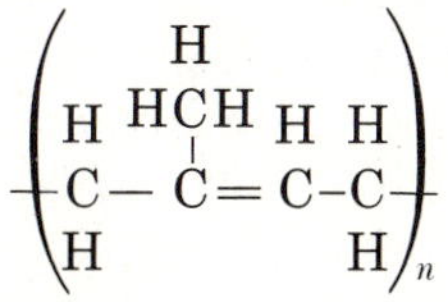

```
 /      H          \
|  H  HCH  H  H     |
|        |          |
-+-C-- C= C-C-+-
|  H          H     |
 \                 /n
```

is vulcanized with 5.0 g of sulfur per 100 g of isoprene. If all the sulfur atoms serve as cross links, what fraction of the potential cross links are bridged?

SOLUTION Basis: 100 amu isoprene + 5 amu sulfur.

(100 amu)/[5(12) + 8(1) amu/mer C_5H_8] = 1.47 mers C_5H_8
(5.0 amu)/(32 amu/S) = 0.156 sulfurs
Sulfurs/mer C_5H_8 = 0.156/1.47 = 0.106.

According to Fig. 7-2.4, up to two sulfur atoms could join every *pair* of isoprene mers. This would mean that there could be 1.47 sulfur atoms per

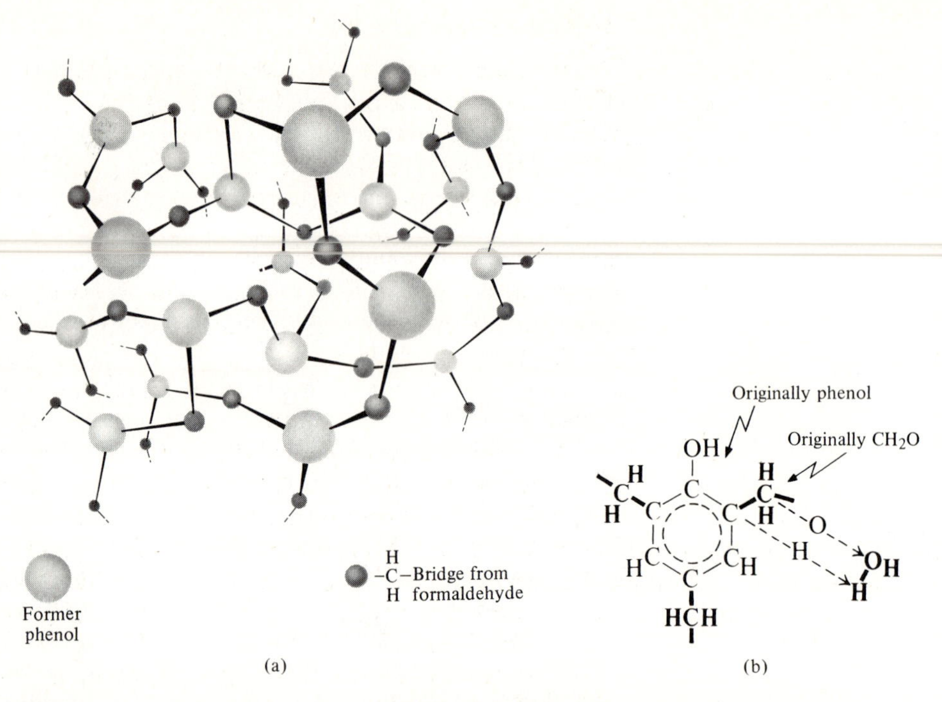

Figure 7-2.5 Network polymer (phenol-formaldehyde). (a) A three-dimensional structure forms because each phenol molecule was trifunctional, that is, it provides three reaction positions. (b) The original phenol and formaldehyde structures are shown in Fig. 7-1.2. Water forms as a by-product.

1.47 isoprene mers.

Fraction = 0.106 of possible sulfur cross links.

ADDED INFORMATION This would give a harder rubber than that found in a tire, in which the sulfur/isoprene ratio is commonly about 0.02.

Oxygen will also cross-link with rubber. We see evidence of this when a rubber article hardens with age. ◄

7-3 CRYSTALLINITY IN POLYMERS

Many polymers, unlike metals—which invariably crystallize when they solidify under normal conditions—solidify without complete crystallization. In this section we shall concentrate on those molecules that do crystallize. In the following section, we shall consider polymers that solidify without crystallizing, and that form solids that may be viewed as having the characteristics of a very viscous liquid.

Molecular crystals

The atoms in metals are approximately spherical in shape. In fact, we spoke of the "hard-ball" atom (Section 2-1). Even in those cases in which atoms may be slightly distorted (cf. information with I.P. 2-3.1), the atoms are sufficiently equiaxed so that we can view a crystal as a collection of individual balls arranged in a repeating long-range pattern.

Very few molecules—and specifically no polymer molecules—have equiaxed shapes with dimensions that are approximately equal in the three directions of space. In fact, the linear molecules of the previous sections are characterized by their extreme elongation in one dimension.

First let us consider the crystallization of the very simple molecule of iodine, I_2. The atoms of this biatomic molecule share a pair of electrons. Consistent with the first statement of Section 7-1, in I_2 there is a strong bond between the two atoms of the molecule but much weaker attraction to other molecules. Thus the I_2 molecule has the shape of a dumbbell (Fig. 7-3.1a) and as such cannot form cubic crystals. The unit cell that results is shown in Fig. 7-3.1(b). Note that the three dimensions are not equal (a = 0.727 nm, b = 0.979 nm, c = 0.479 nm). In this case, however, the unit cell has three 90° angles.

Polyethylene (Table 7-1.3 and Eq. 7-2.4) has long linear molecules. These molecules have strong covalent bonds along the molecular chain. The bonds between chains are much weaker but are still present, so that the chains tend to line up as shown in Fig. 7-3.2. Observe that a single molecule extends through many unit cells. Because the molecules are long, it is difficult to obtain perfect crystallization. For example, all molecules do not end at the same position because each of the many molecules has a different length. Also the tendency to twist and turn, as shown in Fig. 7-1.1, makes perfect alignment difficult to achieve.

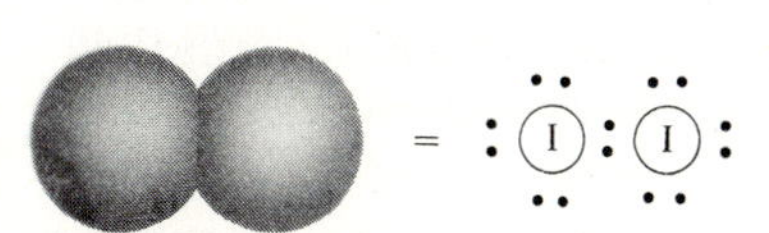

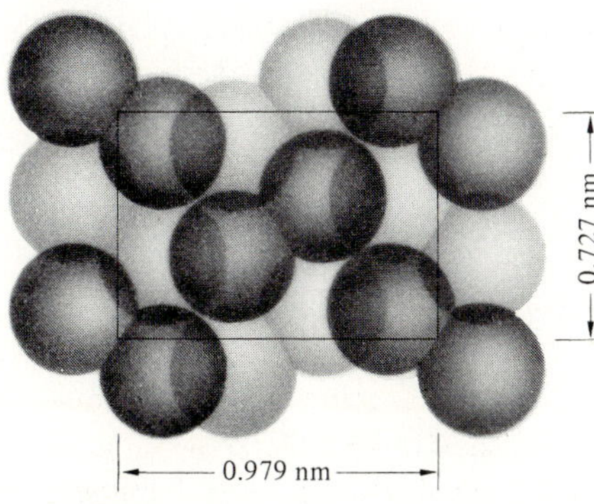

Figure 7-3.1 Molecular crystal (iodine). (a) The iodine molecule, I_2. (b) The unit cell. The c-dimension (not shown) is 0.479 nm. The corner angles of the unit cell are all 90°; however, $a \neq b \neq c$. Therefore the cell is not cubic, but orthorhombic.

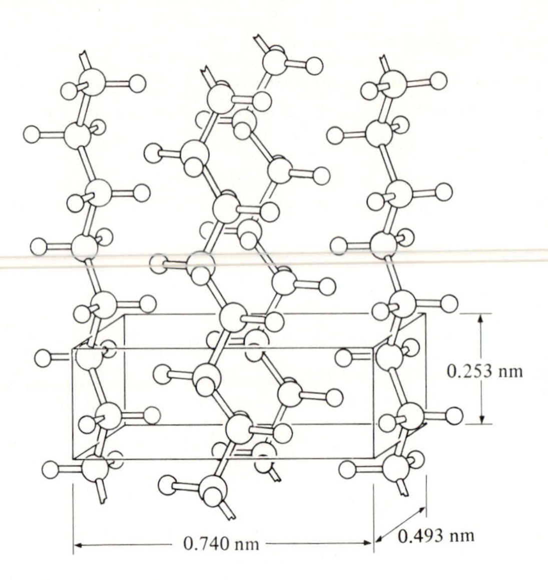

Figure 7-3.2 Polyethylene unit cell. The molecules extend from one unit cell through the next. (M. Gordon, *High Polymers*, London: Iliffe Books, and Reading, Mass.: Addison-Wesley. After C. W. Bunn, *Chemical Crystallography*, London: Oxford University Press.)

Crystallization shrinkage

When iodine is heated, the individual molecules shown in Fig. 7-3.1(b) are more and more agitated by the heat. This produces thermal expansion. More important, there comes a point, the *melting temperature,* at which the agitation is too great for the weak bonds *between* neighboring molecules to be maintained. At this temperature the iodine suddenly loses its long-range crystalline pattern. (The bonds *within* the molecules remain intact.) All crystals that are close-packed (fcc and hcp) undergo marked expansion in volume at this point, as indicated in Fig. 2-4.1. In fact, only in exceptional cases does expansion fail to take place when a material melts. Figure 7-3.3 shows the volume change for iodine, I_2, at the melting (or freezing) point.

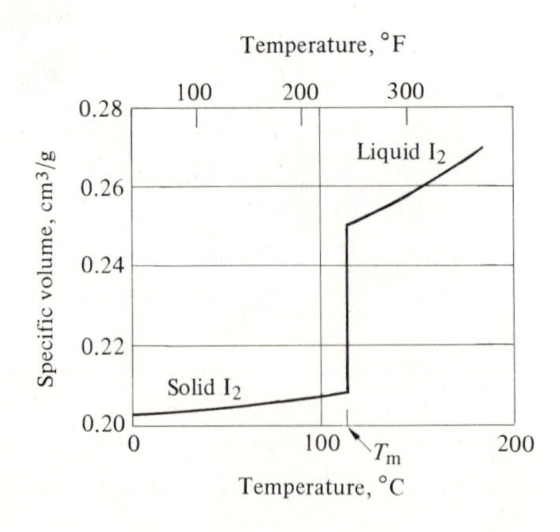

Figure 7-3.3 Volume versus temperature (iodine). The volume change occurs at the melting temperature (T_m = 114°C) because the molecules are packed more tightly in the solid than in the liquid (cf. Fig. 2-4.1).

Starting at 200°C with molten iodine, there is a shrinkage in volume as the temperature decreases. The discontinuous *crystallization shrinkage* at 114°C occurs as the iodine molecules pack themselves more efficiently into the crystal lattice. Below 114°C, shrinkage continues, with thermal agitation reduced, but with less change per degree because the molecules must vibrate at fixed crystal locations.

Illustrative problem 7-3.1 Calculate the density of crystalline iodine.

SOLUTION Inspect Fig. 7-3.1(b) and observe that an I_2 molecule is centered at the corner of each unit cell, and also at the center of each face. Therefore, in this unit cell, there are 4 molecules (8 atoms):

$$\text{Density} = \frac{8 \text{ atoms } (126.9 \text{ g}/0.6 \times 10^{24} \text{ atoms})}{(0.727 \times 10^{-9} \text{ m})(0.979 \times 10^{-9} \text{ m})(0.479 \times 10^{-9} \text{ m})}$$
$$= 4.96 \times 10^{6} \text{ g/m}^3 \qquad (\text{or } 4.96 \text{ g/cm}^3).$$

ADDED INFORMATION The experimental measured density is 4.94 g/cm^3.

A crystal is *orthorhombic* when $a \neq b \neq c$ and the three principal angles are each 90°. ◄

7-4 AMORPHOUS POLYMERS

With small molecules, such as the two-atom iodine molecule, crystals form easily; but the crystallization process with large molecules such as polyethylene is appreciably more difficult. As stated in the previous section, the long, $(C_2H_4)_n$ molecules tend to become twisted and kinked. Entanglements result. Furthermore, one molecule may end and another may not be immediately available to continue into the next unit cell; this makes for poor formation of crystals. Branching also hinders crystallization.

In view of the above complications, it is not surprising that polyethylene may be cooled past the melting temperature (~135°C, or ~275°F) and on down to room temperature without the molecules aligning themselves into a nice crystalline pattern. Supercooling is common among polymers, especially if there are structural irregularities along the molecular chain. For example, the isotactic chain shown in Fig. 7-2.1(a) crystallizes much more readily than the atactic chain of Fig. 7-2.1(b). The more regular the chains of molecules, the more readily they mesh together into crystals, rather like a zipper closing, to give a familiar analogy.

Glass temperature

Even though the molecules of the polymer are entangled when there is thermal agitation, there is a continuous rearrangement of the atoms and molecules within a polymer liquid. And because of this, there has to be extra

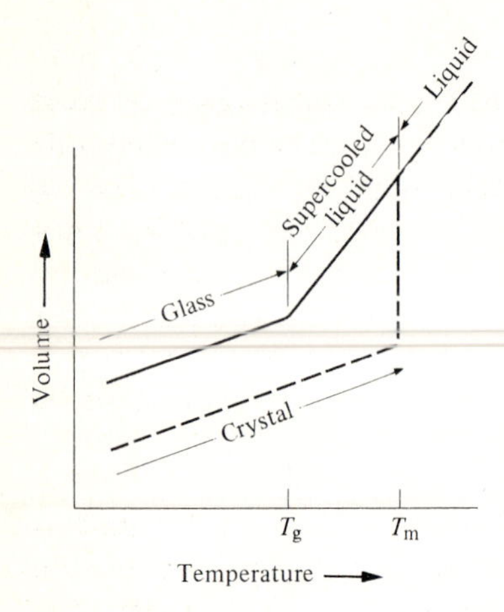

Figure 7-4.1 Volume versus temperature. Most liquids have an abrupt volume contraction when they crystallize. (Cf. Fig. 7-3.3.) However, a liquid may be supercooled below the melting temperature, T_m, if time is not available for the necessary rearrangements into an ordered crystal. This is particularly true for polymer liquids with their large, gangling molecules. They become supercooled. This liquid is still mobile, so decreasing temperatures allow more efficient packing (but not ordering into a crystal). Eventually any rearrangments are precluded and the amorphous (noncrystalline) material becomes a solid, called a *glass*. Continued contraction is solely a result of a decreasing amplitude of atom vibrations. The transition from supercooled liquid to an amorphous solid is called the glass-transition temperature, T_g.

space in the liquid, just as a milling crowd cannot be as close packed as, say, a crowd of stationary people jammed into a football stadium. As the temperature decreases, the thermal agitation lessens, and there is a decrease in volume. This decrease in volume continues below the freezing point into the supercooled liquid range (Fig. 7-4.1). The liquid structure is retained. As with liquid at a higher temperature, flow can occur; however, naturally flow is more difficult (the viscosity increases) as the temperature drops and the excess space between molecules decreases.

Those polymers that are cooled without crystallizing eventually reach a point at which the thermal agitation is not sufficient to allow for rearrangement of the molecules. Although not crystalline, the polymer becomes markedly more rigid. It also becomes more brittle. When there cannot be continued rearrangement of molecules so that there is better packing, additional decreases in volume mean only that the vibrations of the molecules are of smaller amplitude. Thus the slope of the curve in Fig. 7-4.1 becomes much less steep. This point of change in the slope is called the *glass point,* or the *glass-transition temperature,* because this phenomenon is typical of all glasses. In fact, below this temperature, just as with normal silicate glasses, a noncrystalline polymer is a *glass* (although admittedly an organic one). Conversely, we shall observe in Unit 13 that a normal glass is an inorganic polymer.

The glass-transition temperature, or more simply the glass temperature, T_g, is as important to polymers as the melting (or freezing) temperature, T_m, is. The glass temperature of polystyrene (Table 7-1.3) is at ~100°C (212°F). Therefore it is glassy and brittle at room temperature. In contrast, a rubber whose T_g is at −73°C (−100°F) is flexible even in the most severe winter temperatures.

Partially crystalline polymers

Some plastics are partially crystalline. Polyethylene is a good example. As it forms a solid, it is possible for a few unit cells to form relatively good crystals locally. The long molecules, however, extend beyond these crystalline regions into amorphous regions. As a result, various levels of crystallinity are possible. The volume–temperature curves for such polymers lie between the two curves of Fig. 7–4.1. Both amorphous and partially crystalline polymers possess volume in excess of the crystalline packing. This excess volume is called "free-space" and permits easier deformation when external stress is applied (Section 8–3).

Illustrative problem 7–4.1 A polyethylene with no evidence of crystallinity has a density of 0.90 g/cm³. Commercial grades of polyethylene include "low-density" LDPE, with a density of about 0.92 g/cm³, and "high-density" HDPE, with a density of about 0.96 g/cm³. Estimate the fraction of crystallinity in each case.

SOLUTION Referring to Fig. 7–3.2, we may calculate the density of a polyethylene crystal. Basis: 1 unit cell = $\frac{4}{2}$ mers of $\mathord{+}C_2H_4\mathord{+}$.

$$\text{Density} = \frac{(2 \text{ mers})(28 \text{ g}/0.6 \times 10^{24} \text{ mers})}{(0.740 \times 10^{-9} \text{ m})(0.493 \times 10^{-9} \text{ m})(0.253 \times 10^{-9} \text{ m})}$$
$$= 1.0 \times 10^{6} \text{ g/m}^3 \qquad (\text{or } 1.0 \text{ g/cm}^3).$$

Assuming that the change in density is proportional to the crystallization, the crystallization fraction C is

$$C_{\text{HDPE}} = (0.96 - 0.90)/(1.0 - 0.90) = 0.6,$$
$$C_{\text{LDPE}} = (0.92 - 0.90)/(1.0 - 0.90) = 0.2.$$

ADDED INFORMATION The low-density polyethylene softens more between T_g and T_m than the high-density polyethylene does. With less packing, the molecules can rearrange themselves more readily in response to applied stresses.

If polyethylene were 100% crystalline, there would be no glass temperature and the solid would remain essentially rigid up to its melting temperature before suddenly softening into a liquid. ◂

7–5 COPOLYMERS

Each of the vinyl-type mers of Table 7–1.3 has two carbon atoms. Each therefore produces a polymer with a backbone of carbon atoms. It is only logical to ask the question, Can a polymer have more than one type of mer? The answer not only is yes, but also that such combinations are so common that the word *copolymer* has been coined to describe them. Further, the properties of copolymers are often desirable and are specifically prescribed

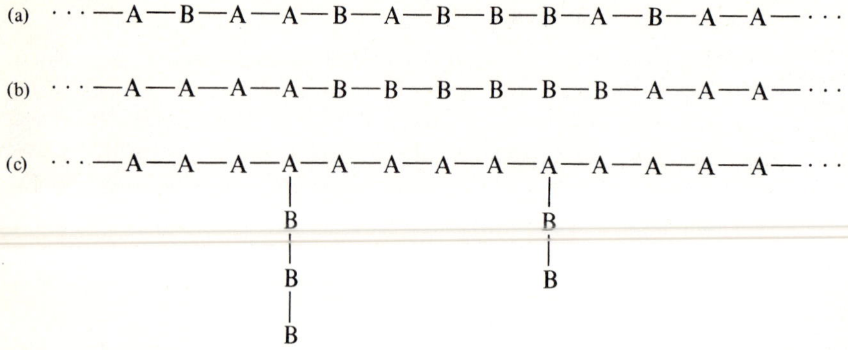

(a) · · · —A—B—A—A—B—A—B—B—B—A—B—A—A— · · ·

(b) · · · —A—A—A—A—B—B—B—B—B—B—A—A—A— · · ·

(c) · · · —A—A—A—A—A—A—A—A—A—A—A—A—A— · · ·
(with side branches —B—B—B from the fourth A and —B—B from the ninth A)

Figure 7-5.1 Copolymers (schematic of A and B mers). (a) Random copolymer. (b) Block copolymer. (c) Graft copolymer. The random and graft copolymers seldom crystallize. Local regions of crystal order can occur in a block copolymer.

for engineering components. In principle, they are solid solution "alloys" of polymers.

Copolymers may be of several types. Figure 7-5.1(a) shows the most general form, in which the chain of molecules contains a *random* sequence of mers. One of the early artificial rubbers, Buna-S, is of this type. Its chain contains mers of styrene, $-(C_2H_3\bigcirc)-$ and butadiene, (C_4H_6). Styrene by itself is a hard, glassy polymer ($T_g \cong 100°C$); butadiene is a soft rubberlike compound.* Styrene lends hardness and a certain amount of mechanical stability to products made from it. Butadiene lends stretch and a degree of flexibility.

The properties of the copolymer may be modified by varying the ratio of the two component mers. Table 7-5.2 shows data for copolymers of vinyl chloride, (C_2H_3Cl), and vinyl acetate, $-(C_2H_3Ac)-$. In general, copolymers with random sequences of mers in their chains are amorphous, because, when the succeeding mers vary, there is scant likelihood that the meshing required for crystallization will take place.

Block copolymers are sketched in Fig. 7-5.1(b). Because there are longer units of a given species of mer in a block copolymer, the crystallization fraction may be higher than it would be in a comparable random copolymer. The resulting plastic may have the characteristic of a two-phase crystalline microstructure, but with molecular chains that tie the adjacent "grains" together. The ABS plastics are triple copolymers of this type; the letters stand for ***a***crylonitrile $-(C_2H_3C{\equiv}N)-$, ***b***utadiene $-(C_4H_6)-$, and ***s***tyrene $-(C_2H_3\bigcirc)-$, respectively.

---

* Butadiene has the mer

$$-\!\!\left[\begin{array}{cccc} H & H & H & H \\ C- & C= & C- & C \\ H & & & H \end{array}\right]_n\!\!-\,, \qquad (7\text{-}5.1)$$

which may be compared directly with isoprene, Eq. (7-2.1). (See Table 7-5.1.)

Table 7-5.1
Butadiene-type molecules

$$\begin{pmatrix} \text{H} & \text{R} & \text{H} & \text{H} \\ \text{C} = & \text{C} - & \text{C} = & \text{C} \\ \text{H} & & & \text{H} \end{pmatrix}$$

	R
Butadiene	–H
Chloroprene	–Cl
Isoprene	$-CH_3$

Table 7-5.2
Vinyl chloride-acetate copolymers*

Item	w/o of vinyl chloride	m/o of vinyl chloride	Range of average mol. wt.	Typical applications
Straight polyvinyl acetate	0	0	4,800–15,000	Limited chiefly to adhesives
Chloride-acetate copolymers	85–87	90	8,500–8,500	Lacquer for lining food cans; sufficiently soluble in ketone solvents for surface-coating purposes
	85–87	90	9,500–10,500	Plastics of good strength and solvent resistance; molded by injection
	88–90	92	16,000–23,000	Synthetic fibers made by dry spinning; excellent solvent and salt resistance
	95	96	20,000–22,000	Substitute rubber for electrical-wire coating; must be externally plasticized; extrusion-molded
Straight polyvinyl chloride	100	100	—	Limited, if any, commercial applications per se; nonflammable substitute for rubber when externally plasticized

* Adapted from A. Schmidt and C. A. Marlies, *Principles of High Polymer Theory and Practice.* New York: McGraw-Hill.

The third type of copolymer (Fig. 7–5.1c) is obtained by grafting side branches of a second polymer onto the main chain. A *graft copolymer* exhibits very little, if any, crystallization.

Illustrative problem 7–5.1 A copolymer contains 10 m/o vinyl acetate and 90 m/o vinyl chloride (m/o = mer percent). (a) What is the w/o vinyl acetate? (b) What is the w/o chlorine?

SOLUTION From Table 7–1.3:

Vinyl acetate mer = $\left(\begin{matrix} H & H \\ C- & C \\ H & Ac \end{matrix}\right)$

where **Ac** is $-OCOCH_3$

Vinyl chloride mer = $\left(\begin{matrix} H & H \\ C- & C \\ H & Cl \end{matrix}\right)$

$$\begin{aligned} \text{Basis: 100 mers} &= \text{10 mers VAc } (= 40C + 20O + 60H) \\ &= \text{90 mers VC } (= 180C + 270H + 90Cl) \\ \text{10 mers VAc} &= (12\text{ amu})(40) + (16\text{ amu})(20) + (1\text{ amu})(60) \\ &= 480\text{ amu C} + 320\text{ amu O} + 60\text{ amu H} \\ &= 860\text{ amu.} \\ \text{90 mers VC} &= (12\text{ amu})(180) + (1\text{ amu})(270) + (35.5\text{ amu})(90) \\ &= 2160\text{ amu C} + 270\text{ amu H} + 3195\text{ amu Cl} \\ &= 5625\text{ amu.} \end{aligned}$$

a) w/o vinyl acetate = 860/(5625 + 860) = 13.3 w/o.
b) w/o chlorine = 3195/(5625 + 860) = 49.3 w/o.

ADDED INFORMATION A copolymer may be viewed as a solid solution of the contributing mers. Just as with the solid solutions in Section 2–5, the overall structural pattern exists with one, two, or several types of components. ◀

Review and Study

SUMMARY

1. The reason that the atoms within a molecule are so tightly bonded is that they share pairs of electrons. Bonds *between* molecules are much weaker. The molecules that go to make up plastics, or polymers, are very large, with thousands of atoms in each molecule.
2. Giant molecules commonly involve chains of carbon atoms. We can understand the structure of these molecules readily if we pay attention to the mer, or

repeating unit. The vinyl units of

$$-\!\left(\begin{matrix} H & H \\ C & - & C \\ H & R \end{matrix}\right)\!-$$

are the most widely encountered mers, where **R** is one of several possible side radicals such as $-H$, $-OH$, $-Cl$, $-CH_3$, etc. (See Table 7-1.3.)

3. If the contributing units have more than two connecting points, a network polymer can result. In Unit 8 we shall see the chain polymers and network polymers have markedly different properties, which presents the engineer with a variety of options.
4. The configuration of a polymer also affects its properties. Variations such as stereoisomers, branching polymers, cross-linking of molecules, and copolymers influence the crystallinity and therefore the behavior of the polymers in service.
5. The glass temperature is a basic thermal characteristic of amorphous materials, just as the melting temperature is a basic thermal characteristic of crystalline materials.

TECHNICAL TERMS

Amorphous Noncrystalline and without long-range order.

Atactic Lack of long-range repetition in a polymer (as contrasted to isotactic).

Bifunctional Molecule with two reaction sites for joining with adjacent molecules.

Branching Bifurcation of a polymer chain.

Configuration Arrangment of mers along a polymer chain. (Rearrangements require bond breaking.)

Conformation Twisting and/or kinking of a polymer chain. (Changes require bond rotation only.)

Copolymer Polymers with more than one type of mer.

Cross-linking The tying together of adjacent polymer chains.

Crystallinity (polymers) Volume fraction of a solid that has a crystalline (as contrasted to an amorphous) structure.

Degree of polymerization (n) Mers per average molecule. Also, molecular mass/mer mass.

Glass An amorphous solid below its transition temperature. A glass lacks long-range crystalline order but normally has short-range order.

Glass-transition temperature (T_g) Transition temperature between a supercooled liquid and its glassy solid.

Isomer Molecules with the same composition but different structures.

Isotactic Long-range repetition in a polymer chain (in contrast to atactic).

Macromolecule Molecules made up of hundreds to thousands of atoms.

Mer, $-\!(\ )\!-$ The smallest repetitive unit in a polymer.

Molecular crystals Crystals with molecules as basic units (as contrasted to atoms).

Molecular size (M) Mass of one molecule (expressed in amu), or mass of 0.6×10^{24} molecules (expressed in grams).

Polyfunctional Molecule with three or more sites at which there can be joining reactions with adjacent molecules.

Polymer Nonmetallic material consisting of (large) macromolecules composed of many repeating units; the technical term for plastics.

Polymer, linear Polymer of bifunctional mers. Chain polymer.

Polymer, network Polymers containing polyfunctional mers that form a 3-dimensional structure.

Rubber A polymeric material with high elastic strains. The strain arises from the unkinking of molecular chains.

Thermal agitation Thermally induced movements of atoms and molecules.

Vinyl compounds *See* Table 7-1.3.

Vulcanization Treatment of rubber with sulfur to cross-link the elastomer chains.

CHECKS

7A _***_-molecules may contain thousands of atoms made up of repeating units called _***_. A molecule with many such units is called a _***_.

7B The large molecules described in 7A are held together primarily by _***_ bonds.

7C If the _***_ polymerization is 750, there are _***_ mers per _***_.

7D The end-to-end length of a kinked molecular chain is proportional to the square root of the number of _***_ in the chain.

7E The kinked _***_ of a rubber molecule provides a high value of elastic _***_ for rubber.

7F Polyvinyl compounds have the composition shown in Eq. (7-1.9). The **R** of polyvinyl chloride, PVC, is _***_; in polystyrene, PS, the **R** is _***_; in _***_, it is –OH; and in polypropylene, it is _***_.

7G During the polymerization of polyvinyl chloride, the additon of each mer destroys one double bond, C = C, and introduces _***_ new single bonds, C–C, per mer. The polymerization of polystyrene eliminates _***_ double bond and forms _***_ new single bonds per mer. For polyisoprene, the numbers are _***_ and _***_ per mer, respectively.

7H _***_ are molecules that have the same composition, but different _***_.

7I A _***_ stereoisomer has greater regularity than does a _***_ stereoisomer.

7J _***_ causes additional entanglements in linear polymers that interfere with plastic deformation.

7K The vulcanization of rubber involves a _***_ of adjacent molecular chains, thus resulting in a harder rubber.

7L For maximum ___*** by sulfur, the sulfur-to-butadiene ratio is ___*** to one.

7M A polyfunctional molecule is required for a ___*** polymer, while a bifunctional molecule will produce a ___*** polymer.

7N Phenol-formaldehyde forms a ___*** polymer because the ___*** molecule is trifunctional.

7O The rate of crystallization for polymers is ___*** than for metal because the rearrangement involves groups of atoms rather than individual atoms.

7P Many polymers are ___***, that is, noncrystalline.

7Q At the ___*** temperature of a polymer, there is a change in the ___*** coefficient.

7R A ___*** is sometimes referred to as a rigid liquid because it is a noncrystalline solid.

7S When compared to a crystalline solid, a glass has excess volume called ___***.

7T A crystal changes to a liquid at the ___***; a glass changes to a ___*** at the glass-transition temperature.

7U A ___*** may be compared to an alloy in that it has one molecular structure with two components.

7V An ABS copolymer contains ___***, ___***, and ___*** mers; it is thus a ___*** polymer.

7W Polybutadiene is shown in Fig. 7-2.4(a). In ___***, one hydrogen of each mer is replaced by a chlorine atom; in isoprene, one hydrogen atom in each mer is replaced by ___***. In each case the hydrogen that is replaced was connected to a carbon atom that is bonded with ___*** other hydrogen atoms.

7X ___*** copolymers possess short side ___*** of a second component along the main chain.

7Y ___*** copolymers crystallize somewhat better than do random copolymers.

STUDY PROBLEMS

7-1.1. What is the molecular size of isoprene (Fig. 7-1.2)?

Answer: 68 amu (or 68 g/mole)

7-1.2. The degree of polymerization of a polyisoprene molecule is 125. (a) How many carbon atoms does it have? (b) What is its molecular size?

7-1.3. A calculation shows that there are 10^{20} polypropylene molecules per gram. (a) What is the average molecular size? (b) What is the degree of polymerization?

Answer: (a) 6000 amu (or 6000 g/mole) (b) 143

7-1.4. The average end-to-end distance of a molecule of polyvinyl alcohol, C_2H_3OH, is approximately 10 nm. (a) What is the degree of polymerization? (b) What is the average molecular size? (c) Change to polyvinyl chloride, and answer parts (a) and (b).

7-1.5. The degree of polymerization for polyacrylonitrile (Table 7-1.3) is 170. (a) What is the molecular size? (b) How many molecules per gram?

Answer: (a) 9000 amu (or 9000 g/0.6×10^{24} molecules)
(b) 0.7×10^{20} molecules/g

7-1.6. (a) What is the mer mass of polyvinyl acetate? (b) Its end-to-end length if the molecular mass is 43,000 amu?

7-1.7. Polytetrafluoroethylene has the structure of ethylene but with all four hydrogen atoms per mer replaced with fluorine atoms. Its molecular size is 21,700 g/mole, that is, 21,700 amu/molecule. How many mers per molecule?

Answer: 217 mers/molecule

7-1.8. Which molecule has greater mass, a polystyrene (PS) molecule with 80 mers or a polyvinyl chloride (PVC) molecule with 130 mers?

7-1.9. The molecular weight of noncrystalline polyvinyl chloride (PVC) is 24,000 g/mole. What is the average end-to-end distance?

Answer: 4.3 nm

7-1.10. Vinylidenes are a class of compounds similar to the vinyls (Table 7-1.3) *except* that there are 2 **R**'s per mer, that is, $H_2CC\mathbf{R}_2$. Common vinylidenes include vinylidene chloride, H_2CCCl_2, and isobutylene, $H_2CC(CH_3)_2$. Which will have the longer molecules if their polymeric molecular sizes are 25,000 g/mole?

7-2.1. Refer to I.P. 7-2.1. (a) What weight percent sulfur would be present if all the possible cross links of polyisoprene are occupied by sulfur? (b) If 50% of the cross links were occupied by sulfur?

Answer: (a) 32%

7-2.2. The formula $\left(\begin{array}{cccc} H & \mathbf{R} & H & H \\ C & C & C & C \\ H & & & H \end{array}\right)$ is the basic formula for several rubbers. (a) In butadiene, **R** is H; (b) in isoprene, **R** is $-CH_3$; (c) in chloroprene, **R** is Cl. Compare their mer weights.

7-2.3. Raw polyisoprene (that is, unvulcanized) gains 2.3 w/o in air by being cross-linked with oxygen. What fraction of the possible cross links are established?

Answer: 0.10

7-2.4. Refer to S.P. 7-2.2. Which rubber will possess the largest percentage of sulfur if 5% of the cross links are vulcanized?

7-2.5. Show the bond changes required for isoprene to polymerize.

7-3.1. Based on Fig. 7-3.3, determine the percent of solidification shrinkage in iodine.

Answer: 17 v/o

7-3.2. The density of polyethylene is 1.01 g/cm^3 when it is fully crystalline. [See Fig. 7-3.2 for the dimensions of a unit cell.] (a) Calculate the number of $\leftarrow C_2H_4 \rightarrow$ mers per unit cell. (b) Check your answer by examining Fig. 7-3.2.

7-3.3. Sketch the three orthogonal views of the atoms in the polyethylene unit cell.

7-4.1. The density of supercooled liquid iodine is 4.3 g/cm^3 at 20°C. What is the fraction crystallinity of some iodine with a density of 4.87 g/cm^3?

Answer: 0.86

7-4.2. What percent "free-space" is there in polyethylene with a density of 0.94 g/cm^3?

7-5.1. A copolymer contains 20 mers of polystyrene (PS) for every 80 mers of polyvinyl chloride (PVC). What is the weight percent chlorine?

Answer: 40 w/o Cl

7-5.2. The mer ratio in a styrene-butadiene copolymer is 1 to 3. What is the carbon content (w/o)?

7-5.3. What mass ratio of PVC and PE is required to give a 1/4 mer ratio?

Answer: 36% PVC and 64% PE

7-5.4. What is the ratio of ethylene mers $\leftarrow C_2H_4 \rightarrow$ to vinyl chloride mers $\leftarrow C_2H_3Cl \rightarrow$ in a copolymer of the two, in which there is 20 w/o chlorine?

CHECK OUTS

7A	macro- mers polymer
7B	covalent (C–C)
7C	degree of 750, molecule
7D	bonds
7E	conformation strain
7F	chlorine a benzene ring polyvinyl alcohol $-CH_3$
7G	two one, two one, two
7H	isomers structures
7I	isotactic atactic
7J	branching
7K	cross-linking
7L	cross-linking (vulcanization) one
7M	network linear
7N	network phenol
7O	slower
7P	amorphous
7Q	glass transition thermal expansion
7R	glass
7S	free space

7T melting temperature
supercooled liquid

7U copolymer

7V acrylonitrile, butadiene, styrene
linear

7W chloroprene (neoprene)
$-CH_3$
no

7X graft
branches (chains)

7Y block

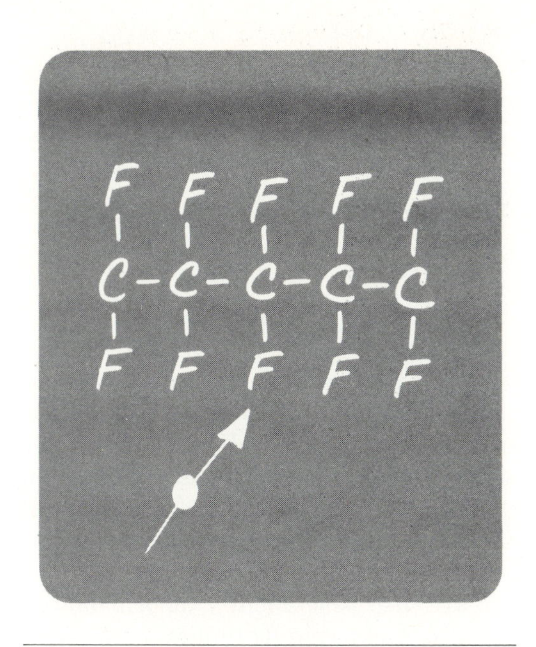

UNIT EIGHT

DEFORMATION OF POLYMERS

Rubbers are polymers that can undergo major elastic strains. Other polymers have larger elastic moduli but still have higher strains than metals. These characteristics must be considered in product design. Polymers are called "plastics" because they are readily deformed to meet production requirements. They may also slowly deform (creep) under adverse service conditions. It thus becomes necessary to look closely at the deformation of these viscoelastic materials.

Contents

Prerequisites Units 3 and 7.

From Unit 8, the engineer should

1) Understand the basis for the large elastic strain in polymers and know why the elastic modulus of linear polymers increases with increased strain and increased temperature.
2) Differentiate between elastic strain, plastic slip in metals, and viscous flow in amorphous solids.
3) • Know how the viscoelastic modulus varies with time and temperature.
4) Relate the property changes in polymers to the glass-transition temperature.

5) Relate the crystallinity of low-density and high-density polyethylenes to their properties.
6) Understand the indicated mechanisms of degradation.
7) Be familiar with technical terms that relate to polymer deformation.

8-1 "PLASTICS"

The general public uses the term "plastics" to describe any of the large class of nonmetallic materials that were the subjects of the past chapter. These materials have polymeric molecules. They typically contain a significant fraction of carbon atoms. As a result, their chemistry is closely related to organic chemistry. Also, covalent bonds are prevalent, since metallic elements are commonly absent and only nonmetallic elements are present.

From one viewpoint, the use of the term "plastics" is unfortunate, since the term is not definitive. Metals are also permanently deformable and, therefore, plastic (Section 3–5). How else could we roll aluminum into foil for kitchen use, or draw tungsten wire into a filament for an incandescent light, or forge a 100-ton ingot of steel into a rotor for a generator (Fig. 4–4.1)? Likewise, glasses, which contain compounds of metals *and* nonmetals, can be permanently shaped at higher temperatures. These are structural cousins to polymers but are not considered to be plastics.

The attachment of the term "plastics" to polymeric materials occurred because these materials are "capable of being molded or modeled as clay or plaster" (Webster). Art potters use wet clay to form their art objects. However, we do not call these plastics. Thus, there is a contradiction in the use of this term. Many engineers prefer to be more specific and use the technical term *polymers*. Even so, the engineer utilizes plastic deformation to shape these materials during processing. In addition, the design engineer must anticipate possible plastic deformation of these materials during their use. Engineers also depend extensively on elastic characteristics of various polymers in their designs, be they for automobile tires or a product such as cross-country skis. We need to understand as much as possible about the deformation of polymers.

Just as with metals, the deformation of polymers includes elastic (recoverable) deformation and permanent deformation. We shall look at elastic deformation first (Section 8–2). We will find that it is not unusual for the elastic deformation to far exceed the 1% strain that is the normal elastic limit in metals, simply because the structure is different. Also we will find that permanent deformation will be much more time dependent than with metals (Section 8–3).

8-2 ELASTICITY OF POLYMERS

There are three structural features that modify the elastic behavior of polymers from that observed in metals (Section 3-4): (a) the majority of polymers are *amorphous* (Section 7-4); (b) many polymers are molecular with *weak intermolecular bonds* (Section 7-1); and (c) a single bond can be *rotated;* for example, C⟲C.

Glassy polymers

The large, gangling molecules of polymers do not lend themselves to the long-range ordering that is required for crystallization. Therefore, we saw in Fig. 7-4.1 that polymeric liquids are readily supercooled below the melting temperature, T_m. Because of its importance, that figure is reproduced as Fig. 8-2.1. As the temperature of a liquid polymer is decreased, the volume of the supercooled melt contracts, partly because the molecular vibrations decrease, but also because the molecules rearrange themselves into denser packing with less "free-space" as they lose their thermal energy. There comes a temperature, however, where molecular movements are too sluggish for further rearrangements. Below this temperature the contraction is solely the result of reduced thermal vibration. The inflection in the V-vs.-T curve of Fig. 8-2.1 is at the *glass-transition temperature,* T_g. This inflection was first observed in commercial silicate glasses. Rigid products of supercooled melts are called *glass* below this temperature. That term can be used whether we are discussing inorganic silicate glasses or polymeric glasses. Recall that these glasses are *amorphous solids*—they are noncrystalline solids.

Of course, glassy polymers respond elastically to applied stresses. A compression stress will push atoms closer together and a tension stress will

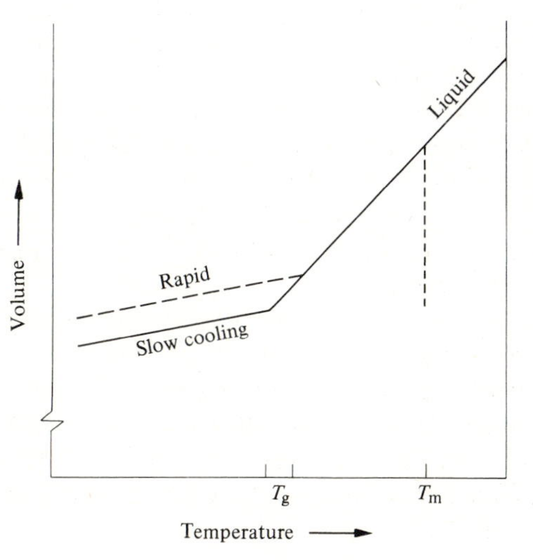

Figure 8-2.1 Volume versus temperature (from Fig. 7-4.1). The supercooled liquid becomes a rigid glass at the glass-transition temperature, T_g. However, that temperature varies slightly with the available time. It is somewhat lower with slow cooling.

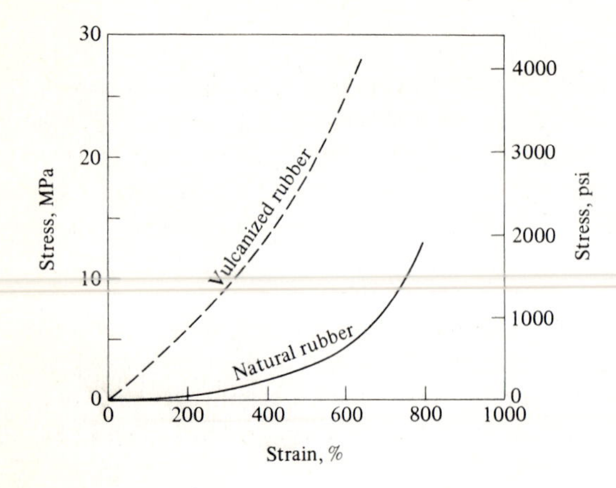

Figure 8-2.2 Stress-strain curves for rubber (isoprene). The elastic modulus, *ds/de*, is low in natural rubber (unvulcanized) until the molecules become aligned with the direction of stress. The vulcanized rubber is cross-linked with less opportunity for extending the molecules. Therefore, it has a higher modulus. (Adapted from Schmidt and Marlies, *High-Polymers*, New York: McGraw-Hill.)

pull them farther apart, just as in crystalline metals. In addition, the bonds across the carbon atoms can be bent to other than the 109° cited in Section 7-1, a situation that does not have a direct counterpart in metals. This allows additional strain for a given stress and leads to significantly lower Young's moduli than for metals. Typically, the elastic moduli for glassy linear polymers (for example, polyvinyl chloride and polystyrene) are less than 5,000 MPa (<750,000 psi). This is in contrast to ~15,000 MPa (~2,000,000 psi) for lead and >100,000 MPa (~16,000,000 psi) for copper. Even the 3-dimensional network polymers, such as phenol-formaldehyde (Section 7-2), have Young's moduli of less than that of lead, the least rigid metal that is included in Appendix A. Although the covalent polymers have strong bonds, their structures may be comparatively flexible.

Oriented polymers

Consider a fiber or thread of nylon-66—a common artificial polymer. As an unoriented glassy polymer, its modulus of elasticity is ~2000 MPa (~300,000 psi). Above the glass-transition temperature, its elastic modulus drops even lower because small stresses will readily straighten the kinked molecular chains. However, once extended and with the molecules oriented in the direction of the stress, larger stresses are required to produce added strain. The elastic modulus increases (Fig. 8-2.2). Now, cool the nylon below the glass-transition temperature without removing the stress, retaining the molecular orientation. The nylon becomes rigid with a much higher elastic modulus in the tension direction (15,000–20,000 MPa or 2,000,000–3,000,000 psi). This is nearly 20 times the elastic modulus of the unoriented glassy polymer. The stress for any elastic extension must work against the rigid backbone of the nylon molecule and not simply unkink molecules.

The above procedure is commonly used in the commercial production of "man-made" fibers (Section 9-5).

Elastomers

Very large elastic strains are possible with minimal stress in the category of polymers that we call *elastomers*. These are *rubbers*. Elastomers have two specific characteristics: (a) their glass-transition temperature, T_g, is below the temperature at which they are commonly used, and (b) their molecules are kinked very highly. The latter is illustrated by the isoprene mer (Eq. 7–2.1) in natural rubber. Each individual mer is arc-shaped (Fig. 8–2.3) because of the rigid (nonrotatable) double bond. The single bond can rotate, however; so the chain becomes highly kinked. When a stress is applied, the molecular chain uncoils and the end-to-end length can be extended several hundred percent with minimal stresses. Some rubbers have initial Young's moduli of less than 10 MPa (<1,500 psi). Of course, once the molecules are extended, the moduli increase (Fig. 8–2.2).

The elastic moduli of metals decrease with increases in temperature (Fig. 3–4.1). In stretched elastomers, the opposite is true because at higher temperatures there is increasingly vigorous thermal agitation in the molecule. Therefore, the molecules more strongly resist the tension forces that attempt to uncoil them. It requires greater stress per unit strain—the *elastic modulus increases with temperature* (Fig. 8–2.4).

When stretched into molecular alignment, many rubbers can form *crystals,* an impossibility when they are relaxed into the highly kinked confor-

(a)

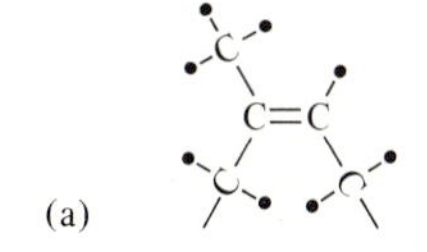

(b)

Figure 8–2.3 Elastic strain in linear polymers (isoprene rubber). Since the single bond rotates, the rubber molecules are normally in a highly kinked conformation. They can be straightened with a small stress. They will rekink through thermal agitation when the stress is released.

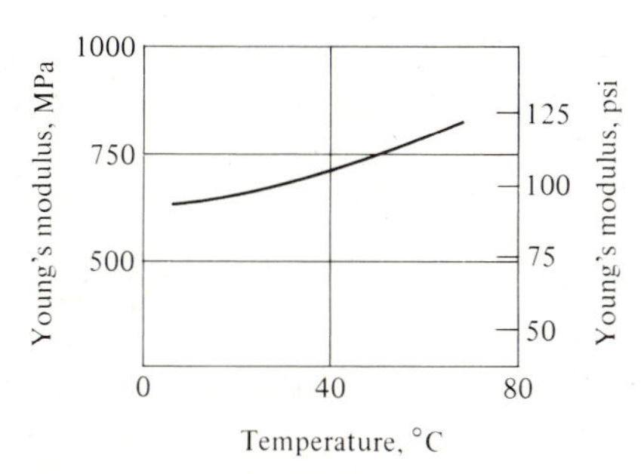

Figure 8–2.4 Young's modulus versus temperature (stretched rubber). The retractive forces of kinking are greater with more thermal agitation. Therefore, the elastic modulus increases with temperature. (Cf. Fig. 3–4.1.)

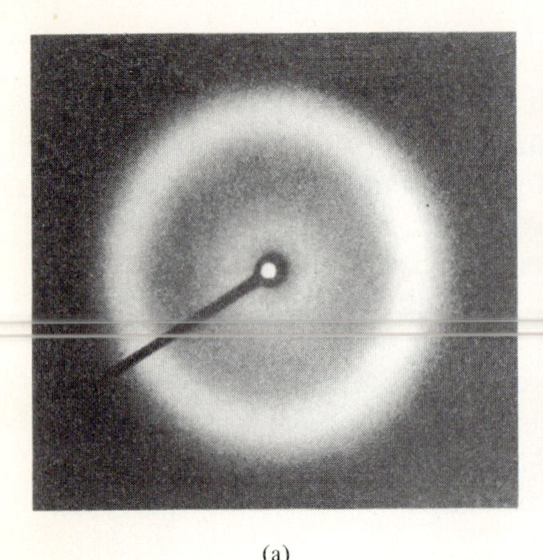

(a)

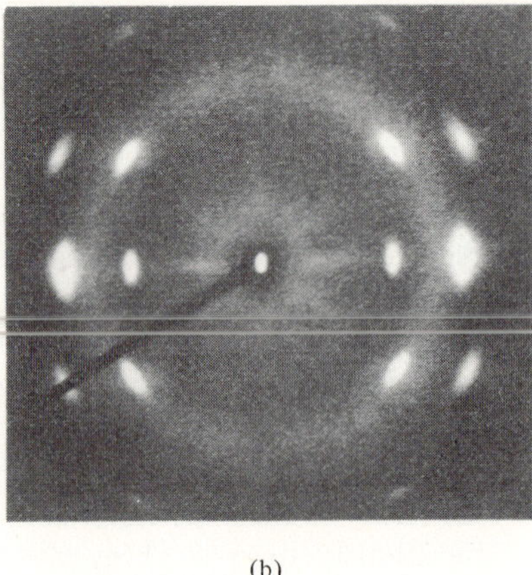

(b)

Figure 8–2.5 Deformation crystallization of natural rubber (polyisoprene) revealed by x-ray diffraction. (a) Unstretched. (b) Stretched. (S. D. Gehman, *Chemical Reviews*, Vol. 26, 1940, page 203.)

mation. This is verified by the information in Fig. 8–2.5. The x-ray diffraction pattern of relaxed isoprene rubber is shown in part (a). The diffraction spots (part b) permit the crystallographer to interpret how the molecules of rubber mesh together when stretched to form molecular crystals (Section 7–3).

8–3 PERMANENT DEFORMATION OF POLYMERS

Viscous polymer melts

Molten polymers can be supercooled below their freezing temperatures without crystallization. These supercooled melts have high viscosities because of their very large molecules. That is, it takes considerable shear stress, τ, to produce flow (or more accurately, a flow gradient, v/y). *Viscosity,* η, may be defined as the ratio of these two (Fig. 8–3.1):

$$\eta = \frac{\tau}{v/y}. \tag{8–3.1}$$

From this, the units for viscosity are Pa·s.* *Fluidity, f,* is the reciprocal of viscosity.

The viscosity increases as the temperature of the supercooled melt decreases. General laboratory experience has shown that the glass-transition temperature, T_g, coincides with a viscosity of approximately 10^{12} Pa·s. Molecular rearrangements by thermal motions are precluded at lower temperatures, simply because the melt is too viscous.

* 1 Pa·s = 10 poises in cgs units.

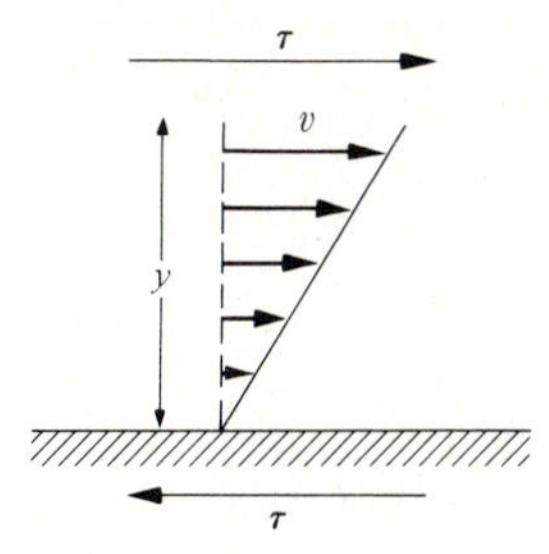

Figure 8-3.1 Viscosity, η, is the ratio of shear stress, τ, to velocity gradient, v/y. Since it is the reciprocal of fluidity, f, it decreases with increased temperature (Eq. 8-3.2).

Table 8-3.1
Viscosities of noncrystalline materials (20°C)

Material	Viscosity, η Pa·s*	Fluidity, f $Pa^{-1}\cdot s^{-1}$
Air	0.000018	5.6×10^4
Pentane, C_5H_{12}	0.00025	4×10^3
Water	0.001	1000
Phenol, C_6H_5OH	0.01	100
Syrup (60% sugar)	0.055	18
Oil, machine	0.1 to 0.6	1.5 to 10
Glycerin	0.9	1.1
Sulfur (120°C)	10^2	10^{-2}
Window glass (515°C)	10^{12}	10^{-12}
Window glass (800°C)	10^4	10^{-4}
Polymers, T_g†	$\sim 10^{12}$	$\sim 10^{-12}$
Polymers, T_g + 15°C	$\sim 10^8$	$\sim 10^{-8}$
Polymers, T_g + 35°C	$\sim 10^5$	$\sim 10^{-5}$

* 1 Pa·s = 0.1 poises.
† T_g = glass-transition temperature.

The viscosity decreases above T_g. The empirical relationship,

$$\log \eta = 12 - (17.5\, \Delta T)/(52 + \Delta T), \qquad (8\text{-}3.2)^*$$

holds for a number of linear polymers, where ΔT is the °C above T_g, that is, $(T - T_g)$. This leads to the last three values of Table 8-3.1, where viscosities of polymeric melts are compared to viscosities of various fluids. The major effect of temperature is evident.

* The more orthodox equation for viscosity that applies for Newtonian fluids is

$$\ln \eta = A + B/T, \qquad (8\text{-}3.3)$$

where B can be related to an activation energy for diffusion. (Cf. Eq. 4-4.1). However, this equation is not the best for polymeric melts, because the long molecules become entangled and do not move past their neighbors as a unit. As a result, we turn to the empirical equation (8-3.2).

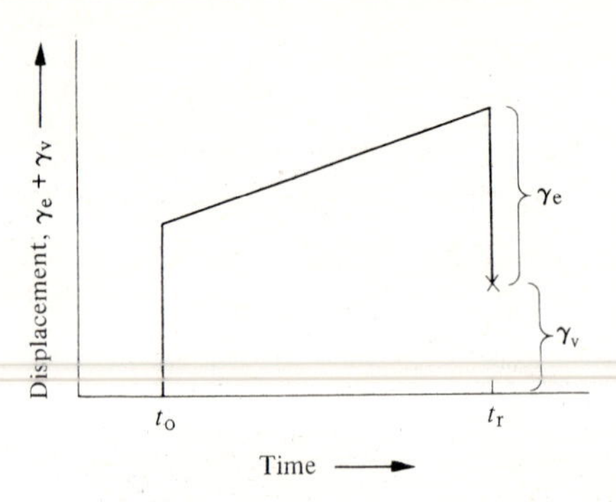

Figure 8-3.2 Viscoelasticity (simplified). When the shear stress, τ, is applied, the elastic strain, γ_e, is immediate. The viscous flow, γ_v, proceeds with time according to Eq. (8-3.4b). When the load is removed at t_r, only the elastic component of the total displacement is recovered.

Polymers must be above their glass-transition temperature to be molded into "plastic" products. The specific temperature depends upon the pressures to be used. These vary from process to process (Section 9-4). Temperature control is critical since only a few degrees up or down doubles the fluidity or viscosity, respectively.

From Fig. 8-3.1, we can define a *viscous flow,* γ_v, as equal to x/y. With $x = vt$, and bringing in Eq. (8-3.1),

$$\gamma_v = vt/y = \tau t/\eta. \tag{8-3.4a}$$

The *rate* of viscous flow, γ_v/t, is therefore equal to τ/η:

$$\gamma_v/t = \tau/\eta. \tag{8-3.4b}$$

That is, the rate increases with shear stress, τ, and is inversely related to viscosity, η.

• Viscoelasticity

When the glass-transition temperature is below service conditions, we can have both elastic shear strain, γ_e, *and* viscous flow, γ_v. We speak of this as *viscoelasticity*. In the simplest situation,* the total shear displacement is

$$\gamma = \gamma_e + \gamma_v \tag{8-3.5a}$$
$$= \tau/G + (\tau/\eta)t. \tag{8-3.5b}$$

The first, γ_e, is immediate and also reversible when the load is removed (Eq. 3-4.3). The second, γ_v, is initially zero but increases continuously with time if the load is held constant. It is not reversible. Together, they give the displacement time curve shown in Fig. 8-3.2 where the stress is applied at t_o and removed at t_r. A higher stress will give a higher curve and a steeper curve. Likewise, higher temperatures introduce more displacement, since this lowers the viscosity and, thereby, increases the flow rate.

---

* Called the *Maxwell model.* There are other more complex models that are presented in texts on polymers.

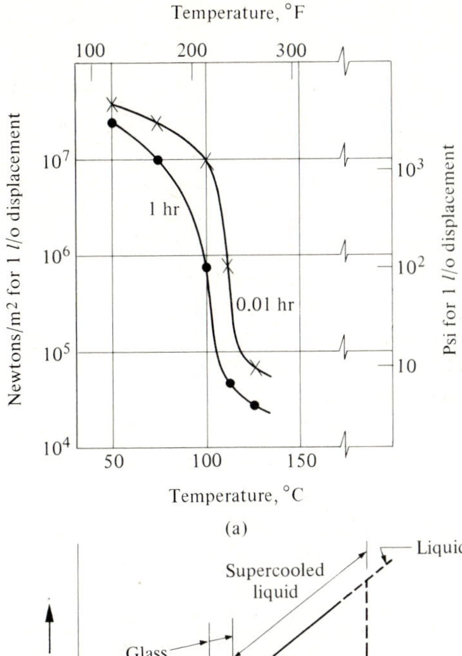

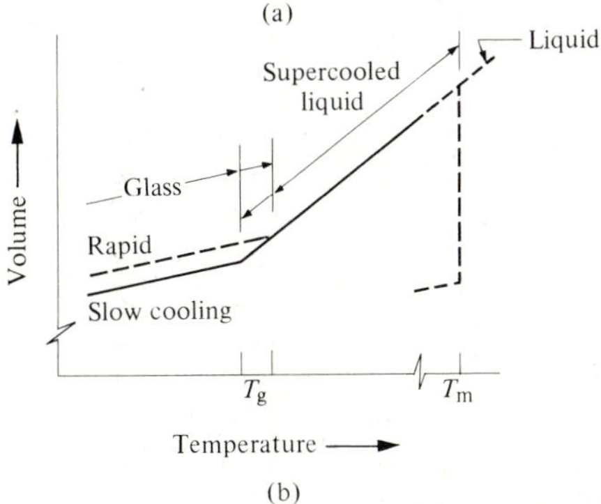

Figure 8-3.3 (a) Softening (polymethyl methacrylate). The shear stress required for unit displacement drops markedly at the glass temperature. The required stresses are also lower when the time is lengthened. (b) Glass-transition temperature, T_g, varies with the cooling and heating rates. When slowly cooled, the molecules of the supercooled liquid have opportunities for rearrangements as the temperature continues to drop.

It is convenient to speak of a *viscoelastic modulus*, M_{ve}, which is the ratio of shear stress to total shear displacement:

$$M_{ve} = \tau/(\gamma_e + \gamma_v) = G\eta/(\eta + Gt). \tag{8-3.6}$$

The viscoelastic moduli for the polymer PMM* are shown in Fig. 8-3.3. This modulus is very sensitive to temperature, but in a rational way. Below the glass-transition temperature, T_g, this modulus is high because there is negligible flow, so the total strain is elastic. In fact, the viscoelastic modulus is close to the shear modulus G for purely elastic responses. Above the glass-transition temperature, which is near 100°C for this polymer, the viscous flow far exceeds the purely elastic strain. As a result, the value of $\tau/(\gamma_e + \gamma_v)$ drops by 2 or 3 orders of magnitude. Below T_g, plastic molding would be an impossible manufacturing process; above T_g, the polymer may be forced into

* Polymethyl methacrylate, also known as lucite, plexiglas, and by other trade names.

a mold. Below T_g, the product is not permanently deformed; above T_g, a slow flow can occur during service.

Since Eq. (8–3.6) has a time factor in it, we have separate curves in Fig. 8–3.3 for the two time spans. They start out together at the left where viscous flow does not occur. However, with 1 hr rather than 36 sec (0.01 hr), the glass-transition temperature is lowered, simply because molecular rearrangements have more time and can occur at a slightly lower temperature ($\sim$10°C). Also with more time, the curve drops lower because viscous flow proceeds longer to give higher values of $(\gamma_e + \gamma_v)$.

• Creep and stress relaxation

A fixed but long-term load on a viscoelastic polymer can lead to *creep*. Conversely, a fixed but long-term strain can lead to *stress relaxation*.

Creep is a slow strain, sometimes as little as 10^{-3}% per hour, but if it is continued for days, weeks, or months, it leads to excessive distortions of the polymeric material. Numerous examples can be cited where creep is to be avoided if possible. An automobile tire that creeps while the car stands over a summer weekend in an airport parking lot develops a "flat spot" on the ground side. Although dimensional changes are minor, the driver is aware of the "thump" that persists until recovery can occur. Cloth fibers that creep can lead to baggy knees in pants. Wooden shelves sag elastically between supports when loaded with books. With time, they can also take on a permanent "set" because the cellulose molecules move by one another under the shear loads in the deflected shelf. The reader can supply additional examples of time-delayed strains.

We can illustrate creep in a rubber schematically (Fig. 8–3.4a). Recall that the initial elastic shear strain, γ_e, is by molecular extension, but that additional viscous flow, γ_v, is by a slow displacement of molecules with respect to their neighbors. After the constant stress is removed, the molecules rekink among their newly obtained neighbors, retaining the permanent viscous displacement.

Stress relaxation is the decay of stress by creep, but with the displacement remaining constant. This is also a common situation. A simple example is a rubber band that has been stretched tightly over a book and then left on a shelf for a year or more. Unless the rubber has been sufficiently cross-linked, the molecules will slowly move with respect to their neighbors, relaxing the stress. The mechanism for stress relaxation and creep are closely related (Fig. 8–3.4b) except that as the value of γ_v increases, γ_e of Eq. (8–3.5a) decreases to maintain the constant total displacement. The molecules of rubber gradually recoil. This means, of course, that when the rubber band is eventually removed from the book, the band will not return to its original dimensions.

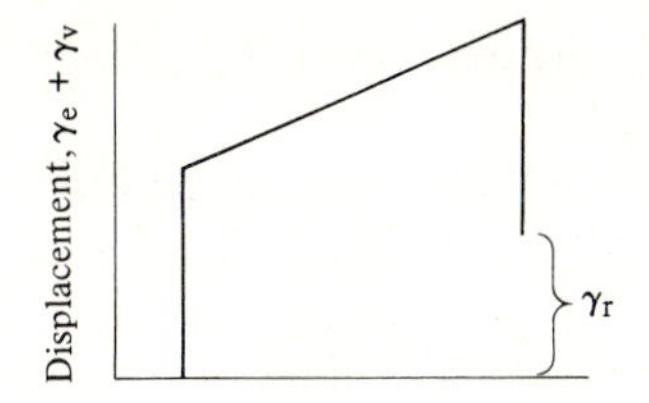

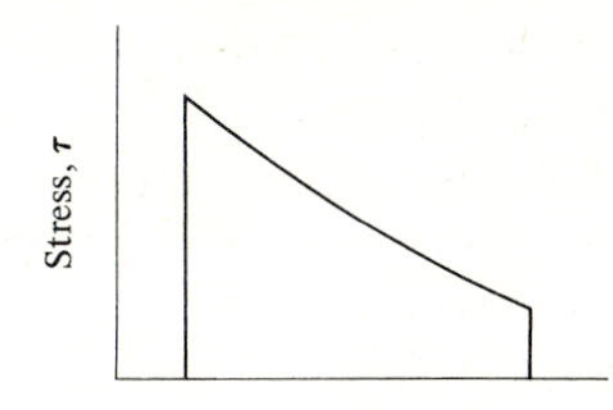

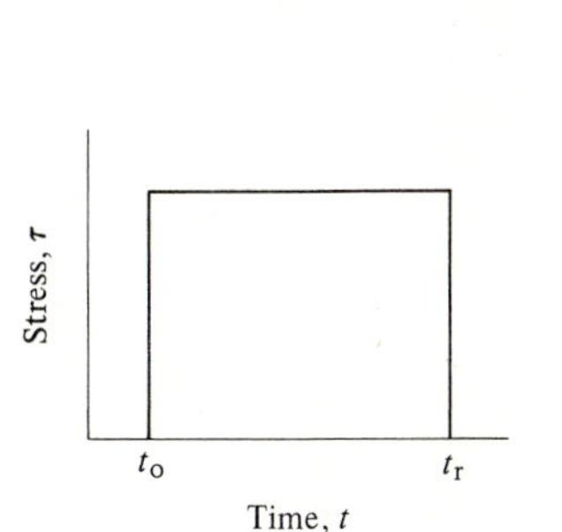

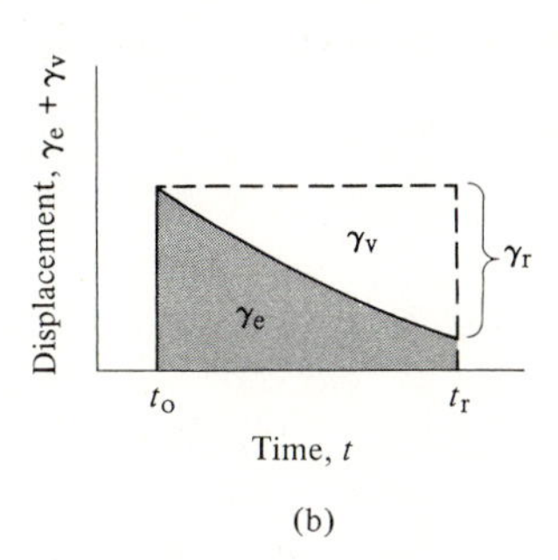

Figure 8–3.4 (a) Creep under constant shear stress. (b) Stress relaxation under constant displacement (simplified). In each case the initial stress produces an elastic strain. Creep proceeds with constant stress by molecular movements. These displacements remain after the load is removed. With a fixed displacement, creep permits the stresses to relax. Only part of the displacement is recovered after the load is removed. (γ_r—displacement after the load is removed.)

If the strain does not change with time, that is, if $\gamma/t = 0$, then from Eq. (8–3.5b) we can write

$$\tau/t = -G\tau/\eta, \tag{8–3.7a}$$

or on a differential basis,

$$d\tau/dt = -\tau/(\eta/G); \tag{8–3.7b}$$

the rate of stress relaxation, $d\tau/dt$, is proportional to the remaining stress, τ. Thus by mathematics*

$$\tau = \tau_o\, e^{-t/\lambda}. \tag{8–3.8}$$

where λ is the *relaxation time* ($= \eta/G$). It is a function of the material and temperature. Its significance is that when $t = \lambda$, $\tau = \tau_o/e$, as shown in Fig. 8–3.5. Since $\lambda = \eta/G$, and since G does not vary rapidly with temperature, λ has the same sensitivity to temperature as does the viscosity, η, (Eq. 8–3.2).

---

* The derivation is

$$d\tau/\tau = -dt/(\eta/G), \tag{8–3.9a}$$

or

$$\ln\tau - \ln\tau_o = -t/(\eta/G), \tag{8–3.9b}$$

where τ_o is the initial stress, and τ is the stress at time, t. The ratio of viscosity to shear modulus, η/G, is called the *relaxation time*, λ.

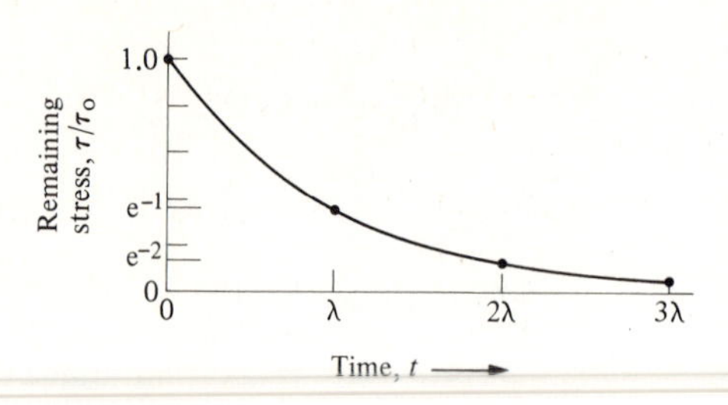

Figure 8-3.5 Stress relaxation. The stress decays at a rate proportional to the remaining stress. After each succeeding period, λ, the stress drops by a factor of $1/e$.

• **Illustrative problem 8-3.1** A rubber band is stretched 1 cm. The initial stress of 1.4 MPa (200 psi) relaxes to 1.0 MPa in 90 days. (a) What is the relaxation time, λ? (b) How long will it take to relax another 0.4 MPa to 0.6 MPa?

SOLUTION From Eq. (8-3.8),

$$\ln(\tau/\tau_0) = -t/\lambda. \tag{8--3.10}$$

a) $\ln(1.0/1.4) = -90 \text{ da}/\lambda,$

$\lambda = 267 \text{ da}.$

b) $\ln(0.6/1.4) = -t/267 \text{ da},$

$t = 226 \text{ da}.$

ADDED INFORMATION An alternate calculation procedure uses 1.0 MPa as the starting stress.

$$\ln(0.6/1.0) = -t/267 \text{ da}$$
$$t = 136 \text{ da}.$$

Thus, added to the initial 90 days, the total time is 226 days. ◀

8-4 PROPERTIES OF POLYMERS FOR SERVICE

Many factors affect the choice of polymers for use in products. Some of these arise from the service conditions—applied stresses, service temperatures, duration of loads, load dynamics, chemical environment, and radiation exposure. The durability of polymers in service is also a function of the structure of the polymer. It is important to know not only the composition of the polymer but also the molecular weight, crystallinity, cross-linking, branching, configuration, etc. When everything is factored in, the picture is complex. We can, however, make some generalized observations that guide the engineer into more detailed considerations.

Simplified correlations

We can use Table 8-4.1 as a starting point. It indicates how the properties of thermoplastic (linear) polymers change with changes in the structure of the molecular chain. The properties are evaluated in the direction of orienta-

Table 8–4.1
Molecular structure and polymeric properties*
(Oriented linear polymers)

Property of Polymer	Increased length of chains	Increased regularity along chains	Increased alignment of chains	Increased rigidity of chains
Strength	Increases	Increases	Increases	Increases
% Elongation	Increases	Decreases	Decreases	Decreases
Elastic modulus	Increases	Increases	Increases	Increases
Flexibility	Decreases	Decreases	Decreases	Decreases
Brittleness at low temperature	Increases	Decreases	Decreases	Decreases
Heat resistance	Increases	Increases	Increases	Increases
Hardness	Increases	Increases	Increases	Increases

* Adapted from R. M. Kell and P. B. Stickney, *Materials in Design Engineering,* Reinhold.

tion. The reader is asked to rationalize these effects on the basis of the discussions in previous sections.

When we try to be more specific, we should turn first to polyethylene. It is probably the least complex of all polymers with its sequence of $\text{-}(CH_2)\text{-}$ segments. Increasing the *molecular weight* increases the melting temperature (Fig. 8–4.1) with the shift from gases to liquids to soft solids (paraffins) and, finally, to polymeric solids that have molecules in excess of 10,000 g/mole. The empirical relationship between the sizes of these molecules and their melting temperature (T_m as K) is

$$T_m \cong [0.0024 + 0.017/m]^{-1}. \tag{8–4.1}$$

In this equation, m is the number of segments, $\text{-}(CH_2)_m\text{-}$, in the paraffin molecule.

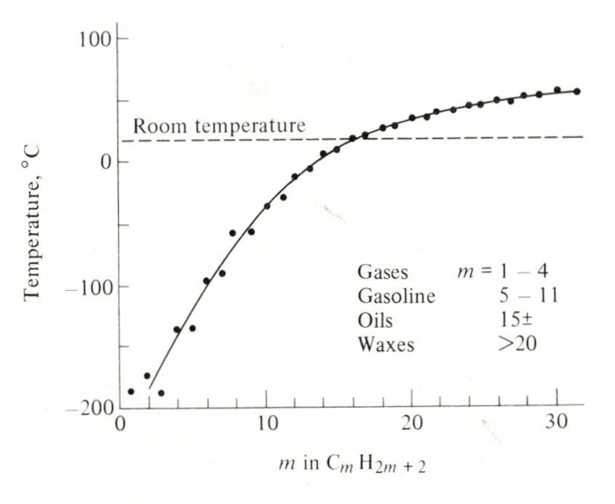

Figure 8–4.1 Melting temperatures versus molecule size in the C_mH_{2m+2} hydrocarbon series.

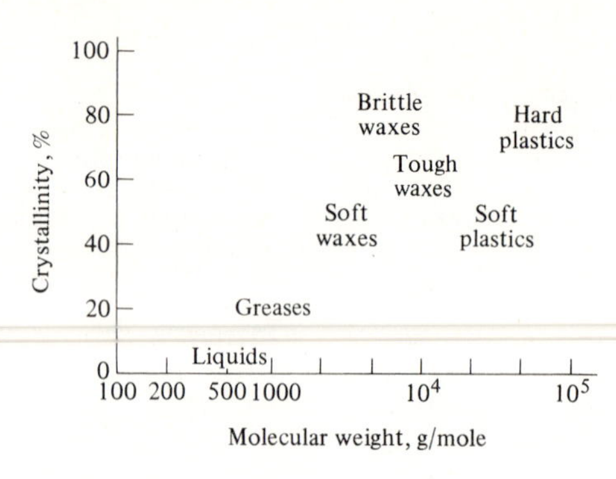

Figure 8–4.2 Characteristics of solid hydrocarbons. This is a generalized extrapolation of Fig. 8–4.1. Increased molecular size and increased crystallinity lead to more rigid solids.

Figure 8–4.2 shows the characteristics of paraffin when the percent crystallinity is considered. Of course, the way these respond in service depends on the loading conditions.

In I.P. 7–4.1, reference was made to low-density polyethylene, LDPE, and to high-density polyethylene, HDPE. These two polymers* have a relatively small difference in density (0.92 vs. 0.96 g/cm³). That difference comes about by the difference in *crystallinity* (~20% for LDPE and ~60% for HDPE). As shown in Table 8–4.2, however, the LDPE has 50% greater thermal expansion coefficient than HDPE; and HDPE has 50% greater thermal conductivity than LDPE. More important is the fact that HDPE has greater heat resistance than LDPE. Specifically, HDPE will withstand 100°C for 10 minutes and can, therefore, be sterilized for use as a food container. This is impossible with LDPE. HDPE, which is more crystalline, softens less and creeps less rapidly at that temperature than does LDPE.

• The *viscoelastic behavior* described in Fig. 8–3.3 is generalized in Fig. 8–4.3(a), where we observe that either high temperatures or sustained loads provide low viscoelastic moduli. Both lead to greater viscous flow under constant stress, or to relaxed stresses with constant load, thus, reducing the $\tau/(\gamma_e + \gamma_v)$ ratio. Conversely, low temperatures and impact loading produce an elastic-type behavior. Under these conditions, the material is relatively rigid, brittle, and hard. A clear plastic triangle that is used by a draftsman is an example.

In the range of the glass temperature, the material is *leathery*; it can be deformed and even folded, but it does not spring back quickly to its original shape. In the *rubbery plateau,* polymers deform readily but quickly regain their previous shape if the stress is removed. A rubber ball and a polyethylene "squeeze" bottle serve as excellent examples of this behavior because they are soft and quickly elastic. At still higher temperatures, or under sustained loads, the polymer deforms extensively by *viscous flow*.

* LDPE is the older product. HDPE, made by a different process, brought its developer a Nobel prize, because (a) it involved some excellent chemistry, and (b) the product had markedly improved heat tolerances for consumer uses.

Table 8-4.2
Characteristics of polyethylenes

	Low-density polyethylene (LDPE)	High-density polyethylene (HDPE)
Density, Mg/m^3 (= g/cm^3)	0.92	0.96
Crystallinity, v/o	~20	~60
Thermal expansion, $°C^{-1}$	180×10^{-6}	120×10^{-6}
Thermal conductivity (watt/m^2)/(°C/m)	0.34	0.52
Heat resistance for continuous use, °C	55–80	80–120
10-min. temp. exposure, °C	80–85	120–125

The curve of Fig. 8–4.3(a) is typical of many amorphous linear polymers. It can be varied, however, with structural modifications that are available for polymeric materials. As sketched in Fig. 8–4.3(b), a highly *crystalline* polymeric material (curve 2) does not have a glass temperature. Therefore it softens more gradually as the temperature increases until the melting temperature is approached, at which point fluid flow becomes significant. The

Figure 8-4.3 Viscoelastic modulus versus structure. (1) Amorphous linear polymer. (2) Crystalline polymer. (3) Cross-linked polymer. (4) Elastomer (rubber).

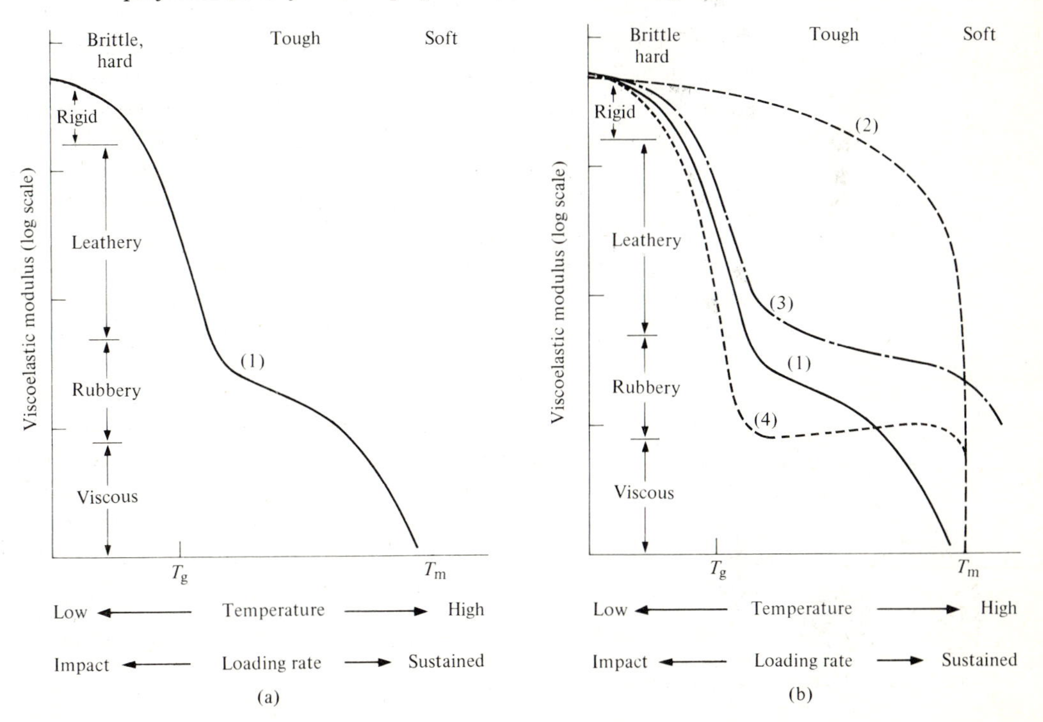

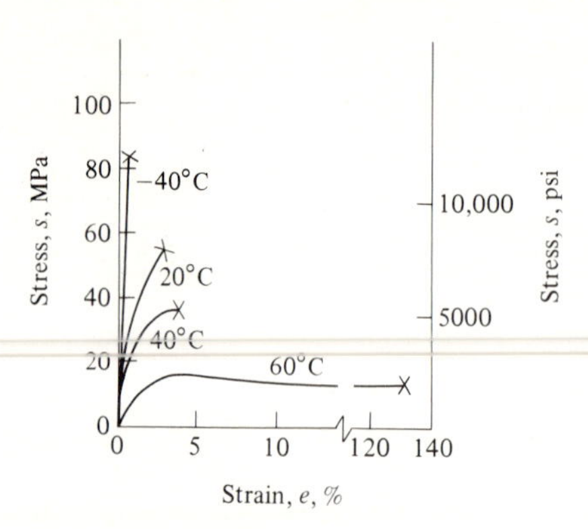

Figure 8–4.4 Stress-strain (PMM with $T_g = 100°C$). Tension fracture is brittle at −40°C. As T_g is approached, molecular movements are facilitated by the applied stress, so that deformation occurs prior to fracture (X). If, however, this deformed plastic is unloaded and heated above T_g, the molecules will rekink. Most of the preceding strain will be removed. (Adapted from Alfrey and Gurnee, *Organic Polymers*, Englewood Cliffs, N.J.: Prentice-Hall.)

higher-density polyethylenes (Table 8–4.2) lie between curves (1) and (2) of Fig. 8–4.3(b) because they possess approximately 50% crystallinity.

The behavior of *cross-linked* polymers is represented by curve (3) of Fig. 8–4.3. A vulcanized rubber, for example, is harder than a nonvulcanized one. Curve (3) is raised more and more as a larger fraction of the possible cross links are connected. Note that the effects of cross-linking carry beyond the melting point into the true liquid. In this respect, a network polymer like phenol-formaldehyde (Fig. 7–2.5) may be considered as an extreme example of cross-linking, which gains its thermoset characteristics by the fact that the three-dimensional amorphous structure carries well beyond an imaginable melting temperature.

Once the glass temperatue is exceeded, *elastomeric* molecules can be rotated and unkinked to produce considerable strain. If the stress is removed, the molecules quickly snap back to their kinked conformations (Fig. 7–1.1). This rekinking tendency increases with the greater thermal agitation at higher temperatures. Therefore the behavior curve (4) increases slightly to the right across the rubbery plateau (Fig. 8–4.3b). Of course, the elastomer finally reaches the temperature at which it becomes a true liquid, and then flow proceeds rapidly.

Failure of polymers

The stress-strain curves of polymers are more varied than for metals. A nonductile polymer, like a nonductile metal, may fracture without any permanent deformation. This is the case for the *s-e* curve of noncrystalline PPM* when the temperature is far below its glass temperature (−40°C vs. 100°C). At higher temperatures, still below T_g, we may observe plastic deformation (Fig. 8–4.4). Although the temperature is too low to permit molecular rearrangement by thermal energy, the molecules can slide by one another when external stresses are applied. In fact, strains of more than

* See footnote following Eq. (8–3.6).

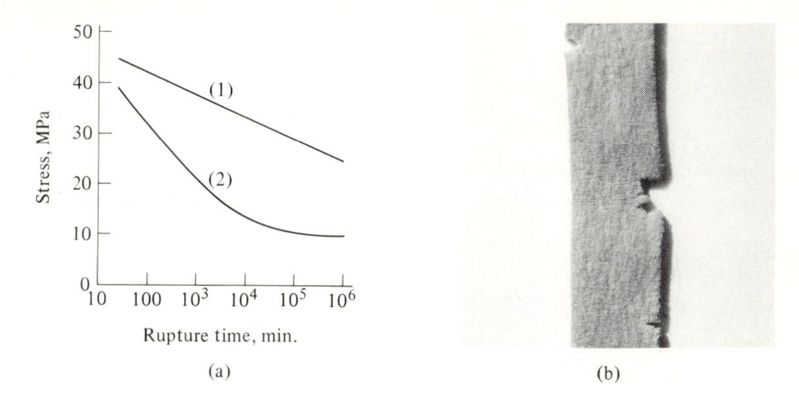

Figure 8-4.5 Accelerated failure. (a) Rupture time (typical). Curve 1: rupture time in a dry environment. Curve 2: rupture time in an organic liquid. (Adapted from Alfrey and Gurnee, *Organic Polymers*, Englewood Cliffs, N.J.: Prentice-Hall.) (b) Oxidation of a stretched rubber band (width, 5mm). Stretched C–C bonds are more readily broken than relaxed bonds. The crack penetrates deeper from the stress concentration at its tip. The same rubber will not fail in the absence of either oxygen or stresses.

100% may be encountered at 60°C. The applied stresses have unkinked the molecules and oriented them into the direction of stressing. The unkinked molecules do not recoil into their amorphous structure at 60°C, since T_g equals 100°C. However, if this deformed plastic is heated in boiling water, the molecules can recoil to their original kinked and coiled conformation.*

The stress-strain curves of Fig. 8-4.4 cannot be projected above T_g, since time becomes a factor and introduces *creep* as a contributor to failure (Section 8-3). Polymers (as well as metals and ceramics) are more sensitive to creep failure when they are exposed to a reactive environment. For example, creep will be accelerated and the time to failure shortened in the presence of a solvent containing small molecules that are able to diffuse among the larger molecules (Fig. 8-4.5a). Also, the stressed bonds of a stretched rubber are more subject to *oxidation* than are the relaxed bonds of unstretched rubber. This commonly leads to the failure of rubber by polymer *scission,* or bond rupture. The reader may recall this in rubber bands that have been over a bundle of papers for months or longer (Fig. 8-4.5b). "Cracks" are commonly visible even though fracture has not yet occurred.

Cyclic loading lowers the maximum design stresses. This is shown in Fig. 8-4.6 and is called fatigue. A static strength of 70 MPa (~10,000 psi) must be decreased to 10 MPa if there are 10^4 reversals of the load, and to still lower values at commonly encountered service conditions of 10^6 to 10^8

---

* Since the reheating above T_g need not be immediate, a time-delayed reshaping is possible. This is sometimes called a *memory effect* because the original geometry of the part can be regenerated.

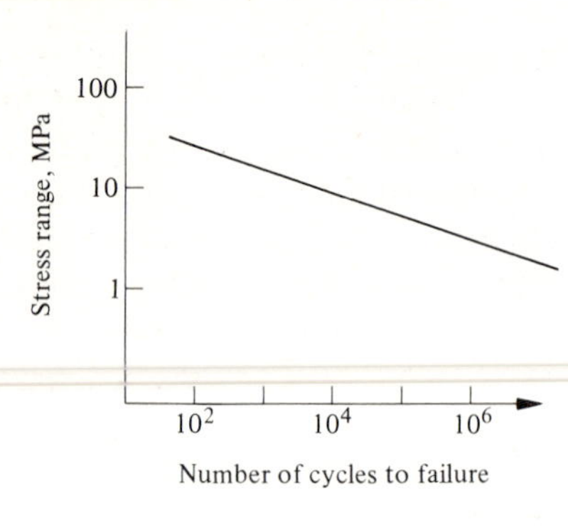

Figure 8-4.6 Accelerated failure (cyclic loading). With cyclic stresses, allowances must be made in design calculations. (Adapted from Alfrey and Gurnee, *Organic Polymers*, Englewood Cliffs, N.J.: Prentice-Hall.)

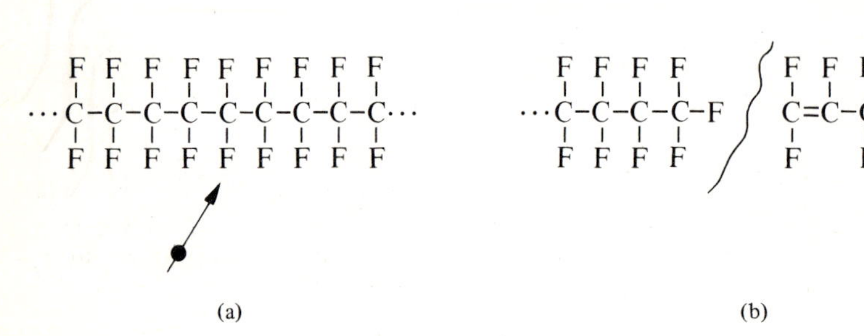

Figure 8-4.7 Scission by neutron radiation (PTFE). Radiation damage occurs when the average size of molecules decreases.

cycles. Each cycle introduces incremental strains that are not reversible. The cummulative effect leads to microscopic flaws and stress concentrations that eventually promote fracture. (See Section 14-3.)

Degradation

Almost all polymers are compounds in which carbon is the main element. Other atoms are bonded along the side of a carbon chain. Most commonly these include hydrogen, oxygen, nitrogen, and chlorine. At high temperatures the bonds that hold these atoms within the polymer may be broken. This can lead to a charring or *carbonization* in which only the carbon remains and the other elements have been volatilized. The most common example of such carbonization is the toasting of bread. Here the cellulose molecule is degraded with the release of H_2O and CO_2 and the retention of the carbonaceous material. It also occurs during the charring of wood in a lumber fire. Of course, the properties change markedly and the wood is no longer useful for structural purposes.

Bonds can also be broken by radiation because a high energy impact may be absorbed at very local points within the material. For example, a photon of ultraviolet light (λ = 300 nm) possesses 6.6×10^{-19} joules of energy, slightly more than the strength of a C–C bond (6.1×10^{-19} J). This can break that bond of a vinyl to form two new molecules of approximately half the original molecular weight. This *scission* is normally not desirable.

Visible light is only about half as energetic as ultraviolet light; however, as it hits a C–C bond, there is still a possibility that the bond will be broken because thermal energy is also present. When the polymer is stressed so that the bond also possesses strain energy, there is even higher probability that visible light can produce scission. This leads to the failure of the stressed rubber cited several paragraphs previously. Neutron radiation, like electromagnetic radiation, can also produce scission. This is illustrated in Fig. 8–4.7 where PTFE (Teflon), normally a very stable molecule, is ruptured.

Review and Study

SUMMARY

Deformation is an important attribute of polymers, so much so that they are commonly called plastics. Plastic deformation is the basis for their final processing. Polymers are also subject to creep in service. Elastic deformation is obvious in many polymers and even serves design purposes in the rubbers.

1. Many polymers are amorphous (noncrystalline) and therefore have a glass-transition temperature. Below T_g, polymers are rigid, brittle glass; above T_g, their molecules can be rearranged in response to stresses or by thermal energy. On heating, the polymer changes from a brittle solid to a leathery solid to a rubbery material and, finally, to a viscous melt.
2. The glass temperature varies with conditions. With unlimited time, T_g is somewhat lower; under impact loading, this transition temperature is higher because there is not time for molecular rearrangements. When a steady stress is applied, T_g is lowered because the external forces facilitate the molecular movements.
3. • Above the T_g of a polymer, shear stresses simultaneously introduce elastic strain, γ_e, and viscous flow, γ_v. The ratio of stress, τ, to total displacement, $\gamma_e + \gamma_v$, is called the viscoelastic modulus. This modulus, which drops precipitously at the glass-transition temperature, is time dependent.
4. • Two simplified approaches to consider viscoelasticity are (a) to evaluate creep under constant stress, and (b) to determine the stress relaxation under constant strain. (We did not consider the more common service condition that involves a combination of the two.)
5. Some polymers are partially crystalline. In general, this improves mechanical properties. Crystallinity is favored when the polymer chain is very regular. Stretching will align the molecules, and therefore increase the crystallinity.
6. In service, polymers can be subject to softening at elevated temperatures, to degradation that produces charring, and to scission or depolymerization in radiation environments—either ultraviolet or neutron. Of course, these potential conditions must be considered for design applications.

TECHNICAL TERMS

Amorphous Noncrystalline and without long-range order.

Carbonization Charring, arising from the loss of side components, leaving only the carbon backbone of the polymer.

Creep Slow permanent deformation. In polymers, this is by viscous flow above the glass transition temperature.

Degradation Reduction of polymers to smaller molecules.

Elastomer Polymer with a large elastic strain. This strain arises from the unkinking of the polymer chain.

Fluidity (f) Coefficient of flowability; reciprocal of viscosity.

Glass *See* Technical terms of Unit 7.

Glass-transition temperature (T_g) *See* Technical Terms of Unit 7.

Orientation (polymers) Strain process by which molecules are elongated into a preferred alignment.

Paraffin hydrocarbons [C_mH_{2m+2}] Molecular chains with only single-bonded carbons.

Polyethylene (HDPE and LDPE) High-density polyethylene is ~60% crystalline. Low-density polyethylene has only minor crystallinity. The properties are significantly different (Table 8–4.2).

• **Relaxation time (λ)** Time required to decay an exponentially dependent value to 37%, that is, $1/e$, of the original value.

• **Rubbery plateau** For a plastic, the range of temperature between the glass temperature and melting temperature, which has a viscoelastic modulus that is relatively constant.

Scission Degradation of polymers by splitting molecules.

• **Stress relaxation** Decay of stress at constant strain by molecular rearrangement.

• **Viscoelastic modulus (M_{ve})** Ratio of shear stress to the sum of elastic deformation, γ_e and viscous flow, γ_v.

Viscoelasticity Combination of viscous flow and elastic behavior.

Viscosity (η) The ratio of shear stress to velocity gradient.

Viscous flow Slow flow in fluids and noncrystalline solids.

CHECKS

8A As with metals, the __***__ and __***__ deformation of polymers refer respectively to the recoverable and permanent deformation.

8B There is a __***__ in the V versus T curve at the __***__-transition temperature of an amorphous polymer.

8C Amorphous polymers have a low elastic modulus because there is __***__ bending as well as __***__ stretching.

8D Since polymer molecules unkink under tension, the elastic modulus __***__ as the strain increases.

8E The elastic modulus of an oriented polymer is ___***___ than the elastic modulus of an amorphous polymer.

8F The elastic modulus of a linear elastomer increases as the temperature is ___***___; the elastic modulus of a metal ___***___ as the temperature is increased.

8G Viscosity, which is the ratio of ___***___ to the flow gradient, is the reciprocal of ___***___.

8H Viscosity increases as the temperature is ___***___, becoming approximately 10^{12} Pa·s at the ___***___ temperature.

• 8I The viscoelastic modulus calculation includes both ___***___ and ___***___ displacement; of these two, the ___***___ displacement is time dependent.

• 8J There is a major drop in the viscoelastic modulus at the ___***___ temperature. This temperature is slightly lower when ___***___ time is available.

• 8K A long-term load above the glass-transition temperature of a polymer produces ___***___; a fixed strain can, with time, lead to ___***___.

• 8L After a period of time equal to the relaxation time ($t = \lambda$), the remaining stress equals ~___***___% of the original stress at t_o.

8M With longer and longer C_mH_{2m+2} molecules, the melting point approaches ___***___ °C.

8N With greater crystallinity, the density of polyethylene is ___***___ because ___***___ · · · . With greater crystallinity, the thermal expansion coefficient is ___***___; based on Fig. 7-4.1, this is because ___***___ · · · .

8O With greater crystallinity, the thermal conductivity of polyethylene is ___***___, and the heat resistance is ___***___.

8P Rubber oxidizes more rapidly when it is ___***___; creep occurs more ___***___ in a solvent that contains small molecules.

8Q In ___***___, the side atoms of a polymer are lost and only the carbon remains.

8R Radiation exposure can lead to ___***___ in polymers in which large molecules are broken into two or more smaller molecules.

STUDY PROBLEMS

8-2.1. Refer to S.P. 5-5.6. Which polymer of Appendix A will permit minimum weight for a given deflection?

8-2.2. Take a thin, "stretchy" rubber band from your desk. Hook it over a door knob, and rig up a scheme so that you may use the rubber band as a "spring balance." By adding ever-increasing incremental loads, plot the load versus stretch and calculate the values of E versus e. Continue incremental additions until fracture. [Therefore, it may not be advisable to use water as your load.]

8-3.1. A linear polymer is raised 10°C above its glass temperature. Estimate its viscosity.

Answer: 1,500 MPa·s

8-3.2. A linear polymer must have a viscosity of 0.4 MPa·s for processing. To what temperature must it be raised?

• 8–3.3. The relaxation time of a rubber is 97 days. (a) How long will it take for the stress to relax to 50% of the original value? (b) To 10%?

Answer: (a) 67 days

• 8–3.4. The stress in a stretched rubber relaxes from 0.20 MPa to 0.15 MPa in 24 hours. How much more time will it take to relax to 0.10 MPa?

• 8–3.5. The relaxation time for creep in plastic thread is 80 hr. How much time is required for the thread to lose half of its stress at a fixed strain?

Answer: 55 hr.

• 8–3.6. Other factors equal, which of the following combinations will show the highest creep rate in a polymer: (a) a long relaxation time and a high stress? (b) a long relaxation time and a low stress? (c) a short relaxation time and a high stress? (d) a short relaxation time and a low stress?

8–4.1. Estimate the melting temperature of polyethylene with a degree of polymerization (a) of 18. (b) Of 1000.

Answer: (a) 75°C (b) 142°C

8–4.2. Cite examples of plastics with the (a) rigid, (b) leathery, (c) rubbery, and (d) viscous characteristics cited in Fig. 8–4.3.

CHECK OUTS

8A elastic
plastic

8B inflection (break)
glass

8C bond
bond

8D increases

8E greater

8F raised
decreases

8G shear stress
fluidity

8H decreased
glass transition

8I elastic, flow
flow

8J glass transition
more

8K creep (viscous flow)
stress relaxation

8L 37, ($1/e$)

8M 143

8N greater, of more efficient packing
decreased, amorphous polymers develop free space above T_g

8O higher
greater

8P stretched
readily (rapidly)

8Q charring (carbonization)

8R scission

UNIT NINE

• MAKING AND SHAPING OF POLYMERS

Although the synthesis of polymers involves organic chemistry, many of the production processes can be understood on the basis of general chemistry and our knowledge of polymer deformation (Unit 8). The engineer is encouraged to examine these because the options and limitations of the use of polymers in design can be better understood and specified. The two main polymerization reactions are addition and condensation. Additives are commonly used. Shaping generally involves three processing steps—softening, molding, and hardening. Processing steps are taken to introduce and retain molecular and orientation during the molding and hardening steps.

Contents

Prerequisites Units 7 and 8.

From Unit 9, the engineer should

1) Know the bonding changes that occur in the two principal polymerization reactions, addition and condensation.
2) Understand the effects of the common polymer additives on polymer properties.
3) Have a descriptive knowledge of the half-dozen principal forming processes, and the factors that favor one over another.

4) Be familiar with methods of introducing molecular orientation into fiber and sheet products.
5) Understand the meaning of the pertinent technical terms.

9-1 RAW MATERIALS

The earliest types of polymeric materials to be used—the naturally occurring materials such as wood, leather, cotton, wool, etc.—still have widespread applications. These materials are used more or less directly except for modifying their shapes. They too contain large molecules, molecules that in principle are similar to those we encountered in the last unit.*

There are several advantages in using materials that have naturally occurring molecules. For one thing, they are already formed and do not have to be manufactured, except for having geometric changes performed on them. They are usually relatively cheap. And they typically have high strength-to-weight ratios. However, these naturally occurring materials also have certain disadvantages: (1) many absorb water, (2) they are combustible, (3) their strength varies with the direction of their structures, (i.e., they have strength anisotropies).

Not surprisingly, therefore, attempts have been made to improve on nature. Plywood, for example, has less anisotropy than wood cut directly from the tree because the "grain" is laid in two directions (Unit 10). It is also possible for plywood to be produced as large sheets, an impossibility with naturally occurring wood. Wood may also be impregnated with other materials to reduce its permeability to moisture (e.g., railroad ties are treated with creosote), and resins may be used to bond fabrics into products that are impervious to water.

Another advance in the state of technology came about years ago when means became available to extract cellulose from cotton and wood, to modify it in various ways, and to regenerate it as a polymeric raw material. Rayon (as a fiber) and celluloid (as a plastic) originated with the invention of a method to extract cellulose. The basic structure of cellulose, cellulose acetate, and cellulose nitrate are shown in Fig. 9-1.1, where the radicals **R** are $-OH$, $-OCOCH_3$, and $-ONO_2$, respectively. The three materials have a very similar structure, except for the added radical. This difference in radical, however, greatly affects their properties. The products made from these materials range from lacquers, to plastics, to explosives (nitrocellulose), depending on what fraction of the $-OH$ units of cellulose are replaced by $-ONO_2$. If $-OCOCH_3$ (that is, acetate radicals) are used as replacements

---

* Although the molecules are similar to those of the man-made materials (for example, Table 7-1.3), they are more complex; in Unit 7 we intentionally limited our discussions to the simpler macromolecules.

```
     R    R
     |    |
     C————C
  H/ H    H\
 -C          C-O-
   \H      /H
    C————O
    |
    HC-R
    H
```
n

Figure 9–1.1 Cellulose, a naturally occurring polymer. In nature the radical, **R**, is -OH. Chemists have learned how to modify cellulose by replacing the -OH's with $-OCOCH_3$ to give cellulose acetate, or with $-ONO_2$ to give cellulose nitrate. The properties depend on the modification. Cellulose acetate is the basis of rayon; cellulose nitrate is guncotton (if all of the -OH's are replaced by $-ONO_2$). Other cellulose derivatives—that is, modifications—are possible.

rather than $-ONO_2$ radicals, the product is nonflammable and is widely used as "safety" film in photography; millions of pounds of it are used to make fibers for the manufacturer of textiles, as well.

Raw materials for many of the newer-type plastics are extracted from coal and/or petroleum products. As an example, coke, the carbonaceous part of coal, plus the methane that is the chief component of natural gas together yield acetylene, C_2H_2:

$$3C + CH_4 \rightarrow 2(H-C\equiv C-H). \qquad (9\text{–}1.1)$$

Acetylene, in turn, may be converted to various vinyl compounds, as follows:

Ethylene	$C_2H_2 + H_2 \rightarrow C_2H_4$	(9–1.2)
Vinyl chloride	$C_2H_2 + HCl \rightarrow C_2H_3Cl$	(9–1.3)
Styrene	$C_2H_2 + C_6H_6 \rightarrow C_2H_3(C_6H_5)$	(9–1.4)
Vinylidene chloride	$C_2H_2 + Cl_2 \rightarrow C_2H_2Cl_2$	(9–1.5)

Other sources of raw material require chemical reactions that are somewhat more complex. We shall not consider them here, except to note that a tremendous industry has developed for the production of raw materials from which plastics can be made. The expansion of this production through the 1970s is plotted in Fig. 9–1.2.

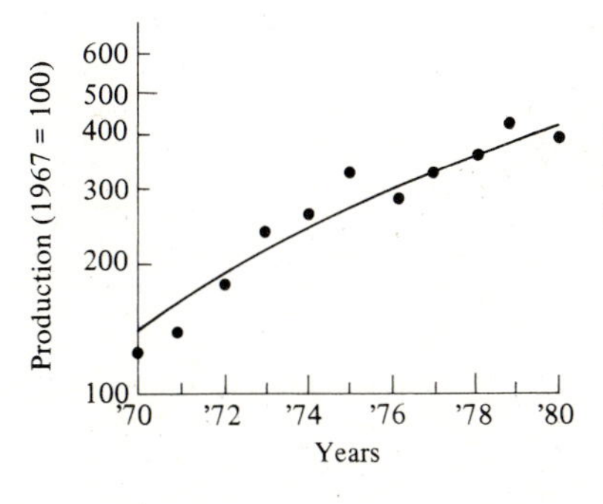

Figure 9–1.2 Production of raw materials for plastics in the 1970s. The trend continues upward as new and improved plastics are developed that require these raw materials.

Illustrative problem 9-1.1 One pound of cellulose is two-thirds converted to cellulose acetate, that is, two of the three radicals of Fig. 9-1.1 are changed from the $-OH$ of cellulose to $-OCOCH_3$. What is the percent gain in weight?

SOLUTION

$$\underset{\text{cellulose}}{C_6H_7(OH)_3O_2} + \underset{\text{acetic acid}}{2HOCOCH_3} \rightarrow \underset{\text{cellulose diacetate}}{C_6H_7(OH)(OCOCH_3)_2O_2} + \underset{\text{water}}{2H_2O}$$

$$[(6)(12) + (7)(1) + (3)(17) + (2)(16)] + \cdots$$
$$\rightarrow [(6)(12) + (7)(1) + 17 + (2)(59) + (2)(16) + \cdots$$
$$162 + \cdots \rightarrow 246 + \cdots$$
$$\frac{246}{162} = 152\%, \text{ or } 52\% \text{ gain in weight.}$$

ADDED INFORMATION Rayon normally requires about a two-thirds replacement of hydroxyls by acetate radicals in cellulose, as indicated above. A triacetate, in which all the **R**'s of Fig. 9-1.1 are acetate, $C_6H_7(OCOCH_3)_3O_2$, was harder to develop because the triacetate was not deformable enough to make fiber production feasible. Fortunately these difficulties have been overcome. Now fibers of triacetate can be made, and their resistance to deformation can be used advantageously because they retain their shape in "permanent-press" textiles and other crease-resistant fabrics. ◄

9-2 BIG MOLECULES OUT OF SMALL MOLECULES

The real breakthrough in the development of plastic materials came when the chemist and the engineer learned how to make large molecules out of small ones in a commercially feasible way. This achievement opened the way to vast new developments. People looking for raw materials were no longer limited to natural polymers such as cellulose or isoprene rubber. Today a wide range of polymers is made by *synthesis* of macromolecules from micromolecules. There are two chief procedures for growing large molecules: (1) *chain-reaction polymerization,* sometimes called *addition* polymerization, and (2) *step-reaction polymerization,* sometimes called *condensation* polymerization.

Chain-reaction (addition) polymerization

The reaction of Eq. (7-2.4) for ethylene that forms polyethylene may be rewritten as a general equation for all vinyl polymers:

$$n\begin{pmatrix} H & H \\ C{=} & C \\ H & \mathbf{R} \end{pmatrix} \longrightarrow \left(\begin{matrix} H & H \\ -C- & C- \\ H & \mathbf{R} \end{matrix}\right)_n \qquad (9\text{-}2.1)$$

As in Table 7–1.3, **R** stands for any of a variety of radical groups. The characteristic of this reaction is that each small original molecule, called a *monomer* (a single unit), must be at least *bifunctional*. That is, it must have reactive sites that may be connected with at least two adjacent molecules, much as a railroad freight car needs two coupling links to become part of a train.

A chain reaction actually involves three steps: (1) initiation, (2) propagation, and (3) termination. We shall focus our attention on the second step, *propagation*. Assume that the reaction has been initiated and that we have a growing vinyl chain already containing a large number of mers, plus an additional vinyl monomer:

$$\begin{array}{c}
\qquad\qquad\qquad\downarrow \\
\begin{array}{ccccccccccc}
 & \mathrm{H} & & \mathrm{H} & & \mathrm{H} & & \mathrm{H} & & & \mathrm{H}\ \ \mathrm{H} \\
\cdots - & \mathrm{C} & - & \mathrm{C} & - & \mathrm{C} & - & \mathrm{C}\bullet & + & & \mathrm{C{=}C} \\
 & \mathrm{H} & & \mathbf{R} & & \mathrm{H} & & \mathbf{R} & & & \mathrm{H}\ \ \mathbf{R}
\end{array}
\end{array} \qquad (9\text{–}2.2)$$

Note that the carbon below the arrow has only three bonds rather than the more stable four (Table 7–1.1). It therefore possesses a *reactive site* •, and will combine readily with the neighboring vinyl monomer:

$$\begin{array}{c}
\qquad\qquad\downarrow \\
\begin{array}{ccccccccccccc}
 & \mathrm{H} & & \mathrm{H} & & \mathrm{H} & & \mathrm{H} & & \mathrm{H} & & \mathrm{H} \\
\cdots - & \mathrm{C} & - & \mathrm{C} & - & \mathrm{C} & - & \mathrm{C} & - & \mathrm{C} & - & \mathrm{C}\bullet \\
 & \mathrm{H} & & \mathbf{R} & & \mathrm{H} & & \mathbf{R} & & \mathrm{H} & & \mathbf{R}
\end{array}
\end{array} \qquad (9\text{–}2.3)$$

Although the reacting carbon gains the required fourth bond, such a reaction obviously does not eliminate the reaction tendencies at the end of the chain, because the new end carbon now has only three bonds and it is now reactive. Sequentially another—and yet another—reactive site is produced, with propagation proceeding as long as the small C_2H_4 molecules are available.

The process of *termination* occurs when the reactive ends of two molecules join:

$$\begin{array}{ccccccccc}
 & \mathbf{R} & & \mathrm{H} & & \mathbf{R} & & \mathrm{H} \\
\cdots - & \mathrm{C} & - & \mathrm{C} & - & \mathrm{C} & - & \mathrm{C}\bullet \\
 & \mathrm{H} & & \mathrm{H} & & \mathrm{H} & & \mathrm{H}
\end{array}
\qquad
\begin{array}{cccccccccccc}
\mathrm{H} & & \mathrm{H} & & \mathrm{H} & & \mathrm{H} & & \mathrm{H} & & \mathrm{H} & \\
\bullet\mathrm{C} & - & \mathrm{C} & - & \mathrm{C} & - & \mathrm{C} & - & \mathrm{C} & - & \mathrm{C} & - \cdots \\
\mathrm{H} & & \mathbf{R} & & \mathrm{H} & & \mathbf{R} & & \mathrm{H} & & \mathbf{R} &
\end{array} \qquad (9\text{–}2.4)$$

$$\begin{array}{cccccccccccccccccccccc}
 & \mathbf{R} & & \mathrm{H} & & \mathbf{R} & & \mathrm{H} & & \mathrm{H} & & \mathrm{H} & & \mathrm{H} & & \mathrm{H} & & \mathrm{H} & & \mathrm{H} & \\
\cdots - & \mathrm{C} & - & \mathrm{C} & - & \mathrm{C} & - & \mathrm{C} & - & \mathrm{C} & - & \mathrm{C} & - & \mathrm{C} & - & \mathrm{C} & - & \mathrm{C} & - & \mathrm{C} & - \cdots \\
 & \mathrm{H} & & \mathrm{H} & & \mathrm{H} & & \mathrm{H} & & \mathrm{H} & & \mathbf{R} & & \mathrm{H} & & \mathbf{R} & & \mathrm{H} & & \mathbf{R} &
\end{array} \qquad (9\text{–}2.5)$$

This termination can occur whenever the active ends of growing molecules of any size happen to encounter one another. Thus few, if any, of the final molecules are the same size. This emphasizes our comments of I.P. 7–1.2 about average molecular weights. Since we usually want materials that have high molecular weights, the polymer technologist strives to develop ways of postponing reactions (9–2.4) to (9–2.5) until the molecules have grown to large sizes. However, the ways of achieving that are beyond the scope of this text.

Step-reaction (condensation) polymerization

The production of dacron (called terylene, trevira, and tergol in other countries) requires a somewhat different reaction. In a very simplified form, this reaction is

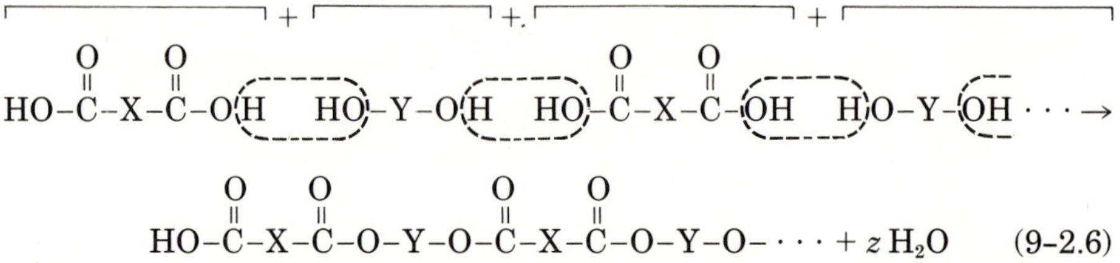

$$\mathrm{HO{-}\overset{O}{\overset{\|}{C}}{-}X{-}\overset{O}{\overset{\|}{C}}{-}OH} + \mathrm{HO{-}Y{-}OH} + \mathrm{HO{-}\overset{O}{\overset{\|}{C}}{-}X{-}\overset{O}{\overset{\|}{C}}{-}OH} + \mathrm{HO{-}Y{-}OH} \cdots \rightarrow$$

$$\mathrm{HO{-}\overset{O}{\overset{\|}{C}}{-}X{-}\overset{O}{\overset{\|}{C}}{-}O{-}Y{-}O{-}\overset{O}{\overset{\|}{C}}{-}X{-}\overset{O}{\overset{\|}{C}}{-}O{-}Y{-}O{-}} \cdots + z\,\mathrm{H_2O} \qquad (9\text{–}2.6)$$

The X and Y may be various groups of atoms, such as $\left(CH_2\right)_n$, in the centers of the initial molecules. Those atoms do not enter directly into the reaction. In each joining step, however, a small *by-product* molecule such as H_2O is released; hence the adjective "condensation." Those large molecules that have

$$\mathrm{-\overset{O}{\overset{\|}{C}}{-}O{-}}$$

units along their chains are called *polyesters,* a term often used to identify this class of plastics.

We need not remember the chemistry of Eq. (9–2.6), but let us compare some features of this reaction with the addition or chain reactions of Eq. (9–2.2). The earlier chain reaction required only one type of monomer and produced no by-product. The step-reaction polymerization typically involves two types of initial micromolecules (or a molecule with dissimilar ends), and produces a by-product (in this case H_2O). Both reactions may involve bifunctional molecules and therefore produce linear polymers. In addition, both reactions may connect polyfunctional molecules into network polymers.

Illustrative problem 9-2.1 Melamine, $C_3N_6H_6$, and formaldehyde, CH_2O, have the following structures:

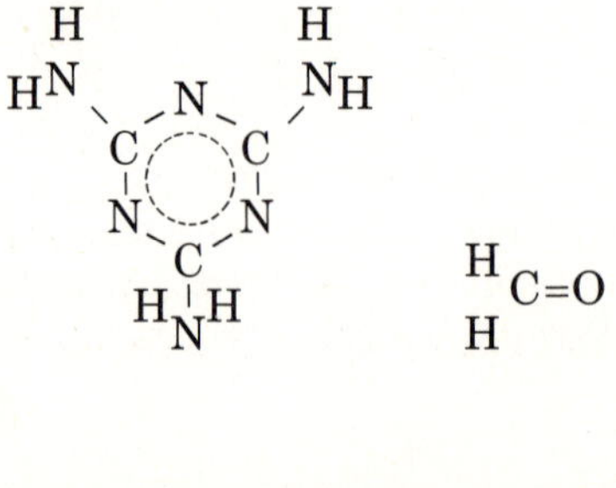

[Melamine structure: triazine ring of alternating C and N atoms, each C bearing an NH₂ group (H–N–H)]

$$\mathrm{\overset{H}{\underset{H}{}}C{=}O}$$

Each CH_2O can react with $-NH_2$ radicals, that is,

$$-N\begin{matrix}H\\ \\ H\end{matrix}$$

joining two melamine molecules and producing a by-product molecule of water (H_2O) as follows:

$$-N\begin{matrix}\mathbf{H}\\ H\end{matrix} + \begin{matrix}H\\ H\end{matrix}C{=}\mathbf{O} + \begin{matrix}H\\ \mathbf{H}\end{matrix}N- \longrightarrow \begin{matrix} & H & H & \\ -N- & & -C- & -N- \\ & H & H & \end{matrix} + \mathbf{H_2O} \qquad (9\text{-}2.7)$$

Show how this condensation (step-reaction) polymerization can produce a framework structure.

SOLUTION

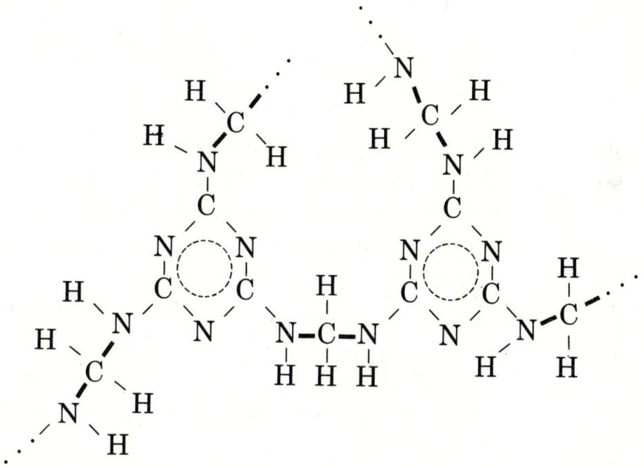

```
                                    .
                                     .
                   .                  \
                 .                     N
        H      .                   H /   \   H
          \   /                          C /
      H     C                      H /    \   H
        \ /   \                            N /
         N      H                          |
         |                                 C
         C                               /   \
       /   \                            N     N           H
      N     N                           |     |           |
      |     |         H                 C     C           C  . .
  H   C     C         |               /   \  /   \      / |
    \ /  \ /  \     N-C-N           /     N     N-C       H
     N    N     \  |  |  |        /          /   
  H /             H  H  H                   H
    C
   /  \
  N     H
 / \
.    H                                               +5H2O
```

ADDED INFORMATION This melamine-formaldehyde polymer, which goes by various trade names, remains rigid at dishwashing temperatures and therefore is widely used in home and restaurant dishes; probably the most familiar trade name is Melmac. It is moderately expensive.

The second H of each $-NH_2$ group is much less readily removed for another $-CH_2-$ bridge because the space available for the new connecting units is limited. As a result, each melamine unit is generally restricted to receiving three connecting $-CH_2-$ bridges to adjacent molecules. ◀

9-3 ADDITIVES

The major component of the plastics made and/or used by engineers are polymeric molecules. Almost invariably, however, these products have additional components that are added to attain specific desired properties.

The materials added to plastic products may be added for purposes of reinforcing and strengthening the product; or they may be added to introduce more flexibility. Thus polyvinyl chloride (PVC) is often used for floor

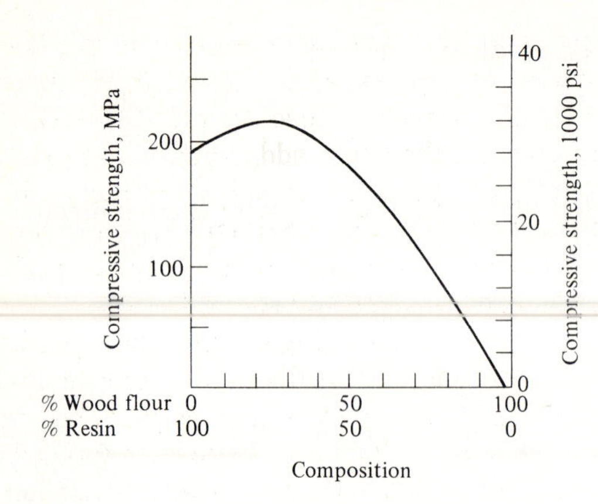

Figure 9-3.1 Addition of a filler to a plastic (wood flour added to phenol-formaldehyde). The mixture of the two is stronger than either alone.

tile; for this usage, an abrasion-resistant *filler* is added. The same polymer is used for raincoats, but with a *plasticizer* added to attain flexibility and to give it characteristics that are markedly different from those of floor tiles. Additives may also be used as *stabilizers,* to keep the polymer from deteriorating, or as *flame retardants.* Finally, additives may be included as *colorants,* for aesthetic purposes. Sometimes certain materials are added to reduce the cost of the product, if at the same time other properties can be improved. For example, by adding a filler, a plastic becomes more rigid and costs less per unit volume than it would otherwise. Additives frequently serve multiple purposes. For example, carbon black strengthens rubber and also absorbs ultraviolet light; this means that the rubber offers more resistance to deterioration during service.

Fillers

We shall discuss those additives that are used in larger proportions than others. Most of them are added to give strength or toughness to plastics. Thus wood flour (a very fine sawdust) is commonly added to a PF plastic (phenol-formaldehyde, Fig. 7-2.5) to increase its strength (Fig. 9-3.1). More important is the fact that 35 v/o wood flour more than doubles the toughness of a PF plastic. As further fringe benefits, wood flour is a replaceable resource that costs less than half what an equal volume of PF would cost. Thus the service properties of the product are improved at the same time the cost of it is being reduced. This is real engineering! Note from Fig. 9-3.1 that the improvement in strength is not a result of the adding of the wood flour alone, because 100 v/o wood flour has nil strength. Rather it is a result of the mutual interaction of the two components, just as a steel made of ferrite plus carbide is stronger than either one of them alone (Fig. 9-3.2).

Fillers may be of various types. We have already cited wood flour, which in reality is a polymer itself, one containing cellulose (Section 9-1). Wood flour has the advantage of being essentially the same density as PF; therefore the product retains a low specific gravity. Silica flour (finely ground SiO_2 made from quartz sand or quartzite rock) is also used (I.P. 9–3.1). It is appreciably harder than wood flour and therefore adds abrasion resistance to the plastic product. Furthermore, it neither burns nor softens at high temperatures; as a result, it adds thermal stability to the product. Of course, it does increase the product's total density because the specific gravity of the SiO_2 is about twice that of common resins. (See Appendix A.)

Fibrous fillers are especially effective for adding strength to a product. These fillers may be glass fibers, organic textile fibers, or mineral fibers such as asbestos. They are often chopped into short lengths so that they may be mixed with the polymeric materials that are subsequently molded as fiber-reinforced plastics (FRP). In Unit 14 we shall give specific attention to composites in which a high volume fraction of continuous fibers reinforce polymers, often with considerable enhancement of properties.

Plasticizers

At normal temperatures, the small molecules of Section 7–1 and Fig. 7–1.2 are generally liquids or gases. In contrast, the macromolecules that we have been describing here are solids because they involve long chains or networks. When small molecules are intimately mixed with macromolecules, they reduce the rigidity of the macromolecular—or polymeric—product. When small molecules surround large ones, the large molecules move more readily, in response to either thermal agitation or external forces. In brief, the small molecules *plasticize* the larger ones. Plastics that are normally stiff, such as polyvinyl chloride, can be made flexible, an obvious requirement if the product is to be used in film or sheet form. A plasticizer, in effect, lowers the glass transition temperature T_g, of which we spoke in Section 7–4 and Fig. 7–4.1, so that molecular movements and rearrangements can occur at room temperature in polymers that would otherwise be rigid.

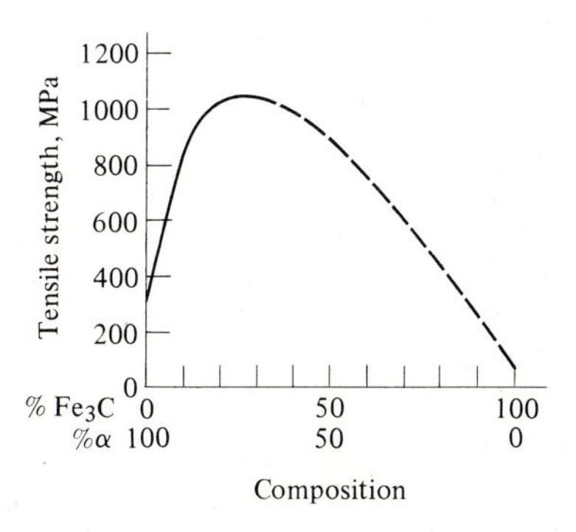

Figure 9–3.2 Addition of carbide to ferrite. The more rigid carbide prevents slip in the soft ferrite matrix; however, since carbide is brittle, 100% carbide is weak.

A plasticizer must have certain characteristics. It should have a high boiling temperature (low vapor pressure) so that it will not readily evaporate. You can understand that a plasticized raincoat would become useless if the plastic were to become stiff and brittle with time; this, of course, could happen if the plasticizer gradually evaporated, leaving a plastic that would be below its glass temperature.

The plasticizer must not be soluble in the liquids with which it comes in contact. Although some plasticizers are nonvolatile, solvents such as petroleum products can dissolve a large variety of micromolecules. Thus the inner surface layer of a plastic container may be embrittled as the plasticizer is depleted by solvents that are stored in the container.

A plasticizer must be "compatible" with the polymer. Although we leave the details to the polymer scientist, let us note that, in the case of some small molecules, the molecule may have greater attraction to another of its kind than to the surface of large molecules. Likewise, strong attraction between two adjacent large molecules may prevent the entry of the small molecules between them for plasticizing purposes. For a plasticizer to be suitable, the small molecules should be attracted to the surfaces of the large molecules; the small molecules should not segregate within the plastic. The choice of the most suitable plasticizer depends on the details of the characteristics of the molecular structure and usually involves extended testing on the part of the manufacturer.

Stabilizers

Polymers deteriorate in two general ways: by degradation and by oxidation (Section 8–4). *Degradation* occurs when the large molecules break up into smaller molecules, thus causing the plastic product to lose strength. Light radiation, particularly by ultraviolet light, can break the molecules into fragments because the light ray, or photon, may supply a powerful "kick" of energy locally onto the bond along the backbone of the molecule. Carbon black is commonly used as a stabilizer because it absorbs the light radiation before the radiation has a chance to degrade the molecule of polymer or rubber. A typical recipe for rubber for a tire to be used on a passenger car includes about 50 lb of carbon black (soot from an oxygen-deficient flame) for every 100 lb of isoprene, for the double purpose of filler and stabilizer.

Deterioration by *oxidation* results when oxygen reacts with the polymeric molecule. In the case of rubber, a change is first noticed when the rubber becomes harder and more brittle. Just as with vulcanization by sulfur (Fig. 7–2.4), oxygen cross-links the rubber (Fig. 9–3.3). As we would expect, therefore, the molecules are no longer independent and their movements are restricted. For cross-linking to occur readily by oxidation, the polymer molecule must contain double bonds. Thus we encounter this type of oxidation much more commonly in rubbers than in other plastics. Ozone

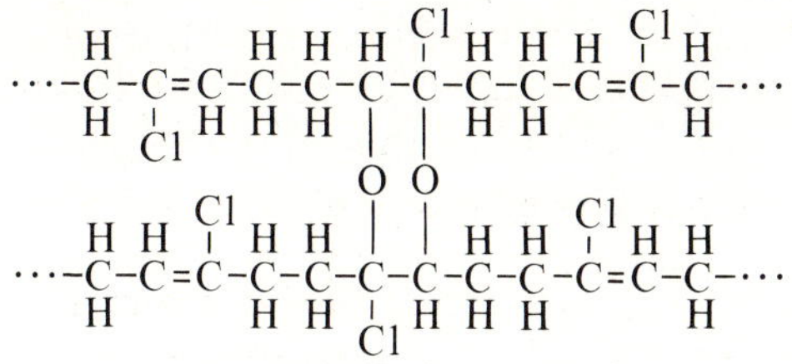

```
                    Cl         Cl
   H      H H H  |  H H H  |  H
····-C-C=C-C-C-C-C-C-C-C=C-C-····
   H  |  H H H  |  |  H H     H
      Cl        O  O
         Cl                Cl
   H H  |  H H  |  |  H H  |  H H
····-C-C=C-C-C-C-C-C-C-C=C-C-····
   H       H H  |  H H H      H
                Cl
```

Figure 9-3.3 Vulcanization of rubber in the presence of oxygen. Like sulfur (Fig. 7-2.4), oxygen from the air can cross-link rubber molecules (chloroprene). This hardens the rubber. Still further oxidation would degrade it to micromolecules.

(which is O_3 rather than O_2) provides a very reactive source of oxygen because it breaks down to O_2 and a single reactive oxygen:

$$O_3 \rightarrow O_2 + O\,\bullet. \tag{9-3.1}$$

Thus rubbers and plastics are usually tested in atmospheres that have controlled additions of ozone as an accelerated test to check their resistance to oxidation. Ultraviolet light also speeds the oxidation of rubber (Fig. 9-3.3) by supplying energy locally to start that reaction.

Some plastics are flammable; however, in general those plastics that contain significant amounts of chlorine and/or fluorine have reduced flammability. Polyvinyl chloride (PVC), polytetrafluoroethylene (Teflon), and chloroprene rubber (the C_4H_5Cl of Table 7-5.1) fall in this category. Not only do the chlorine and fluorine not support combustion, but their presence interferes with the access of air to a hot plastic, so that oxidation is retarded or completely precluded. (However, excessive heating may still soften or char these plastics.) *Flame retardants* are added to other plastics that are to be used where flammability must be avoided. The polymer chemist usually includes antimony trioxide (Sb_2O_3) or compounds containing either chlorine or bromine as flame retardants.

Colorants

The purposes of adding colorants are self-evident. The specifications with respect to colorants are stringent because consumers have specific color preferences; also even slight alterations in color are readily apparent to the human eye. The *dyes* and *pigments** used for colorants must be evenly distributed throughout the plastic. For example, if the dye of a textile fiber is only superficial, slight wear or abrasion produces a color change. A colorant must also be extremely stable against deterioration by oxidation or ultraviolet light, because changes in color usually precede other evidences of deterioration. The technical field of colors and dyes is highly specialized and is beyond the scope of this text.

---

* *Dyes* are molecules that *dissolve* into the structure of the plastic. *Pigments* are *fillers* that retain their identity as a separate phase.

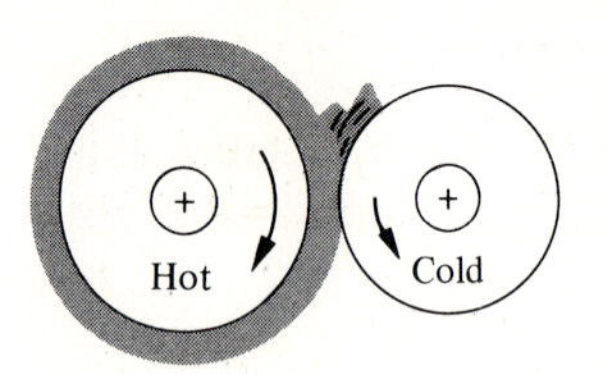

Figure 9-3.4 Open-roll mill (rubber mill). The components of the plastic or rubber are blended by a shearing action in the nip region between the rolls. The two rolls rotate at slightly different speeds. They are also maintained at different temperatures.

Mixing

Uniformity is a necessity for quality materials. Even materials that are composed of multiple phases must have a "homogeneity in their heterogeneity." For example, in a steel it is necessary to have the hard carbides uniformly distributed throughout the tough ferrite matrix (Fig. 6-3.3b). Likewise, it is necessary to have an equal distribution of additives to a polymer. Slight gradients in colorants lead to obvious variations in products. Also, excess plasticizers or fillers at one point and deficiencies a millimeter away lead to suboptimal properties at both locations.

Mixing of additives into a plastic product is far from simple. Commonly, we are mixing unlikes. A silica flour filler has twice the density of many of the polymers with which they are combined. Some additives are in the form of liquids, others are solids, while polymers are either viscous melts or semisolids during the mixing. This latter characteristic is critical. Simple stirring does not suffice.

Most polymer mixing, or *compounding,* is performed as a batch process on an *open-roll mill* (also called a rubber mill). This is illustrated in Fig. 9-3.4. It has two rolls that can be internally heated or cooled. The *nip* of the rolls can be adjusted to different size gaps as required. They operate at different temperatures and speeds (3–4 sec/revolution). The velocity and viscosity gradients that result introduce a kneading action that proves to be quite effective. Also, the temperature difference causes the sheet compound to invest one of the rolls, that is, the plastic mixture blankets one of the rolls and not the other. Typically, rubbers will coat the hot roll. Several hundred passes through the nip achieve a uniform mixing if the operator cuts the blanket and peels it from the coated roll several times, each time feeding it back into the nip to form a new blanket.

Polymers that are subject to oxidation at the mixing temperature must be blended in an internal mixer that is enclosed and excludes air. Rotors and blades perform the kneading action.

Illustrative problem 9-3.1 Forty-five kilograms of silica flour (finely powdered SiO_2) are mixed with each 100 kilograms of melamine formaldehyde (MF). The specific gravities of each are 2.65 and 1.5, respectively. What fraction of the volume is filler?

SOLUTION Basis: 100 g MF and 45 g SiO_2.

$$100\text{ g MF}/(1.5\text{ g/cm}^3) = 67\text{ cm}^3\text{ MF} = 0.8\text{ MF}$$
$$45\text{ g SiO}_2/(2.65\text{ g/cm}^3) = \underline{17}\text{ cm}^3\text{ SiO}_2 = 0.2\text{ SiO}_2$$
$$\text{Total} = 84\text{ cm}^3$$

ADDED INFORMATION A filler is used not only because it is an inexpensive diluent for a moderately expensive polymer, but also because it gives added thermal and dimensional stability to the plastic. ◂

9-4 FORMING PROCESSES

Introduction

There are three general processing steps between the polymer mixture and the final product: (1) softening, (2) molding, and (3) hardening. The *softening* is commonly achieved by heating but may involve plasticization by solvents. The *molding* commonly involves the application of pressure, although there are certain exceptions. The *hardening* may occur by cooling, by a chemical reaction, or by the volatilization of a fugitive plasticizer. Typically, these three steps are sequenced within one process. Thus, as we examine a number of the forming processes, it will be convenient to identify these three steps.

A significant majority of polymers may have their thermal behavior categorized as (1) thermoplastic, or (2) thermosetting. The *thermoplasts* are *linear polymers* (Eqs. 9-2.3 and 9-2.6) with only limited, if any, cross-linking or branching; therefore, they soften at elevated temperatures. As the temperature rises, the molecules can respond to pressure by sliding by one another. Most of the vinyl compounds of Table 7-1.3 are linear and fall into this category. Of course, for thermoplasticity to be effective, the temperatures must be above the glass-transition temperature, T_g (Fig. 7-4.1).* Further, the normal temperatures of use must be in the range in which the plastic retains its shape. The *thermosets* are altered both chemically and structurally during thermal processing. They develop a 3-dimensional structure, either a network structure (Fig. 7-2.5) or a cross-linked structure (Fig. 7-2.4). These structures are only partially completed before forming but become "one big 3-D molecule" when they are processed in the presence of heat and pressure. The *curing* within the heated mold completes the formation of the network. Therefore, a thermosetting polymer gains rigidity before the pressure and added temperature are removed. Thermoplastic polymers must be cooled in the mold (or upon exit).

Finally, before describing the various forming processes, we should examine polymer viscosity more closely than we did in Section 8-3. When a

* T_g decreases somewhat when shear stresses are applied, since rearrangements of molecules do not depend solely on thermal energy.

polymer is under pressure, the relationship between viscosity, η, and the rate of strain, γ/t, is less ideal than that shown in Eq. (8–3.4b).* Several factors contribute to this variance; among them is the "free space" that is present in the supercooled liquid (Section 7–4). This leads to a high compressibility that becomes evident as die swell (a volume expansion at the exit of the die) when the viscous polymer is extruded out of the die. Of course, this is undesirable when specific dimensions are required. It also leads to difficulties in predicting the rate of flow through feed channels (*sprues*) of closed dies. This can mean there will be incomplete filling of the dies during injection molding unless allowances are made.

Extrusion

The extrusion process is sketched in Fig. 9–4.1(a). Starting materials are commonly granules of a thermoplastic polymer that had been sized for easy flow. An *auger* (or screw) feeds the granules into a heated zone where the thermoplastic pellets soften. However, the heating does not come entirely from external heaters. There is considerable energy that goes into the polymer melt from the work of the auger as it compresses the melt and forces it through the die. For example, the motor turning the auger uses as much as 50 kw (~70 Hp) to drive a 10-cm screw. This power is absorbed by the plastic in the form of heat.

In order to control the temperature as closely as possible, the heater is segmented. The temperature of each zone is measured by thermocouples and controlled separately so that the desired temperature profile is obtained along the barrel. This is important because various thermoplastics have different softening characteristics. For example, low-density polyethylene, which is amorphous, softens progressively as the temperature increases. In contrast, nylon softens rather abruptly at a higher temperature because it possesses high crystallinity. This leads to the need for modifications of the screw design of the auger depending on the product. As shown in Fig. 9–4.1, the feed into the heated zone and the compression to eliminate the pores is simultaneous and continuous for polyethylene (Fig. 9–4.1b). The compression is abrupt in Fig. 9–4.1(c) for nylon after the granules have passed well into the heated zones.

The die of an extruder may have a variety of geometries. Of course, the cross section along the length of the product must remain constant. A circular orifice produces a solid rod, and a slit produces sheet or film products. It is also possible to manufacture longitudinal moldings of irregular cross sections. Tubing and pipe must be extruded by mounting a *mandrel* (or "torpedo") in the center of the hole so that extrusion is through the annulus. Alternatively, wire can be continuously fed into the center of the orifice to

* That is one reason we used an empirical η versus T relationship in Eq. (8–3.2) rather than the more orthodox viscosity Eq. (8–3.3).

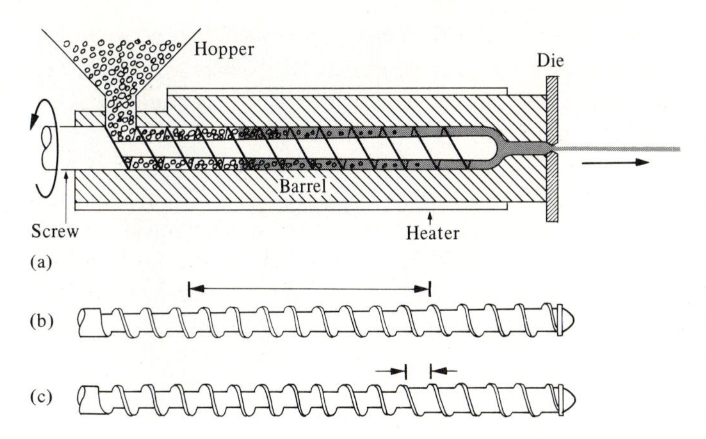

Figure 9-4.1 Extrusion (auger). (a) Schematic. Solid pellets are fed into the heated zone by an auger. The melted thermoplastic is extruded through an open-end die as bars, pipe, or sheet product. (b) Auger for polyethylene, which softens gradually. (c) Auger for nylon, which softens abruptly. The compression and metering section (arrows) is much shorter in the latter.

be coated with rubber insulation. Rates may approach a km/min. The extrusion process is very adaptable to large-scale production.

The plastic must flow as it emerges through the die but must harden immediately upon exit so as to retain its desired shape. The surface-to-volume ratio and the amount of mass to be cooled per second dictate the choice of air or water for cooling.

Injection molding

The open die of the extrusion process may be replaced with a closed die as shown schematically in Fig. 9–4.2(a). The die is a split mold that contains the negative contours of the product to be made. The hot plastic is forced (injected) into the mold by either an auger or a hydraulic plunger. The injection process provides more flexibility in product geometry than does extrusion, because the cross section is not fixed longitudinally. Examples of injection-molded products are numerous and range from plastic ice cream spoons to telephone receivers (Fig. 9–4.2).

The injection molding of *thermoplastic* polymers generally utilizes water-cooled molds for hardening the product. This facilitates production because the product becomes rigid almost immediately and may be removed so there can be a sequel injection. However, this presents complications since the viscous melt must enter through chilled sprues (feed channels) and mold cavities. This is illustrated in Fig. 9–4.3 where we see a rigid shell of chilled polymer that constricts the channel. Higher injection pressures are required in order to feed the far cavities and to avoid porosity from solidification shrinkage.

Thermosetting polymers may be injection molded more readily than they can be extruded, because the injection cycle may include time for *curing,* that is, for the completion of the polymerization reaction. It is not necessary to cool the mold inasmuch as the product "sets" rigidly while hot. Close control is required in the heating and feeding part of the cycle. The

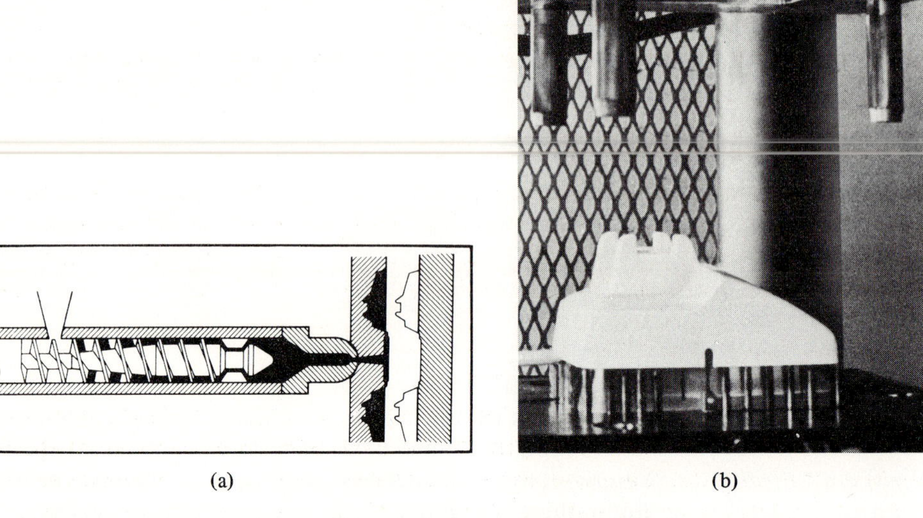

Figure 9-4.2 Injection molding (telephone receivers). (a) Extrustion into closed dies (schematic). The softened thermoplastic is forced through sprues to the cooled die cavity, where it hardens. (b) Formed product in the opened die (rotated 90° since the large auger in part (a) is operated horizontally). (Courtesy of Western Electric Co.)

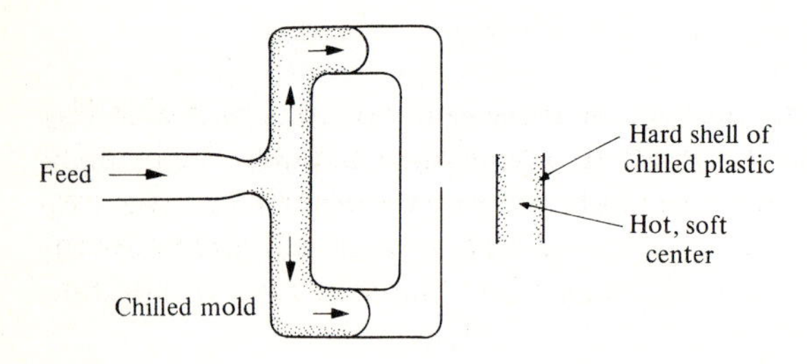

Figure 9-4.3 Flow in chilled mold (schematic). The mold is chilled to harden the plastic for stripping. However this produces a rigid shell that constricts the flow channels. High injection pressures are required.

granular feed must be only partially polymerized with an average of approximately two bonds per mer. This gives it thermoplastic characteristics for molding. Further polymerization reactions occur during curing in the mold to produce a rigid network or cross-linked structure. If that second stage polymerization were to start during the feeding step, the mold would not fill and the auger (or plunger) would bind.

Compression molding

Because of the control problems just cited, thermoplastic resins are more commonly compression molded. An auger or a plunger is not used; rather, a metered amount of pellets or granules are fed directly into the bottom half of

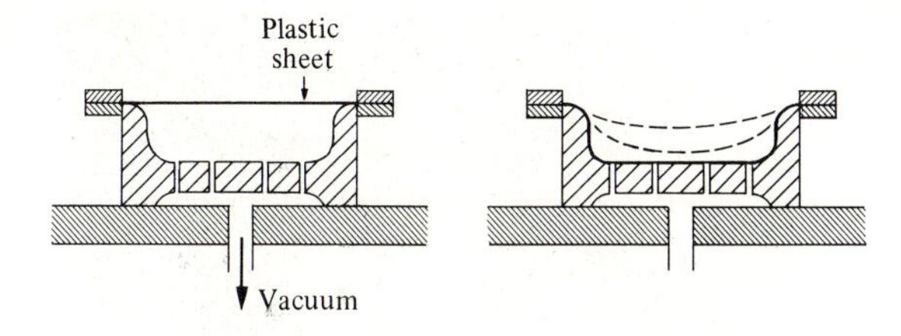

Figure 9–4.4 Sheet molding (vacuum molding). The molding pressure may also be from air pressure or from a mating mold in the upper side.

the mold. The mating half is pressed from above with a hydrostatic pressure of 10 MPa–20 MPa (1500 psi–3000 psi). Compression and reaction proceed during the curing part of the cycle. The product can be hot when it is removed from the mold, and a new cycle can be started immediately.

Sheet molding

In principle, sheet molding is the simplest forming process. An extruded sheet of thermoplastic material is clamped over the edges of a mold. The sheet is heated by infrared heaters. Gravity or a vacuum sags the sheet into the contour of the mold (Fig. 9–4.4). Variants of the process use air pressure or may even incorporate a mating mold on the opposite side.

Sheet molding is an inexpensive process for products with suitable geometries. Applications range from raised-letter signs to the housing for automobile instrument panels.

Blow molding

Bottles and related products having a constricted neck cannot be molded by any of the previous processes. There would be no way to remove the interior mold. In the blow-molding process, which is an adaptation from the container glass industry (Section 13–1), a soft plastic tube is extruded and cut free from the die head (Fig. 9–4.5). The resulting "hollow drop," or *parison,*

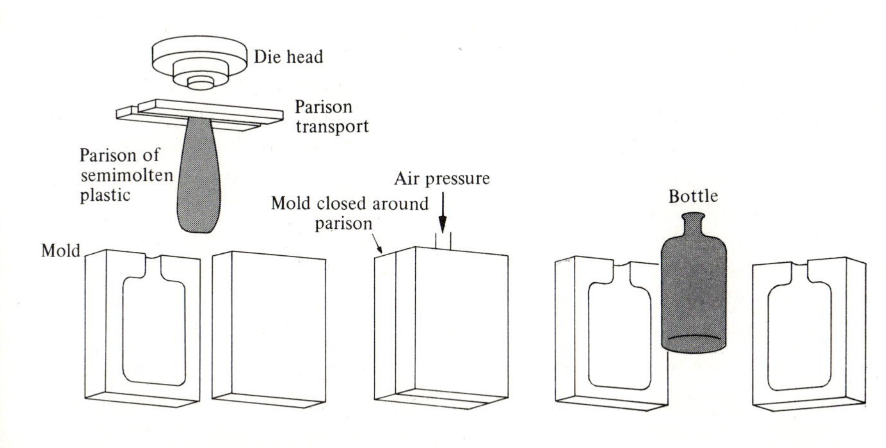

Figure 9–4.5 Blow molding. Air expands the viscous parison of plastic into the shape of the mold. Precise time and temperature controls are required in this process that originated in the glass industry.

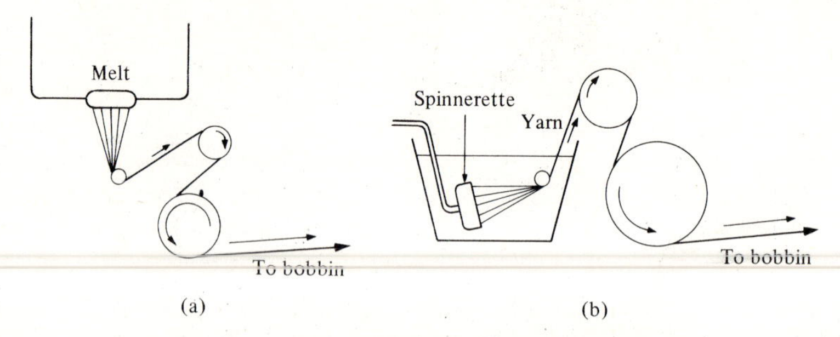

Fig. 9-4.6 Fiber spinning. (a) Melt spinning (polyester fibers). The thermoplastic filaments solidify as they quickly cool. (b) Wet spinning (cellulose fibers). The extruded filaments complete their polymerization in the bath. By moving at different surface speeds (arrow lengths), filaments are stretched between the two take-up rolls to extend and orient the molecules (Section 9-5).

of soft plastic is surrounded by a mold. Air pressure is used to expand the parison until it takes on the shape of the mold. (Cf. Fig. 13-1.4.) Blow molding, more than any other plastic forming process, requires a knowledge of the relationships between viscosity, temperature, and viscoelastic behavior. And not to be ignored is the accumulated empirical experience of the operator.

Spinning

Fibers are made by forcing the plastic through a multiple orifice *spinnerette*. These contain as many as 50 or 100 holes, each less than 0.2 mm in diameter. In *melt spinning*, the thermoplastic polymers, such as nylon and polyesters, are heated to low viscosity. The slender, emerging fibers (Fig. 9-4.6a) cool quickly in a current of air before they travel over a take-up roll.

Rayon is made by a *dry-spinning* operation. It is dissolved in acetone to produce a thick solution that is extruded through the spinnerette. The acetone evaporates* to permit the fibers to dry before going over the take-up roll. In each of the above spinning operations, the filaments travel at about 15 m/s (3000 ft/min).

A third, slower process is *wet spinning*. It is used when the fibers must react with the bath to complete the polymerization reaction after they leave the spinnerette (Fig. 9-4.6b).

* It is collected and reused.

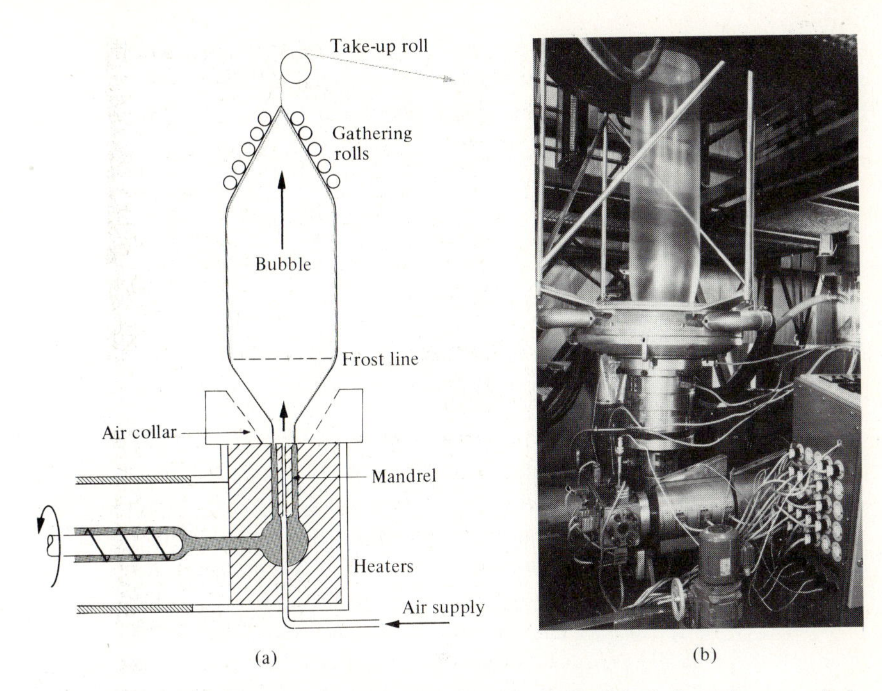

Figure 9–5.1 Bubble forming (plastic film). The cylindrical sheet is expanded simultaneously in two directions as it cools below the glass-transition temperature. Therefore it develops biaxial strength in the two dimensions of the film. (a) Schematic. (b) Production of polyvinylidene chloride $(C_2H_2Cl_2)_n$. (Courtesy of Dow Chemical U.S.A.)

9–5 PROCESSING FOR MOLECULAR ORIENTATION

On several occasions (Sections 7–1 and 8–2) it has been stated that polymers may be uncoiled and aligned by tension and/or shear stresses. When this is done, the strength and elastic moduli are significantly higher. Fibers can easily be processed to take advantage of molecular orientation. Figure 9–4.6 shows the procedure. The fibers are wrapped 360+° around each roll; the second roll rotates appreciably faster than the first. Therefore, the fibers can be stretched several hundred percent, inducing crystallization. This crystalline alignment is maintained at temperatures below the melting temperature.

The same principle could be used for a film product. Here, however, an undesirable anisotropy would be introduced. Specifically, the film would be strong in one direction and weak at right angles.* A more desirable process includes simultaneously stretching in the two coordinate directions so that the molecular orientation is biaxial. The *bubble method* (Fig. 9–5.1) is one

* This is demonstrated by analogy with newsprint, in which the wood fibers are preferentially aligned during the paper-making process. As a result, newsprint tears straight and easy in one direction and with more difficulty at 90°.

such process. A cylindrical film is extruded through an annular die. Air is blown through the mandrel and expands the sleeve into a larger cylinder.* There are both circumferential and longitudinal stretching to provide the biaxial orientation. The film may be slit if desired, or it may be heated sealed into plastic bags.

Review and Study

SUMMARY

Natural polymeric materials have been widely used through history. Using technology, chemists and engineers have developed processes to improve many of the raw materials.

1. Chemical processing may be used to modify the natural molecule, for example, exchanging –OH's of cellulose for acetate radicals to obtain cellulose acetate. Properties are also modified. In contrast the monomers of vinyls are synthesized from simpler molecules, for example, acetylene (C_2H_2).
2. Polymerization of small molecules (monomers) into extended molecules (polymers) is accomplished by two principal types of reactions. These are (a) by chain reaction (also called addition polymerization), and (b) by step reaction (also called condensation polymerization). Chain reactions add monomers at a reactive site on the end of the growing molecule. In turn, each added mer reacts, until two growing chains meet. There is no by-product from the reaction. The step reaction releases a small molecule as a by-product. Polyesters exemplify these step reactions; vinyls illustrate the chain reactions.
3. Just as metallic alloys contain several components, engineered plastics contain additives. These include fillers, plasticizers, stabilizers, and solvents. Their purposes are to add strength, flexibility, toughness, etc., or to provide durability against degradation and deterioration by service conditions.
4. The majority of polymer forming processes utilize their deformability. Therefore, most processing involves sequential steps of (a) softening, (b) molding, and (c) hardening. Thermoplastic polymers are linear polymers that soften with heating and harden on cooling. Thermosetting polymers react while heated to become network polymers and therefore take on a permanent set.
5. During extrusion, softened plastic is forced through an open die. Injection molding forces soft plastic into a closed die. These two methods are primarily applicable to thermoplasts. Thermosets more commonly are formed by compression molding in which metered amounts of partially polymerized resin is compacted in a hot mold where reaction is completed. Other processing modes include sheet molding and blow molding.
6. Deformation is included as a processing step in fiber and film products because it provides molecular orientation and enhanced mechanical properties. Spinning and the bubble method illustrate two procedures for providing orientation (and even crystallization) in fiber and film products, respectively.

* Expansion ceases when the film approaches the glass-transition temperature.

TECHNICAL TERMS

Additives Nonpolymeric materials incorporated into plastics for purposes of filling, stabilizing, plasticizing, coloring, etc.

Blow molding Processing by expanding a parison into a mold by air pressure. *Also see* Glass-bottle making (Section 13–1).

Bubble processing Process for biaxial straining of plastic film by air pressure inside a cylindrical extrusion.

Colorants Dyes (soluble) or pigments (particles) used as additives for coloring purposes.

Compression molding Product formation by compaction and heating of resin powders.

Compounding Formulation and mixing of materials for plastic products.

Degradation Deterioration of polymers through breaking bonds.

Extrusion Mechanical forming by compression through open-ended dies. (Cf., Fig. 9–4.1.)

Fiber-reinforced plastics (FRP) Composite of glass fibers and plastics.

Filler Additive used for the purpose of strengthening and/or extending the basic polymer.

Film Two-dimensional product; by definition, less than 0.5-mm thick.

Flame retardant Additives that reduce combustion, commonly by providing an oxygen-smothering gas.

Injection molding Process of molding a material in a closed die. For thermoplasts, the die is appropriately cooled. For thermosets, the die is maintained at the curing temperature for the plastic.

Plasticizer Micromolecules added among macromolecules to induce deformation and flexibility.

Polymerization, addition (chain-reaction) Polymerization by sequential addition of monomers.

Polymerization, condensation (step-reaction) Polymerization by a reaction that also produces a small by-product molecule.

Reactive site (•) Open end of a free radical, in which there are not enough bonds to meet the requirement of Table 7–1.1. *Also see* Eq. (9–2.2).

Spinning (polymers) Fiber-making process by filament extrusion.

Stabilizer Additive introduced into plastics for the purpose of retarding chemical reactions while the article is in service.

Thermoplasts Plastics that soften and are moldable due to the effect of heat. They are hardened by cooling but soften again during subsequent heating cycles.

Thermosets Plastics that polymerize further on heating; therefore heat causes them to take on an additional set. They do not soften with subsequent heating.

CHECKS

9A Natural polymeric materials include __***__, __***__, and __***__.

9B Many natural polymeric materials are __***__, that is, their properties vary with direction.

9C Several vinyl compounds use ___*** as a starting material, which is reacted with H_2, HCl, benzene, etc., to give ___***, ___***, and ___***, respectively.

9D ___*** polymerization (also called chain reaction) proceeds because there is a ___*** at the end of the molecule. This is transferred to the new end with each monomer addition.

9E Chain-reaction polymerization can be ___*** when the ___*** on the ends of two propagating molecules meet.

9F ___*** polymerization does not produce a by-product; however, ___*** polymerization does produce a small molecule such as ___*** with each reaction step.

9G Polyvinyl chloride (PVC) used for a raincoat has a ___*** as an additive, while PVC used for a floor tile generally has a ___*** as an additive.

9H Two advantages of using a filler in plastics are (1) ___*** , and (2) ___***.

9I In comparing properties of polymers after (1) the addition of 30 v/o glass powder and (2) the addition of 30 v/o glass fibers, there is negligible difference of ___***, but a significant difference in ___***, with the glass ___*** providing the higher value.

9J The introduction of small molecules will ___*** a polymer and in effect lower the ___*** so that molecular rearrangements can occur more readily.

9K Carbon black stabilizes a rubber because it ___*** visible and ultraviolet radiation and therefore avoids ___*** of the polymer molecules.

9L For coloring purposes, a ___*** is present as a separate phase within the polymer product, while a ___*** dissolves into the polymer product.

9M Oxidation of rubbers and similar polymers is particularly rapid in any atmosphere that contains ___***. In fact, it is intentionally introduced into atmospheres for ___*** oxidation tests.

9N By having the two rolls of a "rubber mill" operate at slightly different ___***, mixing is achieved by a shearing action.

9O Processing generally requires three steps ___***, ___***, and ___***. The first step commonly involves ___***; the second, ___***.

9P For a polymer to be thermoplastic, the molecule must be ___***; thus most of the ___***-type polymers fall in this category.

9Q Thermosetting polymers develop a ___*** structure, either because the original molecules are ___***-functional, or through ___***.

9R ___*** harden by cooling; ___*** harden, or cure, by continued reaction to complete the formation of a ___***.

9S ___*** is an example of a polymer product that is formed by ___*** through an open-end die.

9T The auger of an extruder for amorphous polymers has a ___*** compression zone than does the auger for an extruder for polymers that have greater crystallinity.

9U A closed die is used for ___*** molding; the die is a ___*** mold that must be ___*** so that the product is rigid before opening and removal.

9V Spinning is a process for making ___***. There are three distinct types of spinning operations—__***__, __***__, and __***__—that lead to hardening by cooling, evaporation, and a chemical reaction, respectively.

9W One-dimensional stretching is sufficient to provide __***__ in a fiber product; however, two-dimensional stretching is required for a __***__ product.

STUDY PROBLEMS

9–1.1. How many metric tons of acetylene, C_2H_2, and hydrochloric acid, HCl, would it take to produce 7,000,000 tons of vinyl chloride? [This is the approximate world production of polyvinyl chloride predicted for 1985.]

Answer: 2,900,000 tons C_2H_2

9–1.2. Cellulose has a specific gravity of about 1.5 and is the chief component of dry wood. White pine weighs 380 kg/m^3. Estimate the fraction of dry pine that is pore space.

9–1.3. (a) What is the mer weight of cellulose? (b) Of cellulose triacetate?

Answer: (a) 162 amu

9–1.4. How many grams of acetylene (C_2H_2) are required to produce 100 g of each of the vinyl products of Eqs. (9–1.2, 9–1.3, 9–1.4, and 9–1.5)?

9–2.1. Show how the double bonds change during the addition of polymerization of isoprene to polyisoprene.

9–2.2. Show how phenol-formaldehyde can form a network polymer. [*Comment:* There is room for $-CH_2-$ bridges to connect to only three of the six carbons.]

9–2.3. Nylon is a condensation polymer of molecules such as $HO{\cdot}CO{\cdot}(CH_2)_4{\cdot}CO{\cdot}OH$ and $H_2N(CH_2)_6NH_2$. (a) Sketch the structure of the two molecules. (b) Show how polymerization can occur. (c) What is the by-product? (d) What is the percent weight loss?

Answer: (d) −14 w/o

9–2.4. Urea, $H_2N{\cdot}CO{\cdot}NH_2$, and formaldehyde, CH_2O, can react to form a condensation polymer, urea-formaldehyde. (a) Sketch the structure of the two molecules. (b) Show how polymerization can occur.

9–2.5. Determine the weight change in the polymerization of Eq. (9–2.6), if X equals $(CH_2)_4$ and Y equals $(CH_2)_2$.

Answer: −0.17

9–3.1. A 3.2-g sample of polyvinyl chloride-based plastic floor tile is heated to 500°C so that only 1.3 g of silica-flour ash remains. What fraction of the volume was filler? The specific gravities of PVC and silica flour (SiO_2) are 1.3 and 2.65, respectively.

Answer: 25 v/o filler

9–3.2. As a maximum, a pair of isoprene (C_5H_8) mers may be joined by two oxygen atoms (Fig. 9–3.3). By this reaction, 1 kg of isoprene rubber could increase its mass to ______ kgs.

9–3.3. Suppose that you manufacture plastic ball bearings, and you want to have 50 v/o of finely chopped glass fibers (ρ = 2.4 g/cm^3) in the nylon (ρ = 1.15 g/cm^3)

that is to be used for these bearings. How much glass fiber should be batched with each kilogram of nylon in making the plastic–glass composite?

Answer: 2.08 kg

9-3.4. A 2.0-g/cm³ density is required for a plastic product. A silica flour (quartz) filled phenol-formaldehyde is satisfactory. What weight ratios are required? [See Appendix A.]

9-4.1. A 5-kW motor is used to power an auger extruding polymethylmethacrylate. Assuming 80% of the power appears as a temperature rise of the PMM, how much will the temperature rise if the through-put is 120 kg/hr? [c_p of PMM is 1.5 J/g·°C.]

Answer: 80°C

9-4.2. An outdoor sign, 163-cm in diameter, is vacuum-formed into its final shape. What force would be required behind a die to provide the same force in a die-stamping operation?

CHECK OUTS

9A	wood, leather, fibers
9B	anisotropic
9C	acetylene (C_2H_2) ethylene, vinyl chloride, styrene
9D	addition free radical (reactive site)
9E	terminated reactive sites (free radicals)
9F	addition (chain-reaction) condensation (step-reaction) H_2O
9G	plasticizer filler
9H	wear resistance, cost, rigidity
9I	density strength fibers
9J	plasticize glass-transition temperature
9K	absorbs scission (degradation)
9L	pigment dye
9M	ozone accelerated
9N	speeds (rpm)
9O	softening, molding, hardening, heating pressure
9P	linear vinyl
9Q	network poly- cross-linking
9R	thermoplasts thermosets network
9S	pipe (sheet) (rods) extrusion
9T	longer
9U	injection split water-cooled
9V	fibers melt, dry wet
9W	orientation film

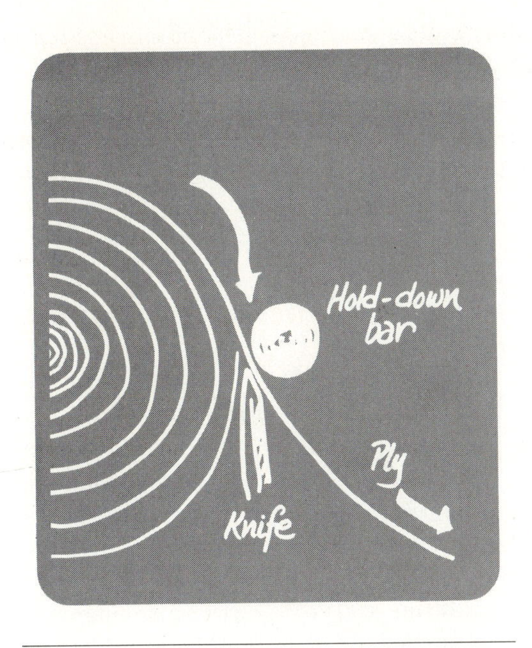

UNIT TEN

• WOOD AND WOOD PRODUCTS

Wood is the most widely used of all materials of construction; it is a replaceable resource and can receive final processing on the job. These factors favor the wide use of wood. The reader is aware that wood has an internal structure that affects its properties but probably is not acquainted with details. Finally, wood and wood products can be modified to enhance their properties for specific applications. These are the reasons for the engineer to look more closely at this common material.

Contents

Prerequisites Units 7 and 8, and selected sections of Unit 3.

From Unit 10, the engineer should

1) Become familiar with the various levels of structure—growth patterns, fiber, and molecular—and their contribution to properties.
2) Identify the principal directions in wood and their relationships to the commonly encountered properties.
3) Understand the effects of moisture in wood, particularly on dimensional changes and durability.

4) Be acquainted with the common processing procedures for wood and wood products so that their use may be optimized.
5) Understand the mechanism of the more common causes of wood deterioration, so that appropriate prevention measures may be specified.
6) Have the vocabulary to understand technical discussions on wood specifications and usage.

10-1 WOOD AS A MATERIAL

Wood is a complex material; thus, we have delayed consideration until certain principles were presented in the preceding chapters. However, wood could justifiably have been the first material considered in this book, since wood is prime. (a) Mankind used wood prior to the Stone Age. (b) More wood is used than any other solid material. (c) Its properties obviously relate to its structure. (d) Wood is easily processed, at least in its more widely used forms. (e) Furthermore, wood will gain in future importance, because it is a renewable resource and requires low net energy for processing.

The structure of wood, while obvious on a macroscale, is intricate on a microscale. Its composition includes a number of components that combine into complex molecules and fibers. Because of its fibrous structure, its properties are directional. Thus, wood is not elementary.

Characteristics of wood

Wood is of botanical origin. Therefore, its constitution, molecular species, and cellular geometry are fixed by nature. Because of this, there are both consistancy and variety in the structure of wood.

The structure of wood forms recognizable patterns related to the tree species that produced it. Since there are literally thousands of tree species in the world, these patterns are useful for identification and diagnostic purposes, often providing a first approximation of properties.

Most wood exhibits a gross annual structure on the cross section that arises from the repeating growth cycles during the seasons, (Fig. 10-1. 1a). All wood contains *biological cells* that commonly possess an elongated spindle shape (Fig. 10-1.1b). The microscopic structure of wood consists of *fibers* (Fig. 10-1.1c) with diameters of micrometer dimensions (and millimeter lengths). The principal molecular species is *cellulose* (Fig. 10-1.1d) that forms structures at the submicroscopic level. As with any natural material, each of these structural levels contains discontinuities, ranging from readily visible knots to amorphous regions in otherwise crystalline cellulose.

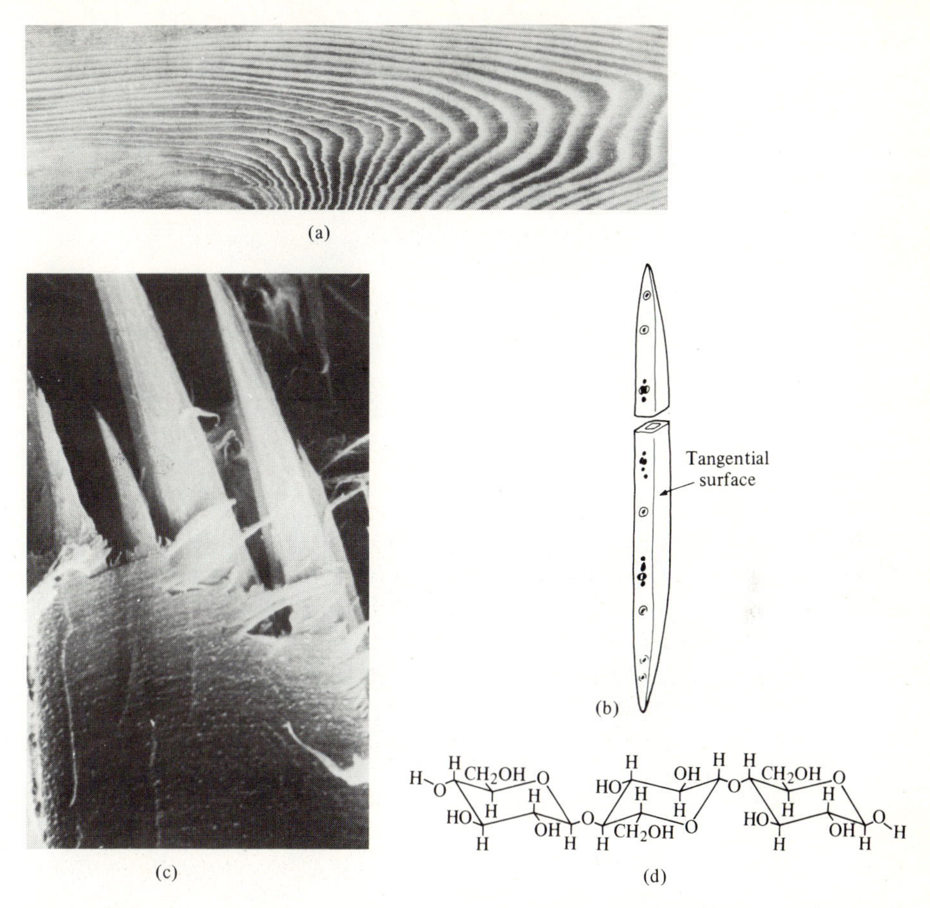

Figure 10-1.1 Structures of wood. (a) Surface pattern formed by annual growth rings. (Courtesy of Forest Products Laboratory, Forest Service, USDA.) (b) Biological cell, or fiber (longitudinal section). (After Hower and Manwiller, *Wood Science*.) (c) Microfibrils from the cell wall. ×2500. (Courtesy of Prof. Hiroshi Harada, Kyoto University.) (d) Molecule (a portion of a long-chain cellulose molecule). Cf. Fig. 9-1.1.

Wood contains considerable *porosity,* which in turn affects the properties. Wood is *hydroscopic,* absorbing water in its pores and adsorbing it on the fibers in its cell walls, where it induces volume changes. Since the properties of wood (Section 10-3) are *directional,* the volume changes just cited are not uniform, often leading to internal stresses and to warpage (Fig. 10-1.2a). Fracture paths are usually affected by the growth pattern (Fig. 10-1.2b). Being botanical in nature, wood is subject to biological deterioration (Fig. 10-1.3a) as well as to mechanical, chemical, and thermal degradation (Fig. 10-1.3b). However, wood will not deteriorate if maintained in the dry state under normal conditions.

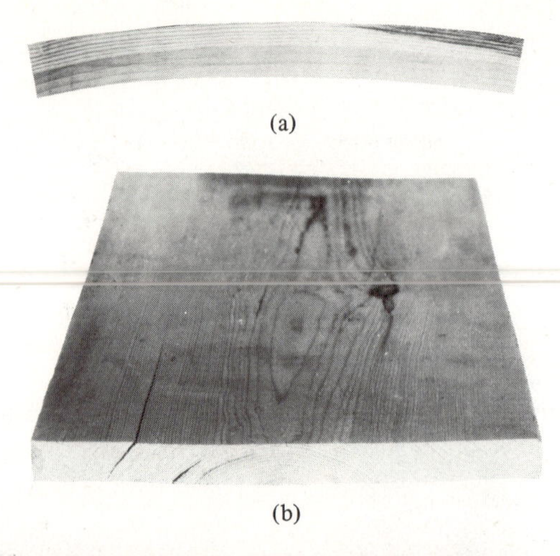

(a)

(b)

Figure 10-1.2 Drying shrinkage. (a) Warpage from nonuniform drying. (b) Splitting along growth rings (shakes.) (Courtesy of Forest Products Laboratory, Forest Service, USDA.)

(a)

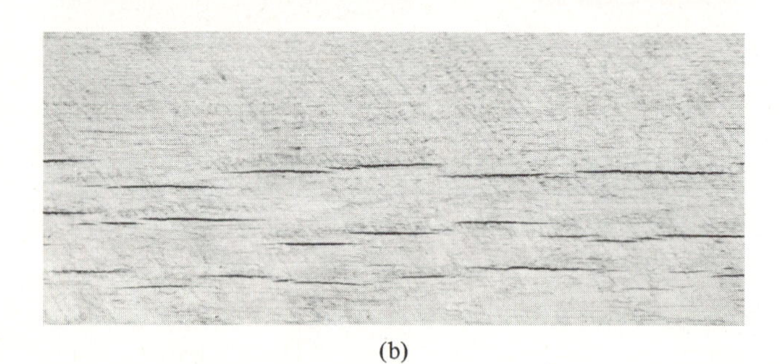

(b)

Figure 10-1.3 Wood degradation. (a) Biological deterioration (fungus growth on a dead limb). (b) Surface cracking (checking of a board) from moisture cycling. (Courtesy of Forest Products Laboratory, Forest Service, USDA.)

Although wood is a natural product, this does not preclude modification and control by man. The most direct processing is the sawing operation that provides construction lumber, which is the most widely used wood product (Table 10-1.1). More complex modifications are performed on products such as plywood, particle board, paper, and cellulose derivatives (Section 10-4). These involve taking wood apart and reassembling it into the desired predetermined forms.

Table 10-1.1
Wood and wood products
(Thousand m^3.)

	Softwoods	Hardwoods
Sawed lumber	138,000	38,400
Veneer and ply	28,300	3,600
Pulpwood	78,500	30,200
Misc. Industry	6,500	5,500
Fuelwood	3,000	12,200
	254,300	89,900

Adapted from Marra, *Journal of Educational Modules for Materials Science and Engineering.*

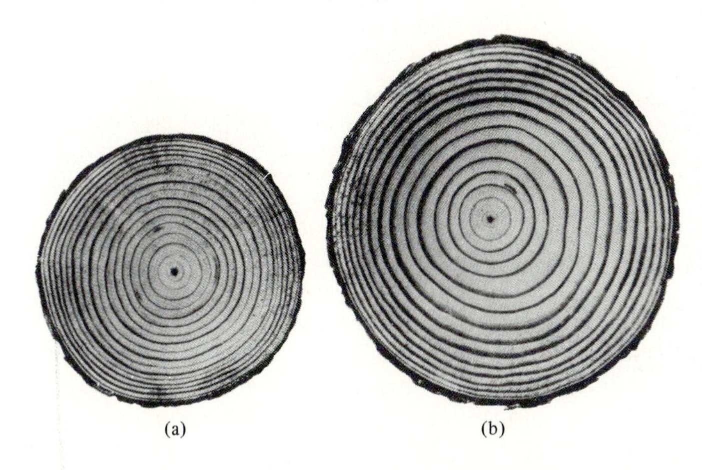

Figure 10-1.4 Growth rates by forest management. Although the larger tree is not as old as the smaller tree, it contains twice as much commercial wood. (Cf. $d^2 \times h$.) The smaller one grew in an unmanaged, overcrowded stand where it had to compete for sunlight and moisture. The other is from a grove that was thinned to give the best trees room to thrive. Because growth rate affects cell structure, the properties of wood from the two trees will not be identical. (American Forest Institute.)

Although it will not be discussed in this book, note should be made that the wood of the future may be modified through genetic control and intensive "forest farming" practices. Changes on an experimental basis have already been accomplished in rates of growth (Fig. 10-1.4), density, straightness of grain, and fiber length. Future possibilities include ratios of cellulose to other constituents, and the crystallinity of cellulose, among other modifications.

10-2 STRUCTURE OF WOOD

Growth patterns

The trunk of a tree grows by adding another ring every season. Usually this is in the annual sequence of spring and summer growing seasons; however, in tropical regions, the rings may reflect a sequence of wet and dry periods. When the growth is fast, the biological cells, called *tracheids* in softwood, are hollow, thin-walled, spindle-shaped cylinders (Fig. 10-1.1b). This

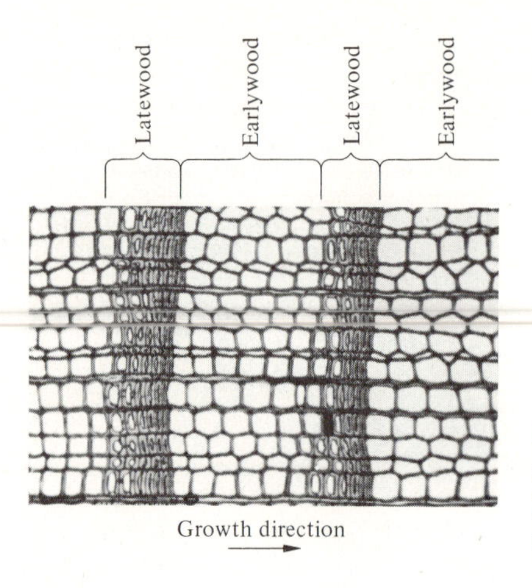

Figure 10-2.1 Biomaterials (wood). The cells that form with early growth are larger but have thinner walls than those that form with later growth.

growth is called the *earlywood* (Fig. 10-2.1). Later in the growing season, when the growth rate slows down, the cells have thicker walls and are denser. This is *latewood*. Because of its higher density, it is stronger than the earlywood. The cyclic pattern of the earlywood and latewood produces the typical ring pattern as seen in the cross section of the tree (Fig. 10-1.4).

These alternating layers of earlywood and latewood and the biological cells that are oriented longitudinally in the tree are primarily responsible for the *anisotropy* of wood. They account for the easy longitudinal splitting, variation of elastic moduli, thermal expansion, conductivity, etc., with direction (Section 10-3). Because of these variations wood technologists discuss properties in terms of the *longitudinal* direction, l, the *radial* direction, r, and the *tangential* direction, t, (Fig. 10-2.2). Also, the lumbermen will distinguish between *quartersawed* wood and *plainsawed* wood (Fig. 10-2.3). Of course, the cut of the lumber may have some intermediate orientation;

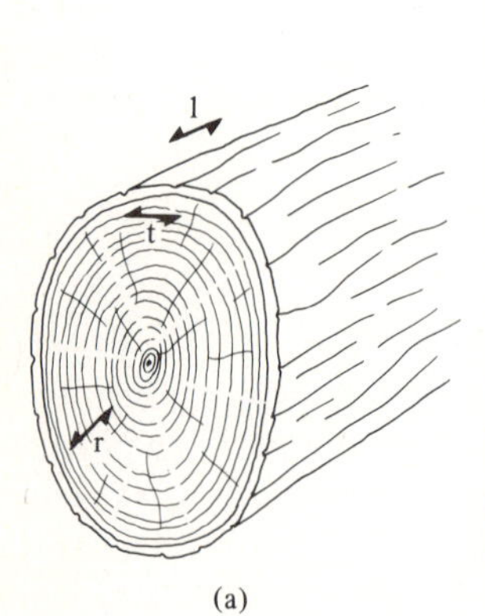

(a)

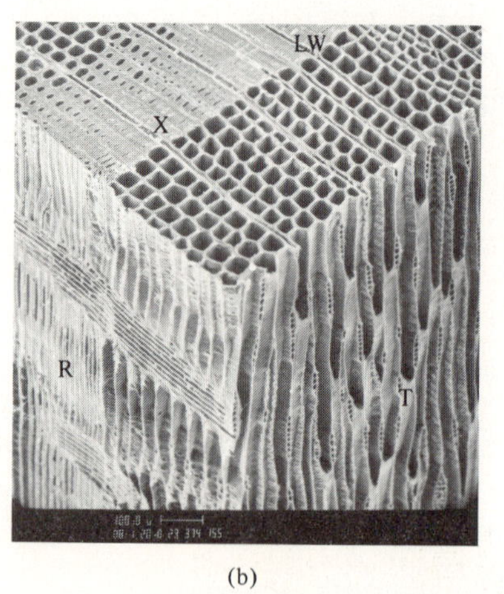

(b)

Figure 10-2.2 Directions in wood. Longitudinal, l; radial, r; tangential, t. (a) Log (sketch). (b) Scanning photomicrograph (Douglas fir). (Courtesy of W.A. Côtè, Jr., S.U.N.Y. College of Environmental Science and Forestry, Syracuse.) Cross section, X; radial section, R; tangential, T; late wood, LW.

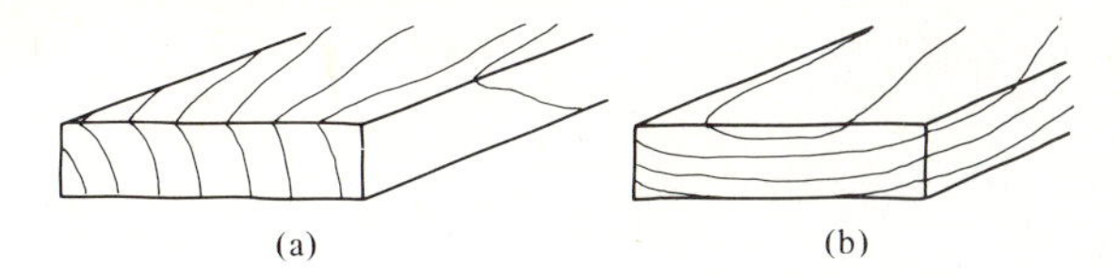

Figure 10–2.3 Saw cuts. (a) Quartersawed. The width of the board is approximately radial. (b) Plainsawed. The width of the board is approximately tangential.

however, it becomes sufficiently obvious to the carpenter that the properties of these two cuts differ, so that intentional selection is sometimes made between the two as they are used in construction and in cabinetmaking.

The new tracheids (biological cells) form in the *cambium layer,* which is located at the inner boundary of the bark. This is where the growth is initiated; however, the *sapwood,* which includes only the wood added during several recent seasons, aids in the growth process. It conducts water and the dissolved minerals that are required for cell development upward to the leaves and branches. The sapwood also accumulates and stores cell nutrients produced by photosynthesis. The center portion of the tree is called the *heartwood.* This serves no biological function other than contributing to the structural stability of the tree. Commonly, but not always, the wood darkens in color when it ceases being sapwood and becomes heartwood (Fig. 10–2.4). Not only does the darkening provide a color contrast that is valued for its visual appeal in many woods, but it commonly makes the wood more durable.

Softwoods and hardwoods

The *softwoods* include the pines, spruces, firs, etc.,—in brief, the cone-bearing trees, or *conifers.* The *hardwoods* include the oaks, ashes, birches,

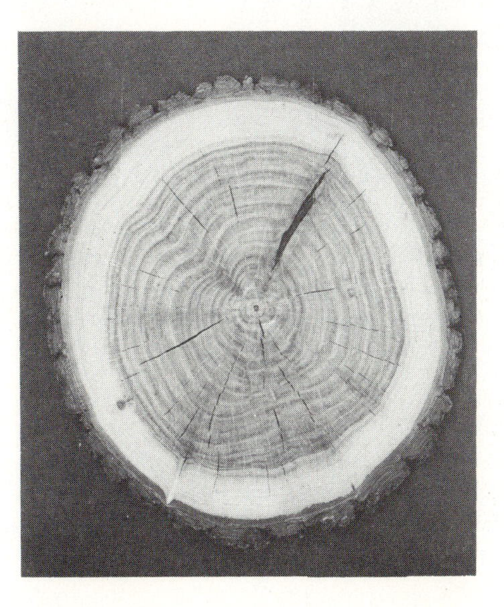

Figure 10–2.4 Heartwood (dark) and sapwood (light). The sapwood is functional in the living tree as a conductor of water and dissolved minerals. The heartwood is dead but it adds stability to the tree. The color change varies from species to species. (Courtesy of R. J. Thomas, North Carolina State University.)

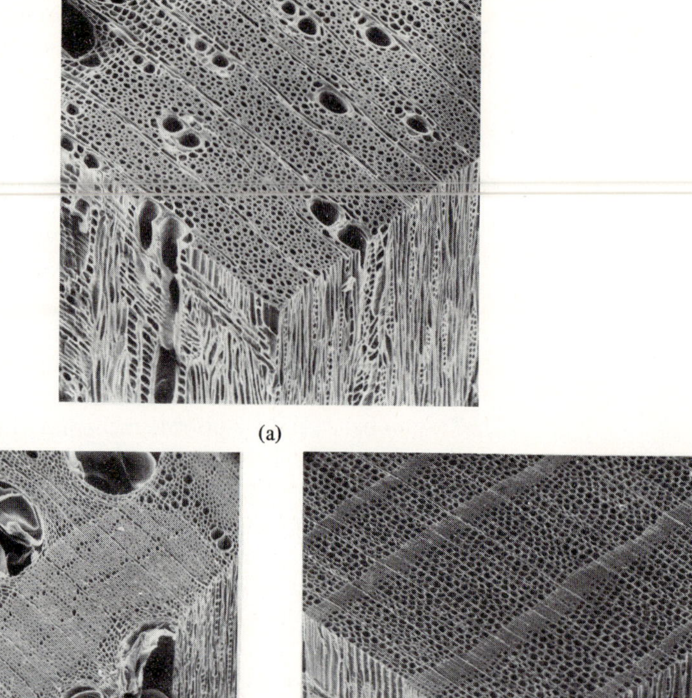

(a)

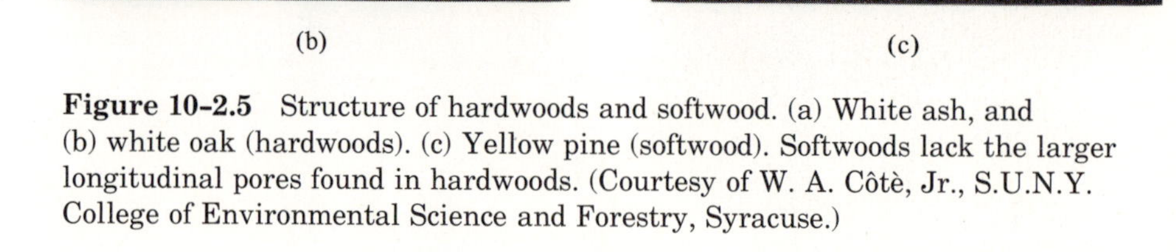

(b) (c)

Figure 10-2.5 Structure of hardwoods and softwood. (a) White ash, and (b) white oak (hardwoods). (c) Yellow pine (softwood). Softwoods lack the larger longitudinal pores found in hardwoods. (Courtesy of W. A. Côtè, Jr., S.U.N.Y. College of Environmental Science and Forestry, Syracuse.)

maples, walnuts, etc., which are the broad-leafed or *deciduous* trees. These two labels are less than satisfactory for engineering purposes because some softwoods, (for example, yews) are harder than some hardwoods (for example, balsa). However, it is true that on the average, the softwoods are softer, less dense, and more easily cut and worked than are the hardwoods.

There are some distinct microstructural differences between softwoods and hardwoods. Softwoods are typically much simpler in structure than are the hardwoods. Softwoods are composed predominantly of one type of cell, the tracheid, which serves all mechanical and biological functions except storage of cell nutrients. Hardwoods are composed of a variety of cell types to carry out the various functions, including distinctive longitudinal vessels for conduction purposes. These can be seen on the cross section as openings or pores (Fig. 10-2.5b), and on the longitudinal section as minute grooves. In some hardwoods, such as oak, the pores are very large in the early wood. The cabinetmaker is aware of their presence since these pores readily absorb stain and produce a pleasing visual texture. Other hardwoods, such as ash (Fig. 10-2.5a), have smaller pores that are more uniformly distributed across the growth rings. These woods receive a finish that is much more uniform in appearance. Softwoods lack these pores (Fig. 10-2.5c). In commercial practice, softwoods are most commonly used for construction and hardwoods for furniture and decorative effects—for reasons obviously other than strength.

Rays and resin canals

Another macrostructural feature of woods is the development of *rays* that extend from the center of the tree to the bark at right angles to the growth rings. These are most obvious in oak wood (Fig. 10-2.6). Wood rays consist of a ribbon-like collection of food-storing cells.

Figure 10-2.6 Rays (oak). These radial structures, which serve a storage function, add stability to the structure of wood but also provide a discontinuity that can initiate fracture. (Courtesy of Forest Products Laboratory, Forest Service, USDA.)

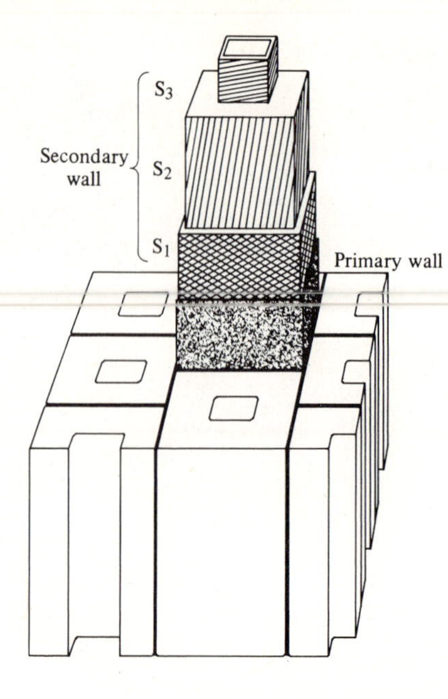

Figure 10-2.7 Structure of wood cells. The cell walls in wood contain three layers—S_1, S_2, and S_3. Each contains microfibrils aligned at an angle to the longitudinal axis. The center cavity is called the lumen. (Courtesy of R. J. Thomas, North Carolina State University.)

Finally, softwoods, such as pine and spruce, possess longitudinal *resin canals*. These intercellular spaces transport resin to an exposed surface when the tree is cut or is otherwise injured. The resin effectively seals the exposed surface. This resin can also be converted to a commercial product, for example, turpentine.

Finer structures

The wood cells that show in Figs. 10-1.1(b) and 10-2.5 possess a structure within themselves. They are made up of several layers of *microfibrils* that spiral at various angles around the hollow cavity called the *lumen*. A typical microfibril has a cross section of 4×10 nm. They are sketched schematically in Fig. 10-2.7. There are three secondary layers that account for 99% of the cell wall:

S_1	10%–22% of wall thickness	50°–70° from longitudinal;
S_2	70%–90% of wall thickness	10°–30° from longitudinal;
S_3	2%– 8% of wall thickness	60°–90° from longitudinal.

The multiple layers and the angles of pitch are nature's way of simultaneously giving strength and flexibility to the wood cell.

Molecular composition of wood

Cellulose is the principal molecular species in wood. It is a polymer with mers of $nC_6H_{10}O_5$ (Fig. 10–1.1d). These molecules make up the microfibrils of Fig. 10–2.7. Polycellulose possesses a high degree of crystallinity as a result of the presence of both –H and –OH attachments to the molecule (Fig. 10–1.1d). The degree of polymerization, n, sometimes exceeds 30,000. Cellulose accounts for ~42 percent (dry basis) of the mass of most wood.

Molecules of *hemicellulose* (a family of molecules that are distinct from, but related to, cellulose) account for another 25%–33% of the mass of the wood. These molecules have only a few hundred mers. The specific hemicellulose differs in hardwoods from those in softwoods.

A third type of molecule common to wood is *lignin*. It is a hard amorphous polymer. Lignin is complex and differs in softwoods and hardwoods. The amount of lignin present is complementary to the amount of hemicellulose, that is, it is high (~33%) when the hemicellulose is low (~25%), and low when the latter is high. As a generality, hardwoods contain less lignin and more hemicellulose than do softwoods. Lignin serves to bind the fibers together.

Irregularities in wood

The most obvious structural irregularity in wood is a *knot*. A knot was initially a limb of a tree. As the tree grew, the former limb was surrounded by the wood of the trunk. The direction of the grain in the knot is, therefore, at approximately right angles to the grain in the bulk of the wood. If the branch is alive at the time the trunk surrounds it, the two become *intergrown* so that fibers are continuous from the knot to the wood of the trunk. If the branch is dead when the trunk of the tree surrounds it, the knot becomes *encased* (Fig. 10–2.8) with no cellular connection to the grain of the wood

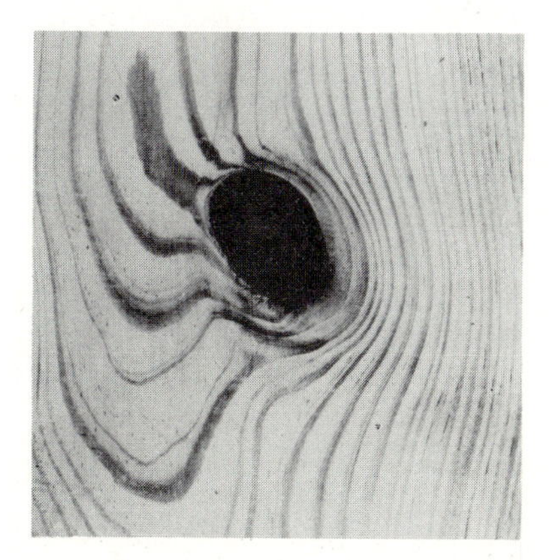

Figure 10–2.8 Knots. These are former branches that have been surrounded by wood growth on the trunk. A knot from a dead branch does not have cellular connection to the surrounding wood. (Courtesy of Forest Products Laboratory, Forest Service, USDA.)

proper. This type of knot may drop out when the lumber is dried, creating the typical knot hole. The shape of the knot varies due to the angle of cut, and depending on its location in a board, may have a pronounced effect on the strength of the board.

Reaction wood is the tree's response to unbalanced growth situations. For example, a tilted tree, or a crooked trunk, results in one side being underneath the normally vertical trunk and the other side in the upper position. The response in softwood is to produce an abnormal wood tissue on the lower side of the trunk (Fig. 10-2.9). It is called *compression* wood because it is formed in a zone containing compression stresses. In hardwood trees, the adjustment is made on the upper side; the abnormal tissue is called *tension* wood because it is the reverse condition. In general, these abnormal cells are more dense than regular wood—30% to 40% greater than normal in softwoods and usually 5% to 10% more dense in the hardwoods. However, the fibril angle in the cell wall is greater, which lowers the strength and increases the shrinkage ratio in the longitudinal direction. Since this invariably introduces excessive warpage during drying, reaction wood is considered undesirable.

Softwoods contain pitch, a protective resin that can accumulate within any crack or opening in the wood. *Pitch pockets* are openings in the wood that extend parallel to the annual rings and are usually lens-shaped. Most pitch pockets are small and because of their shape they produce little deviation of the fibers around them. Hence, there is little or no effect on strength. They are sometimes undesirable because they affect both the appearance of the bare wood and the adherence of paint or adhesives. If very large, they can lead to structural weaknesses.

Figure 10-2.9 Reaction wood. Eccentric growth occurs in wood that supports tension or additional compression during growth. Reaction wood is denser, not strong on a weight basis, and subject to abnormal shrinkage and greater warpage. (Courtesy of Forest Products Laboratory, Forest Service USDA.)

10-3 PROPERTIES OF WOOD

Wood in a tree contains a high percentage of water, and even seasoned wood contains moisture. The amount of moisture affects all properties. In extreme cases, it can more than double the density. Up to a certain point, it increases the volume and conductivity and lowers the strength and elasticity. The moisture content varies with the surrounding environment. The interior woodwork of a house may contain as little as 6% moisture in midwinter, to 15% moisture after a humid season; a wood piling on a boat dock can possess more than 100% water (oven-dry basis) at the water line. Thus, before considering properties, it will be helpful to describe how the water is contained within the wood.

Moisture in wood

The reference point for moisture content is oven-dried wood with 0% H_2O. Where wood is exposed to air, which always contains some water vapor, wood gains or loses moisture depending on the relative humidity of the air and the existing moisture status of the wood. It is always seeking an equilibrium. The equilibrium amount during midwinter is the 6% cited above. This *ab*sorption into the wood is really an *ad*sorption onto the surface of the microfibrils described in Section 10-2. Warm humid summer air carries more moisture; therefore, more moisture is adsorbed onto the microfibrils. The microfibrils become saturated with moisture, and expansion ceases when about 30%–35% H_2O (dry basis) is present in the wood. This is called the *fiber-saturation point,* FSP. This is an important point, because shrinkage occurs and all strength properties increase as the moisture drops below this level. Moisture in excess of the FSP enters the lumen within the cells as free water (Fig. 10-2.7). That does not occur from atmospheric moisture during normal lumber usage. The wood would need to be submerged in water or exposed to a continuous presence of water on its surface to fill these cavities.

Each *one percent* of moisture (dry basis) below the FPS changes the dimensions of typical wood:

radial	~0.15 l/o per % moisture,
tangential	~0.25 l/o per % moisture,
longitudinal	~0.01 l/o per % moisture.

Thus, there are constant but minor volume changes in finished wood in service related to prolonged moisture fluctuations of the atmosphere.

Density

Common woods in the United States have densities in the range of 0.4 to 0.7 g/cm³. Among these woods, redwood and oak are near the limits (Table 10-3.1). However, other woods are more extreme; for example, balsa has a density of only 0.13 g/cm³, and a few tropical woods have a density greater than water, and therefore sink in water. The data of Table 10-3.1 are on a 12% moisture basis.

The differences in density of the woods in Table 10-3.1 are primarily a result of the differences of volume of the cell cavities within the wood, since the wood substances—cellulose, hemicellulose, and lignin—have a density very close to 1.5 g/cm³. An increase in moisture content below the fiber-saturation point does not increase the density appreciably in thin, unrestrained cross sections of wood since the swelling volume almost exactly equals the weight or volume of the water absorbed by the wood substance. However, in larger pieces, the density increases. For example, with 10% moisture, the mass increases 10% and the volume increases about 4% (from the figures earlier in this section) to give a 5%–6% increase in density. Just below the fiber-saturation point, and with 30% moisture, the density is 15%–18% greater. Once the fiber-saturation point is exceeded, excess water fills the cell cavities and the density increases rapidly, eventually exceeding the density of water. The wood becomes *water-logged* and would sink if placed in water.

Table 10-3.1
Properties of common woods (12% moisture)*

	White oak	Paper birch	Douglas fir	White pine	Redwood (old rough growth)
Density, g/cm³	0.68	0.55	0.50	0.38	0.40
Hardness (radial), N†	6,000	4,000	2,900	1,900	2,100
Young's modulus, MPa	12,300	11,000	12,500	10,100	9,200
longitudinal‡ (10^6 psi)	(1.78)	(1.60)	(1.81)	(1.46)	(1.33)
Strength, MPa					
(psi)					
tension (radial)§	5.4	—	2.4	—	1.7
	(780)		(350)		(250)
compression (long.)	51	39	51	35	42
	(7400)	(5700)	(7400)	(5100)	(6100)
compression (radial)	7.4	4.1	5.2	3.2	4.8
	(1100)	(590)	(750)	(460)	(700)
modulus of rupture	105	85	87	67	69
	(15,200)	(12,300)	(12,600)	(9,700)	(10,000)

* Data extracted from Wood Handbook USDA, Forest Service, *Agric. Hdbk.*, No. 72.
† Force required to embed 0.444-in. (11.3-mm) ball to one half of its diameter.
‡ See text for ranges of E_r and E_t.
§ Longitudinal strength in tension $S_{t(l)}$ is commonly 20 $S_{t(r)}$.

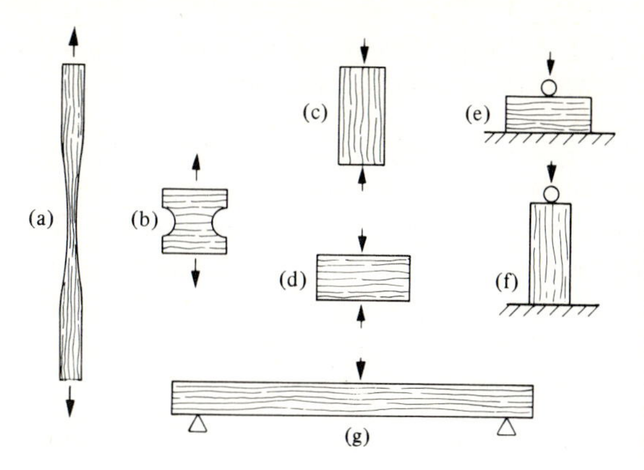

Figure 10-3.1 Test specimens of wood. (a, b) Tension. (c, d) Compression. (e, f) Hardness. (g) Modulus of rupture. The modulus of rupture is the calculated stress at lower (tension) surface.

Elasticity and strength

Selected mechanical properties of several woods are given in Table 10–3.1. In general, their Young's moduli lie near the center of the following ranges:

longitudinal	7500 MPa–16,000 MPa	(1,100,000 psi–2,400,000 psi);
radial	500 MPa– 1,000 MPa	(75,000 psi– 150,000 psi);
tangential	400 MPa– 700 MPa	(60,000 psi– 100,000 psi).

The high values for longitudinal stressing is expected since that is the alignment of the wood cells. The other two orientations are low because the cell cavity permits easy strain. The radial direction is stiffened slightly by the rays (Fig. 10–2.6).

The longitudinal strength in tension, $S_{t(l)}$, is commonly 20 times (and sometimes as much as 40 times) the radial strength in tension, $S_{t(r)}$. This contrast will not be surprising to those readers who have worked with wood. Test specimens for mechanical properties must be specifically cut for each orientation (Fig. 10–3.1). Unlike ductile metals, the strength in compression of wood does not closely match the strength in tension. Longitudinal compression strengths, $S_{c(l)}$, are lower than in tension, $S_{t(l)}$, because the denser latewood buckles and permits shear to occur. Radial strengths in compression, $S_{c(r)}$, are higher than $S_{t(r)}$ (Table 10–3.1) because the tension failure is by splitting.

Hardness can be measured by indenting the wood with a ball that is larger than the thickness of the growth rings (Fig. 10–3.1 e,f). This is similar to the procedure used in the Brinell hardness test (Section 3–3), except that the loads are lower. The difference between indentations in longitudinal and radial directions is not as great as for strength values. This is because the stresses under an indentor never remain unidirectional but become three-dimensional.*

Thermal properties

The approximate linear *thermal expansion coefficients* of wood are dependent upon the density, ρ, as follows:

$$\alpha_r \simeq [6\rho + 2] \times 10^{-5}/°C;$$
$$\alpha_t \simeq [6\rho + 3] \times 10^{-5}/°C;$$
$$\alpha_l \simeq 0.4 \times 10^{-5}/°C.$$

In general, these lead to smaller dimensional changes than those encountered with moisture variations (I.P. 10–3.1). Thus, it is necessary to include thermal expansion design calculations only in special situations.

The *thermal conductivity* of wood is low as compared with other materials. This constitutes an advantage for wood when used in building construction, because wood permits lower heat losses than masonry or metal. Like other properties, thermal conductivity is anisotropic, being 2 to 3 times as great longitudinally as radially (and tangentially).

The low thermal conductivity of wood has an unexpected benefit in case of fire. Although wood will burn and/or char, the interior regions of wooden beams are sufficiently insulated so that strength losses are delayed, sometimes avoiding a structural collapse that occurs with exposed metal beams that become hot and therefore weak.

As wood is heated to elevated temperatures, it first dries to lose absorbed water. It will ignite spontaneously in air if the temperature exceeds ~400°C (~750°F). In the absence of air, wood chars by losing volatiles and dissociating the cellulose:

$$C_6H_{10}O_5 \rightarrow \text{carbon} + \text{water vapor}. \tag{10–3.1}$$

The heating value of dry wood is 17,000 kJ/kg to 21,000 kJ/kg (4 kcal/g to 5 kcal/g) depending on the resin content. Of course, moisture content greatly affects this value because water not only dilutes the mass but also requires heat to be vaporized.

Durability of wood

Under favorable conditions, wood may retain its characteristics for centuries. However, wood can be a victim of weather, decay, insects, and fire. Engineers will benefit in their design activities from knowledge about these degradation mechanisms.

* Among the common woods listed in Table 10–3.1, the softwoods are softer than the hardwoods. Although typical, this contrast cannot be generalized. Balsa, a hardwood, has a hardness value of less than 500 N.)

Figure 10–3.2 Blistered paint. Moisture that migrates from the building interior is concentrated behind an impermeable layer of paint. (Courtesy of Forest Products Laboratory, Forest Service, USDA.)

Weathering progresses most rapidly by causing *checking* of the wood surface. If the surface of unprotected dry wood becomes wet, it swells and introduces compressive stresses in the surface layer. Since cellulose, like all polymers, is viscoelastic (Section 8–3), it creeps slightly to relieve those stresses. With subsequent drying, the wood in this surface layer contracts, opening up microcracks. Repeated wetting and dryings accentuate this cracking into deep checking and eventual loss of surface fibers (Fig. 10--1.3b).

Weathering by ultraviolet *radiation* leads to scission of the cellulose in the surface fibers and to loss of mechanical properties. (Cf. Fig. 8–4.7.) These changes in the wood are usually accompanied by changes in color.

The weathering of wood is most easily avoided by painting. In order to be effective, a paint must be opaque to light and must repel moisture. Of course, it is also necessary for a tight bond to form between the paint and the surface of the wood. This implies that the paint must penetrate the structure of the wood to establish that bond. Also, many paints must be applied over a dry surface to assure maximum anchorage. The bond of the paint to the surface is most readily lost if moisture migrates to the exterior of a building from the inside, forming blisters (Fig. 10–3.2). For example, moisture in the air inside a house migrates through the wall to the colder, outer surface if there is no moisture barrier in the wall. In turn, in many cases the paint on the outside surface is a barrier so that critical amounts of the moisture are easily condensed just underneath the paint, causing a separation of the paint from the wood, that is, *blistering*.

Decay occurs by biological action, most commonly by the growth of *fungi,* a primitive form of plant life. The fungi release enzymes that depolymerize the cellulose, hemicellulose, and lignin. It is impossible to avoid the threat of fungi. Although sterilization of the wood by heat destroys any fungi existing in the wood, their spores are invariably present in the surrounding air. Fortunately, fungi spores cannot germinate and grow in wood if the moisture content is below the 20% that is typical in most structures. Rapid growth does not occur until the moisture content approaches the fiber saturation point (~30%). For these reasons, wood framing in houses may remain sound for historical periods of time. However, wood below the ground line, or on sills that retain water, can deteriorate fast. Once deterioration starts, the wood may also be invaded by insects of various kinds.

The key to avoiding decay is to maintain the wood in a "dry" condition. Alternatively, wood may be impregnated with toxic chemicals that preclude fungus growth. Currently, considerable attention is being given to the choice of preservatives that are effective in the wood but do not become a contaminant to the environment.

Some woods contain their own natural preservatives in their heartwood (for example, redwood, cedar, cypress, and black locust) because they possess natural toxic chemicals. Other woods, plus the sapwood of the above woods, do not have this natural protection.

Insects* also thrive in moisture saturated wood. The presence of many of these organisms follow the decay by fungi. There are, however, insects and borers that will invade sound wood, particularly if there is a connection to their habitat in adjacent ground. Damage caused by termites, carpenter ants, and certain beetles is most widespread. Control is most effective if it eliminates possibilities for subterranean contact and also utilizes chemicals that are toxic to these organisms. Impregnation with creosote, tars, and oils is also used, specifically for piles, docks, and other submarine structures where marine borers are a problem.

There is a technical problem encountered in any impregnation treatment that requires filling or coating the internal pores of the wood. The preservative must wet the wood to enter the capillary pores. Also, in pressure processes, the air must be removed to avoid its entrapment and consequent exclusion of the preservative. Heating expands the air, thus facilitating its removal. Tar and creosote also enter more readily when hot. A vacuum process must be used, however, to assure the most thorough impregnation. Of course, such a heating-evacuation-impregnation-cooling cycle is expensive but warranted for certain applications.

Treatment for *fire retardation* by chemicals such as ammonium phosphate, zinc chloride, boric acid, and ammonium sulfate is based primarily upon their effect in preventing formation of volatiles that ignite in the presence of air. Fire-retardant paints can be used; they blister and foam to temporarily insulate the underlying wood, keeping it below the temperature of spontaneous ignition. Those paints must not contain volatile components that are combustible.

Illustrative problem 10-3.1 A quartersawed board was trimmed to 18 mm × 91 mm (nominal 1 × 4) when it contained 21% moisture. (a) What cross-sectional dimensions will it have if it loses two thirds of its moisture? (b) Repeat for a plainsawed board.

---

* Although biologically incorrect, we will include all small organisms in our present discussion.

SOLUTION Moisture change = −14% (dry basis). Using the data in Section 10–3 for typical wood:

a) Width = $[1 + (-14\%)(0.0015/\%)]$ 91 mm = 89 mm;
Thickness = $[1 + (-14\%)(0.0025/\%)]$ 18 mm = 17.4 mm.
Dimensions = 17.4 mm × 89 mm.

b) Width = $(1 - 0.035)$(91 mm) = 88 mm;
Thickness = $(1 - 0.021)$(18 mm) = 17.6 mm.

ADDED INFORMATION If the wood were clamped by nailing, or otherwise, the above shrinkages of 2.1% and 3.5% would introduce tension stresses. Cracking results if the stresses exceed the corresponding strength in tension ($S_{t(r)}$ or $S_{t(t)}$) of the wood. ◀

Illustrative problem 10–3.2 The bulk density of some dry wood is ~ 0.54 g/cm³. (a) Estimate the percent porosity of dry wood if the true density of the cellulose, hemicellulose, and lignin averages 1.5 g/cm³. (b) This wood absorbs 12% water (dry basis). What is the volume increase? The density increase?

SOLUTION Basis: 1 cm³ dry wood = 0.54 g.

a) Volume solids = $(0.54\ \text{g/cm}^3)/(1.5\ \text{g/cm}^3)$
= $0.36\ \text{cm}^3$.
Porosity = $(1\ \text{cm}^3 - 0.36\ \text{cm}^3)/1\ \text{cm}^3$
= 0.64 (or 64%).

b) From data in Section 10–3:

$$\text{Volume (moist wood)} = [1 + 12(0.0015)] \times [1 + 12(0.0025)] \times [1 + 12(0.0001)]\ \text{cm}^3 = 1.05\ \text{cm}^3.$$

$$\text{Increase} = (1.05\ \text{cm}^3 - 1.0\ \text{cm}^3)/1.0\ \text{cm}^3 = 5\ \text{v/o}.$$

$$\text{Mass (moist wood)} = (0.54\ \text{g})(1.12) = 0.605\ \text{g}.$$

$$\text{Density} = 0.605\ \text{g}/1.05\ \text{cm}^3 = 0.58\ \text{g/cm}^3.$$ ◀

10–4 PROCESSED WOOD AND WOOD PRODUCTS

The most direct use of wood is in roundwood products—utility poles, posts, pilings, and some mine timbers. The only processing is trimming, debarking, and commonly a preservation treatment.

Construction lumber

Most wood is used as boards with rectangular cross section. They vary over a wide range of dimensions. Although strictly speaking, quartersawed lumber is to be preferred over plainsawed lumber for critical uses (Fig. 10–4.1), to

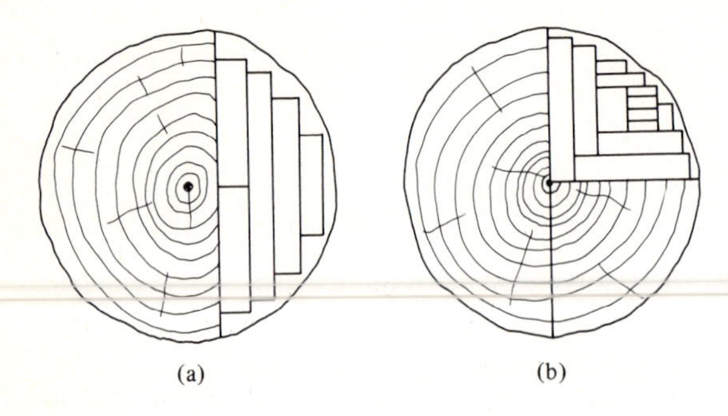

Figure 10-4.1 Lumber cuts. (a) Plainsawed. (b) Quartersawed. The former is more rapid and cheaper. The latter produces the maximum amount of prime cuts.

Table 10-4.1
Construction lumber sizes

Nominal size*	Typical size (dry) inches	mm
1 × 2	$\frac{3}{4} \times 1\frac{1}{2}$	20 × 40
1 × 4	$\times 3\frac{1}{2}$	20 × 90
1 × 8	$\times 7\frac{1}{4}$	20 × 185
2 × 4	$1\frac{1}{2} \times 3\frac{1}{2}$	40 × 90
2 × 6	$\times 5\frac{1}{4}$	40 × 140
2 × 8	$\times 7\frac{1}{4}$	40 × 185
2 × 10	$\times 9\frac{1}{4}$	40 × 235

*Dimensional units do not need to be used, since the nominal size is only a label.

restrict cutting to quartersawing would not only be very expensive, but quite unnecessary for most uses of lumber. Therefore, "breakdowns" are made that maximize other values in the log (Fig. 10-4.1). The sawyer uses both trade-offs in the choice of cuts and a knowledge of the company's order file to optimize the value from the log.

The *nominal sizes* of boards do not coincide with the actual sizes (Table 10-4.1). The reason becomes apparent when examining Fig. 10-4.1(b). The spacing of the saw blade must be based on a common multiple, or else there will be excessive losses. The saw cut consumes 2 mm–5 mm, depending on the log size and the type and size of saw. Then the rough-cut lumber must be planed to the finished dimension. This consumes 1 mm–3 mm on each side of the board. Also the wood is commonly cut before it is completely dried; therefore there is shrinkage. Thus, the final dimensions differ measurably from the original saw spacing that identifies the nominal size.*

* Since the nominal size is a code, or label, and not the actual dimension, there is no need to require a change to the SI units. For example, a label of 5-by-10 (centimeters), rather than 2-by-4 (inches), would not add any technical consistency when identifying a wall stud. On the other hand, it can be desirable to use metric dimensions when designing laminated wooden beams for structures that require engineering calculations (Fig. 10-4.5).

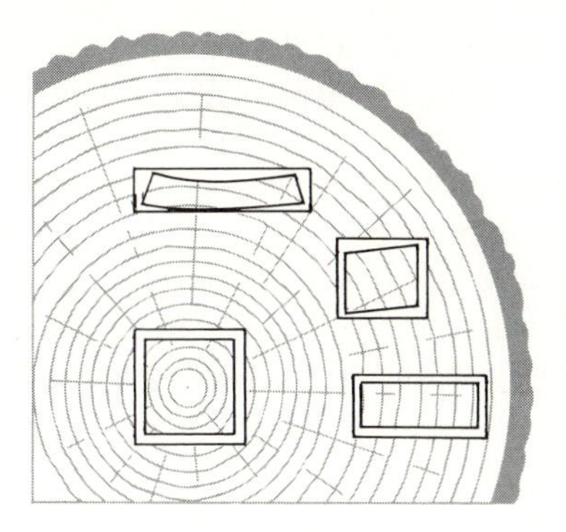

Figure 10–4.2 Shrinkage of wood. Since the structure is anisotropic (that is, it varies with direction), shrinkage is nonuniform. This may cause warpage if the wood is cut before it has dried completely. (Shrinkage is exaggerated in this sketch.)

Drying of lumber

The wood of a living tree continuously receives moisture from the roots and transports it upward to the leaves. The moisture content of the wood in a tree ranges from 30% to more than 200% (oven-dry basis), depending on the species. In general, the moisture content of softwoods is more than of hardwoods. Also, the sapwood of the conifers (the softwood trees) always has more moisture than does the heartwood. This last generalization does not apply to hardwoods.

When the tree is cut and the root supply of moisture removed, the moisture content starts to drop to that which is in equilibrium with the surrounding air. We saw earlier (Section 10–3) that this ranges from 6% to 15%, depending upon humidity and temperature. By losing moisture, the wood becomes more durable, being less subject to biological degradation. However, unless carefully controlled, drying also produces volume changes that can lead to warping and checking.

The first moisture to be removed from the wood vacates the cell cavities and increases void spaces. This does not introduce significant shrinkage. However, once the moisture content is below the fiber saturation point (Section 10–3), shrinkage occurs with further drying. We saw this in the data of the previous section. Since the shrinkage differs radially, tangentially, and longitudinally, warpage may result (Fig. 10–4.2). *Twists* can also occur if the two ends of the board have a different relative position in the original log. (This is possible because the tree varies in diameter along its length. Also, some of the original logs may not have been straight.) Since the surface dries ahead of the interior, *checks** may occur unless care is taken to control the drying rate.

The drying process may be either in the open air or in kilns. In both cases, the cut lumber must be stacked for easy air movement. *Open-air drying* does not require energy costs but is slower and less uniform. *Kiln drying* uses forced air, and considerable heat input. It can be comparatively

* *Checks* are longitudinal cracks across the growth rings. *Shakes* are cracks parallel to the curvature of the growth rings.

rapid. Note, however, that when heat is used, it is desirable to avoid any drying until the lumber is fully heated. This is done by "steaming." In that manner, the moisture can diffuse to the surface as fast as it is removed from the surface, and checking and warping are less likely to occur. Kiln drying may be in the 100°C range, which is sufficient to kill insects and most microorganisms. Kiln drying will reduce moisture contents of 2-cm (1-in.), air-dried hardwood boards from 20% to 6% in 7–14 days. Up to 50 days are required to air-dry green hardwood lumber to the 20% level.

Grading of lumber

Variations are found from piece to piece of lumber. This has led wood producers and industrial wood consumers to establish a series of lumber grades. They enable the user to specify the quality that most nearly meets the intended purpose. For a given type of lumber, the grade is based on (a) the dimensions, and (b) the features that may affect quality, for example, strength and durability. The more obvious features include knots, cracks, pitch pockets, and stain.

Hardwood that is sold to industrial consumers who manufacture wood products is graded to indicate the proportion of the boards that can be cut into useful pieces, called "cuttings." The highest grades must have one face clear and the reverse side sound, without defects such as cracks, rot, or pith. The lowest grades specify only that the wood of the cuttings be sound. Grade steps include "firsts," "seconds," "selects," and three grades of "common." The "firsts" must have more than 91% usable clear face cuttings. At the other extreme, "No. 3 common" may have as little as 25%.

Softwood is used much more extensively for construction. There are three principal grading categories: (a) stress-graded, (b) nonstress-graded, and (c) appearance lumber. The initial category applies to framing lumber for joists, rafters, etc., which are nominally 2 inches (5 cm) or greater in thickness. The grade labels are "select," No. 1, No. 2, and No. 3. With 2 × 4's, the grade labels are "construction," "standard," and "utility." Nonstress-grade lumber refers to thinner boards, usually 1-in. nominal thickness. Grades are No. 1, No. 2, and No. 3, (or "construction," "standard," and "utility"). These are illustrated in Fig. 10–4.3 for white pine. Appearance lumber is used for siding, for door casings, and in many other visible locations. The grades for appearance lumber are closely associated to the type of wood. For example, "clear heart" and "select" apply to cedar, and "clear, all-heart," "clear," and "select" are redwood grades. "Vertical grain" is sometimes part of the grade designation.

Since the grading of lumber is not simple, most lumber is graded under the supervision of inspection bureaus. The details are too extensive to present in the space available here. Therefore, the engineer who specifies wood for technical purposes is urged to contact the appropriate standards organizations for specifics.

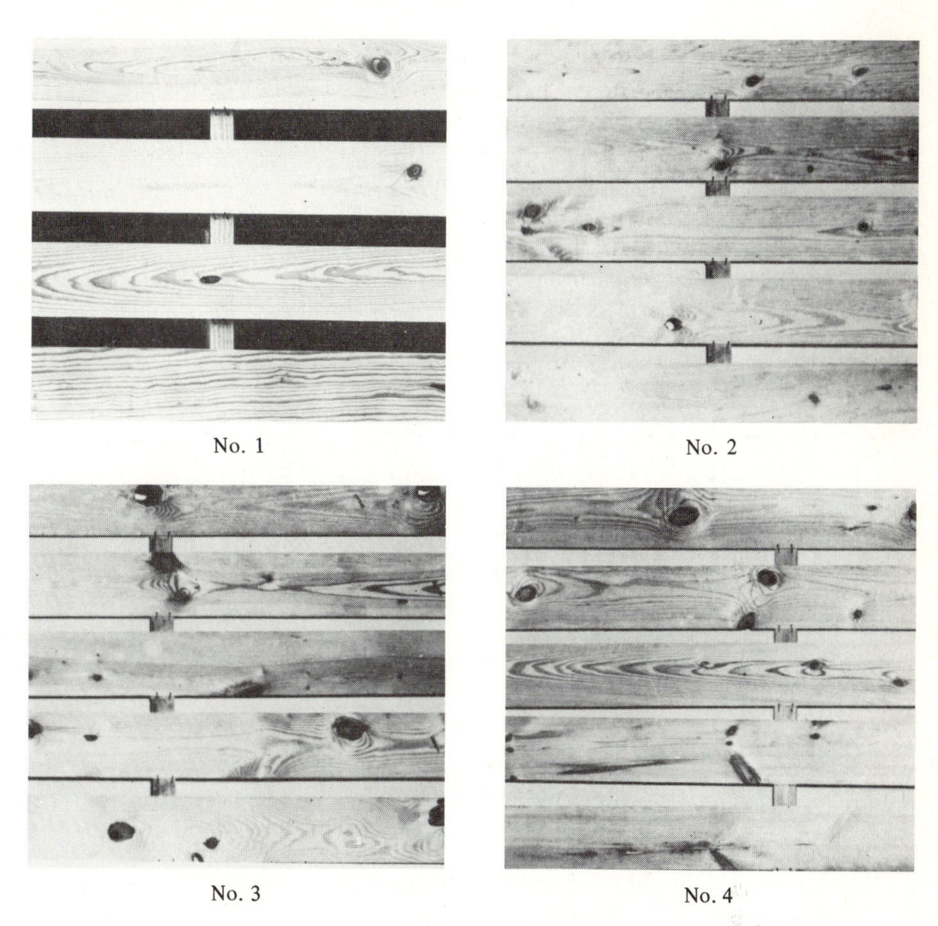

Figure 10-4.3 Grades of lumber for nonstress applications (1 × 4 pine). The No. 1, No. 2, and No. 3 grades are also called "construction," "standard," and "utility," respectively. (Courtesy of Forest Products Laboratory, Forest Service, USDA.

Modified wood

Plywood, laminated wood, particle board, and fiber board are widely used wood products. Each utilizes wood that is restructured into panels. They fulfill many construction needs not obtainable from the unaltered wood.

Plywood is made by laminating separate plies alternately at right angles to one another. That is, the longitudinal directions of the grain structure are at 90°. This produces a biaxial panel—one that has comparable properties in two planar directions. (The properties are not identical in the thin, third direction; however, this does not matter since panels are generally used in planar applications.) The plies are generally produced by rotating a log against a large knife blade. By first softening the wood with steam

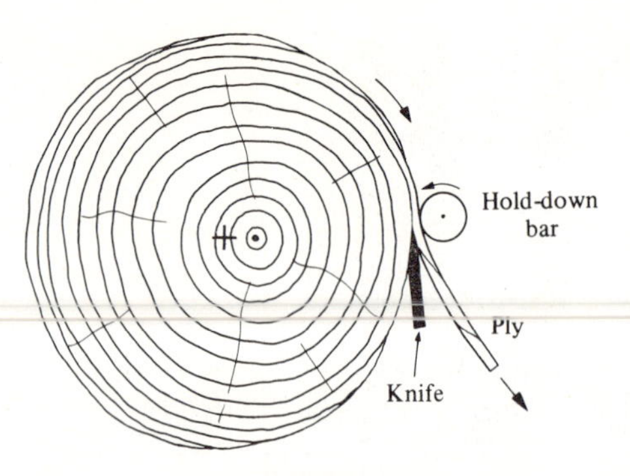

Figure 10-4.4. Cutting veneer. The wood is initially softened by steam or hot water. Since the longitudinal and tangential shrinkages differ, drying must precede the laminating step.

or hot water, and using an appropriate hold-down bar (Fig. 10-4.4), uniform veneers can be cut thinner than 1 mm when necessary.

Compared to solid wood, the plywood has added resistance to splitting and improved dimensional stability because of its cross-ply construction. This utilizes the anisotropy of wood advantageously (Section 10-3). Glues for interior plywood can sometimes be simpler ones that develop strength by drying. Exterior plywood is commonly bonded with phenol-formaldehyde or similar resins that are water resistant. Since these must be thermally set at high temperatures, exterior plywood is a more expensive product.

Laminated wood is a composite of a number of smaller boards bonded into a larger unit. Thus, it is possible to build up large beams (Fig. 10-4.5), arches, and columns, and to make special shapes, such as chair frames and aircraft propellers. The grain of the contributing boards is laid parallel. The boards can be selected to avoid knots and dried to avoid severe checking that would result in the drying of large timbers. The result is a superior product.

Figure 10-4.5 Laminated wood beam. Greater uniformity and larger sizes are achieved than with nonlaminated beams because the effects of knots and checks can be reduced. (Courtesy of Forest Products Laboratory, Forest Service, USDA.)

Figure 10-4.6 Fibers of pulped wood (paper). Fibers are released by mechanical and/or chemical means. (×100.) (Courtesy of A. Nisson, Westvaco Corporation.)

Fiberboard contains wood that was first pulped and then reconstituted into a panel. The product is porous and with low density when it is to be used for acoustical tile or for insulation sheathing for house construction. In contrast, hardboard panels may be made that are denser than wood and are therefore highly wear-resistant.

All panel products are made to specific sizes since additional trimming and planing are not necessary.

Paper

Most paper is made from specially prepared wood fibers (Fig. 10-4.6). Both softwoods and hardwoods are used. The former gives long fibers that add strength and tear-resistance to the paper. Hardwood fibers introduce greater opacity to the paper; therefore, paper from hardwood is preferred for printing stock. The majority of printing papers (other than newsprint) also contain clay additives to further increase the opacity and to enhance the surface smoothness.

In order to obtain wood fibers as shown in Fig. 10-4.6, it is necessary to *pulp* the wood. *Mechanical pulping* is used for newsprint and other papers that do not have strength and longevity as major requirements. *Chemical pulping* releases the fibers by dissolving away the lignin and hemicellulose. Then the fibers are mechanically "beaten" to cause *fibrillation,* which loosens some of the microfibrils from the surface of the fibers (Fig. 10-4.7). This increases the surface area, and when combined with swelling, enhances the cohesion between fibers.

Paper has directional properties* because the paper-making process deposits a suspension of fibers (~4% fibers and 96% water) onto a rapidly

* Try two simple experiments. (a) Tear a newspaper lengthwise and crosswise. In which direction does it tear more easily? Straighter? (b) Use the cardboard off the back of a tablet and cut it into an 8-in. square (~200 mm square). First place an easily stretched rubber band around it in one direction. Remove the rubber band and replace it at 90°. Contrast the two responses of the cardboard. (It may be necessary to try a second rubber band if the first one is too weak or too strong.) Rationalize the results that you observed for these two experiments.

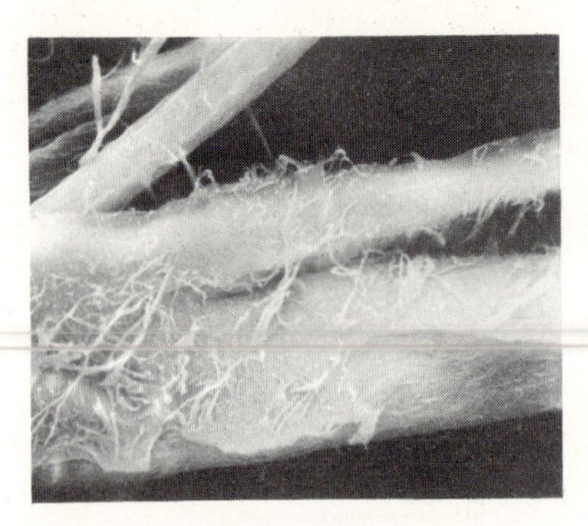

Figure 10–4.7 Microfibrils (paper). Additional mechanical beating subdivides the fibers. The strength of the paper is enhanced with this fibrillation because there are more opportunities for cohesion by hydrogen-bonding at the fiber surfaces. (Courtesy of A. Nisson, Westvaco Corporation.)

moving wire mesh belt of a paper-making machine. The fibers fall into partial alignment as they are retained on the moving mesh while most of the water runs through. The balance of the water is removed as the mesh and the paper pass over vacuum and heating stations in the paper-making machine.

The *recycling* of paper receives considerable popular attention. Obviously, this should be encouraged; however, it should also be noted that technical problems exist. Repulping invariably leads to shorter fibers and, therefore, weaker paper. Secondly, fillers, such as clay, that were previously added for opacity and surface smoothness must first be removed and then reapplied. This is a more expensive process than when working with virgin pulp. For these reasons, recycled magazines should be segregated from newsprint. Finally, inks, both black and colored, lead to a discoloration of the recycled product unless steps are incorporated to remove them. Typically, recycled paper must be used to produce lower grade products than initially.

Review and Study

SUMMARY

The properties of wood are structurally controlled. Therefore its processing and use must be closely related to its structure.

1. Wood has a lamallar macrostructure arising from its annual growth pattern. Its microstructure is determined by its biological cells, which are elongated and therefore introduce directional properties. The biological cells contain three layers of microfibrils that provide strength and flexibility. The principal molecular species is cellulose, $(C_6H_{10}O_5)_n$.
2. Earlywood and latewood refer to the fast-growing and slow-growing portions of the growth ring. In the latter, the cell walls are thicker and the center cavity

smaller. Softwoods include wood from conifers; hardwoods, the broadleaf trees. *There are exceptions* within the two groups; however, in general, the hardwoods are somewhat harder, slightly more dense, and commonly stronger than softwoods. Their fibers are more opaque and shorter than fibers from softwoods.

3. There are three principal directions in wood—longitudinal (l), radial (r), and tangential (t). Mechanical properties vary in these three orientations, being highest in the longitudinal direction.
4. Wood swells and contracts with moisture variations that arise from the humidity of the air, adsorbing (and desorbing) on the fiber surfaces. The cell cavities are not filled with water until the fiber saturation point (~30% H_2O on a dry basis) is exceeded. Dimensional changes are greatest in the tangential direction (~0.25%/% H_2O), less in the radial direction (~0.15%/%), and almost nil in the longitudinal direction.
5. Wood is durable under favorable conditions. It can deteriorate by weathering, particularly from cycles of wetting and drying, and from ultraviolet radiation. Weathering can be retarded by coatings of paint. Wood is also subject to biological failure through decay and insect attack. These can be avoided by keeping the wood dry, or by the impregnation of selected preservatives. Fire destruction is well known. Fire retardants are available; however, in general, preventative measures are paramount.
6. Structural defects in wood include knots (former branches), reaction wood (wood grown under tension or with added compression), pitch pockets (in softwoods), checks (radial cracks), and shakes (tangential cracks).
7. Wood can be modified (a) to produce plywood panels with biaxial properties; (b) to construct laminated boards and beams with improved properties; (c) to make fiberboard, either porous or dense; and (d) to serve as the raw material for paper products.

TECHNICAL TERMS

Anisotropic Having different properties in different directions.

Cellulose Natural polymer of $\left(C_6H_{10}O_5\right)_n$.

Checking (wood) Longitudinal cracking (perpendicular to the growth rings), commonly caused by alternate wetting and drying cycles.

Construction lumber Wood cut to nominal dimensions for building purposes.

Decay (wood) Deterioration by microorganisms that produce enzymes that depolymerize the cellulose or the lignin.

Earlywood First wood growth during the growing season. Part of the wood growth ring with larger, thinner-walled cells. (Spring wood.)

Fiber Biological cells of wood. Typically, micrometer diameters and millimeter lengths. These contain microfibrils in the cell wall.

Fiber-saturation point (FSP) Upper limit of moisture adsorption onto the wood fibers. The limit of volume expansion by moisture absorption into the wood. (Excess water enters the wood pores.)

Grading (lumber) Quality specifications of lumber established by product/trade organizations.

Hardwood Wood from deciduous (broad-leaved) trees.

Heartwood Wood from the interior (nonfunctional) part of the tree trunk. In most trees it is darker in color.

Hydroscopic Absorbs moisture (usually as a function of the humidity).

Knot Defect in lumber originating from a former branch.

Laminated wood Wood product built up from thinner boards.

Latewood Part of the wood growth ring with smaller, thicker-walled cells. Formed late in the growing season (summer wood).

Lignin Important constituent of many woods; denser than cellulose.

Microfibrils Discrete filamentary units of the cell wall.

Nominal size (wood) Size label, approximating the "as-cut" cross-sectional dimensions in inches. Always larger than the actual finished dimensions.

Plainsawed Wood cut with the width of the board approximately tangential to the growth rings (thickness approximately radial).

Plywood Panels made from laminated plies; each ply cut is a longitudinal-tangential sheet; alternate plies are laid at 90°.

Pulping Disintegration of wood (mechanically or chemically) into its fibrous components.

Quartersawed Wood cut with the width of the board approximately radial to the growth rings (thickness approximately tangential).

Rays Radial structure of wood. (See Fig. 10–2.6.)

Reaction wood Nonuniform wood arising from abnormal stresses during growth.

Sapwood Wood in the outer (still functional) part of the tree trunk.

Shakes Cracks that follow the growth rings (tangential).

Softwood Wood from conifers. It is commonly used for construction lumber.

Veneer Thinly cut wood (usually premium) applied as a surface laminate to other wood.

Weathering Environmental deterioration of wood. Moisture cycling and ultraviolet radiation are the prime causes of degradation.

CHECKS

10A The structure of wood that arises from repeating ___***___ pattern is called growth rings.

10B The seasonal growth just cited can involve spring and summer growth; in some regions of the earth, it involves ___***___ and ___***___ periods.

10C The ___* wood___ has fast growth and thinner-walled cells. The ___* wood___ is denser.

10D The three principal directions of anisotropy in wood are ***, ***, and ***.

10E The cross section of a * sawed 1 × 4 board will show more growth rings than a * sawed board of the same nominal size.

10F Typically, * wood trees are cone-bearing, while * wood trees are deciduous.

10G A familiar exception to the generalization that hardwoods are harder than softwoods is ***.

10H Unlike the * wood, the heartwood exhibits no biological function.

10I The biological cells of most wood contain three layers of *** that spiral around a central cavity.

10J The principal molecular species of wood is *** in which the degree of polymerization may be as high as 30,000.

10K A *** is a former limb of the tree that has been surrounded by the wood of the tree trunk.

10L *** wood is wood that was subjected to stresses during growth, commonly as a result of the tree being nonvertical.

10M The wood in 10L is typically more dense than normal wood. It also has significant *** shrinkage.

10N Oven-dried wood with *** % moisture is the reference for moisture content.

10O Water starts to fill the pores within wood when the ***-*** point is exceeded. Below that moisture level, water is adsorbed onto the ***.

10P Typical summer and winter moisture contents of wood trim within a house are *** % and *** %, respectively.

10Q The least dimensional change during the drying of wood is in the *** direction; the most is in the *** direction.

10R Wood with 10% moisture content will not be 10% more dense (on a g/cm^3 basis) than oven-dried wood because *** . . .

10S The highest elastic modulus of wood is in the *** direction; the lowest is in the *** direction.

10T A denser wood has a * er thermal expansion coefficient (volume).

10U Wood ignites spontaneously in air at temperatures in excess of *** °C (or *** °F).

10V *** results when wet wood swells and viscously deforms the cellulose, followed by drying.

10W Fungi growth in wood is rapid if the moisture content exceeds ***.

10X Warpage occurs during the drying of a board since the shrinkage differs in the *** and *** directions.

10Y Cracking can occur during the drying of plywood following lamination because shrinkage differs in the *** and *** directions.

10Z After pulping, wood fibers are "beaten" to cause fibrillation. This increases the *** between fibers in the paper product.

STUDY PROBLEMS

10–1.1 Calculate the mass of a mer in a polycellulose molecule.

Answer: 162 amu (or 162 g/mole)

10–1.2. Assume the height of the two trees in Fig. 10–1.4 are proportional to the trunk diameter. How do the annual growth rates, G, compare (a) on a volume basis? (b) On a mass basis? [The dry density of the wood in part (b) of Fig. 10–1.4 is 95% of that in part (a).]

Answer: (a) $(G_b)_V = 2.5\ (G_a)_V$

10–3.1. What is the typical volume expansion as wood increases from 6% to 12% moisture?

Answer: 2.4 v/o

10–3.2. Refer to S.P. 10–3.1. The wood is dried from 12% moisture to 6% moisture. What is the density change?

10–3.3. The equilibrium moisture content for wood in houses can vary from 6% in winter to 15% in the summer. Calculate the biannual dimensional change of a solid Douglas fir table top, 900 mm × 1500 mm × 20 mm, (a) when it is quartersawed (width is radial); (b) when it is plainsawed (width is tangential).

Answer: (a) Δw = 12 mm, Δt = 0.4 mm, Δl = 1 mm

10–3.4. Water absorption into wood from moist air is initially among the wood fibers (and not into the pores). The fiber-saturation point for most wood is 30 w/o (of oven-dry wood). What is the density increase of birch when it is changed from 10% moisture to 20% moisture?

10–3.5. A solid wood door was made by gluing together edges of quartersawed boards of Douglas fir. That is, the width of the board is in the radial direction, and the thickness is in the tangential direction (and, of course, the length is the longitudinal direction of the wood). The door was initially trimmed to 761 mm by 2035 mm (to fit a 765 × 2050 mm opening) in the late fall when the moisture content of the wood was 9 percent. Will the door "stick" when the moisture content increases to 14 percent?

10–3.6. The trimmed door in S.P. 13–3.5 was shut when its moisture content increased from 9 to 14 percent. What force would develop if the door is 30 mm thick? Assume that the doorway casing is rigid.

Answer: ~10^5N (or 22,000 lb_f)

10–3.7. A 1-in. cube (25.4-mm cube) of wood is cut with the longitudinal, transverse, and radial orientations parallel to the cube edges. What force can be supported in the three orientations without exceeding 1% elastic strain? [Assume mid-range values of the Young's moduli.]

Answer: $\|_L$ = 17,000 lb_f (or 76,000 N)

10–3.8. Refer to S.P. 5–5.6. How do the woods of Table 10–3.1 compare with the materials of Appendix A regarding minimum weight for a given deflection in the longitudinal direction?

10–3.9. Douglas fir has Young's moduli and shrinkage coefficients in the three directions that are typical for most woods. Its radial strength is 2.4 MPa. A quartersawed (1 × 8) board is clamped when it contains 11% moisture. Estimate the moisture level at which radial cracking (checking) will occur.

Answer: 9% H_2O

10–3.10. (a) Estimate the compressive stress that is necessary in the radial direction of wood to prevent expansion as wood absorbs moisture from the 7% level to 11%. (b) In the tangential direction.

10–3.11. Refer to S.P. 10–3.10 (a). Will any of the woods of Table 10–3.1 fail by radial compression? [Assume E_r to be 750 MPa.]

Answer: Birch and pine

10–3.12. Birchwood veneer is impregnated with phenol-formaldehyde (Fig. 7–2.5) to ensure resistance to water and to increase the hardness of the final product. Although dry birch weighs only 0.55 g/cm^3, the true specific gravity of the cellulose-lignin combination is 1.52. (a) How many grams of phenol-formaldehyde (PF) are required to impregnate 10,000 mm^3 (0.6 in.^{3}) of dry birchwood? (b) What is the final density?

Answer: (a) 8.3 g PF

CHECK OUTS

10A seasonal

10B wet
dry

10C early
late

10D longitudinal, tangential, radial

10E quarter
plain

10F soft
hard

10G balsa

10H sap

10I microfibrils

10J cellulose

10K knot

10L reaction

10M longitudinal

10N zero

10O fiber-saturation
microfibrils

10P 15, 6

10Q longitudinal
tangential

10R the volume expands (but not 10%)

10S longitudinal
tangential

10T higher

10U 400°C (750°F)

10V checking

10W the fiber-saturation point

10X tangential, radial

10Y longitudinal
tangential

10Z cohesion

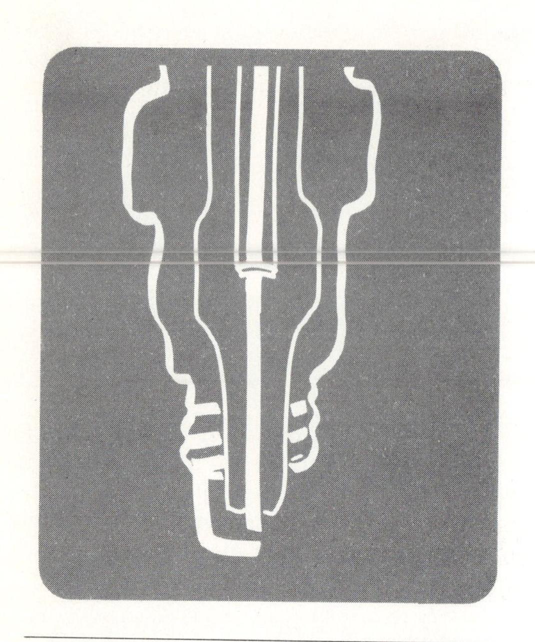

UNIT ELEVEN

CERAMIC MATERIALS

The lay public thinks of ceramics as artware and knows that the products of the earliest artisans were ceramics. The engineer also knows that ceramic materials are used in a wide range of technical products extending from high-speed cutting tools, to piezoelectric transducers, to high-frequency magnets, and to optic fibers. Inherently, ceramic materials are complex, because they are compounds of metals and nonmetals, rather than only metallic atoms. As a result, we shall limit our consideration to the simpler ceramics and the coordination of unlike (metal-nonmetal) neighbors. However, this approach will let us see the origin of the various properties that are characteristic of ceramics.

Contents

Prerequisites General chemistry; Unit 2; and Sections 1-3, 5-3, and 5-4.

From Unit 11, the engineer should

1) Be able to calculate lattice constants and densities of the three basic AX structures.
2) Possess the 3-D picture of the ZnS structure, because this will be basic to the principal semiconducting compounds in Unit 18.
3) Envisage the locations of 6-f and 4-f interstitial sites in an fcc lattice of anions, because they are important in magnetic materials (Unit 19).
4) Be familiar with the similarities and the differences between the structure (and properties) of ceramic and nonceramic phases.
5) Possess the ability to understand technical terms when discussing ceramic compounds with engineering peers.

11-1 CERAMIC COMPOUNDS (AX)

Ceramic materials contain *compounds* of *metallic and nonmetallic elements*. Thus, they include simple *binary* compounds such as MgO and LiF, which have equal numbers of positive and negative ions. Ceramic* compounds also include complex materials such as clay, with use dating back to prehistoric times. In this chapter, we shall direct our principal attention to the simpler compounds that contain only two or three types of elements. They will illustrate many of the structure–property relationships that are utilized by the design engineer. However, we shall not ignore clays, if for no other reason than that they are widely used by artisans, both ancient and modern.

The metallic elements are those that readily lose their valence electrons and become positive ions. They reside in the lower half and/or on the left side of the periodic table as it is invariably drawn (Fig. 1-3.2). The nonmetallic elements readily accept (or share) electrons to become negative ions. They are located in the upper-right section of the periodic table.

Compounds of two elements do not form unless there is an energy preference for atoms to have unlike neighbors. Thus, energy† is lowest when magnesium atoms are surrounded by oxygen atoms (or in this case, ions). This is in contrast to gold, which has an energy preference for additional gold atoms and not for surrounding oxygen atoms as neighbors. Therefore,

---

* The term *ceramic* originated from *keramikos,* the Greek word for clay products.

† More specifically, the "free energy."

gold does not oxidize under normal conditions. Since the energy preference between unlike ions is commonly very great, each ion will coordinate with as many unlike neighbors as possible. Therefore, the structures of many ceramic compounds are dictated by the atom or ion sizes. Of course, there must also be a charge balance.

Ionic radii

The atomic radii of atoms in pure metals are simply half of the interatomic distances.* Each atom of the neighboring pairs is identical. In a compound the interatomic distance† involves two unlike atoms or ions. Almost always they differ in size. It is not possible to partition the interatomic distance between the two ions as precisely as we can measure the interatomic distance. The latter is easily measured to four or even five significant figures. By indirect measurement, it is possible to assign the radii of ions to only two or occasionally three significant figures. Thus, we assign 0.068 nm and 0.133 nm as the radii of Li^+ and F^-, respectively, although we know that the Li^+–F^- distance† is 0.20086 nm. Also, $r_{Mg^{2+}} \cong 0.066$ nm and $R_{O^{2-}} \cong 0.14$ nm; but the Mg^{2+}–O^{2-} distance† is 0.2106 nm. Part of the uncertainty arises because the size of an ion will vary somewhat as it pairs up with different neighbors.

Table 11–1.1 lists the more widely accepted radii for various common ions. Also see Appendix B. Let us note several points. First, the cations are smaller than the comparable anions, for example, $r_{K^+} = 0.133$ nm and $R_{Cl^-} = 0.181$ nm, although their masses are 39.1 amu and 35.5 amu, respectively. This is to be expected since the K^+ has lost its valence electron, so the remaining electrons are drawn closer to the nucleus. In contrast, the anions have more electrons than protons. There is mutual repulsion among these electrons, so that the space commanded by the negative ion is increased. Also observe that a ferric ion, Fe^{3+}, is smaller than a ferrous ion, Fe^{2+}. This is consistent with expectations.

We may further observe from Table 11–1.1 that the ions are larger when they have a higher coordination number, that is, more neighbors. With more neighbors, the packing factor is higher and more electrons are present per unit volume. This increases the electronic repulsion and therefore increases the interatomic distances and ionic radii.

AX structures (CN = 8)

One type of binary compound has the structure shown in Fig. 11–1.1. It is cubic; each ion is coordinated with *eight* unlike neighbors (CN = 8); there

* Determined by x-ray diffraction measurements.

† Atom center to atom center.

Table 11-1.1
Radii of selected ions

Ion	Approximate* radius, nm			Ion	Approximate* radius, nm		
	CN = 4†	CN = 6	CN = 8		CN = 4	CN = 6	CN = 8
Al^{3+}	0.046	0.051		Mn^{2+}	0.073	0.080	
Ba^{2+}		0.134	0.138	Na^{+}		0.097	
Be^{2+}	0.032	0.035		Ni^{2+}	0.063	0.069	
Ca^{2+}		0.099	0.102	O^{2-}	0.128	0.140	0.144
Cl^{-}		0.181	0.187	$(OH)^{-}$		0.140	
Cr^{3+}		0.063		P^{5+}		0.035	
Cs^{+}		0.167	0.172	S^{2-}	0.168	0.184	0.190
F^{-}	0.121	0.133		Si^{4+}	0.038	0.042	
Fe^{2+}	0.067	0.074		Sr^{2+}		0.112	
Fe^{3+}	0.058	0.064		Ti^{4+}		0.068	
K^{+}		0.133		Zn^{2+}	0.067	0.074	
Li^{+}		0.068		Zr^{4+}		0.079	0.082
Mg^{2+}	0.060	0.066	0.068				

* The radii are approximate. In LiF, for example, the center-to-center distance between Li^+ and F^- ions can be accurately measured as 0.20086 nm; this equals $(r_{Li^+} + R_{F^-})$. Beyond this it is necessary to approximate what fraction of that distance is in each ion. That approximation leaves a small probable error to the radii values. (Patterned after Ahrens.)

† CN = coordination number, that is, the number of contacting neighbors in the compounds. For ions, $1.1\ R_{CN=4} \approx R_{CN=6} \approx 0.97\ R_{CN=8}$. See also Appendix B.

Figure 11-1.1 AX structure (CN = 8, CsCl-type). (a) 3-D. (b) Plan. Each ion, + and −, is coordinated with eight neighbors. This structure requires $r/R \geq 0.73$ to permit CN = 8. It is *not* body-centered cubic since the center of the unit cell is different from the corner of the unit cell.

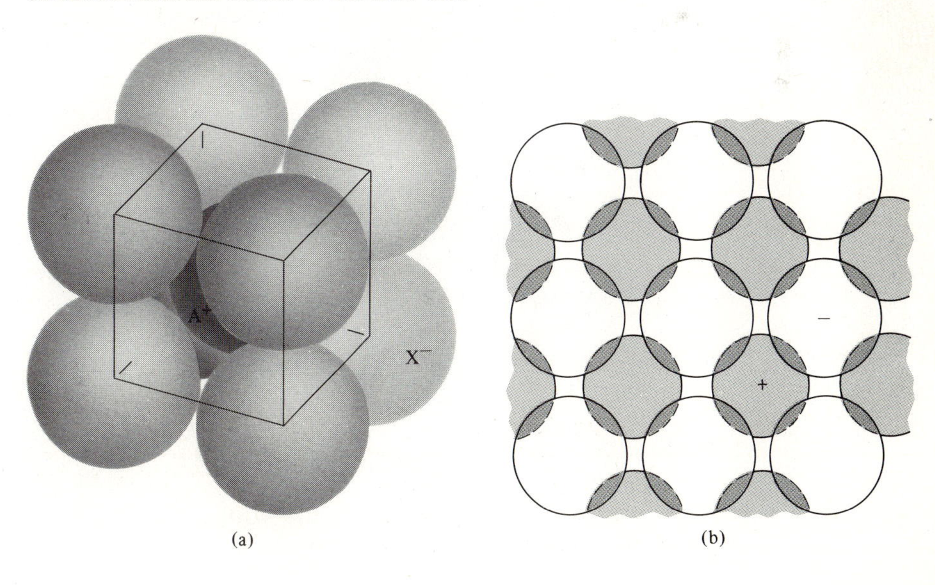

are equal numbers of positive and negative ions (A = X); and each unit cell contains just one ion of each type. The prototype for this structure is CsCl, a compound in which the size ratio of the ions is $r_{Cs^+}/R_{Cl^-} = 0.9$. (Data are from Table 11-1.1.) With this ratio, eight Cl^- ions can be coordinated with each Cs^+ ion without interferring with other Cl^- ions.

Observe that the unit cell is *not* bcc, since the center atom of the cube is not identical with the corner. Also observe that this structure would be impossible if r/R were less than 0.73. Below that ratio, the negative ions would be in contact and repel each other, and the cations could not simultaneously coordinate with all eight anions.

AX structures (CN = 6)

A second type of binary compound is presented in Fig. 11-1.2. This is the most common structure for ceramics, being the structure for nearly 300 compounds. It is probably the simplest binary structure since it is easily described as an alternating sequence of unlike atoms along the three axial directions. It is cubic; each ion is coordinated with *six* unlike neighbors (CN = 6); there are equal numbers of the two elements (A = X); and each unit cell contains four of each type.

Figure 11-1.2 AX structure (CN = 6, NaCl-type). (a) 3-D. (b) Plan. Each positive and negative ion is coordinated with six neighbors. This structure requires $r/R \geq 0.41$ to permit CN = 6. It *is* face-centered cubic since the center of each face is identical in every respect to the corner of the unit cell.

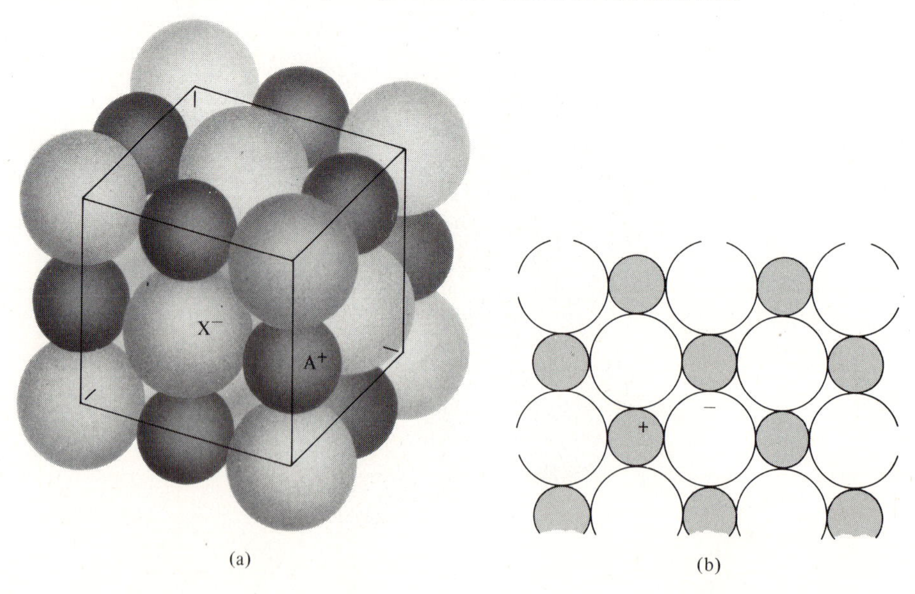

The prototype for this structure is NaCl, a compound in which the size ratio of the ions is r_{Na^+}/R_{Cl^-} = 0.097 nm/0.181 nm = 0.54 (Table 11–1.1). With this ratio, six Cl^- ions can be coordinated with each Na^+ ion without interferring with other negative ions. However, note that this structure would be impossible if r/R were less than 0.41 (I.P. 11–1.3). Below that ratio, the negative ions would be in contact and repel each other, and the positive ions could not simultaneously coordinate with all six neighboring anions.

We can observe that the NaCl unit cell is fcc, since the face-centered positions are identical in every respect with the corner positions. Common ceramic compounds that have this structure include MgO, FeO, TiC, MnS, LiF, NiO, UC, and CaS, to name a few. Although they cannot have this structure if $r/R < 0.41$, there is no restriction against this NaCl-type structure if $r/R > 0.73$. However, with larger ratios, ionic compounds generally assume the CsCl-type structure with CN = 8, since each ion has additional unlike ions as neighbors and, therefore, becomes more stable.

AX structures (CN = 4)

Figure 11–1.3 presents a third binary compound. It is cubic; each atom is coordinated with *four* unlike neighbors (CN = 4); there are equal numbers

Figure 11–1.3 AX structure (CN = 4, ZnS-type). (a) 3-D. (b) Plan. The dashed ions lie at lower (or higher) levels. This structure requires $r/R \geq 0.22$ to permit CN = 4. It *is* face-centered cubic since the center of each face is identical in every respect to the corner of the unit cell.

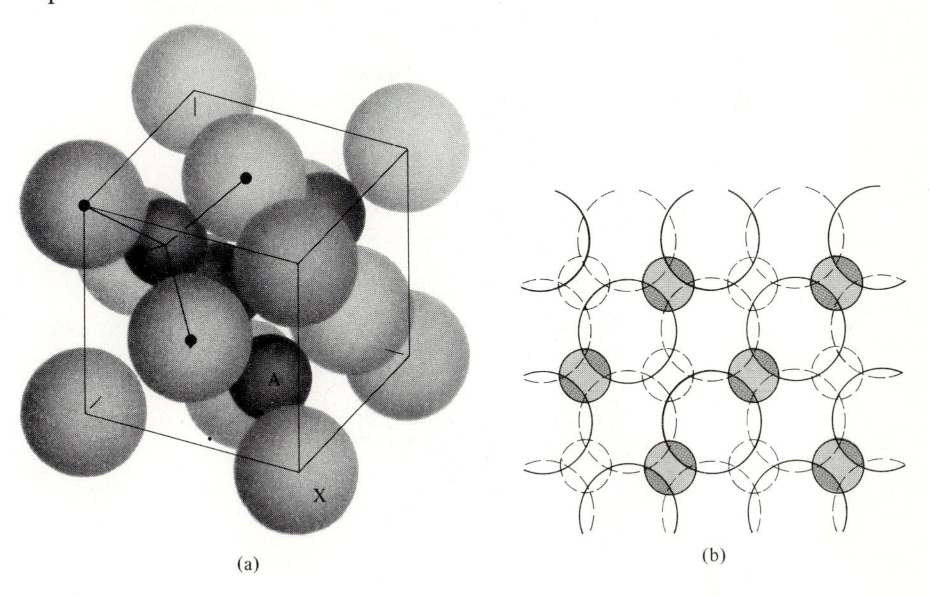

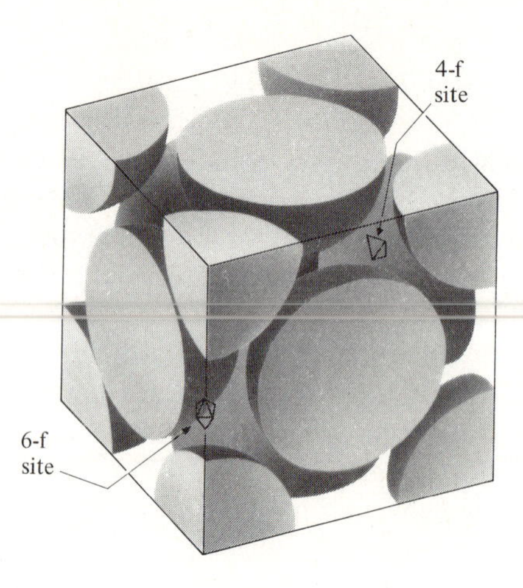

Figure 11-1.4 Interstitial sites (fcc lattice of anions). The cations reside in the spaces (interstices) among the anions. Each six-fold (6-f) site has six neighboring anions. The number of 6-f sites equals the number of anions. Each four-fold (4-f) site has four neighboring anions. There are twice as many 4-f sites as the number of anions. Thus only half of them are occupied in an AX compound.

of the two elements (A = X); and each unit cell contains four of each type. The prototype for this structure is one of the polymorphs of ZnS. Its unit cell is fcc, since face-centered positions are identical in every respect with the corner positions.*

Ionic compounds will choose this structure if r^+/R^- is small. (If r^+/R^- is greater than 0.41, the NaCl structure is favored for ionic compounds, since each ion would have more unlike neighbors.) However, there are few fully ionic compounds with A = X, where r^+/R^- is less than 0.4. Even so, there are a number of compounds with this structure. Included are many of the semiconducting compounds such as GaAs, AlP, InSb (Unit 18). These compounds require CN = 4 for covalent bonding—as does diamond, which has this same basic structure. This ZnS-type structure is an important one for electronic materials.

Interstitial sites

The cations in the CsCl-type structure occupy 8-*fold sites*, that is, within the structure there are *interstices*, or holes, among eight larger anions. All such 8-f sites are occupied by positive ions.

Cations occupy all of the 6-fold sites among the negative ions in the NaCl-type structure. Figure 11-1.4 locates one of these 6-f sites. If the fcc unit cell is drawn in the usual manner, the 6-f sites are located at the center of all 12 edges and at the center of the unit cell for a total of 4 per unit cell.

Figure 11-1.4 also identifies a 4-fold site among an fcc array of anions. There are eight such 4-f sites within the unit cell. Thus, only half of them

* Both NaCl-type and ZnS-type structures are fcc. However, they are not the same; nor are they identical with fcc metals. (Cf, Fig. 11-1.3 with Fig. 2-2.2.)

Table 11-1.2
Selected AX structures

Prototype compound.	Lattice of A (or X)	CN of A (or X) sites	Sites filled	Minimum r_A/R_X	Other compounds
CsCl	Simple cubic	8	All	0.73	CsI
NaCl	fcc	6	All	0.41	MgO, MnS, LiF
ZnS	fcc	4	$\frac{1}{2}$	0.22	β-SiC, CdS, AlP

are occupied in a ZnS-type structure. (Cf. Fig. 11-1.3.) Table 11-1.2 summarizes these patterns for the three basic structures. (In addition, there are other, more complicated AX patterns that we have not described here.)

Illustrative problem 11-1.1 X-ray data show that the unit cell dimensions of cubic MgO are 0.4211 nm. It has a density of 3.6 g/cm³. How many Mg^{2+} ions and O^{2-} ions are there per unit cell? [The structure of MgO is that of NaCl (Fig. 11-1.2).]

SOLUTION

$$\rho = \frac{\text{mass/u.c.}}{\text{volume/u.c.}}.$$

$$3.6\ \text{g/cm}^3 = \frac{n(24.3\ \text{g}/0.6 \times 10^{24}\ \text{Mg}^{2+}) + n(16.0\ \text{g}/0.6 \times 10^{24}\ \text{O}^{2-})}{(0.4211 \times 10^{-7}\ \text{cm})^3/\text{u.c.}}.$$

$$n = 4\ \text{Mg}^{2+}\ \text{ions/u.c.} = 4\ \text{O}^{2-}\ \text{ions/u.c.}$$

NOTE Since MgO is an AX compound, the number of Mg^{2+} ions must equal the number of O^{2-} ions. ◀

Illustrative problem 11-1.2 (a) Based on the ionic radii of Table 11-1.1, what is the distance between the *centers* of nearest neighboring Cl^- ions in NaCl? In CsCl?
b) Assume that the "hard-ball" model applies in ionic materials such as NaCl and CsCl. What is the closest approach of the Cl^- ion "surfaces"?

SOLUTION (a) Examine the structure of Fig. 11-1.2, where A is Na^+ and X is Cl^-. The closest approach of *centers* is $(r_{Na^+} + R_{Cl^-})\sqrt{2}$. Therefore,

$$D_{Cl^- - Cl^-} = (0.097\ \text{nm} + 0.181\ \text{nm})\sqrt{2} = 0.393\ \text{nm}.$$

Repeat for CsCl in Fig. 11-1.1; use the CN = 8 radii of Table 11-1.1:

$$D_{Cl^- - Cl^-} = (0.172\ \text{nm} + 0.187\ \text{nm})(2)/\sqrt{3} = 0.414\ \text{nm}.$$

b) Distance between *surfaces*: In NaCl,

$$d_{Cl^- - Cl^-} = 0.393\ \text{nm} - 2(0.181\ \text{nm}) = 0.031\ \text{nm}.$$

In CsCl,

$$d_{Cl^--Cl^-} = 0.414 \text{ nm} - 2(0.187 \text{ nm}) = 0.040 \text{ nm}.$$

ADDED INFORMATION The ionic radius ratios are

$$r_{Na^+}/R_{Cl^-} = 0.097 \text{ nm}/0.181 \text{ nm} = 0.54,$$
$$r_{Cs^+}/R_{Cl^-} = 0.172 \text{ nm}/0.187 \text{ nm} = 0.92.$$

The Cs^+ ions can have eight neighbors because the ionic radius ratio is >0.73. ◀

Illustrative problem 11-1.3 Determine the minimum r/R for a compound with CN = 6.

SOLUTION Refer to Fig. 11-1.2 and I.P. 11-1.2. The minimum ratio occurs if the anions, X, have radii such that they just touch one another. In such a case,

$$R + R = (r + R)\sqrt{2},$$
$$2R = r\sqrt{2} + R\sqrt{2},$$
$$r/R = 0.41.$$

ADDED INFORMATION If the radius ratio, r/R, is less than 0.41, a 4-fold coordination is mandatory. If it is greater than 0.41, however, 6-fold coordination—that is, CN = 6—is not mandatory. If some other factor such as covalent bonds (shared electrons) becomes important, as it does in ZnS (Fig. 11-1.3), there may be 4-fold coordination with $r/R > 0.41$. ◀

11-2 CERAMIC COMPOUNDS (A_mX_p)

If the valences of the two ions are not equal, the above AX structures are not applicable. Of course, many compounds exist with valence ratios other than 1 : 1. For example, Al_2O_3 has a 2 : 3 ratio; each aluminum ion is +3 and each oxygen ion is −2. Likewise in ZrO_2, another ceramic compound, the valences are +4 and −2. The list could be extended to cover many other compounds in many other ceramic materials.

AX_2 structures (CaF_2-type)

The prototype for the most common of these structures is CaF_2. In addition, ZrO_2 and UO_2 are commercial ceramic compounds of this type. The Ca^{2+} ion has a radius of about 0.1 nm. The F^- is not very large, since it is in the first row of the periodic table (Fig. 1-3.2) and has only one excess electron. Therefore the radius ratio, r/R, is 0.8, and each positive ion can be surrounded by eight negative ions. This determines the structure (Fig. 11-2.1), and as in CsCl, the cations of CaF_2 have a coordination number CN = 8. However, there are only one half as many Ca^{2+} ions as F^- ions, since electrical neu-

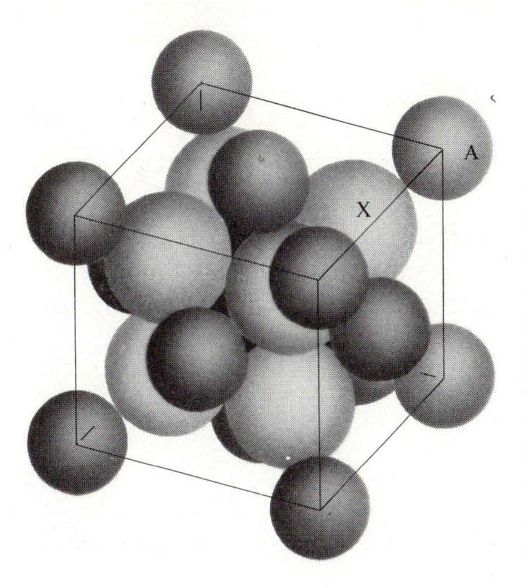

Figure 11–2.1 AX_2 structure (CaF_2). Since the radius ratio is about 0.8, each positive ion can have eight F^- neighbors. (Observe the atom on the front face, which has four additional neighbors in the next unit cell.) However, there must be only one half as many positive ions, Ca^{2+}, as negative ions, F^-; therefore half the 8-fold sites are vacant, that is, the site at the center of the unit cell and the sites at the center of each edge.

trality must be maintained; one half of the cation sites are therefore vacant. This happens to be an important feature of UO_2, a nuclear fuel with this structure, because these unoccupied sites can accumulate fission fragments from nuclear reactions without introducing excessive strains in the solid.

A_2X_3 structures (Cr_2O_3-type)

Aluminum oxide, Al_2O_3, is the most widely used compound in ceramic materials of the nonclay variety. The prime constituent of the spark plug insulator in Fig. 11–2.2, for example, is Al_2O_3. This compound is also used in many other electrical ceramics and abrasion-resistant materials. The prototype for the Al_2O_3 crystal is Cr_2O_3. We may describe Al_2O_3 and Cr_2O_3 crystals as follows. (1) The O^{2-} ions form a hexagonal pattern (cf. Fig. 2–3.1). (2) Cr^{3+} ions (or Al^{3+} ions) are in 6-fold sites among the O^{2-} ions; that is, these cations have CN = 6. (3) Since the A : X ratio is 2 : 3, only two thirds of the 6-f sites are occupied. (4) Finally, since one third of those sites are vacant, the structure adjusts slightly. (This latter adjustment is noticeable to the crystallographer, but will not affect the properties of interest to us.)

Illustrative problem 11–2.1 What is the packing factor of the ions in CaF_2 if we assume spherical atoms with radii as shown in Table 11–1.1? Assume hard-ball ions.

SOLUTION There are $8F^-$ ions (CN = 4) and $4Ca^{2+}$ ions (CN = 8) per unit cell. The distance ($r_{Ca^{2+}} + R_{F^-}$) is one fourth of the cube diagonal:

$$a = \frac{4(0.102 + 0.121)}{\sqrt{3}} = 0.515 \text{ nm},$$

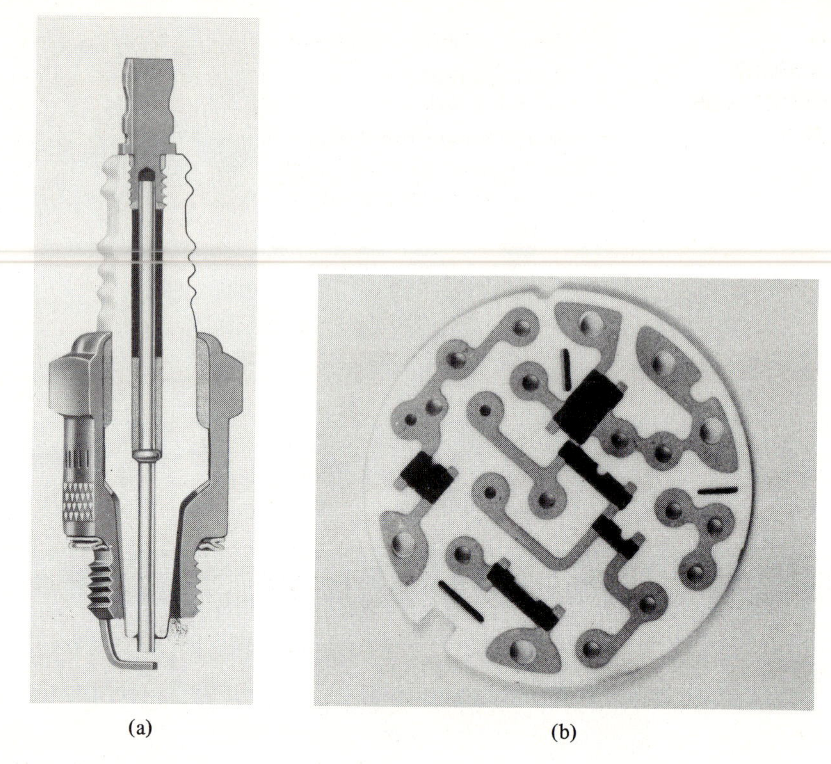
(a) (b)

Figure 11-2.2 Al_2O_3 ceramics. (a) Sparkplug insulator. (Courtesy of Champion Spark Plug Co.) (b) Substrate for a printed circuit. (Courtesy of AC Spark Plug Division, General Motors.) The principal material in these electrical insulators is Al_2O_3. Each Al^{3+} ion has lost its three valence electrons, and each O^{2-} ion contains two extra electrons, which it holds tightly, so that they are not available for electrical or thermal conduction. The cations (3+) and anions (2−) are strongly attracted to one another, as evidenced by the high melting temperature (~2020°C, or 3670°F), and great hardness of Al_2O_3. (One name for Al_2O_3 is emery.)

(cont.)

$$\text{PF} = \frac{4(4\pi/3)(0.102\text{ nm})^3 + 8(4\pi/3)(0.121\text{ nm})^3}{(0.515\text{ nm})^3}$$

$$= \frac{(17.7 + 59.4)}{136.5} = 0.56.$$

ADDED INFORMATION The actual lattice dimension, a, is 0.547 nm because the eight F^- ions around the center vacant site have some repulsion for each other, and this causes a decrease in the packing factor. ◀

11–3 CERAMIC COMPOUNDS ($A_mB_nX_p$)

Ceramic compounds often contain two distinct kinds of metal atoms. For example, materials of the $BaTiO_3$-type are used for phonograph cartridges and for pressure transducers; these devices require materials that interchange mechanical and electrical forces. Also materials of the $NiFe_2O_4$-type are used for magnetic deflection yokes of TV tubes.

Ionic materials of the $A_mB_nX_p$-type must have positive and negative charges that are balanced; the two examples above have

$$(Ba^{2+} + Ti^{4+} = 3O^{2-}) \quad \text{and} \quad (Ni^{2+} + 2Fe^{3+} = 4O^{2-}),$$

respectively. This places another restriction on the structure in addition to the one of radii ratios.

$BaTiO_3$-type structures

Figure 11–3.1 shows this structure. Above 120°C it is cubic, with the Ba^{2+} ions at the corner of each unit cell (CN = 12). The Ti^{4+} ions are at the center of each unit cell (CN = 6). The oxygen ions are at the center of each face. They are closely coordinated with two Ti^{4+} ions and somewhat less closely associated with four Ba^{2+} ions. We shall see in Section 19–3 that below 120°C there is a slight displacement in these ions, a shift that has important consequences for dielectric behavior. The high-temperature form of Fig. 11–3.1 will serve our present purposes.

$NiFe_2O_4$-type structures

Magnetic ceramics can best be described by Fig. 11–1.4, which shows only the oxygen ions. Among these O^{2-} ions are 6-fold sites (cf. Fig.11–1.2). The number of these sites equals the number of O^{2-} ions; however, in $NiFe_2O_4$

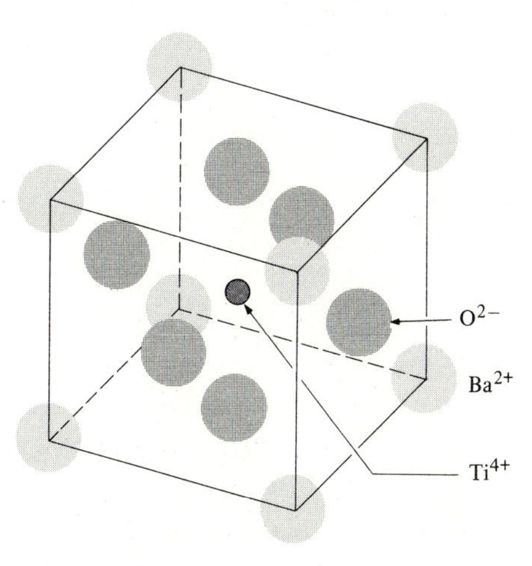

Figure 11–3.1 Ternary compound ($BaTiO_3$ above 120°C). Each unit cell cube has a Ba^{2+} ion at the corner, a Ti^{4+} ion at its center, and O^{2-} ions on each face. Of course, we could have located the cube to corner on the Ti^{4+} ions (I.P. 11–3.1).

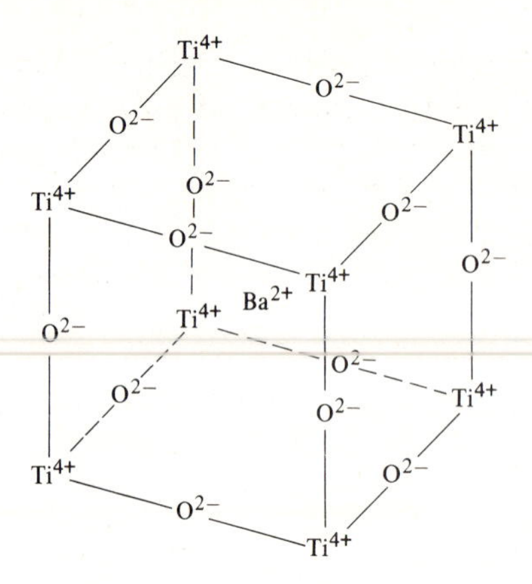

Figure 11-3.2 $BaTiO_3$ with unit cell corners at the Ti^{4+} ion. See I.P. 11-3.1 and compare with Fig. 11-3.1

only half are occupied. The Ni^{2+} ions and half the Fe^{3+} ions sit in these 6-fold sites (CN = 6). Among the O^{2-} ions, there are also 4-fold sites (cf. Fig. 11-1.4). There are twice as many of these sites as there are O^{2-} ions. In $NiFe_2O_4$, one eighth of these 4-fold sites (CN = 4) are occupied by the remaining half of the Fe^{3+} ions.

Illustrative problem 11-3.1 Redraw Fig. 11-3.1 so that the Ti^{4+} ion is at the corner of the unit cell.

SOLUTION See Fig. 11-3.2. We may describe this compound as having Ba^{2+} ions at the cell center, Ti^{4+} ions at the cell corners, and O^{2-} ions at the center of each cell edge.

ADDED INFORMATION Figures 11-3.1 and 11-3.2 are equal alternatives. Each presentation provides the same 1/1/3 ion ratio for Ba/Ti/O.

	Fig. 11-3.1	Fig. 11-3.2
Ba^{2+}	8/8	1
Ti^{4+}	1	8/8
O^{2-}	6/2	12/4 ◀

11-4 POLYMORPHS

The numbers of polymorphic structures (Section 2-6) can increase significantly as additional types of atoms are introduced into the crystal structure. Most single-component metals have only one stable structure; some have two (for example, Co, Ti, and Zr); a few have three or more.* Of course,

---

* Iron is among this group. Included are γ (fcc), α and δ (bcc), and an hcp polymorph that forms under high-pressure conditions at ambient temperatures.

there are numerous compounds with only one polymorph; however, we also find compounds with a large number of polymorphs, particularly if the bonds are covalent and the structure is not closely packed. Silica and SiC, for example, have more than twenty each.

As a general rule, high temperatures favor those structures with greater symmetry and less density because the atoms possess greater thermal agitation. Also, high pressures favor the more dense polymorphs. For example, at 10,000 atmospheres, solid H_2O no longer has the hexagonal structure that we see in snowflakes but is tetragonal with a density of 1.53 g/cm^3.*

Names of ceramic compounds

It is convenient to name various ceramic phases; just as it is convenient to use ferrite as a label for an "iron-rich bcc phase that is stable below the eutectoid temperature." This becomes particularly useful for ceramics since there often are more polymorphs for ceramic compounds than for metallic compositions, for example, silica, with its 22. In general, we will not need to learn many phase names; however, some of the common ones such as quartz (SiO_2 of Fig. 11-6.3), mullite (of Fig. 11-7.3), and corundum (Al_2O_3), will be encountered in this and the next unit. Of greater importance is the need to appreciate that these phases have temperature and compositional areas of stability in the phase diagrams, for example, mullite in the SiO_2–Al_2O_3 system.

11-5 SOLID SOLUTIONS

Substitutional solutions

Just as in the case of metals, ceramic compounds may have interstitial or substitutional solid solutions (Section 2-5). For ceramic materials, we shall consider only substitutional solid solutions. Both size *and* charge of the replacement atoms must be comparable for substitutional solid solutions. Thus, from Table 11-1.1, we may expect extensive replacement of Mg^{2+} ions (r = 0.066 nm) by Ni^{2+}, Fe^{2+}, and Zn^{2+} ions, since they are all divalent and have radii of 0.069 nm, 0.074 nm, and 0.074 nm, respectively. In contrast Li^{+} (r = 0.068 nm) cannot replace Mg^{2+} unless some adjustment in charges is made.

One of the better-known ceramic solid solutions is ruby, in which a fraction of one percent of Cr^{3+} ions replace Al^{3+} ions in Al_2O_3. This solid solution is used not only as a gem but also in lasers.

---

* Give some thought to the consequences, if this were the normal solid polymorph of H_2O. The "ice" would sink to the bottom of the lakes; no freezing expansion; etc.

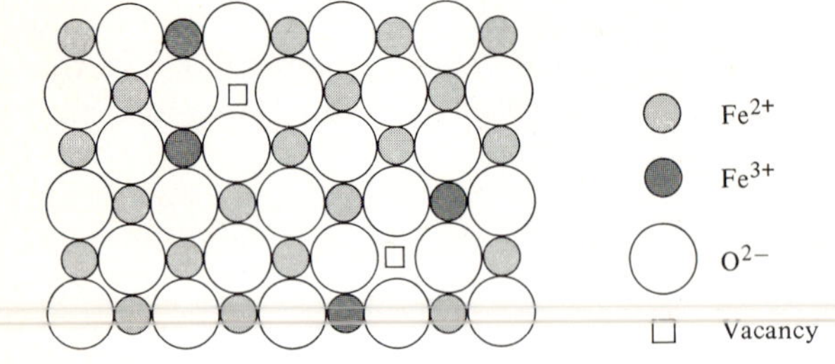

Figure 11-5.1 Defect structure ($Fe_{1-x}O$). This structure is the same as NaCl (Fig. 11-1.2), except for some iron ion vacancies. Since a fraction of the iron ions are Fe^{3+}, rather than Fe^{2+}, the vacancies are necessary to balance the charge. The value of x ranges from 0.05 to 0.16, depending on temperature and the amount of available oxygen.

Fe^{2+} O^{2-} Fe^{3+} O^{2-} Fe^{2+} O^{2-}
O^{2-} Fe^{2+} O^{2-} O^{2-} Fe^{2+}
Fe^{2+} O^{2-} Fe^{2+} O^{2-} Fe^{3+} O^{2-}

Figure 11-5.2 Electronic conduction (ceramics). Compounds with multiple valence ions can conduct charge by electron hopping. In general, such conduction is found only among the transition elements. (Limited conductivity is also possible by ionic diffusion.)

Defect structures

Compounds also have a type of compositional variation not found in metals: *nonstoichiometric,* or *defect structures.* This is best described by iron oxide (Fig. 11-5.1). Iron oxide, $Fe_{1-x}O$, in which the predominant positive ion is Fe^{2+}, has the same basic structure as MgO or NaCl (Fig. 11-1.2). Almost invariably, however, some ferric (Fe^{3+}) ions are present. Since two Fe^{3+} ions have six charges, they replace three Fe^{2+} ions and, in the process, leave an ion vacancy, □. These vacancies affect diffusion rates. More important for electrical considerations, Fe^{3+} ions permit electronic conduction. An electron can hop from an Fe^{2+} ion to an Fe^{3+} ion, transporting charge along an electric field. This is the basis of the conductivity through ceramic semiconductors (Fig. 11-5.2).

Illustrative problem 11-5.1 The phase diagram of Fig. 11-5.3 shows a solid solution of MgO–FeO. What is the Mg^{2+}/Fe^{2+} ion ratio in the solid (Mg, Fe)O that is in equilibrium with the liquid oxide at 1600°C?

SOLUTION Basis: 100 g solid = 35 g MgO and 65 g FeO at 1600°C. From Fig. 1-3.2, the molecular weights are

MgO: 40.3 amu = 40.3 g/mole,
FeO: 71.8 amu = 71.8 g/mole.

Moles MgO = 35 g/(40.3 g/mole) =	0.87 moles	or 49% of solid is MgO
Moles FeO = 65 g/(71.8 g/mole) =	0.91 moles	or 51% of solid is FeO
	1.78 moles	or 100% solid.

Thus 0.49 MgO/0.51 FeO = 0.49 Mg^{2+}/0.51 Fe^{2+} = 0.96 ion ratio.

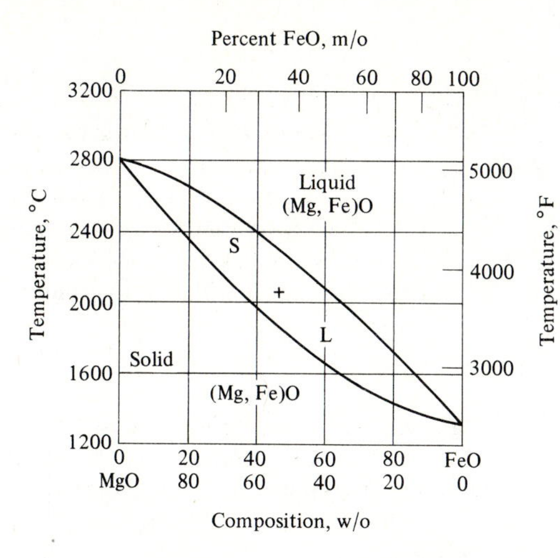

Figure 11-5.3 Oxide solutions (FeO-MgO). In both solid and liquid, Fe^{2+} and Mg^{2+} ions can substitute for each other. This is the same as saying that FeO and MgO replace each other in solution. (See I.P. 11-5.1.)

ADDED INFORMATION Phase diagrams for ceramics are used in the same way phase diagrams are used for metals (Unit 5). No new principles are involved. ◀

Illustrative problem 11-5.2 Figure 11-5.4 shows the ceramic part of the Fe-O diagram. The compounds $Fe_{1-x}O$ (or ϵ), Fe_3O_4 (or ζ), and Fe_2O_3 (or η) are present. Note from the top abscissa that the $Fe_{1-x}O$ composition range has somewhat more than 50 a/o O^{2-}. This occurs because some Fe^{3+} ions are present. What is the Fe^{2+}/Fe^{3+} ion ratio when ϵ contains 52 a/o O^{2-}?

SOLUTION Basis: 100 ions = 52 O^{2-} + y Fe^{2+} + z Fe^{3+};

$$y + z = 48.$$

Based on a charge balance (− equal +),

$$2(52) = 2y + 3z.$$

Solving simultaneously, we obtain

$$104 = 2(48 - z) + 3z = 96 + z,$$
$$z = 8 \quad \text{and} \quad y = 40;$$
$$y/z = 40/8 = 5/1 = Fe^{2+}/Fe^{3+}.$$

ADDED INFORMATION Note another feature of the Fe–O phase diagram. The ϵ phase, that is, $Fe_{1-x}O$, decomposes on cooling by a eutectoid reaction at 560°C:

$$\epsilon(76.74\ \text{Fe}) \underset{\text{heating}}{\overset{\text{cooling}}{\rightleftharpoons}} \alpha(100\ \text{Fe}) + \zeta(72.36\ \text{Fe}). \tag{11-5.1}$$

Magnetite (Fe_3O_4, or ζ) and hematite (Fe_2O_3, or η) are stable at normal temperatures. ◀

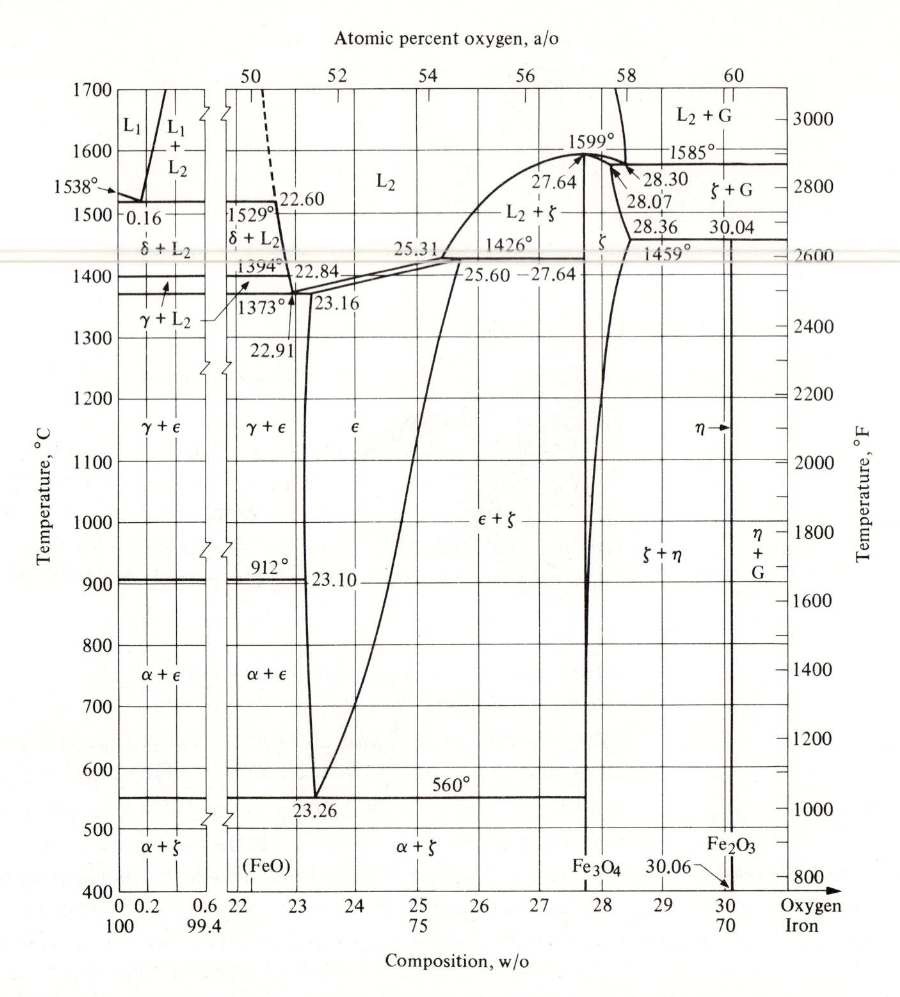

Figure 11-5.4 The FeO system. Rockhounds will recognize Fe_2O_3 (η) as hematite, and Fe_3O_4 (ζ) as magnetite. The ϵ phase, which is not normally found as a mineral, has a defect structure, with the composition $Fe_{1-x}O$. (See Fig. 11-5.1.) (ASM *Handbook of Metals*, Metals Park, Ohio: American Society for Metals.)

• 11-6 SILICA, SILICATES, AND SILICONES

Oxygen and silicon are the two most plentiful elements on the surface of the earth, making up nearly three fourths of the mass (Table 11-6.1). Therefore, it is to be expected that a significant fraction of nature's materials are silica (SiO_2) or silicates. The most common sand is quartz, a compound of silicon and oxygen, SiO_2. Clays are alumino-silicates, and the granitic backbone of the continents is complex silicates.

Table 11-6.1
Composition of the earth's continents

O	~47%	Na	~3%
Si	~27%	K	~2%
Al	8%	Mg	~2%
Fe	5%	Ti	~1%
Ca	4%	All others	<0.1%

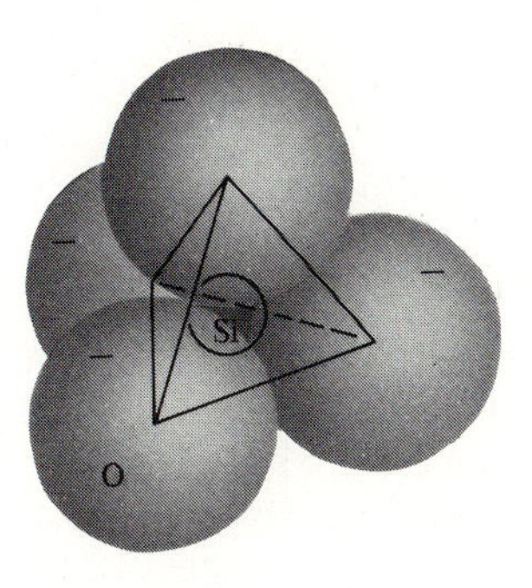

Figure 11-6.1 SiO_4 tetrahedron. This is the basic structural unit of all silicates. The Si–O bonds are very strong. Depending on the number of positive ions present, this tetrahedron may be a tetravalent ion, SiO_4^{4-}, or it may polymerize with other tetrahedra.

SiO_4 tetrahedra

In every silicon-containing rock—sandstone, clays, granites, etc.,—silicon is coordinated with four oxygen atoms (Fig. 11-6.1). These SiO_4 *tetrahedra* are also found in the majority of technical ceramic materials—brick, portland cement, window glass, dinnerware, electrical insulators, and drain tile—as well as in art pottery, artificial teeth, and optical fibers.

This tetrahedral unit* of Si + O is stable on two counts. First, the silicon is partially ionic (Si^{4+}). In the absence of four valence electrons, the silicon ion is so small that each Si can have only four oxygen neighbors. The size ratio, r/R is ~0.3, appreciably less than the 0.41 lower limit for CN = 6. Secondly, the silicon is partially covalent. As with carbon, which is immediately above it in the periodic table, silicon requires four bonds. Not only is CN = 4 favored, also the combination of ionic and covalent bonds leads to a very strong bond. As a result, silicates have been durable over long geological periods. Likewise, silicates provide durable engineering products, if they are designed and processed correctly.

Silica

The compound SiO_2 contains twice as many oxygen atoms as silicon atoms; thus, with $CN_{Si} = 4$, each oxygen atom will be coordinated with two

* This term is used because a four-apexed shape (Fig. 11-6.1) is a tetrahedron, or literally has *four sides.*

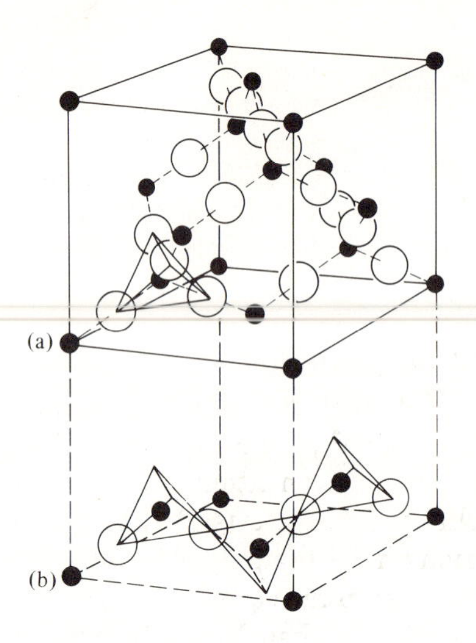

Figure 11-6.2 Silica (SiO_2). (a) Unit cell. This polymorph (cristobalite) is fcc and stable at high temperatures. (b) Sketch showing that each oxygen atom joins two tetrahedra.

Figure 11-6.3 Silica (quartz, the principal phase of most sands and sandstone). (a) Natural crystals. (b) Atomic structure. The basis for the hexagonal structure can be seen. (c) Crystals of quartz (artificial, made for frequency controls of radio circuits). (Courtesy of Western Electric.)

(a)

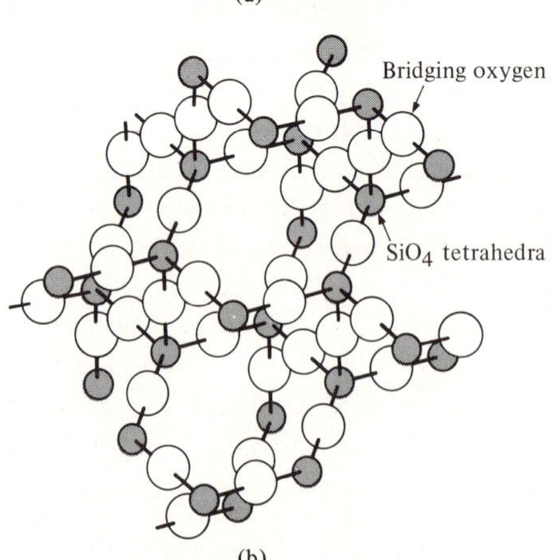

(b)

(c)

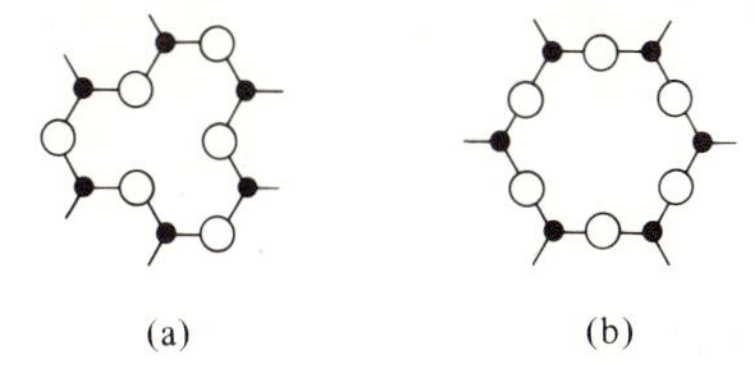

Figure 11-6.4 Bond straightening (quartz). (a) Low temperature (<573°C). The bonds across the oxygen (○) atom form an angle. (b) High temperature (>573°C). The bonds straighten and expand the crystal. This leads to cracking, since quartz (SiO_2) is brittle.

neighbors, $CN_O = 2$.* The resulting structure for the simplest polymorph of SiO_2 is shown in Fig. 11-6.2(a). It is cubic; is least dense of the crystalline SiO_2 phases (ρ = 2.3 g/cm^3); and hence is the highest temperature polymorph. Note that the oxygen atom forms the apex of two tetrahedra (Fig. 11-6.2b). This phase is called cristobalite and is found in brick that is used as liners for various high-temperature furnaces.

The more common silica phase *quartz,* which is found in the majority of sandstones and in most of the beach sands, also has tetrahedra that are joined through their corner oxygens. A double three-dimensional, helical structure forms (Fig. 11-6.3b), which leads to hexagonal crystals. Quartz crystals are found in nature (Fig. 11-6.3a) but may also be made industrially for resonant frequency applications (Fig. 11-6.3c).

The structure of Fig. 11-6.3(b) consists of a rigid three-dimensional network of strongly bonded atoms. Thus, even though the packing factor is low (I.P. 11-6.1), quartz is very stable. It is hard and therefore is used for abrasives on sandpaper; it is extremely resistant to solution in water, as evidenced by its long life as beachsands; and it is refractory, that is, it will not melt readily. Unfortunately, it has one characteristic that limits its use as a high-temperature material. Bonds across the oxygen atoms, Si-O-Si, change angles at 573°C. This is shown schematically in Fig. 11-6.4. An abrupt volume change accompanies this *bond-bending*. Since SiO_2 is a rigid compound, cracking can result unless the rate of heating or cooling is extremely slow.

Vitreous silica

Crystalline silica melts at ~1710°C (~3100°F) where the structure of Fig. 11-6.2 loses its regular, repeating pattern. However, a network structure persists into the liquid because of the strong Si-O bonds that tie each SiO_4-tetrahedral unit to four directions (tetrafunctional). Such a structure is more difficult to sketch in three dimensions than the symmetrical crystalline structure of Fig. 11-6.2. Therefore, let us use a two-dimensional analogy. Figure 11-6.5(a) shows the repetitious array in a crystal where each oxygen is a bridge between adjacent tetrahedra. Each silicon is among four

---

* We've seen this pattern before. In CaF_2, $CN_{Ca^{2+}} = 8$ and $CN_{F^-} = 4$. Since the bonds are between unlike atoms, there must be as many bonds for Si as for O. Hence, with twice as many O atoms as Si atoms, each Si atom will have double the coordination number of each O atom.

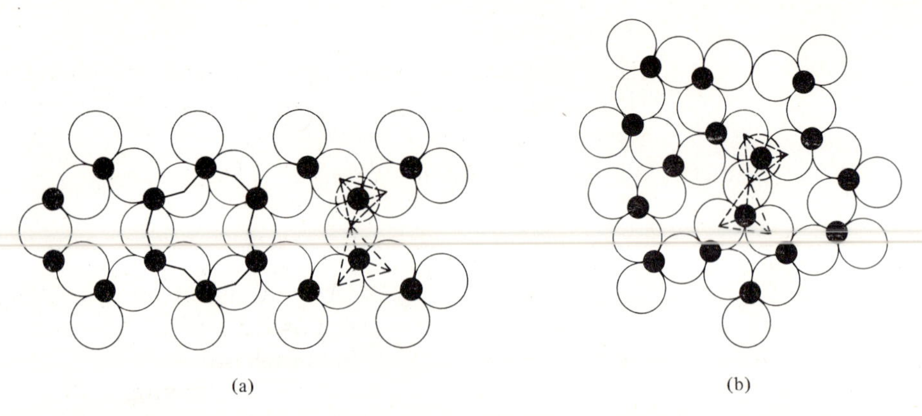

Figure 11-6.5 (a) Crystalline and (b) amorphous silica. (The fourth oxygen of each tetrahedra is deleted for clarity. They lie above, or below, the plane of the sketch.) When supercooled, the liquid silica becomes a solid but amorphous glass. In glassy (vitreous) silica, each oxygen is a bridge between adjacent tetrahedra.

oxygens (if we count one above, or one below the plane of the paper). Figure 11-6.5(b) shows these *bridging* oxygens and SiO_4 tetrahedra in the network without the long-range repetition. This amorphous structure is that of liquid silica.

During cooling, liquid silica should crystallize at 1710°C; however, bond strengths are high and the three-dimensional network is rather immobile. As a result, the time requirement for crystallization is very long. It is easy to supercool the silica to the glass-transition temperature and below, where only elastic movements are possible. (Cf. Section 7-4.) The result is a *vitreous silica,* which is an amorphous silica glass.

The density and packing factor of vitreous silica are low. Furthermore, the bonds across the oxygen may be rotated. Therefore, as the temperature is increased, the increased thermal agitation of the atoms can be accommodated without significant volume expansion. Its expansion coefficient is only $0.5 \times 10^{-6}/°C$! (It is $\sim 22 \times 10^{-6}/°C$ for aluminum.) Vitreous silica is called *fused silica** and is produced and sold for technical and scientific applications that are exposed to abrupt temperature changes, or that require negligible dimensional changes on heating and/or cooling.

We shall see in Unit 13 that the three-dimensional network of bridging oxygen atoms in vitreous silica makes it one of the more viscous glasses for high temperature use. From property considerations, this is good; however, it also means that vitreous silica is very difficult to process into required glass shapes. Thus, vitreous silica is not inexpensive, even though its raw materials are the most widely available of all natural resources.

* Also called "quartz" as a trade name by one producer. This is misleading since it is amorphous and does not have the structure of Fig. 11-6.3.

Silicates

These are compounds of metal ions with silicon and oxygen. Consequently, they have at least three types of elements, and commonly more. We shall consider only two of many silicates—$MgSiO_3$ and one of the feldspars, $KAlSi_3O_8$. Although it contains four elements, the latter is simpler in concept so we shall describe its structure first.

The above feldspar, which is the pink (or tan) phase in granite, may be compared with the silica of Fig. 11–6.2(a), in which one quarter of the silicon atoms have been replaced by aluminum atoms. This is possible because an Al^{3+} ion is small and can substitute for the silicon atom among four oxygen atoms; however, it leaves an imbalance in the charges because the replaced silicon had four charges. The accompanying K^+ provides the charge balance. We may view this feldspar as a substitution of K^+Al^{3+} for Si^{4+} to give $KAlSi_3O_8$ for $4SiO_2$. Note, however, that the substitution would not have been possible were it not for the fact that the packing factor of silica is low, and there is space to accommodate the large K^+ ion. (This is not a solid solution, since both the K^+ and the Al^{3+} always take the same location to form a new crystal lattice.) Sodium feldspar, $NaAlSi_3O_8$, which is used extensively in high-grade porcelains, is analogous to the above potassium feldspar.

Magnesium silicate, $MgSiO_3$, contains Mg^{2+} ions; therefore each $SiO_3{}^{2-}$ unit must possess two negative charges. This occurs as shown schematically in Fig. 11–6.6(a). In three dimensions (Fig. 11–6.6b), the silicon and oxygen atoms form a chain $(SiO_3)_n^{2-}$ in which one third of the oxygen atoms are *bridging* from one tetrahedral unit to the next. The other two thirds of the oxygens are *nonbridging*. Each of the latter carries a negative charge that is attracted to the adjacent Mg^{2+} ions. A material such as this is very *anisotropic,* that is, its properties vary with direction. This is easily appreciated, since within the chain, we have strong Si–O bonds that are partially covalent and between specific atoms. Between the chains, the bonds are coulombic, that is, ionic. An Mg^{2+} ion will attract any and all negative charges that are in the vicinity. The two kinds of bonding respond differently to deformation, fracture, electric fields, etc. For example, cleavage occurs more readily parallel to the chains than across the chains.*

Molten silicates are amorphous and, like vitreous or fused silica, crystallize slowly. Therefore, they can form glasses. However, structures and properties differ from fused silica. An $MgSiO_3$ composition (or Na_2SiO_3, or $CaSiO_3$) forms a linear structure of tetrahedral units (Fig. 11–6.6). When such compositions are melted, the liquid is much less rigid than vitreous silica with its four bridging oxygens. Like linear plastics, these liquids flow readily above their glass temperatures. Commercial glasses are intermediate in compositions with approximately 2 to 3 bridging oxygens per tetrahedral unit. Their properties permit their use in many technical products (Unit 13).

---

* We saw this same principle of anisotropy in the properties of wood (Unit 10).

O⁻ O⁻ O⁻ O⁻ O⁻ O⁻ O⁻ O⁻

–Si–O–Si–O–Si–O–Si–O–Si–O–Si–O–Si–O–Si–O–

O⁻ O⁻ O⁻ O⁻ O⁻ O⁻ O⁻ O⁻

Mg^{2+} Mg^{2+} Mg^{2+} Mg^{2+} Mg^{2+} Mg^{2+} Mg^{2+} Mg^{2+}

O⁻ O⁻ O⁻ O⁻ O⁻ O⁻ O⁻ O⁻

–Si–O–Si–O–Si–O–Si–O–Si–O–Si–O–Si–O–Si–

O⁻ O⁻ O⁻ O⁻ O⁻ O⁻ O⁻ O⁻

(a)

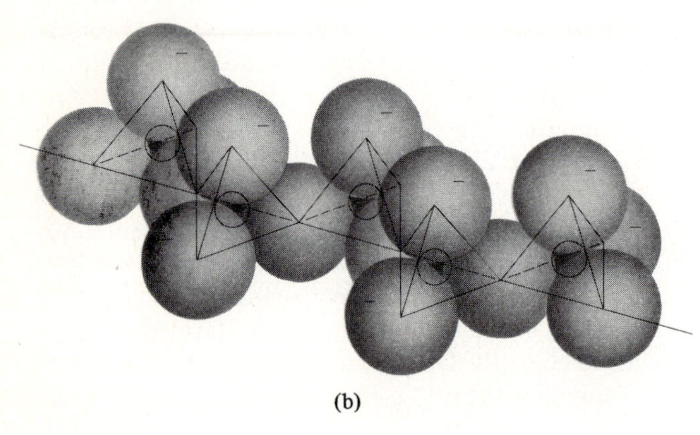

(b)

Figure 11-6.6 Silicate chain $(SiO_3)_n^{2-}$. (a) Schematic. (b) 3-D sketch. Crystallization occurs, but slowly, to form hard, rigid crystals at ~1550°C (~2800°F). The molten liquids of these silicates are less viscous than is liquid silica with its network structure (Fig. 11-6.5b).

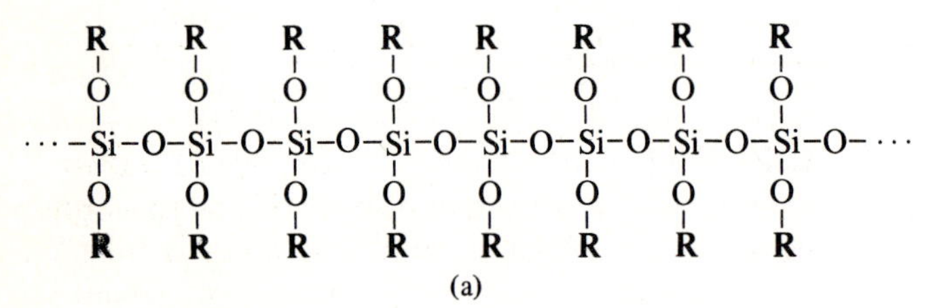

(a)

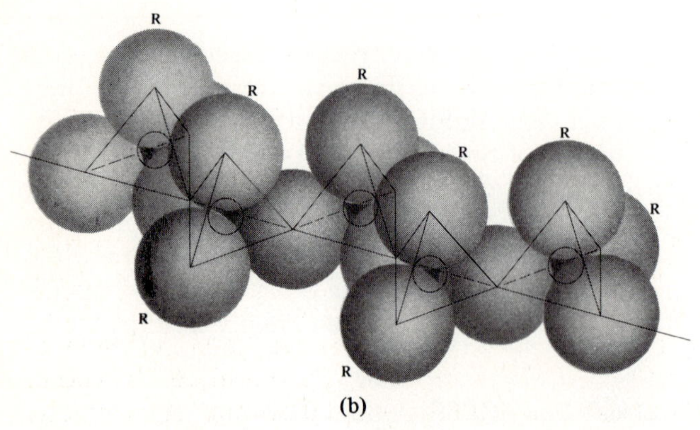

(b)

Figure 11-6.7 Silicones (siloxane). (a) Schematic with **R** indicating -H, $-CH_3$, etc. (b) Polymer chain $(SiO_3\mathbf{R}_2)_n$. Adjacent chains have little attraction to each other. Therefore, these silicones are liquids or soft amorphous solids at ambient temperatures.

Silicones

This family of materials lies between ceramics and polymers, inasmuch as they have some of the characteristics of each. There are a large number of silicones. We shall look at only one variety, shown schematically in Fig. 11-6.7, which may be contrasted with Fig. 11-6.6. In this silicone, the $-(SiO_3)_n-$ backbone does not receive extra electrons to become a long polyatomic ion. Rather, each nonbridging oxygen has combined with an -H, a $-CH_3$, a $-C_2H_5$, or some other side radical. The effect is to give a linear molecule. It is not rigid, since adjacent molecules are not tied to each other by strong bonds. Therefore, individual molecules can rotate independently and will slide by one another with small shear forces. The silicones of Fig. 11-6.7 form viscous liquids that are very stable.

Illustrative problem 11-6.1 Quartz (SiO_2) has a density of 2.65 g/cm³. (a) How many silicon atoms (and oxygen atoms) are there per cm³? (b) What is the packing factor, given that the radii of silicon and oxygen are 0.038 nm and 0.117 nm, respectively?

SOLUTION

$$\text{a) } SiO_2/\text{cm}^3 = \frac{2.65 \text{ g/cm}^3}{(28.1 + 32.0) \text{ g}/(0.6 \times 10^{24} \text{ SiO}_2)}$$
$$= 2.64 \times 10^{22} \text{ SiO}_2/\text{cm}^3$$
$$= 2.64 \times 10^{22} \text{ Si/cm}^3$$
$$= 5.28 \times 10^{22} \text{ O/cm}^3.$$

$$\text{b) } V_{Si}/\text{cm}^3 = (4\pi/3)(2.64 \times 10^{22}/\text{cm}^3)(0.038 \times 10^{-7} \text{ cm})^3 = 0.006$$
$$V_O/\text{cm}^3 = (4\pi/3)(5.28 \times 10^{22}/\text{cm}^3)(0.117 \times 10^{-7} \text{ cm})^3 = 0.354$$
$$\text{Packing factor} = 0.36$$

ADDED INFORMATION Although there is considerable open space within this structure, most single atoms (except for helium) must diffuse through SiO_2 as ions. Thus their charges prohibit measurable movements. ◀

11-7 PHASE DIAGRAMS OF CERAMICS

Binary systems

Phase diagrams for ceramic materials are not different from phase diagrams for alloys, except that the components are compounds rather than elemental. We may see this by comparing binary (two-component) diagrams. The MgO–FeO diagram (Fig. 11-5.3) may be compared directly with the Cu–Ni diagram (Fig. 5-2.4). The two solid compounds, MgO and FeO, have full solubility in each other. This is not surprising since (a) each has O^{2-} as the anion; (b) both cations carry two charges, Mg^{2+} and Fe^{2+}; and (c) both compounds possess the NaCl-type structure. Furthermore, the cation radii are reasonably close together—0.066 nm and 0.074 nm, respectively.

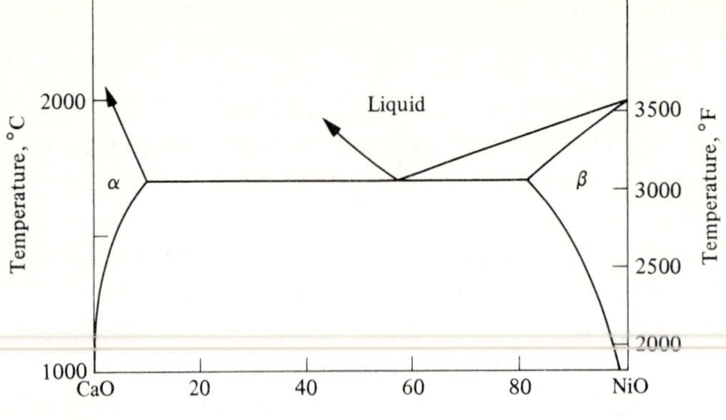

Figure 11-7.1 CaO-NiO system. This system, which contains a eutectic, may be compared directly with the Pb-Sn and the Ag-Cu diagrams of Unit 5. The same principles apply for determining (1) what phase(s)? (2) phase composition(s)? and (3) amounts of phase(s)? (Section 5-4.)

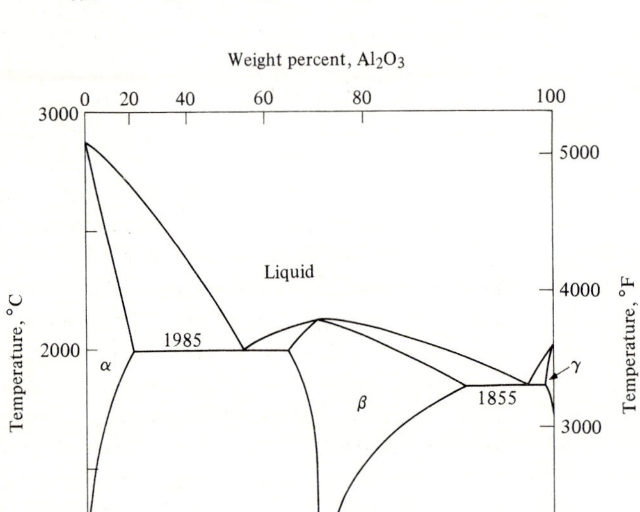

Figure 11-7.2 MgO-Al_2O_3 system. The ternary compound ($MgAl_2O_4$, and labeled β) possesses a solid-solution range at elevated temperature. Its structure is closely related to that of $NiFe_2O_4$ in Section 11-3.

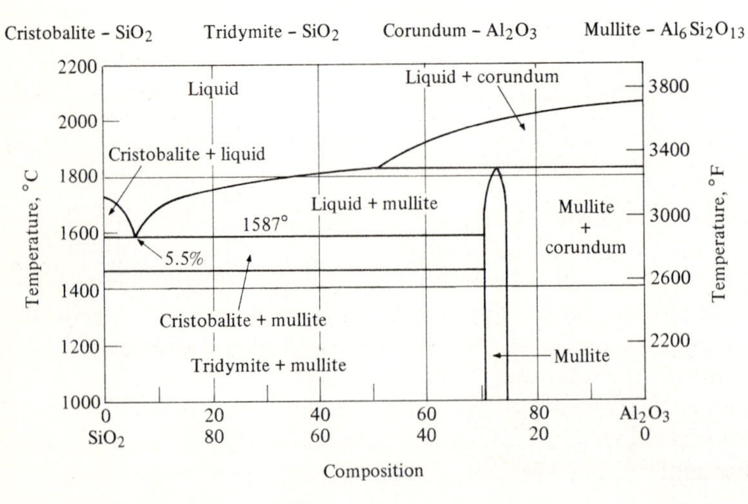

Figure 11-7.3 Al_2O_3-SiO_2 system. (Adapted from Aksay and Pask, *Science*.)

As a result, the lattice constant, a, of MgO is 2(0.066 + 0.140 nm) = 0.41 nm, and that of FeO is 2(0.074 + 0.140 nm) = 0.43 nm (based on the NaCl structure as shown in Fig. 11-1.2). This difference (<5%) permits substitution without excessive lattice strain.

The CaO-NiO phase diagram (Fig. 11-7.1) may be compared directly with the Ag-Cu diagram (Fig. 5-3.3). Each has a eutectic; each has limited solubility of the end members. Silver and copper have the same structure (fcc) but differ sufficiently in size to preclude full solubility (a_{Ag} = 0.408 nm versus a_{Cu} = 0.361 nm, or >10% difference). CaO and NiO have the same structure (NaCl) but differ in size sufficiently to preclude full solubility (a_{CaO} = 0.48 nm versus a_{NiO} = 0.42, or >10% difference).

Some metallic alloys contain intermetallic compounds, for example, $CuAl_2$ in Al-Cu alloys (Fig. 5-5.2). Likewise, ceramic materials may contain compounds of the components; Fig. 11-7.2 shows β, which is $MgO \cdot Al_2O_3$ (or more appropriately $MgAl_2O_4$) in the MgO-Al_2O_3 system. Such compounds, like their intermetallic cousins, are not exactly stoichiometric but possess a solid solution range, particularly at higher temperatures.

As a general rule, the liquidus and solidus of ceramic systems are at higher temperatures than in their metallic counterparts. We see this by comparing the SiO_2-Al_2O_3 system (Fig. 11-7.3) with the Si-Al system (Fig. 5-4.8), or the MgO-Al_2O_3 system (Fig. 11-7.2) with the Mg-Al system (Fig. 5-4.5). The reason is simply that the ceramic compounds usually possess much stronger bonds than the comparable metallic phases. It takes more thermal agitation in order to disarrange the crystal structures into amorphous liquids. Because of this greater thermal stability, ceramic materials find extensive use as *refractories.* These are materials that resist melting and are, therefore, suitable for metal-melting furnaces, ovens to produce coke, glass tanks, cement kilns, power-generating plants, rocket exhaust parts, etc., (Fig. 11-7.4).

Figure 11-7.4 Refractories (roof of electric steel-melting furnace). The roof has just been removed after tapping steel at 1500°C (2900°F). We see the underside. In general, ceramics possess higher melting temperature than do metals; furthermore, they react less with the surrounding air. Thus the higher melting ceramics (refractories) are used for high-temperature applications. (Courtesy of Harbison-Walker Refractories.)

• Three-component ceramics

Ceramics more commonly involve three components than do metals. A common electrical porcelain (called *steatite*) is made by mixing clay with talc, and then firing the product. The two raw materials are somewhat complex but change on calcining to the basic components.

$$\text{Clay:} \quad Al_2Si_2O_5(OH)_4 \xrightarrow{\text{heating}} Al_2O_3 + 2SiO_2 + 2H_2O\uparrow. \tag{11-7.1}$$

$$\text{Talc:} \quad Mg_3Si_4O_{10}(OH)_2 \xrightarrow{\text{heating}} 3MgO + 4SiO_2 + H_2O\uparrow. \tag{11-7.2}$$

Calcining, that is, heating to dissociate the structure and drive off a gas (H_2O), produces an Al_2O_3–SiO_2 composition and an MgO–SiO_2 composition. We can locate these on two sides of a triaxial diagram shown in Fig. 11-7.5. The $Al_2O_3 + 2SiO_2$ composition is at 54 w/o SiO_2–46 w/o Al_2O_3. (See I.P. 11-7.1.) The $3MgO + 4SiO_2$ composition is at 67 w/o SiO_2–33 w/o MgO. Any mixture of calcined talc and calcined clay must lie on a line between these

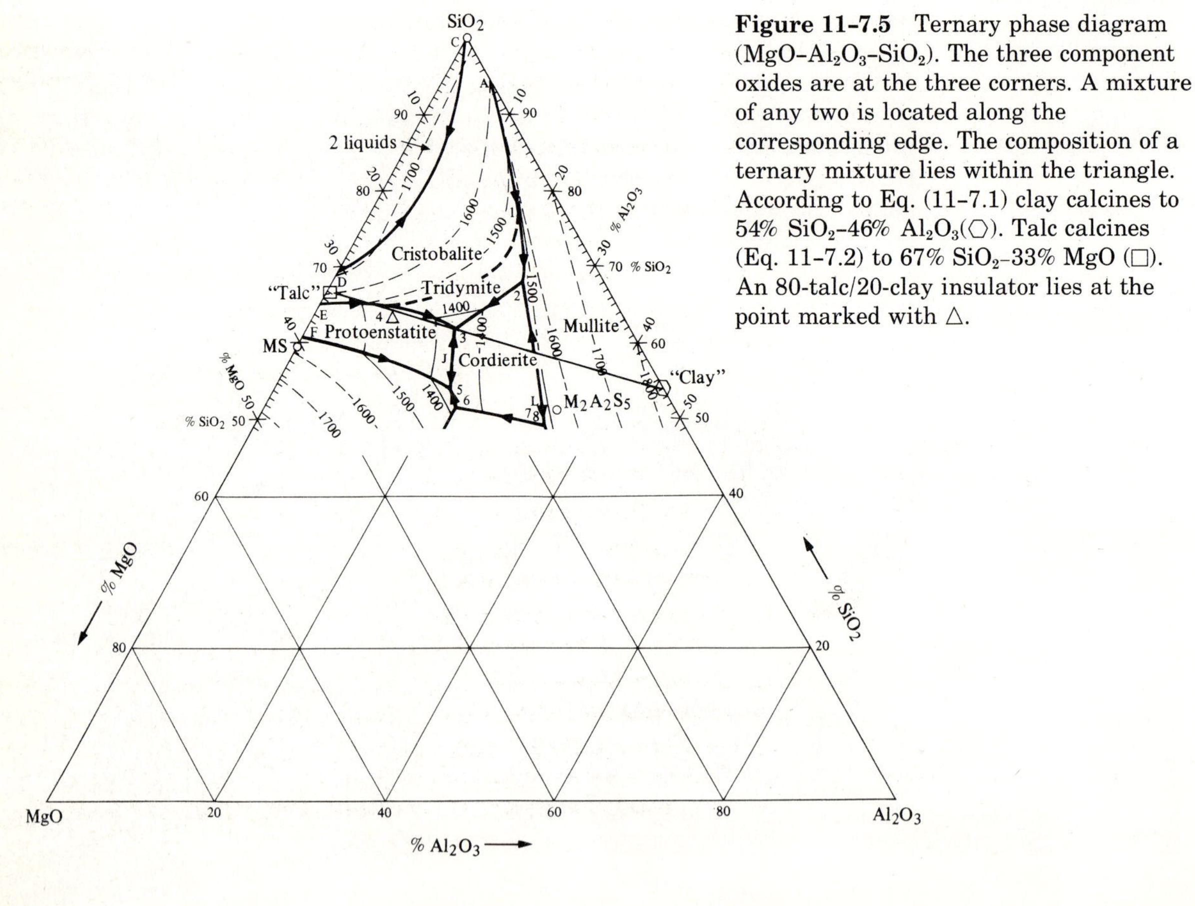

Figure 11-7.5 Ternary phase diagram (MgO-Al_2O_3-SiO_2). The three component oxides are at the three corners. A mixture of any two is located along the corresponding edge. The composition of a ternary mixture lies within the triangle. According to Eq. (11-7.1) clay calcines to 54% SiO_2–46% Al_2O_3(⬡). Talc calcines (Eq. 11-7.2) to 67% SiO_2–33% MgO (□). An 80-talc/20-clay insulator lies at the point marked with △.

Figure 11-7.6 Steatite insulators (circuit substrates). These are produced by pressing mixtures of talc and clay. They are then calcined to remove water, before firing to sinter them into the final product. (Courtesy of Cahner's Publications, from *Ceramic Industry*.)

two compositions. The above steatite electrical porcelains (Fig. 11-7.6) contain about 80% of calcined talc. Therefore, its composition is at the symbol △ in Fig. 11-7.5. Since that figure shows the liquidus temperature for the MgO-Al_2O_3-SiO_2 system, we see that it would be completely liquid* at ~1460°C (or ~2660°F).

Illustrative problem 11-7.1 A fireclay is a relatively pure clay with a composition of $Al_2Si_2O_5(OH)_4$. When it is used to produce refractory brick it undergoes the dissociation reaction of Eq. (11-7.1). Based on Fig. 11-7.3, and assuming equilibrium, what phases will form at 1500°C (2732°F)? What is the maximum weight percent of each?

SOLUTION The H_2O produces a gas, leaving Al_2O_3 and SiO_2:

$$\begin{array}{lllll} Al_2O_3: & 2(27.0) + 3(16.0) & = 102.0 \text{ amu} & = & 46 \text{ w/o} \\ 2SiO_2: & 2[28.1 + 2(16.0)] & = \underline{120.2 \text{ amu}} & = & \underline{54 \text{ w/o}} \\ & & 222.2 \text{ amu} & & 100 \text{ w/o} \end{array}$$

---

* These insulators (Fig. 11-7.6) are manufactured by pressing mixtures of talc and clay into the desired shape. They are then *fired,* that is, heated to first calcine and then to sinter the oxide into a coherent, nonporous product. Obviously, the manufacturer would not want to heat the product above the liquidus temperature, or it would melt and lose its shape. Figure 11-7.5 does point out, however, that properties of 3-component systems can be plotted as a function of composition. For example, the index of refraction of K_2O-Al_2O_3-SiO_2 glasses are given in Fig. 11-7.7 as a function of composition.

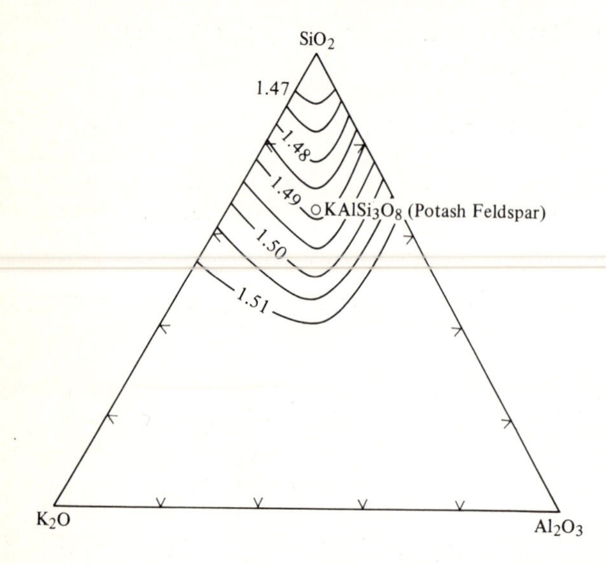

Figure 11-7.7 Ternary refraction diagram (K_2O-Al_2O_3-SiO_2 glasses). The optical indices of refraction are shown as a function of composition. All possible compositions containing these three components can be located on this diagram. Their locations are interpolated in a manner similar to our procedure for phase diagrams in Unit 5 (except that we have a third component).

At 1500°C, we see from the equilibrium diagram that the phases are cristobalite (SiO_2) and mullite ($\sim Al_6Si_2O_{13}$, or $3Al_2O_3 \cdot 2SiO_2$).

$$\text{Fraction cristobalite} = (0.71 - 0.46)/(0.71 - 0) = 0.35.$$
$$\text{Fraction mullite} = 0.46/0.71 = 0.65.$$

ADDED INFORMATION The cristobalite crystallizes very slowly because it contains 100% network-forming oxide, SiO_2. The equilibrium value of 35 w/o cristobalite requires several days to form at 1500°C. ◄

• **Illustrative problem 11-7.2** (a) Locate the following compositions on Fig. 11-7.5: (i) 70MgO-30SiO_2; (ii) 70MgO-30Al_2O_2; (iii) 7Al_2O_3-3SiO_2; (iv) 50MgO-20Al_2O_3-30SiO_2; (v) 1MgO-1Al_2O_3-1SiO_2. (b) Determine the composition of the junction labeled 2,

SOLUTION (b) Hold a horizontal straightedge through the point to the right edge;

$$SiO_2 = 68\%.$$

Hold a 60° straightedge through the point to the left edge;

$$MgO = 9.5\%.$$

Hold a 60° straightedge through the point to the bottom edge;

$$Al_2O_3 = 22.5\%.$$

ADDED INFORMATION Although composition ratios are commonly given in percentage, for example, 70–30, they can also be given in other integer terms, such as 7–3 in (iii) above. ◄

• **Illustrative problem 11–7.3** Refer to Fig. 11–7.5. (a) Locate the composition of talc after dissociation and loss of water.

$$Mg_3Si_4O_{10}(OH)_2 \rightarrow 3MgO + 4SiO_2 + H_2O\uparrow. \qquad (11\text{–}7.2)$$

b) Repeat for clay (Eq. 11–7.1). (c) Determine the location on the MgO-Al_2O_3-SiO_2 diagram of an 8-talc/2-clay mixture to be used for an electrical insulator.

SOLUTION Basis: 100 g starting material = 80 g talc + 20 g clay. (a) From Eq. (11–7.2) and using atomic masses,

$$\frac{80 \text{ g talc}}{(3)(24.3) + 4(28.1) + (10)(16.0) + 2(17.0)} = \frac{y \text{ g MgO}}{3(24.3 + 16.0)} = \frac{z \text{ g SiO}_2}{4(28.1 + 32.0)}.$$

$$\left.\begin{aligned} y &= 25.4 \text{ g MgO} = 33.4 \text{ w/o} \\ z &= \underline{50.7} \text{ g SiO}_2 = 66.6 \text{ w/o} \end{aligned}\right\} \quad \text{See the point marked } \square \text{ in Fig. 11–7.5.}$$
$$76.1 \text{ g dissociated talc}$$

b) From Eq. (11–7.1) and using atomic masses,

$$\frac{20 \text{ g clay}}{(2)(27.0) + 2(28.1) + 5(16.0) + 4(17.0)} = \frac{w \text{ g Al}_2\text{O}_3}{2(27.0) + 3(16.0)} = \frac{x \text{ g SiO}_2}{2(28.1 + 32.0)}.$$

$$\left.\begin{aligned} w &= 7.9 \text{ g Al}_2\text{O}_3 = 46 \text{ w/o} \\ x &= \underline{9.3} \text{ g SiO}_2 = 54 \text{ w/o} \end{aligned}\right\} \quad \text{See the point marked } \bigcirc \text{ in Fig. 11–7.5.}$$
$$17.2 \text{ g dissociated clay}$$

c) Total composition:

Al_2O_3:	7.9 g	= 8.5 w/o
MgO:	25.4 g	= 27.2 w/o
SiO_2:	(50.7 + 9.3) g	= 64.3 w/o
Total:	93.3 g	

(See the point marked △ in Fig. 11–7.5.)

ADDED INFORMATION The location of the composition of the electrical insulator is 64.3% of the distance from the MgO-Al_2O_3 side to the SiO_2 apex; 8.5% of the distance from the MgO-SiO_2 side to the Al_2O_3 apex; and a little over one fourth of the distance from the Al_2O_3-SiO_2 side to the MgO apex.

The total composition may also be located by the lever rule. The total composition lies on a line between dissociated talc, □, and dissociated clay, ○, such that 76.1 g of dissociated talc (81.5 w/o) and 17.2 g of dissociated clay (18.5 w/o) balance their lever at △. ◀

11-8 CHARACTERISTICS OF CERAMIC PHASES

Comparisons with nonceramic phases

We may compare ceramics with both metals and polymers. Like metals, many ceramic phases are crystalline. Unlike metals, their structures commonly do not contain significant numbers of free electrons. Either the electrons are shared covalently with adjacent atoms, or they are transferred from one atom to another to produce an ionic bond, in which case the atoms carry a charge as ions.

Ionic and covalent bonds give ceramic materials relatively high stability. They have a much higher melting temperature, on the average, than do either metals or organic materials. Generally speaking, they are also harder and more resistant to chemical alteration.

Like organic materials, solid ceramic phases are usually insulators. Because of this characteristic, ceramics can be used not only as insulators but also as functional dielectric devices (Unit 19). At elevated temperatures where they have more thermal energy, ceramics will conduct an electric charge, but poorly as compared with metals. Due to the absence of free electrons, most ceramics are transparent, at least in thin sections, and are poor thermal conductors.

Although many ceramic phases are crystalline, we find that they do not crystallize as rapidly as metals. Since metallic crystals are commonly relatively simple, it is necessary to cool them from their liquids to ambient temperatures in less than milliseconds to produce a metallic glass. Glass will form from ceramic compositions with much slower cooling rates; in fact, window glass, etc. (Unit 13), must be slow-cooled intentionally to induce crystallization. The times are many kiloseconds for silicates. (Most nonsilicate compounds crystallize in the range of seconds, in contrast to the milliseconds for metals.) These slower reaction rates for ceramic compositions arise because the structures are more complex than the simple metals of Unit 2, and because the bonds that must be broken for atomic rearrangement are commonly stronger than in metals.

Because ceramic liquids crystallize slowly, they can be supercooled below their freezing temperature. They possess a glass-transition temperature as do polymers; in fact, this characteristic (Fig. 7-4.1) was first observed in the inorganic glasses.

Since ceramics that contain silica and the silicates (Section 11-6) are partially covalent, they will polymerize like many organic compounds. They form both linear and network structures, depending upon the functionality of the tetrahedral unit. Thus, we can find parallels in the viscoelastic and the dielectric properties of organics and of silicate ceramics.

Shear resistance

Ceramic materials and metals may be compared most directly when we consider intermetallic compounds (Section 5-4). Both are strong in shear

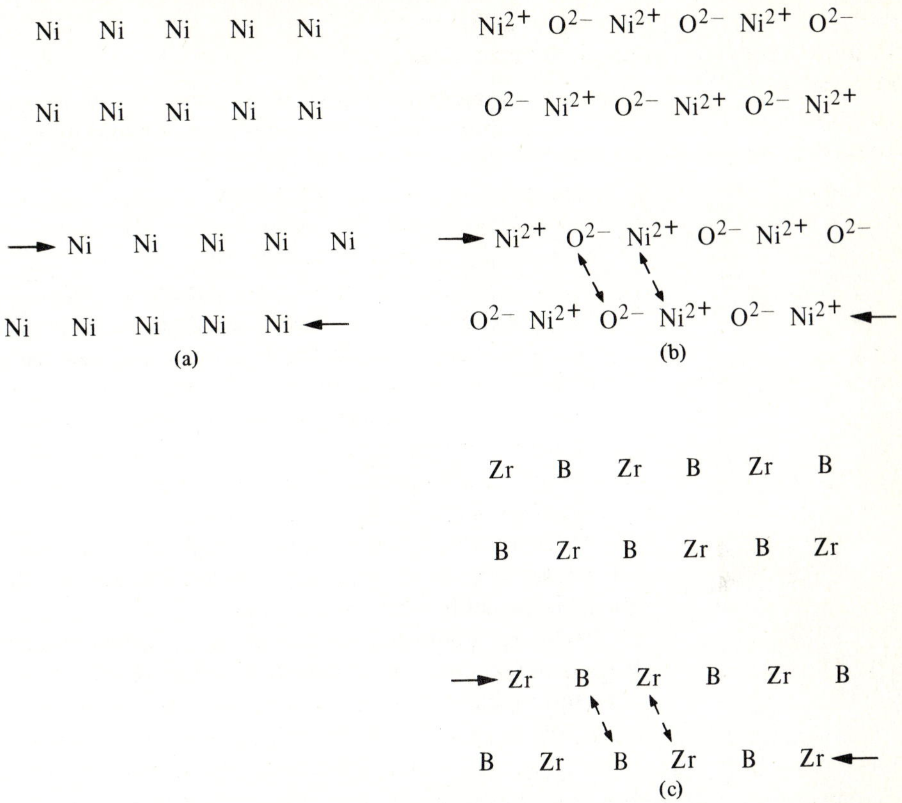

Ni Ni Ni Ni Ni

Ni Ni Ni Ni Ni

→ Ni Ni Ni Ni Ni

Ni Ni Ni Ni Ni ←

(a)

Ni^{2+} O^{2-} Ni^{2+} O^{2-} Ni^{2+} O^{2-}

O^{2-} Ni^{2+} O^{2-} Ni^{2+} O^{2-} Ni^{2+}

→ Ni^{2+} O^{2-} Ni^{2+} O^{2-} Ni^{2+} O^{2-}

O^{2-} Ni^{2+} O^{2-} Ni^{2+} O^{2-} Ni^{2+} ←

(b)

Zr B Zr B Zr B

B Zr B Zr B Zr

→ Zr B Zr B Zr B

B Zr B Zr B Zr ←

(c)

Figure 11-8.1 Comparison of slip processes. (a) Metallic nickel. (b) Nickel oxide (NiO). (c) Zirconium boride (ZrB). Compounds require unlike neighbors. Therefore plastic deformation is resisted on those planes that move like atoms into adjacent positions. This shear resistance does not occur in pure metals (or their nonordered solid solutions). Therefore the compounds, both ceramic and metallic, are harder, stronger, and less ductile than pure metals.

and, as a general rule, both resist plastic deformation. Thus, they are hard, strong in compression, and subject to cracking when submitted to tensile loads.

The shear mechanism that is encountered in single-component metals may be illustrated by Fig. 11-8.1(a). Although the atoms possess new neighbors, the surroundings of each displaced atom are identical to those in the predeformed crystal. In contrast, similar shear in compounds separates *unlike* atoms or ions during the process of slip (Fig. 11-8.1b,c). Thus, the low-energy coordination of *unlike* atoms (or ions) must be altered. This requires an additional shear force and leads to a higher shear stress for plastic deformation. It is especially significant in ionic compounds where ions of like charges develop a very high mutual repulsion.

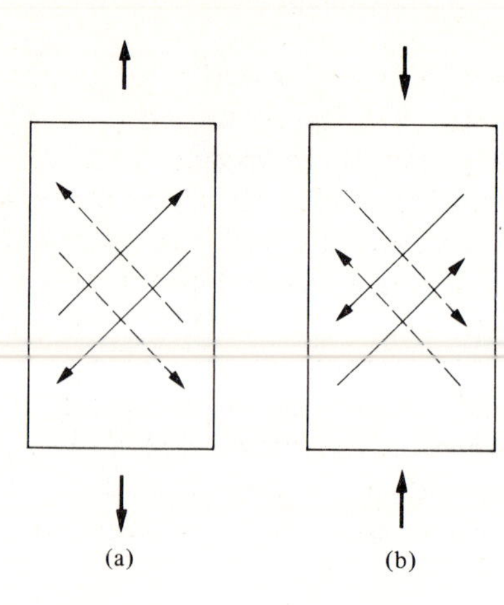

Figure 11-8.2 Resolved shear forces. (a) Tension. (b) Compression. Failure is by shear in ductile materials. Since the highest shear stresses are at 45°, tension strengths and compression strength are comparable when plastic deformation is possible.

Deformation by tension or compression occurs most readily by shear at 45° to the applied forces (Fig. 11-8.2). Thus, in ductile materials where shear is possible, the tension and compression strength are comparable. However, in a ceramic compound or in an intermetallic compound, where the shear resistance is high, other factors come into play. Chief among these are the effects of irregularities such as surface flaws or cracks that are

Figure 11-8.3 Surface flaws (schematic). (a) Microcrack. (b) Compression. The load is transmitted across the crack, so shear resistance determines the strength (Fig. 11-8.2). (c) Tension (ductile material.) The tip of the crack is blunted to relieve the stress concentration. (d) Tension (nonductile material). Since compounds—ceramic or metallic—are strong in shear (Fig. 11-8.1), the tip of the crack does not deform. The stresses are concentrated and the crack length increases (Section 14-2).

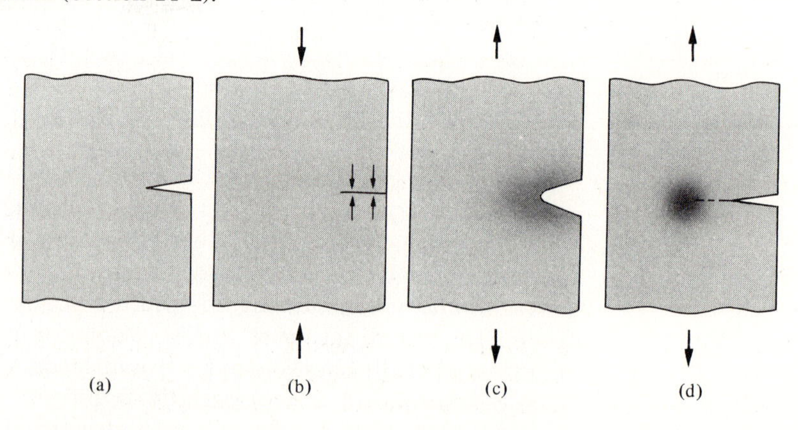

present in almost every material usually on a microscopic scale (Fig. 11–8.3). Under compression, the load is transmitted across the flaw and the tolerable load is determined by the resistance to shear. With a tension load, however, stresses are concentrated at the tip of the crack. If the material is ductile and can undergo shear, the tip of the crack is blunted by plastic deformation, thus lowering the stress concentration. However, if plastic deformation cannot occur because of the high shear resistance, the stress concentration continues to increase as the load increases, even extending the crack and accentuating the stress concentration still further. A *brittle fracture* results. Since these compounds are strong in shear, they are weak in tension. Ceramic materials and metals containing large fractions of compounds (for example, white cast irons) are used most advantageously in compression.

Review and Study

SUMMARY

Ceramic materials contain compounds of metallic and nonmetallic elements. Bearing this in mind, we can relate their structures to metals and to polymers. Similarities and contrasts may be made in the resulting properties.

1. The simpler ceramic compounds, AX, possess one of three structures that have CN = 8, 6, or 4, depending on the ionic radius ratio. Slightly more complex structures, where $A \neq X$, or where there are two cations, may be viewed as variants of the AX structures. In all of these structures, it is advantageous to consider the interstitial sites among the anions, and their occupancy by cations.
2. Ceramic compounds may have a number of polymorphs. Likewise, solid solutions are common. To produce solid solutions, ionic radii must be comparable and the charges must remain balanced. The latter is sometimes achieved by the presence of ion vacancies.
3. • Silica, silicates, and silicones contain SiO_4 tetrahedra as the basic structural unit. These are connected by bridging oxygens that provide a polymeric structure.
4. Phase diagrams serve the same useful role for ceramic materials as they do for metals. In fact, it is desirable to extend their use into three-component systems.
5. Ceramics are generally insulators. They are commonly refractory (high-melting). Some are extremely hard (abrasive). They crystallize more slowly than metals, particularly the silicates. This leads to the formation of glass (Unit 13).
6. Ceramics, like intermetallic compounds, resist shear. Thus, they are nonductile, and strong in compression, but generally weak in tension because stresses cannot be relieved at cracks and flaws.

TECHNICAL TERMS

AX compounds Binary compounds with 1-to-1 ratio of the two elements; commonly ionic.

• **Bridging oxygen** Oxygen atoms shared by two adjacent silica tetrahedra.

Brittle fracture Fracture without plastic deformation, hence with little energy absorption.

Calcine Thermal decomposition of a solid with the release of gas, for example, CO_2 or H_2O.

Ceramics Materials consisting of compounds of metallic and nonmetallic elements.

Clay Fine-grained, hydroplastic raw materials. Formerly, the principal raw material for ceramic products. Sheet-like alumina silicates, $[Al_2Si_2O_5(OH)_4]$.

Compound Material containing two or more types of elements, in established ratios.

Defect structure Compounds with noninteger ratios of atoms or ions. These compounds contain either vacancies or interstitials within the structure.

Glass An amorphous solid below its transition temperature. A glass lacks long-range crystalline order but normally has short-range order.

Ionic radii Assigned radii, so that $(r_{A^+} + R_{X^-})$ equals the interatomic distance.

Interstice Unoccupied space between atoms or ions.

Interstitial site (*n*-fold) An interstice with n (4, 6, or 8) immediate atomic (or ionic) neighbors.

Microcracks Minute surface flaws.

Nonstoichiometric compounds Compounds with noninteger atom (or ion) ratios.

• **Quartz** The most common polymorph of SiO_2.

• **Silicate** Materials containing SiO_4 tetrahedra plus metallic ions.

• **Silicone** "Silicates" with organic side radicals; thus, silicon-based polymeric molecules.

• **SiO_4 tetrahedra** Coordination unit of four oxygens surrounding a silicon atom.

• **Steatite** Insulating materials made from talc + clay.

• **Ternary** Three-components.

Vitreous Glassy, or glasslike.

CHECKS

11A Ceramic materials contain __***__ of metals (which readily __***__ electrons) and nonmetals (which readily accept or __***__ electrons).

11B Of the two, a magnesium atom in the metal and a magnesium ion in its oxide, the radius of the magnesium __***__ can be measured with more precision because __***__ . . . .

11C The radius of a cation (positive) is __***__ than the radius of the corresponding metal; the radius of an anion (negative) is __***__ than the radius of a cation of essentially the same atomic mass.

11D The radius of an oxygen ion with six neighbors is about __***__ % of the radius it has with four neighbors; and about __***__ % of the radius that would be expected with eight neighbors.

11E In order to have CN = 8 (eight neighbors), the radius ratio, r/R, between cations and anions must be __***__ than for CN = 6. The change from CN = 6 to CN = 8 is approximately at r/R = __***__.

11F The prototype AX compound for CN = 6 is __***__; for CN = 4 is __***__. Each of these is __***__ cubic.

11G Since LiF is an AX compound with CN = 6 for both anions and cations, the interatomic distance (which is 0.20 nm atom-center to atom-center) is __***__ % of the lattice constant.

11H The cations in CsCl occupy 8-f interstitial sites among anions. In fcc NaCl, the cations occupy __***__ sites among anions; there are __***__ of these sites per unit cell. In fcc ZnS, the cations occupy __***__ sites; there are __***__ of these sites per unit cell, half of which are occupied.

11I In CaF_2 (and UO_2) there are as many vacant __***__-f sites as there are __***__ ions.

11J In general, higher __***__ favor polymorphs with lower density and greater symmetry; higher __***__ favor polymorphs with greater density.

11K For extensive solid-solution substition of one cation for another, the two must be similar in __***__, and a __***__ must be maintained.

11L Ferrous oxide contains a significant number of ferric, Fe^{3+}, ions. For every 100 Fe^{3+} ions present, there must be 50 __***__ in order to balance charges. A __***__ structure results and the oxide is a __***__ compound.

11M In the oxide just described, electrons can "hop" from an __***__ ion to an __***__ ion to transport charge.

• 11N The __***__ tetrahedron is a very stable form of the two most plentiful elements—__***__ and __***__.

• 11O In quartz, every oxygen atom is at the corners of two adjacent __***__; these may be called __***__ oxygens because of their connecting role.

• 11P Fused silica, also called __***__ silica, has an extremely low __***__ because the __***__ is low. Also, the bonds across the oxygen can bend and __***__ at higher temperatures, thus absorbing thermal energy with minimal expansion.

• 11Q A chain silicate such as $MgSiO_3$ possesses __***__ bridging oxygens and __***__ nonbridging oxygens per silicon.

11R In the calcining process, a phase is dissociated by __***__ with the evolution of a __***__.

• 11S The basic structure of clay, $Al_2Si_2O_5(OH)_4$ calcines to a __***__ % Al_2O_3–__***__ % SiO_2 composition.

• 11T A __***__ diagram can show the composition of a three-component material graphically.

11U Since ceramic phases possess __***__ and __***__ bonds, they possess relatively high stability and generally are insulators. A major exception to the last statement are those compounds that contain ions with more than one __***__.

11V Those ceramic phases without mobile ___***___ are transparent, at least in thin sections.

11W The crystallization rate for ceramics is generally ___***___ than the rate for metals; therefore, ceramics form glasses more readily than metals. The most rapid crystallization rates are measured in ___***___ for ionic ceramics and in ___***___ in those silicates that we use for glasses.

11X As a rule, ceramic phases have a very high ___***___ resistance, since they are compounds with a preference for ___***___ neighboring atoms.

11Y Since ceramics have a high shear strength, they support large ___***___ loads but are sensitive to flaws and imperfections when loaded in ___***___.

STUDY PROBLEMS

11-1.1. Calculate the lattice constant, *a*, for CsCl from data in the Appendix. [Remember that CN = 8.]

Answer: 0.414 nm

11-1.2. Calculate the density of CsCl from the atomic weights and the answer of S.P. 11-1.1.

11-1.3. Calculate the density of nickel oxide, which has the same structure as NaCl.

Answer: 6.8 g/cm^3

11-1.4. The density of LiF (with the NaCl structure) is 2.63 g/cm^3. Does this data give a larger or smaller unit cell size than that calculated from the data in Table 11-1.1?

11-1.5. CdS has a density of 4.8 g/cm^3 and the structure of ZnS. What is its lattice constant, *a*?

Answer: 0.585 nm

11-1.6. (a) From the answer to S.P. 11-1.5 and data in Table 11-1.1, what radius can be assigned to the cadmium in CdS? (b) What is the packing factor of CdS if we assume ions that are spherical?

11-2.1. What is the coordination number of (a) the F^- ions in CaF_2? (b) The Ca^{2+} ions?

Answer: (a) 4

11-2.2. As indicated in the comments following I.P. 11-2.1, the experimental lattice constant for CaF_2 is 0.547 nm. Calculate the density of CaF_2.

11-2.3. Estimate the unit-cell size of ZrO_2, which has the CaF_2 structure at high temperatures.

Answer: 0.485 nm

11-2.4. The uranium oxide used as a nuclear fuel material has the structure of CaF_2. As expected, there are eight O^{2-} ions per unit cell. However, the average uranium ion valence is greater than 4+; therefore, there is an average of only 3.5 metal ions per unit cell. The density of the oxide is 10.9 g/cm³. (a) What is the average distance between the centers of the uranium and oxygen ions? (b) Estimate the radius of the uranium ion.

11–2.5. In Al_2O_3, which is hexagonal, the Al^{3+} ion is in a 6-fold site among six O^{2-} ions. What is the distance between the centers of the closest oxygen ions?

Answer: 0.27 nm

11–3.1. The lattice constant of cubic $BaTiO_3$ is 0.400 nm. (a) Therefore, the center-to-center distance for Ti^{4+} to O^{2-} is _***_ % of that predicted from ionic radii. (b) The center-to-center distance for Ba^{2+} to O^{2-} is _***_ % of that predicted from ionic radii. (A Ba^{2+} ion is listed with R = 0.14 nm for CN = 12.)

Answer: (a) 0.200 nm versus 0.208 nm, or 96%

11–3.2. The center-to-center distance between closest O^{2-} ions in $NiFe_2O_4$ is 0.293 nm. What is the minimum radius (a) of the 6-f interstices? (b) The 4-f interstices? [Refer to Fig. 11–1.4.]

11–4.1. High pressures can force ZnS into the NaCl-type structure from its normal ZnS-type structure. What is the accompanying volume change?

Answer: $V_{NaCl} = 0.86\ V_{ZnS}$

11–4.2. The ion sizes for KCl are such that either the NaCl-type or the CsCl-type structure may form, depending upon the temperature and pressure. (a) Which polymorph will be more dense? (b) Which type will be favored by higher pressures? (c) By higher temperatures? (d) Support your answer for part (a) with calculations.

11–5.1. Ruby is Al_2O_3 with up to 0.5 w/o of the Al_2O_3 replaced by Cr_2O_3. (a) What fraction of the cations are Cr^{3+} ions when 0.5 w/o Cr_2O_3 is present? (b) What fraction of the ions are O^{2-}

Answer: (a) 0.34 a/o (b) 60%

11–5.2. The lattice constant of NiO, which has an NaCl-type structure, is 0.42 nm. Two mole percent Li_2O are in solid solution in NiO. As a result Ni^{3+} forms, and an $Li^+ - Ni^{3+}$ pair replaces an equivalent $Ni^{2+} - Ni^{2+}$ pair in the structure. (a) How many unit cells per Li^+ ion? (b) Per Ni^{3+} ion?

11–5.3. CaO can dissolve into ZrO_2 because Ca^{2+} and Zr^{4+} have nearly the same size. There is a 15/85 ratio of Ca^{2+}/Zr^{4+}. The charge is balanced by O^{2-} ion vacancies. What is the ratio of O^{2-} ion vacancies to O^{2-}?

Answer: $\boxminus/O^{2-} = 0.15/1.85$

• 11–6.1. Use the radii given for silicon and oxygen in I.P. 11–6.1. (a) What is the lattice constant of the cubic polymorph of SiO_2 in Fig. 11–6.2? [*Hint:* The oxygen lies directly between two silicon atoms, and the Si – O – Si bonds parallel the cube-body diagonals.] (b) What is the packing factor, if we assume spherical atoms?

Answer: (a) 0.7 nm (b) 0.3

• 11-6.2. (a) Use the answer(s) of S.P. 11-6.1 to calculate the number of Si atoms per cm^3. (b) Calculate the density of cristobalite, the SiO_2 polymorph shown in Fig. 11-6.2.

11-7.1. Ten grams of a CaO – NiO eutectic composition contain how many grams of α (a) at 1600°C? (b) At 1000°C?

Answer: (a) 3.5 g

11-7.2. At 1500°C, what is the maximum w/o aluminum in the β phase of the $MgO - Al_2O_3$ system?

11-7.3. (a) What is the liquidus temperature for a $60Al_2O_3 - 40SiO_2$ ceramic material? (b) The solidus temperature?

Answer: (a) 1925°C (3500°F)

• 11-7.4. What is the composition of (a) the junction labeled **5** in Fig. 11-7.5? (b) The junction labeled **6?**

Answer: (a) $25MgO - 21Al_2O_3 - 54SiO_2$

• 11-7.5. What is the index of refraction of glass that contains (a) $20Al_2O_3 - 80SiO_2$? (b) $20K_2O - 20Al_2O_3 - 60SiO_2$?

Answer: (a) 1.49

• 11-7.6. An insulator is made with 30 g of talc and 70 g of clay. How many grams of (a) MgO, (b) Al_2O_3, and (c) SiO_2 are there in the calcined product?

Answer: (a) 9.6 g MgO (b) 27.7 g Al_2O_3

• 11-7.7. Locate the composition of the calcined product of S.P. 11-7.6 on Fig. 11-7.5.

Answer: $11MgO - 31Al_2O_3 - 58SiO_2$ (which happens to be at the intersection of the "talc"-"clay" line and the 1600°C isotherm)

CHECK OUTS

11A compounds
release
share

11B atom
adjacent atoms are identical

11C smaller
larger

11D 110
97

11E larger
0.73

11F NaCl
ZnS
face-centered

11G 50

11H 6-f, four
4-f, eight

11I 8-f
positive

11J temperatures
pressures

11K size
charge balance

11L vacancies
defect
nonstoichiometric

11M Fe^{2+}, Fe^{3+}

11N SiO_4
silicon, oxygen

11O tetrahedra
bridging

11P vitreous (glassy)
thermal expansion coefficient
packing factor
rotate

11Q one
two

11R heat
gas

11S 46–54

11T triaxial

11U ionic, covalent
valence

11V electrons

11W slower
seconds
kiloseconds

11X shear
unlike

11Y compression
tension

UNIT TWELVE

• MAKING AND SHAPING OF CERAMICS

As a class, ceramic materials are more resistant to high temperatures than are metals and polymers. Likewise, the deformation of ceramic phases is nearly impossible because of their high shear strengths. As a result, we must use alternative making and shaping processes. Powders are widely used, with shapes obtained by pressing in metal molds, by wet plastic deformation of clay-base products, and by the casting of slips into porous molds that absorb the liquid of suspension. These solid particles must be bonded by sintering, or firing. Exceptions to these processes include glass production (Unit 13), and single-crystal preparations.

Contents

Prerequisites General chemistry; Unit 11; and Sections 1–3, 4–1, and 4–6.

From Unit 12, the engineer should

1) Be acquainted with the principal steps that accompany particulate processing from raw materials through the several shaping options.
2) Identify the various categories of water that exist in presintered products, and relate these to drying and shrinkage.
3) Understand the general principles of sintering, both solid and vitreous.

4) Be able to calculate linear shrinkage from volume shrinkage, or vice versa.
5) Know the principal procedures for preparing single crystals.
6) Understand technical terms that pertain to ceramic processing.

12-1 RAW MATERIALS

Throughout civilization, clay has been a prime raw material for ceramic products. Clay is plastic when it is wet, and it gains strength when it is dried. Therefore, clay can be shaped into desired products and then become usable after the water evaporates. Furthermore, it was discovered in prehistoric times (probably accidentally after fire came into use) that the dried product becomes even stronger if it is heated to a "red" temperature.

Most modern ceramic products contain raw materials other than clay and natural rocks and minerals. These source materials generally must receive initial processing before they are incorporated into the final product.

In general, ceramic raw materials do not melt below 1000°C (and sometimes not even below 2000°C, or 3600°F). Therefore, melting is not a widely used manufacturing process (except for glass, Unit 13). Excluding wet clay, none of the widely encountered ceramic phases has the plasticity common to metals and polymers. Therefore, mechanical forming is not a widely used manufacturing process either. These two limitations promote the use of *particulate* raw materials, that is, powders for direct compaction into the final product shapes.

Natural raw materials

Clay, flint, and feldspar have been the more widely used raw materials that come directly from nature.* In nontechnical terms, *clays* are fine-grained, earthy particles. To the ceramist, a clay is a sheet-structured, layered silicate (Fig. 12-1.1). *Flint* is a rock of microcrystalline silica; while *feldspar* is a mineral of Na_2O, Al_2O_3, and SiO_2, or K_2O, Al_2O_3, and SiO_2 (the tan, or pink, mineral of granite).

There are several clay phases. We shall, however, limit our discussion to kaolinite [$Al_2Si_2O_5(OH)_4$]. Without detailing its structure, we can cite two important features about the clay of Fig. 12-1.1. (a) Its structure is a two-dimensional sheet. In fact, it could be contended that each flake or sheet in that figure is a large 2-D molecule. These sheets readily stack on top of

* Each may require the removal of admixed impurities, but they are generally used without any chemical purification.

Figure 12-1.1 Clay crystals. A very high magnification (×33,000) shows that clay crystals have a sheet structure. When the clay is wet and water molecules are absorbed between the sheets, clay is soft and plastic. ($1\mu = 10^3$nm.) (W. H. East, *J. Amer. Ceram. Soc.*)

others. (b) Small molecules, for example, H_2O, can be adsorbed onto the surface of these sheets, or absorbed between the adjacent sheets. As with polymers, these small molecules plasticize the larger units (Section 9-3). This accounts for the hydroplasticity of wet clay. After the plasticized clay has been dried to remove the *inter*layer water, the adjacent sheets come into contact and develop extensive hydrogen bonding that gives a moderate *dry strength* to the clay.

There are two types of surfaces in clay: the flat surface of the clay sheets and the edges that surround the flake. The latter expose broken Si–O bonds and therefore carry a charge and strongly adsorb ions.

Clay dissociates on heating:

$$Al_2Si_2O_5(OH)_4 \xrightarrow{\text{heat}} Al_2O_3 + 2SiO_2 + 2H_2O\uparrow. \quad (12\text{-}1.1)$$

We commonly label this dissociation that releases a vapor as *calcination.* It occurs at about 500°C (~930°F), and water—usually called *water of hydration*—escapes. However, water does not reside within $Al_2Si_2O_5(OH)_4$ as such. With sufficient heating, the SiO_2 and Al_2O_3 combine to form a glass:

$$Al_2O_3 + 2SiO_2 \longrightarrow \text{glass}. \quad (12\text{-}1.2)$$

This glassy material serves as the bond that gives enhanced strength to all clay-based ceramics, such as brick, chinaware, and porcelains.

Processed raw materials

Alumina, Al_2O_3, is probably the most widely used nonsilicate raw material for ceramic products since it is used for many electrical insulators. It originates as bauxite, the ore of aluminum metal, with the approximate composition of aluminum hydroxide, but with undesirable impurities such as SiO_2 and Fe_2O_3. Chemical purification is required (Table 12-1.1). In brief terms, the process (Bayer) involves (a) grinding, (b) solution of the alumina, then (c) precipitation to give a pure $Al(OH)_3$. That product is calcined to be dissociated to Al_2O_3:

$$2Al(OH)_3 \xrightarrow{\text{heat}} Al_2O_3 + 3H_2O\uparrow . \tag{12-1.3}$$

The above description is overly simplified because it is necessary to control the grain size of the $Al(OH)_3$, which, in turn, affects the particle size of Al_2O_3. Very fine particles of Al_2O_3 ($<1\ \mu m$) generally produce better ceramic products than coarser Al_2O_3.

Oxides, such as BaO, NiO, and PbO, that are to be incorporated into ceramic products do not occur in commercial quantities in nature. Therefore, they must be obtained chemically. It is common to precipitate them as hydroxides [$M(OH)_2$], carbonates [MCO_3], or oxalates [MC_2O_4]. In turn, they are calcined; for example,

$$BaCO_3 \longrightarrow BaO + CO_2\uparrow . \tag{12-1.4}$$

Table 12-1.1
Al_2O_3 production from bauxite (aluminum ore)*

Step	Purpose
1. Dry and calcine bauxite at the mine	Reduces weight for shipping Removes any organic material
2. Grind to − 70 mesh	Accelerates the solution in Step 3
3. Dissolve (leach) in an NaOH solution [2–8 hours at 160°C (320°F) and 60 psi].	Forms soluble Na^+ and AlO_2^- ions, $Al(OH)_3 + NaOH \rightleftarrows AlO_2^- + Na^+ + 2H_2O$
4. Settle any precipitate and insoluble residue, and filter	Removes $Fe(OH)_3$ precipitate and SiO_2
5. Cool to reverse the reaction of Step 3. [Fresh $Al(OH)_3$ particles are added to nucleate further $Al(OH)_3$ precipitation].	Precipitates $Al(OH)_3$
6. Filter and wash. (The NaOH solution is recycled to Step 3.)	Removes $Al(OH)_3$ and salvages NaOH
7. Calcine the $Al(OH)_3$ at 1150°C (2100°F)	Forms Al_2O_3

* Bayer process.

Mixing

The vast majority of ceramic products contain more than one component. However, unlike most metals, ceramic products (other than glass) do not go through a molten stage that facilitates homogenization. Solid-state diffusion is required and this is slow, even slower than in most metals. Because of these factors, mixing is especially important.

To assure thorough mixing, it is necessary to grind the raw materials into very fine particles. Commonly this is done in a ball mill (Fig. 12–1.2). These rotating mills contain a wear-resistant lining and a charge of wear-resistant balls that provide grinding surfaces. *Ball milling* may be either wet or dry. Although wear-resistant, the lining and the balls slowly abrade. Therefore, their composition must be compatible with the composition of the products; for example, an Al_2O_3 lining and balls may be used in grinding porcelain mixes, since the porcelain composition contains some alumina. The particles will have a statistical range of sizes. This is shown in Fig. 12–1.3 for two batches of ball-milled alumina. Dry milling is not as effective as wet grinding because dust becomes suspended within the mill. During wet grinding, the particles are brought between the grinding surfaces more frequently. Of course, wet grinding generally requires a drying step in later processing.

Freeze drying provides a special procedure for intimately mixing two or more components without grinding. To illustrate, it is possible to form an aqueous solution of barium nitrate and ferric nitrate. This solution is frozen, then held in a vacuum until the water is removed. The intimate mixture of

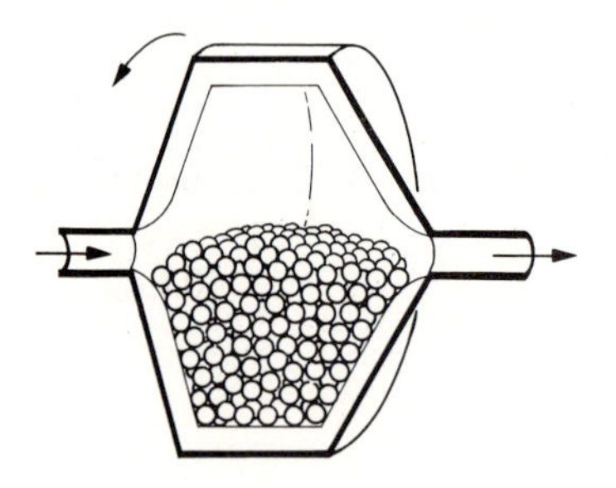

Figure 12–1.2 Ball mill (continuous). This rotating mill grinds the raw material mixture into very fine particles. With fine grinding and intimate mixing, subsequent firing processes are more rapid and the product more uniform.

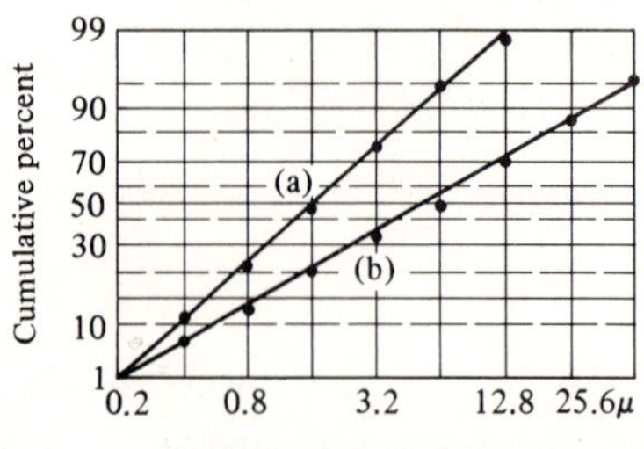

Figure 12–1.3 Diagram of sizing control of ball-milled alumina by grinding: (a) wet grinding; (b) dry grinding. (Adapted from non-gaussian plots by W. D. Kingery, *Ceramic Fabrication Processes*, New York: John Wiley & Sons.)

$Ba(NO_3)_2$ and $Fe(NO_3)_3$ can be calcined to give particles of BaO and Fe_2O_3 on μm-dimension scale. Later firing (Section 12-4) permits fast reactions. In general, freeze drying is more expensive than grinding; however, in many cases shorter firing cycles and more uniform products are possible.

Illustrative problem 12-1.1 A porcelain body is to have the following weight composition: SiO_2, 59; Al_2O_3, 32; K_2O, 3; Na_2O, 1; and CaO, 5. The following are available as raw materials:

1. Quartz	SiO_2		
2. Kaolinite	$(OH)_4Al_2Si_2O_5$	or	$Al_2O_3 \cdot 2SiO_2 \cdot 2H_2O$
3. Potash feldspar	$KAlSi_3O_8$	or	$K_2O \cdot Al_2O_3 \cdot 6SiO_2$
4. Soda feldspar	$NaAlSi_3O_8$	or	$Na_2O \cdot Al_2O_3 \cdot 6SiO_2$
5. Calcite	$CaCO_3$	or	$CaO \cdot CO_2$

Assuming pure raw materials, indicate the number of pounds of each material needed to make 100 kg of body.

SOLUTION

$$CaO = 5 = X_5(56.1/100.1)$$
$$Na_2O = 1 = X_4(62.0/524.6)$$
$$K_2O = 3 = X_3(94.2/556.8)$$
$$Al_2O_3 = 32 = X_4(102/524.6) + X_3(102/556.8) + X_2(101.9/258.2)$$
$$SiO_2 = 59 = X_4(360.6/524.6) + X_3(360.6/556.8) + X_2(120.2/258.2) + X_1$$

Solving, we have

$X_5 = 8.9$ kg calcite
$X_4 = 8.5$ kg soda feldspar
$X_3 = 17.7$ kg potash feldspar
$X_2 = [32 - (8.5)(102/524.6) - (17.7)(102/556.8)](258.2/102)$
$= 68.6$ kg kaolinite
$X_1 = 59 - (8.5)(360.6/524.6) - (17.7)(360.6/556.8) - (68.6)(120.2/258.2)$
$= 9.8$ kg quartz (flint). ◀

12-2 SHAPING PROCESSES

Nonglassy ceramics are made by compacting particles into the desired shape and then sintering them into a coherent mass. Three shaping procedures are common: (1) pressure fabrication, (2) hydroplastic forming, and (3) slip casting.

Pressure fabrication

This procedure is widely used in the preparation of electrical ceramics, magnetic ceramics, and other nonclay products. It also serves as the basis for powder metallurgy (Section 4-6); however, in powder ceramics and pow-

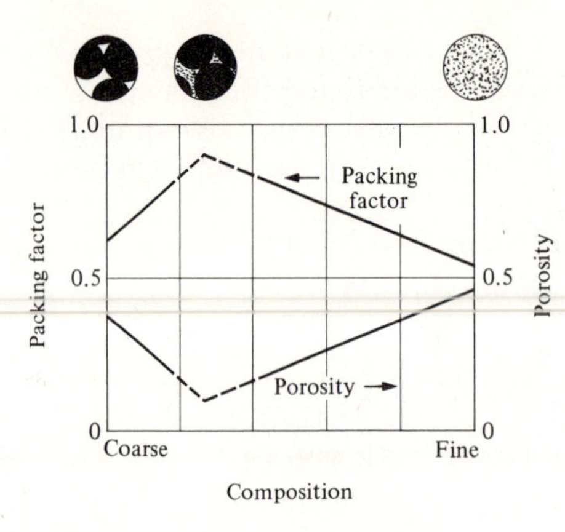

Figure 12-2.1 Mixed sizing. Maximum density occurs when finer particles fill the spaces between coarser particles.

der metallurgy, there is a rather significant difference in processing. In many cases, metal particles making up the powders are deformed as they are compacted so as to densify the product. Except for a limited number of materials at high temperatures, ceramic powders do not compact by pressure alone beyond initial contact. To partially overcome this situation, ceramists employ *sizing*—they use coarse and fine particles to eliminate void space. This is shown schematically in Fig. 12-2.1. Maximum density occurs when there is just enough of the fine fraction to fill the spaces among the coarse particles.

Figures 12-2.2 and 12-2.3 sketch the two main procedures for applying pressure. The directional pressing of Fig. 12-2.2 is simpler in operation than hydrostatic pressing (commonly called *isostatic molding*) but is limited to simple shapes that have equal compaction ratios in all longitudinal sections. The isostatic molding of Fig. 12-2.3 is more complicated but is preferred where possible because there is negligible wall friction to produce pressure gradients.

The importance of wall friction is shown in Fig. 12-2.4, in which bulk densities are shaded to indicate various parts of the compact. Because part

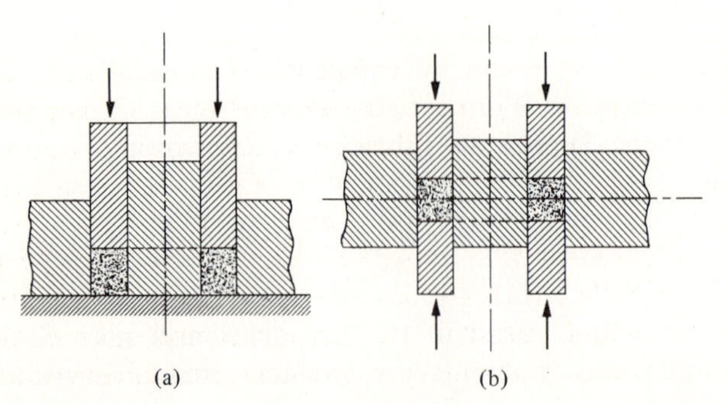

Figure 12-2.2 Pressure fabrication (magnetic toroid). (a) Single-end pressing. (b) Double-end pressing. More uniform density is attained in double-end pressing because wall friction is less with the neutral plane across the center of the product. (See Fig. 12-2.4.)

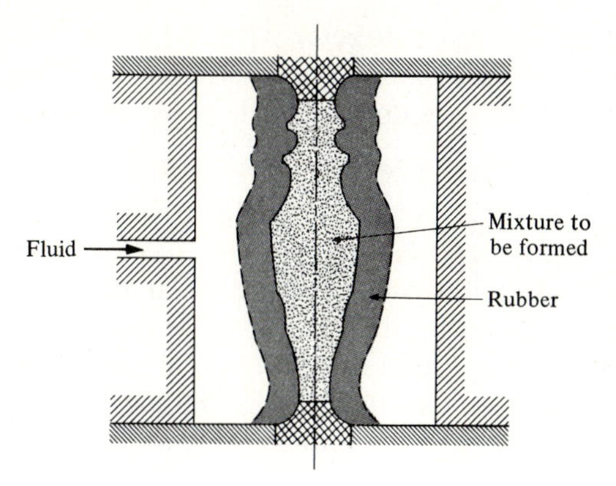

Figure 12-2.3 Isostatic molding (spark plug manufacture). The hydrostatic fluid provides radial pressure to the product. Maximum compaction is possible with this procedure. (Jeffery, U.S. Patent 1,863,854.)

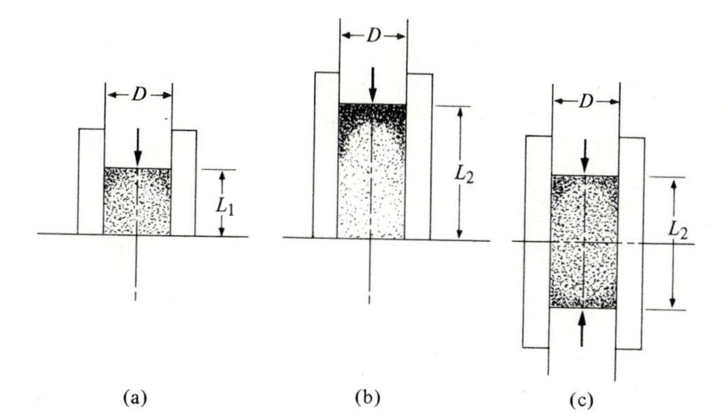

Figure 12-2.4 Bulk density distributions. (a) $L/D = 0.9$. (b) $L/D = 1.8$ (c) $L/D = 1.8$, double-end pressing. More uniform density is obtained with the neutral plane across the center.

of the compressive load is transferred to the mold wall, larger L/D ratios provide greater variations in applied pressures within the mold. Therefore the *double-end* compact of Fig. 12-2.4(c) with a smaller effective L/D ratio produces greater homogeneity and average densities than a single-end compact of the same dimensions (Fig. 12-2.4b).

Hydroplastic forming

Ceramic raw materials with layered structures, specifically clays, are amenable to hydroplastic forming. Their moisture content is adjustable, so that they are plastic enough to be extruded without cracking and also have enough yield strength so that the formed product will maintain its shape during handling and prior to drying. During hydroplastic forming, the layered colloidal particles are oriented by shear stresses. However, a direct comparison of hydroplastic ceramics and the more common metals cannot be made because most metals have cubic crystals and several equivalent planes for slip. The plastic forming of two-dimensional ceramic particles must be compared more directly with the plastic deformation of zinc, magnesium, or other hexagonal metals with only one major slip plane. Because a preferred orientation is developed, it is expected that anisotropic behavior will occur with respect to drying, shrinkage, and various properties.

Slip casting

The third forming process for particulate materials utilizes a suspension, or *slip*, of particles in a liquid (commonly water). The water is absorbed from the suspension into a porous mold leaving a dewatered "shell." When the semirigid shell is thick enough, the balance of the suspension is drained (Fig. 12-2.5). After time for additional dewatering, the mold is disassembled and the piece removed.

The primary advantage of slip casting is that intricate shapes may be formed by this method. Further, it has an economic advantage when production involves only a few items, thus making it a favorite forming process for ceramic artists. It also has industrial applications when thin-walled products are to be made from nonplastic materials, as in the case of high-alumina combustion tubes and thermocouple protection tubes.

Among the important criteria for slip casting is that the suspension remain *deflocculated,* that is, dispersed within a minimum quantity of liquid. It is apparent that if the particles were to settle, the walls of the product

Figure 12-2.5 Slip casting. A slip—a semifluid suspension—is poured into a porous plaster mold. The mold absorbs water, producing a solid shell of the product, which is dried and fired. (F. H. Norton, *Elements of Ceramics*, Reading, Mass.: Addison-Wesley.)

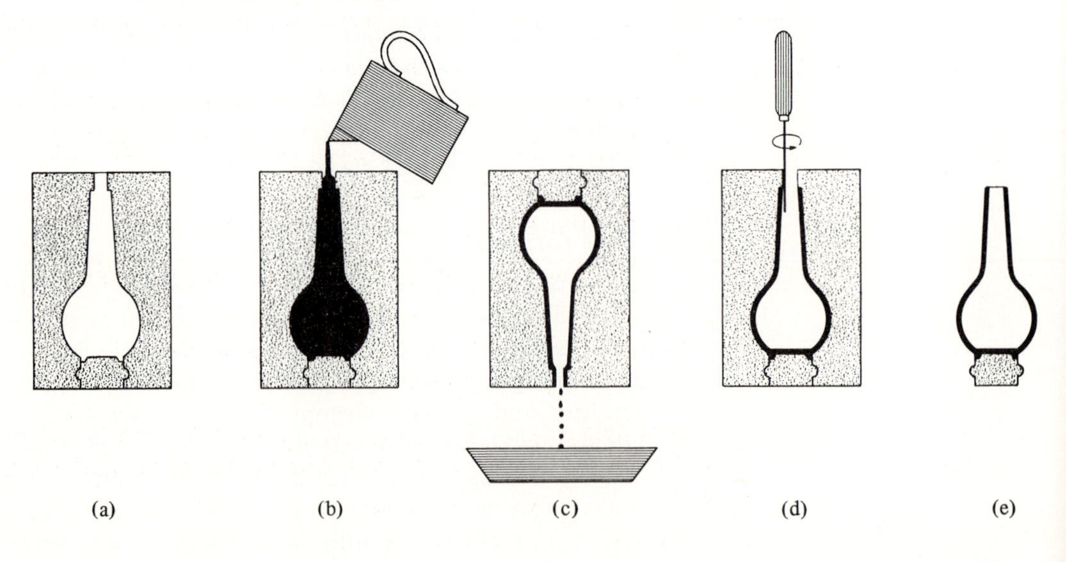

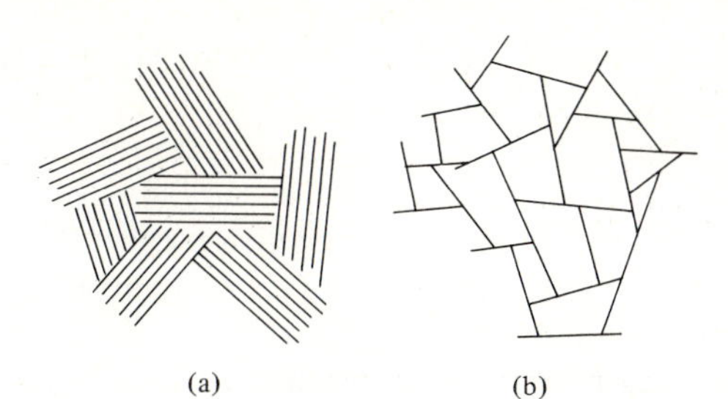

Figure 12-2.6 Flocculation of clay particles. (a) Face-to-face (card-deck) flocculation. (b) Edge-to-face (card-house) flocculation. The former produces a denser product. To obtain this, the ceramist must control particle size and the *pH* of the slip.

would not be uniform. Second, if the water content is high, more time is required for processing, and more shrinkage results. The above criteria are controlled through surface charges on the particles. With adsorption and the broken bonds described in the discussion of clay in the preceding section, it is possible to provide all the surfaces with like electrical charges and introduce mutual repulsion between particles. This promotes deflocculation, or dispersion of particles, and avoids clumping and settling. The ceramists try to do more than this, however. By adjusting the *pH* of the suspension and the adsorbants, they can effect a "card deck" type of structure as the slip is dewatered, rather than a "card house" type of structure (Fig. 12-2.6). The latter retains considerable water, producing a weak shell and permitting excessive shrinkage during drying.

Illustrative problem 12-2.1 Silica sand (~1.0 mm diameter) and silica flour (~0.01 mm) are to be mixed to maximum density. The former has a *bulk* density of 1.6 g/cm³, the latter 1.5 g/cm³. (a) How much of each should be used? (b) What is the maximum density possible? [Both silica sand and silica flour (finely crushed and powdered sand) are quartz, with a true density of 2.65 g/cm³.]

SOLUTION Basis: 1 cm³ of sand = 1.6 g sand.

$$\text{Actual volume sand} = (1.6\text{ g})/(2.65\text{ g/cm}^3) = 0.60\text{ cm}^3.$$
$$\text{Pore space} = 0.40\text{ cm}^3.$$

Fill the pore space of the sand with silica flour.

$$\text{Amount of silica flour} = (0.40\text{ cm}^3)(1.5\text{ g/cm}^3) = 0.6\text{ g}.$$

a) 1.6 g sand to 0.6 g flour.
b) Bulk density = (1.6 g sand + 0.6 g flour)/1 cm³
= 2.2 g/cm³.

ADDED INFORMATION The actual volume of the flour is (0.6 g)/(2.65 g/cm³) or 0.226 cm³. Therefore the overall packing factor is (0.60 + 0.226)/1 cm³, or 0.83. ◄

12-3 DRYING

The initial product after hydroplastic forming, or after slip casting, has a high moisture content. When the product is fabricated by means of pressure, it usually contains small amounts of liquid or fugitive lubricants and binders. Although these need not be water, they must be removed.

To be economical as an engineering process, drying must be rapid. At the same time, it must not be *so* rapid that the product is damaged by cracks or warping as a result of changes in volume. To design an efficient drying process, one must be familiar with the distribution of the liquid within the product, and also with the rate of movements of the liquid. With this knowledge, one can predict various dimensional and property changes with improved accuracy.

Water removal

Water is the liquid most commonly used in ceramic production; therefore we shall give it most of our attention. However, the general principles that apply to water also apply to other liquids, such as oil or carbon tetrachloride. Water may be considered a fugitive vehicle because, after it has been used for purposes of suspension or plasticity, it can readily be removed by evaporation, due to its relatively high vapor pressure.

Water distribution can be categorized into several types. (Not all types are present in all ceramic products prior to drying.) These categories include (1) water of *suspension,* (2) *interparticle* or interlayer water, (3) *pore* water, and (4) *adsorbed* water. We shall not talk further about the water of suspension, because it must be removed by filtration, by decantation, or by the absorption into a mold.

Figure 12-3.1(a) shows interparticle (or interlayer) water in products made of claylike materials. It is in a relatively free state. This interlayer

Figure 12-3.1 Water in wet clay. Shrinkage occurs as the interparticle water is removed (a → b). Further drying removes the pore water and introduces porosity (b → c). The final adsorbed water on the surfaces of the clay particles is tightly held. (After F. H. Norton, *Elements of Ceramics*, Reading, Mass.: Addison-Wesley.)

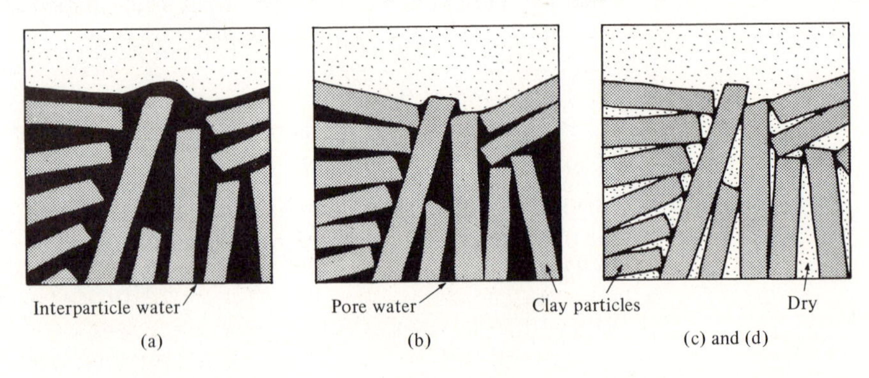

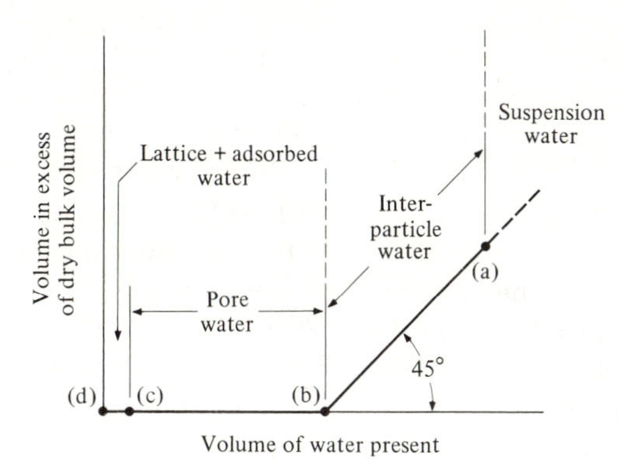

Figure 12–3.2 Shrinkage versus water content (schematic). The letters refer to Fig. 12–3.1. (After F. H. Norton, *Elements of Ceramics*, Reading, Mass.: Addison-Wesley.)

film may be as much as 50 nm thick, which is comparable to the colloidal dimensions of some ceramic raw materials. When the water is removed, the layered particles move closer together (probably by capillary action) and the shrinkage of the previously formed product is readily apparent. During this stage of drying, the volume shrinkage directly equals the volume of interparticle water removed [Fig. 12–3.2(a) to (b)].

Figure 12–3.1(b) shows pore water as the water in the interstices among the particles, after the particles have come into contact. Removal of this water produces very little if any additional shrinkage [Fig. 12–3.2(b) to (c)]. In fact, in special cases, there is even some evidence that a slight expansion occurs because the capillary forces of the liquid disappear.

Hydroplastically formed products generally contain both interparticle and pore water because they usually contain clays or other two-dimensional raw materials. This is also true for slip-cast ceramics. Pressure-fabricated ceramics contain very little (if any) interparticle water, and therefore do not exhibit the large shrinkage depicted in Fig. 12–3.2. This factor constitutes an advantage for pressure molding that partially offsets its other limitations.

Adsorbed water is water that is held by surface forces. It is usually only a few molecules thick; however, its quantity can be significant in ceramic products manufactured from colloidal-sized particles (see I.P. 12–3.2). This water usually requires drying temperatures in excess of 100°C because the molecules are adsorbed to the surface more tightly than they would be bonded to other water molecules of the liquid during normal evaporation.

The drying process requires a movement of liquid from the interior of the shaped product to the surface, where evaporation can occur. The diffusion rate is highly sensitive to temperature; in fact, it increases by a factor of four as the temperature increases from 20°C to 100°C. Let us tie this fact in with the data of Fig. 12–3.2. If the surface is heated and dried while the center is still cool, the surface will shrink [(a) to (b) of Fig. 12–3.2] around the center before the latter shrinks. In contrast, if the entire moist product is heated to 95°C (~200°F) in a humid atmosphere before drying is started, moisture can diffuse outward nearly four times as fast while drying occurs. This procedure, called *humidity drying,* produces a much lower moisture (and dimensional) gradient between surface and center. Cracking and warping are significantly reduced by using this procedure, just as it is for drying wood (Section 10-4).

Illustrative problem 12-3.1 An extruded ceramic insulator that has an initial density of 1.90 g/cm³ shrinks 6.3 l/o (dry basis) during drying. The dry product is 1.75 g/cm³. (a) What was the weight fraction of interparticle water in the extruded shape? (b) What is the volume fraction of open porosity in the final dried product?

SOLUTION Basis: 1 cm³ of dried product = 1.75 g of dried product.

$$\text{Volume before drying} = (1.00\text{ cm} + 0.063\text{ cm})^3 = 1.20\text{ cm}^3.$$
$$\text{Mass before drying} = (1.20\text{ cm}^3)(1.90\text{ g/cm}^3) = 2.28\text{ g}.$$
$$\text{Total water} = (2.28\text{ g} - 1.75\text{ g}) = 0.53\text{ g} = 0.53\text{ cm}^3.$$
$$\text{Interparticle water (from Fig. 12-3.2)} = (1.20\text{ cm}^3 - 1.00\text{ cm}^3) = 0.20\text{ cm}^3 = 0.20\text{ g}.$$

a) Weight fraction interparticle water = 0.20 g/2.28 g = 0.088.
b) Open porosity = (pore volume)/(total volume)
= (0.53 cm³ − 0.20 cm³)/(1.0 cm³) = 0.33.

ADDED INFORMATION The dry weight and dimensions are commonly used as a basis because they are reproducible for sample-to-sample comparison. ◄

• **Illustrative problem 12-3.2** A monolayer of water has adsorbed onto the flat surface of clay. As such, the water molecules are approximately 0.3 nm apart, or approximately 10 H_2O per nm². Estimate the weight percent H_2O (dry basis) if the clay particles are 1-μm plates that are 5-nm thick (average dimension). The true density of clay is 2.6 g/cm³.

SOLUTION

$$\text{Volume of flake} = (\sim 1\,\mu\text{m})^2(0.005\,\mu\text{m}) = \sim 5 \times 10^{-3}\,\mu\text{m}^3 \quad (\text{or } \sim 5 \times 10^{-15}\text{ cm}^3).$$
$$\text{Surface of flake} = 2(\sim 1\,\mu\text{m})^2 \quad (\text{or } \sim 2 \times 10^6\text{ nm}^2).$$
$$H_2O\text{ molecules} = (\sim 10/\text{nm}^2)(2 \times 10^6\text{ nm}^2) = 20 \times 10^6\ H_2O.$$
$$\frac{\text{Mass } H_2O}{\text{Mass dry clay}} = \sim \frac{(20 \times 10^6\ H_2O)(18\text{g}/0.6 \times 10^{24}\ H_2O)}{(5 \times 10^{-15}\text{ cm}^3)(2.6\text{g/cm}^3)} = \sim 0.05 \quad (\text{or } \sim 5\text{ w/o}).$$

ADDED INFORMATION Our estimate assumed 1-μm square flakes. Show why an estimate for hexagonal flakes (Fig. 11-1.1) or for 0.5-μm flakes will give the same answer. In any event, colloidal materials can absorb considerable moisture. [Cf. moisture adsorption in wood (Section 10-3).] ◄

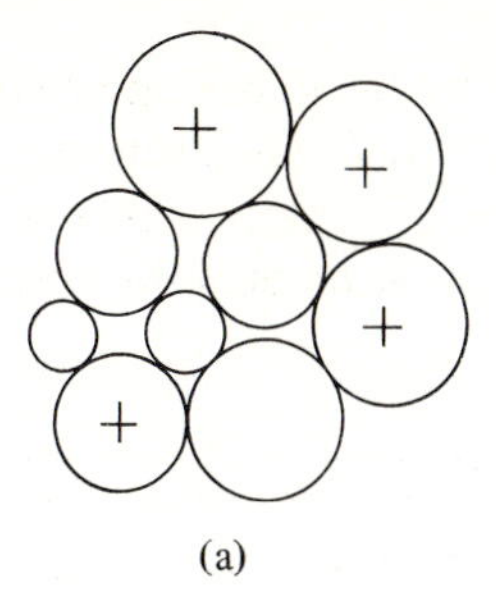

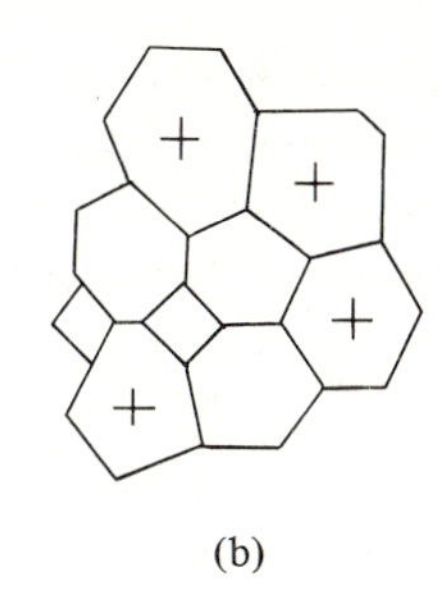

Figure 12–4.1 Solid sintering (schematic). (a) Particles before sintering have two adjacent surfaces, each with high energy. (b) Grains after sintering have one, lower-energy boundary (cf. Fig. 4–6.2).

12–4 BONDING PROCESSES

Since the majority of ceramic products are made by agglomeration, the mechanism of particle bonding is important. Two prime procedures are used: *sintering* and *cementation*. In this section we shall pay attention only to the former. Sintering is performed at elevated temperatures. Cements will be discussed in Unit 15, since they are widely used in concrete, which may be viewed as a composite material.

Solid sintering

The use of heat to bond particles is called *sintering*, or *firing*. In principle the process of sintering is straightforward. Two particles in contact form one boundary rather than two surfaces (Fig. 12–4.1).

The actual mechanism for sintering is somewhat more complicated. Atoms must move from the points of particle contact and diffuse to the surfaces where contact does not exist. The most common route of diffusion is through the crystalline solid. Another route may be along the boundaries and surfaces, which is the path taken by oxygen ions during the sintering of Al_2O_3. We shall not dwell longer on the mechanism of sintering, except to note than many modern ceramics—such as UO_2 for nuclear fuel elements, and Al_2O_3 envelopes for sodium-vapor lights—cannot be produced without an extensive knowledge of the factors that modify sintering.

Firing shrinkage

Extremely high-pressure techniques for compacts of nonductile powders cannot produce full density, that is, zero porosity. We depend on sintering to bring about the final pore removal. The centers of the particles move closer together as atoms or ions move from the points of contact to the surfaces where contact does not occur. (Cf. Fig. 4–6.2.) This produces shrinkage,* which is a "necessary evil" in sintering processes. Allowances must always be made in production for the correct amount of shrinkage. (See I.P. 12–4.1.)

* In a few materials with high vapor pressures—for example, NaCl—sintering can occur by volatilization from the free surfaces and condensation adjacent to the points of contact to enlarge the contact area. This does not produce shrinkage and, furthermore, it does not reduce the total porosity.

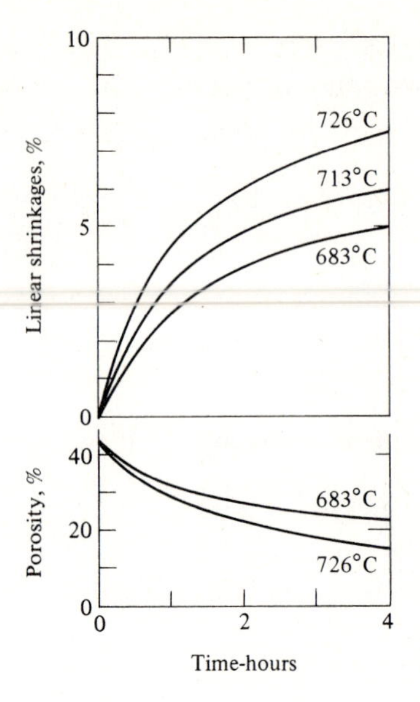

Figure 12-4.2 Sintering shrinkage (NaF). Powdered NaF (−330 mesh) loses porosity as it shrinks during sintering. (Adapted from Allison and Murray, *Acta Metallurgica*.)

Since sintering and, therefore, shrinkage involve diffusion, both shrinkage and pore removal are a function of temperature and time (Fig. 12-4.2). The amount of shrinkage also depends upon the initial porosity. If the initial porosity in Fig. 12-4.2 had been 50% rather than 43%, the final amount of shrinkage would have been greater. This situation may be encountered in many pressed compacts. For example, Fig. 12-4.3 shows a commonly encountered problem. The magnetic toroid was pressed as an open cylinder in which the pressing densification was nonuniform because of the large effective L/D in the cylinder wall. (Cf. Fig. 12-2.4.) Subsequent sintering produced greater shrinkage in the less dense portions of the toroid. In many products, such a variance is not tolerable. Unequal shrinkage and/or warpage may also occur

a) if the distribution of temperature is not equalized, for example, one side of the product is directly exposed to the heat source and the other side shielded;

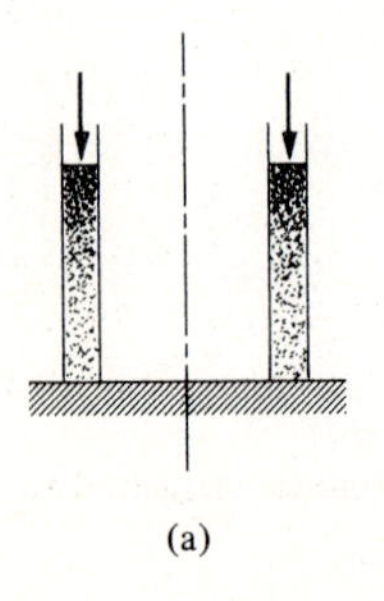

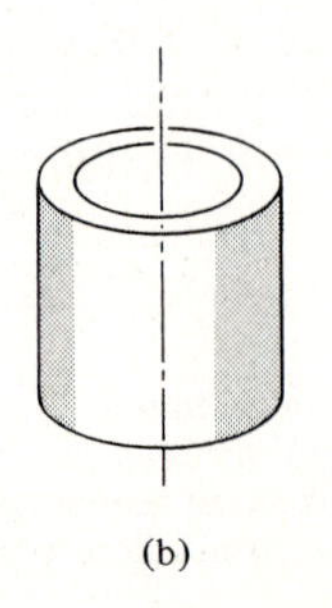

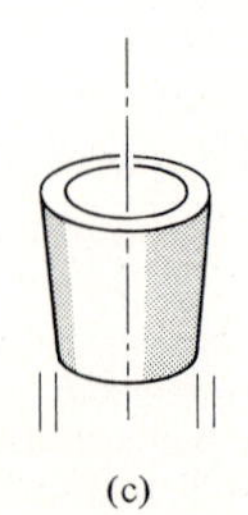

Figure 12-4.3 Dimensional changes in a cylindrical magnet. Unequal shrinkage can result from forming procedures if the packing density is not controlled (Fig. 12-2.4). (a) Single-end pressing. (b) Stripped and dried. (c) Fired.

b) if the time the material is kept at the sintering temperature varies, for example, within thick sections versus thin sections;

c) if there were particle segregation during forming that caused variations in particle size or in compositions;

d) if there is an anisotropy in the structure of the particles, that is, the clay and mica flakes that are pressed to make a steatite insulator (Section 11–7);

e) if flow occurs because of stresses caused by gravity during the high-temperature period of firing; or

f) if there is friction between the kiln base and the shrinking ware.

The engineer finds it mandatory to maintain extremely close control over the processing variables, which include moisture content, particle sizes, drying rates, pressures, and firing temperatures and times.

Vitreous sintering

Siliceous materials normally produce glass during the firing operation. Such materials include the multitude of clay-based ceramics and, as indicated by Eq. (12–1.2), the dissociation products of clay form a glass. We shall see in the next unit that glass responds to viscous deformation. Thus, at firing temperatures, the capillary tension arising from the surface energies gradually shrinks the product, reducing porosity. This action, plus the hydroplastic forming discussed in Section 12–2, made it possible for the early skilled craftsmen to produce intricate, impermeable ceramic products even though they had negligible scientific knowledge of the accompanying structural changes. However, a knowledge of the structural changes can be both useful and interesting to the engineer.

Figure 12–4.4 is a plot of the corner of the $K_2O - Al_2O_3 - SiO_2$ ternary diagram. The older electrical porcelains, called *triaxial porcelains,* are made with three main components: clay [$Al_2Si_2O_5(OH)_4$ calcined to 46 Al_2O_3 – 54 SiO_2], feldspar [$KAlSi_3O_8$], and quartz [SiO_2]. Their compositions are indicated by **C, F,** and **Q,** respectively. The total composition must lie inside a triangle with those compositions at the apexes. The firing reactions can be deciphered from the microstructures shown in Fig. 12–4.5.* (1) The clay has changed completely to glass and small growing mullite ($Al_6Si_2O_{13}$) crystals, the phase we would expect on the basis of the $Al_2O_3 - SiO_2$ diagram (Fig. 11–7.3). (2) The feldspar is a flux that produces considerable glassy liquid. Furthermore the K^+ ions make the glass more fluid than the glass from clay alone. Crystals grow readily in this liquid, to give large mullite needles in original feldspar areas. For this to occur a small fraction of the K^+ ions must

* Since the structural dimensions are in the submicron region, this knowledge had to await modern technology and the development of the electron microscope.

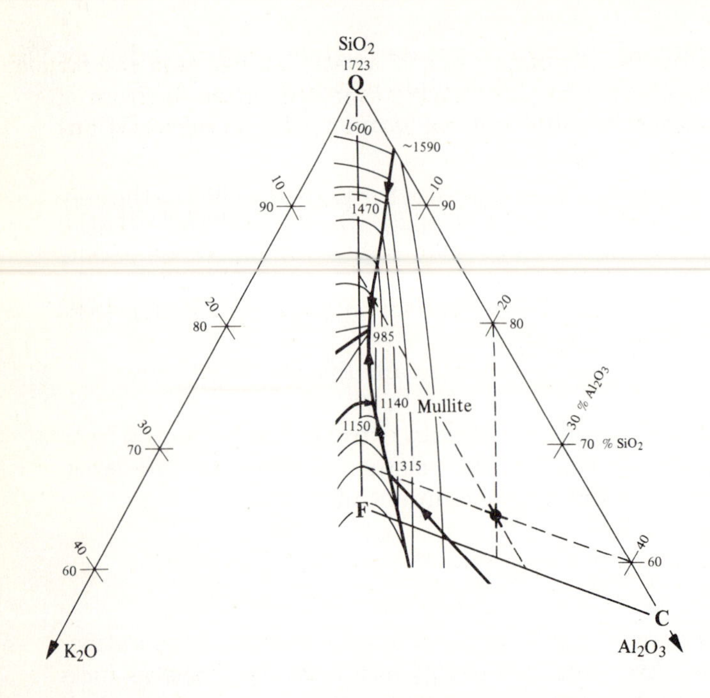

Figure 12-4.4 Portion of the K_2O-Al_2O_3-SiO_2 diagram. Conventional electrical porcelains are made from **C**—clay (Eq. 12-1.1); **F**—feldspar ($KAlSi_3O_8$); and **Q**—quartz (SiO_2). (See I.P. 12-4.2.)

diffuse away from these areas. (3) The quartz reacts only as fast as it is fluxed by the diffusing K^+ ions. Consequently it remains almost intact, providing stability in volume during the firing process. The silica-rich liquid around the quartz grains crystallizes extremely slowly, remaining as a crystal-free glass.

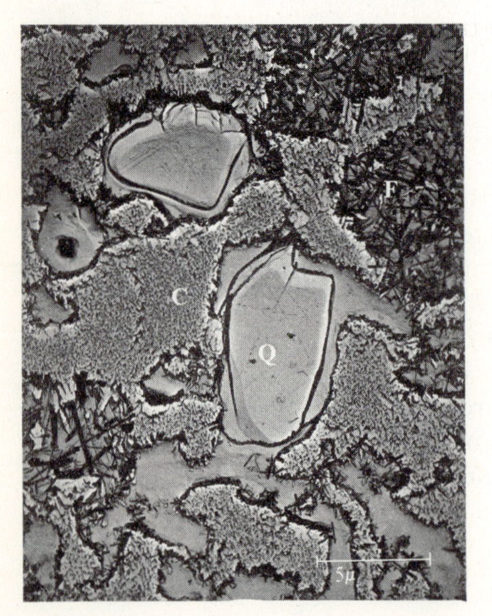

Figure 12-4.5 Electrical porcelain microstructure (×3000). Although not attained, reactions are moving toward the equilibrium. The former clay (**C**) has formed mullite and glass because it contains about $60SiO_2$-$40Al_2O_3$ (Fig. 11-7.3). The former feldspar (**F**) contains more glass and larger mullite crystals because it contains considerable K_2O. The original quartz grains (**Q**) are only slightly altered around the edge by the glassy liquid. Since the original particle sizes were very small (<5 microns) and the glass was semifluid at high temperatures, the pore space was eliminated during firing. (Courtesy of S. T. Lundin, Technological University of Lund, Sweden.)

It takes a long time for the diffusion and crystallization required for complete homogenization to take place. As a result the phase diagram of Fig. 12–4.4, which assumes equilibrium, cannot be used quantitatively but can be used to predict the direction of reaction.

Illustrative problem 12–4.1 The final dimension for a ceramic component must be 17 mm. Laboratory tests indicate that the drying and firing shrinkages are 7.3 l/o and 6.1 l/o, respectively, when based on the dried dimensions. (a) What is the required initial dimension? (b) What was the dried porosity if the final porosity is 2.2%?

SOLUTION

a) $17.0 \text{ mm} = 0.939\, L_d,$

$$L_d = 18.1 \text{ mm};$$

$$L_i = 1.073\,(18.1 \text{ mm}) = 19.4 \text{ mm};$$

b) True volume, $V_{tr} = 0.978\, V_f$;

therefore, $$\frac{V_{tr}}{V_d} = \frac{0.978\, V_f}{V_f/(0.939)^3} = 0.81.$$

$$\text{Porosity} = 100\% - 81\% = 19\%.$$

ADDED INFORMATION This calculation assumes no density changes of the phases. In some cases appropriate corrections would have to be made. ◀

• **Illustrative problem 12–4.2** An electrical porcelain is made into an insulator from 50 w/o clay [$Al_2Si_2O_5(OH)_4$], 40 w/o potassium feldspar [$K_2Al_2Si_6O_{16}$], and 10 w/o silica flour (SiO_2). What is its fired composition? (a) Solve by computation. (b) Solve graphically.

SOLUTION Basis: 100 g = 50 g of clay (which according to Eq. (12–1.1) drops to 43 g with 54.1% SiO_2 and 45.9% Al_2O_3) + 40 g $KAlSi_3O_8$ + 10 g SiO_2.

$$2KAlSi_3O_8 = K_2O{\cdot}Al_2O_3{\cdot}6SiO_2.$$

a) Computation (in grams):

	K_2O, g	Al_2O_3, g	SiO_2, g
Dissociated clay:	0	$0.459 \times 43 = 19.7$	$0.541 \times 43 = 23.3$
Feldspar:	$\left(\frac{94.2}{556.8}\right)(40) = 6.8$	$\left(\frac{102}{556.8}\right)(40) = 7.3$	$\left(\frac{360}{556.8}\right)(40) = 25.9$
Silica flour:	0	0	10.0
Total	6.8	27.0	59.2
Weight percent:	7.3	29.0	63.7

b) Graphically:

Dissociated clay:	43 g	= 46.2 w/o
Feldspar:	40 g	= 43.0 w/o
Silica flour:	10 g	= 10.8 w/o
	93 g	

On Fig. 12–4.4 locate the point within the **F–Q–C** triangular region for the above ratio (intersection of the three dashed lines). This point corresponds to the 7–29–64 point in the overall K_2O–Al_2O_3–SiO_2 triangle, thus checking the computation.

ADDED INFORMATION The clay dissociation follows Eq. (12–1.1), in which 258 amu of $Al_2Si_2O_5(OH)_4$ loses 36 amu of H_2O to give 102 amu of Al_2O_3 and 120.2 amu of SiO_2.

The triaxial diagram is not limited to the corner oxides alone, but can be used for any combination of components that lie within the triangle. ◀

12-5 SINGLE-CRYSTAL PREPARATIONS

Metals are seldom (if ever) used as single crystals; single crystals of polymers are rare, small, and usually imperfect at best. However, there are numerous ceramic phases (and also semiconductors) that find technical usage as single crystals. Figure 11–6.3(c) showed quartz crystals that had been made artifically for use as crystal oscillators to maintain frequency control at radio and microwave wavelengths. Figure 12–5.1(a) shows ruby boules, which are single crystals, and are used as lasers to produce coherent radiation.

There are three common procedures that are available to produce single crystals of ceramic phases: (a) fusion techniques, (b) hydrothermal solutions, and (c) fused salts.

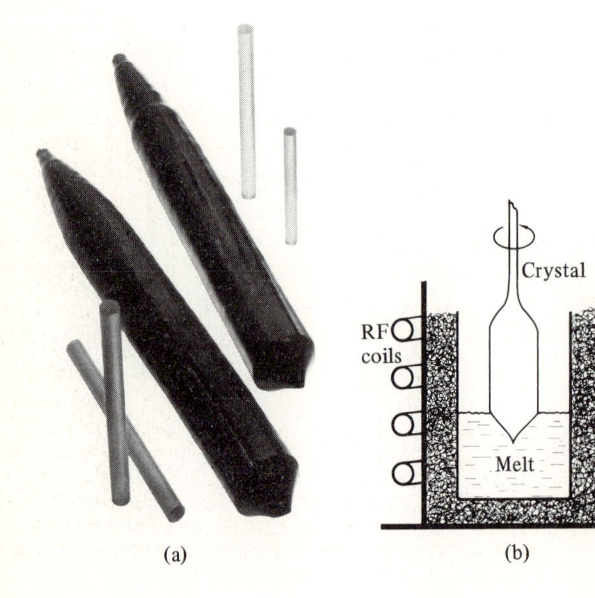

Figure 12–5.1 Single crystals. (a) Boules (up to 100-mm dia.) and laser rods of Al_2O_3. (b) Crystal growth by the Czochralski method (schematic). A seed crystal is lowered to the surface of the melt and rotated as it is slowly raised, growing in the process. (Courtesy of Union Carbide, Electronics Division.)

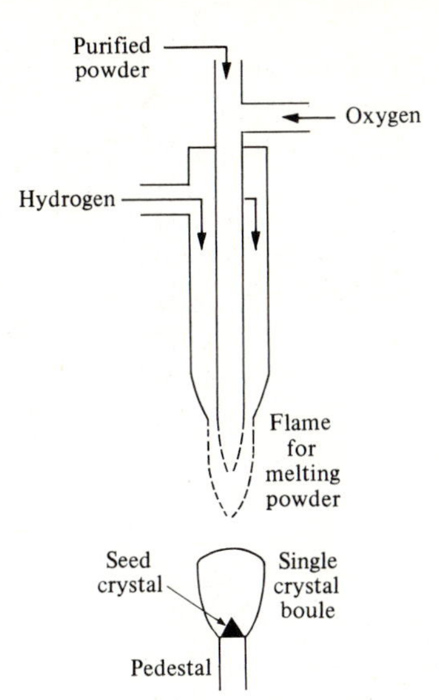

Figure 12-5.2 Single crystal boule (sapphire, that is, Al_2O_3). Al_2O_3 powder is melted in an oxygen-hydrogen flame. The seed grows to a large single crystal.

Flame fusion is shown schematically in Fig. 12-5.2. It is commonly used for crystals that melt without dissociation so that the liquid and the solid have the same composition. In order to grow the single crystal boule, previously prepared powders of the correct composition are melted in an oxygen-hydrogen flame. The small droplets fall to a preselected *seed* crystal that grows into the large, single-crystal boule. Alternatively, the Czochralski method (Fig. 12-5.1b) starts with a seed crystal that is lowered against the surface of a melt and then pulled slowly upward, allowing growth to proceed.

Zone fusion remelts a rod of polycrystalline material by heating one end first and passing the molten zone toward the other end. The local (zone) heating is commonly achieved in semiconductors by induced currents (Section 18-5). Since ceramic phases are generally insulators, heating must be achieved by radiation or conduction. This limits the use of this process. With appropriate controls, the seed crystal continues to grow as the molten zone moves along the rod. In essence, the process in one of recrystallization, in which there is an intermediate molten stage.

Many useful large crystals either do not solidify directly from their melts or have polymorphic changes that preclude direct-fusion techniques. Prime among these is quartz (SiO_2) as pictured in Fig. 11-6.3(c). It must be formed below 573°C (1065°F) in order to avoid dimensional changes such as those described in Fig. 11-6.4. Since the solubility of SiO_2 in an alkaline solution of NaOH and H_2O varies with temperature, quartz sand can be dissolved at T_B and SiO_2 precipitated onto seed crystals at T_A as sketched in Fig. 12-5.3. Since the temperatures that are used are as high as 500°C (930°F), the process must be performed at high pressures (or the water

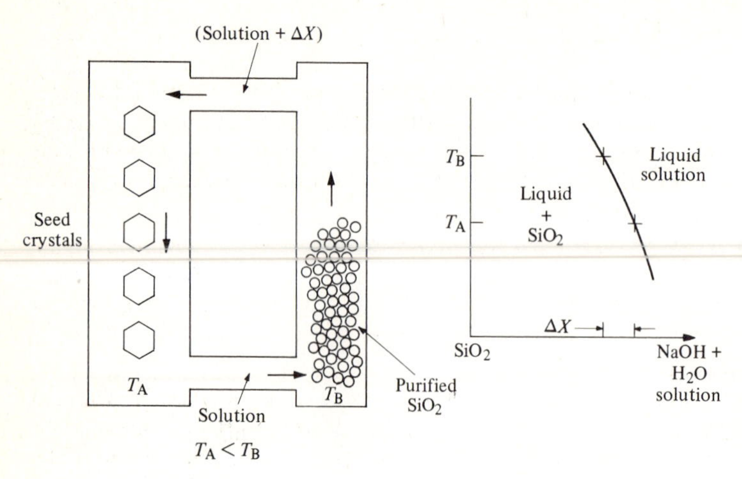

Figure 12-5.3 Schematic drawing of the hydrothermal process for the formation of quartz. The temperature T_B on the solution side is a few degrees higher than the temperature T_A on the precipitation side. Therefore, ΔX of SiO_2 is precipitated from the solution as it circulates from the warmer to the cooler sides. High pressure is required.

would evaporate). Close temperature control is required: (a) T_B should be high to accelerate solution, (b) greater differentials between T_B and T_A increase the chemical efficiency. However, excessive supercooling in A will cause spontaneous nucleation, which must be avoided if we wish to obtain single crystals. The process just described is labeled *hydrothermal* crystallization.

Crystallization from *fused salts* is similar in principle to the hydrothermal process of the previous paragraph. However, use of fused salts rather than water as the solute removes the necessity for high pressures. An example of the phase relationships that permit this type of crystal growth is shown in Fig. 12-5.4. Again, a change in the solubility is utilized to produce controlled supersaturation. Slow cooling is required in order for seed crystals to grow and in order to prevent spontaneous nucleation.

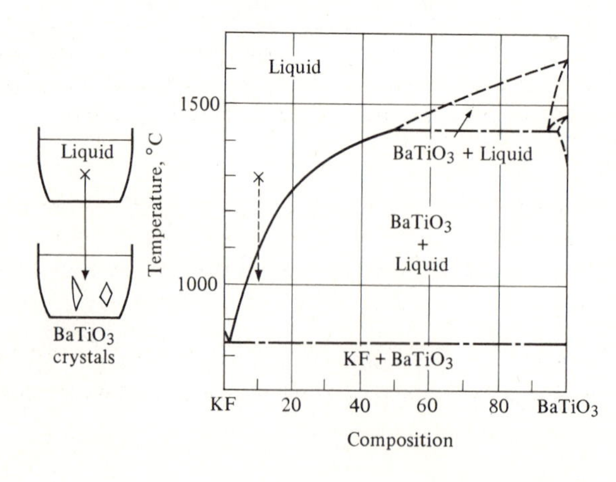

Figure 12-5.4 Growth from fused melts. (Adapted from R. C. DeVries, *J. Amer. Ceram. Soc.*)

Review and Study

SUMMARY

1. Historically, raw materials for ceramic products came directly from their natural sources—clay, flint, and feldspar. Although these continue to be used, present-day products require raw materials that are chemically processed, for example, Al_2O_3, BaO, CaO. Typically, these involve solution and precipitation steps, plus a final calcination.
2. Since they have high melting temperatures and are commonly nondeformable, ceramic products (other than glass) are generally processed from particulate solids. Preliminary steps include extremely fine grinding and mixing, or specialized procedures such as freeze drying. Mixed sizing can lead to higher packing factors and greater densities.
3. Pressure fabrication compresses dry particles into the product shape. Binders are required, for example, $\sim$ 1% wax. Since the uniformity of compaction is a problem, isostatic molding (with hydrostatic pressure) is used where possible. Clay-based products can be shaped by hydroplastic forming, or by slip casting.
4. Drying involves several steps of water removal. Dewatering of slips is achieved by the adsorption of excess water into porous molds (or by filtration). Interparticle water must be removed from hydroplastically formed products by diffusion. This is accompanied by a one-for-one volume shrinkage of the product and can lead to cracking unless (a) the rate is intentionally slow, or (b) the product is preheated before drying to facilitate moisture diffusion. Pore water, adsorbed water, and fugitive binders are removed without shrinkage. Clay and similar materials also lose "water of hydration" through calcination at higher temperatures.
5. The bonding of the powders into a monolithic ceramic product is normally achieved by firing. This is a high-temperature process that sinters the particles together and utilizes atomic diffusion. In many technical ceramics, this diffusion must be achieved within the solid. Clay-based ceramics develop a glassy liquid that facilitates diffusion, and therefore accelerates bonding and pore removal. Shrinkage always accompanies firing. Close process controls are required to maintain uniformity in the final product.
6. Single crystals of ceramic compounds find wide use, particularly in electronic products. Flame fusion, zone fusion, and hydrothermal growth are common processes for their production. In each case, a single seed crystal is used as a nucleus and growth proceeds along a thermal or concentration gradient.

TECHNICAL TERMS

Ball mill A rotating cylinder containing balls or pebbles for grinding and mixing raw materials.

Calcination *See* Technical Terms in Unit 11.

Clay *See* Technical Terms in Unit 11.

Deflocculation Dispersion of colloidal particles within a liquid suspension.

Drying Water-removal step in making hydroplastic and slip-cast ceramic products.

Flocculation Aggregation of particles within a liquid suspension.

Firing The sintering process used to agglomerate ceramic powders into a monolithic solid.

Flame fusion Procedure for growing single crystals within a high-temperature flame.

Flint Natural rock of very fine grained silica.

Hydroplastic Plastic when wet, for example, clays.

Hydrothermal crystallization Procedure for growing single crystals within a heated aqueous solution.

Isostatic molding Pressure fabrication in a rubber mold by hydrostatic pressure.

Packing factor (powders) Ratio of occupied volume to bulk volume. Also (1.0 − porosity).

Pressure fabrication Compacting of particles into a product by pressure.

Shrinkage, drying Shrinkage accompanying drying, in particular, the removal of the interparticle water.

Shrinkage, firing Shrinkage accompanying sintering through the removal of pores by diffusion.

Sintering (solid) Agglomeration of particles by diffusion through the solid.

Sintering (vitreous) Agglomeration of particles by viscous flow of a glass phase.

Slip (liquid) Thick suspension of particles.

Slurry *See* slip (liquid).

Water, absorbed Water contained within a material.

Water, adsorbed Water bonded to a solid surface.

Water, interlayer or interparticle Water among the particles that is responsible for the major volume changes during drying shrinkage.

Water, pore Free water in the pores of unfired ceramics (not adsorbed).

Water of hydration Water released by thermal decomposition. [Generally originates as OH^- ions. *See* Eq. (12–1.1).]

Water of suspension Water supporting colloidal particles in a slip (or slurry).

Zone fusion Procedure for growing single crystals by moving a molten zone along a rod of the material.

CHECKS

12A Most ceramic materials melt well above 1000°C, therefore melting is not a common manufacturing process for ceramics; (an exception is the making of __***__). Also, shaping is generally not attained by __***__; an exception is the making of products from wet clay.

12B Because of the characteristics cited above, most ceramic products are made from __***__.

12C Clays are fine-grained particles but, more specifically, they have a __***__ structure. Feldspars are alkali __***__-silicates.

12D Clays can be plasticized by absorbing ___***___ between the sheets, accounting for the ___***___ of wet clays.

12E Calcination dissociates clay into its components: ___***___, ___***___, and H_2O. The first two combine to form a ___***___ at higher temperatures.

12F Bauxite is the raw material for ___***___, and typically contains ___***___ and ___***___ as impurities, which must be removed in processing.

12G Ball milling serves to both ___***___ and ___***___ the raw materials.

12H ___***___ molding inside a rubber mold provides a nearly hydrostatic pressure and gives more uniform density because ___***___ is negligible.

12I Hydroplastic forming is generally limited to those ceramics with raw materials that have ___***___ structures.

12J A slip, or slurry, is a ___***___ of fine particles in a fluid.

12K The particles remain suspended in a ___***___ slip if the surfaces of the particles possess a ___***___.

12L We can categorize the water that must be removed by drying as ___***___, ___***___, ___***___, and ___***___. Most of the shrinkage occurs during the removal of ___***___ water.

12M Since adsorbed water is held by ___***___ forces, temperatures must be raised significantly above ___***___°C for that water to be removed.

12N Humidity drying requires ___***___ before drying so that the water ___***___ faster.

12O Sintering bonds particles by applying ___***___; atoms must ___***___ away from points of particle contact to other surfaces. The path of diffusion may be along grain boundaries, or through the ___***___.

12P Unequal firing shrinkage arises from (a) ___***___ . . . , (b) ___***___ . . . , or ___***___ . . . .

12Q Since ___***___ is not attained during firing, the phase diagram cannot be used quantitatively to predict the amounts of phases; however, these diagrams are useful to predict the ___***___ of reaction.

12R Flame and zone ___***___, hot aqueous ___***___, and molten ___***___ are used to produce single crystals of ceramic phases.

STUDY PROBLEMS

12–1.1. What weight ratio of $Ba(NO_3)_2$ and $Fe(NO_3)_3$ must be combined to produce a mix for the hard ceramic magnet of $BaFe_{12}O_{19}$ (Table 19–4.2)?

Answer: $Ba(NO_3)_2$-to-$Fe(NO_3)_3$ = 8.3/91.7 (or 1/11)

12–1.2. A bauxite ore contains 90% $Al(OH)_3$ and 10% impurities. If 85% of the $Al(OH)_3$ is recoverable, how many kilograms of Al_2O_3 are available from 100 kg of ore?

12–1.3. (a) What percent weight loss occurs in the calcining (high-temperature dissociation) of clay? (b) Of limestone ($CaCO_3$)? (c) Of $Al(OH)_3$?

Answer: (a) 14%

12–1.4. Nickel oxalate [NiC_2O_4], zinc carbonate [$ZnCO_3$], and Fe_2O_3 are to be mixed to produce a soft ceramic magnet that is a solid solution of $70NiFe_2O_4$–$30ZnFe_2O_4$ (by weight). How much of each of the three raw materials must be mixed to make 100 g of the above composition after calcination in air?

12–1.5. A porcelain body should contain

SiO_2, 61; Al_2O_3, 30; K_2O, 2; Na_2O, 3; CaO, 4.

How many kilograms are required of each of the raw materials that are listed in I.P. 12–1.1 in order to make a 500 kg batch?

Answer: 44.6 Qtz; 289.8 Kao; 59.1 K-spar; 126.9 Na-spar; 35.7 Cal.

12–1.6. What will be the weight loss during the calcination of the batch in S.P. 12–1.5?

Answer: 56 kg (or 10%)

12–1.7. An electrical porcelain is made with a mixture of 72 parts of pyrophyllite ($Al_2Si_4O_{10}[OH]_2$), and 28 parts of magnesite ($MgCO_3$). During processing each component loses all possible H_2O and CO_2. What is the final composition?

Answer: $16MgO$–$25Al_2O_3$–$59SiO_2$

12–1.8. The clay flakes in Fig. 12–1.1 average 0.5 μm in "diameter" and 10 nm in thickness. What is the surface area per gram of the clay, given that the true density is 2.6 g/cm^3?

12–2.1. Equal bulk volumes of the sand and silica flour of I.P. 12–2.1 are mixed together. Estimate the bulk density of the mixture.

Answer: 1.94 g/cm^3

12–2.2. Refer to I.P. 12–2.1. Combinations of (a) 75-sand/25-silica flour, (b) 50/50, and (c) 25/75 are thoroughly mixed and compacted. What is the greatest bulk density possible for each of the combinations? (ρ_{SiO_2} = 2.65 g/cm^3)

12–3.1. A dried, slip-cast product is 6% shorter after drying. What was the v/o shrinkage (based on dried dimension)?

Answer: 19 v/o (dry basis)

12–3.2. A clay drain tile was extruded as a hollow cylinder. The die had an i.d. of 108 mm and an o.d. of 135 mm. The dried dimensions are 99 and 124 millimeters respectively. Wet and dry lengths were 318 mm and 305 mm. How much shrinkage was there (dry basis)?

Answer: 9% radially, 4% longitudinally (and 24 v/o, dry basis)

12–3.3. A cylindrical extrusion should have a dried diameter of 26.7 mm. A die that is 28.4 mm in diameter is required under normal operating conditions of 20 v/o interparticle water (dried basis). What dried diameter will result if 28 v/o water is present rather than the 20 v/o?

Answer: 26.2 mm

12–3.4. Refer to S.P. 12–3.2. A piece is broken from the extruded tile. It weighs 319 g with the initial moisture, 290 g at the end of the drying shrinkage, and 251 g dry. What is the final porosity?

12–4.1. A ceramic insulator is to have a final dimension of 27.0 mm. Its volume shrinkage during sintering is 29.1% (unfired basis). What initial dimension should the insulator have before sintering?

Answer: 30.3 mm

12–4.2. A ceramic magnet for a small motor has a dimension of 11.1 mm before sintering and 10.2 mm after sintering, when its porosity is 4 v/o. (a) What was the volume shrinkage (dry basis)? (b) What was the porosity before sintering?

12–4.3. A ceramic electrical insulator of Al_2O_3 weighs 1.23 g, occupies 0.47 cm^3, and is 0.87 cm long before sintering. After sintering, it is 0.77 cm long. If the true specific gravity is 3.85, that is, 3.85 times as dense as water, what are the porosities (a) before, and (b) after sintering?

Answer: (a) 32 v/o (b) 1.9 v/o

12–4.4. A finished electrical porcelain product has only one v/o porosity, but had 27 v/o porosity after drying and before firing. How much linear shrinkage occurred during firing?

12–4.5. A ceramic wall tile contains 25 w/o mullite ($Al_6Si_2O_{13}$) and 75 w/o glass of an average composition of $7K_2O$–$18Al_2O_3$–$75SiO_2$. How much calcined clay (**C**), feldspar (**F**), and silica flour (**Q**) were required? (Solve graphically on Fig. 12–4.4 by extrapolating as required.)

Answer: 60**C**; 30**F**; 10**Q**

12–4.6. A porcelain was made from 50 w/o clay ($50SiO_2$–$35Al_2O_3$–$15H_2O$), 25 w/o feldspar ($KAlSi_3O_8$), and 25 w/o SiO_2. What is the final composition after firing?

12–4.7. A ceramic product was made with 10 w/o magnesite ($MgCO_3$), 30 w/o talc ($Mg_3Si_4O_{10}[OH]_2$) and 60 w/o clay ($Al_2Si_2O_5[OH]_4$). Determine the final MgO-Al_2O_3-SiO_2 compositions (a) by calculation and (b) graphically.

Answer: 17 MgO–$28Al_2O_3$–55 SiO_2

CHECK OUTS

12A	glass deformation
12B	powders
12C	colloidal alumino-
12D	water hydroplasticity
12E	Al_2O_3, SiO_2 glass
12F	Al_2O_3 Fe_2O_3, SiO_2
12G	grind, mix
12H	isotactic wall friction

12I layered (clay-like)

12J suspension

12K deflocculated
charge

12L water of suspension,
interparticle
pore
adsorbed
interparticle

12M surface
100

12N heating
diffuses

12O heat
diffuse
grains

12P unequal temperatures
unequal pressures
particle segration
unequal times at temperature
friction

12Q equilibrium
direction

12R fusion
solutions
salts

UNIT THIRTEEN

• GLASS AND VITREOUS PRODUCTS

Glass has an unusual role in the technical world. It contains the most abundant elements of the earth; it has a long history; large quantities are made via mass production; yet glass is the critical component of some of our most complicated technical products and systems, for example, lasers and optical fibers for multichannel light communication systems. Although basically a ceramic product, glass is also an inorganic polymer. Its structure, properties, and behavior are related to both categories of materials.

Contents

Prerequisites Units 7, 8, and 11; Sections 9-4, 12-4.

From Unit 13, the engineer should

1) Become familiar with the basic technology of glass production.
2) Understand the general concepts of glass structure, and the role of network formers, network modifiers, and bridging and nonbridging oxygens.
3) Know the significance of glass viscosities as they pertain to production operations.

4) Be familiar with the general factors that affect strength, optical properties, and electrical behavior of glass.
5) Become acquainted with the processing principles that are required for tempered glass, devitrified glasses, and optically active glasses.
6) Know the principal glass-related terms that can be encountered in technical communication.

13-1 GLASS PRODUCTS

Glass, as a product of man's ingenuity, may be as old as bronze. Even older is the use of natural glass (obsidian) for arrowheads and similar artifacts. However, glass beads and glazes, dating back 4500 years, show that humans were manipulating their resources, usually for objects to fulfill aesthetic desires. More than 2200 years ago, glassmakers had become adept at making thin colored glass rods, which were bundled together, sintered, drawn to a smaller cross section, and cut into slices to give multiple mosaics (Fig. 13-1.1a). In turn, these identical plaques were fused together to produce bowls and other simple containers.

Glass achieved a utilitarian role when the blow pipe was invented. This hollow tube of iron permitted the craftsman to use lung power and air pressure to shape hot glass into desired products. It was the first of a long series of glassmaking processes that have now become very sophisticated. The present-day glass-ribbon machine (Fig. 13-1.1b) can produce light-bulb envelopes at the rate of 1,000 per minute of controlled size, thickness, shape, and quality.

In this section we shall summarize the current processes that account for the majority of glass products. It will be necessary to omit a number of equally interesting, but more specialized, procedures.

Flat glass

As indicated by the name, this category includes the glass that is used for automotive and architectural purposes, whether lights for windows in homes or curtain-walls that cover the outside of modern skyscrapers (Fig. 13-1.2). Window glass has undergone a series of developments that date back to blown glass that was rotated onto a disc with a thick blob, or *crown*, in the center. This was followed by blowing large cylinders that were cut and opened into flat sheet. *Drawn* sheet glass was next developed and made by baiting the viscous melt and pulling a wide sheet continuously from the furnace by the thousands of meters, day after day.

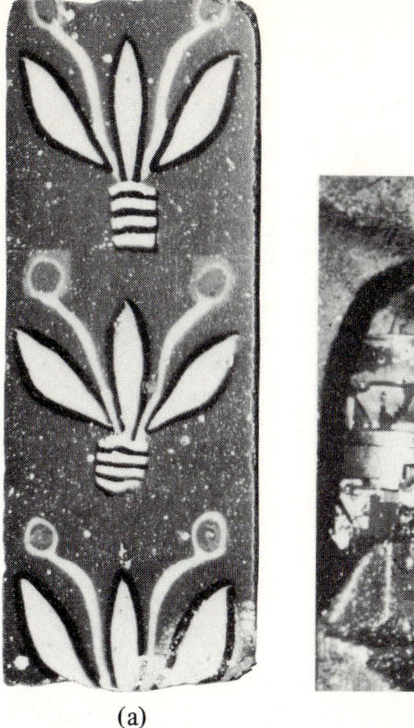

(a)

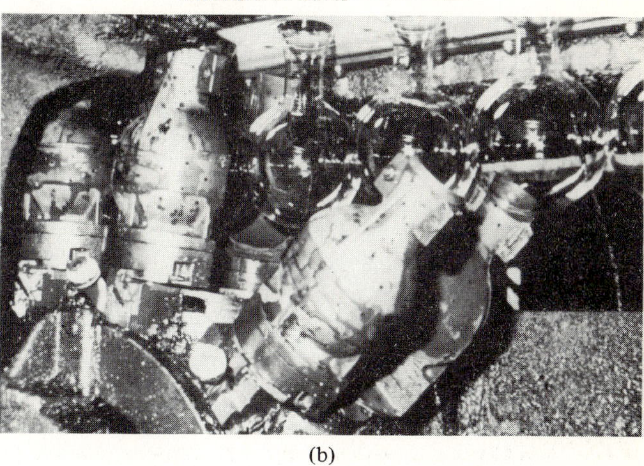

(b)

Figure 13-1.1 Ancient and modern glass. (a) Mosaic glass plaque (Ptolemaic Egypt, 3rd–1st century B.C.). This 3 cm × 7.5 cm floral plaque was made by bundling colored glass rods into the desired pattern, sintering them together, drawing the bundle into a smaller cross section, and cross-cutting them into 5-mm square wafers, which were then sintered edge to edge. (Courtesy of The Corning Museum of Glass.) (b) Modern glass-ribbon machine (light-bulb manufacture). A ribbon of hot, semiviscous glass is fed over a very rapidly moving train of molds and under controlled air jets that blow the glass into the shape of the mold. Hundreds are made per minute. (Courtesy of Lighting Business Group, General Electric Co.)

Figure 13-1.2 Architectural glass (curtain wall, Executive Plaza, Kansas City). Aesthetic appearance, installation costs, and energy considerations enter into the selection of glass as a construction material. (Courtsey of Libbey-Owens-Ford Glass.)

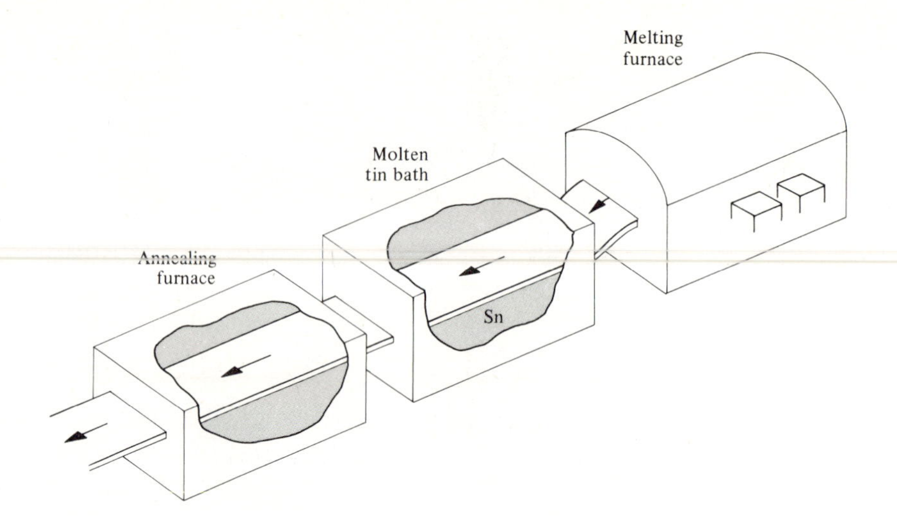

Figure 13-1.3 Float-glass production. Float-glass products are made visually smooth by floating them on a molten tin bath, a continuous process that produces as much as 50m²/min (varying with thickness).

All of the above products had distortions because the two surfaces were not perfectly flat or parallel. Therefore, it became desirable to grind and polish the two surfaces into parallel planes. The product was called *plate glass*. While visually flat, the ground and polished surface was inferior to the original "fire-polished" surface with respect to durability and to mechanical strength. Furthermore, the grinding and polishing operation was labor-intensive and very expensive. As a result, great expense and ingenuity was used to develop a *float process* for making glass (Fig. 13-1.3). A ribbon of glass, up to 4 meters wide, flows from the forehearth of a glass tank into a "float bath," where the glass floats on a bath of molten tin. It moves through this furnace at rates up to 20 cm/s (~40 ft/min). The initial temperature in the float furnace is high enough so that the soft glass forms a uniformly thick layer. The temperature drops along the furnace's length, with the result that the glass is completely rigid with only minor residual stresses at the exit. No polishing is required.

Container glass

Containers possess restricted necks. It is thus impossible to use any process, either casting or pressing, that depends upon matched molds. The alternative is to use air pressure. The reader may be aware of the glass blower who

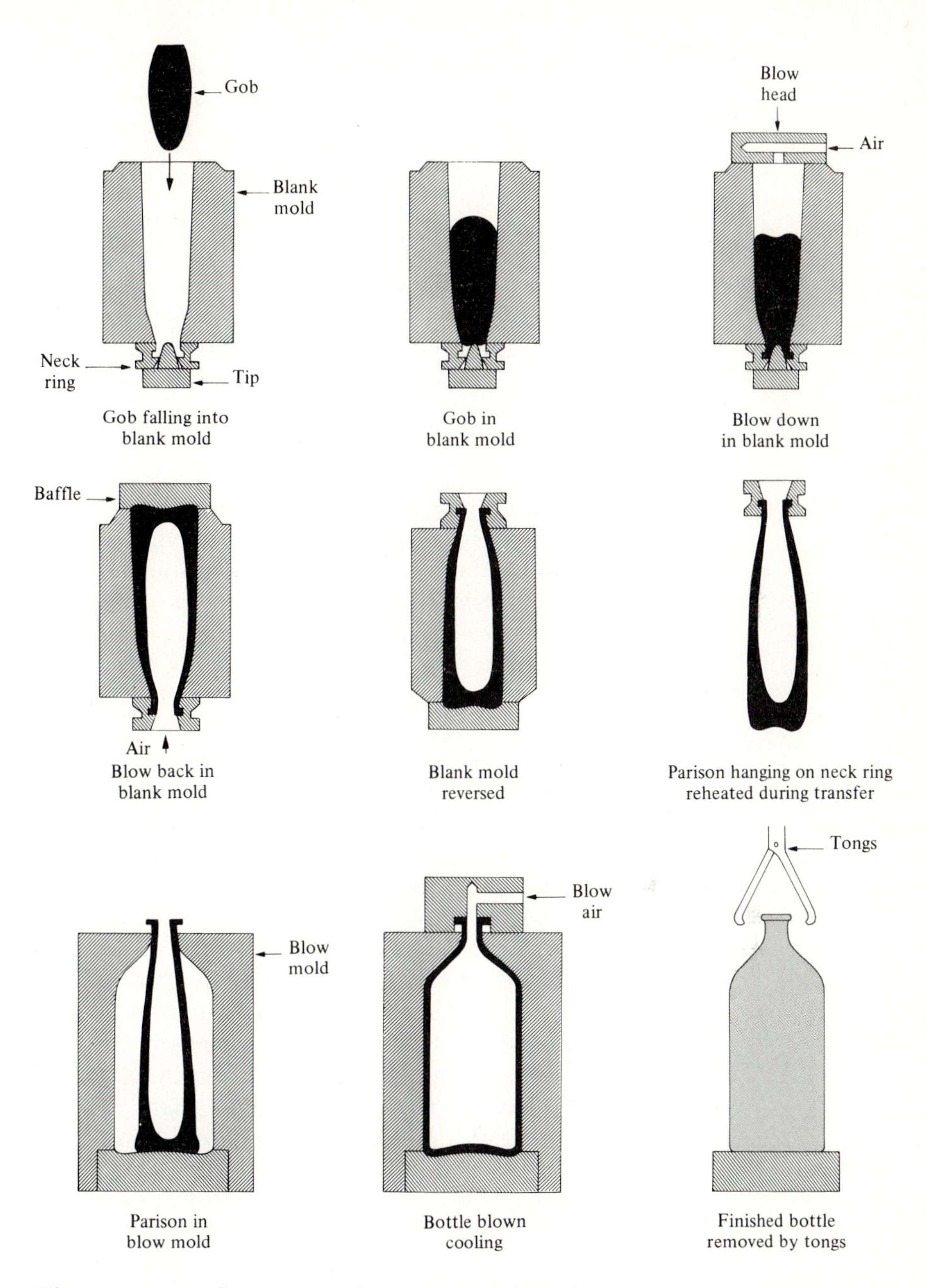

Figure 13-1.4 Steps in the production of a glass bottle. Air is blown into the center of a "gob," or large drop of viscous glass. The blowing and mold manipulations are all automatic mechanical steps. (Cf. Fig. 9-4.7.) (F. H. Norton, *Elements of Ceramics,* Reading, Mass.: Addison-Wesley.)

uses lung power. This procedure is still used by the specialist in the scientific laboratory who makes one-of-a-kind product, or by the artisan at the street-art fair. In industry, however, the production of containers by the thousands, for food, medicine, etc., has been mechanized as shown in Fig. 13–1.4. This process served as the basis for the blow-molding process for plastics (Fig. 9–4.5). The automation of the container-making process requires very close control of the factors that affect viscous flow—composition, temperature, pressures, and timing—so that the final product has uniform wall thickness and surface quality. In particular, the thickness of the bottom of the container can easily be excessively thick, leading to thin and weak walls, and to uneven cooling stresses that are undesirable.

Fiber glass

The history of glass-fiber production is short, although nature is known to produce fibrous glass in connection with vulcanism. In principle, glass fibers are made by thermally softening glass and "pulling fast." There are two industrial versions in use. The first produces a *discontinous,* randomly oriented glass wool that (a) provides a very effective thermal (and acoustical) insulation (Fig. 13–1.5a), and (b) is used in glass-reinforced plastics. Centrifugal forces are utilized to extend the fibers. They are coated with a thermosetting polymer, which not only serves as a binder for the glass mat but also protects the surface of the fibers from mechanical abrasion. The fibers are approximately 10 μm (~0.0004 in.) in diameter. The bulk density of the wool mat is approximately 0.1 g/cm^3.

Continuous glass fibers are made by drawing molten glass through spinnerettes (Fig. 13–1.5b), onto a high-speed, take-up winder. At velocities of 300 m/s-500 m/s, fibers can be drawn as small as 2 μm in diameter. As shown in the upper part of this figure, several hundred filaments are drawn simultaneously and collected into a single strand. These filaments find specific use (a) in glass cloth, (b) in insulation for electrical cord, (c) in communications, and (d) in the reinforcement of high-strength, glass-polymer composites (see Unit 14). The glass fibers must be "sized," that is, they must receive a coating that protects the surface from abrasion, not only from exterior grit, etc., but also from adjacent glass fibers.

Glass coatings

Veneers of glass have extensive use. Glass coatings on ceramic products are called *glazes;* glass coatings on metal are called *vitreous enamels,* or porcelain enamels. These coatings serve several purposes: (a) they prevent moisture and grease access to semiporous ceramic foodware; (b) they prevent oxygen access to heated metals, for example, the liner of a kitchen oven; and (c) they contribute aesthetically. Decorative tile, dinnerware, and artware provide examples of the latter.

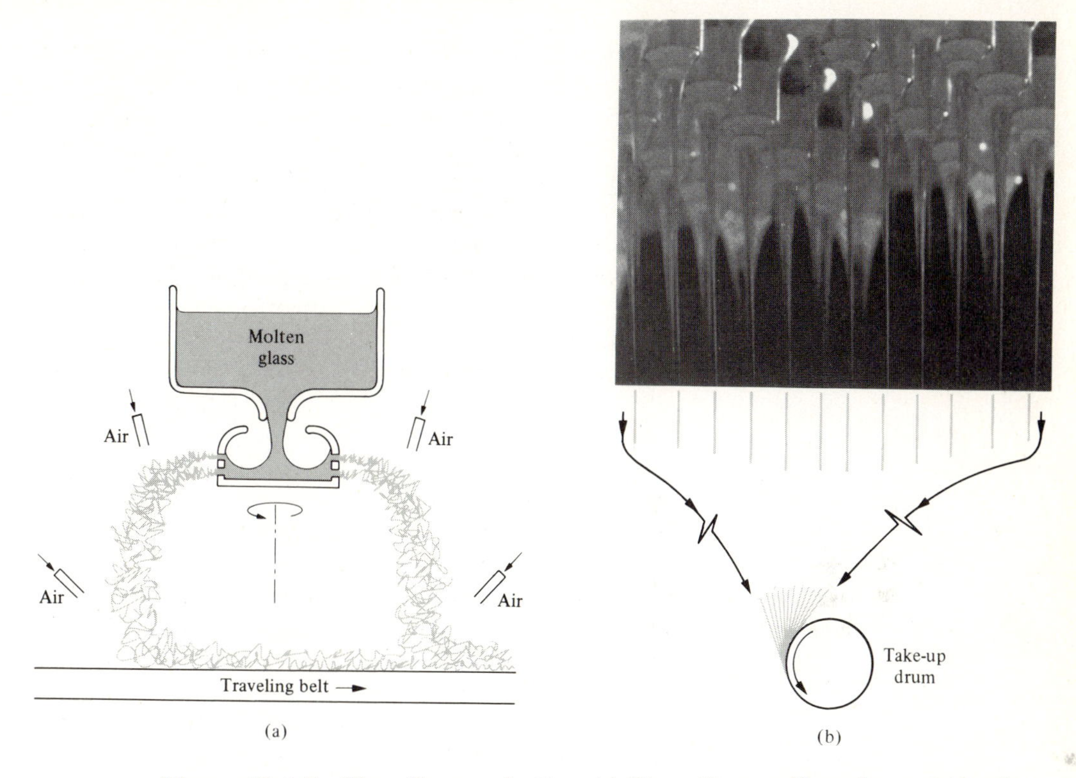

Figure 13-1.5 Glass-fiber production. (a) Discontinuous fibers (by centrifugal spraying). (b) Continuous fibers (by viscous drawing). The former are used for thermal and acoustical insulators. Continuous fibers—used for cloth, electrical insulators, communication, and reinforcement—are drawn through multiple orifices (spinnerettes) at high velocities. (Courtesy of Owens-Corning Fiberglas Co.)

13-2 COMPOSITION AND STRUCTURE OF GLASS

Silicate glasses

With minor exceptions, glasses of technology contain more than 50% SiO_2 (Table 13-2.1). Container glass and flat glass (structural glass) possess 72%-74% SiO_2. These two are the major glasses on a production basis.

In order to understand the structure and composition of these glasses better, let us review vitreous silica as it was introduced in Section 11-6. *Vitreous,* (glassy, or fused) *silica* is noncrystalline and made up of SiO_4 tetrahedra (Fig. 11-6.1). Each of the corner oxygens of the tetrahedra serve as a bridge between adjacent tetrahedra (Fig. 11-6.5). The four bridging oxygens with each tetrahedron produce a rigid structure. As a result, vitreous silica is very viscous even at high temperatures.

Table 13-2.1
Commercial glass types*

Type	Major components, %								Comments
	SiO_2	Al_2O_3	CaO	Na_2O	B_2O_3	MgO	PbO	Other	
Fused silica	99								Very low thermal expansion, very high viscosity
96% silica (vycor)	96				4				Very low thermal expansion, high viscosity
Borosilicate (pyrex)	81	2		4	12				Low thermal expansion, low ion exchange
Containers	74	1	5	15		4			Easy workability, high durability
Flat (float)	73		9	14		4			High durability
Lamp bulbs	74	1	5	16		4			Easy workability
Lamp stems	55	1		12			32		High resistivity
Fiber (E-glass)	54	14	16		10	4			Low alkali
Thermometer	73	6		10	10				Dimensional stability
Lead glass tableware	67			6			17	K_2O 10	High index of refraction
Optical flint	50			1			19	BaO 13; K_2O 8; ZnO 8	Specific index and dispersion values
Optical crown	70			8	10			BaO 2; K_2O 8	Specific index and dispersion values

* Data adapted from various sources, but primarily from A. K. Lyle, "Glass Compositions," *Handbook of Glass Mfg.* (F. V. Tooley, editor). New York: Ogden Publishing Co.

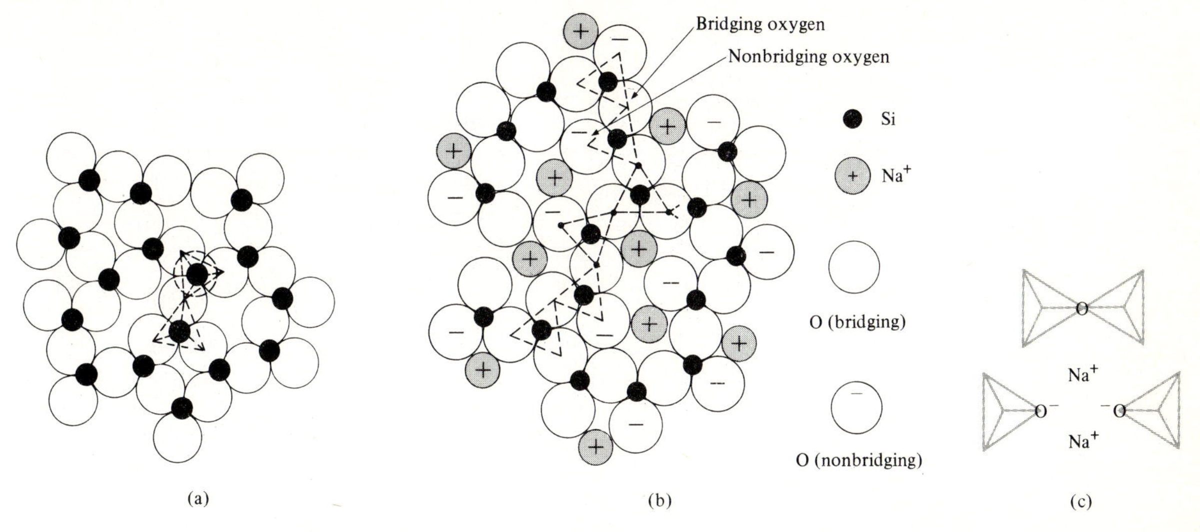

Figure 13-2.1 Structure of glass. (a) Vitreous SiO_2. (b) Na_2O-SiO_2 glass. (c) Oxygen bridging. (The fourth oxygen of each tetrahedron has been deleted for clarity.) In vitreous silica, all oxygen atoms are bridging. In a soda-silica glass, there is one nonbridging oxygen for each Na^+ ion. (In a lime-silica glass, there are two nonbridging oxygens for each Ca^{2+} ion.)

Other oxides, such as GeO_2, B_2O_3, and P_2O_5, are also *glass formers,* because as a liquid they also possess bridging oxygen atoms. The GeO_2 is rare and, therefore, not commonly used, but its chemical similarity to SiO_2 (Group IV) makes it duplicate the atomic coordinations of SiO_2. A glass of B_2O_3 has a trianglular coordination with each boron among three oxygens, and each oxygen atom bridging two B-O triangles.* In a glass, P_2O_5 is coordinated with an average of five oxygens (probably as a mixture of 4 and 6 immediate neighbors).

The introduction of Na_2O and CaO (as well as other Group I and II oxides) modifies the glassy structure of silica by disconnecting some of the oxygen bridges. They are *network modifiers.* This is shown schematically in Fig. 13-2.1 for an Na_2O-SiO_2 glass. For each Na_2O added, two Na^+ ions are formed and two nonbridging oxygens are obtained—one from the oxygen just added, and the second, a former bridging oxygen (Fig. 13-2.1c). More simply, there is *one* nonbridging oxygen for each Na^+ added. Each nonbridging oxygen carries a single minus charge. A CaO addition does the same thing, except that each CaO added produces one Ca^{2+} and *two* nonbridging oxygens: -O^- Ca^{2+} ^-O-.

---

* Refer to Fig. 13-2.1(a), but ignore the oxygen atom above or below the plane of the paper.

The effect of the nonbridging oxygens is to "depolymerize" the structure of the glass. Without CaO or Na_2O, the tetrahedra are *tetra*functional in the polymer terms of Section 7–2. With CaO and Na_2O, the tetrahedra become *tri*functional and in some cases may even be *bi*functional, losing their 3-D network structure. Experience has shown that the soda-lime-silica glass composition used for containers and window glass (Table 13–2.1) is the optimum between a rigid 3-D structure and a thermoplastic product. It is workable at high temperatures; it becomes rigid below its glass-transition temperature; and it is sufficiently siliceous to maintain chemical durability.*

The liquidus temperatures for soda-lime-silica glasses are shown in Fig. 13–2.2. With Na_2O, CaO, and SiO_2, this temperature is very sensitive to composition in the typical container and window-glass composition range. The presence of 1 or 2 percent Al_2O_3 facilitates the melting without decreasing the SiO_2 content by shifting the liquidus contours. Therefore, we may encounter small amounts of Al_2O_3 in soda-lime glasses.

Fiber glasses have special requirements. The Na^+ ions of any glass can react with moisture of the air:

$$Na^+_{(glass)} + H_2O \longrightarrow H^+_{(glass)} + NaOH. \qquad (13\text{–}2.1)$$

Since fiber glass has a large surface area per unit volume, and therefore greater opportunity for an H^+-for-Na^+ exchange, it is made without Na_2O (Table 13–2.1). This requires two other changes to facilitate workability of the molten glass: (a) Some of the SiO_2 is replaced by B_2O_3 and Al_2O_3 as glass formers. The SiO_2 is tetrafunctional; the latter are trifunctional. Therefore, the total network-modifier requirements are reduced. (b) The CaO and MgO contents are increased to meet the network-modifier needs that remain with the absence of Na_2O.

Nonsiliceous glasses

Although the industrial volume of nonsiliceous glasses is minor with respect to silicate glasses, nonsiliceous glasses have technical importance. These are commonly based on the glass-forming oxides—B_2O_3, P_2O_5, and Al_2O_3—in combination with other oxides. For example, borate glasses containing rare earth oxides have high refractive indices with low optical dispersion. These are useful in optical lenses. Phosphate glass that contains 2%–3% FeO is transparent to visible light but absorbs infrared radiation and, therefore, has useful heat-shielding applications. Many other examples are available.

* Still higher Na_2O and CaO contents can lead to reaction of the glass with acids of canned food, or with the surrounding atmosphere.

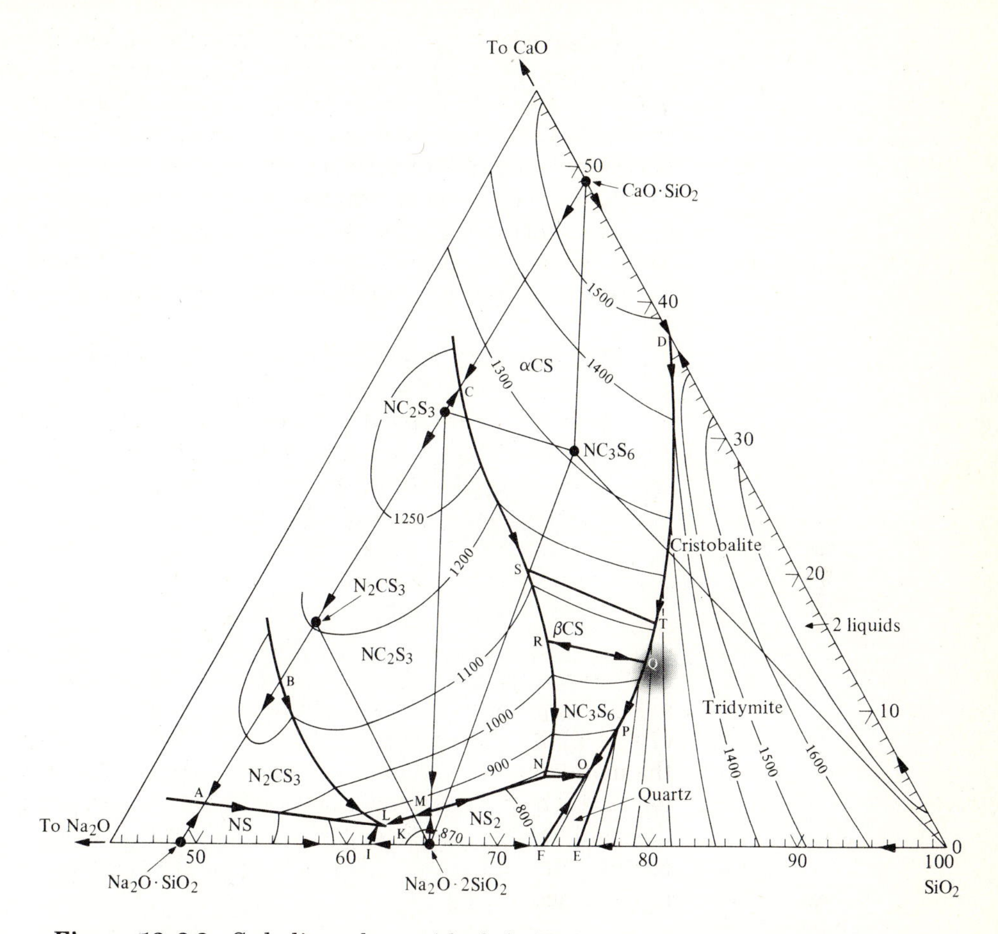

Figure 13-2.2 Soda-lime glasses (shaded). Higher SiO_2 would provide a more durable glass but, as shown by this Na_2O-CaO-Al_2O_3 diagram, would require excessively high melting temperatures. Additional Na_2O (to replace CaO) would provide a lower-melting glass but would lead to deterioration by ion exchange (Eq. 13-2.1). MgO may be substituted for part of the CaO. (Phase diagram based on Morey and Bowen, *Journ. Glass Tech.*)

Glasses do not have to be oxides but may be based on other nonmetallic elements, such as sulfur, selenium, and tellurium, even carbon.* Each of the above elements may be bonded covalently and, therefore, have their crystallization delayed during cooling to retain the amorphous structure of the

* It is even possible to make metallic glass by cooling liquid metal in microseconds. These new engineering materials have fascinating engineering potentials but will not be considered in this unit.

liquid. Chalcogenide glasses of S, Se, and Te have incited considerable interest for their potential use in electrical circuits.

Illustrative problem 13-2.1 A glass contains 80 w/o SiO_2 and 20 w/o Na_2O. What fraction of the oxygens is nonbridging?

SOLUTION Basis: 100 g = 80 g SiO_2 + 20 g Na_2O.

$$80 \text{ g } SiO_2/[(28.1 + 2 \times 16.0) \text{ g/mole}] = 1.33 \text{ mole } SiO_2 \text{ or } 80.6 \text{ m/o } SiO_2.$$

$$20 \text{ g } Na_2O/[(2 \times 23.0 + 16.0) \text{ g/mole}] = 0.32 \text{ mole } Na_2O \text{ or } 19.4 \text{ m/o } Na_2O.$$

$$\begin{aligned} 80.6\, SiO_2 &= 80.6\, Si + 161.2\, O \\ 19.4\, Na_2O &= \underline{\qquad\qquad\qquad 19.4\, O + 38.8\, Na^+} \\ & \quad\ 80.6\, Si + 180.6\, O + 38.8\, Na^+ \end{aligned}$$

Note from Fig. 13-2.1 that there is one nonbridging oxygen for each Na^+ added.

Fraction nonbridging oxygens = 38.8/180.6 = 0.215.

ADDED INFORMATION Calcium forms a double charge and provides electrons to two nonbridging oxygens. Therefore each Ca^{2+} plus the accompanying oxygen of CaO breaks open a bridging oxygen and forms two nonbridging, charge-carrying oxygens. ◄

Illustrative problem 13-2.2 A soda-lime glass contains 13 w/o Na_2O, 13 w/o CaO, and 74 w/o SiO_2 (cf. a float glass of Table 13-2.1). Soda ash (Na_2CO_3) and limestone ($CaCO_3$) are used as sources of the Na_2O and CaO, respectively. As they are heated, CO_2 is evolved, leaving Na_2O and CaO. How many pounds of each are required for 1000 kg of SiO_2?

SOLUTION Based on the above composition, 1000 kg SiO_2 = 175 kg Na_2O and 175 kg CaO.

$$Na_2CO_3 \rightarrow Na_2O + CO_2 \tag{13-2.2}$$

$$\frac{x}{2(23) + 12 + 3(16)} = \frac{175}{2(23) + 16}$$

$$x = 300 \text{ kg } Na_2CO_3$$

$$CaCO_3 \rightarrow CaO + CO_2 \tag{13-2.3}$$

$$\frac{y}{40.1 + 12 + 3(16)} = \frac{175}{40.1 + 16.0}$$

$$y = 312 \text{ kg } CaCO_3$$

In the process, 125 + 137, or 262, kg of CO_2 will be evolved per 1000 kg of SiO_2. ◄

Table 13–3.1
Viscosity constants for common glasses*

Glass type	A	B
Fused silica	−7.9	28,200
Vycor (96% SiO_2)	−3.9	18,800
Borosilicate (pyrex)	−6.0	13,900
Soda-lime (container)	−5.6	11,200

* Equation 13–3.1, with temperature in K and viscosity in Pa·s. Applicable for temperatures at the annealing point (10^{12} Pa·s) and above.

13–3 PHYSICAL PROPERTIES OF GLASS

It is not an exaggeration to state that there are more than 10,000 kinds of glass. The number is so high because incremental compositional changes may be made in one or more of the several dozen available glass components. Each change affects one or more properties. It is possible to adjust a set of properties to precise specifications that are distinct from those of adjacent compositions. These properties include, among others, viscosity, mechanical properties, optical properties, and electrical behavior.

Viscosity

The technical importance of glass viscosity is greatest in its manufacturing processes. The glass-transition temperature, T_g, is sufficiently high so that creep is normally not encountered in service. (Cf. Section 8–3 for polymers.) To a first approximation, the viscosity, η, of glass behaves in the orthodox (Newtonian) manner at elevated temperatures:

$$\log \eta = A + B/T. \qquad (13\text{–}3.1)$$

The constants A and B are unique for each glass (Table 13–3.1). This differs somewhat from the empirical relationship given in Eq. (8–3.2) for linear polymers.*

Figure 13–3.1 (a) shows the viscosity of container glass. For *melting* and *fining*† within the hottest part of the furnace, the viscosity must be as low as 10 Pa·s (100 poises). This compares with the viscosity of a heavy motor oil.

The temperature must be reduced and the viscosity is increased for forming processes. The *working range* is defined as the temperatures between viscosity values of $10^{2.5}$ Pa·s and 10^6 Pa·s ($10^{3.5}$ and 10^7 poises). As shown in Fig. 13–3.1(b), this is a wide range because working procedures

* The linear polymers involve extremely long molecules. Although most glasses are depolymerized from their network structures, they do not become one-dimensional molecules.

† The removal of gas bubbles.

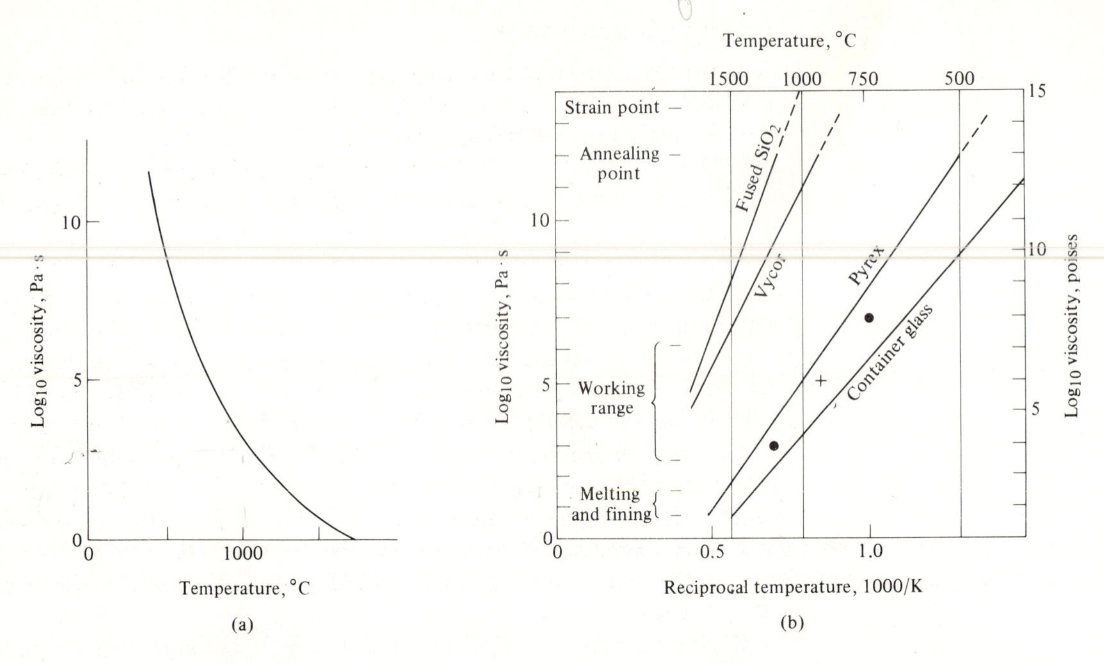

Figure 13-3.1 Viscosity versus temperature. (a) Container glass (soda-lime). (b) Equation (13-3.1). This equation is not valid at the glass-transition temperature and below.

vary from *hot-pressing,* used to make products such as automobile headlights that require a viscous glass ($\eta \cong 10^6$ Pa·s, or 10^7 poises), to the *blowing* of thin glass envelopes for light bulbs at a rate of thousand per minute ($\eta \cong 10^3$ Pa·s).

The *annealing point* is the temperature at which the glass has 10^{12} Pa·s (10^{13} poises). This is in the range of the glass-transition temperature, T_g. The movements of atoms are very sluggish; however, the atoms do move so that stresses can be reduced by a factor of two every few minutes. Therefore, the residual stresses are essentially eliminated in a 15-minute anneal. Almost all glass products are annealed.

The temperature of glass with a viscosity of $10^{13.5}$ Pa·s ($10^{14.5}$ poises) is called the *strain point.* This is on the lower side of the glass transition temperature range. The atomic movements are so sluggish here that rapid cooling does not induce residual stresses. After forming, most glass products are heated at the annealing point to reduce the residual stresses, then cooled slowly to the strain point to avoid new stresses. Subsequent cooling to ambient temperatures can be rapid.

Mechanical properties

Glass is always used in its elastic range (below T_g). *Young's elastic moduli* of all silicate-based glasses are dictated by the Si–O bond and have values of ~70,000 MPa (~10,000,000 psi). If a major part of the SiO_2 is replaced by B_2O_3 (to make Si/B < 1.5), the elastic modulus drops to ~50,000 MPa.

The *strength* of glass can be very high because it cannot undergo shear (Section 11-8). However, as discussed in that section, its tensile behavior is severely affected by flaws or surface imperfections. Of course, we are familiar with this when we "cut" window glass by scoring it with a sharp tool and then bending it to introduce tension. The subtle effects of surface imperfections may be illustrated by measuring the strength of a freshly drawn (pristine) glass fiber. It is possible to get values as high as 5,000 MPa–10,000 MPa (700,000 psi–1,400,000 psi). However, if the glass fiber is touched by one's hand, by another glass fiber, or allowed to reside in moisture-containing air a few hours, the strength drops drastically to less than 500 MPa. Ordinary window glass becomes so flawed before it is used that it can break under a tension stress of less than 200 MPa (<30,000 psi). This is because tension stresses can be concentrated at the roots of flaws when the glass is loaded and deflected.

If a well-defined elliptical hole is placed through a plate of a brittle material, the concentrated stress, s_c, is increased by a *stress concentration factor* of $2\sqrt{c/r}$ over its nominal stress, s_n:

$$s_c = s_n(2\sqrt{c/r}), \tag{13-3.2}$$

where c is the depth of the notch (from the edge), or half the span of an internal hole with a tip radius of r (Fig. 13-3.2). Thus, a cylindrical hole (with $r = c$) doubles the nominal stress at its edge. However, a crack has an exceedingly small tip radius, which we must approximate as being of atomic dimensions, that is, ~0.1 nm. It means that if a crack is large enough

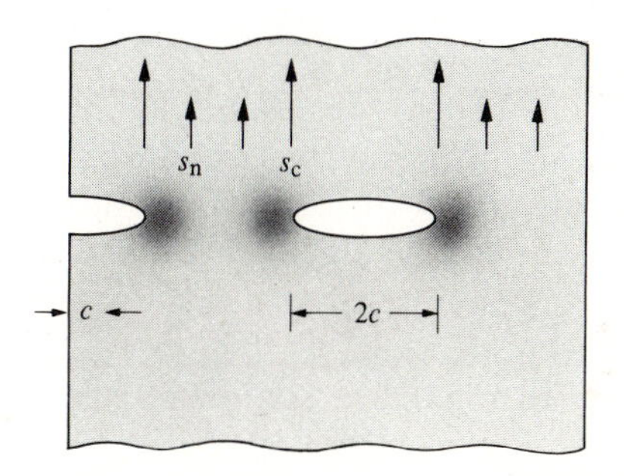

Figure 13-3.2 Stress concentrations (schematic). Stresses are concentrated at the edge of a hole, or crack. Equation (13-3.2) applies for discontinuities through plates.

Table 13-3.2
Characteristic ionic colors

Ion	Atomic number	Characteristic color	
		Oxidized	Reduced
Ti	22	Yellow	None
Cr	24	Yellow-green	Dark green
Mn	25	Light purple	None
Fe	26	Light yellow	Light blue
Co	27	Blue-violet	Blue-violet
Ni	28	Violet to brown	Violet to brown
Cu	29	Green-blue	Red
U	92	Yellow	Green

(~1,000 nm) to be seen in an optical microscope, the stress concentration, $2\sqrt{c/r} = s_c/s_n$, is 200. Thus, the 200 MPa cited in the previous paragraph for window glass can be concentrated to values of 5,000 MPa–10,000 MPa by flaws that cannot be seen optically. These stresses exceed the inherent strength of glass. In Section 13–4, we will examine ways of counteracting these imperfections by introducing surface compression.

Optical properties

The prime optical property for glass lenses is the *index of refraction*. Denser glasses, or more specifically glasses with more electrons per mm^3, have higher indices of refraction. Thus, vitreous silica has a relatively low index ($n_D \cong 1.46$) because its packing factor is low, and both silicon and oxygen have relatively low atomic numbers, hence relatively few electrons. In contrast, PbO- and BaO-containing glasses have higher indices of refraction ($n_D > 1.56$). The lense manufacturer pays attention not only to the refractive index but also to the *dispersion*, or difference between the indices of the shorter (blue) and longer (red) wavelengths. In doing so, the glassmaker not only draws upon oxides, such as SiO_2, Al_2O_3, Na_2O, CaO, BaO, and PbO, but also has developed fluoride, fluoroborate, rare-earth borate, fluorogermanate, and phosphate-type glasses.

Silica, Na_2O, and CaO do not absorb any of the visible colors; therefore, normal plate and container glasses transmit white light. However, transition elements within glass absorb selected wavelengths of light. The unabsorbed colors are transmitted. Table 13–3.2 shows not only that the transition metals will color glass but that the color is generally dependent upon the oxidation state of the metal. Thus, Fe^{3+} provides a light yellow color to glass, while Fe^{2+} ions introduce a light blue color. Of course, if colorless glasses are required, the amount of impurities must be eliminated.

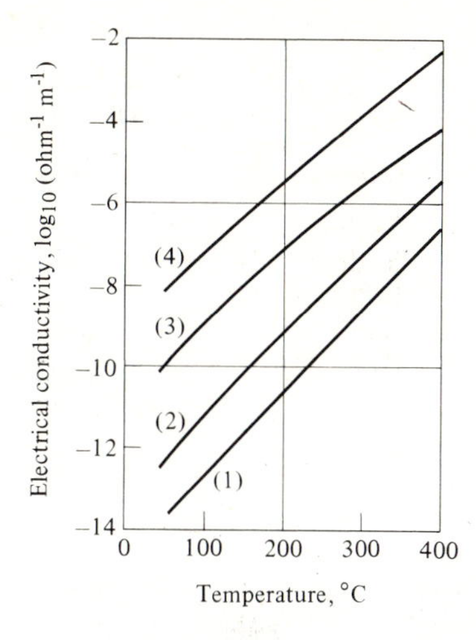

Figure 13-3.3 Electrical conductivities of principal glasses. (1) Fused silica. (2) Vycor glass (96% SiO_2). (3) Borosilicate glass (Pyrex). (4) Container glass (soda-lime).

Electrical behavior

Basically, glass is an insulator. The valence electrons from Na^+, Ca^{2+}, and other positive ions have been transferred and permanently attached to nonbridging oxygen atoms. At higher temperatures, however, *ionic conduction* becomes possible. In particular, Na^+ ions can diffuse through glass toward the negative electrode. They move more readily than Ca^{2+} or O^{2-} ions because the sodium ions are smaller and carry only a single charge. (The silicon atoms do not readily move because they are firmly caged inside tetrahedra of four oxygen atoms each.) The data of Fig. 13-3.3 show the electrical conductivity for four glass types. A comparison with Table 13-2.1 reveals the role of the Na_2O content.

The ionic conductivity just described is never great, but it is sufficient to pass a current through molten glass and produce resistive heating. Small glass furnaces can operate in this manner; also, resistive heating can be used to make glass-to-glass seals, for example, in the joining of the face plate onto the cone of a television tube.

The relative *dielectric constant* of glass will be discussed in Unit 19 where related properties are presented.

Illustrative problem 13-3.1 The viscosity of a glass is 10^7 Pa·s at 727°C and 10^3 Pa·s at 1156°C. At what temperature is it 10^5 Pa·s?

SOLUTION Based on Eq. (13-3.1),

$$\log 10^7 = A + B/1000\ K = 7;$$
$$\log 10^3 = A + B/1429\ K = 3.$$

Solving simultaneously, we obtain,

$$A = 7 - B/1000\ \text{K} = 3 - B/1429\ \text{K},$$
$$B = 13{,}300\ \text{K}.$$
$$A = 7 - 13{,}300/1000 = -6.3.$$
$$\log 10^5 = -6.3 + 13{,}300/T = 5.$$
$$T = 1177\ \text{K}$$
$$= {\sim}900^\circ\text{C} \qquad (\text{or } {\sim}1650^\circ\text{F}).$$

ADDED INFORMATION A graphical solution is possible, simpler, and just as accurate if one uses a straight line on Fig. 13-3.1. (See dots between the two lower curves.) ◄

13-4 ENGINEERED GLASSES

A number of glass products require more than melting, forming, cooling, and annealing in their production histories. This section will review a few of the many special processing procedures and applications available—processes that introduce greater strength, low thermal sensitivity, communication possibilities, and radiation responses.

Tempered glass

Residual stresses in glass are commonly to be minimized. Thus, we anneal most glass products as the final manufacturing step. However, if residual stresses are compressive at the glass surface, and if they are correctly distributed, the engineer can use them to advantage. Consider the rear window of an automobile. As the final step of production, it is heated above the annealing point, but not high enough to start deformation (${\sim}10^7$ Pa·s, or ${\sim}10^8$ poises). Then the glass is cooled rapidly. As the glass surface contracts, the interior is still hot enough for dimensional adjustments (Fig. 13-4.1). However, the surface soon drops below the strain point; therefore, it becomes rigid and cannot adjust to the contraction in the center as it cools more slowly. Consequently, the surface is placed under compression (and the center under tension). The product of this heat treatment is called *tempered glass*. Figure 13-4.2 shows the stress patterns across the thickness of the glass. With the surface in *compression,* the glass is very strong because any tension stress must overcome this compression before the surface is sufficiently stressed in tension to initiate and/or propagate a crack. The purpose of the annealing that was cited in the previous section is to *avoid* the presence of *tension* stresses at the surface where flaws are present.* Here, we intentionally introduce residual stresses.

* If a crack penetrates through the compression skin of tempered glass (for example, by scratching) into the tension zone shown in Fig. 13-4.2, the crack will become rapidly self-propagating. The aftermath of this effect can be observed in a broken rear window of a car, where the crack pattern (Fig. 13-4.3) is a mosaic rather than spear-like fragments.

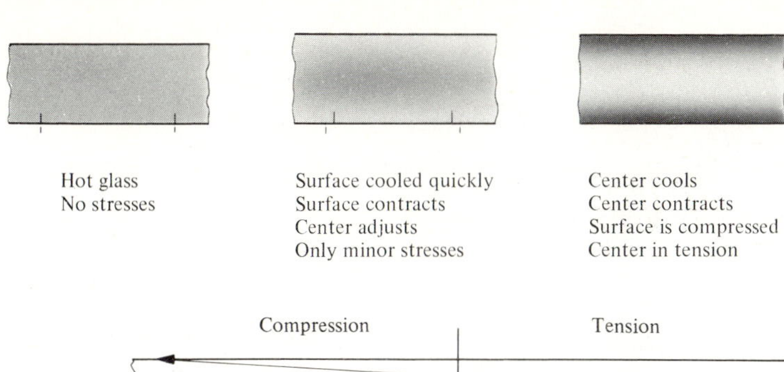

Figure 13-4.1 Dimensional changes in tempered glass (schematic).

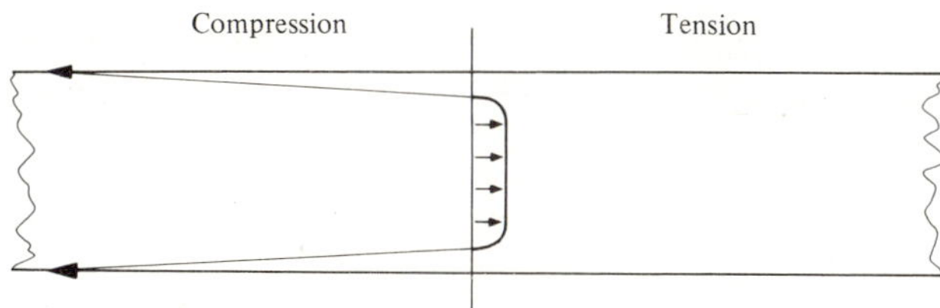

Figure 13-4.2 Tempered glass (residual stresses). The surface is in compression. (Cf Fig. 13-4.1.) These stresses must be exceeded by tension before microcracks can propagate and so fracture will occur.

Figure 13-4.3 Impact fracture modes in glass (intentionally fractured). (a) Tempered glass (typical uses: store doors, ophthalmic lenses, rear automobile windows). (b) Laminated glass (typical uses: automobile windshields, safety shields). (Courtsey of LOF Glass.)

Tempered glass can be given a very high strength through its residual compressive stresses (Fig. 13-4.2); however, it may be sensitive to impact failure. If cracks penetrate the surface zone, the release of the residual tension stresses produce a mosaic fracture. (This fracture mode seldom produces penetrating injuries.)

Since the fracture mode of nontempered glass is obviously different, specifications may call for lamination with plastic in order to hold the sharp fragments intact and limit injuries.

(a)

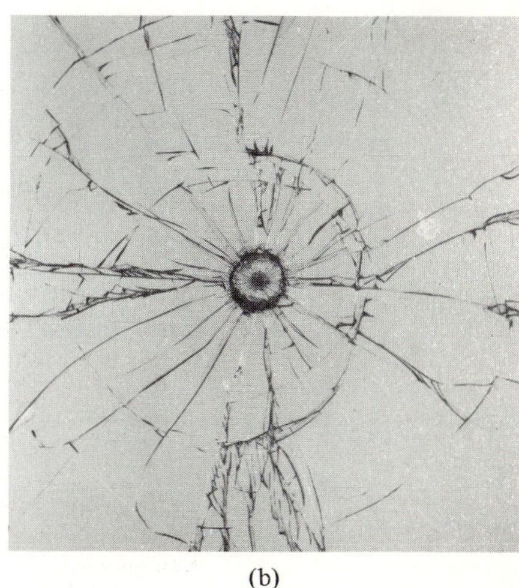

(b)

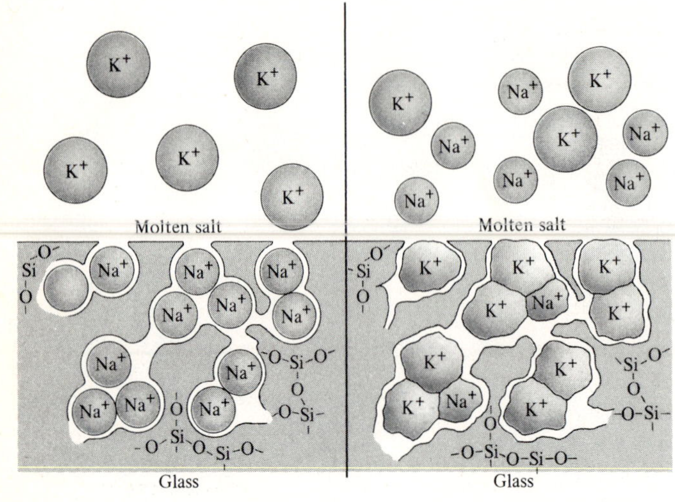

Figure 13-4.4 Chemical tempering (schematic). If a soda-lime glass is heated in molten K_2SO_4, part of the Na^+ ions of the surface zone of the glass are exchanged with larger K^+ ions. These larger ions place the surface under compression, thus strengthening the glass against tensile failure. (Courtesy of Corning Glass Co.)

A variant of the above principle is available with *chemical "tempering,"* which originates through *ion exchange.* A normal glass contains 12 w/o Na_2O to 15 w/o Na_2O (along with CaO and SiO_2, Table 13-2.1). If such a glass is heated in a bath of molten K_2SO_4, there can be an exchange of some of the Na^+ ions in the glass for K^+ ions of the melt:

$$Na^+_{glass} + K^+_{melt} \rightarrow K^+_{glass} + Na^+_{melt}. \tag{13-4.1}$$

However, the K^+ ions are larger than the Na^+ ions. ($r_{K^+} = 0.133$ nm; $r_{Na^+} = 0.097$ nm.) Although it is not possible for all of the Na^+ ions to be replaced by K^+ ions, simply because space is not available, enough K^+ ions move into the glass so that the rigid silicate network of the glass (Section 11-6) is subjected to a compression stress as great as 350 MPa–400 MPa (50,000 psi–60,000 psi). Here, as with the thermally tempered glass, bending or other service loading must overcome the induced compression stresses resulting from *ion stuffing* (Fig. 13-4.4) before the surface encounters tension.

Devitrified glass

Glass is a solid; at the same time it has the structure of a liquid. It has been supercooled; but given a chance it will transform to a crystalline solid, that is, it will *devitrify*. However, the rates are so extremely slow under ambient conditions that certain glass artifacts remain from prehistoric times without

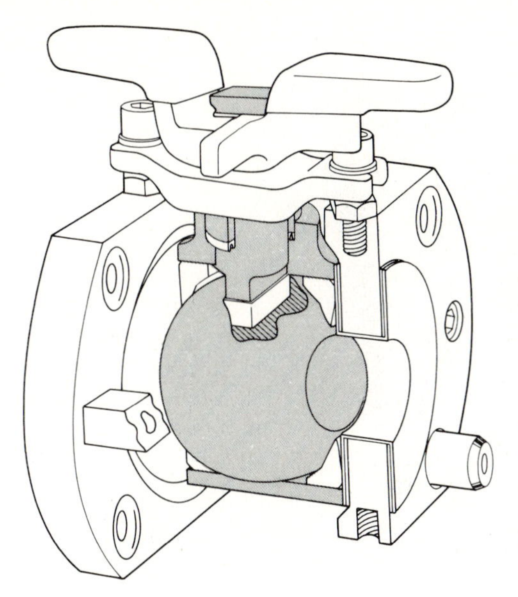

Figure 13-4.5 Devitrified glass product (ball valve in an acid line). Nonporous glass is shaped by conventional glass-forming techniques. It is then heated (1) to an appropriate intermediate temperature, to nucleate crystallization, that is, devitrification. (2) This is followed by a higher-temperature treatment to complete crystallization. The final product is stronger than glass and is less porous (0%) than most sintered ceramic products. (Courtesy of O-I/Schott.)

devitrification.* The rate of devitrification is more rapid at elevated temperatures. The maximum rate usually occurs between 0.75 T_m and 0.90 T_m. At temperatures nearer to the melting temperature, T_m, the supercooling is not sufficiently great to break bonds and nucleate the more stable crystals. In fact, oxides, such as TiO_2 and/or ZrO_2, are normally incorporated into the glass composition to provide nucleants for commercially devitrified glasses.

Devitrified glass has commercial attractions. For example, many products can be readily formed into the desired shape as a glass. They have zero porosity. If the glass can then be devitrified into a crystalline product, it can possess the stability and strengths that are common to crystalline solids. One such example is the ball valve of the acid line of Fig. 13-4.5. Other examples include cookware and similar products.

The prime technical problem in glass devitrification is to ensure that the volume changes during the process are negligible. If a metal shrinks during solidification, stresses can be immediately relieved by easy atom diffusion and by plastic deformation. If there is a volume change in a brittle glass product, cracking can result. Because of this, many devitrified glasses are based on lithium aluminosilicates ($Li_2O \cdot Al_2O_3 \cdot {\sim}8SiO_2$), since they maintain nearly constant volume during crystallization and have very

* These early glasses match closely the compositions of the more durable glasses of present day. This suggests that the earliest glassmakers had mastered the compositions of optimum glass chemistry. However, we now know that many products were made of nondurable composition. They have not remained to tell their story.

Figure 13-4.6 Optical glass (prisms, lenses, and mirrors). Stringent requirements for composition and homogeneity are mandatory to assure clarity and precise optical paths for high resolution. (Courtesy of American Optical.)

small thermal expansion coefficients. Because of the latter property, *glass-ceramics,* as devitrified glasses are sometimes called, find specific use where thermal expansion is critical, for example, telescope mirrors and the cookware cited earlier.

Optically functional glasses

Glass can be designed to serve technical functions as it transmits light. Thus, glass can do more than passively transmit light through a window or a food container. The most familiar technical uses are in *lenses* where *optical* glass can be ground and polished to the correct curvature to aid vision—in the microscopic, in daily sight, and into the far distance (Fig. 13-4.6). The technical specifications include clarity, index of refraction, and dispersion (Section 13-3).

Photosensitive glasses can record a light pattern just as a photographic film does. There are several available mechanisms. In one, the glass contains Ce^{3+} ions and dissolved metal ions such as Au^{+}, Ag^{+}, or Cu^{+}. On exposure to radiation, the Ce^{3+} ion loses additional electrons to become Ce^{5+}. The electrons that were released reduce the gold to metal. By heating the glass to the correct temperature and for an appropriate time, the gold (or copper or silver) collects into submicroscopic colloidal particles that absorb some of the wavelengths of light, thus coloring the glass (for example, ruby glass is colored by gold particles). It is possible to use any optical scheme available to expose the glass and record the desired image.*

Those photosensitive glasses of the above type that contain lithium silicate can be *chemically machined.* A mask exposes the glass with the selected pattern. The exposed areas develop colloid metal particles as de-

* There is no "graininess" in these images as there is in photographic film because the metal particles are colloidal and of submicroscopic dimensions.

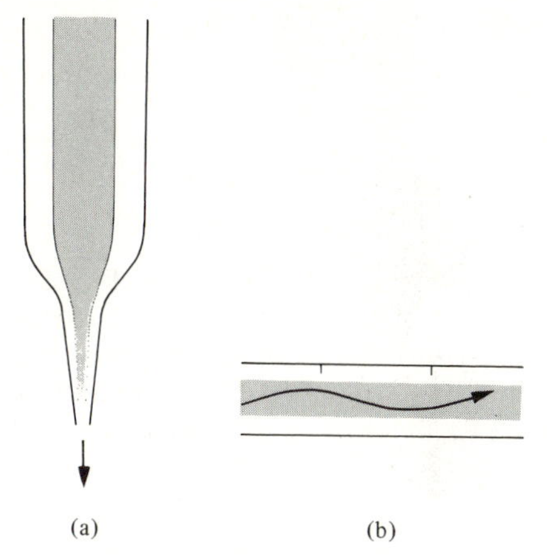

Figure 13-4.7 Optical wave guide (dual fiber). (a) Drawing. A low-index glass (outer) and a normal-index glass (center) are not drawn together. (b) By internal refraction, the low-index glass keeps the light path away from the surface and its flaws.

scribed above. These submicroscopic particles serve as nuclei on which devitrification occurs. In this case, the crystalline areas (which were the areas originally exposed to light) are more readily dissolved by hydrofluoric acid than are the unexposed and still vitreous regions. Very high resolution is possible through photographic projection of the mask. As many as 500 holes can be "drilled" per mm^2, if necessary.

Photochromic glass is photosensitive glass with reversible coloration at ambient temperatures. Silver halides are incorporated into the molten glass but are subsequently precipitated as colloidal particles within the glass. These halide particles are less than 10^{-6} as large as those in photographic emulsions. Thus, while the darkening reaction is the same as in photographic paper, the reduced silver atom and the halide atom are never physically separated far enough but that they can reunite when the excess light is eliminated. The recombination time is a relatively few seconds. Sunglasses of these compositions are the photochromic product that is best known by the general public.

Optical fibers can be made from any continuous glass fiber if the lengths are reasonably short and if light losses are tolerable. However, new multichannel, modulated-light communication systems require optical fibers that will transmit light great distances. Thus, special attention must be given to the composition, structure, and processing of glass fibers. The fibers must be free of all oxides that absorb light—in particular, all transition metal oxides, such as FeO. We depend on total internal reflections to keep the light within the optical waveguide. In principle this is simple, since the critical angle can be calculated from the index of refraction. However, sharp curvatures can lead to "leakages." More importantly, surface flaws, no matter how small, can lead to a scattering of light and, of course, any handling of the fiber will introduce flaws. The effects of these flaws can be avoided if the light-transmission glass is coated with a second glass of lower index. The light is totally reflected within the core glass and does not enter the outer glass, which is exposed to the degrading environment. Such a dual fiber is made by jacketing the higher-index glass rod with a lower-index glass tube and simultaneously drawing them into a composite filament (Fig. 13–4.7).

Lasers can be made of glass since glass will dissolve almost any type of element as an oxide. In glass, neodymium and other rare earth oxides generally replace Cr^{3+}, the ion used in ruby lasers. The latter functions on the basis of its 3-*d* electrons, which are immediately subvalent, and therefore behave differently in noncrystalline glass than in crystalline ruby. The neodymium ions, Nd^{3+}, provide laser* action by more deeply seated electrons that are not influenced by the bonds between adjacent atoms. Glass has an advantage over crystalline lasers in that large crystals are difficult to make, especially with the required structural uniformity. Glass lasers must be made to optical specifications.

Review and Study

SUMMARY

1. Glass has a long history as a man-made material for society. The principal products include (a) flat glass for architectural and automotive purposes, (b) container glass for food and beverages, (c) fiber glass for thermal insulation and for plastic reinforcement, and (d) glass coatings for glazes (coatings on ceramics) and vitreous enamels (coatings on metal).
2. Silica is the glass-forming oxide used in all common glasses. The SiO_4 tetrahedron provides the structural unit. It is tetrafunctional with oxygen atoms serving as connecting bridges. The network structure is modified by soda (Na_2O) and lime (CaO), which introduce nonbridging oxygens and reduce the tetrahedral connections from four to less than three. They lower the viscosity of the glass and increase its workability for producing glass products.
3. The viscosity, η, of glass depends not only upon the addition of network modifiers but also upon temperature. To a first approximation, $\log \eta$ is inversely related to absolute temperature (Eq. 13–3.1). The viscosity must be low for melting and fining (comparable to a heavy motor oil with $\log \eta = \sim 1$). The working range extends from $\log \eta = 2.5$ to $\log \eta = 6$, depending on the product. Annealing is achieved at $\log \eta = \sim 12$, since that permits the reduction of residual stresses in 10 to 15 minutes. Residual stresses are not reintroduced when $\log \eta > 13.5$ because the temperature has dropped below the glass transition temperature. [The above values for $\log \eta$ use Pa·s as units for viscosity.]
4. Modern technology has improved and expanded the use of glass. Examples include (a) tempered glass to which the engineer intentionally introduces residual stresses for strength, (b) devitrified glass that is processed as a formable glass and then crystallized into a nonporous ceramic product, (c) photosensitive

* *L*ight *a*mplification by *s*timulated *e*mission of *r*adiation.

and photochromic glasses that vary their optical transmission with light intensity, and (d) optical fibers that can be used for multichannel communication through the use of ultrapure raw materials and coatings that produce internal reflections. Such a list is incomplete and continues to expand.

TECHNICAL TERMS

Annealing point (glass) Stress-relief heat treatment. The temperature should provide a viscosity of $\sim 10^{12}$ Pa·s (10^{13} poises).

Bridging oxygens Oxygen atoms shared by two adjacent silica tetrahedra.

Devitrified glass Glass with controlled crystallization.

Dispersion Difference in index of refraction, red light versus blue light.

Enamel (vitreous) Glass coating on metals.

Fining Removal of gas bubbles from molten glass.

Flat glass Architectural (and window) glass products.

Glass An amorphous solid below its transition temperature. A glass lacks long-range crystalline order but normally has short-range order.

Glaze Glass coating on ceramics.

Index of refraction Ratio of light velocity in a vacuum to the velocity within a material. (Produces a "bending" of the light at the surface of a material.)

Ion exchange Exchange of ions in a solid solution.

Ion stuffing Ion exchange that produces compressive forces because the new ions are larger than the original ions that were replaced.

Network formers Glass-forming oxides, such as SiO_2 and B_2O_3. These have polyfunctional coordination units that produce network structures through the presence of bridging oxygens.

Network modifiers Oxides that depolymerize the silicate network of glass.

Nonbridging oxygens Oxygens attached to one SiO_4 tetrahedron only, thus not tying tetrahedra together.

Photochromic Reversible light-sensitivity—darkening with increased intensity, and fading with decreased intensity.

SiO_4 tetrahedron Coordination unit of four oxygens surrounding a silicon atom.

Strain point Temperature at which the viscosity, η, of a glass is $10^{13.5}$ Pa·s ($10^{14.5}$ poises).

Stress concentration factor (s_c/s_n) Increase in stress at a notch.

Tempered glass Glass with surface compressive stresses induced by heat treatment.

Vitreous Glassy, or glasslike.

Working range Temperature range for glass-shaping operations, lying between the viscosities of $10^{2.5}$ Pa·s and 10^{6} Pa·s ($10^{3.5}$ to 10^{7} poises).

CHECKS

13A Three of the major product categories for glass include flat glass, __***__ glass, and __***__ glass.

13B Molten __***__ is used in a float bath to __***__ . . . .

13C The effects of __***__, __***__, and __***__ must be controlled for the automatic production of glass containers.

13D Discontinuous glass fibers are produced by __***__ forces; continuous fibers depend upon rapidly drawing the glass through __***__.

13E Glass coatings are called either __***__ or __***__, depending on whether they cover ceramic products or metal products.

13F The principal component of container and flat glass products is __***__.

13G In glass, the silicon atom is coordinated with __***__ oxygen atoms to produce __***__ tetrahedra.

13H A __***__ oxygen connects two tetrahedra within the glass structure.

13I A __***__ former is an oxide that provides bridging oxygens to the structure of glass.

13J A network __***__ reduces the number of bridging oxygen atoms in glass and, therefore, makes hot glass less __***__.

13K Fiber glass must be low in Na^+ ions because they react with the __***__ of the air.

13L The three main components of container glass and flat glass are __***__, __***__, and silica.

13M The viscosity of glass at its __***__ point is such that residual stresses drop by a factor of two every few minutes.

13N The viscosity of glass at its __***__ point is such that rapid cooling will not introduce residual stresses.

13O Read 13M and 13N. The __***__ point is slightly above the glass-transition temperature, and the other one slightly below it.

13P Since glass does not undergo __***__ when loaded, it has a high strength when no flaws are present.

13Q Surface damage of glass can occur from handling, and from __***__.

13R The __***__ is a function of flaw depth and the curvature at the tip of the flaw.

13S Ionic conduction in glass is principally by __*** ions__ because they are the smaller species that are not covalently bonded.

13T The surface stresses of tempered glass are in __***__.

13U For "chemical tempering" by ion exchange, the incoming ion must have a __*** er__ radius than the previous ion.

13V The term __***__ as applied to glass means to crystallize the glass. The rate is greatest at about __***__% of the melting temperature, T_m.

13W Glass-ceramics, as devitrified glasses are sometimes called, must maintain essentially constant __***__ during the devitrification process.

13X Photosensitive glass contains metal ions that have two valences, for example, Ce^{5+} and Ce^{3+}. Of these two, the _***_ ion is produced in the presence of light or other radiation.

13Y The critical property of glass for fiber optics is its _***_.

13Z The core glass of an optic fiber should have a higher _***_ than the surface glass.

STUDY PROBLEMS

13-2.1. A glass contains 80 w/o SiO_2 and 20 w/o CaO. What percent of the oxygen atoms bridge adjacent tetrahedra?

Answer: 76% bridging

13-2.2. A glass contains 70 w/o SiO_2 and 30 w/o K_2O. What fraction of the oxygen atoms are nonbridging?

13-2.3. A soda-lime glass contains 13 w/o Na_2O, 13 w/o CaO, and 74 w/o SiO_2. What fraction of the oxygens are nonbridging?

Answer: 0.30

13-2.4. Soda ash (Na_2CO_3) is the only carbonate added to the batch for a thermometer glass (Table 13-2.1). How many grams of CO_2 will be evolved per 100 g final glass product?

13-2.5. Assume that 44 g of CO_2 (that is, 1 mole) produce 100,000 cm^3 of gas at glass-melting temperatures. How many cm^3 of CO_2 will be evolved per cm^3 of the glass in S.P. 13-2.3? [The density of glass is 2.5 g/cm^3.]

Answer: 1100 cm^3

13-2.6. Approximately 9000 glass bottles (2000 kg) are to be recycled through a glass plant. How much quartz sand ($99SiO_2$-$1Al_2O_3$), soda ash (Na_2CO_3), limestone ($CaCO_3$), and dolomite ($CaMgC_2O_6$) must be added to produce a charge batch for 4600 kilograms of glass for lamp bulbs? [The analyses in Table 13-2.1 are all ±0.3%.]

13-3.1. Refer to the glass in I.P. 13-3.1. (a) What is its viscosity at 1186°C? (b) At what temperature will its viscosity be 10^8 Pa·s?

Answer: (a) 655 Pa·s (b) 657°C

13-3.2. Estimate the temperature of the annealing point of the soda-lime glass in Table 13-3.1.

13-3.3. Calculate the viscosity constants, A and B, for a glass that has a viscosity of 10^{11} Pa·s at 500°C and $10^{3.5}$ Pa·s at 1000°C.

Answer: A = −8.1 B = 14,800 K

13-3.4. What is the working range, °C, for the glass in S.P. 13-3.3?

13–3.5. Assume the radius of a crack tip in glass is 1 nm. How deep must the crack be to concentrate a 25-MPa stress to 5,000 MPa?

Answer: 10^4 nm (or 0.01 mm)

13–3.6. The index of refraction increases in proportion to the linear density of electrons along the path of the light. Vitreous SiO_2, with a mass density of 2.2 g/cm^3, has an index of refraction of 1.46. Estimate the index of refraction for quartz (also SiO_2) with 2.65 g/cm^3.

Answer: $\sim$1.55

13–4.1. Cite two reasons why the stronger temper glass is generally not specified for use in automobile windshields.

13–4.2. Explain the radiating mode of fracture in the glass of Fig. 13–4.3.

CHECK OUTS

13A container
fiber

13B tin
produce parallel glass surfaces

13C time, temperature, composition

13D centrifugal
orifices

13E glazes
enamels

13F silica

13G four
SiO_4

13H bridging

13I glass (network)

13J modifier
viscous

13K moisture (H_2O)

13L soda (Na_2O), lime (CaO)

13M annealing

13N strain

13O annealing

13P plastic deformation

13Q moisture

13R stress-concentration factor

13S Na^+ (positive)

13T compression

13U larger

13V devitrify
70–90

13W volume

13X more positive (Ce^{5+})

13Y transparency

13Z index of refraction

UNIT FOURTEEN

STRONG, TOUGH, AND HARD MATERIALS

Innumerable engineering applications require materials that must support massive loads, absorb large amounts of energy, and/or resist wear. These requirements call for strong, tough, and hard materials. Unlike the forces discussed in Unit 3, the stresses are seldom static; rather, impact loading and cyclic loading are often involved, so that it is necessary to consider fracture and fatigue. Furthermore, we can now consider microstructures, which were not familiar to us in Unit 3. All of these factors must be understood by the designer of products subject to severe loads.

Contents

Prerequisites: Units 2 through 6 (including Section 6–5), and Section 11–8. (Also reference is made to the subsections on mechanical properties of glass in Sections 13–3 and 13–4.)

From Unit 14 the engineer should

1) Know the principal mechanisms of strengthening.

2) Distinguish between the two modes of fracture and their implications in design and in materials failure.
3) Be able to use the stress-intensity factor to calculate stress limitations of steels possessing cracks.
4) Understand composites sufficiently well to make the type of calculations shown at the end of Section 14-5.
5) Be familiar with those abrasives that are more widely used in manufacturing processes.
6) Have sufficient knowledge of the related technical terms for effective communication.

14-1 REVIEW OF STRENGTHENING MECHANISMS

Hard and strong materials resist plastic deformation. In atomic terms, they are strong if dislocations are inhibited from moving. In practical terms, materials resist deformation for a variety of reasons related to their structure. It may be well to review those structural effects, since previous discussions are dispersed among a number of previous units.

Hard phases

The hardest metallic and ceramic phases are compounds—Al_2O_3, TiC, BN, WC, SiC, etc. These materials are often called *abrasives*. These are all refractory compounds with high melting points, and all have high elastic moduli; thus, we may conclude that all of them have very strong bonding forces. However, unlike metals, the primary bonding forces are between unlike elements. As discussed in Section 5-4 and sketched in Fig. 11-8.1, these compounds have a high shear resistance, because the slip mechanism tends to bring like atoms, rather than unlike atoms, into neighboring positions.

Of course, not all compounds have equally high shear strengths; nevertheless, this shear resistance is sufficiently high within compounds so that we can expect that these binary phases will be harder than the phases of the separate components. As an example, $CuAl_2$, CuAl, and Cu_3Al of the Cu-Al system are all harder than either copper or aluminum (Fig. 5-5.2).

Hard phases are typically more brittle than ductile phases. This is because any structural flaw leads to a stress concentration that cannot be relieved by plastic deformation (Section 14-2).

Strain hardening

Ductile materials progressively lose their ductility and gain strength when they are plastically deformed (Fig. 14–1.1a). The terms *cold working* and *strain hardening* are used to describe the deformation process and the property changes, respectively (Section 4–2).

Strain hardening occurs because dislocations that are generated during plastic deformation become entangled and, therefore, lose their mobility (Fig. 4–2.3). Thus, although dislocations are the basis for plastic deformation, too many dislocations restrict plastic deformation and introduce a strengthening that can be used for design purposes in wire and in sheet products.

Strain hardening is removed at higher temperatures by *recrystallization* (Section 4–3). This means that strain-hardened materials cannot be used at elevated temperatures to meet strength requirements. Conversely, recrystallization provides a way to reintroduce ductility into the cold-worked metal for further processing, or to simultaneously anneal and *hot work* a metal (Section 4–4).

Solution hardening

Pure materials are never as hard and strong as are solid solutions of two or more components (Fig. 14–1.1b). This leads not only to single-phase alloys such as brass, bronze, and sterling silver (Section 5–2) but also to multicomponent alloys that are used in sophisticated engineering applications, such as jet engines, since solution hardening is effective at high temperatures.* Most solid-solution alloys can also be strain-hardened for a combined hardening that is greater than either solution hardening or strain hardening alone.

Solution hardening, like strain hardening, arises because dislocation movements are restricted. As sketched in Fig. 5–2.3, impurity atoms anchor dislocations; added energy and greater shear forces are required to free the dislocations for further slip movements.

Grain refinement

At lower temperatures (only), materials with finer grain sizes are stronger† (Fig. 14–1.1c). The explanation is straightforward; namely, grain boundaries interfere with plastic deformation because the slip planes of adjacent

---

* Of course, alloy solutions are not as strong at high temperatures as at ambient temperatures; however, that softening also occurs in pure metals. Therefore, the solid-solution alloy retains its strength advantage.

† This statement applies to low temperatures *only*. At elevated temperatures, where there are significant atom movements, grain boundaries facilitate creep (to be discussed in Section 16–5) and therefore weaken a metal or a ceramic.

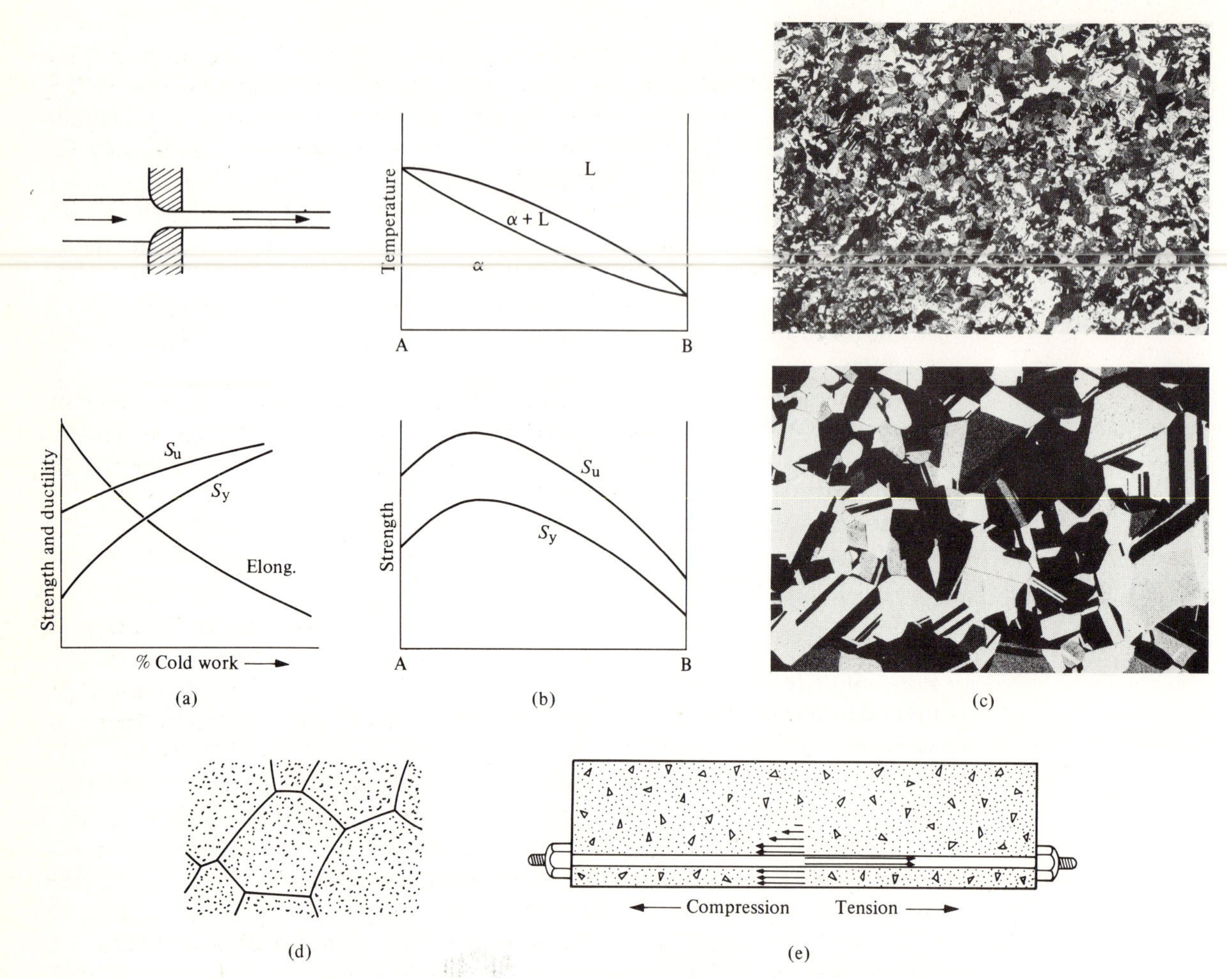

Figure 14-4.1 Strengthening mechanisms. (a) Strain hardening. Plastic deformation in the form of cold work increases the strength of crystalline materials (and decreases the ductility). (b) Solution hardening. Alloys of two components are stronger and harder than either component alone. (c) Grain size and hardness (brass annealed at 580°C, 8 sec. and 1 hr., respectively). At low temperatures (only), a material with a smaller grain size is harder (60 R_f and 40 R_f, respectively) (× 40). (Courtesy of J. E. Burke.) (d) Dispersion strengthening. Very fine, embryonic particles restrict slip. This accounts for the strengths of precipitation-hardened alloys and of tempered martensite. (e) Induced strength. The nonductile concrete is compressed. These stresses must be exceeded before the loaded beam will fail in tension. Induced strength may also be achieved by thermal tempering and chemical tempering.

grains do not have the identical orientations. Grains grow coarser (and fewer) with extended heating, growing faster at higher temperatures. The average grain size *cannot* grow smaller; however, a material may be processed to possess smaller grain sizes by *grain refinement.* This is done by starting a new "crop" of grains. For example, after cold work, an anneal produces new grains by recrystallization. If this anneal is brief, the "refined" grains will be small. In other cases, grain refinement can be achieved by phase changes. For example, austenite of a 1020 steel changes to ferrite during cooling; the new ferrite grains are initially small. With normal cooling, a 1020 steel can have smaller ferrite grains than the austenite previously had.

Also, grain growth can be retarded if a *very* fine dispersion of a second phase is present. For example, in Section 14-4 we will see that the ferrite in a 1020 steel will remain fine if a small amount (~0.03%) of niobium is present to form NbC particles within the ferrite.

Dispersion hardening

In Sections 5-6 and 6-5, attention was given to *precipitation hardening* and to *tempered martensite*. In each of these, the alloy contains a *dispersion* of very small particles of a hard minor phase (Fig. 14-1.1d). In our example of precipitation hardening, $CuAl_2$ had *started* to cluster into particles that are dispersed within a matrix of tough κ—the fcc, Al-rich phase. All precipitation-hardened alloys have this combination of many, miniscule hard particles and a tough matrix. In our example of tempered martensite, the hard, brittle martensite decomposed to a dispersion of many, very small hard carbide particles within a tough ferrite matrix, $M \rightarrow [\alpha + \overline{C}]$. In the early stages while the carbide particles are still submicroscopic ($<<1\ \mu m$), the steel retains most of its hardness but gains usable toughness. This is an ideal situation and is in contrast to strain hardening where a loss of ductility accompanies the gain in strength.

Dispersion-hardened materials have greatest strength when the dispersed particles are smallest—at the very inception of their growth. Because they are very small, their numbers are greatest, and the interparticle distances are least. In brief, there is greater interference to dislocation movements at this stage. Slip cannot pass through these particles that have a high shear resistance, and there is little space for dislocations to move significant distances between the particles. Higher temperatures or extended heating bring about a *coalescence* into larger but fewer particles, and softening accompanies this microstructural change. We speak of *over-aging* and *overtempering* when the softening goes beyond the desired levels.

Over-aging and overtempering are usually avoided in manufacturing processes by limiting the temperature and the heating time. These phenomena will, however, produce restrictions in service conditions. For example, a precipitation-hardened aluminum alloy cannot be used in the leading edge of a supersonic aircraft wing, because the frictional heating will over-age the alloy and reduce the strength to below the specified value. Likewise, a tool steel that readily overtempers can soften and fail if it is used in fast cutting operations. High-speed steels contain alloy additions of carbide formers—Cr, W, V, Mo, etc.—that slow down the carbide coalescence. These steels tolerate higher temperatures that accompany fast cutting.

Induced strength

A means of inducing strength was described in the previous chapter, whereby a surface of a nonductile material such as glass is placed under compression, either by thermal treatments (Fig. 13–4.2), or by chemical treatments (Eq. 13–4.1). In neither case will the surface encounter tension until the induced compressive stresses are exceeded. This provides significant amounts of extra strength, since it is possible to induce high residual compression stresses. This procedure is primarily useful for nonductile materials since they are sensitive to tension failure. However, metals that are shot-peened to introduce surface cold work and a compression layer have improved resistance to fatigue cracking (Section 14–3).

Materials may also be placed under compression by *prestressing*. A simple example is a concrete beam in which the tensile side of a beam is compressed by tightening reinforcing rods before installation (Fig. 14–1.1e).

14–2 FRACTURE

The ultimate mechanical failure is fracture. In a ductile material, fracture is proceeded by plastic deformation, and we speak of *ductile fracture,* or rupture. Brittle materials, such as glass and numerous metals, break by *brittle fracture.* Such a failure undergoes little or no plastic deformation (Fig. 14–2.1).

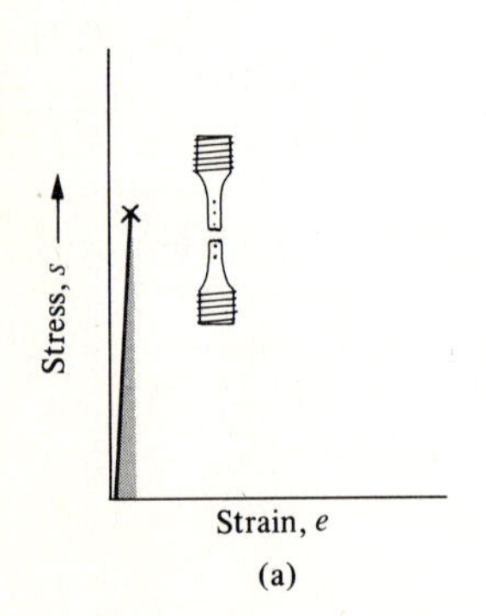

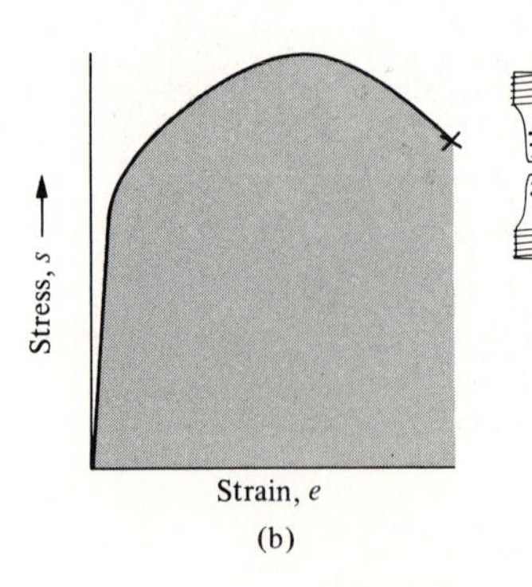

Figure 14–2.1 Fracture. (a) Brittle fracture involves little or no plastic deformation. (b) Ductile fracture requires energy for plastic deformation. Toughness, the energy requirement, is equal to the area under the *s–e* curve.

Toughness is a measure of the energy required for failure by fracture. Brittle fracture requires minimal energy, since the only energy consumption is the separation of atoms along the fracture path, that is, the energy required to form two new surfaces. Ductile fracture requires appreciably more energy, since not only must the fracture produce new surface, but it also plastically deforms significant amounts of adjacent metal. In brief, the amount of deformation energy that is required per unit volume is equal to the integrated area under the stress-strain curve (Fig. 14-2.1b).

Toughness tests

The tensile test of Fig. 14-2.1 is designed to calculate stress and strain from measurements of load and displacement. Although it would be possible to calculate energy of failure from these data, the energy requirements can be measured more accurately if we use the impact tester of Fig. 14-2.2. The

Figure 14-2.2 Toughness test. The notched test specimens—arrow in (a) and sketched in (b)—is broken by the impact of the swinging pendulum (c). The amount of energy absorbed is calculated from the arc of the follow-through swing (I. P. 14-2.1). (Courtesy of U.S. Steel Corp.)

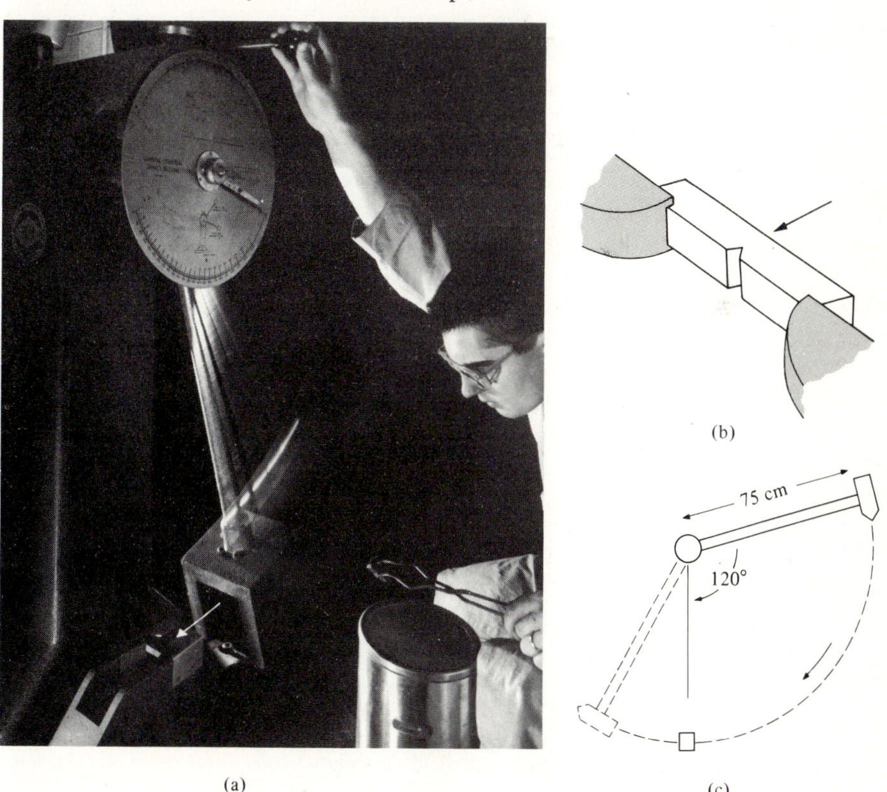

energy of fracture is calculated from the height of follow-through swing of the heavy pendulum (I.P. 14–2.1).

The consumption of energy necessary to cause fracture is concentrated adjacent to the fracture path and is not uniformly distributed throughout the test sample. Therefore, the amount of energy depends considerably on the size and shape of the test specimen and the rate of impact loading. As a result, standardized test specimens and procedures are prescribed for toughness or impact testing. A commonly used test specimen is the Charpy sample. It may have either a V-notch (Fig. 14–2.2b) or a keyhole slot as a means of standardizing the conditions to initiate the fracture. The keyhole specimen has the advantage of being more reproducible.* However, the V-notch relates somewhat better to structural design, in that brittle failures are seldom observed in service if the V-notch specimens show any ductility.

Ductility-transition temperature

Many materials possess an abrupt drop in ductility and toughness as the temperature is lowered. In glass and other amorphous materials, this change corresponds to the glass-transition temperature (Fig. 8–3.3). Of course, metals are crystalline and do not have a glass-transition temperature. However, they have a *ductility-transition temperature,* T_{dt}, (Fig. 14–2.3) that divides a lower temperature regime where the fracture is said to be nonductile from a higher temperature range where considerable plastic deformation accompanies failure.

Figure 14–2.4 shows the energy absorbed during impact fracture by two different steels as a function of temperature. As with the glass-transition temperature (Sections 7–4 and 8–3), the ductility-transition temperature is not sharp, but it depends on the rate of loading, presence of impurities, etc. However, the engineer will be quick to choose Steel C over Steel B of Fig. 14–2.4 if it is to be used in a welded ship in the North Atlantic winter waters. A crack, once started in the steel with a high ductility transition temperature, could continue to propagate with a low energy fracture until the ship is broken apart. There were several unfortunate naval catastrophies that occurred before the design engineer learned how to make appropriate corrections by design redundancies, and metallurgists found that fine-grained steels had lower ductility transition temperatures than coarse-grained steels.

Fortunately, fcc metals do not possess an abrupt change in ductility as the temperature is lowered. Thus, aluminum, copper, and some stainless steels can be used in cryogenic applications (Section 16–7). Unfortunately these metals cannot replace steels in large structures, pipelines, etc.

---

* It is made by drilling a 2-mm hole and then cutting a slot that connects the hole to the surface.

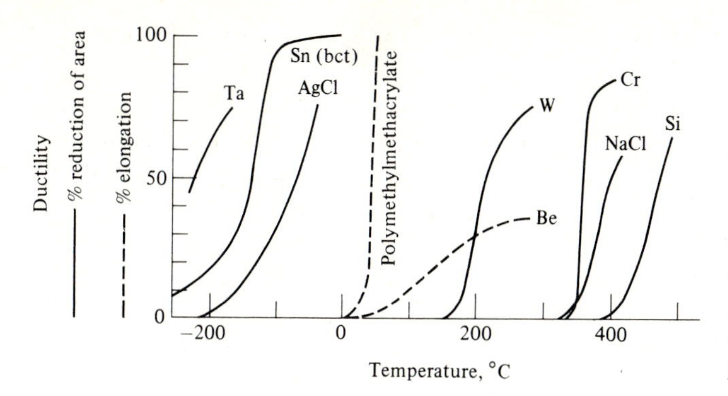

Figure 14–2.3 Ductility versus temperature (tensile tests). Except for fcc metals, most materials abruptly lose ductility at decreasing temperatures. For a given material the transition temperature is higher for higher strain rates, for example, impact loading. (After data by A. H. Cottrell, *The Mechanical Properties of Matter,* New York: Wiley.)

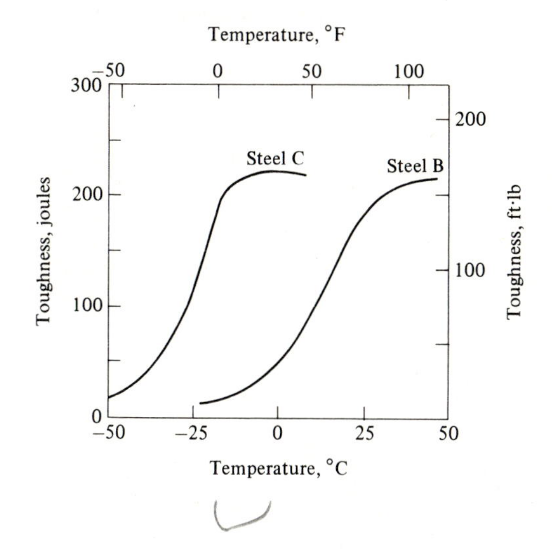

Figure 14–2.4 Toughness transitions. For each steel, there is a marked decrease in toughness at lower temperatures. The transition temperature is significantly lower for Steel C (fine-grained) than for Steel B (rimmed). (Adapted from Leslie, Rickett, and Lafferty, *Trans. AIME.*)

- **Stress-intensity factor**

Failure by fracturing always starts at some geometric irregularity that introduces a stress concentration. It may be a keyway in a shaft (Fig. 20–4.5), a corner of a hatchway on a ship, or a scored line on a piece of window glass. For that reason, test specimens include standardized notches so that the stress states are comparable from sample to sample.

Any notch or crack will concentrate stresses ahead of it. This was indicated in Fig. 13–3.2 and Eq. (13–3.2). Thus, a crack may propagate even though the nominal stress, s_n, is below the strength of the material (as calculated in Section 3–1). In many experiments, metallurgists have measured the nominal stress to propagate a crack of known length, c*. For a given steel and under fixed conditions, the results provide the relationship:

$$s_n\sqrt{\pi c} = \text{constant} = K_{Ic} \tag{14–2.1}$$

* Or half-length for an internal crack. (Cf. Fig. 13–3.2.)

The constant, K_{Ic}, is called the *stress-intensity factor* and the s_n is the design limit for the nominal stress. Thus, a steel with a K_{Ic} value of 100 $MPa \cdot m^{1/2}$ and a 2-mm crack could assume a tolerable nominal stress of 1250 MPa;* with a 5-mm crack, only 800 MPa. Steels with the above K_{Ic} value typically have yield strengths in excess of 1400 MPa (>200,000 psi). This means that if these steels have a crack in excess of approximately 1.6 mm in length, they will fail by fracturing rather than on the basis of their yield strength. Design considerations and inspection must take this into account.

Illustrative problem 14-2.1 An impact pendulum on a testing machine weighs 10 kg and has a center of mass 75 cm from the fulcrum. It is raised 120° and released. After the test specimen is broken, the follow-through swing is 90° on the opposite side. How much energy did the test material absorb?

SOLUTION Refer to Fig. 14-2.2.

$$\Delta E = 10 \text{ kg}(9.8 \text{ m/s}^2)(0.75 \text{ m})[\cos(-120°) - \cos 90°]$$
$$= -36.8 \text{ J} \qquad (\textit{lost} \text{ by the pendulum})$$
$$= +36.8 \text{ J} \qquad (\textit{absorbed} \text{ by the sample}).$$

ADDED INFORMATION In general, materials have more toughness at high than at low temperatures. In fact, there is an abrupt decrease in the toughness of many steels as they are cooled below ambient temperatures. The temperature at which this discontinuous decrease occurs is called the *transition temperature*. ◀

• **Illustrative problem 14-2.2** A steel with a K_{Ic} value of 150 $MPa \cdot m^{1/2}$ also has a yield strength of 1200 MPa. How deep may a crack be before the steel is subject to fracture as a mode of failure?

SOLUTION

$$s_n = 1200 \text{ MPa} = (150 \text{ MPa} \cdot \text{m}^{1/2})/\sqrt{\pi c}$$
$$c = \sim 0.005 \text{ m} \qquad (\text{or } 5 \text{ mm}). ◀$$

14-3 FATIGUE

People have long been aware that certain materials fail if they remain in service a long time. Laymen assumed that the metal got "tired," so they spoke of fatigue failure.

The most familiar type of fatigue occurs with *cyclic stresses*.† We are aware that a thin metal sheet can be broken if we repeatedly bend it back

---

* $s_n = (100 \text{ MPa} \cdot \text{m}^{1/2})/\sqrt{\pi(0.002 \text{ m})} = 1250$ MPa. In English units, 100 $MPa \cdot m^{1/2}$ = 91,000 $psi \cdot in.^{1/2}$; $s_n = (91{,}000 \text{ psi} \cdot \text{in.}^{1/2})/\sqrt{\pi(0.08 \text{ in.})} = 180{,}000$ psi.

† Static fatigue can occur with constant stresses in glass and in metals by *stress corrosion*, where there is a slow destructive reaction with the surrounding environment.

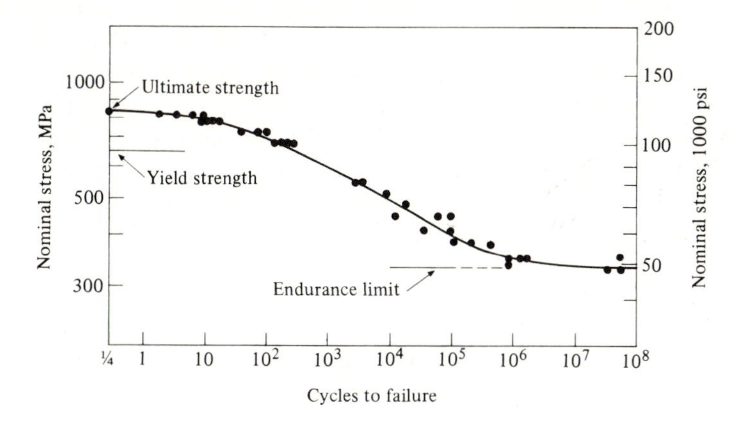

Figure 14-3.1 *S-N* curve (SAE 4140 normalized steel). *S-N* = cyclic *stress* versus *number* of cycles to failure. At the endurance limit, the number of cycles becomes indeterminately large. (Adapted from R. E. Peterson, *ASTM Materials Research and Standards.*)

and forth. More subtle but equally important stress cycles occur in a loaded rotating shaft or a vibrating plate. In one position the surface of the shaft is in tension; 180° later (0.017 sec later at 1800 RPM), it is on the other side and in compression. Under such conditions, 10^6 to 10^8 stress cycles can be accumulated during a relatively short period of service time.

The stress that a material can tolerate under cyclic loading is much less than it is under static loading. The ultimate strength can be used only as a guide because, as Fig. 14-3.1 shows, the tolerable stress decreases as the required number of cycles increases. Were it not for the fact that an *endurance limit* (the stress below which fatigue does not occur in reasonable times) exists for many materials, the uncertainty in design would be great.

Fatigue arises because each half-cycle produces minute strains that are not fully recoverable. When these strains are added together, they produce local plastic strains that are sufficient to reduce ductility in the cold-worked areas (cf. Fig. 4-2.2c) so that submicroscopic cracks are formed. A crack

Table 14-3.1
Surface finish versus endurance limit
(SAE 4063 steel, quenched and tempered to $44R_c$)*

	Surface roughness		Endurance limit	
Type of finish	**μm**	**μin.**	**MPa**	**psi**
Circumferential grind	0.4–0.6	16–25	630	91,300
Machine lapped	0.3–0.5	12–20	720	104,700
Longitudinal grind	0.2–0.3	8–12	770	112,000
Superfinished (polished)	0.08–0.15	3–6	785	114,000
Superfinished (polished)	0.01–0.05	0.5–2	805	116,750

* Adapted from M.F. Garwood, H.H. Zurburg, and M.A. Erickson, "Correlation of Laboratory Tests and Service Performance," *Interpretation of Tests and Correlation with Service,* Amer. Soc. Metals.

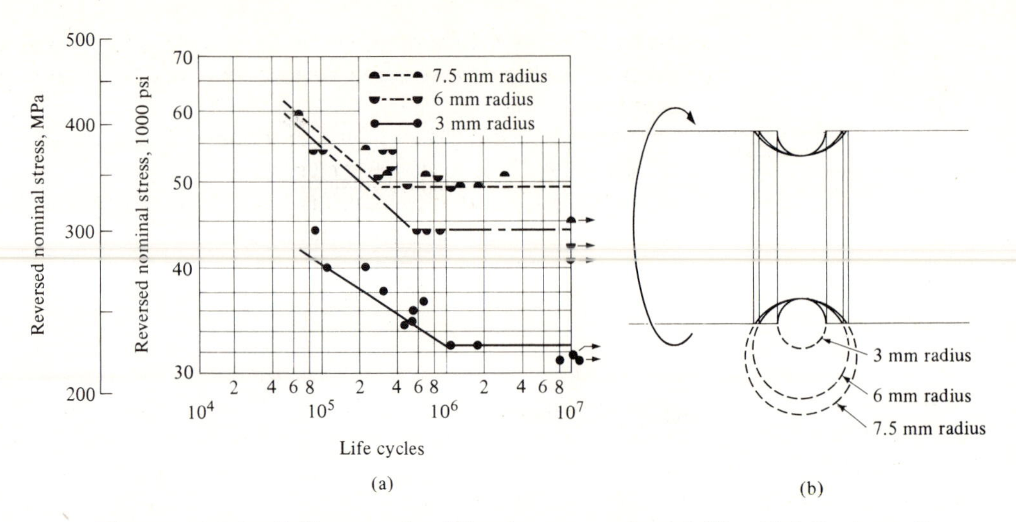

Figure 14-3.2 *S-N* curves for filleted test samples (cf. Fig. 14-3.3). A smaller radius of curvature produces a high stress concentration and therefore a lower endurance limit. (Adapted from M. F. Garwood, H. H. Zurburg, and M. A. Erickson, "Correlation of Laboratory Tests and Service Performance," *Interpretation of Tests and Correlation with Service,* Amer. Soc. Metals.)

serves as a notch that concentrates stresses until finally complete fracture occurs.

These strains are localized along slip planes, at grain boundaries, and around surface irregularities caused by compositional or geometric defects. The influence of geometric irregularities (notches) is illustrated in Fig. 14-3.2 and Table 14-3.1. The three sets of data in Fig. 14-3.2 are all for identical steels; however, a notch of 3 mm serves to reduce the endurance

Figure 14-3.3 Design of fillet. The use of generous fillets is recommended in mechanical engineering design. It should be observed that (c) is a better design than (a), even with some additional material removed. Of course, if too much metal is removed, failure may occur by mechanisms other than fatigue.

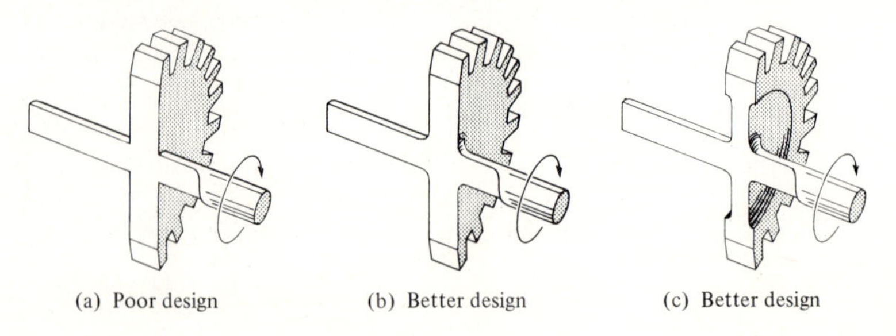

limit by 25%–30%. Likewise, better surface finishes increase the endurance limit because there are fewer surface irregularities on which stresses can concentrate.

Any design that introduces stress concentrations can lead to premature fatigue failure. This is illustrated schematically in Fig. 14–3.3, which shows that the use of fillets is recommended by the mechanical engineer to avoid sharp corners, and to improve the design.

• 14–4 HIGH-STRENGTH, LOW-ALLOY (HSLA) STEELS

For several reasons, steels for construction, steels for automobile bodies, and steels for similar large-scale applications are not amenable to the classic quenching-and-tempering treatments discussed in Section 6–5. For example, many beams are simply too big. Also, a quenched-and-tempered steel cannot be welded without a major alteration of the properties adjacent to the weld. Finally, the thousands of such structural steel products manufactured per day would make quenched-and-tempering costs very high. As a result, millions of tons of "as-hot-rolled" steels have been used over the years with yield strengths of less than 250 MPa (<36,000 psi), because they are the cheapest of steels *and can* be welded.

The above shortcomings have provided incentives to develop stronger, weldable steels with minimal increases in cost. There has been considerable success. With appropriate minor alloy additions and with very close thermal control during the rolling process, the metallurgist has developed weldable structural steels that have yield strengths greater than 500 MPa (>70,000 psi). As a group they are called *high-strength, low-alloy steels* (HSLA steels). Actually they include a number of subgroups. In any event, their availability has permitted the design engineer to reduce weights of automobiles, and to construct large structures at lower costs than would have been possible with previous steels.

Some HSLA steels gain their strength from an extremely small grain size. This grain refinement (Section 14–1) is achieved by the presence of very fine precipitates in the steel. In this case, the precipitates do not form by decreasing the solubility limit for one phase in another, as is the situation for precipitation hardening (Section 5–6); rather, because they are soluble in austenite, γ, but *not* soluble in ferrite, α. Specifically, alloying elements such as vanadium and niobium are added to the steel in very small amounts (<0.05%)* to steel with ~0.2% carbon. These alloying elements are soluble in austenite; however, since they are very powerful *carbide formers,* they react with the immediately adjacent carbon in the steel as the steel changes from γ to α while cooling *during the rolling process.* The total

* This is sometimes called *microalloying.*

volume of VC or NbC particles* is small; but their numbers are astronomical since they are extremely minute (~5 nm). As a result, they are very close together. In effect, they serve as a "fence" across which the ferrite grain boundaries cannot move, with the result that the average ferrite grain dimension is about 5 μm. This would lead to a ferrite grain boundary area of 500–1000 mm^2/mm^3. (Cf. Fig. 2–7.4 and I.P. 2–7.1).†

These HSLA steels that have just been described can be welded, because the grain refinement will occur during the cooling that follows welding, as well as during the cooling that accompanies rolling. As a result, it is possible to have welded structural steel with yield strengths as high as 400 MPa (~60,000 psi).

HSLA steels can develop a combination of austenite plus proeutectoid ferrite while annealing above the eutectoid temperature, since they contain 0.05%–0.25% carbon. (Cf. I.P. 6–3.1.) As expected, they are then *dual phase,* ($\alpha + \gamma$). The austenite will readily change to pearlite during cooling, unless the $\gamma \rightarrow [\alpha + \overline{C}]$ reaction is suppressed by rapid cooling—in which case it forms martensite (Section 6–5). Through appropriate (and very careful) compositional control, it is possible to have a dual-phase steel that will produce ferrite + martensite (α + M), and subsequently tempered martensite. This combination is desirable since it produces a low-yield strength from the presence of ferrite and a relatively high ultimate strength from the tempered martensite. The steel can readily be cold-shaped into a product with considerable ultimate strength before fracture.

Other HSLA steels make use of one or both of the above strengthening mechanisms, plus some strain hardening that occurs in the ferrite during rolling (Section 4–2). The combined effects give greater strength than the individual procedures alone. Of course if an elevated temperature is used, the strain-hardened steel must be cooled sufficiently rapidly after deformation to avoid recrystallization of the ferrite (Section 4–3). However, the cooling rate does not introduce a major problem in sheet and wire products because they have a large surface area per unit volume. These combined effects are less applicable on heavier products where the strain hardening could overload the rolling mill.

There are still other procedures for making high-strength steels with low-alloy contents. Many of these require rather sophisticated metallurgy but are being pursued in an effort to increase strength and decrease weight, and to achieve these goals without the use of expensive alloys that require critical raw materials. It is an area of metallurgy that is receiving much current attention.

---

* These carbides usually contain some nitrogen.

† In some steels, precipitates form from the austenite during hot rolling and restrict the austenite grain growth; this grain refinement is inherited by the ferrite in subsequent cooling.

14-5 COMPOSITES

The early craftsman attempted to improve available materials by combining two or more into a *composite.* Whether it was an addition of straw to sun-dried clay brick in biblical times or a glaze on an earthenware pot, the purpose was to utilize the merits of two or more distinct materials. Technical interest in composite materials has increased progressively with the sophistication of available materials. Fibers of carbon filaments can now be used to reinforce complex polymers; semiconducting coatings are vapor-plated onto selected areas of chips to provide electrical functions; grains of abrasive are bonded by rubber into an industrial grinding wheel; aluminum is plated onto the surface of a plastic molding of car trim; and steel rods are embedded in a specified pattern within the concrete of a hydroelectric dam. In each case, the characteristics of combined materials perform in a manner that would be impossible by either of the contributing materials alone.

In this section we shall give attention to reinforced materials and composites. In Unit 16, surface-coated composites will be presented. The two will illustrate many of the characteristics required in designing composites.

Reinforced materials

Materials are reinforced to make them stronger. When improved strength is the major goal, the reinforcing component must have a large *aspect ratio,* that is, its length/diameter ratio must be high, so that the load is transferred across potential points of fracture. Thus, we place steel rods in a concrete structure (Fig. 14-5.1). And we combine glass fibers and polymers for fiber-reinforced plastics (FRP).

It is obvious that the reinforcement must be the stronger component if it is to carry the load. It may be less obvious that the reinforcement must have the higher elastic modulus. Likewise, it may not be immediately apparent that the bond between the matrix and the reinforcement is critical, since it is generally necessary to transfer the load from the matrix to the fibers or rods if the reinforcement is to serve its purpose.

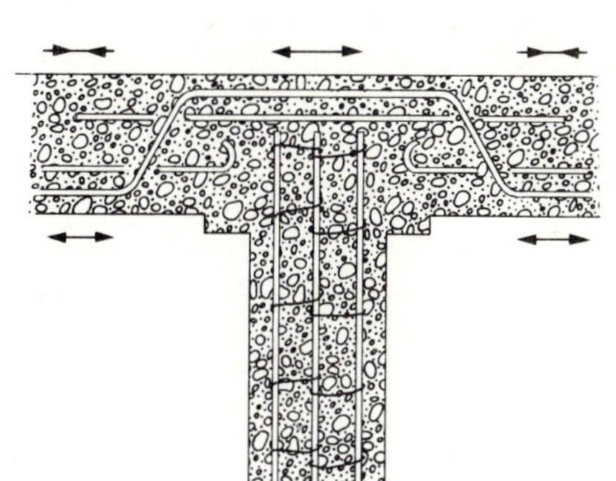

Figure 14-5.1 Reinforced materials (concrete). The materials are selected to match the design requirements. The steel is used for tension; the concrete for compression and for rigidity.

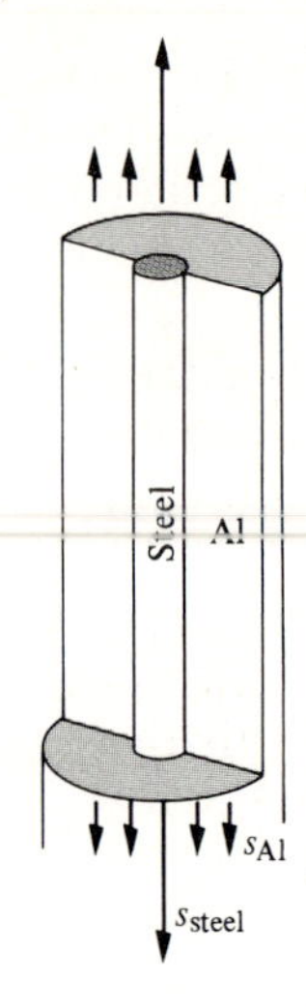

Figure 14-5.2 Stress in composites (steel-reinforced aluminum wire). The strain in the two must be equal. Therefore Eq. (14-5.2) applies.

For the reinforcing material to carry most of the load, *the reinforcement must have a higher Young's modulus than the matrix.* Consider Fig. 14-5.2 in which a steel core reinforces an aluminum wire. When loaded in tension, the two metals must deform together. Assume a strain, e, of 0.001. With the two moduli being ~205,000 MPa (~30,000,000 psi) and ~70,000 MPa (~10,000,000 psi), respectively, and using Eq. (3-1.3), the steel develops a stress of ~205 MPa versus ~70 MPa in the aluminum:

$$e = 0.001 = (s_{steel}/205{,}000 \text{ MPa}) = (s_{Al}/70{,}000 \text{ MPa}). \tag{14-5.1}$$

Because E_{steel} is $\sim 3(E_{Al})$, the stress in the steel is $\sim 3(s_{Al})$. Generally, for a two-component composite,

$$s_1/s_2 = E_1/E_2. \tag{14-5.2}$$

The reinforcing rods of Fig. 14-5.1 are *located in the tensile side* of the beam—on the lower side in mid-span, and on the upper side where the beam passes over the supporting column. Thus, the steel carries the load where the brittle material would have been subjected to tensile cracking. Conversely, the beam would gain negligible additional strength if similar steel rods were placed in the compressive regions. First, concrete is strong in compression (Section 11-8). Also, a long rod (or fiber) will buckle if loaded in compression. Of course, rods with higher moduli are stiffer, but in compressive loading we must generally depend upon larger dimensions* to give rigidity. Concrete provides the volume necessary for rigidity relatively cheaply.

* Actually, larger moments of inertia, which include Young's modulus *and* cross-sectional dimensions.

The modulus of elasticity (in tension) of a fiber-reinforced plastic, E_{FRP}, may be estimated as the volume fraction average if all of the fibers are aligned parallel to the direction of loading:

$$E_{FRP} = \Sigma f_i E_i, \tag{14–5.3}$$

where f_i and E_i are the volume fractions and Young's moduli, respectively, for the components. Consider a plastic reinforced with 50 v/o of parallel glass fibers ($E = 70{,}000$ MPa, or 10^7 psi) and the same fraction of a plastic that has a Young's modulus of 4000 MPa (580,000 psi). The resulting composite has a Young's modulus in its longitudinal direction of ~37,000 MPa (or 5.3×10^6 psi). (Of course, the elastic moduli in the other two coordinate directions will be low and close to that of the plastic, because they are at right angles to the reinforcement.)

If the same amount of glass is incorporated as a woven fabric into the above plastic, the composite gains two-way reinforcement. However, the values drop below that of the above mixture rule (Eq. 14–5.3). Furthermore, the 45° modulus is low (Fig. 14–5.3). It is now common to use matted glass fibers to avoid this anisotropy in composite sheet products. The glass fibers of the reinforcing mats possess a sufficiently random distribution to give uniform elastic moduli and therefore uniform load distributions in two dimensions.

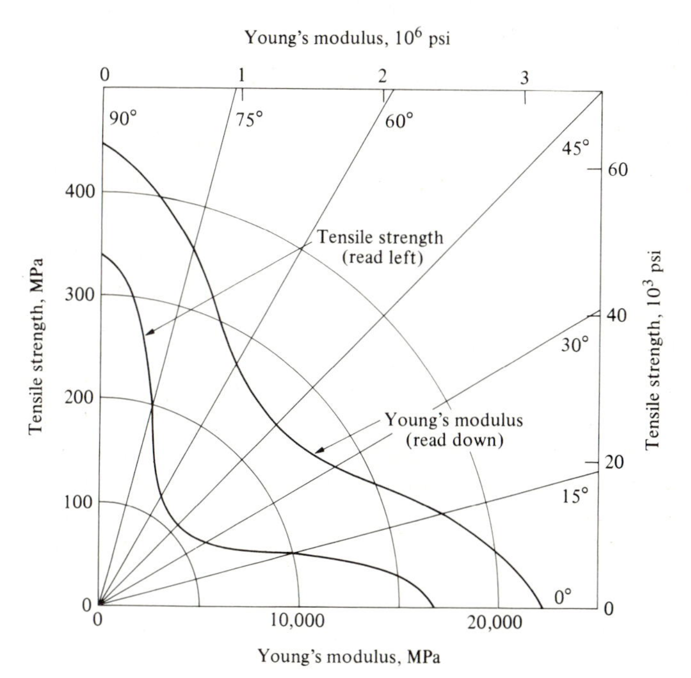

Figure 14–5.3 Directional properties (cross-laminated, glass-reinforced epoxy). The reinforcing fibers are woven in two right-angle directions. Therefore, there is high strength and good rigidity in these directions. The corresponding values are low at 45°. (Adapted from Broutman, *Modern Composite Materials,* Reading, Mass.: Addison Wesley.)

- **Interfacial stresses**

Specifications (ASTM, A-305) for reinforcing bars that are to be used in concrete structure must be *merloned,* that is, they must have a roughened surface pattern to help anchor the rod into the concrete. However, it is not simply a matter of holding the rod in position; the load on the structure must be transferred as shear stresses across the interface between the concrete and the steel. Comparable shear stresses are encountered at the interface between glass fibers and the surrounding plastic matrix. There, however, the shear stresses must be transferred through chemical bonds rather than by surface roughness.

Interfacial shear stresses become particularly important if the fibers are not continuous. This is illustrated in Fig. 14–5.4, where s_f represents the stress to be carried by a continuous fiber—the 205 MPa (30,000 psi) available for Eq. (14–5.1). If the fiber is broken, however, its stress automatically drops to zero at the end of the fiber, and the load is transferred into the matrix (Fig. 14–5.4b). This transfer is by shear stresses across the interface (Fig. 14–5.4c). The shear stress is very high near the fiber ends, and the weaker matrix must carry an overload. This places a premium on long continuous fibers in load-bearing composites. It also favors greater numbers of small diameter fibers rather than fewer larger fibers, since there is more interfacial area to support the shear loads, and less chance that one broken fiber will introduce damaging flaws in the matrix. Finally, a ductile matrix will adapt more readily to the stress concentrations at fiber ends than will a matrix of brittle plastic.

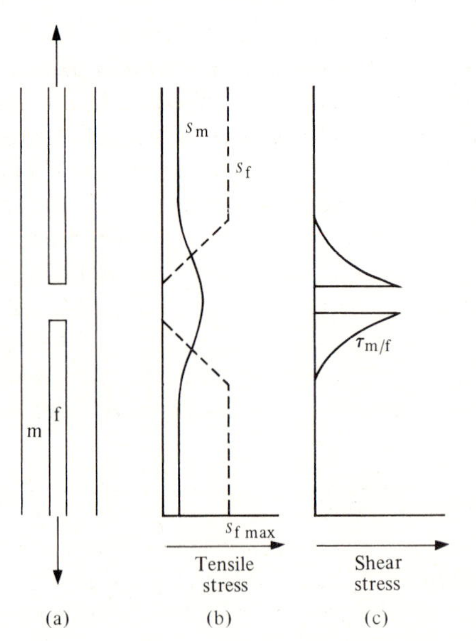

Figure 14–5.4 Stress distribution (at a break in a reinforcing fiber). The fiber stress, s_f, drops from its maximum value to zero. The load must be transferred across the interface from the fiber to the matrix by shear stresses, $\tau_{m/f}$. The matrix has to carry a higher stress, s_m, in the vicinity of the break.

Table 14–5.1
Modulus/density (E/ρ) ratios of common materials

	Density, Mg/m^3	Young's modulus, MPa	E/ρ, N·m/g
Aluminum	2.7	70,000	26,000
Iron and steel	7.8	205,000	26,000
Magnesium	1.7	45,000	26,000
Glass (soda-lime)	2.5	70,000	28,000
Wood (spruce)	0.43	11,000	26,000
Wood (birch)	0.61	16,500	27,000
Polystyrene	1.05	2,800	2,700
Polyvinyl chloride	1.3	<4,500	<3,500
50 v/o glass-plastic*	~1.9	~37,000	~20,000
70 v/o glass-plastic*	~2.1	~50,000	~24,000

* Values for these composites will vary, depending on the plastic; however, these values are typical of most glass reinforced plastics.

- **High-stiffness composites**

Because of their lighter weight, many polymeric composites are attractive to design engineers. Although the composities may have lower ultimate tensile strength, S_u, than many metals, their strength-to-density ratios, S_u/ρ, are high and very desirable. However, as indicated earlier, reinforcements function differently in tension than in compression. In compression, the elastic modulus-to-density ratio, E/ρ, provides a better design criterion where weight is important. Table 14–5.1 shows the E/ρ ratios for a number of common structural materials, including several of those in Appendix A. It is noteworthy that the majority have comparable values. The 50-50 glass-plastic composite immediately below Eq. (14–5.3) has an E/ρ ratio of only 20,000 N·m/g. These lower values are typical for most artificial composites. A glass-reinforced plastic would have to contain more than 70 v/o glass to approach the E/ρ of metals. That percentage is difficult to produce cheaply because special efforts are required to align the fibers perfectly. Thus, attention has been paid to materials with extremely high moduli for reinforcement. Some possibilities are listed in Table 14–5.2. In general, however, these materials have not been amenable to fiber production. Boron and carbon fibers show the most potential; but processing costs are high and engineering applications will be limited until these materials are more generally available.

Illustrative problem 14–5.1 A 1060 steel wire (1 mm in diameter) is coated with copper (combined diameter = 2 mm). The two materials have elastic moduli listed in Appendix A and yield strengths of 280 MPa (40,000 psi) and 140 MPa (20,000 psi), respectively. (a) If this composite is loaded in tension, which metal will yield first? (b) How much load, F, can the compos-

Table 14-5.2
Modulus/density (E/ρ) ratios of materials potentially available* to reinforce composites

	Density, Mg/m^3	Young's modulus, MPa	E/ρ, N·m/g
Alumina	3.9	400,000	100,000
Boron	2.3	400,000	170,000
Beryllium	1.9	300,000	160,000
BeO	3.0	400,000	130,000
Carbon	2.3	700,000	300,000
Silicon carbide	3.2	500,000	160,000
Silicon nitride	3.2	400,000	120,000

* Each of these will have to be made available in fiber form at competitive costs, before widespread use may be anticipated.

ite carry in tension without plastic deformation? (c) What is the modulus of elasticity, $\overline{E}$, of this composite material?

SOLUTION

$$\text{Area}_{st} = (\pi/4)(0.001 \text{ m})^2 = 0.8 \times 10^{-6} \text{ m}^2.$$
$$\text{Area}_{Cu} = (\pi/4)(0.002 \text{ m})^2 - 0.8 \times 10^{-6} \text{ m}^2 = 2.4 \times 10^{-6} \text{ m}^2.$$

For elastic strain (Eq. 3-1.3), and using data from Appendix A,

$$(s/E)_{st} = e_{st} = e_{Cu} = (s/E)_{Cu}$$
$$s_{st} = s_{Cu}\,(205{,}000 \text{ MPa})/(110{,}000 \text{ psi}) = 1.86\, s_{Cu}.$$

a) With a 1.86 stress ratio, the steel is stressed 260 MPa when the copper is stressed 140 MPa. Therefore, the copper yields first.

b)
$$\begin{aligned} F_{total} &= F_{Cu} + F_{st} \\ &= (140 \times 10^6 \text{ N/m}^2)(2.4 \times 10^{-6} \text{ m}^2) \\ &\quad + (260 \times 10^6 \text{ N/m}^2)(0.8 \times 10^{-6} \text{ m}^2) \\ &= 540 \text{ N} \qquad \text{(or 55 kg on earth).} \end{aligned}$$

c) From Eq. (14-5.3).

$$\begin{aligned} \overline{E} &= (fE)_{st} + (fE)_{Cu} \\ &= 0.25\,(205{,}000 \text{ MPa}) + 0.75\,(110{,}000 \text{ MPa}) \\ &= 130{,}000 \text{ MPa.} \end{aligned}$$

Alternatively,

$$\begin{aligned} e &= (s/E)_{st} = [F/(A_{st} + A_{Cu})]/\overline{E}. \\ \overline{E} &= (E/s)_{st}\,[540 \text{ N}/(3.2 \times 10^{-6} \text{ m}^2)] \\ &= \left[\frac{205{,}000 \times 10^6 \text{ N/m}^2}{260 \times 10^6 \text{ N/m}^2}\right]\left[\frac{540 \text{ N}}{3.2 \times 10^{-6} \text{ m}^2}\right] \\ &= 130 \times 10^9 \text{ N/m}^2 \qquad \text{(or 130,000 MPa).} \blacktriangleleft \end{aligned}$$

Illustrative problem 14-5.2 What is the thermal expansion coefficient of the Cu-coated, 1060 steel wire of I.P. 14-5.1?

SOLUTION In the composite, $(\Delta L/L)_{st} = (\Delta L/L)_{Cu}$, and in the absence of an external load, $F_{Cu} = -F_{st}$. Basis for calculation, $\Delta T = 1°C$, and the data of Appendix A:

$$(\Delta L/L)_{st} = (\Delta L/L)_{Cu},$$

$$\alpha_{st}\Delta T + \left(\frac{F/A}{E}\right)_{st} = \alpha_{Cu}\Delta T + \left(\frac{F/A}{E}\right)_{Cu},$$

$$(\sim 11 \times 10^{-6})(1) + \frac{F_{st}/0.8 \times 10^{-6}\ \text{m}^2}{205 \times 10^9\ \text{N/m}^2} = (\sim 17 \times 10^{-6})(1) + \frac{-F_{st}/2.4 \times 10^{-6}\ \text{m}^2}{110 \times 10^9\ \text{N/m}^2};$$

$F_{st} \cong +0.61$ N (tension on heating);

$F_{Cu} \cong -0.61$ N (compression on heating).

$$(\Delta L/L)_{Cu} = (17 \times 10^{-6}/°\text{C})(1°\text{C}) + \frac{-0.61\ \text{N}/(2.4 \times 10^{-6}\ \text{m}^2)}{110 \times 10^9\ \text{N/m}^2};$$

$$\bar{\alpha} = \sim 15 \times 10^{-6}/°\text{C}.$$

ADDED INFORMATION The thermal expansion varies slightly with the exact type of steel. ◀

• 14-6 ABRASIVE PRODUCTS

Ultrahard materials such as Al_2O_3 (emery) and SiC are commonly called *abrasives*. They are used for grinding, either as grit or in a bonded form as abrasive wheels. Although hardness, per se, is an important requirement for abrasives, a degree of toughness is also important, especially when the abrasive is used for grinding high-strength, ductile materials. Moreover, since a grinding action is essentially a cutting operation (Fig. 14-6.1), high temperatures develop locally, and a high melting temperature is required. Fortunately, this does not present a major problem, because strong interatomic bonds provide both hardness and refractoriness, that is, the ability to stand high temperatures.

Properties

The *hardnesses* of abrasives are commonly indexed by one of two scales. The *Mohs scale*, which was originated by mineralogists, measures hardness by scratching (cutting). The *Knoop scale* measures the degree of penetration into a material by a diamond indenter. This is a static test. The Mohs scale

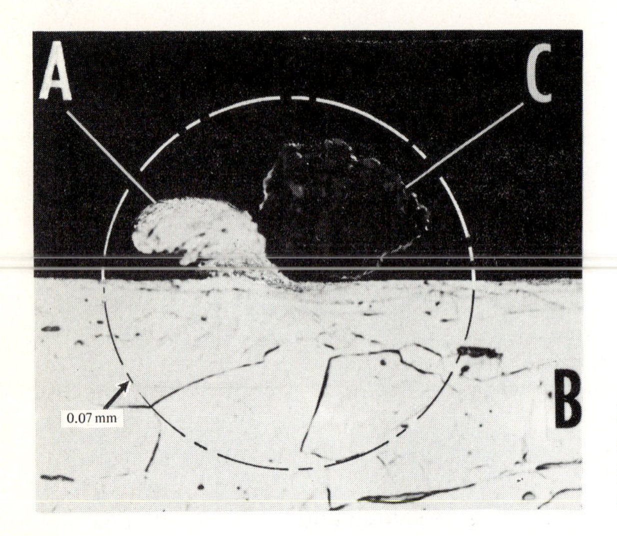

Figure 14-6.1 Photograph (×500) of grinding action. The steel chip (A) is shown being cut from the workpiece (B) by grinding grit (C). (Courtesy of Cincinnati Milacron.)

is simple and widely used, but it is not satisfactory for measuring the hardnesses of abrasives, because all abrasives have values of greater than 9 (Table 14-6.1). The Knoop scale is somewhat more quantitative since figures are available for gradations of hardness (Table 3-3.1).

The relative toughness and friability of different abrasive materials depends on a number of variables. Let us take Al_2O_3 as an example. When we have pure Al_2O_3, the abrasive grains (white) are extremely friable and fracture readily under the stresses of grinding. In contrast, an Al_2O_3 that contains Fe_2O_3 and/or TiO_2 in solid solution (red or brown) is appreciably tougher and more resistant to wear and fracture. The technical significance of this difference between the pure Al_2O_3 and its variants is that the engineer can make a choice of the correct abrasive for the job. For hard materials, the friable grain is best, since it maintains its angular cutting edges

Table 14-6.1
Hardness of abrasive materials

Material	Composition	Mohs hardness number	Approximate Knoop hardness	Melting temperature, °C
Diamond	C	10	8000	>3500
Boron carbide	B_4C	—	3500	2450
Silicon carbide	SiC	—	3000	>2700
Titanium carbide	TiC	—	2800	3190
Tungsten carbide	WC	—	2100	2770
Corundum	Al_2O_2	9	2000	2050
Topaz	$SiAl_2F_2O_4$	8	1500	Dissociates
Quartz	SiO_2	7	1000	Transforms

by fracturing, even though it implies a more rapid consumption of the abrasive. Tougher abrasives are used for high-strength ductile materials because considerable energy is absorbed in the grinding operations.

Bonded abrasives

Abrasive grit may be bonded into grinding wheels and other products. The identification codes of Fig. 14-6.2 are for manufacturer-user specifications. The terms "abrasive type," "grain size," and "bond" are self explanatory, but "grade" and "structure" require further explanation.

The *grade* refers to the friability of the product. The grain is readily torn from a "soft" grinding wheel. The wheel resists attrition during use in a "hard" grinding product. Grade is controlled in manufacture by (a) the nature of the grain (see above), (b) the nature of the bond, and (c) the relative amount of bond and grain. A soft wheel is self-cleaning, but short-lived.

The *structure* refers to the pore space among the grains within the bonded abrasive. Surface grinding, for example, requires an open structure to allow ample clearance of grinding chips. On the other hand, in precision grinding of bearing surfaces to prescribed dimensions and contours, a dense structure is required; hence, grinding is slower.

Vitrified alumina example

27* A 120 G 3 V 7†

Abrasive type	Grain size		Grade		Structure	Bond
A, Alumina	Coarse		"Soft"		Dense	V, Vitrified
C, Silicon carbide	4		A		0	S, Silicate
	6		B		1	R, Rubber
	8		C		2	B, Resinoid
	10	70	D		3	E, Shellac
	12	80	E	N	4	O, Oxychloride
	14	90	F	O	5	
	16	100	G	P	6	
	20	120	H	Q	7	
	24	150	I	R	8	
	30	180	J	S	9	
	36	220	K	T	10	
	46	240	L	U	11	
	50	280	M	V	12	
	60	320		W	13	
		400		X	14	
		500		Y	15	
		600		Z	16	
		Fine	"Hard"		Open	

C 36 R 10 R

Silicon carbide—rubber example

*The manufacturer's code for abrasive mix is optional with each manufacturer.
†Optional code for bond.

Figure 14-6.2 Designations for bonded abrasives.

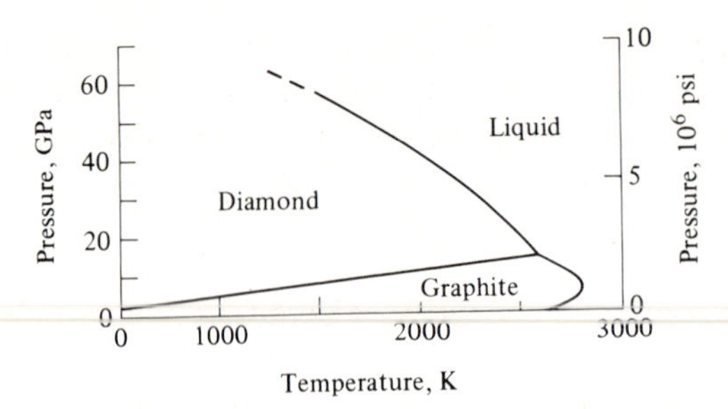

Figure 14-6.3 Part of the carbon-T-P phase diagram. High pressures are required to form diamond. High temperature facilitates the polymorphic changes between diamond and graphite.

Industrial diamonds

The word diamond brings sparkling gems to the mind of the average person. Diamonds have even greater value as a technical material, because of their extreme hardness. It is the hardest of all natural substances and was found only in nature until after the middle of the present century. Several tons per month are now made artificially from graphite through the combined use of high pressure and high temperature (Fig. 14-6.3).

Diamond is hard because of the extremely strong C–C covalent bonds that tie every carbon into a network structure with four neighboring carbon atoms. Diamond will revert to graphite if given the opportunity (Fig. 14-

Figure 14-6.4 Industrial diamonds. (a) High-impact strength. Used for sawing more than 2 million tons of granite per year. (b) Medium-impact strength. Used for grinding ~1 billion eyeglasses per year. (c) Low-impact strength. Used for sharpening millions of cemented-carbide tools per year. (Courtesy of General Electric Co., from *Materials Technology*.)

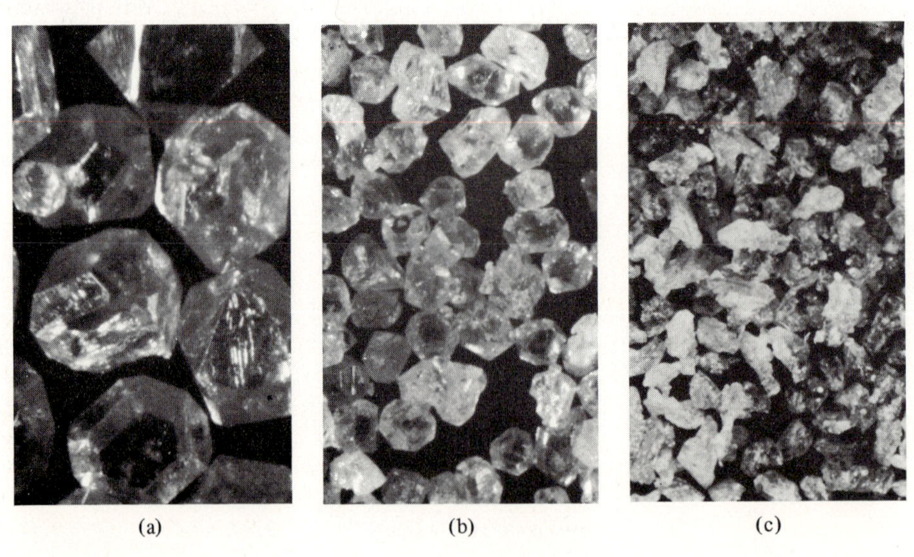

6.3). Higher temperatures facilitate this polymorphic change, because more thermal energy is available to help break the bonds. Thus, even if the cost should warrant it, diamond is not satisfactory for high-temperature applications as are many other abrasive materials.

Synthetic diamonds can be made with variations in toughness (or impact strength) as shown in Fig. 14-6.4. This option is very restricted in natural diamonds. Thus, artificial diamonds have reduced costs by 95% in processes such as the dressing (sharpening) of cemented-carbide tools. This is important because there are thousands of tons of tungsten carbide used each year in these tools in the industrial plants of the world. Of course, there are also many other abrasive uses for diamond.

Review and Study

SUMMARY

1. Strengthening mechanisms include (a) strain hardening, (b) solution hardening, (c) grain refinement, and (d) dispersion hardening—all of which restrict slip by interferring with dislocation movements. A fifth method of strengthening, induced residual stresses, is particularly applicable for nonductile materials that do not deform by slip. The induced compressive stresses must be exceeded before the material will crack in tension.
2. Engineers categorize fracture as ductile or brittle, depending upon the amount of accompanying plastic deformation. The ductile fracture absorbs appreciably more energy than the brittle (nonductile) fracture and will thus lessen the probability of product failure. The energy requirement depends not only on the material but also on the geometry, the rate of strain, and temperature. Many materials possess a ductility transition temperature, below which fracture becomes relatively nonductile.
3. • Stresses are intensified at geometric irregularities. Stress intensity factors. K_{Ic}, relate the design limit for the nominal stress to the depth of a propagating crack. This permits the engineer to predict whether a steel will fracture or fail by yielding in tension.
4. Materials fail by fatigue from cyclic loading. The tolerable stresses decrease as the number of cycles increases. However, an endurance limit does exist for many materials. This permits design calculations for an indefinitely large number of cycles. Endurance limits are lowered by anything that concentrates stress—sharp corners, cracks, tool marks, and rough-ground surfaces.
5. • The development of high-strength, low-alloy (HSLA) steels has nearly doubled design strength for structural steels. These microalloyed steels obtain their strengths by several mechanisms. Some are extremely fine-grained as a result of very finely dispersed carbides; some are dual-phase with ferrite and martensite; some combine strain hardening with other strengthening mechanisms. In any event, weight saving and increased materials savings are possible.

6. Composites utilize the merits of two materials to obtain properties that are not available from either component alone. To be effective, the reinforcing component of a composite not only must be strong but must possess an elastic modulus that is greater than that of the matrix component. • The bond between the two must withstand high shear stresses wherever there are fiber ends.

• 7. Abrasives are also refractory (high-melting) as a result of their strong bonding. Requirements differ for grinding high-strength ductile materials and hard, brittle materials; the former require a tough abrasive, the latter a friable one.

TECHNICAL TERMS

Abrasive Hard, mechanically resistant material used for grinding or cutting; commonly made of a ceramic material.

Aspect ratio Length/diameter ratio of reinforcing fibers in a composite.

Composite Material containing two or more distinct materials.

Compound *See* Technical Terms in Units 5 or 11.

Dispersion hardening Strengthening by the presence of many small particles within a tough matrix, for example, precipitation hardening and tempered martensite.

Ductility-transition temperature (T_{dt}) Temperature that separates the regime of brittle fracture from the higher temperature range of ductile fracture.

Endurance limit The maximum stress allowable for unlimited cycling.

Fatigue Fracture arising from cyclic stressing in service.

Fracture, ductile Fracture accompanied by plastic deformation and, therefore, by energy absorption.

Fracture, nonductile Fracture, with negligible plastic deformation and a minimum of energy adsorption. Brittle fracture.

Grain refinement Treatment to reduce (and retain) fine-grain sizes. It involves the nucleation and restricted growth of new crystals.

• **High-strength, low-alloy (HSLA) steels** Weldable structural steels with enhanced strengths.

Induced strength Strength achieved by introducing compressive stresses into the surface of a nonductile material.

• **Interfacial shear stresses** Stresses introduced between matrix and reinforcement in a composite material.

Knoop hardness number (KHN) Hardness index determined with a microindentor. (*See* Table 3-3.1.)

Mohs hardness number Hardness index determined by scratch testing.

Solution hardening *See* Technical Terms in Unit 5.

Strain hardening *See* Technical Terms in Unit 4.

• **Stress intensity factor (K_{Ic})** A constant (for a given steel) that relates the design limit for the nominal stress to the depth of a propagating crack (Eq. 14-2.1).

CHECKS

14A Hardening and __***__ mechanisms include __***__ hardening, __***__ hardening, and __***__ hardening, among others.

14B A plastically deformed metal is __***__, and less __***__ than a comparable annealed metal.

14C The introduction of __***__, which are linear imperfections, into a material by plastic deformation makes further deformation more difficult.

14D Annealing by recrystallization decreases the __***__ and increases the __***__ of a previously cold-worked material.

14E The hardness of a 50/50 solid solution is __***__ than the average hardness of the two components.

14F A __***__-hardened alloy can also be __***__-hardened.

14G A fine-grained material has a large __***__ area per unit volume that interferes with slip at __***__ temperatures, but facilitates deformation at __***__ temperatures.

14H A dispersion-hardened material can have both __***__ and __***__ because the hard particles that restrict slip are in a __***__ that absorbs energy before it fractures.

14I The strength of a dispersion-hardened material decreases when particles __***__, becoming __***__ and __***__.

14J A high-speed tool steel contains alloying elements that reduce the rate of __***__ of the hard carbide particles. These alloying elements are called __***__ because they form very stable carbides.

14K Induced stresses provide an important means of strengthening materials that are __***__. Commonly, induced strength is achieved by producing __***__ stresses in the surface; in reinforced concrete, the induced strength is achieved by introducing __***__ on the __***__ side of the beam.

14L In ductile fracture, __***__ accompanies the rupture. With ductile fracture, energy is required not only to form new surface but also to __***__ . . . .

14M It is necessary to standardize the geometry of test samples for toughness, since the required energy depends on the __***__ and __***__ of the sample.

14N The __***__ temperature separates fractures into a lower-temperature regime that is __***__ from a higher-temperature regime that is __***__.

14O For cryogenic applications, it is advantageous to use __***__ cubic metals because they lack a __***__.

14P Other things equal, a steel with a __***__-grain size will have a higher T_{dt}.

• 14Q A __***__ is a stress-raiser because it focuses stresses at points of small __***__. The stress-intensity factor is proportional to nominal stress and __***__.

14R Cyclic stresses can lead to __***__ if the stress level is above the __***__. The test for ultimate strength (Section 3–2) involves __***__ cycle(s).

• 14S Structural steels are generally not quenched and tempered because (1) they are too __***__, (2) quenched and tempered steels cannot be __***__, and (3) __***__ . . . .

• 14T HSLA steels refer to __***__ - __***__, __***__ - __***__ steels. A common procedure for making these steels is by grain __***__, in which very fine __***__ prevent the growth of the ferrite grains as they form from __***__.

• 14U A __***__-phase steel contains __***__ and __***__, which subsequently changes to __***__. The ferrite gives it a low yield strength that permits __***__; the tempered martensite gives it a high __***__.

• 14V Refer to 14U. __***__ hardening may be combined with that procedure to give greater strengths to the __***__ in the steel.

14W For a composite to show improved strength, the reinforcement must be stronger than the matrix. In addition, it must have a higher __***__.

• 14X Reinforcing rods (or fibers) usually do not add much to the compressive strength of a composite because rods with a long __***__ ratio easily __***__ in compression, unless a large cross section is involved.

14Y To a first approximation, the elastic modulus of an oriented composite is proportional to the __***__ of each component.

14Z The properties of __***__ composites are very directional, or __***__.

• 14AA The interface between the reinforcement and the matrix must be well bonded in order to withstand __***__ stresses.

• 14AB In __***__, the ratio of S_u/ρ serves as a basis for comparison; in __***__, a better basis is E/ρ.

• 14AC Hardness, __***__, and __***__ are factors to be considered for abrasive selection. Friable abrasives provide __***__ cutting edges; __***__ abrasives are better for high strength, ductile metals.

• 14AD Diamond is stable at higher __***__ than is graphite.

STUDY PROBLEMS

14-1.1. Making use of data in previous units, determine (a) how much zinc should be in a cold-worked brass to give a ductility of at least 20% elongation, and a hardness of a least 70 R_b? (b) How much should it be cold-worked? [Your answer should not exceed 36% Zn because of processing complications.]
Answer: (a) 65Cu–35Zn (b) $e_{cw} = 20\%$

14-1.2. An alloy of 95Al–5Cu was solution treated at 550°C, then cooled rapidly to 400°C, where it was held for 24 hours. During that time, it produced a microstructure with 10^6 particles of θ per mm^3. The θ particles ($CuAl_2$) are nearly spherical and have twice the density of the κ matrix. (a) Approximately how far apart are these particles? (b) What is the average particle dimension?

14-1.3. A microstructure has spherical particles of β, with an average dimension, $\overline{d}$, that is 10% of the average distance, $\overline{D}$, between the centers of the adjacent particles. (a) What is the volume percent of β? (b) What is the ratio of $\overline{d}/\overline{D}$ when there is 0.5 v/o β?

Answer: (a) 0.05 v/o (b) $\overline{d}/\overline{D} = 0.2$

14–2.1. A test specimen is expected to absorb half as much energy during fracture as did the specimen in I.P. 14–2.1, that is, 18.4 J. What angle of follow-through swing should be expected?

Answer: 104.5°

14–2.2. What would the follow-through angle have been in I.P. 14–2.1 if the initial angle of the pendulum had been 105° before testing the sample?

14–2.3. The value of K_{Ic} for a steel is 160 MPa·m$^{1/2}$ (146,000 psi·in.$^{1/2}$). What is the maximum tolerable crack when the steel carries a nominal stress of 760 MPa (110,000) psi?

Answer: 14 mm (or 0.55 in.)

14–2.4. The steel of S.P. 14–2.3, which has a 6-mm (0.24-in.) crack, can support what maximum nominal stress without the crack advancing?

14–2.5. Convert K_{Ic} = 60 MPa·m$^{1/2}$ to psi·in$^{1/2}$.

Answer: 55,000 psi·in.$^{1/2}$

14–2.6. A steel has a yield strength of 550 MPa (80,000 psi), and a K_{Ic} value of 40 MPa·m$^{1/2}$ (36,500 psi·in$^{1/2}$). What will be the limiting design stress if the minimum tolerable crack is 1.5 mm (0.06 in.) and no plastic deformation is permitted?

• 14–4.1. A "microalloyed" steel contains 0.20% carbon and 0.05% vanadium. (Assume the resulting VC particles have a density that is not significantly different from the density of ferrite.) (a) How many particles are there per cm³ if the average diameter is 10 nm? (b) What is a typical distance between particles?

Answer: (a) $\sim 10^{15}/\text{cm}^3$ (b) $\sim 10^{-5}$ cm (or 100 nm)

• 14–4.2. The steel in S.P. 14–4.1 is heated and the particles are allowed to coalesce until their average diameter is 55 nm. How many particles are there per cm³?

• 14–4.3. A "dual-phase" steel contains 0.15% carbon, is quenched from 750°C (1380°F), and is tempered. (The density of the tempered martensite is essentially the same as the proeutectoid ferrite.) (a) What fraction of the metal is ductile proeutectoid ferrite? (b) What is the carbon content of the tempered martensite?

Answer: (a) 0.8 (b) 0.65

• 14–5.1. Refer to I.P. 14–5.1 and use appendix data. What is the mean density of the composite?

Answer: 8.64 g/cm³

14–5.2. A steel-reinforced aluminum wire has the same dimension as the steel-reinforced copper wires in I.P. 14–5.1. What fraction of tension loads will be carried by the steel?

14–5.3. Glass fibers provide longitudinal reinforcement for a nylon. The fiber diameter is 20 μm and the volume fraction is 0.45. (a) What fraction of the load is carried by the glass? (b) What is the stress in glass when the average stress in the composite is 14 MPa (2000 psi)?

Answer: (a) 95% (b) s_g = 30 MPa (s_n = 1.2 MPa)

14–5.4. What is the thermal expansion coefficient of the composite in S.P. 14–5.3?

14–5.5. What is the thermal expansion coefficient of the composite in S.P. 14–5.2?

Answer 16.8×10^{-6}/°C

14–5.6. A glass-reinforced plastic rod (fishing pole) is made of 67 v/o borosilicate glass fibers in a polystyrene matrix. What is the modulus of elasticity of the composite?

14–5.7. A 2.5 mm steel wire coated with 0.5 mm of copper (total dia. = 3.5 mm) is loaded with 900 kg. (a) What is the strain? (b) How much strain would occur with a 3.5 mm copper-free steel wire? (c) A 3.5 mm copper wire (assuming the yield strength is not exceeded)?

Answer: (a) 0.006

14–5.8. The composite wire of S.P. 14–5.7 is stress-relieved at 400°C (750°F) and cooled rapidly to 10°C (50°F). (a) Which metal is in tension? (b) What is the stress?

• 14–5.9. What is the glass-nylon interface area (mm^2) per millimeter of the composite of S.P. 14–5.3 (dia. = 1 mm)?

Answer: 70 mm^2/mm

• 14–5.10. The composite of S.P. 14–5.9 develops an interfacial shear stress of 7 MPa at the end of a broken glass fiber. With the same loading, what shear stress develops in a composite with 15 μm fibers? (Other factors remain the same.)

Answer: 5.25 MPa

14–5.11. (a) What volume fraction of aligned glass fibers is required in polystyrene to give a composite the same specific stiffness, E/ρ, that birch has (Table 14–5.1)? (b) Comment on its practicality.

Answer: (b) The volume fraction is close to the limit of cylinder packing—0.9. Therefore, care is required to assure exact alignment.

CHECK OUTS

14A strengthening
strain
solution
precipitation (age)

14B stronger (harder)
ductile (tough)

14C dislocations

14D strength (hardness)
ductility

14E greater

14F solution (precipitation)
strain

14G grain-boundary
low
high

14H strength, toughness
matrix

14I coalesce
larger, fewer

14J growth (coalesence)
carbide-formers

14K brittle (nonductile)
compression
reinforcement
tension

14L deformation
plastically deform adjacent material

14M size, shape

14N ductility-transition
nonductile (brittle)
ductile (tough)

14O face-centered
ductility-transition temperature

14P coarse

14Q crack (flaw) (notch)
radii of curvature
$\sqrt{c}$

14R fatigue (fracture)
endurance limit
1/4

14S large
welded
costs become prohibitive

14T high-strength, low-alloy
refinement
particles
austenite

14U dual-
ferrite, martensite
tempered martensite
deformation
ultimate strength

14V strain
ferrite

14W elastic modulus

14X aspect
buckle

14Y volume fraction

14Z woven
anisotropic

14AA shear

14AB tension
compression

14AC toughness, friability
sharp
tough

14AD pressures

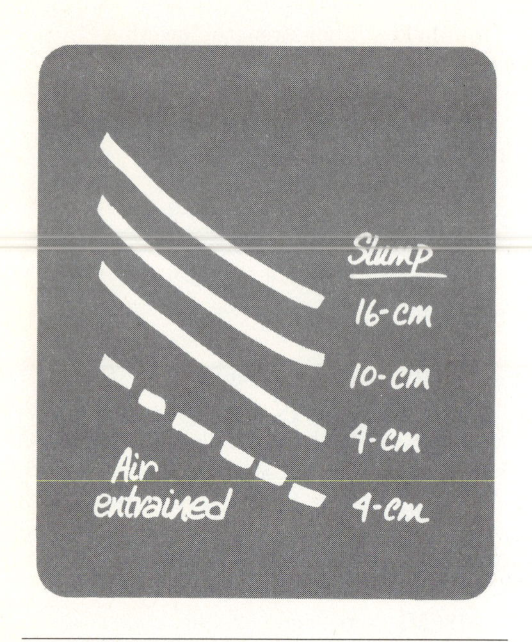

UNIT FIFTEEN

• CONCRETE AND RELATED MATERIALS OF CONSTRUCTION

These familiar materials are very widely used. In fact, their total tonnage of products exceeds that of all other solid engineering materials combined (excluding wood). Moreover, their production and use requires significant technology if maximum design potentials are to be achieved. Concrete is a monolithic product of aggregates bonded with a cement, such as portland cement or asphalt. In this context, brick may be viewed as preshaped aggregate. The technology of concrete differs from that of most of the other materials we have examined, since a significant step of its production is on the job site. This places an added responsibility on the engineer.

Contents

Prerequisites: Unit 2; Sections 11-8, 14-2, and 14-5.

From Unit 15 the engineer should

1) Be familiar with the major types of aggregates, and the measures of their size distributions.
2) Distinguish among the several methods for calculating density and porosity.
3) Know that the hardening process for cement and concrete is one of hydration, rather than one of drying.

4) Understand the requirements of a good concrete mix, and be able to calculate the volume of a unit mix.
5) Become acquainted with related types of construction materials.
6) Know the technical terms in this topic area.

15-1 AGGREGATES

Crushed rock, gravel, sand, and other aggregates are widely available and inexpensive when compared with most other types of materials of construction. It is to be expected that they are widely used. *Aggregates* are defined as granular materials that can be bonded into a monolithic product such as concrete*; however, large volumes of aggregate are also used without bonding for roads, fills, etc. Specifications for construction aggregates involve durability, size distribution, and other properties.

Types of aggregates

Aggregates may be either natural or processed (Table 15-1.1). In practice, the distinction between sand and gravel for portland cement concrete is placed at approximately a No. 4 sieve, which has a mesh opening of nearly 5 mm (Table 15-1.2). A somewhat finer sand is used for asphalt concrete.

Excluding *sizing*, which is required for all aggregate selection, the most common processing step is that of *crushing*. Crushed aggregates possess angular particles in contrast to rounded particles for most natural aggregates. Angular material is generally required as the coarse aggregate for asphalt concrete (Section 15-4).

Lightweight aggregates may be porous. Although volcanic cinders (scoria) and pumice are available in a few localities, most light-weight structural aggregates are manufactured from low-melting clays or shales. Gas that is evolved when the clays are heated form bubbles that expand the vitrified clay and lower the apparent density.

Quality of aggregates

All natural aggregates have been durable within their original environments. They have been there for centuries. Furthermore, most aggregates are stronger than the cement that bonds them into a monolithic product. Therefore, we may conclude that they generally have sufficient durability.

---

* *Concrete* is the bonded product; the term *cement* refers to the bonding material. Unless specifically stated otherwise, our reference to *concrete* will be aggregate bonded with portland cement (Section 15-2). An *asphalt concrete* is aggregate bonded with asphalt (Section 15-4).

Table 15-1.1
Types of aggregates

Type	Requirement	Uses
Natural		
Gravel	>No. 4 sieve*	Portland cement concrete
Sand	<No. 4 sieve*	Concrete (portland cement and asphalt), mortar
Heavyweight aggregates	Dense rocks, for example, barite, magnetite, etc.	Radiation shielding
Processed (crushed)		
Crushed rock	>No. 4 sieve*	Concrete (portland cement and asphalt)
Sand (fine)	<No. 4 sieve*	Concrete, mortar
Processed (lightweight)		
Expanded shale (heated to vitrify and evolve gases)	<15 mm	Lightweight structural concrete
Vermiculite (expanded mica)	<12 mm	Insulating concrete (low strength)

* See Table 15-1.2. The No. 4 seive (~5 mm) is considered the dividing point between sand and gravel for portland cement concrete. A finer sand (<No. 8) is used for asphalt concrete.

Table 15-1.2
Sieve series used in aggregate analyses (openings)*

ASTM Series		
1.5	38 mm	1.50 in.
1	25.4	1.00
$\frac{3}{4}$	19.1	0.75
$\frac{1}{2}$	12.7	0.50
$\frac{3}{8}$	9.5	0.375
No. 4	4.75	0.187
8	2.38	0.0937
16	1.19	0.0469
30	0.59	0.0232
50	0.297	0.0117
100	0.149	0.0059
200	0.074	0.0029

* Another sieve series (Tyler) differs slightly from the ASTM Series. It has half steps ($\sqrt{2}$) between the principal sizes and extends to smaller openings. It is more widely used in ceramics and powder metallurgy.

There are exceptions, however. Stratified and porous aggregates, such as slate, some sandstones, and gneisses, are vulnerable to freezing and thawing damage in a concrete; their original environment did not have a similar combination of water saturation and temperature fluctuation. The freezing of absorbed water in the pores of individual aggregate particles can lead to cracking from expansion pressures in the surfaces of concrete pavement. The chief test to anticipate freezing–thawing resistance is to measure the *water absorption* (ASTM-C127).

Abrasion wear on highways is not comparable to natural forces of deterioration. Therefore, tests have been developed to evaluate impact resistance of aggregates (ASTM-C131).

Aggregates must be free of organic materials and other surface coatings, because such material precludes the development of a good bond between the cement and the sand or rock surface. Again, there are standardized tests for the presence of organic impurities (ASTM-C40 and C87).

Size distributions

The engineer must have a measure of the size of the aggregate. For example, Table 15-1.1 uses the 4-sieve to differentiate between gravel and sand (or between crushed rock and "processed" sand). However, sizes vary, even within these two major categories, and there is need to know the extent of variation. Therefore, it is common to perform a *sieve analysis* on aggregates. Figure 15-1.1 shows the size distribution of two common types of sand—

Figure 15-1.1 Aggregate size distributions. (a) Narrowly graded; for example, beach sand. (b) Broadly graded; for example, river sand. (c) Cumulative sieve analyses.

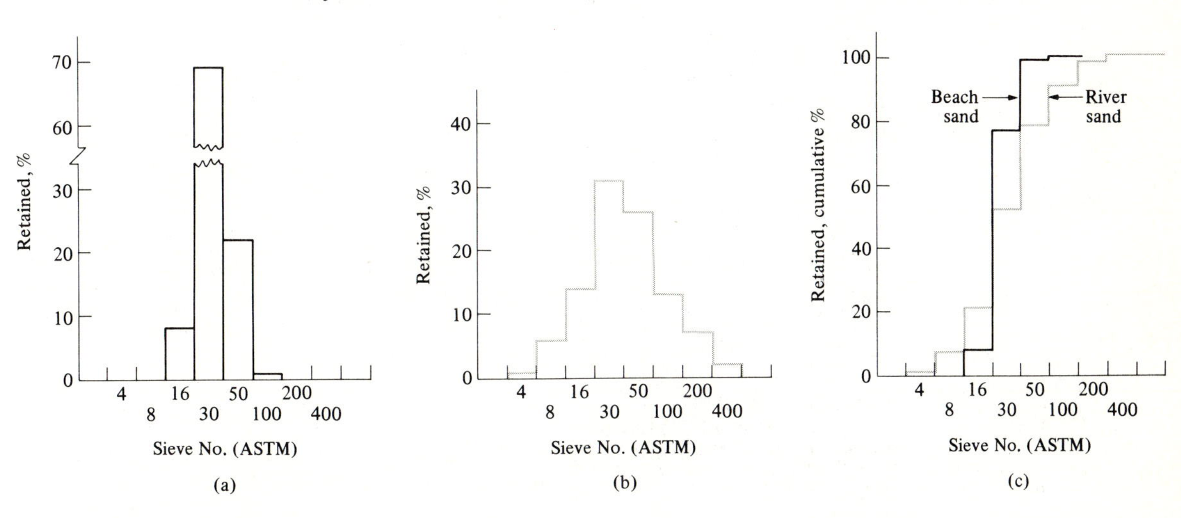

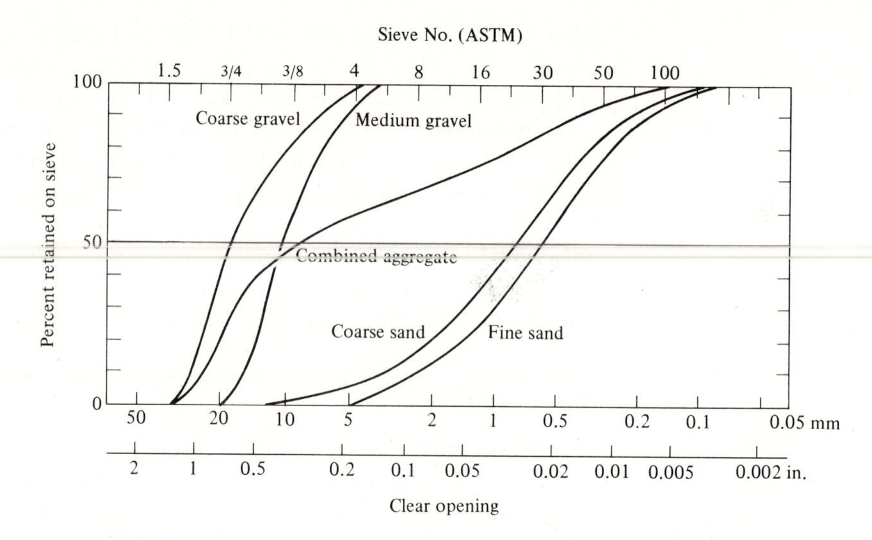

Figure 15-1.2 Analyses of typical aggregates used for concrete mixes. The combined aggregate is a mixture of coarse gravel (60%) and coarse sand (40%). The combination fills space more completely than gravel alone or sand alone. [Note that the ordinate is plotted as percent retained, which equals (100 − percent passed).] (Data adapted from Troxell, Davis, and Kelly, *Composition and Properties of Concrete*.)

beach and river. In this case, they both possess the same *mean size*—30 mesh, or approximately 0.6 mm (Table 15-1.2). However, the former is very *narrowly sized*. The latter is *broadly sized*. Figure 15-1.2 shows the size distributions of four typical aggregates, plus a 60/40 gravel/sand combination that could be used for the total aggregate in a typical concrete mix.

Packing factor

Ideally, for use in construction, an aggregate should possess a high *packing factor*. Higher strength and lower water permeability are realized. Also, less of the more expensive portland cement or asphalt is required for bonding into concrete.

Unlike the packing of atoms (I.P. 2-1.2), no two pieces of aggregate have the same size and shape. Therefore, we cannot calculate density directly from their sizes as we did for metals. However, experiments have shown that maximum densities are obtained for mixed aggregates when their size distributions follow the Weymouth curve:

$$p_i = (d_i/D)^{0.45}, \tag{15-1.1}$$

where p_i is the cumulative fraction *passing* any chosen sieve size with a clear opening of d_i. The maximum sieve opening in the aggregate mix is D.

Thus, if the maximum size just passes the 3/4 sieve (19.1 mm), 29% should pass the No. 16 sieve (1.19 mm), that is, $(1.19\text{mm}/19.1\text{ mm})^{0.45} = 0.29$. Likewise, 53% should pass the No. 4 (4.75 mm) sieve.

When plotted, Weymouth's curve (also called Fuller's maximum density curve) follows the dashed lines of Fig. 15-1.3 (starting at the top from the maximum size, *D*.) Neither the coarse sand nor the coarse gravel of Fig. 15-1.2 approximate these maximum density curves on Fig. 15-1.3. However, there is reasonable approximation to maximum density for the 60-40 mix of coarse gravel and coarse sand. This is why concretes always contain at least two distinctly different aggregates.

Aggregate porosity

Some of the porosity of sand and gravel aggregates is permeable to water, asphalt, etc. These *permeable* (open) pores are important since absorbed water can cause freezing expansion in cold climates, and lead to cracking. In addition, the permeable pores are commonly associated with capillaries along the rock planes that are weaker. In asphalt concrete, the permeable pores contain extra asphalt that does not lead directly to increased bonding. In general, the open, permeable porosity should be kept to a minimum.

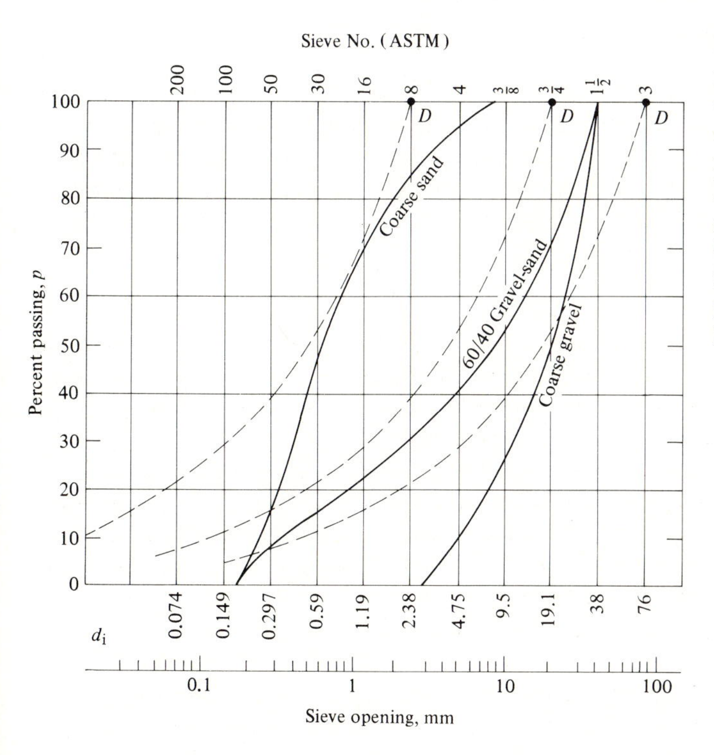

Figure 15-1.3 Size distributions for maximum packing (dashed lines). The coarse gravel and the coarse sand distributions (from Fig. 15-1.2) do not match the Weymouth curves (Eq. 15-1.2); however, the 60/40 gravel/sand mixture gives a close approximation for sizes above 0.3 mm. In effect, the sand fills the interstices among the gravel; thus less portland cement or asphalt is required for bonding.

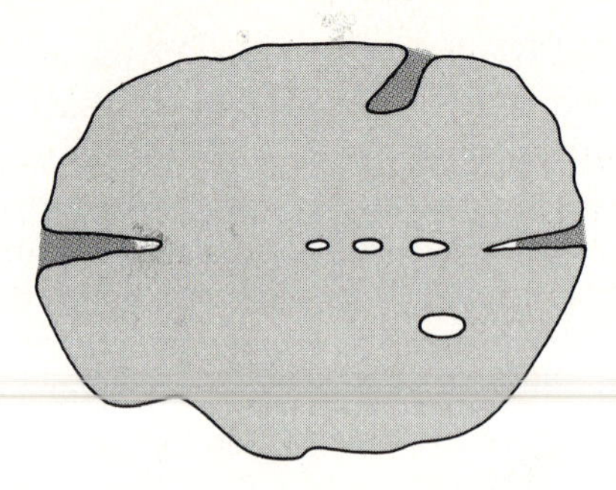

Figure 15-1.4 Aggregate porosity (schematic). Permeable pores are open to fluids. Impermeable porosity includes closed pores, and capillaries too small for fluid access. (Dark tint: permeable pores; unshaded: impermeable pores.)

Impermeable pores include isolated closed pores, and also capillaries that are too small to be filled, either because the air cannot escape or because asphalt does not wet the capillary surfaces (Fig. 15-1.4). This porosity generally is not as critical; in fact, aggregate for lightweight concrete is intentionally chosen to contain a significant volume fraction of closed pores.

The *total*, or *bulk volume*, V_b, of the aggregate pieces includes the *true volume*, V_t, the volume of the impermeable pores, V_{ip}, and the volume of the permeable pores, V_{pp}:

$$V_b = V_t + V_{ip} + V_{pp}. \tag{15-1.2}$$

The *apparent volume*, V_a, excludes the volume of the permeable pores:

$$V_a = V_b - V_{pp} = V_t + V_{ip}. \tag{15-1.3}$$

Densities may be *bulk, apparent,* or *true,* corresponding to the volume selected. Percent porosity is based on the bulk volume and may be either an *apparent porosity,*

$$P_a = V_{pp}/V_b, \tag{15-1.4}$$

or a *true porosity,*

$$P_t = (V_{ip} + V_{pp})/V_b. \tag{15-1.5}$$

Illustrative problem 15-1.1 A sample of crushed basalt weighed 482 g after being completely dried. The water level changed from 452 cm³ to 630 cm³ when it was added to a large graduated cylinder and remained for 24 hours. After the water was drained away and the excess water removed, the sample weighed 487 g. (a) What is the apparent volume? The total (bulk) volume of the crushed fragments? (b) The permeable porosity of the crushed rock? (c) The apparent density?

SOLUTION

a) Apparent volume $= 630\ \text{cm}^3 - 452\ \text{cm}^3 = 178\ \text{cm}^3$
Open (permeable) pores $= (487\ \text{g} - 482\ \text{g})/(1\ \text{g/cm}^3) = 5\ \text{cm}^3$
Total (bulk) volume of aggregate $= 183\ \text{cm}^3$.

b) Open (permeable) porosity = 5 cm^3/183 cm^3 = 0.027, (or 2.7 v/o).
c) Apparent density = 482 g/178 cm^3 = 2.71 g/cm^3.

ADDED INFORMATION In this problem statement, the open-pore volume is found within the crushed fragments and does not include the interparticle space. ◀

15-2 CEMENTS

Cements of various types have been used for centuries. These range from casein glues to modern polymeric adhesives, and from calcined lime mortars to portland cement that is made by the millions of tons per year. In this section, we shall consider only the latter, which is used (a) to bond aggregate into concrete, and (b) as the bond in mortar for use between bricks. *Portland cement* was given that name by its inventor, Joseph Aspdin, because the hardened product resembled a widely used limestone quarried on the Isle of Portland that joins the coast of England.

Portland cement

The most common commercial cement, portland cement, contains compounds such as Ca_2SiO_4, Ca_3SiO_5, and $Ca_3Al_2O_6$. First let us consider the latter: tricalcium aluminate. It hydrates at normal temperatures in the presence of water:

$$Ca_3Al_2O_6 + 6H_2O \rightarrow Ca_3Al_2(OH)_{12}. \tag{15-2.1}$$

This *hydration* reaction is reversible at high temperatures and will dissociate, returning from $Ca_3Al_2(OH)_{12}$ to $Ca_3Al_2O_6$, much as clay dissociates on heating (Eq. 12-1.1). However, that need not concern us at the moment. The hydration of Eq. (15-2.1) is a surface reaction for the cement particles; therefore the reaction occurs more rapidly when the $Ca_3Al_2O_6$ is finely ground. The reaction product adheres tightly to the surface of various types of rock. This enables it to serve as a cement between two adjacent surfaces of sand, gravel, or rock.

The more common portland cement phases are calcium silicates, Ca_2SiO_4 and Ca_3SiO_5. In a greatly simplified form, the hydration reaction of Ca_2SiO_4 may be shown as

$$Ca_2SiO_4 + xH_2O \rightarrow Ca_2SiO_4 \cdot xH_2O. \tag{15-2.2}$$*

---

* This reaction is more correctly stated as

$$2Ca_2SiO_4 + (5 - y + x)H_2O \rightarrow Ca_2[SiO_2(OH)_2]_2 \cdot (CaO)_{y-1} \cdot xH_2O + (3 - y)Ca(OH)_2, \tag{15-2.3}$$

where x varies with the availability of water and y is approximately 2.3. Equations (15-2.2) and (15-2.4) will suffice for our purposes in this book. However, a civil engineer should be aware of Eq. (15-2.3) because it is the basis of the changes in volume of concrete, which vary with humidity.

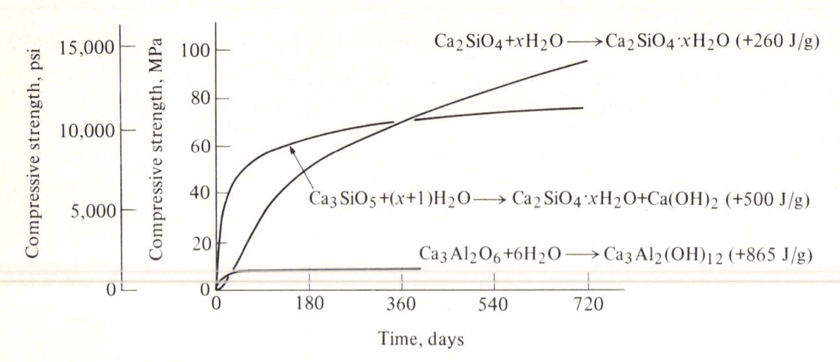

Figure 15-2.1 Cement hydration and strength (portland cement components). For strength to develop, water must be present for hydration. Concrete does *not* set by drying. Heat is given off during hydration as indicated. Its removal may require special provisions in massive structures, such as dams.

The product is almost completely amorphous and therefore provides a stronger bond with the surface of the aggregate than the tricalcium aluminate discussed above. The hydration of dicalcium silicate is slower than the hydration of the tricalcium aluminate of Eq. (15-2.1). See Fig. 15-2.1 for the time required to approach its maximum strength.

Finally, when tricalcium silicate hydrates, it releases free lime, $Ca(OH)_2$:

$$Ca_3SiO_5 + (x + 1)H_2O \rightarrow Ca_2SiO_4{\cdot}xH_2O + Ca(OH)_2. \qquad (15\text{-}2.4)$$

This occurs relatively rapidly, giving a rapid initial *set*, but the eventual maximum strength of the concrete is less than the product of Eq. (15-2.2).

Each of these reactions emphasizes that cement does not harden by drying, but by the chemical reaction of hydration. In fact it is necessary to keep concrete moist to ensure proper setting. The hydration reactions described above release heat, as shown in Fig. 15-2.1. Engineers take advantage of this in cold climates in which it may be necessary to pour concrete at temperatures that are slightly below the freezing temperature of water. Conversely, the *heat of hydration* presents a problem when concrete is poured into massive structures such as dams. In these extreme cases, special cooling is required, and a refrigerant is passed through embedded pipes that are left in place and later act as reinforcement.

The variables we have discussed suggest that the engineer may utilize various kinds of cement to obtain different characteristics. For example, a cement with a larger fraction of tricalcium silicate (Ca_3SiO_5) sets rapidly and thus gains strength early.* In contrast, in cases in which the heat of reaction might be a complication, and more setting time is available, a Ca_2SiO_4-rich cement is used. The American Society for Testing and Materials (ASTM) lists the following types of cement: (1) Type I, used in general concrete construction in which special properties are not required; (2) Type

* A *high, early-strength cement* can also be achieved by grinding the cement more finely, as discussed in connection with Eq. (15-2.1).

III, used when one wants very strong, early-strength concrete; (3) Type IV, used when a low heat of hydration is required. There are also two other types (II and V) for special applications in which sulfate attack is a design consideration.

The manufacture of portland cement

As indicated earlier, portland cement contains Ca_2SiO_4, Ca_3SiO_5, and $Ca_3Al_2O_6$, as the principal compounds and may be obtained by processing naturally occurring rocks. More commonly, the cement manufacturer combines two different materials, limestone ($CaCO_3$) and clay [$Al_2Si_2O_5(OH)_4$] in appropriate proportions. These are heated with the initial reactions of

$$CaCO_3 \rightarrow CaO + CO_2 \uparrow, \tag{15-2.5}$$

and

$$Al_2Si_2O_5(OH)_4 \rightarrow Al_2O_3 + 2SiO_2 + 2H_2O \uparrow. \tag{15-2.6}$$

However, the heating is extended to higher temperatures (Fig. 15-2.2) so that the CaO, SiO_2, and Al_2O_3 react to form a clinker of the above compounds—Ca_2SiO_4, and Ca_3SiO_5, and $Ca_3Al_2O_6$. Invariably, there is some iron oxide present, which reacts with the CaO and Al_2O_3 to give $Ca_4Al_2Fe_2O_{10}$.

Figure 15-2.2 Cross section of a rotary kiln for clinkering portland cement. (After Norton, *Elements of Ceramics*, Reading, Mass.: Addison-Wesley.)

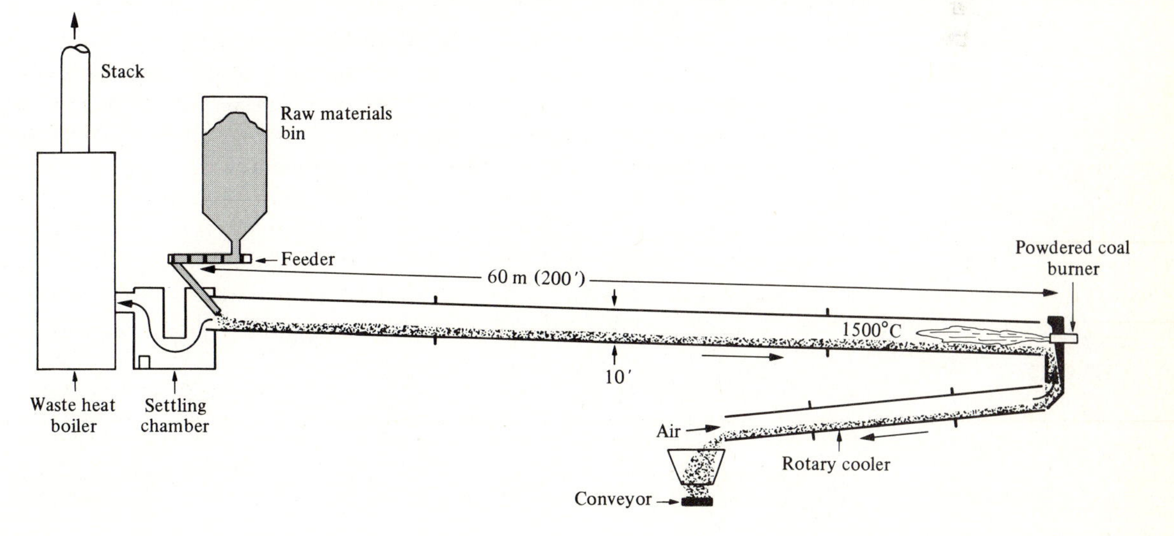

After clinkering, the product is ground very finely to make it react more rapidly when mixed with water for producing concrete. The size is normally reported as surface area per gram since the surface area strongly affects the reaction rates. Finer grinding is used for the high, early-strength cements. Typically, the hydraulic components of the two common types of portland cement are

	Type I General	Type III High, early-strength
Ca_2SiO_4	24	13
Ca_3SiO_5	50	60
$Ca_3Al_2O_6$	11	9
$Ca_4Al_2Fe_2O_{10}$	8	8
Surface area, cm^2/g	1800	2600

Illustrative problem 15-2.1 Two kilograms of finely ground cement are added to 12 kg of aggregate and 1 kg of water. What will be the temperature rise due to the hydration, if half of the heat is lost? (a) If the cement is solely Ca_3SiO_5? (b) Solely Ca_2SiO_4? (c) Solely $Ca_3Al_2O_6$? [The heat capacity of concrete is ~1 J/°C·g.]

SOLUTION Heats of hydration, ΔH_h are shown in Fig. 15-2.1.

$$0.5(2{,}000\text{ g})\,\Delta H_h = (15{,}000\text{ g})(\sim 1\text{ J/°C·g})(\Delta T)$$
$$\Delta T = \Delta H_h/(\sim 15\text{ J/°C·g}).$$

a) $\Delta T_{Ca_3SiO_5} = (500\text{ J/g})/(\sim 15\text{ J/°C·g}) = \sim 33°C.$

b) $\Delta T_{Ca_2SiO_4} = (260\text{ J/g})/(\sim 15\text{ J/°C·g}) = \sim 17°C.$

c) $\Delta T_{Ca_3Al_2O_6} = (865\text{ J/g})/(\sim 15\text{ J/°C·g}) = \sim 58°C.$

ADDED INFORMATION Since the Ca_2SiO_4 reaction is slow (Fig. 15-2.1), 50% or more of the heat can be readily lost during the setting time. However, the temperature of a concrete containing high, early-strength cement, which is finely ground and contains more Ca_3SiO_5, can rise significantly during the initial stage of setting. This becomes a design consideration in large structures. ◂

15-3 CONCRETE

Cement refers to the material used to bond concrete. *Concrete* is the final monolithic product of aggregate and cement.

On an overly simplified basis, portland cement concrete is a composite of gravel (or crushed rock) and an admixture of sand to fill the voids. The space remaining among the sand is filled with a "paste" of cement and water (Fig. 15-3.1). The cement hydrates when the concrete *sets*, or hardens.

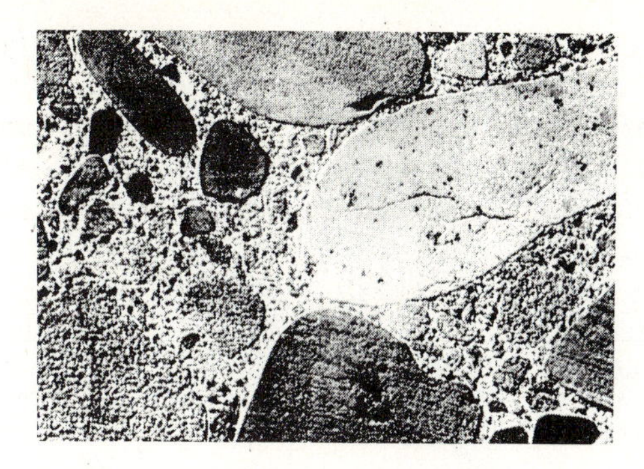

Figure 15-3.1 Concrete (×1.5). This is a composite with sand among the crushed rock or gravel. The pores among the sand are filled with a "paste" of cement and water. The water reacts with the cement to produce bonding. (Courtesy of the Portland Cement Association.)

Concrete mixes

Strength is generally the prime consideration in designing a concrete. This property is usually determined by the water-cement ratio (Fig. 15-3.2).

Workability is another important factor because the concrete must fill the forms and enter between steel reinforcement without introducing voids. This characteristic is dependent upon the size and amount of aggregate and is measured by a *slump* test (Fig. 15-3.3). A slump of 2 cm to 8 cm (1 in.-3 in.) is preferred for foundations, pavements, and similar massive structural units. Beams, columns, and other heavily reinforced structures may require greater *workability* (slumps to at least 12 cm, or 5 in.) in order to ensure complete placement of the concrete and the absence of voids.

The maximum *aggregate size* is also a function of the dimension of the final concrete product. For example, the coarse aggregate should not exceed one third the thickness of slabs, nor three fourths the spacing between reinforcing rods. Otherwise, it is possible to develop "bridging" and accompanying voids in the concrete.

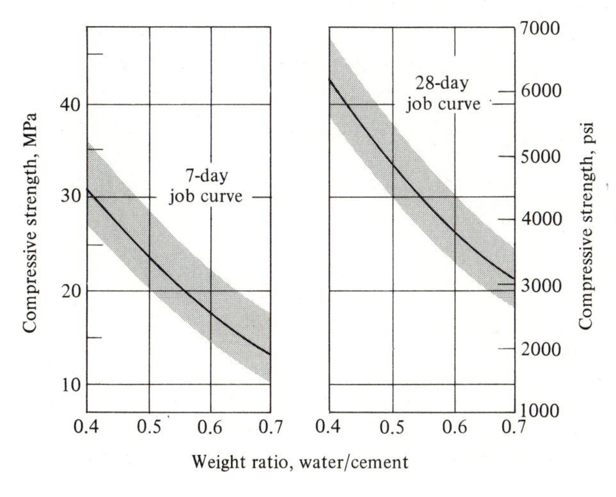

Figure 15-3.2 Strength of concrete versus water-cement ratio (Type 1 portland cement, moist-cured at 20°C (68°F)). There must be enough water to provide workability to the mix and to ensure that the forms are filled without voids. Additional water simply requires extra space and means less bonding. (After Cernica, *Fundamentals of Reinforced Concrete*, Reading, Mass.: Addison-Wesley.)

Figure 15-3.3 Slump test for workability. A concrete mix must be sufficiently "workable" to fill mold and eliminate voids in the corners of the forms or among reinforcing rods. Therefore, specifications are made on the amount of slump a wet mix should possess. However, excess water reduces the strength (Fig 15-3.2). When more water is added for workability, more cement is required to meet the required water-cement ratio. (Courtesy of the Portland Cement Association.)

The *water requirements* of a concrete mix vary with both the required slump and the size of the coarse aggregate. Figure 15-3.4 shows the general pattern in which (a) more water is required for greater slump, and (b) less water is required for coarse aggregate.

The *strength* of the concrete is controlled not only by the amount of cement in the mix but, more specifically, by the water-cement ratio (Fig. 15-3.2). Thus, if a concrete is to have greater workability (greater slump) for casting purposes, it must also have a higher cement content to maintain strength. Conversely, the excess water with a higher *water-cement ratio* decreases the strength. Specifically, *capillary voids* remain in the cement when the concrete is eventually dried of excess water. Of course, these weaken the concrete. As a result, it is preferred to have a water-cement ratio (w/c) of less than 0.55 by weight in almost all structures, and as low as

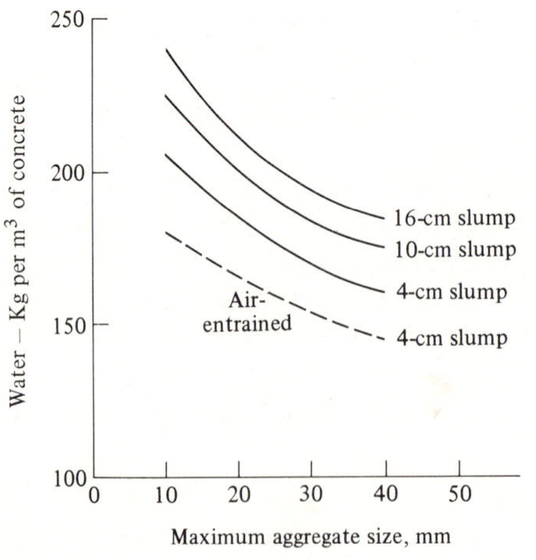

Figure 15-3.4 Water requirements for concrete (dashed line: air-entrained concrete). Less water is required with smaller slump requirements, with coarser aggregate, and with air entrainment. With more water added, more cement must be added to meet the water-cement ratios for strength (Fig 15-3.2).

0.45 (a) for thin sections (<10 cm) in continuously wet locations subject to freezing and thawing, and (b) in structures exposed to sea water or sulfates. Strengths (after 28 days) for these latter applications should exceed 35 MPa (5000 psi).

Air entrainment

A major technical development occurred when it was learned that the incorporation of air-entraining agents into concrete that produced small (~1 mm) bubbles increased the resistance of concrete to freezing and thawing damage. This addition has become standard practice where such exposure is expected. The amount of entrained air should be varied with the same factors that dictate aggregate size [~8 v/o for concretes that require 10-cm aggregate, to ~4 v/o for 50-cm (2-in.) aggregate]. Expectedly, the air entrainment reduces the strength (~1 MPa, or 150 psi per volume percent of entrained air); however, the dramatic (~10-fold) increase in winter durability proves to be a valuable trade-off. The rationale for the benefits of air entrainment is not simple, if indeed it is understood. A probable explanation lies with the "blunting effect" the air bubbles have on cracks as they progress during repeated freezing expansions. It is also assumed that the bubbles serve as an "expansion chamber" to receive water that is forced out of the adjacent concrete during freezing exposure. Whatever the explanation, the benefits of air-entrainment are real and have saved millions of dollars through the extended lives of highways and structures.*

Properties of concrete

The full *strength* of the cement in concrete is approached over a period of time depending on the composition and the water-cement ratio (Figs. 15–2.1 and 15–3.2).

The *permeability* of concrete to moisture varies with the amount of capillary porosity in the cement. Therefore, permeability is a function of the water-cement ratio used in the mix. For example, a mix with 4-cm (1.5-in.) aggregate has relative permeability values of 5 at a w/c ratio = 0.5, 40 at w/c = 0.6, and 150 at w/c = 0.7. This is another incentive to keep the water-cement ratio below 0.5 in foundations and subgrade structures.

The *density* and *porosity* of concrete, like those of aggregates (Section 15–1), have different values depending on the presence of open and closed (permeable and impermeable) pores. Equations (15–1.2 to 15–1.5) may be

* There is a second benefit from air entrainment in addition to the increased freezing and thawing resistance. With air entrainment, the aggregate content may be increased for a given slump requirement. Conversely, less cement is required per m^3, thus reducing costs.

used. However, in the concrete product, the pore space in the cement among the aggregate particles must be included as permeable pores.

Bulk density: $$\rho_b = \frac{m}{V_b} = \frac{m}{(V_t + V_{ip} + V_{pp})}. \quad (15\text{-}3.1)$$

Apparent density: $$\rho_a = \frac{m}{V_a} = \frac{m}{(V_t + V_{ip})} = \frac{m}{(V_b - V_{pp})}. \quad (15\text{-}3.2)$$

True density: $$\rho_t = \frac{m}{V_t} = \frac{m}{(V_b - V_{ip} - V_{pp})}. \quad (15\text{-}3.3)$$

As before, the subscripts are a(apparent), b(bulk), ip(impermeable pores), pp(permeable pores), and t(true).

The *thermal expansion* of concrete is $\sim 12 \times 10^{-6}$/°C, (or 7×10^{-6}/°F). Normally this is tolerable and, in fact, it is possible to pave a continuous slab of concrete for a highway without expansion joints. As with continuously welded railroad rails, the dimensional changes are carried as elastic strain and induced stresses. However, if other expansions occur, simultaneously, or if the stresses become misaligned because of subsoil movements, then pavement buckling can occur. Secondly, unless heat is removed from massive structures, the heat of hydration can raise the temperature of the concrete during setting and introduce cracking from thermal expansion.

Illustrative problem 15-3.1 A *unit mix* of concrete contains 0.15 m³ (5.3 ft³) of gravel, 0.12 m³ (4.2 ft³) of sand, and 100 kg (220 lb) of cement. Using the data given below, calculate the number of unit mixes required for a driveway that is 19 m long, 7.5 m wide, and 13 cm thick (62 ft × 25 ft × 5 in.), if you use 50 liters (13.2 gal) of water per mix.

	Unit weight*		Apparent density	
	Mg/m³*	(lbs/ft³)*	Mg/m³	(lbs/ft³)
Gravel	1.73	(108)	2.62	(163)
Sand	1.76	(110)	2.65	(165)
Cement	1.51	(94)	3.15	(196)
Water	1.00	(62.4)	1.00	(62.4)

* Mass per unit volume (m³, or ft³) including interparticle pores. This volume is the volume required for storing, shipping, and measuring materials before mixing.

SOLUTION Apparent volume per unit mix:

Gravel	(1.73 Mg/m³)(0.15 m³)/(2.62 Mg/m³)	= 0.099 m³
Sand	(1.76 Mg/m³)(0.12 m³)/(2.65 Mg/m³)	= 0.080
Cement	(100 kg)/(3150 kg/m³)	= 0.032
Water	(50 l)(1 kg/l)/(1000 kg/m³)	= 0.050
Volume of unit mix		= 0.261 m³

Unit mixes = (19 m × 7.5 m × 0.13 m)/(0.261 m³) = 71.

ADDED INFORMATION There is sand in excess of the pore space among the gravel ($0.15\ m^3$–$0.099\ m^3$). This is necessary because it is essentially impossible to distribute the sand exactly right for maximum packing. Likewise some excess water is necessary (beyond hydration requirements) to ensure workability.

If an air-entraining agent that introduces 5 v/o air bubbles had been added, each unit mix would be $0.275\ m^3$, and 68 unit mixes would be required. More importantly, (a) the concrete would resist freezing/thawing damage better, or (b) it would be possible to have the same strength with additional aggregate in the unit mix, thus requiring less of the more expensive cement. ◀

15-4 ASPHALT CONCRETE

In principle, the term concrete applies to all bonded aggregates including asphaltic, and not just to that concrete bonded by portland cement (Section 15-3). In this section we shall consider concrete bonded with asphalt, and call it *asphalt concrete.*

Asphalt

Asphalt is a product of oil production. Like coal tar, *asphalt* is a bitumen or residue of higher-melting organic compounds that remain after distillation removes gaseous and fluid fractions. It does not have a specific composition; rather it is a mixture of many molecular species in which hydrocarbons predominate. In effect, asphalts are naturally occurring linear polymers that become thermoplastic on heating (Section 9-4). Typically, the glass transition temperature is $\sim -25°C$ ($\sim -10°F$). Thus, asphalt tends to be brittle in the winter, or under impact loading, and behaves viscously in hot, summer exposures (Unit 8).

Modified asphalt

Since asphalt is a petroleum residue, its properties can be varied in the refinery process. This does not imply that the product and properties are uncontrolled; rather, the refinery operator has the option of the amounts of light fractions that are present. Also, the asphalt may be liquified by being remixed with volatile oils (cutbacks) or emulsified with water. The asphalt becomes hard and stable when these fugitive additives evaporate.

Rapid-curing (RC) asphalt. Gasoline, or naphtha, may be added to asphalt. When either or both of these are present, the asphalt is initially fluid but quickly stiffens after placement because the solvents readily volatilize at ambient temperatures to provide a rapidly curing product.

Medium-curing (MC) asphalt. Kerosene, which is less volatile and somewhat cheaper than gasoline, produces a medium-curing asphalt.

Slow-curing (SC) asphalt. Heavy oils are sometimes left with the asphalt residue on refining. These produce a slower curing asphalt cement because the oils have very slow evaporation rates.

Emulsified asphalt. Mechanical treatments can be used to produce a water-based emulsion. In general, these are applied as surface coatings that harden when the emulsion "breaks" and the water evaporates.

There is a trend away from the cutbacks, that is, the use of petroleum solvents, because (1) their evaporation adds undesirable hydrocarbons to the atmosphere, and (2) it takes less petroleum product to liquify the asphalt by heating it than by using a solvent.

Asphalt-bonded aggregates

Asphalt concretes are widely used to pave roads (Fig. 15–4.1), parking lots, driveways, and airports. In hot mixes, the asphalt is heated until it becomes fluid and mixed with heated aggregates at a "hot plant." It is delivered, spread, and compacted before it cools and becomes stiff. The technical variables of the product include (a) stability, (b) durability, and (c) skid resistance.

The *stability* of an asphalt concrete requires that it will not flow under load. This is achieved by quantity and shape of the crushed aggregate. The maximum amount of *crushed* rock should be present, with grading that provides smaller sized particles in the spaces among the coarser particles (Fig. 15–1.3). Excess asphalt permits the aggregate to "float" and shift in the asphalt. Of course, this is undesirable. Figure 15–4.2 shows a schematic sketch of (a) an idealized asphalt concrete, (b) an unsatisfactory asphalt concrete with rounded aggregate, and (c) an unsatisfactory concrete with too much asphalt. In the first example with crushed rock, there are numerous points of contact that give "aggregate interlock" to promote mechanical stability. In both (b) and (c), the concrete can gradually flow as it is repeatedly loaded with vehicle acceleration, or with continuous traffic.

Asphalts can "age" by hardening and becoming embrittled. This can lead to cracking and to loss of cohesion in an asphalt concrete. Since this loss of *durability* is associated with evaporation and with oxidation of the asphalt, a thin-film oven test (TFOT) is commonly required (ASTM-D-1754). A thin layer (3.2 mm, ⅛ in.) of asphalt is heated in air for 5 hours at 163°C (325°F) to determine changes in hardness, ductility, viscosity, weight, etc. Pavements with embrittled asphalts are more sensitive to freezing/thawing damage and to the stripping of the asphalt from the aggregate (loss of cohesion).

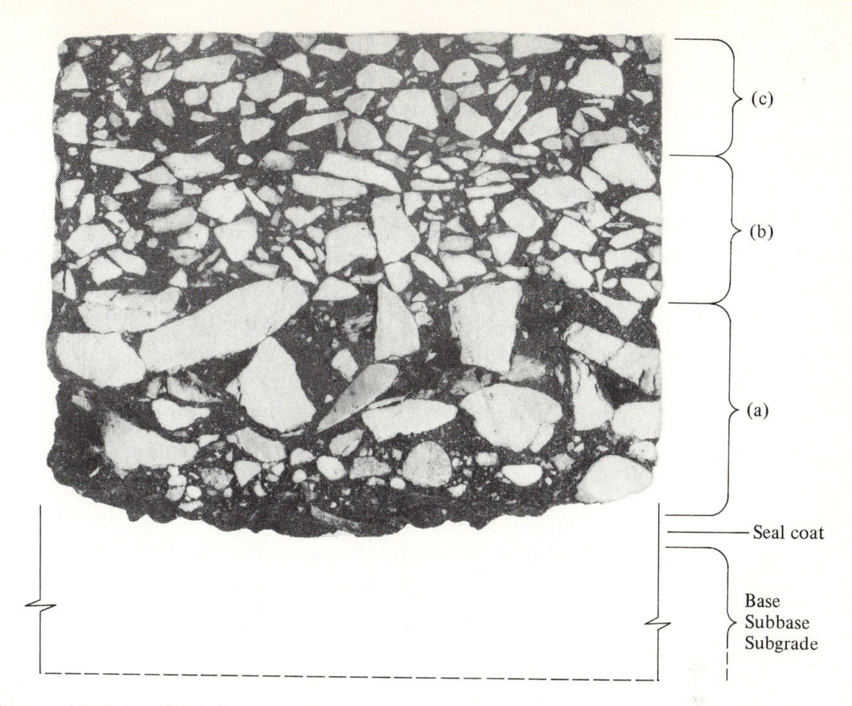

Figure 15-4.1 Core from asphalt pavement (×0.5). Like portland cement concrete, this product is agglomerated; but viscous asphalt, rather than a hydrated silicate, serves as the bond. (a) Bonding course. Sand (30 w/o) and asphalt (5 w/o) fill the interstices among the coarse stone (65 w/o) to give a rigid support above the subgrade aggregate. (b) Leveling course. The stone is smaller; otherwise the composition is similar to the preceding course. (c) Wearing course. More asphalt (6 w/o), more sand (55 w/o), and 4–5 w/o mineral filler (fly-ash) provide an impermeable, tough surface to resist traffic wear. (Core from Michigan Highway Testing Laboratory.)

Figure 15-4.2 Asphalt concrete (schematic). (a) Ideal, with contact of crushed-rock aggregate. (b) Unsatisfactory, since rounded aggregate reduces stability. (c) Unsatisfactory, excess asphalt.

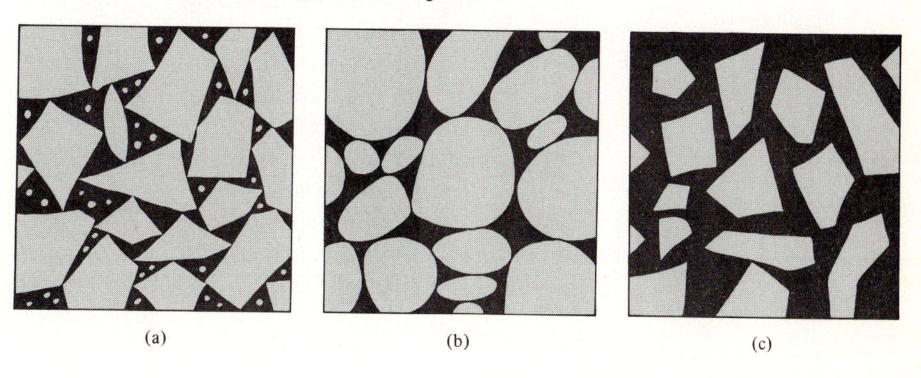

Skid resistance is a quality that is obviously related to asphalt concretes that are primarily used for pavements. Excess asphalt cement permits the aggregate to be forced below the surface of the pavement, thus producing an undesirable surface solely of asphalt. Excess surface asphalt must be avoided by mix control. If encountered, its effect may be partially combated by the addition of aggregate chips to the surface at a later date. The skid resistance of asphalt concrete can also decrease if the aggregate becomes polished from road traffic.

15-5 STRUCTURAL CERAMICS

Brick, like concrete, finds wide use in structural applications. Bricks are the oldest structural ceramic products, first being used as sun-dried, adobe blocks in prehistoric times. We even have early record of their improvement in ancient Egyptian days through the addition of straw as a reinforcing component.*

Present-day brick are vastly superior to these original products and are made at the rates of hundreds per minute by an extrusion auger. They are fired to introduce a glassy or vitrified bond into the product. The firing temperatures and times are such that melting starts but is incomplete. During subsequent cooling, the glass becomes a rigid (but still noncrystalline) solid that gives coherency to the product. In addition, the glass formation reduces porosity and permeability, which provides the product with better resistance to environmentally caused deterioration.

The relationships between properties and intended use can be illustrated by the *ASTM specifications* for ordinary building brick (Table 15-5.1). Three grades are indicated, based on the severity of the winters to be encountered or, more specifically, on the exposure to freezing and thawing cycles. One requirement is *strength*. Obviously a stronger brick will better withstand the stresses that accompany the freezing expansion of absorbed water. The second requirement is for limits on porosity and, therefore, on *water absorption*. This is to limit the origin of frost damage.

Since stronger, less porous bricks require more firing during production, more fuel is used, and the cost is greater. Thus, it would be unwise to specify Grade I quality if the brick are not to be exposed to severe conditions. Conversely, the buyer is willing to pay more if properties are obtained that will preclude early degradation.

The specifications of Table 15-5.1 take into account one other factor: Properties will vary from specimen to specimen. Thus, it is common practice to set specifications on an *average value*. However, care should be taken to assure that certain minima are met because even a few inferior units generally are not compensated for by a good average. Therefore, limits are also set.

---

* Exodus 5:7

Table 15-5.1
Physical specifications of building brick (ASTM C 62)

Grade	Minimum compression strength		Maximum H_2O absorption	
	Average of 5 bricks, MPa (psi)	Of one brick, MPa (psi)	Average of 5 bricks, %	Of one brick, %
I*	20.7 (3000)	17.2 (2500)	17	20
II†	17.2 (2500)	15.2 (2200)	22	25
III‡	10.3 (1500)	8.6 (1250)	No limit	No limit

* I: Brick intended for use where a high degree of resistance to frost action is desired, and the exposure is such that the brick may be cracked when permeated with water that can freeze.

† II: Brick intended for use where exposed to temperatures below freezing, but unlikely to be permeated with water, or where a moderate and somewhat uniform degree of resistance to frost action is needed.

‡ III: Brick intended for use as backup or interior masonry.

Review and Study

SUMMARY

Construction for contemporary society requires massive amounts of materials. The prime requirements are strength, durability, and "on-the-site" fabrication. Thus, use is made of bonded aggregate and of brick (which are preshaped aggregate).

1. Aggregate is most commonly natural sand and rock, either crushed or uncrushed. In practice, sand is smaller than a No. 4 seive opening (4.75 mm, or 0.187 in.). Aggregates may be graded by the mean-sieve size but more commonly are analyzed for their size distribution. Narrowly sized sand, for example, beach sand, is retained by two or three sequential sieves. Broadly sized sand, for example, river sand, has a range of size across six or more sieves, each differing by a factor of two. Prime aggregate has low water absorption (to minimize freezing/thawing damage) and minimal organic or clay contents (to assure good surface bonding).
2. Portland cement is the principal cement of construction to bond aggregates. It hardens by hydration. The rates of reaction and the heat released by the chemical reaction vary sufficiently with the composition and the powder size of the cement so that cements may be tailored for high early-strength (Type III), for massive structures (Type IV), and for general purpose use (Type I).
3. Concrete is composed of aggregate and a hydraulic cement. It is formulated to have the desired slump for workability and the minimum water-cement ratio for maximum strength. The entrainment of small (~1 mm) air bubbles improves workability and freezing/thawing resistance. A high strength and a low water permeability are the principal property requirements. Concrete with a lightweight (porous) aggregate is used where design requirements restrict the weight and/or thermal conductivity.

4. Aggregates and concretes possess both permeable (open) pores and impermeable pores. We thus find it useful to refer to (a) the true volume, which ignores all pore volume, (b) the bulk (total) volume, which includes the true volume and all pore space, and (c) the apparent volume, which includes the impermeable pore space with the true volume but excludes the open-pore space. With these volumes, we can calculate the (a) true density, (b) bulk density, and (c) apparent density; as well as (a) the true (total) porosity, and (b) the apparent (permeable) porosity.
5. Asphaltic concretes utilize bitumen residues of oil production to bond aggregates. Asphalt may be liquified by adding volatile oils as solvents. The curing time is determined by the evaporation time of the additive. "Hot-mix" asphalt concrete is achieved by heating both the asphalt and aggregate. Stability, durability, and skid resistance are the principal requirements for road construction.
6. Structural clay products develop their strength through a glassy bond. Requirements to avoid fracture during freezing and thawing include strength and low water absorptions.

TECHNICAL TERMS

Aggregate Coarse particulate materials such as sand, gravel, and crushed rock used for construction purposes.

Air entrainment (concrete) The incorporation of small ($\sim$1 mm) bubbles of air into concrete to give increased resistance to freezing/thawing damage (and improved workability). The bubbles are formed by the addition of small quantities of surface-active chemicals.

Asphalt Bitumen residues from oil production.

Cement Bonding material that adheres to the two adjacent solid surfaces.

Cement, portland A hydraulic, calcium silicate (and aluminate) cement widely used for structural concrete.

Concrete Agglomerate of aggregate and a hydraulic cement.

Density, apparent Mass divided by apparent volume (material + closed pores).

Density, bulk Mass divided by total volume.

Density, true Mass divided by true (pore-free) volume.

Grading Size distribution of aggregate.

Heat of hydration Energy released as heat during the hydration reaction, for example, of portland cement.

Hydration Chemical reaction consuming water: $Solid_1 + H_2O \rightarrow Solid_2$.

Permeability Coefficient of flow of a fluid through a porous solid.

Pores, impermeable Pores without access to surrounding fluids.

Pores, permeable (open) Pores accessible to fluids.

Porosity, apparent (Permeable-pore volume)/(total volume).

Porosity, total (true) (Total pore volume)/(total volume).

Sand (for concrete) Aggregate that passes a No. 4 sieve (4.75 mm, or 0.187 in.).

Sieve series A sequence of mesh sizes for grading aggregate. The ASTM series is used predominantly for structural aggregates.

Slump test Standardized test for assessing the workability of a concrete mix.

Unit mix Ratio of gravel, sand, portland cement (=1, or 100) and water in a concrete mix. May be based on either weight or volume.

Water absorption, % Test used to measure open porosity.

Water/cement ratio, *w/c* Weight ratio of the water to portland cement used in a unit mix of concrete.

Workability Flowability for a concrete mix to fill concrete forms without the presence of large voids. Measured by slump test.

CHECKS

15A Crushed aggregates possess __***__ particles in contrast to the __***__ particles of the most natural aggregates.

15B Typically, beach sands are more __***__ sized than river sands.

15C __***__ pores are open.

15D The __***__ volume includes the volume of the closed pores, but not of the permeable pores.

15E The value of the apparent density will be greater than of the __***__ density.

15F The hardening of portland cement is by __***__.

15G Type-__***__ cement is used for general concrete construction.

15H A major reason that high, early-strength cement sets faster is that it has been __***__ during processing.

15I The __***__ of a concrete is measured by a slump test.

15J If more water is added to increase the workability of a concrete, it is also necessary to add __***__.

15K The major advantage of air-entrained concrete over regular concrete is its resistance to __***__.

15L The strength of concrete gradually _*_creases as setting time extends from weeks to months, and even years.

15M A concrete with __***__ aggregate requires less water per m^3.

15N Asphalt is a bitumen of higher-melting compounds that are obtained from __***__ production.

15O Typically the glass temperature of asphalt is near __***__.

15P The "curing" of a modified asphalt is by __***__ of the admixed solvents.

15Q An asphalt is considered __***__ if it does not flow under applied loads.

15R For an asphalt to be considered durable, it must resist __***__ and loss of cohesion with the aggregate.

15S The coarse aggregate of asphalt pavement should be __***__ in shape.

15T Modern brick have a __***__, or vitrified, bond.

15U __***__ and __***__ are two factors that affect the resistance of brick to frost damage.

STUDY PROBLEMS

15-1.1. Since the minimum spacing between reinforcing rods in a concrete is 50 mm, the maximum aggregate should be no more than 38 mm (1.5 in.). If maximum aggregate packing is to be realized, what size sieve should pass (a) 30% of the aggregate? (b) 75% of the aggregate?

Answer: (a) 8-sieve (2.5-mm opening) (b) $\frac{3}{4}$-sieve (20-mm opening)

15-1.2. Crushed limestone consists of $CaCO_3$ (ρ = 2.71 g/cm³). A container (30-cm cube) can hold 51.0 kg of dry crushed limestone when level full; 8.0 kg of water can be added to the full container of crushed rock. (a) What is the true volume of the limestone? (b) What is the apparent volume of the crushed rock? (c) The apparent density? (d) What is the volume of the impermeable pores?

15-1.3. A brick weighs 2570 g. After being in water overnight, its suspended (in water) weight is 1520 g. The brick is taken from the water and any excess surface water is removed (but not removed from the permeable pores). The weight is 2593 g. (a) What is the volume of the permeable pores? (b) What is the apparent volume? (c) What is the bulk (total) volume? (d) What is the apparent porosity? (e) The bulk density?

Answer: (a) 23 cm³ (b) 1050 cm³ (c) 1073 cm³ (d) 2.1% (e) 2.4 g/cm³

15-2.1. Sixty cm³ of water and 100 g of $Ca_3Al_2O_6$ are used in a test block (with no aggregate). How much pore space will be present in the product after hydration and subsequent drying because of the excess water that was used?

Answer: 20 cm³ from excess water (40 g of water required for hydration)

15-2.2. A cement contains 30% Ca_2SiO_4, 55% Ca_3SiO_5, and 15% $Ca_3Al_2O_6$. What minimum water-cement ratio is required to hydrate the cement? [Assume x of Eqs. (15-2.2) and (15-2.4) is 5.]

15-2.3. Refer to S.P. 15-2.1. (a) How much heat will be evolved in the test block? (b) What temperature increase would develop if the heat could not escape? (c_p of product = ~1 J/g·°C; c_p of free H_2O = 4.2 J/g·°C.) (c) What other factors should be considered in determining the expected temperature increase in the concrete?

Answer: (a) 86,500 J (b) ΔT = ~400°C!!!

15-2.4. Type I portland cement is hydrated. How much heat is evolved per kilogram of cement? [The heat of hydration of $Ca_4Al_2Fe_2O_{10}$ is comparable to that of Ca_3SiO_5.]

15-3.1. A 4-in. × 8-in. × 16-in. "solid" concrete block weighs 41.1 lbs dry. It is saturated with water and weighs 43.0 lbs. (a) What is the bulk density? (b) The apparent density? (c) The apparent porosity? (ρ_{H_2O} = 62.4 lbs/ft³ = 1 g/cm³.)

Answer: (a) 139 lbs/ft³ (or 2.2 g/cm³) (b) 154 lbs/ft³ (or 2.5 g/cm³)

15-3.2. In order to determine the volume of an irregularly shaped concrete building block, three measurements were made: weight (dry) = 17.57 kg; weight (saturated with water) = 18.93 kg; weight (suspended in water) = 10.43 kg. (a) What is the bulk volume of the block? (b) The apparent volume of the block?

15–3.3. In a certain construction project, the builder uses a concrete mix composed of dry cement, sand, and gravel in volume proportions of 1:2:3.5, respectively, and an 0.53 *w*/*c* ratio (weight basis). The densities are those cited in I.P. 15–3.1. Air-entraining agents are added to give 4.5 v/o closed pore space. Calculate the bulk density of the wet concrete mix.

Answer: 2.31 Mg/m^3 (or 2.31 g/cm^3)

15–3.4. A unit mix of concrete comprises 42.7 kg (94 lb) of cement, 160 kg (350 lb) of gravel, 125 kg (275 lb) of sand, and 21 kg (46 lb) of water. Using the data in I.P. 15–3.1, calculate the number of unit mixes required for a driveway of $20 m^3$ (700 ft^3).

CHECK OUTS

15A	angular rounded
15B	narrowly
15C	permeable
15D	apparent
15E	bulk
15F	hydration
15G	I
15H	more finely ground
15I	workability
15J	more cement
15K	freezing/thawing damage
15L	increases
15M	coarse
15N	petroleum
15O	−25°C (−10°F)
15P	evaporation (volatilization)
15Q	stable
15R	cracking
15S	angular
15T	glassy
15U	porosity strength

UNIT SIXTEEN

MATERIALS IN HOSTILE ENVIRONMENTS

Engineering products cannot remain in their shipping cartons but must encounter adverse service conditions that can lead to deterioration and failure. The costs arising from corrosion and the related repairs and replacements are estimated to amount to 5% of the gross national product of any industrial nation ($50–$150 billion/yr in the United States alone). Efficient energy conversion requires high temperatures and materials that will not fail prematurely by oxidation or creep. To these are added low-temperature applications and radiation damage, with the result that one of today's most active technical areas involves the evaluation of service environments and their effects on materials.

Contents

Prerequisites Units 2, 3, 5, and 6; and Sections 8–4, 14–1, 14–2, and 14–3.

From Unit 16 the engineer should

1) Anticipate the effects of abnormal surroundings upon the structure and properties of common materials.
2) Know the mechanism of corrosion reactions, both anode and cathode.
3) Be acquainted with the several variants of the galvanic cells that produce corrosion.
4) Be familiar with the more direct means of corrosion control.
5) Identify the two general mechanisms of failure at high temperatures, and be aware of the materials that will better withstand those conditions.
6) Understand the simpler mechanisms of radiation damage and damage recovery.

16–1 INTRODUCTION

Hostile environments are those surroundings that bring about accelerated and/or unexpected failure. However, the engineer is supposed to take into account potential service conditions in product design. Therefore, the engineer should expect the unexpected and be prepared for the premature. It is necessary to be acquainted with various adverse service conditions and their effects on materials.

Adverse conditions may involve a chemical attack upon the material by the surroundings—either corrosion or oxidation. Extreme temperatures, both hot and cold, can affect the rates of normal responses to stresses and, therefore, affect properties. And energetic radiation can penetrate a material to induce radiation damage.

There are distinct ways material-property relationships are affected. With corrosion and oxidation, the reactions are at the surface of the material (Sections 16–2,3,4). High temperatures accelerate the atom movements deep within the material (Section 16–5) while cryogenic temperatures preclude normal atom responses to applied forces (Section 16–7). Finally, radiation damage occurs because the normal crystal and molecular structures are degraded atom by atom (Section 16–8).

16-2 CORROSION REACTIONS

Metallic *corrosion* is the electrochemical change of a metal to its positive ions. For example,

$$Zn \rightarrow Zn^{2+} + 2e^-, \tag{16-2.1}$$

or more generally

$$M \rightarrow M^{n+} + ne^-. \tag{16-2.2}$$

The ionic product no longer has its former metallic characteristics that we encountered in Units 2 through 6. For corrosion to proceed, it is necessary to have a medium into which the ions are accumulated. This is the *electrolyte.* Commonly, but by no means always, the electrolyte is an aqueous solution. Corrosion also requires companion reactions that will consume the electrons of Eqs. (16–2.1) and (16–2.2). Examples include

$$4e^- + O_2 + 2H_2O \rightarrow 4\,(OH)^-, \tag{16-2.3}$$

and

$$2e^- + Cu^{2+} \rightarrow Cu. \tag{16-2.4}$$

The chemist calls changes that release electrons *oxidation,* and changes that consume electrons *reduction.* The former occurs at *anodes* within an electrochemical system; reduction reactions occur at *cathodes.*

For metallic corrosion to proceed, there must be an *anode,* an *electrolyte,* and a *cathode*—a combination we call a *galvanic cell.* That is, there must be a metallic electrode that is oxidized and then releases electrons; there must be a solution (liquid or solid) into which the reaction product can enter, and from which the cathode reactants are supplied (the O_2 and H_2O of Eq. (16–2.3), or the Cu^{2+} of Eq. (16–2.4); finally, there must be another metallic electrode that supplies electrons to reduction reactions similar to Eqs. (16–2.3) and (16–2.4).

Neither anodic reactions nor cathodic reactions occur to any significant extent in isolation,simply because the material would build up a negative or positive potential, which would halt further oxidation or reduction, respectively. However, it will pay us to look at each of them independently because it will be less complicated. Furthermore, anodic and cathodic reactions are simply reversals of each other. Thus,

$$Cu \rightarrow Cu^{2+} + 2e^- \tag{16-2.5a}$$

is anodic, and

$$Cu \leftarrow Cu^{2+} + 2e^- \tag{16-2.5b}$$

is cathodic. As discussed before, the former involves corrosion. The latter has engineering significance, because in this case it is an *electroplating* reaction. Metallic corrosion is electroplating in reverse.

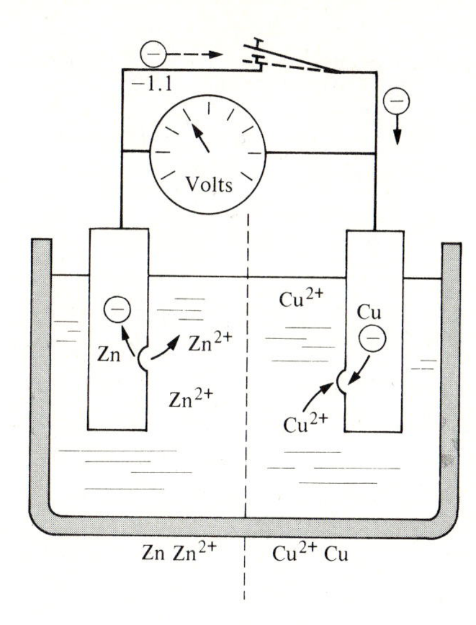

Figure 16–2.1 Galvanic cell (Zn–Cu). With the switch closed, zinc provides electrons through the external circuit to copper. A −1.1-volt potential difference develops when the circuit is opened. (Table 16–2.1, with molar solutions.)

Anode reactions

All metals can serve as anodes and lose electrons to become positive ions. However, all metals do not react equally. For example, the zinc of Eq. (16–2.1) releases electrons more readily than does the copper of Eq. (16–2.5a). This is shown in Fig. 16–2.1. Since the zinc oxidizes (loses electrons) more readily than copper, it serves as a negative electrode and registers a −1.1 volts with respect to copper.*

We can make similar comparisons between other pairs of electrode reactions. In doing so, we can develop a series of electrode potentials (Table 16–2.1). By historical convention, the hydrogen reaction is considered as the reference.† As measured in Fig. 16–2.1, the potential difference between the Zn and Cu anode reactions is 1.1 volts. The convention among electrochemists and corrosion engineers identifies the voltage difference as negative. However, the sign is a matter of choice as long as one is consistent. (See footnote to Table 16–2.1.)

Cathode reactions

Each of the reactions of Table 16–2.1 is a cathode reaction if the direction is *reversed*. In each case, this involves supplying electrons from another

* This is the voltage at 25°C when pure zinc and pure copper are in 1-molar solutions of their salts. Expectedly, the voltage will be different under other conditions at the anode, in the electrolyte, or at the cathode.

† This is an arbitrary, but logical reference, since H^+ ions are present in all aqueous solutions. Other ions are not as universally present.

Table 16-2.1
Electrode potentials (25°C; 1-molar solutions)

Anode half-cell reaction (the arrows are reversed for the cathode half-cell reaction)	Electrode potential used by electrochemists and corrosion engineers,* volts		Electrode potential used by physical chemists and thermodynamists,* volts
$Au \rightarrow Au^{3+} + 3\,e^-$	+1.50	↑ Cathodic (noble)	−1.50
$2\,H_2O \rightarrow O_2 + 4\,H^+ + 4\,e^-$	+1.23		−1.23
$Pt \rightarrow Pt^{4+} + 4\,e^-$	+1.20		−1.20
$Ag \rightarrow Ag^+ + e^-$	+0.80		−0.80
$Fe^{2+} \rightarrow Fe^{3+} + e^-$	+0.77		−0.77
$4(OH)^- \rightarrow O_2 + 2\,H_2O + 4\,e^-$	+0.40		−0.40
$Cu \rightarrow Cu^{2+} + 2\,e^-$	+0.34		−0.34
$H_2 \rightarrow 2\,H^+ + 2\,e^-$	0.000	Reference	0.000
$Pb \rightarrow Pb^{2+} + 2\,e^-$	−0.13	Anodic (active) ↓	+0.13
$Sn \rightarrow Sn^{2+} + 2\,e^-$	−0.14		+0.14
$Ni \rightarrow Ni^{2+} + 2\,e^-$	−0.25		+0.25
$Fe \rightarrow Fe^{2+} + 2\,e^-$	−0.44		+0.44
$Cr \rightarrow Cr^{2+} + 2\,e^-$	−0.74		+0.74
$Zn \rightarrow Zn^{2+} + 2\,e^-$	−0.76		+0.76
$Al \rightarrow Al^{3+} + 3\,e^-$	−1.66		+1.66
$Mg \rightarrow Mg^{2+} + 2\,e^-$	−2.36		+2.36
$Na \rightarrow Na^+ + e^-$	−2.71		+2.71
$K \rightarrow K^+ + e^-$	−2.92		+2.92
$Li \rightarrow Li^+ + e^-$	−2.96		+2.96

* The choice of signs is arbitrary. Since we are concerned with corrosion, we will use the middle column.

source. Of course, this does not guarantee that all of these reactions will occur at the cathode. Obviously, the combination of Ag^+ ions and electrons will not occur unless silver is present in the electrolyte. The common cathode reaction,

$$2H_2O + O_2 + 4e^- \rightarrow 4(OH)^-, \tag{16–2.3}$$

will not occur if no oxygen is present. Likewise, sodium will not plate out of sea water since a significant number of H^+ ions are also present. Specifically, the electrons will react preferentially with the hydrogen ions,

$$2H^+ + 2e^- \rightarrow H_2\uparrow, \tag{16–2.6}$$

since that reaction is more cathodic (higher in the electrode potential series shown in Table 16–2.1). Furthermore, concentration has an effect. For example, copper ions will plate out of a concentrated Cu^{2+} solution more readily than from a dilute Cu^{2+} solution.

This effect of molar concentration, C, upon the electrode potential, $\mathcal{E}$, can be readily calculated as

$$\mathcal{E} = \mathcal{E}_0 + (0.0257/n)\ln C. \tag{16–2.7}$$

The standard electrode potential (Table 16–2.1) is $\mathcal{E}_0$. The term n is the number of electrons in the reaction—2 for Cu^{2+}, Ni^{2+}, Fe^{2+}; 3 for Al^{3+}; one for Ag^+, Na^+, etc.

Concentration cells

From Eq. (16–2.7), we observe that the electrode potential $\mathcal{E}$ is $\mathcal{E}_0$ when the electrolyte concentration is 1-molar. More concentrated solutions make the electrode potential of Table 16–2.1 more cathodic; more dilute solutions make the reactions more anodic. This leads to a *concentration cell* as sketched in Fig. 16–2.2. If the concentration is 0.01-molar on the left side, and 2.2-molar on the right, the resulting voltage would be $(+0.34 - 0.06)$V versus $(0.34 + 0.01)$V, or -0.07 V between the two electrodes. When connected electrically, this induces corrosion on the metal in the dilute electrolyte and plating on the metal in the more concentrated solution.

Oxidation cells

In corrosion, metal atoms are ionized and enter the electrolyte. The released electrons can move toward the cathode where they are consumed. An oxygen-enriched, aqueous electrolyte will utilize these electrons according

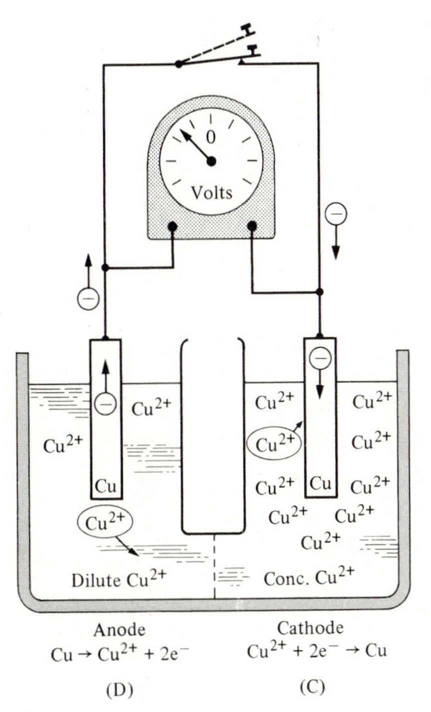

Figure 16–2.2 Concentration cell. When the electrolyte is not homogeneous, the less concentrated area becomes the anode.

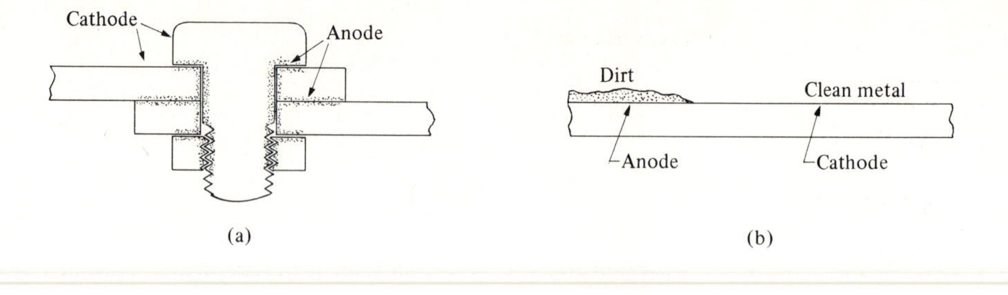

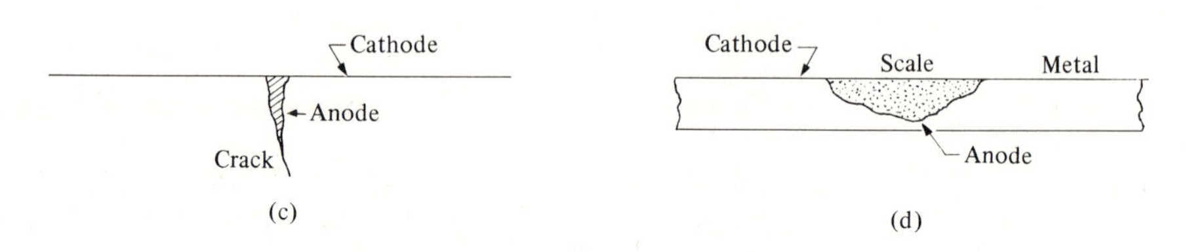

Figure 16-2.3 Oxidation cells. Inaccessible locations with low oxygen concentrations become anodic. This situation arises because the mobility of electrons and metal ions is greater than that of oxygen or oxygen ions.

to Eq. (16-2.3). Of course, if the electrolyte has no oxygen, then that particular cathode reaction cannot occur. In the absence of other cathode reactants, corrosion cannot proceed.*

From the above, we find that oxygen greatly accelerates corrosion; but the oxygen reacts at the cathode where corrosion does not occur. Corrosion occurs at the anode. Thus, *oxygen produces an oxidation cell that accentuates corrosion, but it produces corrosion where the oxygen concentration is lower.*

The above generalization is significant. Corrosion may be accelerated in inaccessible places such as in cracks and crevices and under accumulations of dirt and other surface contaminants (Fig. 16-2.3) because the oxygen-deficient areas serve as anodes. This may lead to pit corrosion because the anode digs deeper and deeper while the surrounding exposed area is cathodic and therefore not corroded.

Localized corrosion

Galvanic couples of Zn–Cu, Fe–Sn, or Fe–($H_2O + O_2$) that can be paired from Table 16-2.1 are obvious candidates for corrosion (of the anodic member of the pair). Corrosion may also occur on a single surface of a piece of metal. It

* Make-up water for steam-power plants is intentionally deaerated so as to preclude Eq. (16-2.3) as a cathode reaction. Without this reaction available, which consumes electrons, the corrosive anodic reactions are greatly inhibited.

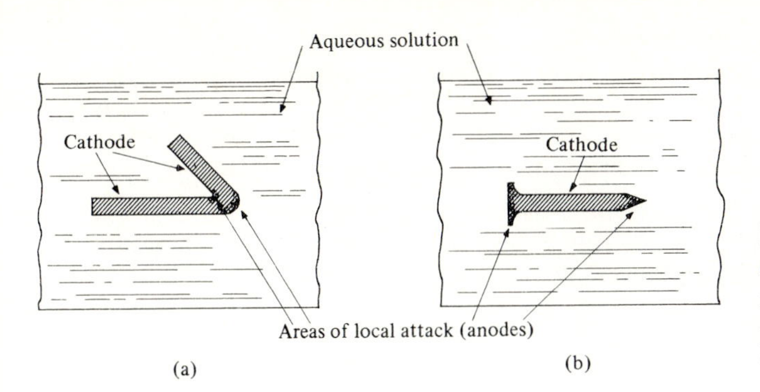

Figure 16–2.4 Stress cells. In these two examples of strain hardening, the anodes are in the cold-worked areas. The electrode potential of a strained metal is higher than that of an annealed metal.

occurs because the surface is not completely homogeneous. There are local anodes and cathodes on a microscale. For example, the carbide and the ferrite areas of the pearlite in Fig. 6–3.1 become anodes and cathodes, respectively. The electrode potential difference is small, but corrosion proceeds with time. The boundary between grains is invariably anodic to the interior of the grain. That is the reason we can observe the grain boundary through a microscope after the surface has been etched (corroded) in a laboratory acid (Fig. 2–7.2). Impurities and solidification segregation also introduce localized galvanic cells. Finally, plastically deformed regions of metals, such as the sheared point or the cold-deformed head of the typical wire nail, are anodic to less deformed regions (Fig. 16–2.4). This is because strain energy is introduced with the dislocations that are generated during plastic deformation. All in all, it is almost impossible to avoid corrosion. It is up to the engineer to minimize it (Section 16–4).

Illustrative problem 16–2.1 Silver and zinc are coupled in a standard galvanic cell (Table 16–2.1). (a) What is the electrode potential difference? (b) Change the electrolyte concentrations to 1.5 molar Ag^+ and 0.021 molar Zn^{2+}. What is the electrode potential difference?

SOLUTION From Table 16–2.1 (where $H_2 \rightarrow 2H^+ + 2e^- = 0.0$ V), the two *anode* reactions are

$$Zn \rightarrow Zn^{2+} + 2e^- \qquad \mathcal{E}_0 = -0.76 \text{ V};$$
$$Ag \rightarrow Ag^+ + e^- \qquad \mathcal{E}_0 = +0.80 \text{ V}.$$

a) $\qquad \text{Difference} = -1.56 \text{ V}.$

b) Using Eq. (16–2.7),

$$\mathcal{E}_{0.021Zn^{2+}} = -0.76 + (0.0257/2) \ln 0.021 = -0.81 \text{ V};$$
$$\mathcal{E}_{1.5Ag^+} = +0.80 + (0.0257/1) \ln 1.5 = +0.81 \text{ V}.$$
$$\text{Difference} = -1.62 \text{ V}.$$

ADDED INFORMATION Since the Ag reaction is "above" the Zn reaction, it becomes the *cathode* reaction, and its direction is reversed to one of consuming electrons.

Illustrative problem 16-2.2 In order to determine the electrode potential of cadmium, Fig. 16-2.1 was duplicated with cadmium for zinc and silver for copper. The voltmeter read −1.20 volts. What is the cadmium electrode potential with respect to hydrogen?

SOLUTION Using the same sign notation as in Fig. 16-2.1,

$$\begin{array}{lr} Cd \rightarrow Cd^{2+} + 2e^- & \mathcal{E}\ V \\ Ag \rightarrow Ag^+ + e^- & +0.80\ V \\ \hline \text{Difference} = & -1.20\ V \end{array}$$

Sum: $\mathcal{E} = -1.20\ V + 0.80\ V = -0.40\ V.$

ALTERNATE SOLUTION

	Cd (anode)	Ag (cathode)	−1.20 V
	H_2 (anode)	Ag (cathode)	−0.80 V
Difference:	Cd (anode)	H_2 (cathode)	−0.40 V.

ADDED INFORMATION Cadmium ionizes to Cd^{2+}.

If cadmium were cathodic to silver [+0.80 V − (−1.20 V) = +2.0 V], it would be more noble (and valuable) than gold. Thus, even with the confusion that can occur with signs, the reader should be able to position cadmium correctly with respect to the other elements. ◄

Illustrative problem 16-2.3 How long does it take to electroplate 450 g (1 lb) of nickel from an Ni^{2+} electrolyte with a current of 160 amperes?

SOLUTION

$$\text{Time} = \frac{(450\ g)(0.6 \times 10^{24} Ni)(2el/Ni^{2+})(0.16 \times 10^{-18} A{\cdot}s/el)}{(58.7\ g\ Ni)(160\ A)}$$

$$= 9200 \text{ seconds} \qquad (2.6 \text{ hr}).$$

ADDED INFORMATION Corrosion is electroplating in reverse. Although the currents are in the milliamp range, the times can be much longer, so that much metal can be lost. ◄

16-3 CORROSION OF METALS IN SERVICE

Corrosion reactions can be predicted under laboratory conditions. For example, the data in Table 16-2.1 are for 25°C and 1-molar solutions. We can even make adjustments for other concentrations (Eq. 16-2.7). A number of other factors come into play in service, some of which cannot be handled quantitatively. However, it is still useful to consider them so that the engineer can anticipate the consequences.

Anode/cathode areas

The current (amperes) from the anode must equal the current to the cathode.* If the anode is small compared to the cathode, the current density, A/mm^2, will be high and corrosion will be rapid. Conversely, a small cathode is not critical, since corrosion does not occur at that electrode.

Polarization

During corrosion, ions are introduced into the electrolyte at the anode surface. Normally, their concentration is low with no effect upon the corrosion rate. At the cathode, however, the corrosion rate can be influenced by the accompanying reactions. Consider Eq. (16-2.3), which supplies oxygen to the cathode surface. As the corrosion proceeds in a stagnant electrolyte, the oxygen may be consumed faster than it diffuses to the cathode. The cathode side of the reaction slows down and hence the rate must slow down at the anode too, since excess electrons cannot "pile-up" in the electrodes. We call this depletion and its consequences *polarization.* †

Polarization is beneficial because it slows down corrosion.‡ Unfortunately, it is an erratic phenomenon. Corrosion may be retarded by polarization, but if the temperature rises and the cathode reactants can diffuse more rapidly to the surface, corrosion is accelerated. Polarization retards corrosion in a pipe when no flow occurs, but the pumping of the liquid (the electrolyte) removes the polarization and corrosion proceeds.

Passivation

According to Table 16-2.1, aluminum should be extremely anodic (−1.66 V with standardized conditions). We would expect it to corrode more readily than zinc or iron. It doesn't—at least under normal conditions. It can be used for boats, kitchen pans, automobile parts, etc. The reason is straightforward. Aluminum oxidizes readily, forming Al_2O_3 on the surface. The Al_2O_3 coating is only a few atomic layers thick; however, it is an electrical insulator and adheres very tightly to the metal surface. As a result, the metal is effectively isolated from further corrosion reactions. It is *passive.* §

Metals such as titanium and stainless steels react similarly in an oxygen-containing environment. An 18-8 stainless steel‖ forms a Cr_2O_3-

---

* Electrons travel from the anode to the cathode in the metallic path of the circuit. By convention, current travels in the opposite direction—from the (+) to the (−). This means that current travels from the anode to the cathode through the electrolyte.

† This is not the same as the polarization we will encounter in Unit 19.

‡ It is not beneficial in a battery where we depend upon the electrochemical reaction to provide a current through the external circuit.

§ By making the aluminum be the anode in an appropriate electrolyte, the engineer can force the oxide film to become a thicker layer and therefore provide greater protection. We call this *anodizing.*

‖ 18% Cr-8% Ni in iron.

Table 16-3.1
Galvanic series of common alloys*

Graphite	Cathodic ↑	Nickel—A
Silver		Tin
12% Ni, 18% Cr, 3% Mo steel—P		Lead
20% Ni, 25% Cr steel—P		Lead-tin solder
23 to 30% Cr steel—P		12% Ni, 18% Cr, 3% Mo steel—A
14% Ni, 23% Cr steel—P		20% Ni, 25% Cr steel—A
8% Ni, 18% Cr steel—P		14% Ni, 23% Cr steel—A
7% Ni, 17% Cr steel—P		8% Ni, 18% Cr steel—A
16 to 18% Cr steel—P		7% Ni, 17% Cr steel—A
12 to 14% Cr steel—P		Ni-resist
80% Ni, 20% Cr—P		23 to 30% Cr steel—A
Inconel—P		16 to 18% Cr steel—A
60% Ni, 15% Cr—P		12 to 14% Cr steel—A
Nickel—P		4 to 6% Cr steel—A
Monel metal		Cast iron
Copper-nickel		Copper steel
Nickel-silver		Carbon steel
Bronzes		Aluminum alloy 2017-T
Copper		Cadmium
Brasses		Aluminum, 1100
80% Ni, 20% Cr—A		Zinc
Inconel—A		Magnesium alloys
60% Ni, 15% Cr—A	↓ Anodic	Magnesium

* Adapted from C. A. Zapffe, *Stainless Steels,* American Society for Metals.
A-active; P-passivated.

containing passive coating, if it is exposed to oxidizing conditions, for example, in the presence of HNO_3. Thus, 18–8 will resist nitric acid. However, if this same steel is immersed in hydrochloric acid (HCl), it loses its passivity and becomes *active*. The HCl is not only nonoxidizing but it also removes any of the Cr_2O_3 coating that previously passivated the surface.

Table 16-3.1 is a *galvanic series* of metals as they are distributed in the cathodic/anodic spectrum. We observed that a number are listed twice—in their passive condition (P) with a protective film, and in their active condition (A) without the surface layer. In using these chromium-containing steels and alloys for their corrosion resistance, assurance must be available that their operating conditions will also be oxidizing.

Stress corrosion

Corrosion is always more severe in the simultaneous presence of *a tensile stress and an electrolyte*. This *stress corrosion* is not a uniform attack; rather, it proceeds as a nonductile crack in an otherwise ductile metal. As a result, eventual failure is by brittle fracture rather than a "wasting-away." Higher stresses shorten the time for fracture (Fig. 16-3.1). The corrosive environ-

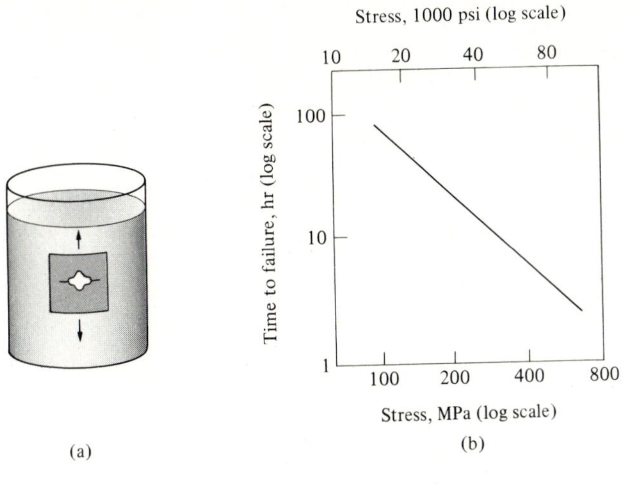

Figure 16–3.1 Stress corrosion. (a) Testing (schematic). A pre-notched sample is stressed in tension within an electrolyte. (b) Higher stresses reduce the time for fracture. (After Guy, *Introduction to Materials Science*, (New York: McGraw), with data from McEvily and Bond.)

ments that lead to stress corrosion are usually highly specific, for example, electrolytes with Cl^-, $NO^-{}_3$, and OH^- ions generally lead to fracture across the grains of the metal (Fig. 16–3.2).

The explanation of stress corrosion is not fully understood. One possibility is that corrosion is extremely rapid at the tip of a flaw where the atoms are in a stressed condition (Sections 13–3 and 14–2). Thus, the crack penetrates faster than the metal can yield plastically, unlike a crack in normally ductile materials.

Design engineers must use stress levels far below the static yield strength measured in Section 3–2 if their products are used in corrosive environments. Of course, the sensitivity to stress corrosion varies from material to material.

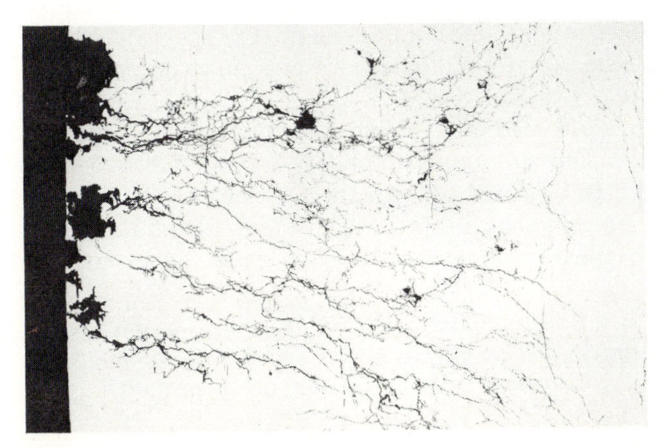

Figure 16–3.2 Stress-corrosion cracking (316 stainless steel, boiling NH_4Cl). This type of failure is common when metals are stressed in the presence of an electrolyte. (×150.) (Courtesy of A. W. Loginow, U.S. Steel Corp.)

Metals are not the only materials subject to failures of this type. Although it is called *static fatigue,* the failure of glass when stressed over a period of time in the presence of water vapor follows a pattern nearly identical to the stress corrosion of metals. Likewise, polymers crack more readily when stressed in the presence of solvents, moisture, various lubricants, and even oxygen (Section 8-4).

16-4 CORROSION CONTROL

Deterioration of metals by corrosion leads to enormous costs in maintenance, in replacement, and in product and personal liability. An estimate places these costs at 5% of a modern country's gross national product. Naturally, major attention must be given to contain these costs. The answer is not to make cars out of gold, or even stainless steels (Fig. 20-4.2), because no country possesses the resources for either. The control of corrosion entails technical approaches.

Corrosion control can fall into several categories. First is the avoidance of galvanic cells. A second, common approach is to isolate the surface. Protective coatings may be applied to the surface, or passive films may be induced to form. Finally, control may be by cathodic protection.

Avoidance of galvanic cells

Naturally, unlike metals should not be in contact if corrosion is a potential problem. Thus, plumbing codes call for a teflon washer in the coupling between copper and steel pipes in a house; iron or steel screws should not be used in bronze marine hardware; etc. The above are obvious; unfortunately, the engineer sometimes forgets the obvious. There are designs where contact between unlike metals is unavoidable; but efforts may be made to eliminate the effects. For example, mechanical engineers specify that the "make-up" water for a steam power plant be deaerated so as to minimize the most common cathodic reaction (Eq. 16-2.3).

Galvanic corrosion should also be avoided on a microscale. Annealed sterling silver (92.5 Ag–7.5 Cu) will have two phases—α and β (Fig. 5-3.3). As a result, corrosion can be established with the Cu-rich β as the anode. However, if the sterling silver is cooled rapidly from the single-phase region of Fig. 5-3.3, it is much more resistant to corrosion because the metal is single phase.

The above example can be supplemented by other alloys. Thus, we can generalize to say that *single-phase alloys* are more corrosion-resistant than are the *same alloys with multiple phases*. Of course, other properties will also be affected by this choice.

Table 16-4.1
Comparisons of inert protective coatings

Type	Example	Advantages	Disadvantages
Organic	Baked "enamel" paints	Flexible Easily applied Cheap	Oxidizes Soft (relatively) Temperature limitations
Metal	Noble metal, electroplates	Deformable Insoluble in organic solutions Thermally conductive	Establishes galvanic cell if ruptured
Ceramic	Vitreous enamel, oxide coatings	Temperature resistant Harder Does not produce cell with base	Brittle Thermal insulators

Protective surfaces

Table 16-4.1 lists the main categories of protective coatings, together with their merits and weaknesses. Since service conditions are so diverse, it cannot be said that one category is superior to another. They are all important.

Protective films also isolate the metal surface from the environment. The *passive* oxide film discussed in the previous section is one such example. Another example is a rust *inhibitor*. In this case, chemicals, such as chromates, are added to the electrolyte. Chromate ions adsorb onto the surface to give a protective film not greatly unlike the passive film that forms on an 18-8 stainless steel (Section 16-3). Certain nitrate and ferric salts act the same way. In contrast, sulfite $(SO_2)^=$ ions and hydrazine (N_2H_2) are oxygen scavengers and inhibit corrosion by removing the principal cathode reactants. Their use should be limited to closed systems where the oxygen will not be restored from the air. Most of us have closest familarity with inhibitors through their presence in automotive antifreezes. By incorporating them with the ethylene glycol, etc., the life of the radiator can be extended significantly.

Cathodic protection

The corrosion process itself may be used to give cathodic protection. For example, iron may be protected by providing a *sacrifical anode* of zinc or magnesium that makes the iron the cathode. This type of protection is used in hot water tanks, and for buried pipelines (Fig. 16-4.1). It is standard practice to use zinc anode bars on ocean-going vessels. The anode, when spent, can be replaced easily and relatively cheaply.

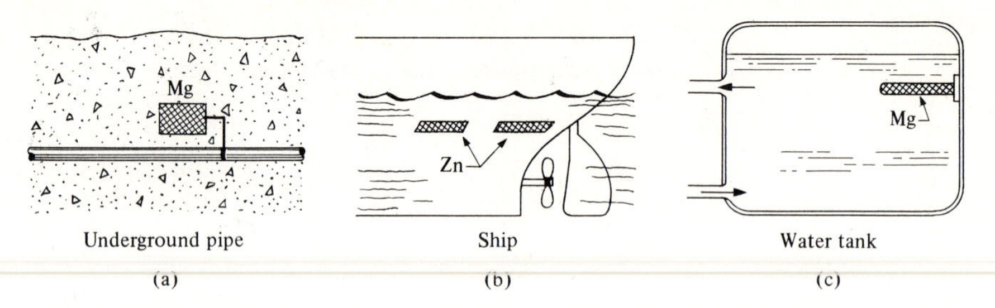

Figure 16-4.1 Sacrificial anodes. (a) Buried magnesium plates along a pipeline. (b) Zinc plates on ship hulls. (c) Magnesium bar in an industrial hot-water tank. Each of these sacrificial anodes may be easily replaced. They cause the *equipment* to become a cathode.

The above type of protection is also widely established for steel sheet and wire products. Essentially all fence wire and sheet metal roofing are *galvanized,* that is, they have been dipped in molten zinc to give them a surface that is a sacrificial anode (Fig. 16-4.2a). Note that the protection mechanism is different from that given by a tin coating (Fig. 16-4.2b). The tin is more noble than iron, therefore it simply serves as a "metallic paint" that isolates the iron from the electrolyte. However, a pinhole or a scratch through the tin coating introduces a galvanic cell with a small anodic area. Localized corrosion can then be very rapid.

The sacrificial anodes described above function by making a cathode out of the steel in the fence, ship, or water tank. The same result may be achieved by an *impressed voltage.* In this case, a d.c. current, rather than

Figure 16-4.2 Protective coatings for steel (galvanic action). (a) Galvanized steel. Zinc serves as the anode if the coating is punctured. The steel becomes the cathode and is protected. (b) Tin plate. The tin protects the iron while the coating is continuous. When the coating is punctured, the steel is corroded because it is anodic to the tin (Table 16-3.1). The small exposed area corrodes rapidly.

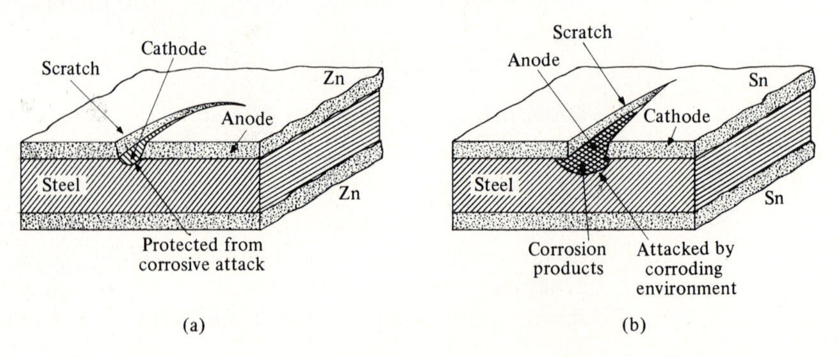

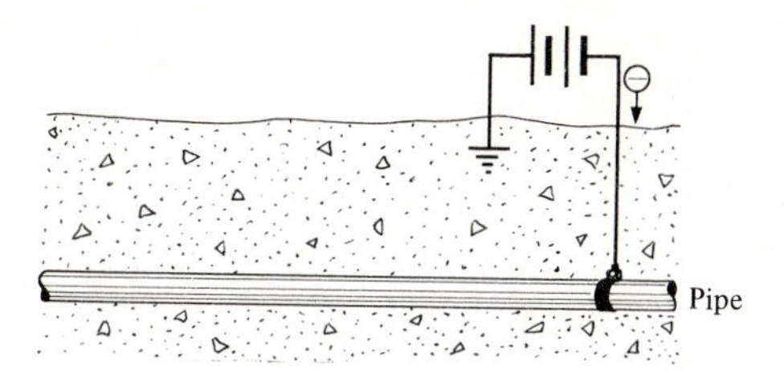

Figure 16-4.3 Impressed voltage. A small d.c. voltage will provide sufficient electrons to make the equipment a cathode.

zinc or magnesium, supplies the electrons to establish the cathode (Fig. 16-4.3).

• Stainless steels

Stainless steels are used extensively to avoid corrosion in hostile environments. They are relatively expensive; however, their cost is warranted where a premium can be paid for extended life, or where food and drug products require a near absence of contamination. Because they are expensive and draw upon scarce raw materials, particularly chromium, the engineer should hesitate specifying stainless steels unless there are no alternatives.

Essentially all stainless steels contain 12% or more of chromium and may have a total of 25% alloying elements other than iron. Thus, these are *high-alloy steels.* The chromium promotes passivation.

Metallurgists commonly separate stainless steels into one of four categories based on their structure or heat treatability (Table 16-4.2). The 300 series are *austenitic,* generally because considerable nickel is present, but sometimes from manganese additions. In view of their high-alloy con-

Table 16-4.2
Stainless steels

Type	General Composition, %	Principal Characteristics
Austenitic	>17Cr–>7Ni, <0.1C (Mn may substitute for part of Ni)	High corrosion resistivity; nonmagnetic.
Ferritic	~12Cr, <0.12C	Least expensive of stainless steels; magnetic, nonhardenable.
Precipitation hardening	~17Cr–7Ni + Al and/or Ti (<0.07C)	Hardenable without carbide formation; precipitate stable to 400°C (750°F).
Martensitic	~12Cr, 0.3 to 1C	Quenched and tempered. >500 BHN; less resistant to corrosion than austenitic or ferritic stainless steels.

tent and solution hardening, most stainless steels are relatively strong, even in their annealed condition. The austenitic steels are the most ductile of the various stainless steels and, therefore, are readily processed into the sheet and tube products that require sterilization for food or drugs. The austenitic stainless steels are also nonmagnetic. The *ferritic* stainless steels are less expensive than the austenitic steels because they have lower total alloy content. As a result, they are used where corrosion resistance is desired but not mandatory, for example automobile trim. Both of these varieties are made with minimal carbon content since (1) carbon combines with chromium as chromium carbides, thereby lowering the passivity of the γ or α phase, and (2) the chromium carbides introduce a second phase that promotes galvanic corrosion. The *precipitation hardenable* stainless steels (17-7 PH) depend on the precipitation of nickel with Al and/or Ti, just as $CuAl_2$ separates from aluminum alloys (Section 5-6). The *martensitic* stainless steels require additional carbon contents to raise the hardness to >500 BHN (>200,000 psi, or >1400 MPa). This is at the expense of corrosion resistance.

Stainless steels are not always inert to corrosion. This may be a consequence of the processing treatments; also, service conditions can alter this susceptibility to corrosion. The following example of one type of stainless steel indicates the relationship between structure and corrosion behavior. Of course, other stainless steels will have other responses.

Example of alternatives

An 18 Cr-8 Ni stainless steel contains austenite that is stable at ambient temperatures. Such a steel is not used primarily for applications requiring

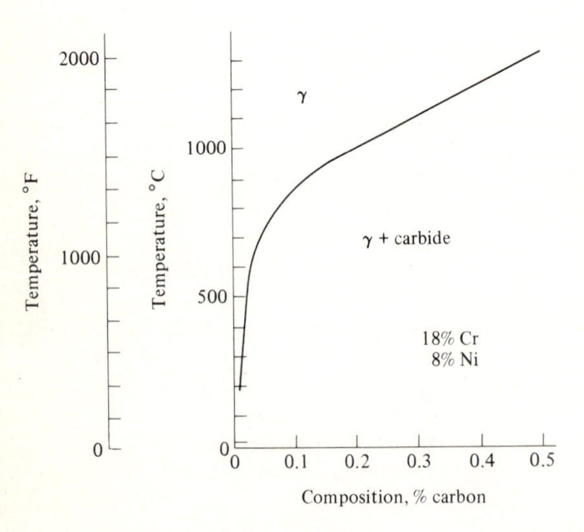

Figure 16-4.4 Carbon solubility in austenitic stainless steel. The carbon solubility in an 18-8 type stainless steel decreases markedly with temperature. Consequently, the carbon will precipitate if cooling is not rapid. The precipitated carbide is rich in chromium. (Adapted from E. E. Thum, *Book of Stainless Steels*, Metals Park, Ohio: American Society for Metals.)

Figure 16-4.5 Carbide precipitation at the grain boundaries (×1500). The small carbon atom readily diffuses to the grain boundary. It will precipitate there as a chromium carbide if sufficient time is available (a few seconds at 650°C). Galvanic cells are then formed. (P. Payson, *Trans. AIME.*)

high hardness, but rather in corrosive applications. Therefore carbon, which is more soluble in austenite at high than at low temperatures (Fig. 16-4.4), is kept to a minimum. If steel containing 0.1% carbon is cooled rapidly from about 1000°C(~1800°F), a separate carbide does not form and galvanic cells are not established. On the other hand, if the same steel is cooled slowly, or held at 650 ± °C (~1200°F) for a short period of time, the carbon precipitates as a chromium carbide, usually in the form of a fine precipitate at the grain boundaries (Fig. 16-4.5). In the latter case, two effects are possible: (1) galvanic cells may be established on a microscopic scale, or (2) the carbon forms chromium carbide (more stable then Fe_3C), which depletes the grain-boundary area of chromium and removes its passivating protection locally (Fig. 16-4.6). Either of these effects accentuates corrosion at the grain boundaries and is to be avoided (Fig. 16-4.7).

Figure 16-4.6 Chromium depletion adjacent to the grain boundary. The carbide precipitation consumes nearly ten times as much chromium as carbon. Since the larger chromium atoms diffuse slowly, the Cr content of the adjacent areas is lowered below protection levels.

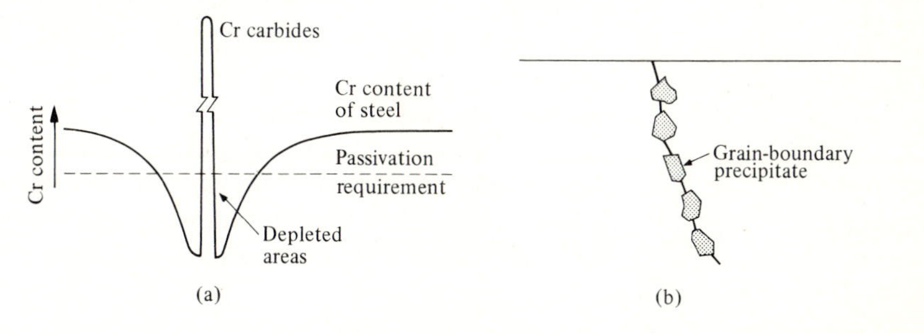

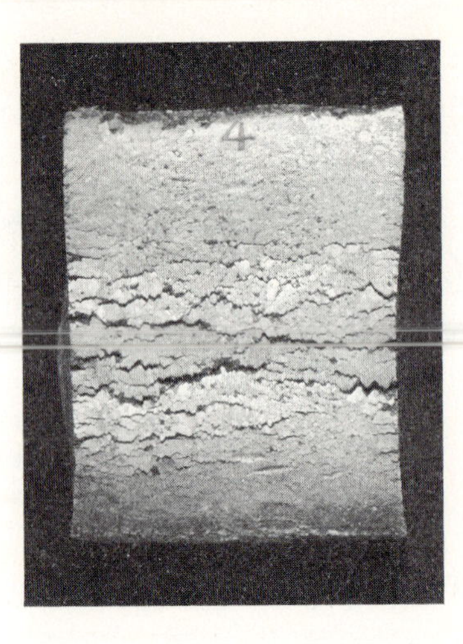

Figure 16–4.7 Intergranular corrosion (bent steel specimen, $\times 2$). This type of corrosion becomes severe if the steel has been heated into the carbide-precipitation range. (W. O. Binder et al., *Corrosion of Metals* (Chapter 3), Metals Park, Ohio: American Society for Metals.)

There are several ways to inhibit intergranular corrosion; the choice, of course, depends on the service conditions:

1. *Quenching to avoid carbide precipitation.* This method is commonly used unless (a) service conditions require temperatures in the precipitation range, or (b) forming, welding, or size prevent such a quenching operation.
2. *Provision for an extremely long anneal in the carbide separation range.* This technique offers some advantage because of (a) agglomeration of the carbides, and (b) homogenization of the chromium content so that there is no deficiency at the grain boundary. However, this procedure is not common because the improvement in corrosion resistance is relatively small.
3. *Selection of a steel with less than 0.3% carbon.* As indicated in Fig. 16–4.4, this would virtually eliminate carbide precipitation. However, such a steel is expensive because of the difficulty of removing enough of the carbon to attain this very low level.
4. *Selection of a steel with high chromium content.* A steel that contains 18% chromium corrodes less readily than a plain carbon steel. The addition of more chromium (and nickel) provides additional protection. This, too, is expensive because of the added alloy costs.
5. *Selection of a steel containing strong carbide-formers.* Such elements include titanium, niobium, and tantalum. In these steels, the carbon does not precipitate at the grain boundary during cooling because it is precipitated earlier as titanium carbide, niobium carbide, or tantalum carbide, at much higher temperatures. These carbides are innocuous because they neither deplete the chromium from the steel nor localize

the galvanic action to the grain boundaries. This technique is used frequently, particularly with stainless steel that must be fabricated by welding.

Although the above examples are somewhat specific, they do indicate methods that are used to reduce the extent of corrosion in metals. The exact choice of procedure depends on the alloy and the service conditions involved.

16-5 METALS AT HIGH TEMPERATURES

Typically, metals are weaker at elevated temperatures than at normal temperatures. In addition, they are subject to rapid oxidation. In brief, metals should not be specified for high-temperature applications without attention being given to necessary technical adjustments. The *elastic modulus* of a metal decreases with increased temperature. We saw this in Fig. 3-4.1. The *strengths* of metals also decrease (Fig. 16-5.1); however, the change is not simply one of lowering the yield or tensile strengths. The tolerable loads must be reduced further because the metal creeps and eventually fails by stress-rupture.

Creep

At high temperatures, strain is time-dependent because even low stresses introduce a slow plastic deformation called *creep*. This is because a significant number of atoms can move to new locations in response to the stress. This is the natural attempt to relieve the stresses. However, if the load remains constant, the creep continues.

The *creep rate* is the strain per unit time, de/dt. It may be only a fraction of a percent per day or per month. However, in a steam boiler or similar application, steel tubing may be in service at 650°C (1200°F) for years. The steam pressures at these temperatures are very high, so that the steel is never free from stresses. Thus, even 0.01% strain per day of service is not tolerable.

Creep has the characteristic of a high initial rate, followed by a long period of slower, but *steady-state creep,* that is, a constant creep rate, which is called "Stage-2" creep (Fig. 16-5.2). Higher temperatures or higher stresses increase the slope of the e-vs.-t curves; conversely slower creep rates accompany lower stresses and lower temperatures (Fig. 16-5.3). Eventually there is necking of the metal and, unless the load is decreased, the rate increases in Stage 3 before final *rupture*.

Creep rates are faster in fine-grained materials than in the same material with less grain boundary area per unit volume. When the temperature is high enough to permit atom movements, the grain boundary facilitates the atom's relocation from sites of high stress to positions of low stress. This

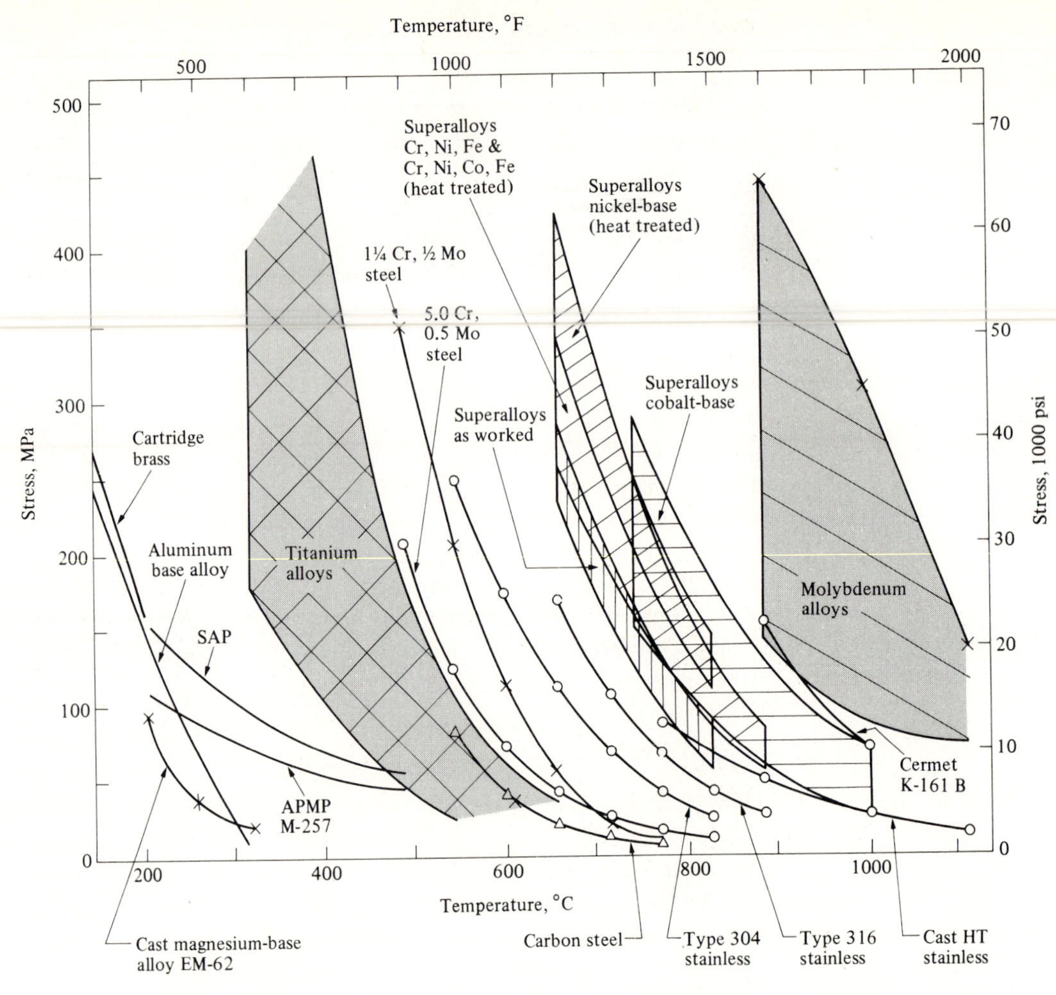

Figure 16-5.1 Superalloys. Stresses to produce rupture in 1000 hours. (Cross and Simmons, *Utilization of Heat-Resistant Alloys*, Metals Park, Ohio: American Society for Metals.)

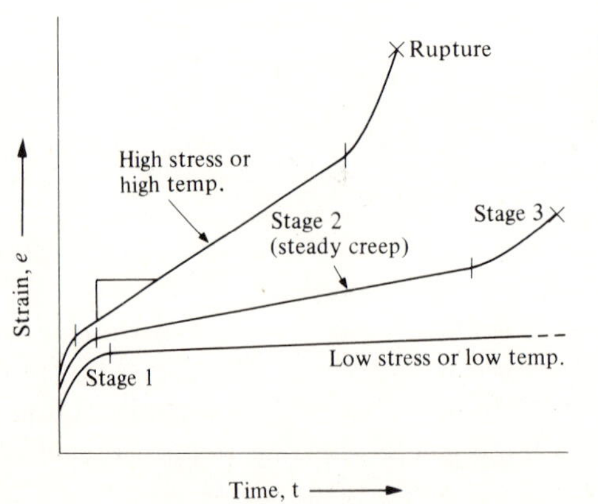

Figure 16-5.2 Creep. The steady rate of creep, de/dt, in the second stage determines the useful life of the material.

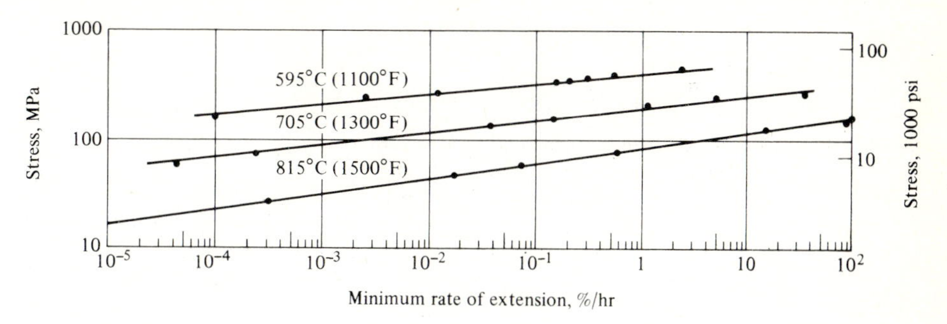

Figure 16-5.3 Creep data (type 316 stainless steel). Higher stresses and higher temperatures increase the creep rate. (From Shank, in McClintock and Argon, *Mechanical Behavior of Metals*, Reading, Mass: Addison-Wesley.)

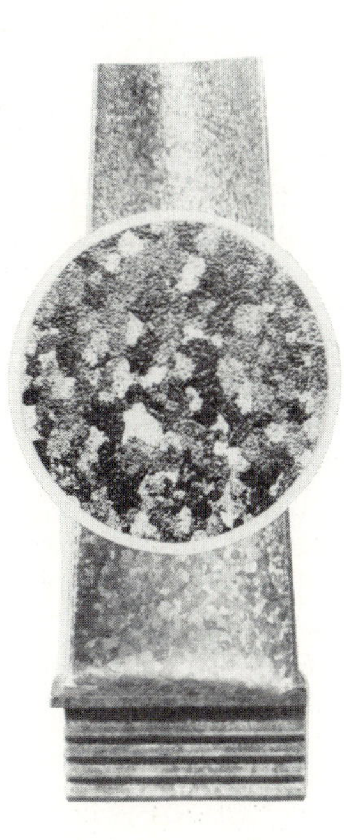

Figure 16-5.4 Coarse-grained metal in turbine blades. These blades are used at high temperatures; therefore the grain-boundary area is minimized by the development of coarse grains in processing. In contrast, grain boundaries increase the strength at low temperatures. (Courtesy of M. E. Shank, Pratt and Whitney Aircraft Group, United Technologies.)

means that *at high temperatures** coarse-grained materials are stronger than fine-grained materials (Fig. 16-5.4).

Oxidation

Although metals oxidize at any temperature, oxidation is particularly important at high temperatures because the rate is significantly higher. In

* This is not true at low temperatures where deformation follows slip planes, and grain boundaries interfere with dislocation movements (Section 14-1).

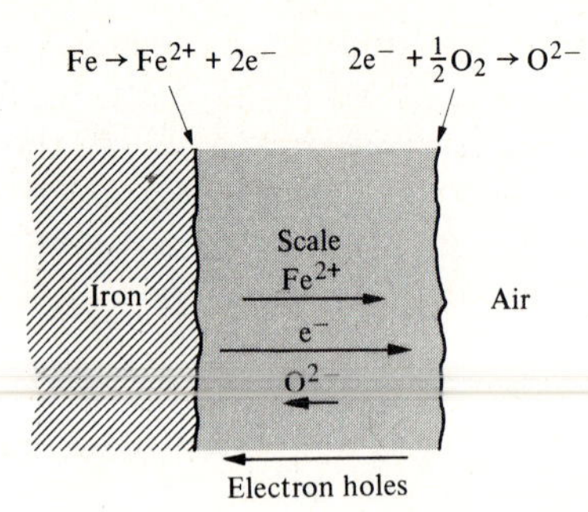

Figure 16-5.5 Scaling mechanism. Fe^{2+} and electrons diffuse more readily through the scale than do O^{2-} ions. As a result, the reaction $Fe^{2+} + O^{2-} \rightarrow FeO$ is predominant at the scale–air interface.

simplest terms, the oxidation reaction is

$$\text{Metal} + O_2 \rightarrow \text{metal oxide.} \tag{16–5.1}$$

The actual reaction is somewhat more complex because an oxide scale soon separates the metal from the air that supplies the oxygen. For oxidation to continue, the metal must either diffuse outward or the oxygen diffuse inward through the scale. This is illustrated in Fig. 16–5.5 for iron. At high temperatures the first oxygen molecules react to give

$$2Fe + O_2 \rightarrow 2FeO. \tag{16–5.2}$$

Additional reaction requires that Fe^{2+} ions diffuse through the scale to meet the oxygen. This occurs much more rapidly than any possible diffusion of oxygen in the opposite direction because (1) the Fe^{2+} ions ($r = 0.074$ nm) are much smaller than O_2 molecules, (or for O^{2-} ions, $R = 0.140$ nm); (2) FeO is iron deficient, specifically $Fe_{1-x}O$, with an Fe/O ratio of 0.96 in the scale adjacent to the metal (Fig. 11–5.2). This deficiency introduces Fe vacancies to permit easy diffusion of the Fe^{2+} ions. Electrons must go through the scale, too. Again, Fig. 11–5.2 reveals the mechanism, which is by electron hops from Fe^{2+} ions to Fe^{3+} ions. The former Fe^{2+} ion is now an Fe^{3+} ion and may receive still another electron to continue the transfer.

Oxidation of the type shown in Fig. 16–5.5 has a parabolic rate, that is, the thickness, x, increases with the square root of time, t.

$$x = k\sqrt{t}. \tag{16–5.3}$$

This relationship is predictable by metallurgists, since the rate of this type of corrosion is controlled by the thickening scale layer. If, however, the scale flakes off, or if the scale occupies less volume than the original metal,* the rate of oxidation will be more rapid than indicated in Eq. (16–5.3) and approaching

$$x = k't. \tag{16–5.4}$$

This more rapid oxidation is particularly characteristic of the alkali metals (Li, Na, etc.), and the adjacent alkaline earth metals (Mg, Ca, etc.).

* See I.P. 16–5.1.

The energies released by aluminum oxidization to Al_2O_3 are very great. However, the oxide layer that forms nearly halts all further oxidation. [The k of Eq. (16-5.3) is extremely small.] As a result, the scale that develops is nearly invisible and provides very excellent protection from further oxidizing. Two factors contribute to this: (1) more so than many other metal oxides, the Al^{3+} and O^{2-} ions are very tightly bonded, so that the aluminum ions cannot readily diffuse through the oxide layer toward the surface, and (2) the crystal structures of Al_2O_3 and aluminum are *coherent*, that is, the crystal structures of the two phases match dimensions. There is, thus, a strong bond between the scale and metal.

Stainless steels were considered in Section 16-4 because of their corrosion resistance to corrosive liquids. Furthermore, those characteristics that make them passive in liquid electrolytes also make them oxidation resistant in hostile gaseous atmospheres. Like the aluminum of the last paragraph, a chromium-containing steel forms a coherent chromium oxide scale. This permits the use of stainless steels at higher temperatures than is possible for either low-alloy or plain-carbon steels.

Decarburization

The two components of an alloy seldom have identical oxidation tendencies. Thus, we see chromium oxidizing preferentially before iron to passivate stainless steel (Section 16-3). Carbon oxidizes readily at the surface of plain-carbon steels. This reaction is not restricted by a scale since the product is carbon monoxide:

$$2C + O_2 \rightarrow 2CO\uparrow. \tag{16-5.5}$$

Carbon also diffuses from the interior of the steel and disappears as a gas. This leaves a decarburized zone behind the surface (Fig. 16-5.6). Of course, a decarburized surface is very soft compared to the balance of the steel.

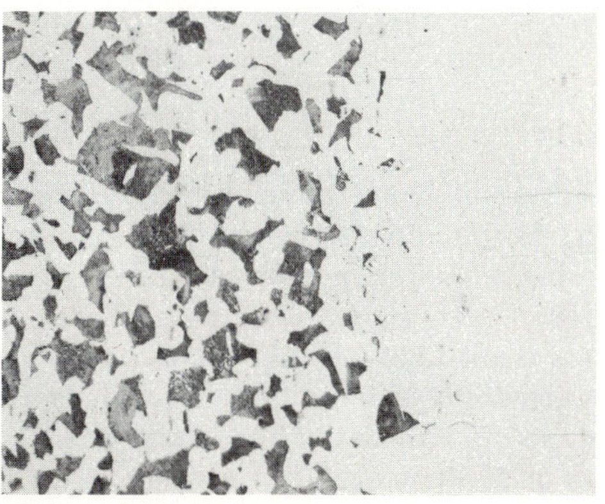

Figure 16-5.6 Decarburization ($\times 100$). The 1040 steel is softened as the carbon is preferentially oxidized from the surface.

Illustrative problem 16-5.1 One cubic centimeter of magnesium (ρ = 1.74 g/cm^3 is oxidized to MgO (ρ = 3.6 g/cm^3). What is the volume of the resulting oxide?

SOLUTION

$$\begin{aligned} 1\ \text{cm}^3 &= 1.74\ \text{g Mg;} \\ \text{g MgO} &= 1.74\ \text{g}\ (40.31\ \text{amu/MgO})/(24.31\ \text{amu/Mg}) \\ &= 2.9\ \text{g MgO.} \\ \text{Volume of MgO} &= 2.9\ \text{g}/(3.6\ \text{g/cm}^3) \\ &= 0.8\ \text{cm}^3\ \text{MgO.} \end{aligned}$$

ADDED INFORMATION The Mg-O bond is sufficiently stronger than the Mg-Mg bond in the metal, so there is a volume contraction with the addition of oxygen. [*Note:* T_{Mg} = 650°C; and T_{MgO} = 2800°C.] The volume contraction cracks the scale and permits more rapid oxidation than found with aluminum and other metals. ◄

Illustrative problem 16-5.2 The distance from the center of one Al^{3+} ion to the centers of six nearest Al^{3+} ions in Al_3O_3 is 0.267 nm. Compare this with the distance between the centers of the closest aluminum atoms in the fcc metal.

SOLUTION Refer to Fig. 2-2.2(a):

$$\text{distance} = 2\,R_{Al}.$$

From Appendix B, $R = 0.1431$ nm;

$$\begin{aligned} \text{distance} &= 2(0.1431\ \text{nm}) = 0.2862\ \text{nm} \\ \Delta d/d &= (0.2862\ \text{nm} - 0.267\ \text{nm})/0.267\ \text{nm} \\ &= 0.07 \qquad \text{(or 7 l/o).} \end{aligned}$$

ADDED INFORMATION With a mismatch of only 7 l/o, the aluminum atoms at the boundary can be associated with both phases. This leads to a coherent structure and a strong adherence of the scale to the underlying metal. ◄

• 16-6 REFRACTORY MATERIALS

Technology demands ever higher temperatures of operation. A major reason for this is that energy converts more efficiently from one form to another when higher temperatures are used. This leads to a demand for gas turbines, and for fusion reactors, to name but two examples. Likewise, temperatures in excess of 1500°C (>2700°F) are required to process many of the metallic and ceramic products that the engineer uses.

Very few metals can withstand temperatures for extended periods of time in excess of 650°C (>1200°F) without rapid losses by oxidation. One group that does, called *superalloys,* makes extensive use of nickel and cobalt. Another group, called *refractory metals,* includes tungsten, molybdenum, tantalum, and niobium. In general, these metals are dense and oxidize catastrophically unless covered with a protective coating. *Refractories* are ceramics with melting temperatures in excess of 1500°C.

Superalloys

The working ranges of several superalloys were shown in Fig. 16-5.1. Both nickel and cobalt are significantly more noble than iron. Therefore, they are the major elements of these alloys. The oxides that form are adherent, so that they provide a measure of protection.

These alloys are widely used in critical regions of jet engines and gas turbines (Fig. 16-6.1). Their load-carrying abilities decrease with increased temperature (Fig. 16-5.1); thus, design considerations must be carefully balanced between the stress requirements and the need for high operating temperatures. The principal phase of these alloys (fcc) contains a number of elements for solution hardening. These multicomponent alloys are stronger than are two-component solid solutions, which have received our previous attention. This is important since solution hardening will not "anneal out"

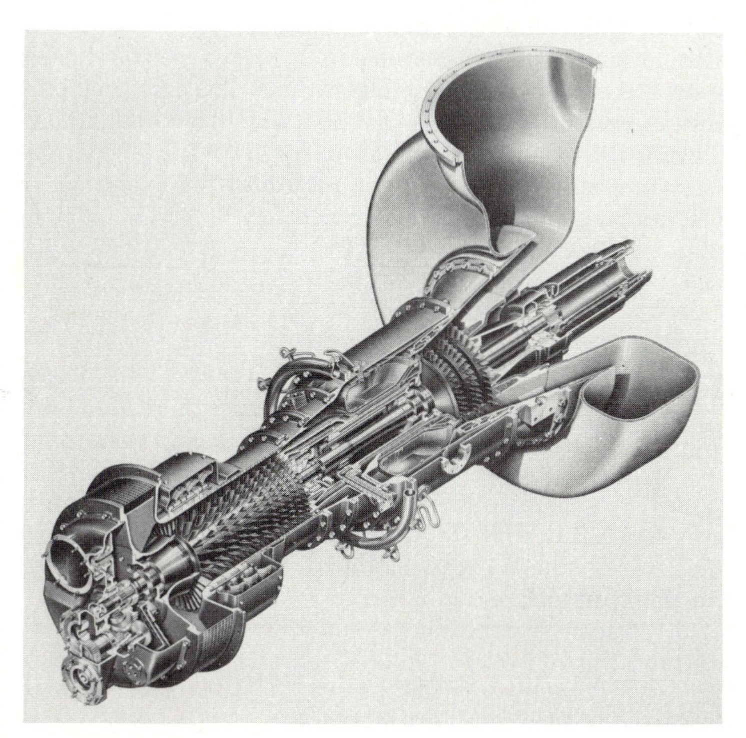

Figure 16-6.1 Gas turbine. The metals of the rotor must withstand the highest possible temperatures in order to obtain maximum fuel efficiency. Nickel and cobalt-based alloys are commonly used; however, these alloys always contain elements for solution hardening and dispersion hardening (Section 14-1). (Courtesy of Solar Turbines International.)

Table 16-6.1
Refractory metals

Metal	Density, g/cm³	Melting temperature		Young's modulus	
		°C	(°F)	GPA	(10^6 psi)
Tungsten	19.3	3410	(6170)	345	(50)
Tantalum	16.6	2996	(5420)	185	(27)
Molybdenum	10.2	2610	(4730)	325	(47)

as will strain hardening or precipitation hardening. Most of these alloys also contain a dispersion of phases such as Ni_3Al or Ni_3Ti, which are hard, ordered structures* and, therefore, alloy strengtheners (Section 14-1). Thus, high-temperature metals are engineered to utilize a combination of strengthening mechanisms—solution hardening, dispersion hardening, and, in this case, coarse grains (Section 16-5).

Refractory metals

The refractory metals cited above—W, Mo, and Ta—melt above 2400°C (>4000°F). They are rigid (high elastic modulus), and they have a high density (Table 16-6.1). Unfortunately, they cannot be used above 1200°C (2200°F) without oxidation protection, because their oxides melt and even vaporize much below the metals' melting temperatures. The oxidation protection that is most successful is coatings that contain silicon and similar elements. These form a glassy surface. Such a surface layer has low permeability for the gases encountered in service and retard further oxidation. A $MoSi_2$ coating (0.1 mm) will protect molybdenum for 30 hrs at 1650°C (3000°F).

Refractories

Temperature-resistant, nonmetallic materials are called *refractories*. These are commonly oxides. The oxides are widely used (1) because there are many naturally occurring oxides; (2) because oxides typically have very strong bonds, to withstand high temperatures, for example Al_2O_3, MgO, SiO_2, ZrO_2; and (3) because oxides are generally stable in air. Nonoxide refractories are also available, for example, SiC, AlN, Ce_2S_3, graphite. These refractories are unaltered by temperatures well in excess of 1500°C (>2700°F) but require some oxygen protection if used directly in air.

Several natural refractories are best examined in terms of the SiO_2–Al_2O_3 phase diagram (Fig. 11-7.3). This is redrawn as Fig. 16-6.2.

* Titanium (or aluminum) at cube corners, and Ni at face centers.

Quartzite, which is predominantly SiO_2, is a widespread rock. It will withstand 1500°C, even in the presence of fluxes such as CaO, FeO, and MgO. It is vulnerable to Al_2O_3 and, therefore, the raw materials must be chosen accordingly. In contrast, fireclay refractories withstand higher temperatures if they are enriched in Al_2O_3. The best grade clays dissociate on heating as follows:

$$Al_2Si_2O_5(OH)_4 \rightarrow Al_2O_3 + 2SiO_2 + 2H_2O\uparrow. \quad (12\text{-}1.1)$$

This *calcination* leads to a 45–55 ratio of Al_2O_3-SiO_2. From Fig. 16-6.2, we see that additional Al_2O_3 produces a progressively higher liquidus temperature. When the Al_2O_3-SiO_2 ratio exceeds ~72–28, the solidus is also above 1800°C (>3300°F).

The predominant refractories for steel and copper production are the basic refractories. These come from calcination of carbonate rocks, for example,

$$CaCO_3 \rightarrow CaO + CO_2\uparrow, \quad (16\text{-}6.1)$$

and

$$MgCO_3 \rightarrow MgO + CO_2\uparrow. \quad (16\text{-}6.2)$$

The resulting oxides (which give basic characteristics to aqueous solutions) possess very high melting temperatures (MgO: 2800°C; CaO: 2550°C). Their usable temperatures, however, are appreciably below these high figures because they react extensively with other oxides; the lowest eutectic of the CaO-Al_2O_3 system is ~1400°C. The ceramist intentionally avoids such combinations, just as a metallurgist avoids the use of certain metals in oxidizing environments.

Typically, refractory materials are pressed into brick shapes and are laid in furnace walls—most generally without mortar. Monolithic walls

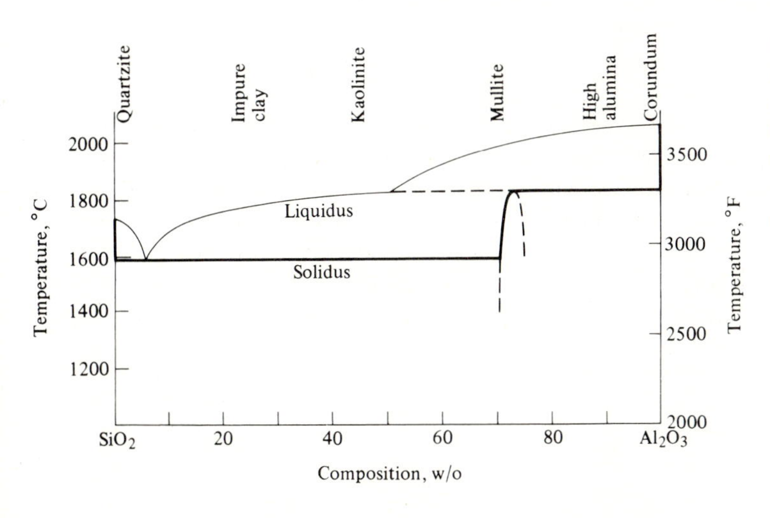

Figure 16-6.2 Alumino-silicate refractories. Silica refractories are readily obtainable from quartzite, but the Al_2O_3 must be held low. Clays and other alumino-silicates are more refractory when enriched in Al_2O_3.

both of rammed mixes and of "gunned" mixes receive increasingly wide usage. The latter may be compared to the spray-gunning of concrete, except that they are applied to the red-hot interior walls of a furnace where they dry and sinter almost immediately. Such spraying applications have the advantage of repairing furnaces without using extended periods of time for cooling and reheating the massive structures. However, sprayed refractories are never as dense as are pressed brick.

In addition to temperature resistance, refractories must resist *spalling*. This is thermal cracking, chiefly from the sharp temperature gradients in the wall of a furnace. Several factors go into the *spall resistance index*, SRI, of a refractory. Chief among them are thermal diffusivity, h; the strength, S; the thermal expansion coefficient, α; and the elastic modulus, E:

$$\text{SRI} = hS/\alpha E. \tag{16-6.3}$$

A low thermal expansion coefficient, α, leads to smaller dimensional changes, and a smaller elastic modulus, E, introduces less stress from the dimension changes that are restrained. Unfortunately, neither of these factors may be altered for a given material. A high thermal diffusivity, h, prevents a sharp temperature gradient but also leads to greater heat losses. This factor is partially controllable in a given refractory by the amount of porosity. The strength factor, S, may also be varied. Ideally, a strong material resists fracture and therefore spalling. In practice, however, high strengths may lead to fewer, but more major, cracks with extensive material losses. If the refractory produces many small, less damaging cracks, it is possible for dimensional adjustments to occur without major spalling. Thus, actual spalling does not correlate perfectly with Eq. (16-6.3).*

Insulating refractories receive added attention with increased energy costs, particularly since all types of furnaces are major fuel consumers. One approach to reduce heat losses is to increase the wall thicknesses and lower the temperature gradient. This is not always possible, either because of materials costs or because of design restrictions on space, etc. A second approach is to introduce porosity within the refractory. This may involve a less dense pressing of the brick; it may also be achieved through the introduction of sawdust or similar combustible materials within the original mix so that voids are introduced when the final product is fired; finally, lowest thermal transfer is realized in fibrous products with very low packing factors. The familiar glass wool is such a product that is amenable to moderate temperature exposures. Fibers of refractory materials are available for temperatures up to ~1600°C (Fig. 16-6.3) to achieve significant fuel savings in industrial furnaces.

* This is not a unique situation. The service behavior of a metal axle generally does not correlate with its ultimate strength because of fatigue (Section 14-3).

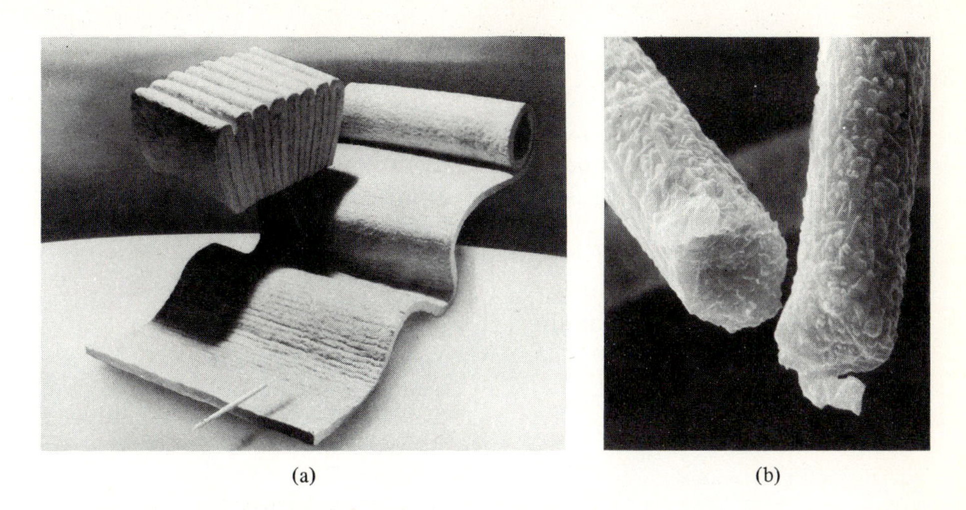

(a) (b)

Figure 16–6.3 High-temperature insulation (fibrous). (a) Insulating blanket that is folded and pinned to the furnace wall with ceramic anchors. (Courtesy of Carborundum Co.) (b) Refractory fibers (7 μm dia.). The more stable high-temperature fibers (1500°C) are crystallized and have minimum glass present, thus differing from normal insulation. (Courtesy of L. E. Olds, Johns-Manville Corp.)

• 16–7 MATERIALS AT SUBNORMAL TEMPERATURES

It is a widely known fact that many materials become brittle at subnormal temperatures. This loss of toughness was described for polymers and rubbers in Section 8–3. Below the *glass-transition temperature*, T_g, molecules of these materials cannot respond readily to stresses through viscous flow past their neighbors. As a result, they do not absorb energy before fracture. The same is true for silicate glasses (Section 13–4). Many metals possess a *ductility-transition temperature*, T_{dt}, below which very little plastic slip occurs; hence, fracture is nonductile (Section 14–2). Of course, these behavioral changes must be considered in materials selection for product design.

Modern technology presses the range of useful temperatures to ever lower levels. For example, it is now feasible to transport and store natural gas as a liquid ($B.P._{CH_4} = -161.5°C$). This reduces the container volume by a factor of more than 500 from atmospheric pressures. Liquid oxygen (−183°C, or 90 K) is a standard raw material in steel processing and related metallurgical industries. Likewise, superconductivity is a highly attractive property for a number of advanced engineering designs ranging from magnetically levitated trains to high-wattage power transmission (Fig. 16–7.1). These require temperatures to within 20° of absolute zero.

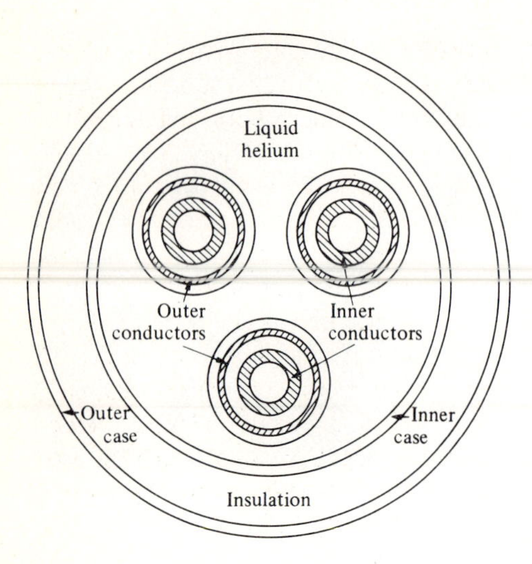

Figure 16-7.1 Superconducting cable for power transmission (cross section). This 138-kVA power route would require liquid helium to maintain superconductivity. Not only must the inner case and the conductors retain toughness and ductility at −265°C, all equipment handling the liquid helium and liquid nitrogen must meet low-temperature requirements. (Adapted from D. P. Snowden, "Superconductors for Power Transmission," *Scientific American*.)

Cryogenic service

Engineers and physicists refer to *cryogenic temperatures* when they speak of technology below −100°C (173 K). This extends to 0 K (−273.15°C). The paramount property for cryogenic design is *toughness*. Strengths are important but not critical, since the strength commonly increases at lower temperatures. *Thermal expansion coefficients* are important because parts of the cryogenic equipment must be cooled several hundred degrees as it goes into service while other parts remain near ambient levels. In general, this does not pose new design problems because similar temperature differentials are realized in elevated temperature equipment. *Thermal conductivity* is more critical in low-temperature design than in high-temperature design, because inward diffusing heat is difficult to remove. In contrast, heat losses from equipment that operate above 300°C can be replaced with additional fuel or with auxiliary heaters.

In general, plain-carbon steels cannot be used at temperatures below −50°C (−60°F), even when they are fine-grained, because of their transition temperature to nonductile fracture (Section 14-2). Austenitic steels, like other fcc metals, do not have a marked ductility transition; therefore, these steels are widely used as cryogenic materials. Nickel is the most available "austenite-former"; hence, steels with ≥8% Ni, or with Ni − Mn combinations, find wide use. Almost invariably, cryogenic steels must be low in carbon (<0.2%), because carbides are not ductile and they accentuate brittle fracture.

16-8 RADIATION DAMAGE

Radiation—by energetic neutrons, by ultraviolet light, by β-rays (accelerated electrons), or by other forms of particles and rays—introduces energy very locally within a material. A single electron, atom, or bond may receive the full impact, while neighbors experience only the aftereffects. An atom may be knocked out of the crystal lattice to introduce a vacancy and an interstitial; a bond may be broken to cut a molecule in two; an electron may be knocked away from an atom to produce an ion; or a proton may be knocked out of a nucleus to transmute an element. All of these modify the structures of the exposed materials and, hence, the properties.

Figure 16-8.1 shows schematically the effects of a high-velocity neutron entering a crystal lattice. It dislodges first one atom, then another until the neutron is captured by the nucleus. In doing so, it may introduce ten or more vacancies (and interstitials). There is also disruption along with paths of the neutron and the ejected atoms. These imperfections inhibit the slip pro-

Figure 16-8.1 Radiation effects in a crystal lattice arising from the neutrons entering at the left. (C. O. Smith, *Nuclear Reactor Materials*, Reading, Mass.: Addison-Wesley.)

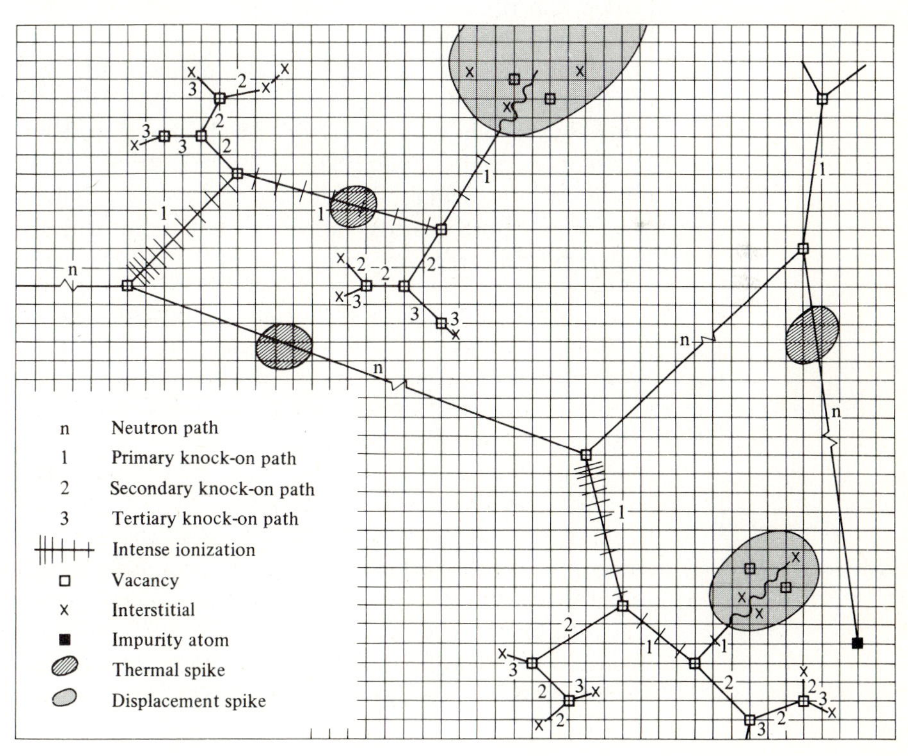

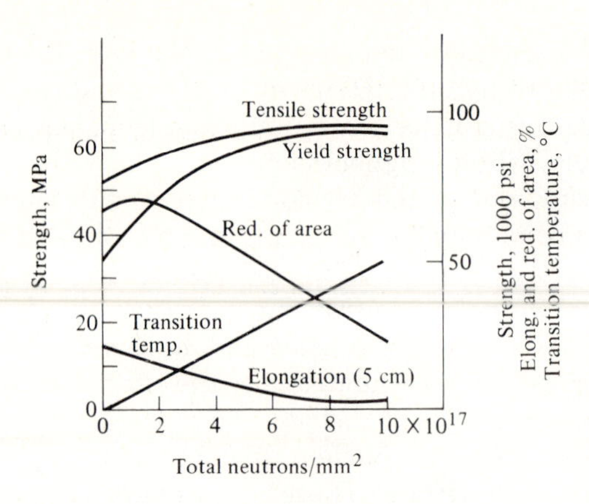

Figure 16-8.2 Radiation damage to steel (ASTM A-212-B carbon-silicon steel). (Adapted from C. O. Smith, ORSORT, Oak Ridge, Tenn.)

cesses with the result that hardness and strength are increased as shown in Fig. 16-8.2. There are accompanying decreases in ductility and in toughness. We call these changes *radiation damage*, since the loss in toughness is generally more critical to the exposed material than is the gain in strength. *Recovery* is possible by heating the material so that the vacancies will be filled and the interstitials disappear.

The above property changes, and the ability to have recovery, lead us to the naive conclusion that radiation damage and recovery may be compared with strain hardening and recrystallization. In each case, imperfections are introduced, which can be removed by heating. However, strain hardening introduces dislocations. These are linear defects involving many coordinated atoms, whereas radiation damage introduces isolated defects. The latter are more readily "annealed out" than the dislocations. For this reason, radiation recovery occurs at lower temperatures than does recrystallization.

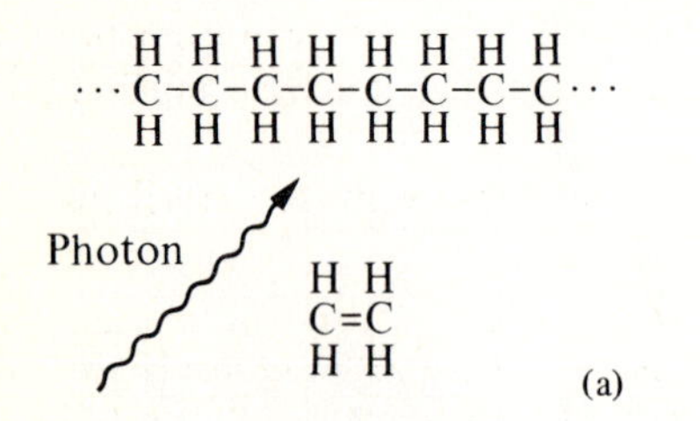

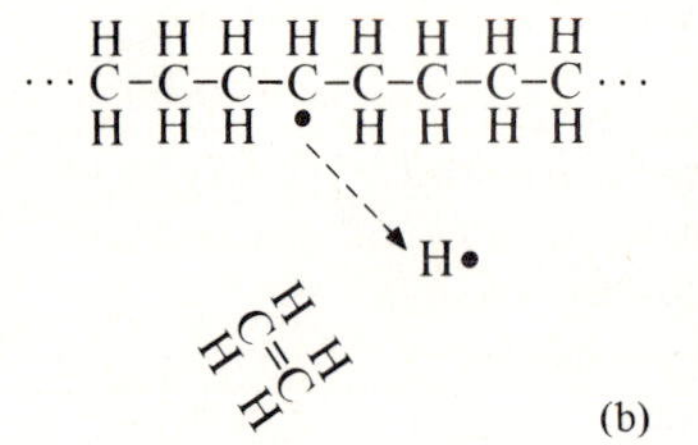

H H H H H H H H
C–C–C–C–C–C–C–C
H H H | H H H H
H H
C–C•
H H
(c)

Figure 16-8.3 Branching by irradiation. A photon can supply the activation energy necessary to cause branching. A neutron can produce the same effect.

Table 16-8.1
Effects of radiation on various materials*

Integrated fast neutron flux, n·cm/cm³ (or nvt)	Effect
10^{14}	Germanium transistor—loss of amplification
	Glass—coloring
	Polytetrafluoroethylene—loss of tensile strength
10^{15}	Polymethyl methacrylate and cellulosics—loss of tensile strength
	Water and least stable organic liquids—gassing
	Natural and butyl rubber—loss of elasticity
10^{16}	Organic liquids—gassing of most stable ones
	Butyl rubber—large change, softening
	Polyethylene—loss of tensile strength
10^{17}	Mineral-filled phenolic polymer—loss of tensile strength
	Natural rubber—large change, hardening
	Hydrocarbon oils—increase in viscosity
10^{18}	Metals—most show appreciable increase in yield strength
	Carbon steel—reduction of notch-impact strength
	Polystyrene—loss of tensile strength
10^{19}	Ceramics—reduced thermal conductivity, density, crystallinity
	All plastics—unusable as structural materials
	Carbon steels—severe loss of ductility, doubled yield strength
10^{20}	Carbon steels—increased fracture-transition temperature
	Stainless steels—yield strength tripled
10^{21}	Aluminum alloys—reduced but not greatly impaired ductility
	Stainless steels—reduced but not greatly impaired ductility

* Indicated exposure levels are approximate. Indicated changes are at least 10%. The table is reprinted from C. O. Smith, *Nuclear Reactor Materials*, Reading, Mass.: Addison-Wesley.

It was shown in Section 8-4 that energies from either neutrons or ultraviolet light produce *scission* and degrade polymers. In other cases (for example, polyethylene) radiation can induce *branching* as shown in Fig. 16-8.3. This may actually increase the temperature resistance of the polymer, since these linear molecules can no longer slide by each other as readily. It is not predictable whether scission or branching will have a more significant effect on the structure and properties.

Electrical properties are also affected by radiation since (a) ionization occurs and electrons are freed, and (b) the structures become less perfect. These factors will be discussed in the next two units.

Table 16-8.1 presents examples of property changes that are attributable to neutron radiation.

Illustrative problem 16-8.1 The energy, E, of a light photon (sometimes described as a small energy "packet" within a light ray) may be calculated as follows:

$$E = h\nu, \tag{16-8.1}$$

Where h is a constant with a value of 6.62×10^{-34} joule·sec and ν is the frequency of light. What is the energy of a photon of ultraviolet light with a wavelength of 300 nm?

SOLUTION Since the frequency of light ν is the velocity c (= 3×10^8 m/sec) divided by the wavelength λ (=300 nm, or 3×10^{-7} m),

$$E = (6.62 \times 10^{-34} \text{ joule·sec})(3 \times 10^8 \text{ m/sec})/(3 \times 10^{-7} \text{ m})$$
$$= 6.6 \times 10^{-19} \text{ joules} \qquad (1.6 \times 10^{-19} \text{ cal}).$$

ADDED INFORMATION The constant h is called *Planck's constant* and enters widely into calculations that involve the wave characteristics of energy. Usually it is expressed as 6.62×10^{-34} J·s, or 1.58×10^{-34} cal·sec. Also, 4.13×10^{-15} eV·sec is used when energies are expressed in electron volts (I.P. 18–2.2). ◂

Review and Study

SUMMARY

Materials deteriorate in corrosive liquids and gases principally by surface attack. Extreme temperatures alter the response time for atomic movements within solids, leading to creep at elevated temperatures and to nonductile fracture at low temperatures. Radiation introduces point imperfections internally.

1. Metallic corrosion is by oxidation to form metallic ions. This anode reaction releases electrons that must be consumed at the cathode. In addition to the anode and cathode, an electrolyte is necessary for corrosion to proceed.
2. Oxygen normally accelerates corrosion because it consumes electrons at the cathode; however, the increased corrosion occurs in the oxygen-deficient region. An exception occurs and oxygen inhibits corrosion when the oxygen passivates the metal by producing a protective surface film. Stainless steel, titanium, and aluminum benefit by such protection.
3. Corrosion may be minimized (a) by the avoidance of galvanic cells; (b) by the presence of protective surfaces, including paints, vitreous enamels, metallic coatings, passivation, and adsorbed ions (rust inhibitors); and (c) by electrochemical action, including impressed voltages and sacrificial anodes.
4. • Stainless steels fall in four categories—austenitic, ferritic, precipitation hardening, and martensitic. Even within these groupings, the engineer has options available for their compositions and processing. However, trade-offs must commonly be made between maximum mechanical properties and optimum corrosion resistance in specifying these steels for technical products.
5. The elastic moduli and the strengths of metals decrease at elevated temperatures. Creep and eventual rupture are typically more critical than yield and tensile strengths in designing for high-temperature service. All metals are subject to accelerated oxidation as temperatures rise. The thickness of an adhering scale is proportional to the square root of time. However, if the scale cracks, or "flakes-off," the oxidation rate proceeds linearly.

- 6. Superalloys, refractory metals, and oxide refractories are available for high-temperature use. Each has limitations.
- 7. In general, materials lose ductility abruptly at low temperatures. Except in some fcc metals, this increased brittleness must be a factor in engineering design for cryogenic applications.

8. Radiation damage occurs because atoms (and electrons) are knocked out of their normal positions, thus altering the basic properties. Recovery is facilitated by annealing.

TECHNICAL TERMS

Anode The electrode that supplies electrons to an external circuit. (Corrodes.)

Cathode The electrode that receives electrons from an external circuit.

Cell, concentration Galvanic cell arising from nonequal electrolyte concentrations (The more dilute solution produces the anode.)

Cell, oxidation Galvanic cell arising from nonequal oxygen potentials. (The oxygen-deficient area becomes the anode.)

Corrosion Deterioration and removal by chemical attack.

Creep A slow deformation by stresses below the normal yield strength (commonly occurring at elevated temperatures).

Creep rate Creep strain per unit of time.

Cryogenic Low temperatures, generally $<-100°C$.

Decarburization Removal of carbon from the surface zone of steel by elevated temperature oxidation.

Ductility transition temperature (T_{dt}) *See* Unit 14.

Electrode potential ($\mathcal{E}$) Voltage developed at an electrode (as compared with a standard reference electrode).

Electroplating Cathodic reduction process, opposite of corrosion. Electrons are supplied by external circuit.

Galvanic cell A cell containing two dissimilar metals and an electrolyte.

Galvanic series Sequence (cathodic to anodic) of corrosion susceptability for common metals. Varies with passivity and with electrolyte composition. (Cf. Electrode potentials, which are for standard electrolytes.)

Galvanization The process of coating steel with zinc to give galvanic protection.

Impressed voltage D.c. voltage applied to make a metal cathodic during service.

Inhibitor An additive to the electrolyte that promotes passivation.

Oxidation (general) The raising of the valence level of an element.

Passivation The condition in which normal corrosion is impeded by an adsorbed surface film on the electrode.

Polarization (electrochemical) Depletion of reactants at a cathode surface, thus reducing the corrosion current.

Radiation damage Structural defects arising from exposure to radiation.

Recovery (radiation) Return of normal properties following radiation damage by annealing to eliminate vacancies and interstitials.

Reduction Removal of oxygen from an oxide; the lowering of the valence level of an element.

• **Refractories** Materials capable of withstanding extremely high temperatures.

• **Refractory metals** Metallic elements with melting temperatures in excess of 2400°C.

Sacrificial anode Expendable metal that is anodic to the product it is to protect.

Scale Surface layer of oxidized metal.

Scission Degradation of polymers by radiation that splits the molecule in two.

Spalling Cracking from thermal stresses.

• **Stainless steel** High-alloy steel (usually containing Cr, or Cr + Ni) designed for resistance to corrosion and/or oxidation.

Static fatigue Delayed fracture arising from stress corrosion.

Stress corrosion Corrosion accentuated by stresses.

• **Superalloys** Multicomponent alloys designed for high-temperature service.

CHECKS

16A Hostile environments include adverse _***_, _***_, and/or _***_ conditions.

16B Corrosion includes an _***_ reaction in which the atom loses valence electrons. Of the two, anode and cathode, corrosion occurs at the _***_.

16C A reduction reaction occurs at the _***_ where the atoms, or ions _***_ electrons.

16D For corrosion to occur, there must be an _***_, a _***_, and an _***_.

16E The commercial process of _***_ has reactions that are opposite to those of corrosion.

16F Of the two, copper and iron, _***_ is more anodic; of the two, zinc and iron, _***_ is more anodic.

16G The oxidation of _***_ is used as an arbitrary reference for other electrode reactions.

16H The most common cathode reaction that leads to the rusting of iron in moist air is _***_.

16I Reference electrolyte solutions have a _***_ concentration; more _***_ solutions are more anodic.

16J When the electrolyte concentration varies, corrosion occurs more readily where the concentration is more _***_.

16K In general, oxygen accelerates the rate of corrosion with corrosion occurring most rapidly where the oxygen content is ***.

16L On a microscale, a grain boundary becomes the *** and the interior of the grain the ***; likewise, annealed metal is cathodic to *** metal because *** . . . .

16M Electrons move from the *** to the *** through the metallic part of the circuit; current flows from the *** to the *** through the electrolyte.

16N Depletion of cathode reactants leads to *** and therefore slower corrosion rates. The rates can be increased by increasing the *** or ***.

16O A surface of a metal becomes *** if it develops a protective oxide film; it becomes *** if the film is destroyed.

16P The combination of *** and an electrolyte leads to more rapid corrosion and failure by *** fracture. This type of failure is related to *** fatigue for glass and polymers.

16Q Protective surfaces include coatings of ***, ***, and ***, and also *** films. *** are effective because they supply oxygen-containing ions that are adsorbed by the surface.

16R Zinc bars bolted to the side of ships provide corrosion protection because they become ***. D.c. currents also provide protection if connected so that electrons move *** the metal to be protected (the current flowing the other direction).

16S Galvanized steel is protected because *** . . . ; anodized aluminum is protected because *** . . . ; stainless steel does not corrode readily because *** . . . .

• 16T Austenitic stainless steels contain *** as the principal alloying element(s), that is, more than 5%. Ferritic stainless steels contain *** as the principal alloying element(s).

• 16U High-grade stainless steels contain an absolute minimum of ***, because it reduces the effective amount of chromium.

• 16V Strong carbide-formers such as *** and *** can be added to stainless steels to tie up the carbon that would normally react with chromium.

16W Stage-*** creep has the most uniform rate of the three stages. Its creep rate increases with an increase in *** and *** and with a decrease in ***.

16X For oxidation to proceed and for a scale to form, there must be *** through the scale. If the scale is adherent and doesn't flake off, the scale requires *** times as long to become 2 mm thick as it took to become 1 mm thick.

16Y When *** occurs at the surface of a steel, the microstructure becomes more ferritic.

• 16Z Two of the principal metals used in superalloys are *** and ***. Since these alloys are designed for high-temperature use, *** hardening cannot be used; rather strengthening involves *** hardening.

- 16AA Silicide coatings on refractory metals help retard oxidation because the silicon forms a ___***___ that restricts diffusion.
- 16AB Some widely occurring oxides that can serve as refractories include ___***___ and ___***___. Less common, but more refractory, oxides include ___***___ and ___***___.
- 16AC Spalling resistance is greater for refractories that have high values of ___***___ and ___***___, and low values of ___***___ and ___***___.
- 16AD One of the most important properties for low-temperature service is ___***___; this is because there is little or no ___***___ to accompany fracturing.

16AE Radiation damage involves the introduction of many ___***___ defects; in contrast, ___***___ involves the introduction of dislocations, which are linear defects.

16AF Because of the contrast in 16AE, ___***___ from radiation damage occurs at lower temperatures than does ___***___ after strain hardening.

16AG ___***___, caused by radiation, degrades a polymer by reducing the ___***___ molecular size.

STUDY PROBLEMS

16–2.1. A galvanic couple includes iron in a 0.002 molar solution of Fe^{2+} ions, and magnesium in a 0.3 molar solution of Mg^{2+} ions. What is the electrode potential difference?

Answer: -1.86 V

16–2.2. In a stagnant solution, the copper ion concentration is 0.012-molar at point A of a copper surface; it is 0.002-molar at point B. What potential difference develops for possible corrosion?

16–2.3. What ratio of Zn^{2+} concentrations is required in electrolytes to introduce an electrode potential difference of 35 mV?

Answer: 15/1

16–2.4. The electrode potential difference between cobalt and copper in 1-molar (1-M) solutions is -0.62 volts. What is the electrode potential of cobalt with respect to zinc within 0.1-M Co^{2+} and 0.1-M Zn^{2+} solutions?

16–2.5. What current density, A/cm², is required to plate 20 μm of chromium onto a surface in 20 min from a Cr^{2+} solution?

Answer: 0.044 A/cm²

16–2.6. A dry cell with a zinc anode (cell wall) provides 15 amperes of current from its 300 cm² of surface. How long will it take for 10% of the 1-mm thick wall to corrode (if one assumes uniform corrosion)?

16–5.1. Alumina (Al_2O_3) has a density of 3.85 g/cm³. What is the volume change as the metal oxidizes to Al_2O_3?

Answer: +32%

16–5.2. Scale of NiO on nickel is 200 μm thick. The corresponding thickness of metal was _***_ μm. [NiO has the NaCl-type structure shown in Fig. 11–1.2; nickel is an fcc metal.]

• 16–5.3. Refer to S.P. 16–5.2. How do the closest Ni-to-Ni distances compare in NiO and in the metal?

Answer: 0.296 nm vs. 0.249 nm, or 19% greater in NiO. (This is one reason that NiO is less protective to its metal than is Al_2O_3. See I.P. 16–5.2.)

16–8.1. (a) What frequency must a photon have to supply an additional electron volt of energy to an atom? (b) What is its wavelength?

Answer: (a) 0.24×10^{15} Hz (b) 1240 nm (This is in the infrared range.)

16–8.2. Photons of blue light ($\lambda = 480$ nm) lead to extensive scission (bond breaking) in a polymer. How much energy is involved with each photon?

CHECK OUTS

16A temperature
chemical
radiation

16B oxidation (electrochemical)
anode

16C cathode
gain

16D anode, cathode, electrolyte

16E electroplating

16F iron
zinc

16G H_2

16H $2H_2O + O_2 + 4e^- \rightarrow 4(OH)^-$

16I 1-molar
dilute

16J dilute

16K lower

16L anode
cathode
deformed
of the residual strain energy

16M anode
cathode
anode
cathode

16N polarization
temperature, agitation

16O passivated
activated

16P stress
brittle
static

16Q glasses (ceramics)
paints (polymers)
metals
oxide
inhibitors

16R anodic
into

16S zinc is anodic to iron
Al_2O_3 forms a protective film
the surface becomes passivated

16T chromium, nickel
chromium

16U carbon

16V titanium, niobium

16W 2
temperature, stress
grain size

16X	diffusion four
16Y	decarburization
16Z	nickel, cobalt strain solution
16AA	glass (oxide)
16AB	SiO_2, Al_2O_3 ZrO_2, Cr_2O_3
16AC	strength, thermal conductivity thermal expansion coefficient elastic modulus
16AD	toughness, ductility
16AE	point plastic deformation
16AF	recovery annealing (softening)
16AG	scission average

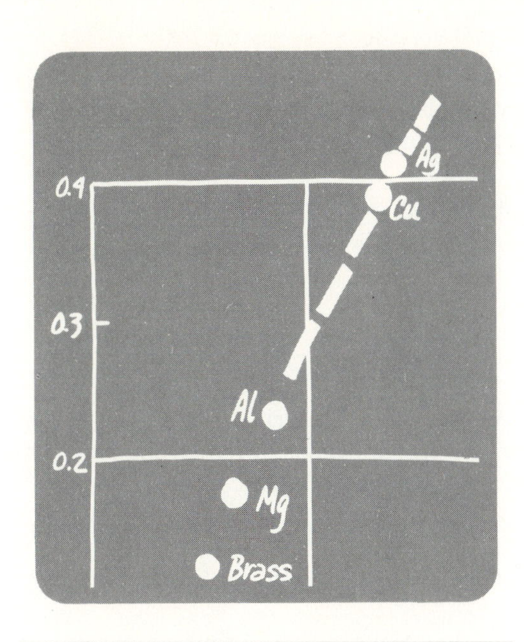

UNIT SEVENTEEN

METALLIC CONDUCTORS

Highly conductive materials are sometimes categorized as metals. However, not all metals conduct with equal facility; and, in fact, the conductivity of a given metal varies with temperature, purity, deformation, etc. Therefore, we shall look more closely at the factors that affect conductivity (and resistivity). Finally, attention will be given to the differences that separate metals from semiconductors and insulators.

Contents

Prerequisites Unit 2; Sections 4–2, 4–3, 5–1, 5–2, and 16–8; and general physics.

From Unit 17, the engineer should

1) Know the units of conductivity, resistivity, resistance, mobility, electric field, and drift velocity.
2) Relate the mean free path for electron movements to temperature, solid solutions, strain hardening, radiation damage, etc.; and, in turn, relate these to drift velocity, mobility, conductivity, and resistivity.

3) Know how thermal and electrical conductivities are related.
4) Develop the concept of energy bands and energy gaps in conductors, semiconductors, and insulators.
5) Become acquainted with the technical terms that pertain to electrical conductivity.

17-1 ELECTRICAL CONDUCTIVITY (AND RESISTIVITY)

Electrical *resistivity,* ρ, is a property of a material. *Resistance* (ohms), which is commonly used in circuit calculations, differs from resistivity (ohm·m) in that resistance possesses geometric factors:

$$R = \rho\,(L/A). \tag{17-1.1}$$

This assumes a uniform cross-section area, A, along the length of L, of the conductor.

Conductivity, σ, is the reciprocal of resistivity and thus possesses units of $\text{ohm}^{-1}\cdot\text{m}^{-1}$;

$$\sigma = 1/\rho. \tag{17-1.2}$$

It is also a property of a material and may be expressed either (a) as a ratio of the *current density* (amp/m^2) and the *electric field* (volts/m),

$$\sigma = i/\mathcal{E} = (\text{A/m}^2)/(\text{V/m}), \tag{17-1.3}$$

or (b) as the product of the *number* of charge carriers, n, per unit volume; the *charge* of each, q; and the charge *mobility,* μ;

$$\sigma = nq\mu. \tag{17-1.4a}$$

Since the charge mobility is the mean or net velocity (m/s) that is attained within an electric field (V/m), the units of mobility are (m/s)/(V/m), or ($\text{m}^2/\text{V}\cdot\text{s}$), and for Eq. (17-1.4a) are

$$\sigma = (\text{m}^{-3})(\text{A}\cdot\text{s})(\text{m}^2/\text{V}\cdot\text{s}) = \text{ohm}^{-1}\cdot\text{m}^{-1}. \tag{17-1.4b}$$

The charge per carrier, q, is the charge per electron (or its multiple) for all types of electrical conductivity. That value is 0.16×10^{-18} coulombs, or 0.16×10^{-18} A·s.

In *metals,* conductivity occurs by electron transport. However, metals are not conductors simply because they possess electrons. All atoms have electrons, but not all materials conduct equally readily. For example, both rubber and copper possess $\sim 0.3 \times 10^{24}$ electrons/g. Rubber is an insulator ($<10^{-12}\text{ohm}^{-1}\cdot\text{m}^{-1}$); copper is a metallic conductor ($\sim 10^{8}\ \text{ohm}^{-1}\cdot\text{m}^{-1}$)—more

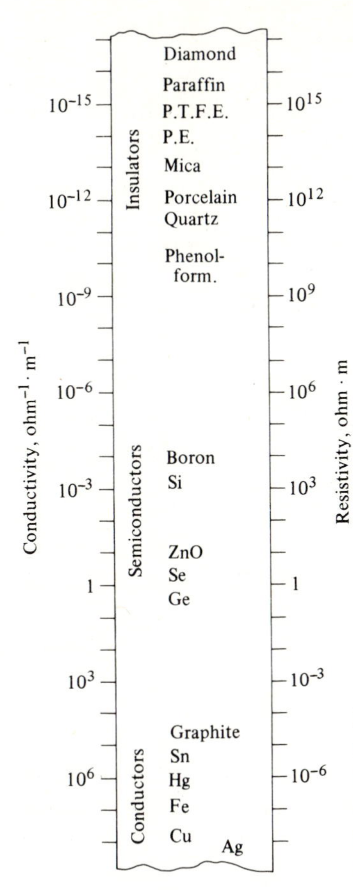

Figure 17-1.1 Range of conductivities. Metals normally have conductivities in excess of 10^5 ohm^{-1} · m^{-1}. Typical commercial semiconductors and insulators are shown. Other materials fill the remaining gaps to give a full spectrum of conductivities (and resistivities).

than 20 orders of magnitude greater. Therefore, we will need to look more closely at the difference between conductors and insulators (Fig. 17-1.1). That will be done in Section 17-6. In the meantime, we may point out that the conduction electrons of metals tend to be accelerated and gain more energy (and momentum) as they move toward the positive electrode. This gain in velocity continues until the electron is deflected by some irregularity in the structure that interferes with the wavelike motions of the electrons. When deflected into a direction that carries it toward the negative electrode, the electron loses velocity until its direction is reversed once again. Thus, there is a net movement of the electrons—more toward the positive electrode than with the field. This leads to a mean, or *drift velocity*, $\bar{v}$, that was cited in connection with mobility and with Eq. (17-1.4). The distance traveled between deflections, or reversals in direction, is called the *mean free path*.

Illustrative problem 17-1.1 The electrical conductivity of copper is $59 \times 10^6\ \text{ohm}^{-1}\cdot\text{m}^{-1}$ at 20°C. (a) What is the resistance of a wire with a 1-mm diameter that is 2 km in length? (b) If it carries a current of 1.3 A, what will be the voltage drop?

SOLUTION

a)
$$\begin{aligned}\rho &= 1/59 \times 10^6\ \text{ohm}^{-1}\cdot\text{m}^{-1}\\ &= 16.9 \times 10^{-9}\ \text{ohm}\cdot\text{m};\\ R &= (16.9 \times 10^{-9}\ \text{ohm}\cdot\text{m})(2000\ \text{m})/(\pi/4)(0.001\ \text{m})^2\\ &= 43\ \text{ohms}.\end{aligned}$$

b) From Eq. (17-1.3),

$$\begin{aligned}\mathcal{E} &= [1.3\ \text{A}/(\pi/4)(0.001\ \text{m})^2]/(59 \times 10^6\ \text{ohm}^{-1}\cdot\text{m}^{-1})\\ &= 0.028\ \text{V/m} = \Delta E/2000;\\ \Delta E &= 56\ \text{V}.\end{aligned}$$

Alternatively,

$$\begin{aligned}E &= IR\\ &= (1.3\ \text{A})(43\ \Omega) = 56\ \text{V}. \blacktriangleleft\end{aligned}$$

Illustrative problem 17-1.2 There are 10^{19} electrons/m^3 to serve as carriers in a material that has a conductivity of $0.01\ \text{ohm}^{-1}\cdot\text{m}^{-1}$. What is the drift velocity, $\bar{v}$, of those carriers when 0.17 volts is placed across a 0.27 mm distance within the material?

SOLUTION From Eq. (17-1.4),

$$\begin{aligned}\mu &= 0.01\ \text{ohm}^{-1}\cdot\text{m}^{-1}/(10^{19}/\text{m}^3)(0.16 \times 10^{-18}\ \text{A}\cdot\text{s})\\ &= 6.25 \times 10^{-3}\ \text{m}^2/\text{V}\cdot\text{s}.\\ \bar{v} &= (6.25 \times 10^{-3}\ \text{m}^2/\text{V}\cdot\text{s})(0.17\ \text{V}/0.00027\ \text{m})\\ &= 4\ \text{m/s}.\end{aligned}$$

ADDED INFORMATION Observe that 4 m/s is not the actual velocity, but a net velocity of the electrons as they migrate toward the positive electrode. The actual velocity varies from zero to high values depending upon the time in the mean free path between deflections. ◀

17-2 ELECTRICAL CONDUCTIVITY (AND RESISTIVITY) VERSUS TEMPERATURE

As shown in Fig. 17-2.1, the resistivities of the pure metals increase nearly directly with temperature. The rationale for this increase is as follows: (a) As the temperature increases, there is more thermal agitation and, therefore, a less perfect crystal lattice through which the electrons must move. (b) This decreases the mean free paths of the electrons and reduces their net or drift velocity. (c) The lower net velocity decreases the mobility and, hence, lowers the conductivity (and raises the resistivity).

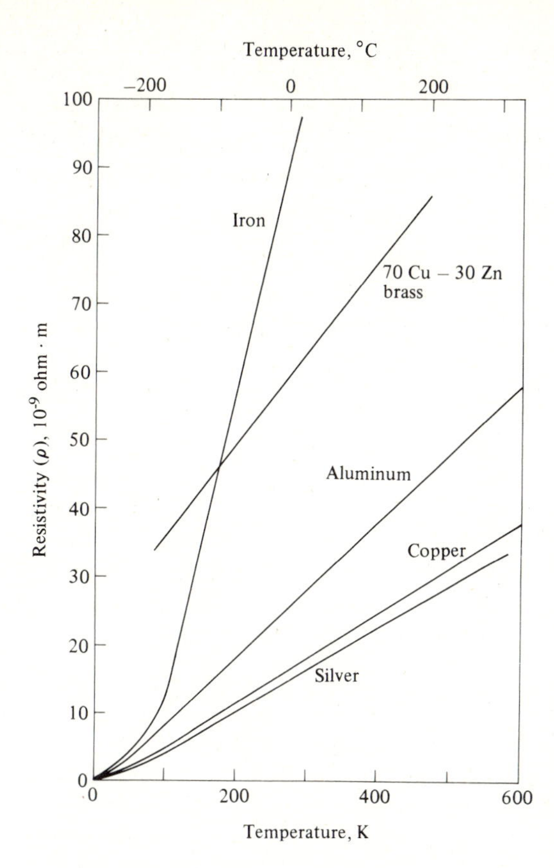

Figure 17-2.1 Resistivity versus temperature (metals). The resistivity of metals is linear with temperature under normally encountered conditions.

Temperature resistivity coefficient

The curves of Fig. 17-2.1 may be fitted empirically to the equation,

$$\rho_T = \rho_{0°C}\,(1 + y_T\,\Delta T), \tag{17-2.1}$$

where $\rho_{0°C}$ is the resistivity at 0°C and ΔT is $(T - 0°C)$; ρ_T is the resistivity at temperature, T. The slope of the curve, called the *temperature resistivity coefficient*, y_T, is ~0.004/°C (Table 17-2.1). This means that the resistivity for pure metals doubles between 0°C and 250°C, and that the mean free path is cut by a factor of two.* Alloys have lower values of y_T because they already have shorter mean free paths and higher resistivities (Section 17-3).

Equation (17-2.1) gives a good approximation of resistivities from the ambient temperature to the melting temperature of metals. However, that equation cannot be used at cryogenic temperatures because the curves of Fig. 17-2.1 do not intercept the origin as straight lines.

---

* This assumes that q and n of Eq. (17-1.4) are constant. Of course, q is, since it's the charge per electron. In a metal, our assumption for a constant n is valid, since there are always electrons available for conductivity. This latter assumption is not valid for semiconductors (Unit 18).

Table 17-2.1
Temperature resistivity coefficients

Metal		Resistivity at 0°C*, ohm·nm	Temperature resistivity coefficient, y_T, °C^{-1}
Aluminum		27	0.0039
Copper		16	0.0039
Gold		23	0.0034
Iron		90	0.0045
Lead		190	0.0039
Magnesium		42	0.004
Nickel		69	0.006
Silver		15	0.0038
Tungsten		50	0.0045
Zinc		53	0.0037
Brass	(Cu-Zn)	~60	0.002
Bronze	(Cu-Sn)	~100	0.001
Constantan	(Cu-Ni)	~500	0.00001
Monel	(Ni-Cu)	~450	0.002
Nichrome	(Ni-Cr)	~1000	0.0004

* These values will not agree with those in Appendix A, since they are based on different reference temperatures.

• Superconductivity

Certain metals and a large number of intermetallic compounds possess *superconductivity,* having zero resistivity and undetectable magnetic permeability. At a very low temperature they lose their resistivity completely.

The transition from normal conductivity to superconductivity is abrupt and occurs as a function of temperature and magnetic field (Fig. 17-2.2). The critical values of temperature, T_c, and magnetic field, H_0 at 0 K for several superconductors are shown in Table 17-2.2. Since the H-T curve of Fig. 17-2.2 is essentially parabolic for various superconductors, the data of

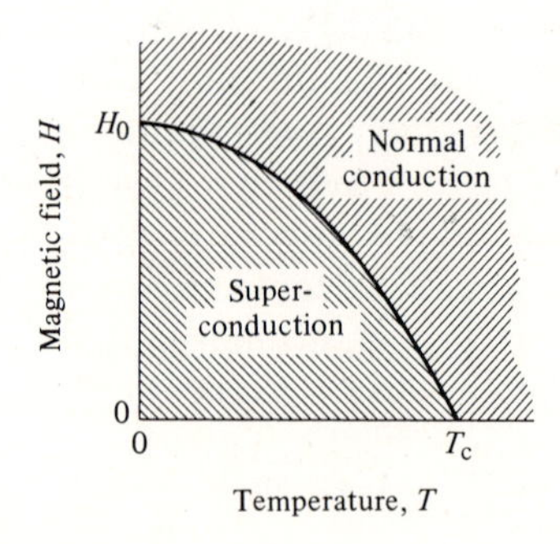

Figure 17-2.2 Conditions for superconduction. Zero resistivity and negligible magnetic permeability occur when superconducting metals are in low magnetic fields at low temperatures (Eq. 17-2.2).

Table 17-2.2
Values of T_c and H_0 for selected superconductors (See Fig. 14-2.2.)

Material	Magnetic field H_0 at 0 K, MA/m (oersteds)	Transition temperature T_c in zero field, K
Al	8.4 (106)	1.2
Hg	32.9 (413)	4.2
Nb	166 (2000)	9.2
Sn	24.3 (305)	3.7
Ti	1.6 (20)	0.4
V	104 (1310)	5.0
Nb_3Sn	400 (5000)	18.1
V_3Si		17.1
NbN		16.0
MoC		8.0
CuS		1.6

Table 17-2.2 permit a calculation of the critical magnetic field, H_c, for selected temperatures, T:

$$H_c = H_0\,(1 - T/T_c)^2. \tag{17-2.2}$$

Approximately half of the metallic elements and a large number of intermetallic compounds are known to be superconductive. Others, including all of the alkali metals and the noble metals, have been observed down to <0.1 K with no evidence of a transition to superconductivity. General and empirical rules reveal that superconductivity occurs most readily (a) among those metals with low conductivities of the normal type, and (b) among those with 3, 5, or 7 valence electrons. These generalizations have led to the formulation of numerous intermetallic superconductors, many with higher critical temperatures than with their pure metals. Of course, appreciably higher temperatures are persistently sought, since that characteristic would expand the application potentials for this attractive behavior.

Illustrative problem 17-2.1 A copper wire has a resistance of 0.5 ohm per 100 m. Consideration is being given to the use of a 75Cu-25Zn brass wire instead of a copper wire. What would the resistance be if the brass wire were the same size?

SOLUTION

$$R_{75-25} = \rho_{75-25}(L/A) = (1/\sigma_{75-25})(L/A),$$
$$R_{\mathrm{Cu}} = \rho_{\mathrm{Cu}}(L/A) = (1/\sigma_{\mathrm{Cu}})(L/A).$$

Based on Fig. 17-3.1, where $\sigma_{75-25} = 0.3\sigma_{Cu}$;

$$\frac{R_{75-25}}{R_{Cu}} = \frac{\sigma_{Cu}}{0.3\sigma_{Cu}},$$

$$R_{75-25} = R_{Cu}/0.3 = \frac{(0.5 \text{ ohm}/100 \text{ m})}{0.3} = 1.7 \text{ ohm}/100 \text{ m}.$$

ADDED INFORMATION If a wire with low resistance were the only specification, one could afford to pay more than three times as much for copper wire as for brass wire. ◀

Illustrative problem 17-2.2 At what temperature will iron and brass have the same resistivity?

SOLUTION From Eq. (17-2.1) and Table 17-2.1,

$$[\rho_{0°C}(1 + y_T \, \Delta T)]_{Fe} = [\rho_{0°C}(1 + y_T \, \Delta T)]_{Brass}$$

$$(90 \text{ ohm·nm})[1 + (0.0045/°C)\Delta T] = (60 \text{ ohm·nm})[1 + (0.002/°C)\Delta T].$$

$$\Delta T = -100°C = T - 0°C;$$

$$T = -100°C. ◀$$

17-3 ELECTRICAL CONDUCTIVITY (AND RESISTIVITY) OF ALLOY PHASES

The conductivities of solid solutions are always less than those of the pure metal. For example, as nickel dissolves in fcc copper,

pure copper	$\sigma = 60 \times 10^6$ ohm^{-1}·m^{-1}	$\rho = 0.017 \times 10^{-6}$ ohm·m;
99 a/o Cu–1 a/o Ni	35×10^6 ohm^{-1}·m^{-1}	0.029×10^{-6} ohm·m.

Also, as copper dissolves in fcc nickel,

pure nickel	$\sigma = 14 \times 10^6$ ohm^{-1}·m^{-1}	$\rho = 0.07 \times 10^{-6}$ ohm·m;
99 a/o Ni–1 a/o Cu	13×10^6 ohm^{-1}·m^{-1}	0.08×10^{-6} ohm·m.

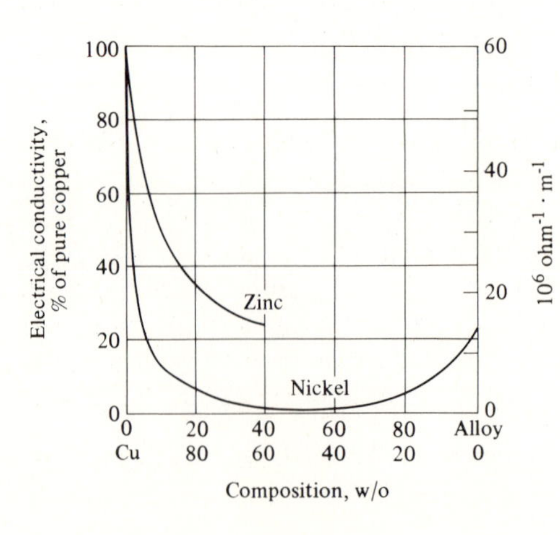

Figure 17-3.1 Electrical conductivity of solid solution alloys. The conductivity is reduced when nickel is added to copper, and when copper is added to nickel, because the impurity atoms shorten the paths of the electrons between deflections.

Table 17–3.1
Solution resistivity coefficients (in copper at 20°C)*

Solute	Coefficient, ohm·m
Ag	$y_x = 0.2 \times 10^{-6}$
Al	$y_x = 0.8 \times 10^{-6}$
Ni	$y_x = 1.2 \times 10^{-6}$
Si	$y_x = 2.0 \times 10^{-6}$
Sn	$y_x = 2.9 \times 10^{-6}$
Zn	$y_x = 0.2 \times 10^{-6}$

* For dilute solutions only (<10 a/o).

These data at 20°C for conductivities and resistivities, respectively, are shown graphically on a weight-percent basis in Fig. 17–3.1, along with similar data for brass.*

Observe that we cannot interpolate linearly between the conductivities of the two component metals since the conductivities of both decrease as they are alloyed. Also, observe that while brass is stronger and cheaper than copper (Fig. 5–2.2), the brass is unsuitable for most electrical wiring. According to Fig. 17–3.1, a 70–30 brass has only 27% of the conductivity (and ~3.7 times the resistivity) of pure copper. Thus, it would take nearly four times as much brass to maintain the same resistance in a conductor as with pure copper.

Solution resistivity coefficient

The increased resistivity (decreased conductivity) that arises from solid solution may be expressed empirically as

$$\rho_x = y_x x\,(1-x), \qquad (17\text{–}3.1)$$

where x is the atom fraction of solute (and $(1-x)$ is the atom fraction of solid solvent). The *solution resistivity coefficient,* y_x, is specific for each binary alloy. Table 17–3.1 lists values for several solutes in solid copper.

The solid solution resistivity, ρ_x, originates from the same mechanism that varies the resistivity with temperature. (a) Each foreign atom reduces the perfection of the crystal lattice through which the electrons must move. [For example, the local electric field around a nickel atom (28 protons) is different from that around a copper atom (29 protons).] (b) This decreases

* The brass data end at ~40% Zn because that is the solubility limit for zinc in fcc copper (Fig. 5–4.7).

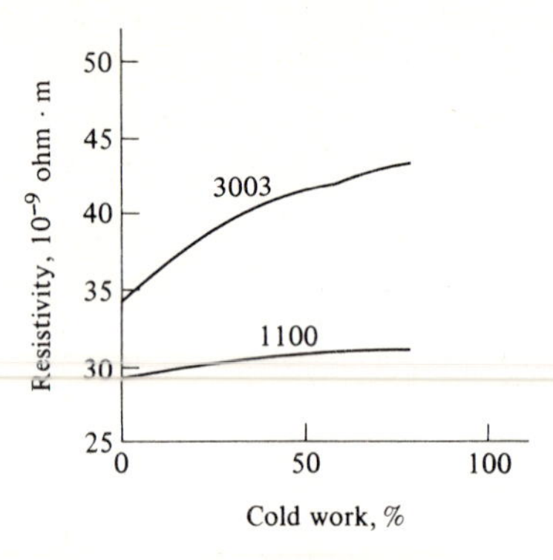

Figure 17-3.2 Electric resistivity versus cold work (wrought aluminum alloys); 1100 = 99.9% Al; 3003 = 1.2% Mn, balance Al.

the mean free path of the electrons and reduces the net, or drift, velocity. (c) The lower net velocity decreases the mobility and, hence, lowers the conductivity (and raises the resistivity).

The total resistivity, ρ, of an annealed solid solution is the sum of the two resistivities, temperature and solution,

$$\rho = \rho_T + \rho_x, \tag{17-3.2}$$

which are obtained from Eqs. (17-2.1) and (17-3.1).

Radiation damage and strain damage

Radiation exposure increases the electrical resistivity of metals. As sketched in Fig. 16-8.1, atoms are displaced. The local electric field is altered both where the displaced atom resides and where the vacancy remains. Each imperfection can deflect electrons and shorten the mean free path. Recovery from radiation damage is relatively easily obtained at moderate temperatures, since the imperfections are eliminated by a sequence of individual atom movements (Section 16-8).

Cold work also increases the electrical resistivity (Fig. 17-3.2). The increase arises in part from the dislocations that are generated by plastic deformation (Section 4-2). They deflect the electrons in their wavelike movements to shorten the mean free path and decrease the mobility. Cold work also displaces individual atoms and introduces vacancies and interstitials. Recovery and recrystallization after cold work restore the original conductivity to the annealed product.*

* The recovery of conductivity begins ahead of the softening that accompanies recrystallization (Section 4-3). This is because part of the added resistivity occurs from the vacancies and interstitials that are easily eliminated. The dislocations that account for the hardening require more extensive rearrangements. (Cf. recovery from radiation damage in Section 16-8.)

Illustrative problem 17-3.1 Estimate the resistivity of an 99Cu–1Al alloy at −100°C.

SOLUTION First, convert the alloy to atom percent.

Alloy:	100 amu		
Cu :	99 amu/(63.5 amu/Cu)	= 1.56 Cu	(or 97.7 a/o)
Al :	1 amu/(27.0 amu/Al)	= 0.037 Al	(or 2.3 a/o).
		1.60	

Using Eq. (17-3.2) and data from Tables 17-2.1 and 17-3.1,

$$\begin{aligned}\rho &= \rho_T + \rho_x \\ &= (16 \times 10^{-9}\ \text{ohm·m})(1 + (0.0039/°\text{C})(-100°\text{C})) \\ &\quad + (0.8 \times 10^{-6}\ \text{ohm·m})(0.023)(0.977) \\ &= 28 \times 10^{-9}\ \text{ohm·m};\\ \sigma &= 0.036 \times 10^{9}\ \text{ohm}^{-1}\text{·m}^{-1}. \blacktriangleleft\end{aligned}$$

Illustrative problem 17-3.2 A brass alloy is to be used in an application that must have an ultimate strength of more than 275 MPa (40,000 psi) and an electrical resistivity of less than 50×10^{-9} ohm·m (resistivity of Cu = 17×10^{-9} ohm·m). What percent zinc should the brass have?

SOLUTION Data from Figs. 5-2.2 and 17-3.1.

	0 10 20 30 % Zn
$S_u \geq 275$ MPa;	0 to 15%
$1/\rho = \sigma \geq 0.02 \times 10^{9}\ \text{ohm}^{-1}\text{·m}^{-1}$.	21% and up
Use 15% to 21% Zn.	OK ⟷

ADDED INFORMATION Alloys falling anywhere in the 15%–21% Zn will meet the specification. However, since zinc is cheaper than copper, there is reason to move to the high side of the range. Thus, an 80-20 brass, if available, would be preferred. One would hesitate specifying a 79-21 brass since any small variation in composition or behavior could miss the specification limits. ◀

17-4 THERMAL CONDUCTIVITY IN METALS

Thermal energy may be transported (a) by atomic vibrations and (b) by moving electrons. The latter mechanism is predominant in metals and absent in insulators. Therefore, in metals there is a correspondence between electrical conductivity and thermal conductivity. This is shown by comparing the thermal conductivities in Fig. 17-4.1 for brass and for Cu–Ni alloys, with their electrical conductivities in Fig. 17-3.1.

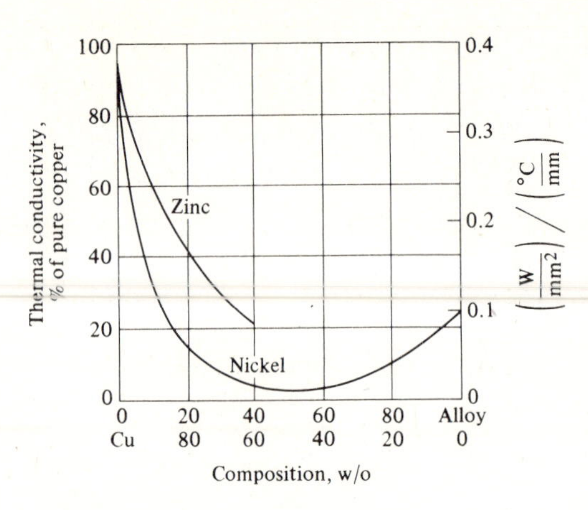

Figure 17-4.1 Thermal conductivity of solid solutions. These data correlate with the electrical conductivity of the same solid solutions (Fig. 17-3.1). In each case the solute atoms reduce the mean free path of the electron movements to decrease the conductivities.

Conductivity ratios

For metals (only), it is possible to generalize and state that the ratio between thermal conductivity, k, and electrical conductivity, σ, is a constant. This *conductivity ratio* is sometimes called the Wiedemann-Franz ratio, W-F:

$$\text{W-F} = k/\sigma. \tag{17-4.1}$$

The conductivity ratio may be obtained from the plot in Fig. 17-4.2. It is $\sim 7 \times 10^{-6}$ watt·ohm/°C at 20°C when the two conductivities are expressed as (watts/m²)/(°C/m) and (ohm^{-1}·m^{-1}), respectively. This is a convenient ratio since thermal conductivity is a difficult property to measure, whereas electrical conductivity is more simple. Not uncommonly, thermal conductivity may be estimated with greater accuracy by measuring the electrical conductivity and calculating k from the conductivity ratio, 7×10^{-6} watt·ohm/°C, in Eq. (17-4.1) than by measuring it directly.

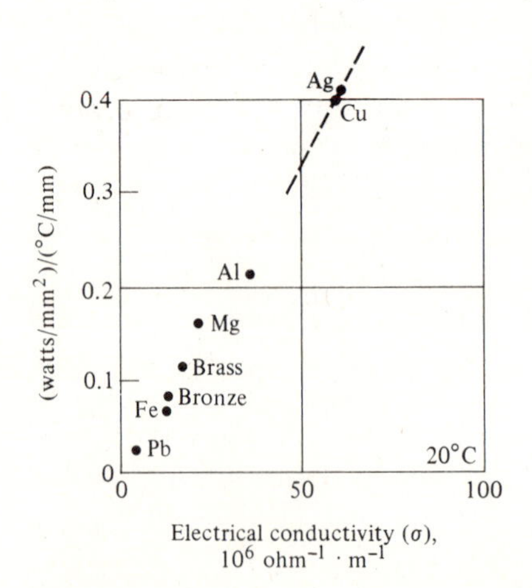

Figure 17-4.2 Conductivity ratio, k/σ. Since thermal energy may be carried by electrons, those metals with high electrical conductivity also have high thermal conductivities:

$$\text{W-F} = k/\sigma = \sim 7 \times 10^{-6} \text{ watt·ohm/°C}.$$

Illustrative problem 17-4.1 An annealed bar of 99.9% aluminum is cold worked from a diameter of 12 mm to 8 mm. This book contains no applicable thermal conductivity data. You make an estimate.

SOLUTION We have resistivity data for 1100 aluminum in Fig. 17-3.2.

$$e_{cw} = [(12\text{ mm})^2 - (8\text{ mm})^2]/(12\text{ mm})^2 = 56\%.$$
$$\rho = \sim 32 \times 10^{-9}\text{ ohm}\cdot\text{m};$$
$$\sigma = 31 \times 10^{6}\text{ ohm}^{-1}\cdot\text{m}^{-1}.$$
$$k = \sigma\,(\text{W-F}) = (31 \times 10^{6}\text{ ohm}^{-1}\cdot\text{m}^{-1})(7 \times 10^{-6}\text{ W}\cdot\text{ohm/°C}) = 220\text{ W/m}\cdot\text{°C} \qquad (\text{or } 0.22\text{ (W/mm}^2)/(\text{°C/mm}).$$

ADDED INFORMATION Laboratory measurements give 0.2 (watts/mm·°C). ◀

• 17-5 CONDUCTIVITIES (AND RESISTIVITIES) OF MULTIPHASE MATERIALS

A material that contains a mixture of two or more phases does not follow Eq. (17-3.1). That equation gave the resistivity of a solid solution within a single alloy phase. The conductivity of a multiphase mixture depends not only upon the amounts of the several phases but also upon their distribution. As a result, a calculation could become complex. However, let us examine three easily described mixtures as shown in Fig. 17-5.1. *Mixture rules* can be formulated.

The special case of Fig. 17-5.1(a) has mixtures in *parallel*. This leads to a mixture rule that is linear with volume fraction, f, of the two phases. For thermal conductivity,

$$k = f_1 k_1 + f_2 k_2; \tag{17-5.1a}$$

or for electrical conductivity,

$$\sigma = f_1 \sigma_1 + f_2 \sigma_2. \tag{17-5.1b)*}$$

The second special case is for mixtures in *series* (Fig. 17-5.1b). This leads to a mixture rule that relates the reciprocals of the conductivities. For thermal conductivity,

$$1/k = f_1/k_1 + f_2/k_2; \tag{17-5.3a}$$

* We can derive this equation from the physics law for parallel resistors,

$$(1/R) = (1/R_1) + (1/R_2); \tag{17-5.2a}$$

or in terms of resistivity (from Eq. 17-1.1),

$$(A/\rho L) = (A_1/\rho_1 L) + (A_2/\rho_2 L). \tag{17-5.2b}$$

Since $A_1 = f_1 A$, and $A_2 = f_2 A$,

$$1/\rho = f_1/\rho_1 + f_2/\rho_2, \tag{17-5.1c}$$

which is equivalent to Eq. (17-5.1b).

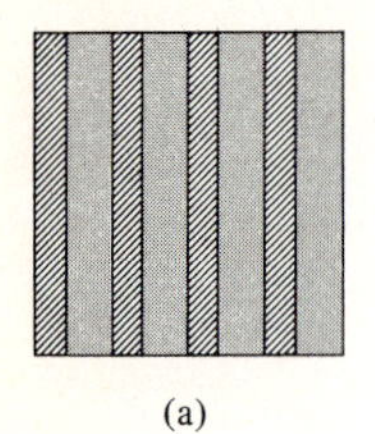
(a)

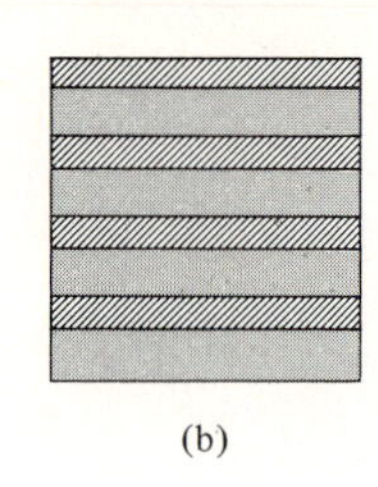
(b)

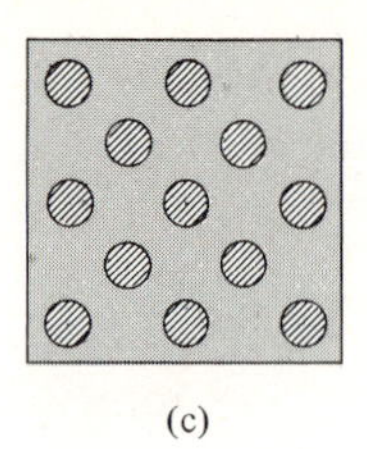

(c)

Figure 17-5.1 Conductivity versus phase distribution (idealized). (a) Parallel conductivity (Eq. 17-5.1). (b) Series conductivity (Eq. 17-5.3). (c) Conductivity through a material with a dispersed phase (Eq. 17-5.5).

or for electrical conductivity,

$$1/\sigma = f_1/\sigma_1 + f_2/\sigma_2. \qquad (17\text{-}5.3b)^*$$

Although the two mixture rules just described have special applications, neither is typical of microstructures of materials. The sketch in Fig. 17-5.1(c) represents an idealized form of microstructure containing *dispersed particles*. There are many such microstructures—for example, spheroidite in which equiaxed carbides are dispersed in a matrix of ferrite (Fig. 6-3.3b). The corresponding mixture rule takes on two empirical forms, depending on whether the continuous (matrix) phase, c, is markedly more, or markedly less, conductive than the dispersed phase, d (Fig. 17-5.2). When $k_c > 10\ k_d$,

$$k \cong k_c\,(1 - f_d)/(1 + f_d/2). \qquad (17\text{-}5.5a)$$

When $k_c < 0.1\ k_d$,

$$k \cong k_c\,(1 + 2f_d)/(1 - f_d). \qquad (17\text{-}5.5b)$$

When the two conductivities are not markedly different or when the dispersed particles develop contact, the conductivity of the mixture is approximately linear (on a volume fraction basis) between the conductivities of the two phases.

Equations (17-5.5a and b) are approximate. They were derived for spheres of uniform size and spacing. We do not find such exact geometries in any real microstructure. Nonspherical phase distributions, such as pearlite (Fig. 5-3.3a), should require a different, and more complex, mixture rule. Fortunately, the less regular the microstructural geometry is, the more appropriate a linear mixture rule becomes. Thus, for many multiphase microstructures,

$$k \cong \Sigma\, f_i k_i. \qquad (17\text{-}5.6)$$

The relationships of this section apply to the conductivities of both metallic and nonmetallic mixtures since they do not depend on the mechanism of conductivity.

---

* The reader is asked to derive this equation from the physics law for series resistors,

$$R = R_1 + R_2 + \cdots. \qquad (17\text{-}5.4)$$

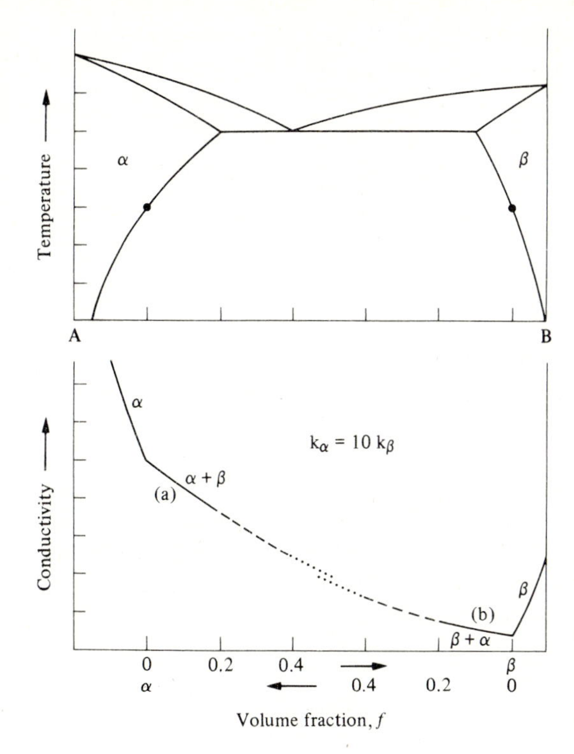

Figure 17-5.2 Conductivity of two-phase microstructures (dispersed phases in a continuous matrix). Curve (a): β is dispersed in a matrix of α. Curve (b): α is dispersed in a matrix of β. (The curves are not extended beyond $f_d = 0.4$, because grains of the second phase commonly develop contact.)

Illustrative problem 17-5.1 A 95/5 bronze bearing that was made of powdered metal contains 16 v/o porosity as dispersed pores. Estimate (a) its thermal conductivity from the data in Fig. 17–4.2; (b) its electrical conductivity from the data in Appendix A.

SOLUTION Consider this a two-phase material of bronze and air, the latter with near-zero conductivity.

a) $k_c = \sim 0.08\ (\text{watts/mm}^2)/(°\text{C/mm})$
$k = 0.08\ (1 - 0.16)/(1 + 0.08)$
$= \sim 0.06\ \text{watts/°C·mm}$ (or 62 W/°C·m).

b) $\sigma_c = \sim 10^7\ \text{ohm}^{-1}\cdot\text{m}^{-1}$
$\sigma = 10^7\ (0.84/1.08)$
$= \sim 8 \times 10^6\ \text{ohm}^{-1}\cdot\text{m}^{-1}$.

ADDED INFORMATION The pores are somewhat spherical in the final sintered product. The conductivities would have been much lower prior to sintering, because the pores would have extended far between the metal particles, and the bronze would have had only limited continuity. ◀

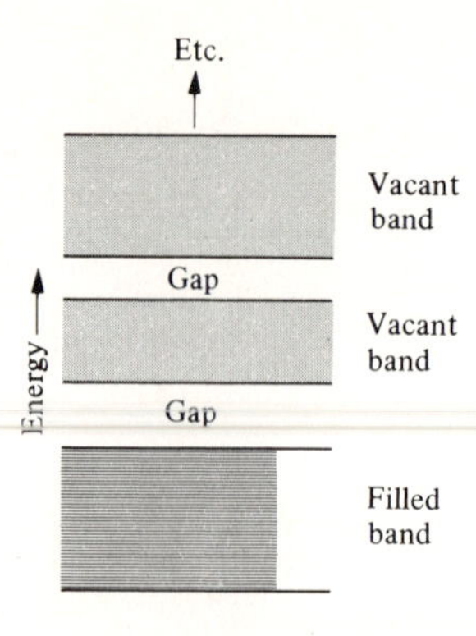

Figure 17-6.1 Energy bands and gaps (schematic of insulators). Valence electrons completely fill the lower energy bands in an insulator. They cannot receive the increased incremental energy to produce a drift velocity because there is an overlying forbidden energy gap. (The crystal lattice cannot support wave motions with those energies.)

17-6 CONDUCTORS, SEMI-CONDUCTORS, AND INSULATORS

As observed in Section 17-1, the difference between a metallic conductor and an insulator is not the number of electrons. Although both rubber and copper possess $\sim 0.3 \times 10^{24}$ electrons per gram, their conductivities differ by ~20 orders of magnitude. Electronic conductivity and resistivity originate with the constraints placed on the valence electrons.

In any solid, the energies of electrons are grouped into *bands*. In an *insulator*, the valence electron band is filled. A wide forbidden *energy gap* lies above the valence band, separating it from the next available energy band (Fig. 17-6.1).* We also noted in Section 17-1 that valence electrons in metals achieve a net, or drift, velocity as they are accelerated in their movements toward the positive electrodes (and decelerated in the opposite direction). Gaining velocity, they also gain energy; however, these increases in energy are not possible in an insulator with its filled energy band and overlying forbidden energy gap. Therefore, it is impossible to provide them with a drift velocity, and no charge is transported.

Metals differ from insulators in that the valence band is *not* filled (Fig. 17-6.2a). Thus, the valence electrons (•) of metals can be energized to higher velocities and remain within the permissible energy range (Fig. 17-6.2b). Also, with some of the electrons energized toward the upper part of the valence band, there are vacant levels deeper in the band. Those electrons that are moving toward the negative electrode can drop to these electron holes (○) at lower energy levels as they are decelerated. These also contribute to the net, or drift, velocity of the charge. This is not possible in the filled valence band of an insulator.

Semiconductors have resistivity values between the extremes of metallic conductors and insulators (Fig. 17-1.1). Like insulators, semiconductors

* An energy gap occurs because the electrons possess the characteristics of a standing wave within a solid material. Certain frequencies, and therefore certain energies, cannot maintain a standing wave in a perfect lattice—just as a perfectly tuned piano will not produce standing waves that do not match the characteristics of the vibrating strings.

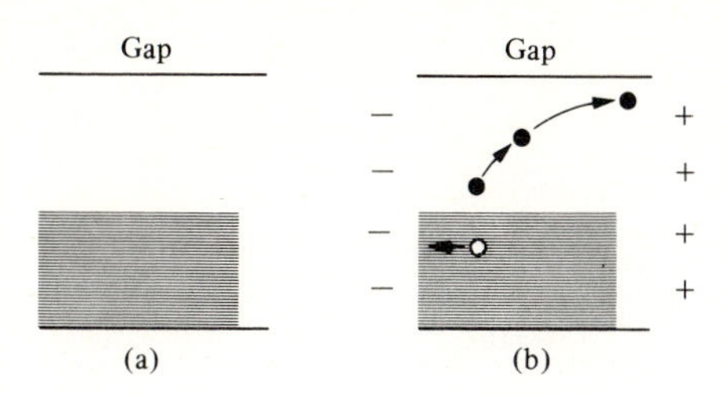

Figure 17-6.2 Valence energy band (schematic of metals). (a) At absolute zero. (b) $T > 0$ K. In a metal the valence band is not filled; therefore electrons (•) can be energized by an electric field to produce net movements toward the positive electrode. Electron holes (○) in the lower part of the valence band move toward the negative electrode.

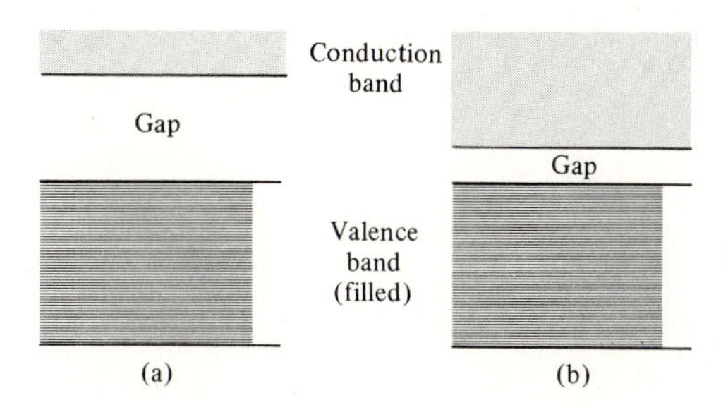

Figure 17-6.3 Energy gaps (schematic). (a) Insulators. (b) Semiconductors. In the latter, the gap is sufficiently narrow so that a limited number of electrons can "jump the gap."

have filled valence bands; unlike insulators, the energy gap between the valence band and the next higher band is narrow (Fig. 17-6.3). Being narrow, some of the electrons can "jump the gap," that is, some electrons can receive enough thermal or photon energy so that they are raised across the forbidden energy gap and "live" in the next energy band. In that band they can be accelerated in their response to the positive electrode and raise the net, or drift, velocity.

Review and Study

SUMMARY

1. Conductivity (and resistivity) are properties of materials independent of their shapes. Conductivity is the product of charge carriers, charge per carrier, and charge mobility. Alternatively, it is the ratio of current density to electric field. Charge mobility is the ratio of drift velocity to electric field.
2. The drift velocity is determined by the mean free path that an electron moves before it is deflected, or reflected, and is greatest in perfect lattices. Increased thermal vibrations (from increased temperatures), foreign atoms (from alloying), and displaced atoms (from plastic deformation, from radiation damage, and even grain boundaries) provide crystalline irregularities that shorten the mean free path and, therefore, increase the resistivity (decrease the conductivity) of a metal.

3. In a metal, the major portion of the thermal conduction is by electron transport. Therefore (in metals), there is a close correlation between electrical and thermal conductivity. We even use the ratio $k/\sigma = \sim 7 \times 10^{-6}$ watt·ohm/°C as a means of estimating thermal conductivities of metals.
4. Mixture rules are available for estimating the conductivities (thermal and electrical) of multiphase materials. Except where the conductivity of the phases differ markedly, for example, in a porous material, the conductivities vary approximately linearly with volume fraction.
5. Metals differ from semiconductors and insulators by possessing an energy band that is only partially filled with valence electrons.

TECHNICAL TERMS

Conductivity (σ) Transfer of thermal or electrical energy along a potential gradient.

Conductivity ratio (W–F) Ratio of thermal conductivity to electrical conductivity—commonly called the Wiedermann–Franz ratio.

Drift velocity ($\bar{v}$) Net velocity of electrons in an electric field.

Electron charge (q) The charge of 0.16×10^{-18} coul (or 0.16×10^{-18} amp·sec) carried by each electron.

Electron hole (p) Electron vacancy in the valence band that serves as a positive charge carrier.

Energy band Permissible energy levels for valence electrons.

Energy gap (E_g) Unoccupied energies between the valence band and the conduction band.

Insulator Nonconductor of (a) electrical energy or (b) thermal energy; in either case, the insulator has significant electronic resistivity. Material with filled valence bands and a large energy gap.

Mean free path Mean distance traveled by electrons (or by elastic waves) between deflections or reflections.

Metal Material with partially filled valence bands. (*Also see* Fig. 1-3.2.)

Mobility (μ) The drift velocity, $\bar{v}$, of an electric charge per unit electric field, $\mathcal{E}$, (m/sec)/(volt/m). Alternatively, the diffusion coefficient of a charge per volt, $(m^2/sec)/volt$.

Resistance (R) Ohms, or volts/amp.

Resistivity (ρ) Reciprocal of conductivity (usually expressed in ohm·m).

Resistivity coefficient, solid solution (y_x) Resistivity arising from additions of solute. Change in resistivity per unit of solute.

Resistivity coefficient, thermal (y_T) Electrical resistivity arising from thermal agitation. Change in resistivity per °C.

Semiconductor A material with controllable conductivities, intermediate between insulators and conductors.

Superconductivity Property of zero resistivity and magnetic permeability near 0 K.

Valence band Highest energy band normally occupied by electrons. It is only partially filled in metals.

CHECKS

17A Conductivity is the __***__ of resistivity; it is also the ratio of __***__ to __***__; furthermore, it is the product of __***__, __***__, and mobility, where mobility is the ratio of __***__ to __***__.

17B The mean free path of an electron is reduced when the atomic structure is less regular. The situation occurs at __***__ temperatures, with __***__ alloy contents, and with __***__.

17C If the conductivity of a metal is 4×10^7 $ohm^{-1} \cdot m^{-1}$ at −100°C and 2×10^7 $ohm^{-1} \cdot m^{-1}$ at 0°C, it will be __***__ $ohm^{-1} \cdot m^{-1}$ at +100°C.

17D Many unalloyed metals double their resistivity between 0°C and approximately __***__ °C.

17E Superconductivity has upper limits for __***__ and __***__. Good superconductors are normally __***__ metallic conductors.

17F The conductivity of brass is less than of copper because the __***__ of electron movements are shortened by a different __***__ around the zinc atoms than that around the copper atoms.

17G The imperfections that decrease conductivity in a strain-hardened material are __***__ and __***__; in a radiation-damaged material, they are primarily __***__.

17H During the heating that follows cold work, the electrical conductivity is recovered ahead of the softening because __***__ . . . .

17I In metals, the thermal energy is transferred primarily by __***__; therefore, the thermal conductivity is proportional to the __***__.

• 17J Other factors equal, the conductivity through a mixture in parallel is __***__ than the conductivity through a mixture in series.

• 17K In calculating the conductivity of a porous material, the pores may be considered to be the second phase, with a conductivity of ~__***__.

17L The energy band of a __***__ is only partially filled; while it is filled for both __***__ and __***__.

17M In a semiconductor, a useful number of electrons can be raised across the energy __***__; extremely few possess that much energy in an __***__.

17N If an electron is raised to a higher energy level, an __***__ is left below.

17O In an insulator, electrons cannot be accelerated for __***__ movement toward the positive electrode because there are no __***__ levels in the valence band.

STUDY PROBLEMS

17–1.1. (I^2R review) A dry cell (1.5 V) is connected across a 10-ohm circuit. (a) What is the resulting current? (b) What is the wattage, or energy rate? (c) How much energy will be used per hour?
Answer: (a) 0.15A (b) 0.225 W, or 0.225 J/s
(c) 810 J, or 0.225 Wh

17–1.2. (I^2R review) For cathodic protection (Section 16–4), 1 kW of d.c. power is fed to a pipeline from 16 V sources. (a) What is the current? (b) What is the effective resistance of the circuit?
Answer: (a) 62 A (b) 0.26 ohms

17–1.3 The resistivity of iron is 10^{-7} ohm·m. What is the resistance per meter of an iron wire that has a 1.5-mm diameter?
Answer: 0.056 ohm/m

17–1.4 A wire must have a diameter of less than 1 mm, and a resistance of less than 0.1 ohm/m. Which of the materials in Appendix A is suitable?

17–1.5. An 18-m copper wire connects two terminals of a dry cell (1.5 V). (a) What is the current density in the wire? (b) How many watts go through the wire, if it has a 3-mm² cross section? (c) What diameter wire must be used to limit the current to one ampere?
Answer: (a) 5×10^6 A/m² (b) 22 W (c) 0.5-mm

17–1.6. Refer to part (c) of the preceding problem. How many electrons pass through the wire per minute?

17–1.7. A material has a resistivity of 0.1 ohm·m. There are 10^{20} electrons/m³ that can serve as charge carriers. (a) What is the conductivity? (b) The charge mobility? (c) What electric field is required for a drift velocity of 1 m/s?
Answer: (a) 10 ohm^{-1}·m^{-1} (b) 0.63 m²/V·s (c) 1.6 V/m

17–1.8. A material with a charge mobility of 0.39 m²/V·s develops a conductivity of 100 ohm^{-1}·m^{-1}. (a) How many charge carriers are involved? (b) What is the current density when the drift velocity of the charge carriers is 10 m/s?

17–2.1. What is the resistivity of a tungsten wire at 800°C?
Answer: 230 ohm·nm (or 230×10^{-9} ohm·m)

17–2.2. (a) At what temperature will the resistivity of copper double its 100°C value? (b) Although its resistivity is high, the cupronickel alloy of constantan has technical merits as a conductor in certain applications. Discuss.

17–2.3. The electric power loss in a low-voltage copper power line is 1.01% per kilometer on a 0°C (32°F) day. What is the power loss on a 30°C (86°F) day?
Answer: 1.13%

17–2.4. A 6% variation (maximum) is permitted in the resistance of a conductor between 0° and 25°C. Which metals of Table 17–2.1 meet the specification?

17–2.5. Nichrome is used as a heating element in a toaster. The toaster draws 550 W (at 110 V) when it is hot (1000°C). (a) What is the current? (b) What is the initial amperage each time it is turned on?
Answer: (a) 5 A (b) 7A

17–3.1. From Tables 17–2.1 and 17–3.1, estimate the resistivity of a 95Cu–5Sn bronze (a) at 0°C; (b) at 100°C.

Answer: (a) 93×10^{-9} ohm·m (b) 102 ohm·nm

17–3.2. Determine the resistivity of a red brass (85Cu–15Zn) at 210°C.

17–3.3. An annealed cupronickel alloy must have an ultimate strength of >350 MPa (>51,000 psi), and an electrical conductivity of $>5 \times 10^6$ $ohm^{-1} \cdot m^{-1}$. Select a composition with (a) a minimum nickel content; (b) a minimum copper content.

Answer: (a) 90Ni–10Cu

17–3.4. A copper alloy (either brass or cupronickel) is required that has an ultimate strength of at least 245 MPa (35,000 psi) and a resistivity of less than 50×10^{-9} ohm·m. (Resistivity of copper = 17×10^{-9} ohm·m.) (a) Select a suitable alloy, bearing in mind that $\$_{Ni} > \$_{Cu} > \$_{Zn}$. (b) What is the ductility of the alloy that you chose?

17–3.5. A brass wire must carry a load of 45 N (10 lb) without yielding and have a resistance of less than 0.033 ohm/m (0.01 ohm per foot). (a) What is the smallest wire that can be used if it is made of 60–40 brass? (b) 80–20 brass? (c) 100% Cu?

Answer: (a) 1.8 mm (0.07 in.) (b) 1.4 mm (0.055 in.) (c) 1.2 mm (0.047 in.)

17–3.6. A wire with a diameter of 1 mm has a resistance of 0.33 ohm/m. It is cold-drawn to a diameter of 0.8 mm and then has a resistance of 0.66 ohm/m. What fraction of the initial conductivity remains after the cold working?

17–4.1. Estimate the thermal conductivity of silver at 0°C from the data in Table 17–2.1.

Answer: 0.47 W/mm·°C

17–4.2. Estimate the thermal conductivity of copper at 40°C.

17–4.3. The design engineer has a choice of either a brass alloy or a Cu–Ni alloy to meet room-temperature specifications of hardness > $80R_f$, $S_u > 275$ MPa (40,000 psi), and thermal conductivity, $k > 40$ W/m·°C (>10% k_{Cu}). Pick a suitable composition.

Answer: Use 85Ni–15Cu

17–4.4. A certain application requires a piece of metal having a yield strength greater than 100 MPa and a thermal conductivity greater than 0.04 W/mm·°C. Specify either an annealed brass or an annealed Cu–Ni alloy that will meet the requirements.

• 17–5.1. Refer to Fig. 6–6.2(b), in which ferrite and graphite have the properties of iron and graphite in Appendix A. (a) Estimate the volume fraction of graphite.* (b) Estimate the thermal conductivity of this cast iron. ($k_{gr} = \sim 0.002$ W/m·°C.)

Answer: (a) $12 \pm 3\%$ v/o (b) 0.06 W/m·°C

* An "eye-ball" estimate is possible. A better estimate is obtained by laying a transparent millimeter scale at random across the photomicrograph several times, and counting the fraction of the ticks that lie on the graphite.

• 17-5.2. Derive Eq. (17-5.3).

• 17-5.3. A laboratory furnace liner is principally Al_2O_3, but it is porous and has a density of only 3.0 g/cm^3 rather than the full 3.85 g/cm^3. Estimate its thermal conductivity from the data in Appendix A.

Answer: 0.022 W/mm·°C

CHECK OUTS

17A reciprocal
current density
electric field
charge concentration
electron charge
drift velocity
electric field

17B higher temperatures
higher alloy contents
radiation exposure
plastic deformation

17C 1.3×10^7

17D 250

17E temperature, magnetic field
poor

17F mean free path
electric field

17G dislocations, point imperfections
point imperfections

17H point imperfections are removed first

17I electrons
electrical conductivity

17J greater

17K zero

17L metal
semiconductor, insulator

17M gap
insulator

17N electron hole

17O net
vacant

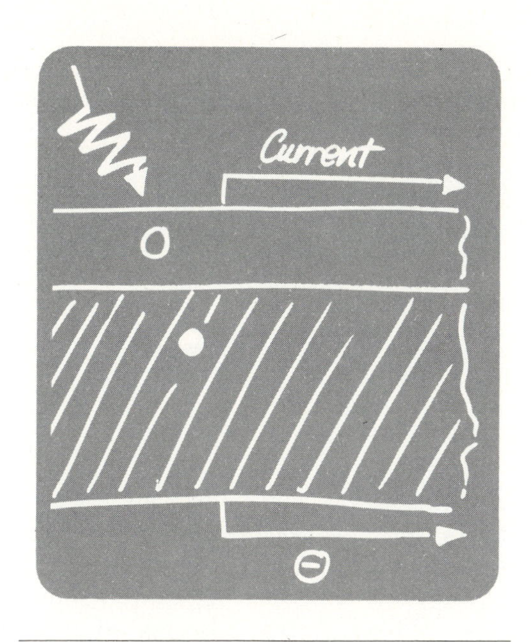

UNIT EIGHTEEN

SEMICONDUCTORS AND SIMPLE ELECTRONIC DEVICES

A technical revolution was initiated when the transistor was invented. It is based on semiconducting materials. The teamwork of materials scientists and electrical engineers has developed a multitude of devices and products that penetrate into every facet of modern society—the outstanding example being the computer. Electronic devices play a particularly important role in the various branches of engineering, so much so that every engineer should know the basic principles of semiconductivity and of junction devices.

Contents

Prerequisites General physics; Units 2 and 17; and Sections 11–1, 11–5 and 12–5.

From Unit 18, the engineer should

1) Develop an understanding of the relationship of the energy gap, and of temperature, to the number of carriers available in an intrinsic semiconductor.

2) Understand the mechanisms of both n-type and p-type semiconduction.
3) Be able to calculate the conductivity of an extrinsic semiconductor from the impurity concentration and the charge mobility.
4) Be able to explain the bases for simple semiconducting devices of both the conduction and the junction categories.
• 5) Be familiar with the elementary steps for device preparation.
6) Possess an understanding of the technical terminology in the semiconduction area.

18-1 SEMICONDUCTIVITY

In the previous unit, semiconductors were described in two distinct ways. Figure 17-1.1 shows semiconductors located between conductors ($\sigma > 10^5$ ohm$^{-1}\cdot$m^{-1}) and insulators ($\sigma < 10^{-10}$ ohm$^{-1}\cdot$m^{-1}, or $\rho > 10^{10}$ ohm·m). In Section 17-6, semiconductors were described as those materials with filled *valence bands,* but with a small *energy gap* between the upper filled band and the overlying vacant energy band. This latter concept is of greater importance to the engineer than the first, because it gives a basis for understanding electronic devices. Thus, it will be desirable to look more closely at the charge carriers in semiconductors and at the energy gap that is involved in semiconduction.

Charge carriers

Negative charge carriers include *electrons* and *anions.* The latter are atoms that possess one or more excess electrons. The charge on an electron is 0.16×10^{-18} A·s; and the charge on a negative ion is an integer multiple of the above value.

Positive charge carriers are *cations* and electron *holes.* The former are easily envisaged. They are on atoms that are deficient by one or more electrons. The charge on a positive ion is also an integer multiple of 0.16×10^{-18} A·s, depending on the number of deficient electrons. Cations migrate with the electric field to the negative electrode; anions move in the opposite direction to the positive electrode.

Just as a cation is an atom that is short one or more of its valence electrons, an electron hole is a material that lacks one or more electrons in its valence band. This is probably a new concept for most readers. Figure 18-1.1 may help to develop that concept. As stated before, a semiconductor has a full *valence band,* VB; a small *energy gap,* E_g; and an unoccupied

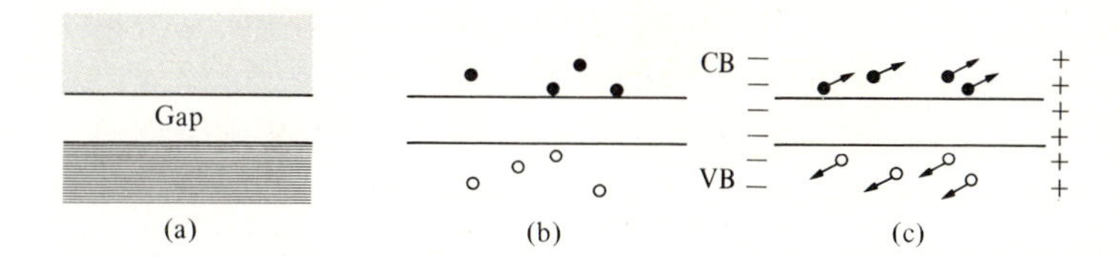

Figure 18–1.1 Semiconductors (schematic). (a) Zero kelvin. The lower band is filled; the gap is narrow; and the upper band is empty. (b) Increased temperature. Some electrons jump the narrow energy gap, leaving electron holes in the valence band. (c) Electric field. The electrons in the conduction band, CB, move toward the positive electrode. The electron holes in the valence band, VB, move toward the negative electrode.

energy band above the gap that we will be calling the *conduction band,* CB. This is in the absence of any thermal energy, that is, at zero Kelvin. As the temperature is raised, a small fraction of the electrons in the valence band will pick up enough thermal energy to "jump the gap." This leaves electron holes in the valence band (Fig. 18–1.1b). Just as an electron-deficient atom (a cation) carries a positive charge, the absence of an electron in the valence band carries a positive charge. This electron hole will move toward the negative electrode—and the electron in the upper, or conduction, band is accelerated toward the positive electrode (Fig. 18–1.1c).

Conductivity calculations

Both electrons and electron holes transport charge. Therefore, the total conductivity in a semiconductor is the sum of the conductivity by negative carriers, σ_n, and the conductivity by positive carriers, σ_p. Modifying Eq. (17–1.4a), we have

$$\sigma = \sigma_n + \sigma_p = n_n q \mu_n + n_p q \mu_p. \tag{18–1.1}^*$$

The two kinds of carriers are (a) the electrons in the conduction band, and (b) the electron holes in the valence band. With the situation we have selected for Fig. 18–1.1, there are equal numbers of electrons, n_n, and electron holes, n_p.

Illustrative problem 18–1.1 A semiconductor has a conductivity of 0.01 $\text{ohm}^{-1}\cdot\text{m}^{-1}$. With mobilities of 0.50 $\text{m}^2/\text{V}\cdot\text{s}$ and 0.02 $\text{m}^2/\text{V}\cdot\text{s}$, respectively, for the electrons and electron holes, calculate (a) the number of electrons in the conduction band, and (b) the number of electron holes in the valence band.

---

* Strictly speaking, the signs of μ_n and μ_p are opposite because the two types of charge carriers are accelerated in opposite directions. Likewise, the signs of the charges are opposite. However, the two minus signs for the negative carriers cancel, relieving us of the necessity of inserting those signs in Eq. (18–1.1).

SOLUTION The numbers n_n and n_p will be the same, since one hole remains for each electron that jumps the gap. From Eq. (18–1.1),

$$\sigma = n(q\mu_n + q\mu_p),$$
$$n = \sigma/q(\mu_n + \mu_p)$$
$$= 0.01/(0.16 \times 10^{-18}\ \text{A·s})^{-1}(0.50 + 0.02\ \text{m}^2/\text{V·s})^{-1}.$$
$$n_n = n_p = 1.2 \times 10^{17}/\text{m}^3 \qquad (\text{or } 1.2 \times 10^8\ /\text{mm}^3).$$

ADDED INFORMATION The number of charge carriers increase exponentially with temperature; $10^{16}/\text{m}^3$ is typical of many semiconductors at 20°C. ◄

18–2 INTRINSIC SEMICONDUCTORS

Those semiconducting materials that do not depend upon impurities for their conductivity are called *intrinsic* semiconductors. They are inherently semiconductive.

Group IV semiconductors

The elements of the fourth group of the periodic table include carbon (diamond), silicon, germanium, and tin (gray). These are shown with their neighbors in Fig. 18–2.1 Each of these elements has four valence electrons and is similar chemically. As a result, they have the same crystal structure (Fig. 18–2.2a). Likewise, they have comparable band structures; each has filled valence bands, and an energy gap, above which there is another energy band that is vacant (at zero kelvin).

The sizes of the energy gaps of these four members of Group IV are not identical—decreasing from ~6 eV (or ~10^{-18} J) for diamond, to 1.1 eV for silicon, 0.7 eV for germanium, and only 0.1 eV (or 0.016×10^{-18} J) for gray tin. With decreasing gap sizes, we find that an increasing fraction of the valence electrons possess the required thermal energy to jump the gap at room temperature (Fig. 18–2.3). These fractions are shown in Table 18–2.1.

II	III	IV	V	VI
Be	B	C	N	O
Mg	Al	Si	P	S
Ca	Ga	Ge	As	Se
Sr	In	Sn	Sb	Te

Figure 18–2.1 Group IV elements. (Cf. Fig. 1–3.2.) There are four valence electrons per atom; therefore the valence bands are exactly filled. There are no electrons in the conduction band, unless additional energy (for example, heat) is supplied.

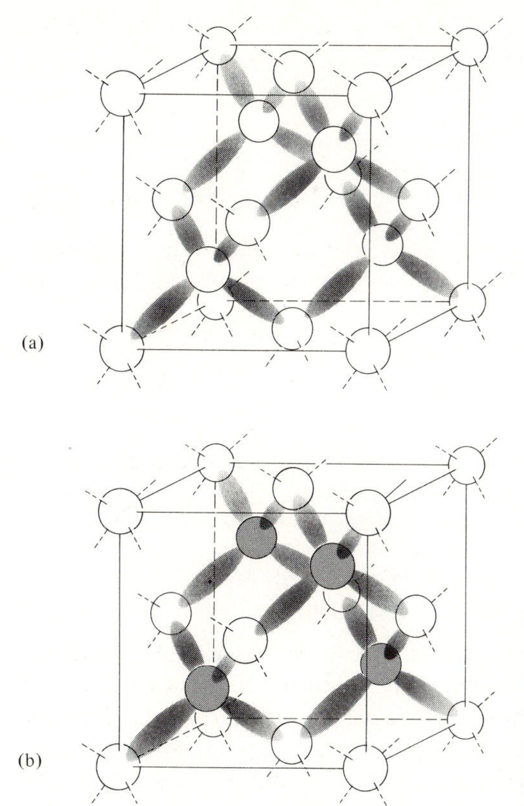

Figure 18-2.2 Crystal structure of familiar semiconductors. (a) Diamond, silicon, germanium, gray tin. (b) ZnS (Fig. 11-1.3), GaP, GaAs, InP, etc. The two are similar, except that two types of atoms are in alternate positions in the semiconducting compounds. All atoms have CN = 4; each material has an average of four valence electrons per atom, and two electrons per bond.

Figure 18-2.3 Energy gaps (Group IV elements). (a) Carbon (diamond). (b) Silicon. (c) Germanium. (d) Tin (gray). Since diamond has a very large energy gap (~6 eV), very few electrons jump the gap; therefore, it has extremely low conductivity. The energy gap of gray tin is ~0.1 eV. At normal temperatures tin has more than 10^{16} electrons and electron holes per mm^3; its conductivity is ~2×10^{16} ohm$^{-1} \cdot$ m^{-1}.

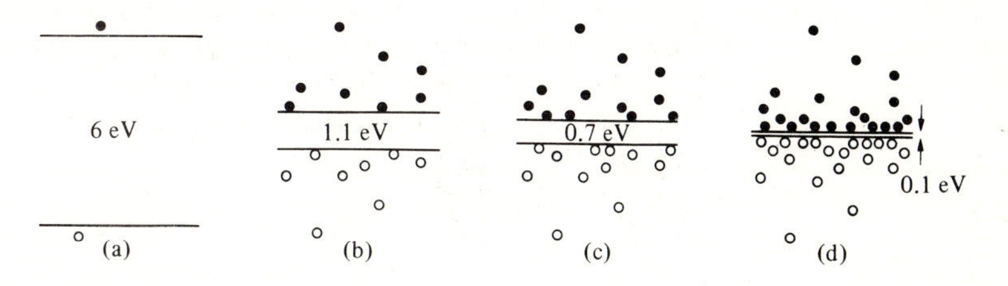

Table 18-2.1
Energy gaps in semiconducting elements

	Energy gap E_g		At 20°C (68°F)	
Element	**10^{-18} J**	**eV**	**Fraction of valence electrons with energy > E_g**	**Conductivity σ, ohm^{-1}·m^{-1}**
C(diamond)	0.96	~6	~$1/30 \times 10^{21}$	$<10^{-16}$
Si	0.176	1.1	~$1/10^{13}$	5×10^{-4}
Ge	0.112	0.7	~$1/10^{10}$	2
Sn(gray)	0.016	0.1	~1/5000	10^6

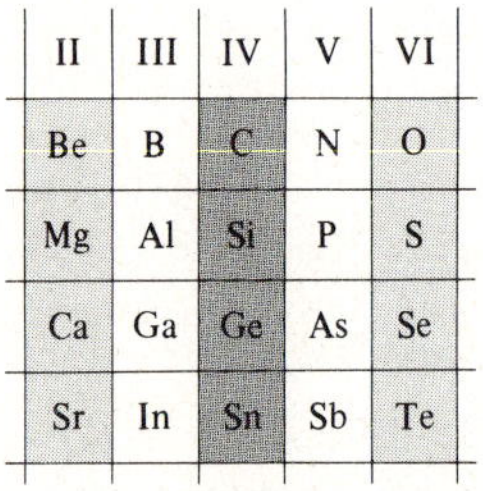

II	III	IV	V	VI
Be	B	C	N	O
Mg	Al	Si	P	S
Ca	Ga	Ge	As	Se
Sr	In	Sn	Sb	Te

Figure 18-2.4 III–V elements. Compounds of Group III and Group V atoms average four valence electrons per atom; therefore, the valence bands are exactly filled, and the crystal structure corresponds to that of the Group IV elements (Fig. 18-2.2).

With more electrons in the conduction band (and also more holes in the valence band), the conductivity also increases. The energy gap of diamond is so wide that its conductivity is less than 10^{-16} ohm^{-1}·m^{-1}. It is an insulator. In contrast, germanium has a conductivity of 2 ohm^{-1}·m^{-1}; and gray tin, 10^6 ohm^{-1}·m^{-1}.*

Semiconducting compounds

In addition to semiconducting Group IV elements, there are a number of closely related semiconducting compounds. Most common among these are the III–V compounds, which have equal numbers of Group III elements and Group V elements, for example, AlP, GaAs, and InSb (Fig. 18-2.4). These III–V compounds (as well as several II–VI compounds) have the same structures as do the Group IV semiconductors except that alternate atoms are different (Figs. 18-2.2a and b). The structure shown in Fig. 18-2.2(b) is also identical with that of Fig. 11-1.3, where we examined ceramic AX compounds with CN = 4.

* Germanium is an excellent semiconductor and was used in the early transistors. However, it is very uncommon and therefore expensive. Silicon is widely available. Tin has two polymorphs, gray and white. Gray tin has the structure of Fig. 18-2.2. White tin is more dense (7.3 g/cm³ versus 5.8 g/cm³) and, therefore, has energy bands that overlap. Thus, white tin is not a semiconductor, but a metallic conductor. (Recall, also, that carbon has two polymorphs with markedly different electrical properties.)

Table 18-2.2
Properties of common semiconductors (20°C)*

Material	Energy gap E_g		Mobilities, m^2/volt·sec		Intrinsic conductivity, $ohm^{-1}·m^{-1}$	Lattice constant, *a*, nm
	10^{-18} J	eV	Electron, μ_n	Hole, μ_p		
Elements						
C(diamond)	0.96	~6	0.17	0.12	$<10^{-16}$	0.357
Silicon	0.176	1.1	0.19	0.0425	5×10^{-4}	0.543
Germanium	0.112	0.7	0.36	0.23	2	0.566
Tin (gray)	0.016	0.1	0.20	0.10	10^6	0.649
Compounds						
AlSb	0.26	1.6	0.02	—	—	0.613
GaP	0.37	2.3	0.019	0.012	—	0.545
GaAs	0.22	1.4	0.88	0.04	10^{-6}	0.565
GaSb	0.11	0.7	0.60	0.08	—	0.612
InP	0.21	1.3	0.47	0.015	500	0.587
InAs	0.058	0.36	2.26	0.026	10^4	0.604
InSb	0.029	0.18	8.2	0.17	—	0.648
ZnS	0.59	3.7	0.014	0.0005	—	—
SiC (hex)	0.48	3	0.01	0.002	—	—

* Revised data collected by B. Mattes.

The III–V and II–VI compounds are semiconductors because they average four valence electrons per atom. Identical with the Group IV elements, the valence bands are completely filled; there is also an energy gap between the valence band and the next energy band. The size of the energy gap varies from compound to compound but, in general, is greater for compounds containing elements higher in the periodic table. (See the sequence (a) of GaP, GaAs, GaSb, or (b) of AlSb, GaSb, InSb in Table 18-2.2 and Fig. 18-2.4).

The electrical engineer also designs electronic devices with *alloys of compounds,* for example, a solid solution of InAs and GaAs to give (In,Ga)As, or a solid solution of InP and InAs to give In(As,P). Furthermore, 4-component compounds of (In,Ga)(As,P) are possible. All of these possess the same structure as the individual III–V compounds (Fig. 18-2.2b). The energy gaps of these III–V alloys are intermediate (but not necessarily linear) between the energy gaps of the end members (Fig. 18-2.5). Likewise, the lattice constants are intermediate. The relationships of Fig. 18-2.5 are useful to the designers of complex devices where it is necessary to match the lattice size with specific energy-gap values.

Semiconductivity (intrinsic) versus temperature

At room temperature, 20°C (68°F), the average energy of the valence electrons is 0.025 eV. This average increases in proportion to the absolute tem-

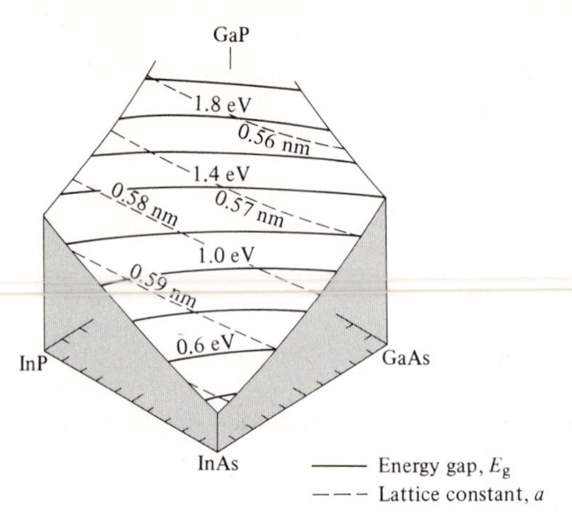

Figure 18-2.5 Alloys of III-V compounds. By adjusting the In–Ga ratio and/or the P–As ratio, alloys may be selected with desired combinations of energy gaps and lattice constants. (Adapted from C. J. Nuese, *J. Ed. Mod. Mat'ls Sci. & Engr.*)

perature, in K. More importantly, the fraction with abnormally high energies increases markedly.

Let us examine the conductivity of germanium as a function of temperature (Fig. 18-2.6). Materials scientists tell us that theory and experiment check each other to give

$$\ln \sigma = A - \frac{E_g}{0.00017T}. \tag{18-2.1}$$

Thus, if we plot log σ versus $1/T$, we get a straight line, and

$$\begin{aligned} \ln \sigma_2/\sigma_1 &= \frac{E_g}{0.00017}\left[\frac{1}{T_1} - \frac{1}{T_2}\right] \\ &= \frac{E_g}{0.17}\left[\frac{1000}{T_1} - \frac{1000}{T_2}\right]. \end{aligned} \tag{18-2.2}$$

Again T must be the absolute temperature in K. The simplest means of measuring the energy gap, E_g, of a semiconductor is to measure its conductivity, σ, at two different temperatures, and then solve Eq. (18-2.2). (This can also be done graphically.)

Intrinsic semiconduction

Figure 18-2.7 reveals that each conduction electron that breaks loose in a semiconductor, such as pure germanium, also produces an electron hole. For this reason an intrinsic semiconductor has an equal number of n-type (electron) and p-type (hole) charge carriers.

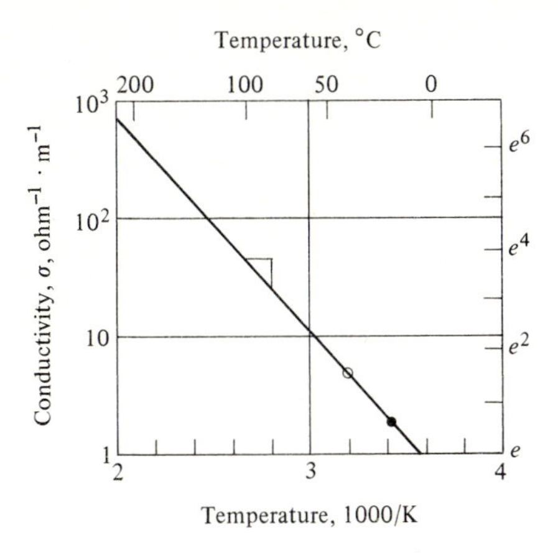

Figure 18–2.6 Semiconduction versus temperature (germanium, E_g = 0.7 eV). The plot is a straight line because it follows Eq. (18–2.1).

Figure 18–2.7 Intrinsic semiconductor (germanium). (a) Schematic presentation showing electrons in their covalent bonds (and their valence bands). (b) Electron-hole pair. (Positive electrode at the left.) (c) Energy gap, across which an electron must be raised to provide conduction. For each conduction electron, there is a hole produced among the valence electrons.

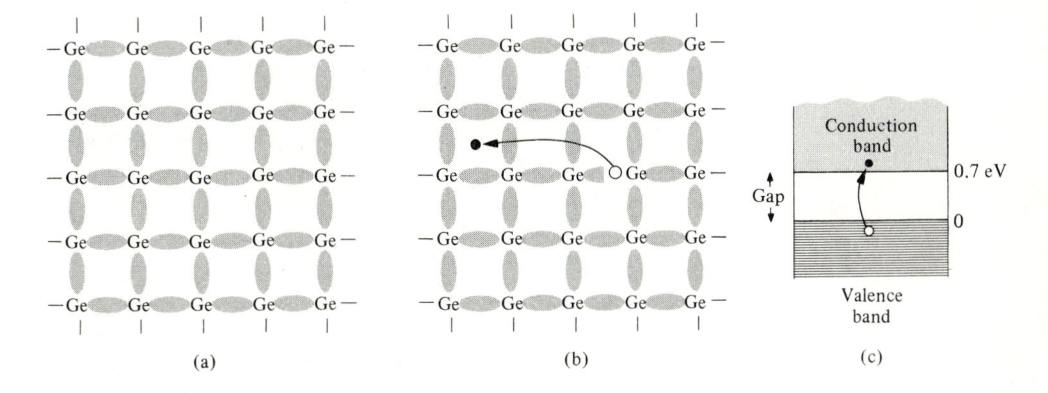

Photoconduction

Thermal energy raises a relatively small fraction of the electrons out of the valence band of silicon into the conduction band (1 out of 10^{13}, according to Table 18–2.1). In contrast, if an electron is hit by a photon of light, it may readily be energized to the conduction level (Fig. 18–2.8). As an example, a photon of red light (wavelength λ = 660 nm) has 1.9 eV of energy, more than enough to cause an electron to jump the 1.1 eV energy gap in silicon. Thus the conductivity of silicon increases markedly when it is exposed to light.

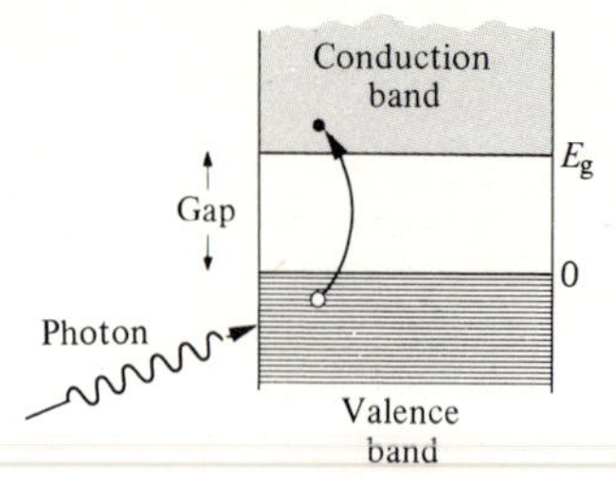

Figure 18-2.8 Photoconduction. A photon (that is, light energy) raises the electron across the energy gap, producing a "conduction electron + valence hole" pair, forming charge carriers. Recombination (Eq. 18-2.3b) occurs when the electron drops back to the valence band.

Recombination

The reaction to produce an *electron-hole pair,* as shown in Fig. 18-2.8, may be written as

$$E \rightarrow n + p, \tag{18-2.3a}$$

where E is energy, n is the conduction electron, and p is the hole in the valence band. In this case the energy came from light.

Since all materials are more stable when they reduce their energies, electron-hole pairs recombine sooner or later:

$$n + p \rightarrow E. \tag{18-2.3b}$$

In effect, the electron drops from the conduction band back to the valence band, just the reverse of Fig. 18-2.7(c). Were it not for the fact that light or some other energy source continually produces additional electron-hole pairs, the conduction band would soon become depleted.

The time required for recombination varies from material to material. However, it follows a regular pattern because within a specific material every conduction electron has the same probability of recombining within the next second (or minute). This leads to the relationship

$$N = N_0 e^{-t/\tau}, \tag{18-2.4a}$$

which we usually rearrange to

$$\ln (N_0/N) = t/\tau. \tag{18-2.4b)*$$

In these equations, N_0 is the number of electrons in the conduction band at a particular moment of time (say, when the source of light is turned off).

---

* Equation (18-2.4) can be derived through calculus (by those who wish to do so) from the information stated above:

$$dN/dt = -N/\tau, \tag{18-2.5a}$$

and

$$dN/N = -dt/\tau; \tag{18-2.5b}$$

and hence Eq. (18-2.4b)

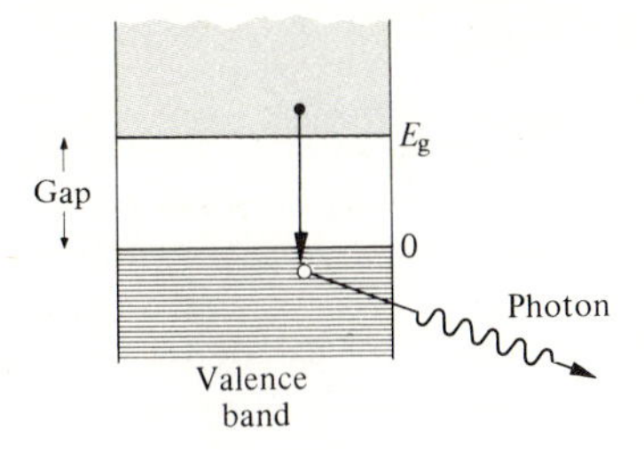

Figure 18-2.9 Luminescence. Each millisecond, a fraction of the electrons energized to the conduction band return to the valence band. As the electron drops across the gap, the energy may be released as a photon of light.

After an additional time t, the number of remaining conduction electrons is N. The term τ is called the *relaxation time,* or recombination time, and is characteristic of the material. (Compare this discussion with that for stress relaxation in Section 8-3.)

Luminescence

The energy released in Eq. (18-2.3b) may appear as heat. It may also appear as light. When it does, we speak of luminescence (Fig. 18-2.9). Sometimes we subdivide luminescence into several categories. *Photoluminescence* is the light emitted after electrons have been activated to the conduction band by light photons. *Chemoluminescence* is the word used when the initial activation is due to chemical reactions. Probably *electroluminescence* is best known, because this is what occurs in a TV tube, in which a stream of electrons scans the screen, activating the electrons in the phosphor to their conduction band. Almost immediately, however, the electrons recombine, emitting energy as visible light.

Since the recombination rate is proportional to the number of activated electrons, the intensity, I, of luminescence also follows Eq. (18-2.4):

$$\ln (I_0/I) = t/\tau. \tag{18-2.6}$$

For a TV tube, the engineer chooses a phosphor with a relaxation time such that light continues to be emitted as the next scan comes across. Thus our eyes do not see a light-dark flickering. However, the light intensity from the previous trace should be weak enough so that it does not compete with the new scan that follows one thirtieth of a second later. (See I.P. 18-2.3.)

Illustrative problem 18-2.1 The resistivity of germanium at 20°C (68°F) is 0.5 ohm·m. What is its resistivity at 40°C (104°F)?

SOLUTION Based on Eq. (18-2.2) and energy gap of 0.7 eV (Table 18-2.2),

$$\ln \sigma_{40^\circ}/\sigma_{20^\circ} = \ln \rho_{20^\circ}/\rho_{40^\circ} = \frac{0.7}{0.17}\left[\frac{1000}{293} - \frac{1000}{313}\right]$$
$$= 4.1(3.41 - 3.19) = 0.90.$$

Therefore, in terms of resistivity,

$$\rho_{40^\circ} = \rho_{20^\circ}\, e^{-0.90} = 0.5 \text{ ohm·m } (0.4)$$
$$= 0.2 \text{ ohm·m.}$$

This problem may also be solved graphically (Fig. 18–2.6).

ADDED INFORMATION It is possible to measure resistance changes (and therefore resistivity changes) of $<0.1\%$. Therefore one can measure temperature changes of a small fraction of a degree. (See S.P. 18–4.2.)

The units of Eq. (18–2.2) are

$$\ln \frac{\text{ohm}^{-1}\cdot\text{m}^{-1}}{\text{ohm}^{-1}\cdot\text{m}^{-1}} = \frac{\text{eV}}{\text{eV/K}}\left[\frac{1}{\text{K}} - \frac{1}{\text{K}}\right]. \blacktriangleleft$$

Illustrative problem 18–2.2 The energy of a light photon is calculated from

$$E = h\nu, \tag{18–2.7a}$$

or alternatively,

$$E = hc/\lambda, \tag{18–2.7b}$$

where c is the velocity of light, 3×10^8 m/sec; λ is the wavelength of light; and h is a constant whose value is 4.13×10^{-15} eV·sec when energy is expressed in electron volts. (a) What is the minimum frequency of light necessary to cause photoconduction in gallium phosphide, GaP, if all the energy comes from the photons? (b) What color light is this?

SOLUTION The energy gap is 2.3 eV (from Table 18–2.2).

a) $\nu = 2.3 \text{ eV}/4.13 \times 10^{-15} \text{ eV·sec)}$
$= 5.57 \times 10^{14}/\text{sec}.$

b) $\lambda = c/\nu = (3 \times 10^8 \text{ m/sec})/(5.57 \times 10^{14}/\text{sec})$
$= 540 \text{ nm}.$

This is in the yellow part of the visible spectrum.

ADDED INFORMATION Longer wavelengths (lower frequencies) can also activate those electrons that possess considerable thermal energy at the moment they are hit. However, the number of electrons that have thermal energy of more than a few tenths of an eV is extremely limited. Therefore, for all intents and purposes, red light does not produce photoconduction in GaP because the red light has longer wavelengths (less energy) than yellow light. ◀

Illustrative problem 18–2.3 The scanning beam of a television tube covers the screen with 30 frames per second. What must the relaxation time for the activated electrons of the phosphor be if only 20% of the intensity is to remain when the following frame is scanned?

SOLUTION Refer to Eq. (18-2.6):

$$\ln(1.00/0.20) = (0.033 \text{ sec})/\tau,$$
$$\tau = 0.02 \text{ sec.}$$

ADDED INFORMATION We use the term *fluorescence* when the relaxation time is short compared to the time of our visual perception. If the luminescence has a noticeable afterglow, we use the term *phosphorescence.* ◀

18-3 EXTRINSIC SEMICONDUCTORS

Those materials whose conductivities are controlled by the presence of impurity atoms are called *extrinsic* semiconductors.

N-Type semiconductors

Consider some silicon that contains an atom of phosphorus. Phosphorus has five valence electrons rather than the four that are found in pure silicon. In Fig. 18-3.1(a), the extra electron is present independently of the electron pairs that serve as bonds between neighboring atoms. This electron can be pulled away from the phosphorus atom and carry a charge toward the positive electrode (Fig. 18-3.1b). Alternatively, in Fig. 18-3.1(c), the extra electron—which cannot reside in the valence band because it is already full—is located near the top of the energy gap. From this position—called a *donor* level, E_d—the extra electron can easily be activated into the conduction band. Regardless of which model is used, Fig. 18-3.1(b) or 18-3.1(c), we

Figure 18-3.1 Extrinsic semiconductors (*n*-type). A Group V atom has an extra valence electron beyond the average of four sketched in Fig. 18-2.2. This fifth electron can be pulled away from its parent atom with very little added energy, and "donated" to the conduction band, to become a charge carrier. We observe the donor energy level, E_d, as being just below the top of the energy gap. (a) An *n*-type impurity, such as phosphorus. (b) Ionized phosphorus atom. (Positive electrode at left.) (c) Band model.

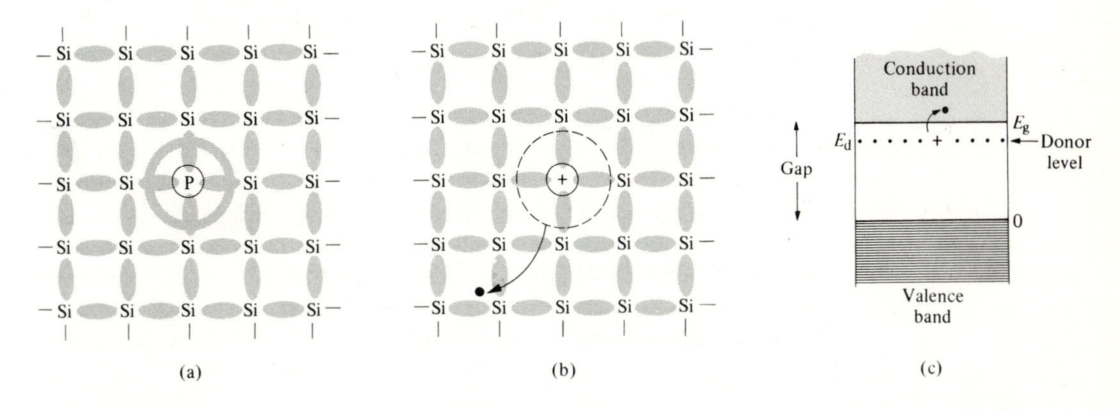

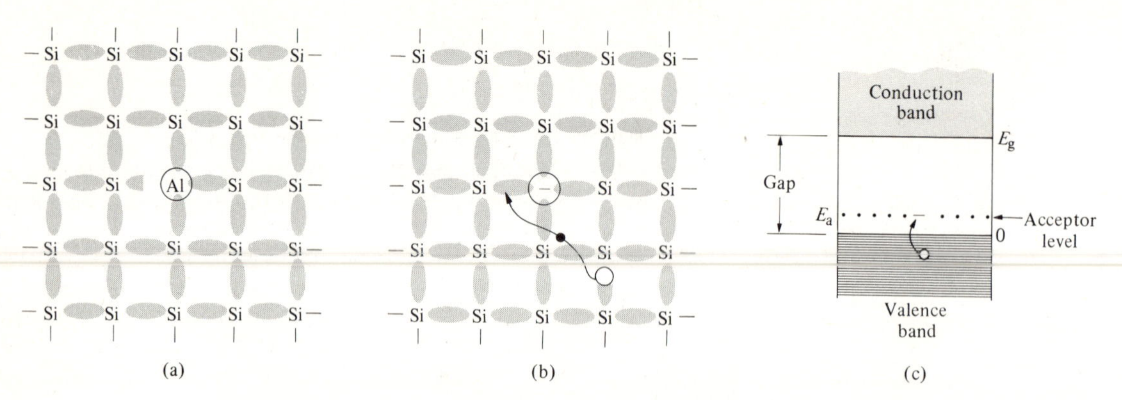

Figure 18-3.2 Extrinsic semiconductors (p-type). A Group III atom has one less valence electron than the average of four sketched in Fig. 18-2.2. This atom can accept an electron from the valence band, thus leaving an electron hole as a charge carrier. The acceptor energy level, E_a, is just above the bottom of the energy gap. (a) A p-type impurity such as aluminum. (b) Ionized aluminum atom. (Negative electrode at right.) (c) Band model.

can see that atoms from Group V of the periodic table (Fig. 1-3.2) can supply negative, or n-type, charge carriers to semiconductors.

P-type semiconductors

Group III elements have only three valence electrons. Therefore, when such elements are added to silicon as impurities, electron holes come into being. As shown in Fig. 18-3.2(a) and (b), each aluminum atom can accept one electron. In the process a positive charge moves toward the negative electrode. Using the band model (Fig. 18-3.2c), we note that the energy difference for electrons from the valence band to the *acceptor* level, E_a, is much less than the full energy gap. The electron holes remaining in the valence band are available as positive carriers for p-type semiconduction.

Donor exhaustion (and acceptor saturation)

If the energy difference for electrons from the donor level to the conduction band is small in comparison with the size of the energy gap, all the electrons in the donor level can be raised to the conduction band, even at room temperatures. This is illustrated in Fig. 18-3.3. At absolute zero, beyond the right edge of the figure, all the electrons are either in the valence band or in the donor levels, and Fig. 18-3.1(a) applies. As the temperature is raised,

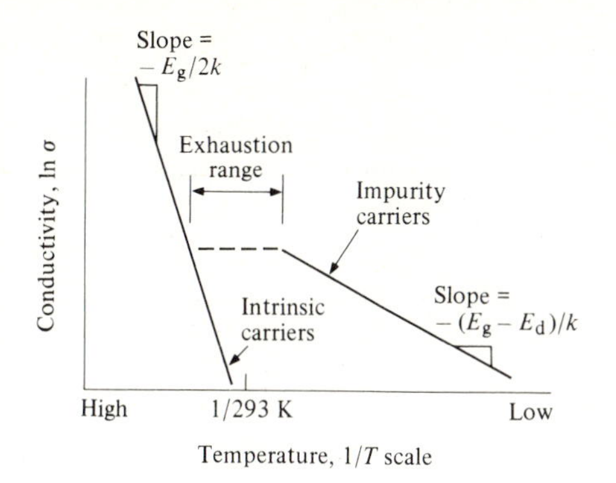

Figure 18-3.3 Donor exhaustion. Intrinsic (left-hand curve) and extrinsic (right-hand curve) conductivity require energies of E_g and $(E_g - E_d)$, respectively, to raise electrons into the conduction band. At lower temperatures, donor electrons provide most of the conductivity. Exhaustion occurs when all of the donor electrons have entered the conduction band, and before the temperature is raised high enough for valence electrons to jump the energy gap. The conductivity is nearly constant in this temperature range.

more and more of the donor electrons jump to the conduction band and are able to carry charge. When the thermal energy of the electrons is approximately equal to this small energy difference $(E_g - E_d)$, essentially all the donor electrons have jumped to the conduction band and this reservoir of charge carriers arising from impurities is exhausted. As a result there is a conductivity plateau; only after the temperature is increased appreciably does the thermal energy become sufficient to cause electrons to jump across the full energy gap from the valence band to the conduction band.

Donor exhaustion of *n*-type semiconductors has its parallel in acceptor saturation of *p*-type semiconductors. The reader is asked to make the various comparisons. Donor exhaustion and acceptor saturation are important to materials and electrical engineers, since they provide a region of essentially constant conductivity. This means that it is less necessary to compensate temperature changes in electrical circuits than it would be if the ln σ versus $1/T$ characteristics followed an ever-ascending line.

Illustrative problem 18-3.1 Silicon, according to Table 18-2.2, has a conductivity of only 5×10^{-4} ohm$^{-1}\cdot$m^{-1} when pure. An engineer wants it to have a saturation conductivity of 200 ohm$^{-1}\cdot$m^{-1} when it contains aluminum as an impurity. How many aluminum atoms are required per m^3?

SOLUTION Since the intrinsic conductivity is negligible compared with 200 ohm$^{-1}\cdot$m^{-1}, assume that all the conductivity comes from the holes:

$$n_p = (200\ \text{ohm}^{-1}\cdot\text{m}^{-1})/(1.6 \times 10^{-19}\ \text{A}\cdot\text{s})(0.0425\ \text{m}^2/\text{volt}\cdot\text{sec})$$
$$= 3 \times 10^{22}/\text{m}^3.$$

ADDED INFORMATION Each aluminum atom contributes one acceptor site; hence, one electron hole. Therefore 3×10^{22} aluminum atoms are required per m^3. This is, of course, a large number; however, it is still small compared with the number of silicon atoms per m^3. (See S.P. 18-3.3a.) ◀

18-4 APPLICATIONS OF SEMICONDUCTORS

Metallic conductors are used primarily to transfer electrical energy from one site to another. Exceptions are few and simple. (a) A high-resistance wire may be used as a heating element, for example, in a toaster. Also, extremely fine wires of tungsten are used for incandescent lights. (b) Conductors are also used for coils to produce magnetic fields.

In contrast, semiconductors are used primarily in device applications and not for energy transfer. The devices fall into two principal categories: (a) conduction and resistance devices, and (b) junction devices.

Conduction and resistance devices

We have already seen that the conductivity of a *photoconductor* will vary directly with the amount of incident light. This capability leads to *light-sensing* devices. The radiation does not have to be visible—it may also be ultraviolet or infrared, providing the photons have energy comparable to or greater than the energy gap in the semiconductor.

A second device is a *thermistor*. It is simply a semiconductor that has had its resistance calibrated against temperature. If the energy gap is large, so that the ln σ versus $1/T$ curve is steep, it is possible to design a thermistor that will detect temperature changes of 10^{-4}°C.*

Because many semiconducting devices have low packing factors, they have a high compressibility. Experiments show that as the volume is compressed, the size of the energy gap is measurably reduced; this, of course, increases the number of electrons that can jump the energy gap. (Cf. Fig. 18–2.3.) Thus pressure can be calibrated against resistance for *pressure gauges*.

A *photomultiplier* device makes use of electron activation, first by photons *and* then by the electrons themselves. Assume, for example, that a very weak light source, even just one photon, were to hit a valence electron. Our eye would not have been able to detect it. However, if that electron is raised to the conduction band, and simultaneously the semiconductor is within a very strong electric field, that electron is accelerated to high velocities and high energies. In turn, it can activate one or more additional electrons that also respond to the very strong field. The multiplying effect may be used to advantage. A very weak light signal may be amplified. With appropriate focusing, the image in nearly complete darkness may be brought into visible display.

---

* In technical practice, thermistors are better than other types of thermometers for measuring small temperature *changes*. However. thermocouples, etc., are more convenient for measuring the temperature itself. The measurement of temperature changes is important in microcalorimetric studies involving chemical or biological reactions.

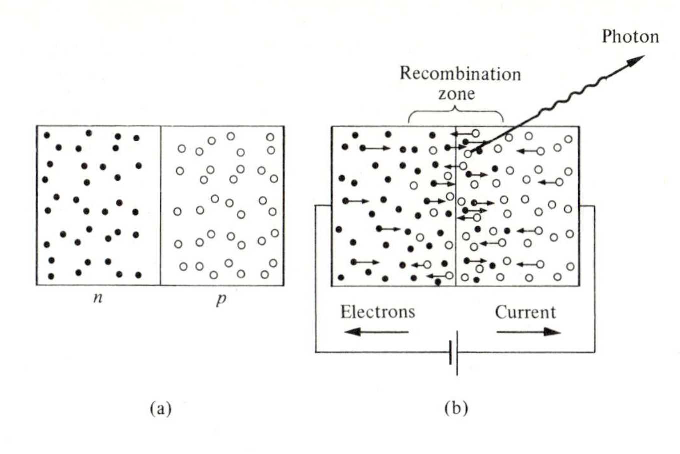

Figure 18–4.1 Light-emitting diode (schematic). (a) An LED is a junction device between n-type and p-type semiconductors. (b) When a forward bias is placed across the junction, carriers of both types have a net movement across the junction where they recombine, emitting photons (Eq. (18–2.3b) and Fig. 18–2.9).

Junction devices (diodes)

A number of devices utilize junctions between n-type and p-type semiconductors. The most familiar of these is the *light-emitting diode* (LED). We see it used in the digital displays (red) that are placed in many hand calculators. An LED operates on the principle shown schematically in Fig. 18–4.1. The charge carriers on the n-side and p-side of the junction are electrons and holes, respectively. Their net movements are zero. If a voltage is placed across the device in the direction shown, the holes of the valence band have a net movement through the junction, into the n-type material; conversely, the electrons of the conduction band cross into the p-type material. Adjacent to the junction, there are excess carriers that recombine and produce luminescence:

$$n + p \rightarrow \text{photon}. \tag{18–4.1}$$

When GaAs is used, the photons emitted in the recombination zone are red; a phosphorus substitution to give Ga(As,P) produces green photons.

The junction of Fig. 18–4.1 can also serve as a *rectifier*, that is, it is an electrical "check valve" that lets current pass one way and not the other. With the *forward bias* of Fig. 18–4.2(a), current can pass because carriers—both electrons and holes—move through the junction. With a *reverse bias* of Fig. 18–4.2(b), the carriers are pulled away from each side of the junction to leave a carrier-depleted "insulating zone" at the junction. If a greater voltage is applied, this *depletion zone* is simply widened. Usable current passes only with the forward bias.

The last statement holds over a wide range of reverse voltages. There is a point, however, where a surge of current is allowed to pass because there is an electrical breakdown in the insulating zone. Specifically, those very few

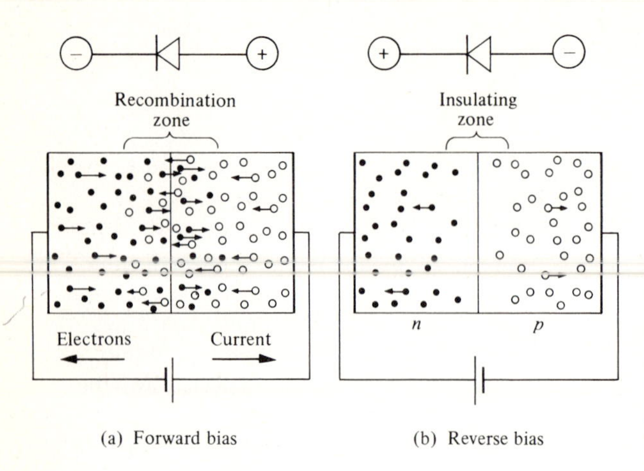

Figure 18–4.2 Rectifier (schematic). (a) Current flows with a forward bias because charge carriers pass the junction. (b) With a reverse bias, charge carriers are depleted farther from the junction region. Extrinsic conductivity disappears from the junction region, and only a small amount of intrinsic conductivity remains.

carriers that are in the depleted zone are accelerated to high velocities by the steep potential drop. As with the photomultiplier device that was previously described, these energetic electrons can knock other electrons loose. An *avalanche* develops that leads to a high current. In effect, we have a "safety valve" that opens at a definite voltage.

Diodes, based on the above principle, may be designed for breakdown voltages ranging from 1 or 2 volts to several hundred volts and with current ratings from milliamperes to a large number of amperes. Called *Zener diodes*, these devices can serve advantageously as filters, gates, and constant voltage controls.

• Solar cells

The photocells of Section 18–2 and Fig. 18–2.8 are *photoconductive*. The II–VI compound, CdS, provides an example. The resistance is lowered by additional carriers that arise from photons of light. Photocells may also be *photovoltaic* in which a potential difference is established across a *p–n* junction. The potential difference is proportional to the intensity of the incident light. This potential can feed an external circuit as a *solar cell*. A large area of *n*-type silicon is required. It is covered with a very thin *p*-type layer by

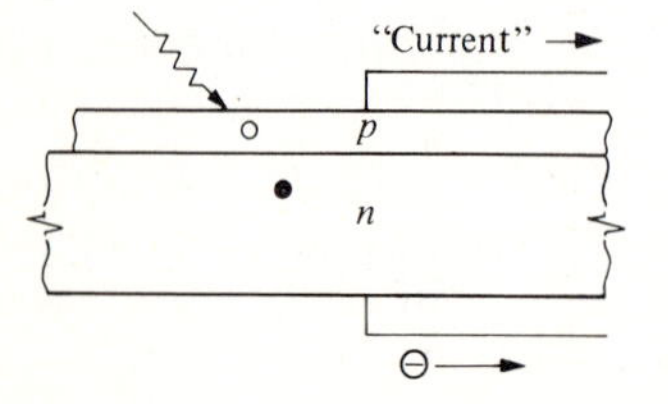

Figure 18–4.3 Solar cell (schematic). A photon produces an *n–p* pair in the *p*-type surface layer. Those electrons that enter the underlying *n*-type material before recombination continue into the external circuit. Unlike the device of Fig. 18–2.8, which is photoconductive, this device is photovoltaic, since it produces a voltage—and therefore power.

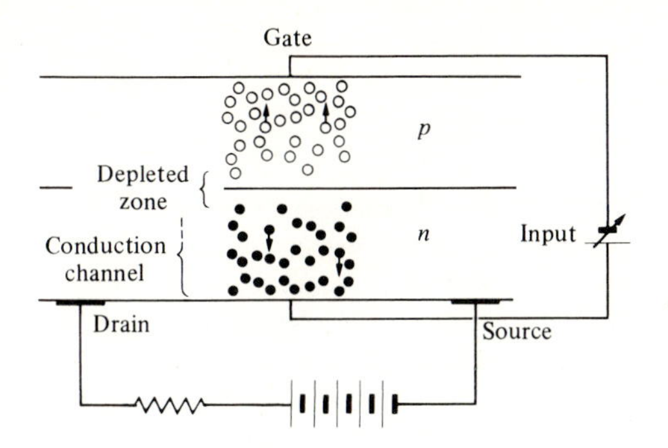

Figure 18–4.4 Transistor (field-effect). A single n–p junction is used. A voltage change across the gate, which is reverse-biased, changes the width of the depleted zone and, hence, the cross section of the conduction channel between the source and the drain. A small variation in the input voltage produces a major change in the resulting current through the conduction channel.

adding Group III elements, such as boron, to the surface. Now, when an n–p pair is formed by sunlight in the surface of the p-type material, a significant number of electrons cross the junction into the n-type material where they can feed an external circuit (Fig. 18–4.3). The p-type surface must be extremely thin because most of the n–p pairs form soon after the photons enter the silicon. If the junction were not nearby, the electrons could recombine with the holes before they enter the n-type silicon and are able to leave the solar cell and deliver power. The maximum open-circuit voltage is only ~0.5 volts, and the short-circuit current is typically less than 0.05 amp/cm^2. As a result, large areas are required for meaningful power production.

• Transistors

Transistors are junction devices that amplify weak signals into stronger usable outputs.

The simplest type of transistor* makes use of a carrier-depleted zone to amplify the output of a circuit. This is called the *field-effect transistor* (FET). In Fig. 18–4.4, the p–n junction is reverse-biased by the input signal. As the signal varies, the carrier-depleted zone varies in size and thus alters the resistivity between the *source* and the *drain*. In turn, the current going to the output varies in a controlled manner. A small signal produces a significant current fluctuation.

A more common transistor has two junctions in series. They may be either p–n–p or n–p–n and are called *junction transistors*. The former has been used somewhat more in the past; however, we shall consider the n–p–n transistor, since it is easier to visualize the movements of electrons, rather than the movements of holes. However, the principles behind each type are the same.

---

* The term *transistor* is a contraction of "transfer resistor," since early rationale for its performance utilized such a concept of operation.

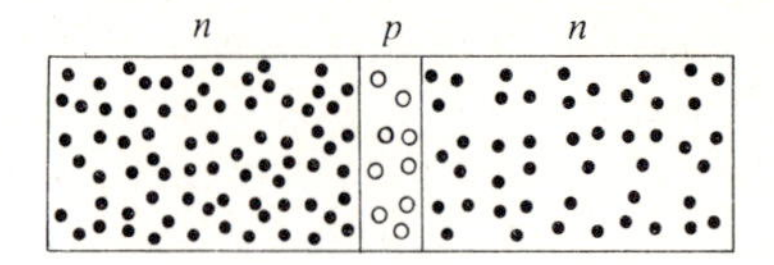

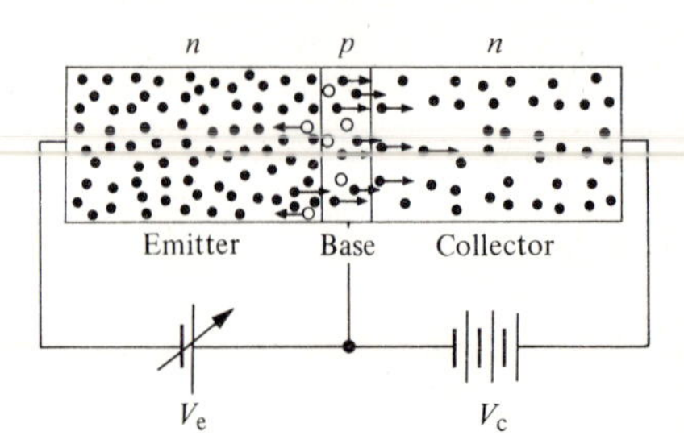

Figure 18-4.5 Transistor (junction, *n–p–n*). The number of electrons crossing from the emitter-base junction is highly sensitive to the emitter voltage. If the base is narrow, these carriers move to the base-collector junction, and beyond, before recombination. The total current flux, emitter to base, is highly magnified, or amplified, by fluctuations in the voltage of the emitter.

Before describing the makeup of the transistor, let us recall that as holes move across the junction with a forward bias (Fig. 18–4.2a), they recombine with the electrons in the *n*-type material according to Eq. (18– 2.3b). Likewise, the electrons combine with the holes as the electrons move beyond the junction and into the *p*-type material. However, the reaction of Eq. (18–2.3b) does not occur immediately. In fact, an excess number of positive and negative carriers may move considerable distances beyond the junction. The number of excess, unrecombined carriers are an exponential function of the applied voltage and become important in transistor operation.

A transistor consists of an *emitter*, a *base*, and a *collector* (Fig. 18–4.5). For the moment, consider only the *emitter junction*, which is biased so that electrons move into the base (and toward the collector). As discussed a moment ago, the number of electrons that cross this junction and move into the *p*-type material is an exponential function of the emitter voltage, V_e. Of course, these electrons start at once to recombine with the holes in the base; however, if the base is narrow, or if the recombination time is long (τ of Eq. 18–2.4), the electrons keep on moving through the thickness of the base. Once they are at the second junction, the *collector junction*, the electrons have free sailing, because the collector is an *n*-type semiconductor. The total current that moves through the collector is controlled by the emitter voltage, V_e. As the emitter voltage fluctuates, the collector current, I_c, changes exponentially. Written logarithmically,

$$\ln I_c \simeq \ln I_0 + V_e/B, \tag{18–4.2a}$$

or

$$I_c = I_0 e^{V_e/B}, \tag{18–4.2b}$$

where I_0 and B are constants for any given temperature. Thus, if the voltage in the emitter is increased even slightly, the amount of current is increased markedly. It is because of these relationships that a transistor serves as an amplifier.

• **Illustrative problem 18-4.1** A solar cell can produce 0.012 A/cm² at 400 millivolts. The voltage can be increased to 40 V by placing 100 cells in series. What area should each cell have to generate 1 kW of energy for a hot-water heater?

SOLUTION

$$\begin{aligned}\text{Required amperage} &= 1000\text{ W}/40\text{ V}\\ &= 25\text{ A}.\\ \text{Area} &= 25\text{ A}/(0.012\text{ A/cm}^2)\\ &= 2100\text{ cm}^2, \qquad (\text{or } 2.25\text{ ft}^2),\end{aligned}$$

that is, 0.21 m² for each of the 100 cells. ◄

Illustrative problem 18-4.2 Zinc sulfide is used as a thermistor. To what fraction sensitivity, δ, must the resistance be measured to detect 0.001°C change at 20°C?

SOLUTION Since the geometry is constant, and with Eq. (18-2.2),

$$\begin{aligned}\delta &= \frac{R_1 - R_2}{R_1} = \frac{\rho_1 - \rho_2}{\rho_1} = 1 - \frac{\rho_2}{\rho_1} = 1 - \frac{\sigma_1}{\sigma_2};\\ \ln \sigma_2/\sigma_1 &= -\ln(1 - \delta)\\ &= \frac{-3.7}{0.00017}\left(\frac{1}{293} - \frac{1}{293 + 0.001}\right).\\ \ln(1 - \delta) &= -2.5 \times 10^{-4};\\ \delta &= 0.00025, \qquad (\text{or } 0.025\%).\end{aligned}$$

ADDED INFORMATION A bridge-type instrument would be required. ◄

• **Illustrative problem 18-4.3** A transistor has a collector current of 4.7 milliamperes when the emitter voltage is 17 millivolts. At 28 millivolts, the current is 27.5 milliamperes. Given that the emitter voltage is 39 millivolts, estimate the current.

SOLUTION Based on Eq. (18-4.2),

$$\begin{aligned}\ln 4.7 &\simeq \ln I_0 + 17/B = 1.55,\\ \ln 27.5 &\simeq \ln I_0 + 28/B = 3.31.\end{aligned}$$

Solving simultaneously, using *milli*units, we have

$$\ln I_0 \simeq -1.17 \quad \text{and} \quad B \simeq 6.25.$$

At 39 millivolts,

$$\ln I_c \simeq -1.17 + 39/6.25 \simeq 5.07,$$
$$I_c \simeq 160 \text{ milliamp.}$$

ADDED INFORMATION The electrical engineer modifies Eq. (18–4.2) to take care of added current effects. These, however, do not change the basic relationship: The variation of the collector current is much greater than the variation of the signal voltage. ◀

• 18–5 PREPARATION OF SEMICONDUCTING MATERIALS AND DEVICES

The major steps of semiconductor processing include material purification, single-crystal growing, and device preparation.

Material purification

The composition of a semiconductor is very critical. Some impurities introduce donors and negative (*n*-type) carriers; others introduce acceptors and positive (*p*-type) carriers. Even when these *dopants* are desired, their amounts must be closely controlled to parts-per-million (ppm) levels, or less. Therefore, it is common practice to purify the silicon (or other semiconductors) to the highest level possible, and then make precise additions of the required dopant.

The purification of silicon is primarily by chemical procedures. Trichlorosilane ($SiHCl_3$) is distilled at its boiling point (32°C) and allowed to decompose on a hot (~950°C) silicon rod in the presence of hydrogen:

$$SiHCl_3 + H_2 \rightarrow Si + 3HCl. \tag{18–5.1}$$

With careful process controls, the silicon that forms on the rod has impurity levels of less than one part in 10^{10} and contains less boron and phosphorus than other types of purification, for example, zone refining. (Zone refining is used for germanium purification.)

Crystal growing

Single crystals are required for the large majority of applications, since grain boundaries reduce carrier mobility and decrease the recombination time of excess carriers. The latter decrease affects the performance of many junction devices. Single-crystal growth generally utilizes one of two techniques in semiconductor technology—crystal-pulling and floating-zone methods (Fig. 18–5.1). (Also see Section 12–5, and Fig. 12–5.1.)

The *crystal-pulling* method first melts the semiconducting material. Then a small single-crystal seed is touched to the surface and slowly pulled away (~1 mm/min) as it is rotated (~1/sec). If the liquid is only slightly

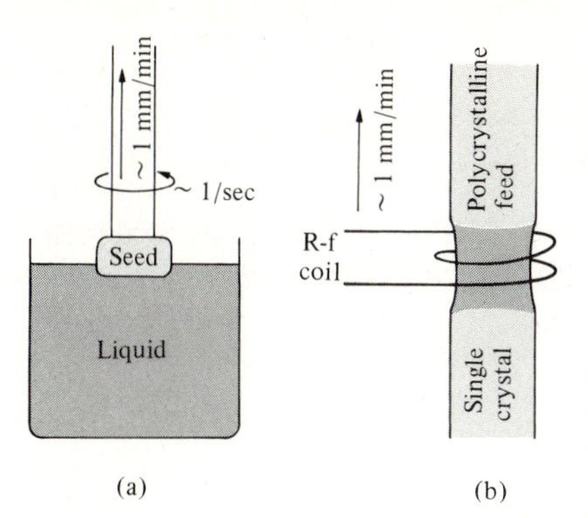

Figure 18–5.1 Single-crystal growth (semiconductors). (a) Crystal-pulling method. The seed—a single crystal—is slowly pulled upward. The liquid crystallizes on its lower surface. (b) Floating-zone procedure. The molten zone is raised along the semiconductor bar, solidifying the lower edge as a single crystal. The liquid is held in position by surface tension and does not come into contact with a container.

above its melting temperature, it will solidify on the seed crystal as the seed is pulled upward. The solidifying atoms continue the crystal structure of the seed. Dopants of Group III or Group V compounds may be added to the liquids in the amounts ($\sim 10^{-6}$ a/o) required to make p- and n-type products.

The above technique (Fig. 18–5.1a) is satisfactory for germanium and other materials that melt below 1000°C. However, it is not as satisfactory for silicon for a couple of reasons. Silicon melts above 1400°C and, therefore, more readily picks up contaminants from the container and furnace walls. Also, the dopants more readily vaporize so that compositional control becomes more difficult. Therefore, a floating-zone process is used.

The *floating-zone* method starts with a rod ($\sim$5 cm dia.) of purified polycrystalline silicon sitting on a disc of a previously prepared single crystal. The two are melted where they are in contact by r-f heating. The r-f coil is then slowly raised (Fig. 18–5.1b) to move the molten zone upward. The polycrystalline solid melts with the upper movement and feeds the molten zone. The initial single crystal grows upward with the movement by crystallization at the lower side of the molten zone. As with the crystal-pulling process, the upward movement of the molten zone advances $\sim$1 mm/min.

Thin ($\sim$0.25 mm) wafers are cut across the bar. These are polished and chemically cleaned; then dopants are added in an *epitaxial* layer. This is a layer that grows onto the surface as a continuation of the underlying crystal. The growth comes from a gaseous mixture whose composition is adjusted to give either an n-type or a p-type layer.*

Device preparation

Figure 18–5.2 shows a typical junction transistor with a *planar* configuration, that is, the preparation and contacts are on one flat surface rather than at the ends and sides as in Fig. 18–4.5. The thicker substrate, an n-type

* $SiCl_4$, H_2, and PH_3 for n-type silicon, and $SiCl_4$, H_2, and B_2H_6 for p-type silicon. (Cf. Eq. 18–5.1.)

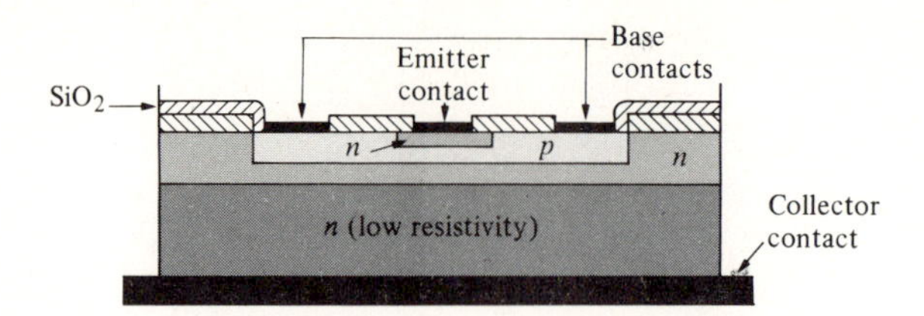

Figure 18-5.2 Planar transistor (junction-type). The substrate is a low-resistivity, n-type, single-crystal wafer of silicon. It is covered with an epitaxial layer of higher resistivity silicon that serves as the collector. Boron is diffused through the openings of an SiO_2 mark to produce the p-type base. A second masking admits phosphorus for the n-type emitter. Aluminum contacts are added.

silicon base that was crystallized in the floating-zone process, is doped to have low resistivity for easy conduction. The overlying (epitaxial) layer was described in the previous paragraph and has higher, but controlled, resistivity. Following a masking procedure, boron is *diffused* from the surrounding gas into the epitaxial silicon, changing the selected regions from n-type to p-type. The latter becomes the base of the transistor. (Cf. Fig. 18-4.5.) A second masking and diffusion step introduces a Group V element to produce the n-type emitter. Final maskings, etching, and coating establish a passive silica (SiO_2) surface and contacts for circuit leads.

The thickness of the base in Fig. 18-5.2 may be only 0.5 μm. With this short distance between the two junctions, any recombination can be minimized as described in Section 18-4. The dimensions and dopant concentrations are controlled by diffusion time and temperature, and by gas composition. This has been developed into a highly reproducible production process.

Ion implantation is a newer process that is also finding wide use in device preparation. In this process, the dopants (Group III and Group V elements) are bombarded into the semiconductor surface with a high energy ion beam ($\sim$10,000 eV). The surface does not have to be heated as is necessary for diffusion; therefore, the heating steps and the possibility of unwanted impurities are avoided. The number and depth of implanted ions are controlled by the time and voltage of the ion beam.

The above processes are amenable to mass production. Hundreds of chips that contain the junctions of the integrated circuit are produced simultaneously on a wafer (Fig. 1-1.1). The finished wafer must be scribed and separated into the individual chips, each with an integrated circuit. The final step is to package the integrated circuit (IC) for protection and for connection into the customer's product.

Review and Study

SUMMARY

1. Electronic semiconductors transport charge (a) by electrons in the conduction band, and (b) by electron holes in the valence band. There are more charge carriers and higher conductivities in those semiconductors with a small energy gap because more electrons are able to jump the gap.
2. Group IV elements and III–V compounds are the most widely used semiconductors. They possess four valence electrons per atom, thus completely filling their valence bands. All have the same structure. In general, the heavier ones have smaller energy gaps and, therefore, higher conductivities.
3. Intrinsic semiconductors possess the same numbers of negative and positive carriers. Extrinsic semiconductors are either *n*-type, in which impurities donate electrons to the conduction band, or *p*-type, in which impurities accept electrons from the valence band and thus produce electron holes.
4. Conductivities of individual semiconductors increase with increased temperature, with increased light, and with increased pressure. This leads to various detection devices such as thermistors, light sensors, and pressure gages. The recombination of electrons and electron holes can produce luminesence.
5. Junction devices place *n*-type and *p*-type semiconductors in contact. A forward bias carries electrons and electron holes across the junction (where they recombine). A reverse bias depletes the carriers from the junction region so that there is negligible current. Thus, a junction can serve as either a rectifier or a light-emitting diode. Other junction devices include Zener diodes, • photovoltaic cells (solar cells), and • transistors—both field-effect and junction.
6. • Processing requires special procedures to chemically purify the semiconducting materials. This is followed by melting and the growth of single crystals, either by a crystal-pulling method or by a floating-zone method. Device preparation usually involves masking for vapor-plating, etching, and/or ion implantation on wafers containing a multitude of integrated circuits.

TECHNICAL TERMS

Acceptor Impurities that accept electrons from the valence band and, therefore, produce an electron hole in the valence band.

Acceptor saturation Filling of acceptor sites. As a consequence, additional thermal activation does not increase the number of extrinsic carriers.

Band, conduction (CB) Energy band of conduction electrons. Electrons must be in this band to be carriers.

Band, valence (VB) Filled energy band below the energy gap. Conduction in this band requires holes.

Charge carriers Electrons in the conduction band provide *n*-type (negative) carriers. Electron holes in the valence band provide *p*-type (positive) carriers.

• **Crystal-pulling** Method of growing single crystals by slowly pulling a seed crystal away from a molten pool.

Depletion zone The region adjacent to an *n–p* junction that lacks charge carriers. (A reverse bias increases the width of the depleted zone.)

Donors Impurities that donate charge carriers to the conduction band.

Donor exhaustion Depletion of donor electrons. Because of it, additional thermal activation does not increase the number of extrinsic carriers.

Electron charge (q) The charge of 1.6×10^{-19} coul (or 1.6×10^{-19} amp·sec) carried by each electron.

Electron holes (p) Electron vacancy in the valence band that serves as a positive charge carrier.

Electron-hole pair A conduction electron in the conduction band and an accompanying electron hole in the valence band, which result when an electron jumps the gap in an intrinsic semiconductor.

Energy gap (E_g) Forbidden energies between the valence band and the conduction band.

• **Floating zone** Method of growing single crystals, which melts an isolated zone within a bar. The material solidifies as a single crystal on the bottom side of the rising molten zone.

Forward bias Potential direction that moves the charge carriers across the *n–p* junction.

• **Ion implantation** The introduction of dopants by bombarding the surface of the semiconductor with a high-energy ion beam.

Junction (*n–p*) Interface between *n*-type and *p*-type semiconductors.

Light emitting diode (LED) A *p–n* junction device designed to produce photons by recombination.

Luminescence Light emitted by the energy released as conduction electrons recombine with electron holes.

Photoconduction Conduction arising from activation of electrons across the energy gap by means of light.

Photomultiplier Device that uses a photon to trigger an electron avalanche in a semiconductor. From this, weak light signals can be amplified.

• **Photovoltaic** Production of an electrical potential from incident light.

Recombination Annihilation of electron-hole pairs.

Recombination time (τ) *See* relaxation time.

Rectifier Electric "valve" that permits forward current and prevents reverse current.

Relaxation time (τ) Time required for electron-hole pairs to recombine and decrease the remaining numbers of 1/e of the original levels.

Semiconductors Materials with controllable conductivities, intermediate between insulators and conductors.

Semiconductors, compound Compounds of two or more elements with an average of four shared electrons per atom.

Semiconductors, extrinsic Semiconduction from impurity sources.

Semiconductors, intrinsic Semiconduction of pure materials. The electrons are excited across the energy gap.

Semiconductors, *n*-type Impurities provide donor electrons to the conduction band. These electrons are the majority charge carriers.

Semiconductors, *p*-type Impurities provide acceptor sites for electrons from the valence band. Electron holes are the majority charge carriers.

• **Solar cell** Photovoltaic device for sun-to-electrical energy conversion.

Thermistor Semiconductor device with a high resistance dependence on temperature. It may be calibrated as a thermometer.

• **Transistor** Semiconductor device for the amplification of current. There are two principal types, field-effect and junction.

Zener diode A *p–n* junction with controlled breakdown voltage under reverse bias.

CHECKS

18A The ___***___ of semiconductors is filled; furthermore, the ___***___ between the valence band and the overlying conduction band is small.

18B Charge carriers include positive ions, ___***___, electrons, and ___***___.

18C An electron hole is the absence of an electron in the ___***___; this "vacancy" will move toward the ___***___ electrode with a same effect as if a negative charge moved toward the ___***___ electrode. Hence, we consider the electron hole as a ___***___ charge carrier.

18D The conductivity of a semiconductor is the sum of the conductivity in the ___***___ band by negative carriers and the conductivity in the ___***___ band by positive carriers.

18E The simplest semiconductors are the elements of Group ___***___ of the periodic table; in this group, the ___***___ elements have narrower energy gaps.

18F The simplest compound semiconductors contain elements of Groups ___***___ and ___***___ with an average of ___***___ valence electrons per atom. Those compounds with ___***___ elements have wider energy gaps.

18G The numbers of electrons jumping the energy gap depend upon the reciprocal of the ___***___ ___***___. In order to make a linear plot, the numbers must be expressed in terms of ___***___.

18H Electrons can be raised across the energy gap, E_g, by ___***___, by ___***___, or by energetic ___***___. In the first case, the numbers are proportional to $e^{-E_g/2kT}$; in the second, the ___***___-conduction depends upon the number of ___***___ hitting the semiconductor.

18I When an electron jumps the gap, an electron-___***___ pair is formed. This is eliminated by the ___***___ of the electron and hole.

18J The rate of recombination is proportional to the ___***___ of electrons in the conduction band; mathematically, this leads to a relaxation time, after which 1/___***___ of the original number of electrons remain in the conduction band (assuming additional electrons are not raised across the gap).

18K The term ___***___ applies to those semiconductors that possess semiconductivities by virtue of impurities; a semiconductor without impurities is called ___***___.

18L Group V elements such as As or Sb, serve as ___***___ that provide an electron to the ___***___ of silicon, making it a ___***___-type semiconductor.

18M Group III elements such as ___***___ and ___***___ serve as ___***___ that receive electrons from the ___***___ band of silicon, making it a ___***___-type semiconductor.

18N If all of the Group III atoms in impure silicon receive electrons, we speak of ___***___ ___***___; the comparable limit for ___***___-type semiconductors containing As or Sb, is called ___***___ ___***___.

18O A light-sensing device may be either photo-___***___ or photo-___***___. Of these, the ___***___ device does not require a p–n junction.

18P A semiconductor device for a ___***___ works on the principle that the energy gap is ___***___ in size with an increase in density, so that more electrons can enter the conduction band.

18Q A ___***___ device depends simply on the fact that more electrons enter the conduction band at higher ___***___.

18R Junction devices require two types of semiconductors, ___***___ and ___***___.

18S ___***___ occurs at the junction of a light-emitting diode to produce ___***___ of light.

18T Current passes through the junction of a ___***___ when a ___***___ bias is applied.

18U In a ___***___ and in a ___***___ the conduction electrons are accelerated to high enough energies to form additional electron-hole pairs.

• 18V One type of ___***___ requires two p–n junctions with reverse orientations. A slight increase in the emitter voltage greatly increases the numbers of carriers traveling through the ___***___, thus giving ___***___.

• 18W In a second type of ___***___, a slight change in the electric ___***___ changes the size of the conduction ___***___ and greatly alters the amount of current flow.

18X Of the two, germanium and gallium, ___***___ would much more markedly affect the conductivity of silicon because ___***___ . . . .

18Y Of the two, Si with 10^{-8} a/o Ga, and Si with 10^{-8} a/o As, the impurity of ___***___ would lead to greater conductivity because ___***___ . . . .

• 18Z With the ___***___ technique for single-crystal growing, a single-crystal ___***___ is slowly rotated and pulled away from the molten surface.

• 18AA With the ___***___ technique for single-crystal growing, purified ___***___ silicon melts and then crystallizes onto a single crystal as the molten region is raised.

• 18AB If a surface layer is formed that matches the crystal pattern of the substrate, we call it an ___***___ layer.

• 18AC ___***___ and ___***___ can be used to produce a thin ___***___ layer on the surface of n-type silicon.

STUDY PROBLEMS

18–1.1. Silicon is at a temperature where there are 10^{17} electrons per m^3 that have risen to the conduction band. The resistivity is 270 ohm·m. The mobility of the electrons in the conduction band is 0.19 $m^2/V\cdot s$; what is the mobility of the holes in the valence band?

Answer: 0.04 $m^2/V\cdot s$

18–1.2. The density of silicon is 2.33 g/cm^3 (or 2.33 Mg/m^3). What fraction of the valence electrons have entered the conduction band in S.P. 18–1.1? [Silicon is a Group IV element; therefore, it has four valence electrons per atom.]

18–2.1. What fraction of the intrinsic conductivity of silicon is from negative carriers?

Answer: 0.82

18–2.2. The intrinsic conductivity in germanium is 2 $ohm^{-1}\cdot m^{-1}$ at 20°C. What is the conductivity from (a) the positive carriers? (b) The negative carriers?

18–2.3. How many conduction electrons (and electron holes) does intrinsic germanium have when its conductivity is 4 $ohm^{-1}\cdot m^{-1}$?

Answer: $4.2 \times 10^{19}/m^3$

18–2.4. The mobility of electrons in silicon is 0.19 $m^2/V\cdot s$. (a) What voltage is required across a 2-mm chip of Si to produce a drift velocity of the electrons of 0.7 m/s? (b) What electron concentration must be in the conduction band to produce a conductivity from negative carriers of 20 $ohm^{-1}\cdot m^{-1}$? (c) What would be the total conductivity for this silicon if no impurities are present?

18–2.5. The intrinsic conductivity of silicon, according to Table 18–2.2, is 5×10^{-4} $ohm^{-1}\cdot m^{-1}$ at 20°C. What is the conductivity at 60°C?

Answer: 0.007 $ohm^{-1}\cdot m^{-1}$

18–2.6. At what temperature will the conductivity of silicon be one half of its conductivity at 30°C?

18–2.7. What energy gap is required to permit a semiconductor to double its conductivity between 15°C and 25°C?

Answer: 1.0 eV

18–2.8. A semiconductor has a resistance of 3.1 ohms at 19°C. The temperature is raised to 21°C and the resistance decreases to 2.9 ohms. What is the size of the energy gap?

18–2.9. At an elevated temperature, one of every 10^{12} valence electrons in intrinsic silicon is in the conduction band. (a) What is the conductivity? (b) What is the temperature?

Answer: (a) 7.4×10^{-3} $ohm^{-1}\cdot m^{-1}$ (b) 61°C

18–2.10. To what temperature must gray tin be reduced to have only one of every 10,000 valence electrons in the conduction band? [Gray tin has the same structure as silicon, but with $a = 0.649$ nm.]

18–2.11. (a) What is the minimum frequency of light required to supply all of the energy for an electron to be raised across the energy gap in gallium arsenide (GaAs)? (b) What is the wavelength? (c) Will visible light supply enough energy?

Answer: (b) 885 nm

18–2.12. Gallium arsenide, GaAs, releases red photons (λ = 650 nm) during the recombination of electrons and holes. How many electron volts of energy do the photons possess?

18–2.13. The relaxation time for a phosphor is 100 milliseconds. (a) How much time is required before the light intensity is 50% of its initial value? (b) What relaxation time would be required to have 50% intensity in 50 ms?

Answer: (a) 69 ms (b) 72 ms

18–2.14. Refer to I.P. 18–2.3. Assume a phosphor is used with a relaxation time of 0.05 sec. What fraction of the light intensity will remain when the next scan is made?

18–3.1. Extrinsic germanium used for transistors has the resistivity of 0.017 ohm·m and an electron hole concentration from impurities of 1.6×10^{21} holes/m^3. (a) What is the mobility of the holes in the germanium? (b) What impurity atoms could be added to the germanium to create holes?

Answer: (a) 0.23 m^2/V·s (b) Any Group III element (Fig. 18–2.1)

18–3.2. Extrinsic germanium is formed by melting 2×10^{-6}g of arsenic with 100 g of germanium. (a) Will the semiconductor be *n*-type or *p*-type? (b) Calculate the concentration of arsenic (in atoms/m^3) in the germanium. [The density of germanium is 5.35 g/m^3.]

18–3.3. Silicon has a density of 2.33 g/cm^3 (or 2.33 Mg/m^3). (a) What is the concentration of silicon atoms per m^3? (b) Phosphorus is added to silicon to make it an *n*-type semiconductor with a conductivity of 100 ohm^{-1}·m^{-1}. What is the concentration of the donor atoms per m^3?

Answer: (a) 5×10^{28} Si/m^3 (b) 3.3×10^{21}/m^3

18–3.4. How many silicon atoms are there for each aluminum atom in I.P. 18–3.1?

18–3.5. Silicon has one atom of gallium replacing one silicon atom per 10^6 unit cells. (a) Will the silicon be *n*-type or *p*-type? (b) On the basis of Table 18–2.2, what is the conductivity?

Answer: (a) *p*-type (b) 42 ohm^{-1}·m^{-1}

18–3.6. Indium phosphide, InP, has one atom of gallium replacing one phosphorus atom per 10^6 unit cells. (a) Will the InP be *n*-type or *p*-type? (b) On the basis of Table 18–2.2, what is the conductivity?

18–3.7. Silicon (0.0001 w/o) is added to intrinsic gallium arsenide. How much is its conductivity changed?

Answer: None. Si is Group IV.

18–3.8. Three grams of n-type silicon that had been doped with phosphorus to produce a conductivity of 600 $ohm^{-1} \cdot m^{-1}$ are melted with three grams of p-type silicon that had been doped with aluminum to produce a conductivity of 600 $ohm^{-1} \cdot m^{-1}$. (a) What is the resulting conductivity? (b) Will it be p-type or n-type?

• 18–4.1. Eight watts are required of a solar cell that produces 60 A/m^2 at 0.45 volts. What area is required?

Answer: 0.3 m^2

18–4.2. Intrinsic silicon is to be used as a thermometer. How sensitive ($\pm$%) should the resistance instrument be to detect a 10^{-2} °C temperature change at the temperature of the human body (37°C)? [*Hint*: Change $[1/T_1 - 1/T_2]$ to $[T_2 - T_1]/T_2T_1$.]

18–4.3. Refer to I.P. 18–4.3. If the emitter voltage is doubled from 17 to 34 millivolts, by what factor is the collector current increased?

Answer: 15

18–4.4. A transistor operates between 10 mV and 100 mV across the emitter. At the lower voltage the collector current is 6 mA; at the higher voltage, 600 mA. Estimate the current when the emitter voltage drops to 50 mV.

• 18–5.1. Assuming fully efficient operation, how many passes by an r-f induction coil are required to zone refine a rod of 90Si–10Al material to "8-nines" purity, that is, 99.999999% Si at the starting end of the zone refined rod? [*Hint*: According to Fig. 5–4.8, the Al ratio of C_S to C_L near the melting temperature of silicon is 0.001.]

Answer: 3 passes

• 18–5.2. What is the conductivity of the above "8-nines" silicon containing 10^{-6} a/o aluminum (S.P. 18–5.1)?

CHECK OUTS

18A	valence band gap	18F	III, V four lighter
18B	anions electron holes	18G	absolute temperature logarithms
18C	valence band negative positive positive	18H	heat, light, radiation photo- photons
18D	conduction valence	18I	hole recombination
18E	IV heavier (more metallic)	18J	number e

18K extrinsic
intrinsic

18L donors
conduction band
n

18M B, Al, Ga, In
acceptors
valence
p

18N acceptor saturation
n
donor exhaustion

18O conductive
voltaic
photoconductive

18P pressure gage
reduced

18Q thermistor
temperature

18R *n*-type, *p*-type

18S recombination
photon

18T rectifier (LED)
forward

18U photomultiplier device
Zener diode

18V transistor
base
amplification

18W transistor
field
channel

18X gallium
Ga introduces acceptors
(Ge is a Group IV element)

18Y As
mobility is higher in the conduction band

18Z crystal-pulling
seed

18AA zone-fusion
polycrystalline

18AB epitaxial

18AC diffusion, ion implantation
p-type (epitaxial)

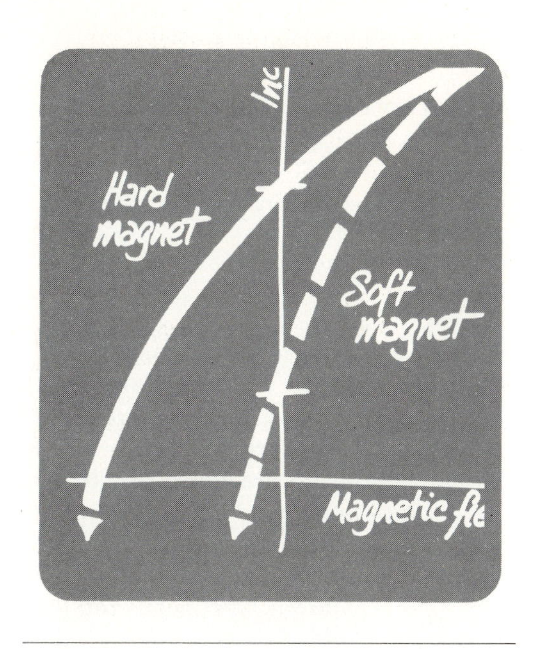

UNIT NINETEEN

NONCONDUCTING CIRCUIT MATERIALS (DIELECTRIC AND MAGNETIC COMPOUNDS)

Conductors transport electrical energy and signals through circuits and systems. Nonconductors also play an important role and, in fact, are as necessary as are the conductors. First, they isolate the electrical potentials of the two or more lines of a circuit. Secondly, they provide means for specifying capacitances and/or inductances for circuit elements. Finally, they can provide functional devices such as transducers and magnetic memory.

Contents

Prerequisites Units 2 and 11; and Sections 3-1, 12-5, 13-3, and 17-6. (Chemistry is required for parts of Section 19-4.)

From Unit 19, the engineer should

1) Understand the origin of electric polarization in terms of charge displacements.
2) Relate the above to the effects of frequency and temperature upon relative dielectric constants.
3) Know the basis of piezoelectricity and its role in transducer devices.
4) • Be able to explain the similarities and differences between piezoelectric and ferroelectric materials.

- • 5) Be able to calculate the magnetization of a ferrimagnetic material.
- • 6) Identify the structural factors that produce hard or soft magnets.
- 7) Understand the meanings of technical terms that pertain to dielectric and magnetic materials.

19-1 ELECTRICAL INSULATORS

An *insulator* has the role of an electrical isolator. Insulators are made of dielectric materials. *Dielectrics* must separate two electrical conductors without conducting charge between them. Metals are obviously excluded. Many, but not all, ceramic and polymeric materials are dielectrics. In terms of the discussion in Section 17-6, dielectrics must have a large energy gap. Alternatively, the valence electrons must be anchored to atoms, or to the bonds between atoms, so they cannot transport charge. Thus, we view polyethylene as a dielectric because the C–C and the C–H bonds retain the electrons covalently; and we view Al_2O_3 as a dielectric because the valence electrons of the aluminum atoms have become anchored to the oxygen atoms. Furthermore, the Al^{3+} and O^{2-} ions that are formed cannot be relocated by the electric field, because these ions are bonded very strongly to each other.

• Dielectric strength

Dielectric strength is the maximum electric field that can be placed across an insulator without electrical breakdown. This is not a strength in terms of Section 3-2, where failure occurred at a critical force per unit area, that is, stress. Rather, the material is subjected to an electrical field, or voltage gradient (volts/meter). Even so, the term dielectric strength is apt since the electrical failure is commonly accompanied by a physical destruction of the material—insulators of plastic film are punctured, and ceramic insulators are cracked.

The dielectric strengths and resistivities for a number of insulators are shown in Table 19-1.1. It should be emphasized, however, that the data given are not fully independent of thickness, because most dielectric materials contain heterogeneities that lead to breakdown paths.

Dielectric failure can originate from several sources. Perhaps the most common cause is the arcing across *external* surfaces (Fig. 19-1.1). Surfaces that have been contaminated by dirt, or even by adsorbed moisture, are specifically subject to such a breakdown. Likewise, connecting pores within

Table 19-1.1
Typical properties of electrical insulators

Material	Resistivity (20°C) ohm·m	Dielectric Strength,* V/mm
Ceramic Materials		
Soda-lime glass	10^{13}	10,000
Pyrex glass	10^{14}	14,000
Fused silica	10^{17}	10,000
Mica	10^{11}	40,000
Steatite porcelain	10^{12}	12,000
Mullite porcelain	10^{11}	12,000
Polymeric Materials		
Polyethylene	10^{13}–10^{16}	20,000
Polystyrene	10^{16}	20,000
Polyvinyl chloride	10^{14}	40,000
Natural rubber	—	16,000–24,000
Polybutadiene	—	16,000–24,000
Phenol-formaldehyde	10^{10}	12,000

* Not constant with thickness

Figure 19-1.1 Surface breakdown (high-voltage insulators). Contamination of an insulator surface can lead to leakages and external shorting. Insulators are glazed to minimize contamination and prevent moisture absorption. (R. Russell, *Brick and Clay Record.*)

the bulk of an insulator provide breakdown paths. Therefore, glazes and other coatings are used to limit moisture absorption.

Internal failure usually originates from those impurities that provide donor electrons. These are rapidly accelerated along potential gradients. In turn, they energize other electrons until an avalanche of electrons cascades through the material and causes failure.

It is difficult to make generalizations about dielectric strength. However, some phenomenological relationships may be cited: (a) dielectric strengths are only broadly related to resistivity; (b) impurities lower dielectric strengths; (c) dielectric strengths decrease as the temperature is raised; and (d) porosity decreases dielectric strengths because the number of possible surface paths for shorting is increased.

19-2 ELECTRICAL POLARIZATION

Types of polarization

Although dielectrics do not transport charge, they can respond to an electric field. For instance, if polyethylene, $\left(C_2H_4\right)_n$, is placed in an electric field, the electrons shift toward the positive electrode, and the protons shift the atomic nuclei slightly toward the negative electrode. In an a.c. field, the above displacements move forth and back in phase with the circuit frequency. Likewise, electrons in atoms such as oxygen, silicon, fluorine, and chlorine "dance in step" with the a.c. field by shifting toward one side of the atoms, then toward the other.

The phenomenon just described is called *electronic polarization* because the shifts of electrons and atomic nuclei produce a negative and positive side to the material. The "centers of gravity" for the positive and negative charges are displaced with respect to each other.

Ionic polarization occurs in a similar manner in ionic solids. The negative ions shift slightly toward the positive electrode, and the positive ions shift slightly toward the negative electrode (Fig. 19-2.1). The ions reverse their displacements with each half cycle of an a.c. field. However, because atomic masses are much greater than the electronic masses, ionic polarization is limited to $\sim 10^{13}$ Hz. Electronic polarization can keep up with an alternating field until $\sim 10^{16}$ Hz.*

Molecular polarization occurs with polar molecules, that is, with molecules that have positive and negative ends. For example, methyl chloride, CH_3Cl, is like methane except a chlorine atom replaces one of the hydrogen atoms. The chlorine atom has a complement of 17 electrons, while each hydrogen atom is simply an exposed proton on the end of a covalent

* The frequency of visible light lies about halfway between these two frequencies. Therefore, incident light can produce electronic polarization, but not ionic polarization.

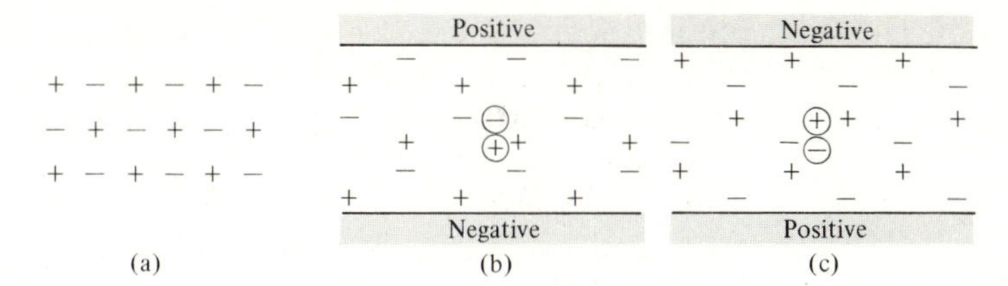

Figure 19-2.1 Ionic polarization (exaggerated). (a) No electric field. The centers of gravity for the positive ions and for the negative ions are identical. (b) Electrical field. The center of gravity for the negative ions, −, is raised; the positive ions, +, lowered. (c) Reversed field. The ionic displacements are reversed. (Center of positive charges, ⊕ ; negative charges, ⊖ .)

bond (Fig. 19-2.2a). The "centers of gravity" for negative and positive charges are separated. Thus, this molecule has a *dipole moment, p*, which is a product of the separation distance, d, and the charge, Zq—in this situation, $26(0.16 \times 10^{-18}$ coul)—since there are 26 electrons (and 26 protons) in CH_3Cl.

$$p = (Zq)d = Qd. \qquad (19\text{-}2.1)$$

In the case of methyl chloride, which is a gas, the whole molecule "flips" with each reversal of the electric field. Molecular solids respond less readily; however, they are not immobile. For example, polyvinyl chloride $\text{-}(C_2H_3Cl)\text{-}_n$ has numerous *polar sites* along its length—one with each chlorine. These will shift orientation with the application of an electric field, if the temperature is above the glass transition temperature. The limiting frequency for molecular polarization in amorphous polymers, such as PVC, is not as sharp

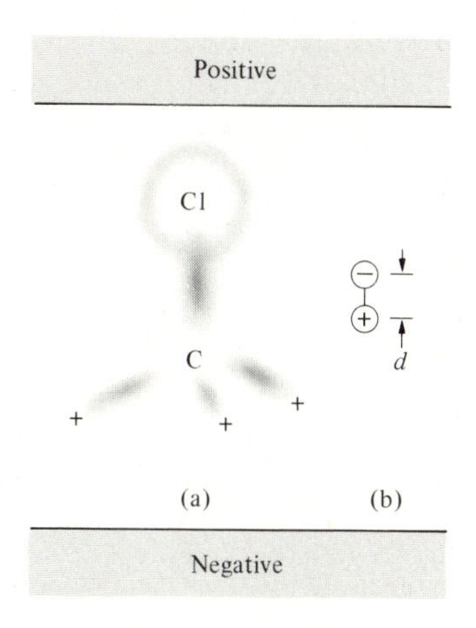

Figure 19-2.2 Molecular polarization (schematic with CH_3Cl). Asymmetric molecules possess a positive and negative end and orient themselves in an electric field. (a) Electron regions are shaded. Hydrogen atoms are protons at the end of a covalent bond. (b) Electric dipole. (Center of positive charges, ⊕ ; negative charges, ⊖ .)

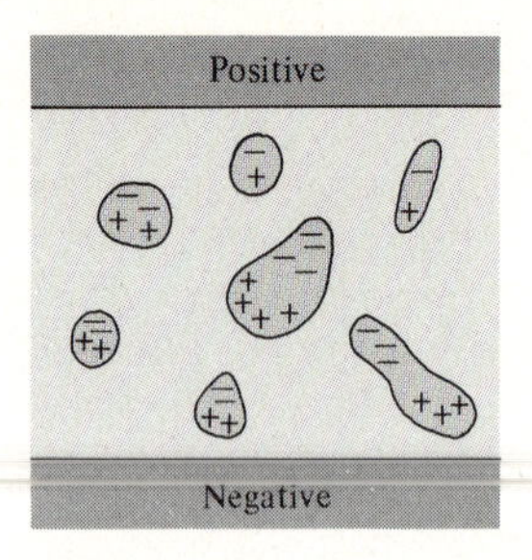

Figure 19-2.3 Space charge. When conducting particles are present within a dielectric, polarization can occur. Each particle develops a positive side and a negative side in an electric field.

as it is for ionic and electronic polarization, because in such a glass no two polar sites have the same surroundings. Some can respond rather rapidly to an alternating field; others are more restricted.

A *space charge* (also called interfacial polarization) develops when there is local conduction within a dielectric. For example, if Al_2O_3 contains small particles of aluminum, the conduction electrons in the latter can move from one side to the other side of each metal particle matching the a.c. frequency that is applied. However, they are trapped within the particle and cannot discharge at the adjacent electrode (Fig. 19-2.3). The Al_2O_3-Al example just cited for illustrative purposes is not typical of engineering products. It is not uncommon, however, to find semiconducting phases within oxide insulators, for example, Ti_2O_3 in TiO_2.

Polarization (quantitative)

Polarization, $\mathcal{P}$, is a property of the material. On a quantitative basis, it is the total of the dipole moments, Σp, per unit volume, V.

$$\begin{aligned}\mathcal{P} &= (\Sigma p)/V = \Sigma(Zqd)/V \\ &= \mathcal{P}_e + \mathcal{P}_i + \mathcal{P}_p + \mathcal{P}_{sc}.\end{aligned} \tag{19-2.2}$$*

Polarization varies with frequency. We saw this earlier where only electronic polarization respond above 10^{13} Hz, and molecular polarization is limited to lower frequencies. Space charge varies significantly with phase composition and microstructural geometry.

Relative dielectric constant

The effects of polarization show up in a capacitor where the two electrodes are separated by a dielectric. First, however, let us consider a pair of electrodes without a dielectric (Fig. 19-2.4a). The *charge density* of $\mathcal{D}_0$ (in coul/m²) that is established on the electrodes of a capacitor is proportional to the electric field, $\mathcal{E}$ (in volts/m);

$$\mathcal{D}_0 = \epsilon_0 \, \mathcal{E} = \epsilon_0 E/d. \tag{19-2.3}$$

* The subscripts refer to electronic polarization (e), ionic polarization (i), the permanent polarization (p), and the space charge (sc).

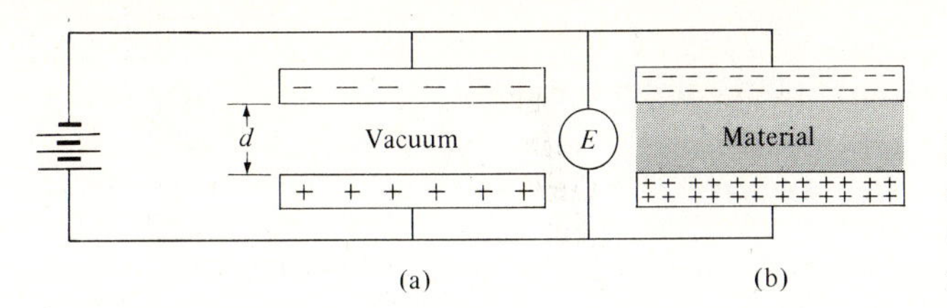

Figure 19-2.4 Charge density. The charge density (coulombs/m^2) on the electrodes depends on the electric field, $\mathcal{E}$, and the polarization of material that is present.

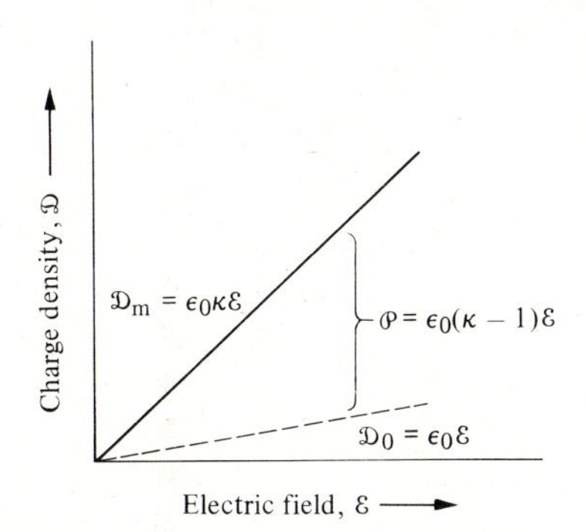

Figure 19-2.5 Charge density versus electric field. Polarization, $\mathcal{P}$, is that part of the charge density resulting from the presence of the material.

The proportionality constant, ϵ_0, is the *capacivity*, or *permittivity*, of vacuum, and has a value of 8.85×10^{-12} C/V·m. This is the slope of the lower curve in Fig. 19-2.5.

Now consider Fig. 19-2.4(b), which shows a dielectric material between the two electrodes. The charge density in this case is $\mathcal{D}_m$, and it is also proportional to the electrical field (the upper curve of Fig. 19-2.5). The *relative dielectric constant,* κ, is the ratio $\mathcal{D}_m/\mathcal{D}_0$, or

$$\kappa = \mathcal{D}_m/\mathcal{D}_0 = \kappa\epsilon_0\,\mathcal{E}/\epsilon_0\,\mathcal{E}. \tag{19-2.4}$$

The relative dielectric constant is dimensionless.

The polarization, $\mathcal{P}$, of the dielectric material that lies between the electrodes produces the increased charge density shown in Fig. 19-2.5.

$$\mathcal{P} = \mathcal{D}_m - \mathcal{D}_0. \tag{19-2.5a}$$

Alternatively,

$$\mathcal{P} = (\kappa - 1)\epsilon_0\,\mathcal{E}. \tag{19-2.5b}$$

The units for this last equation and Eq. (19-2.2) are

$$\frac{\text{coul·m}}{\text{m}^3} = (-)\left(\frac{\text{coul}}{\text{volt·m}}\right)\left(\frac{\text{volt}}{\text{m}}\right). \tag{19-2.5c}$$

Selected values of the relative dielectric constant are given in Table 19-2.1. Note that these are for 20°C, and for the specified frequencies—60 Hz and 10^6 Hz. We can make several observations. Nonpolar polymers, for example, polyethylene $-(C_2H_4)-_n$ and teflon $-(C_2F_4)-_n$, have low dielectric con-

Table 19-2.1
Relative dielectric constants (20°C, unless stated otherwise)

	At 60 Hz	At 10^6 Hz
Soda-lime glass	7	7
Pyrex glass	4.3	4
"E" glass	4.2	4
Fused silica	4	3.8
Porcelain	6	5
Alumina	9	9
Nylon 6/6	4	3.5
Polyethylene $\text{-}(C_2H_4)_n\text{-}$	2.3	2.3
Teflon, $\text{-}(C_2F_4)_n\text{-}$	2.1	2.1
Polystyrene, PS	2.5	2.5
Polyvinyl chloride, PVC		
plasticized ($T_g \approx 0$°C)	7.0	3.4
rigid (T_g = 85°C)	3.4	3.4
Rubber (12 w/o S, $T_g \approx 0$°C)		
−25°C	2.6	2.6
+25°C	4.0	2.7
+50°C	3.8	3.2

stants since they contribute only electronic polarization. Polar polymers have higher dielectric constants that arise from molecular polarization. Typically, ceramic insulators have higher dielectric constants than polymeric insulators; however, numerous exceptions can be cited. Finally, the high dielectric constant of some materials, for example, TiO_2, arises from a space charge that occurs from the presence of a very fine semiconducting precipitate.*

The relative dielectric constant varies with frequency because the polarization varies. The dielectric constant can also vary with temperature when molecular polarization is involved. This is shown in Fig. 19-2.6 for vulcanized rubber. Below the glass transition temperature, T_g, the molecules cannot become oriented with the electric field because the structure is "locked in place." Above T_g, molecular polarization can occur and a higher dielectric constant results. Observe, however, that two factors become important. In the vicinity of T_g, the dielectric constant is less at 10^6 Hz than at 60 Hz (or at 0 Hz, that is, d.c.). In this temperature range, the molecules are not always able to reverse orientations as fast as the a.c. field does. Finally, the maximum polarization (and therefore dielectric constant) decreases at higher temperatures simply because the dipole orientation is destroyed by thermal agitation.

* This high value is attractive; however, such materials also have high dielectric losses, which are generally undesirable.

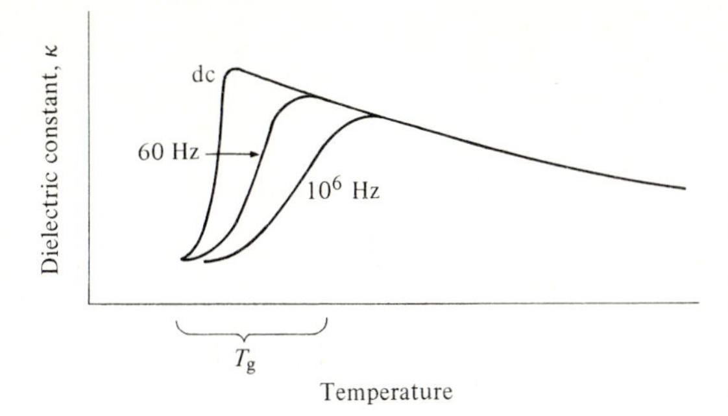

Figure 19-2.6 Dielectric constant versus temperature. Below the glass temperature, T_g, the molecular dipoles cannot respond to the alternating electric fields. (The glass temperature is lower with low frequencies—Fig. 8-3.3b.) Above T_g, thermal agitation destroys the polarization and therefore reduces the dielectric constant κ.

Illustrative problem 19-2.1 A capacitor (Fig. 19-2.4a) with two parallel plates 1 cm × 2 cm each, receives a 2.25-volt potential difference between the electrodes. (a) How far apart must they be to produce a charge density of 10^{-8} coul/m²? (b) How many electrons accumulate on the negative plate under these conditions? No dielectric insulator is placed between these plates.

SOLUTION
a) From the first paragraph of the preceding subsection,

$$\mathfrak{D}_0 = (8.85 \times 10^{-12})\mathcal{E}, \qquad (19\text{-}2.3)$$
$$10^{-8}\text{ coul/m}^2 = (8.85 \times 10^{-12}\text{ coul/volt·m})(2.25\text{ volt}/d),$$
$$d = 0.002\text{ m} \qquad (\text{or } 2\text{ mm}).$$

b) $$\text{No. of electrons} = \frac{1\text{ cm} \times 2\text{ cm} \times 10^{-12}\text{ coul/cm}^2}{1.6 \times 10^{-19}\text{ coul/electron}} = 1.25 \times 10^7\text{ electrons.} \blacktriangleleft$$

Illustrative problem 19-2.2 The relative dielectric constants κ of a glass and of a plastic are 3.9 and 2.1, respectively. What voltage should be applied between electrodes that are separated with 0.13 cm of glass if one wants the same charge density $\mathfrak{D}_m$ as would develop on another set of electrodes separated by 0.42 cm of plastic and 210 volts?

SOLUTION

$$\kappa_{gl}\mathfrak{D}_0 = \mathfrak{D}_{gl} = \mathfrak{D}_{pl} = \kappa_{pl}\mathfrak{D}_0.$$

Since

$$\mathfrak{D}_0 = (8.85 \times 10^{-12}\text{ coul/volt·m})\mathcal{E},$$
$$[\kappa(8.85 \times 10^{-12})(E/d)]_{gl} = [\kappa(8.85 \times 10^{-12})(E/d)]_{pl},$$
$$3.9\,E/0.13\text{ cm} = 2.1(210\text{ volts})/(0.42\text{ cm}),$$
$$E = 35\text{ volts.} \blacktriangleleft$$

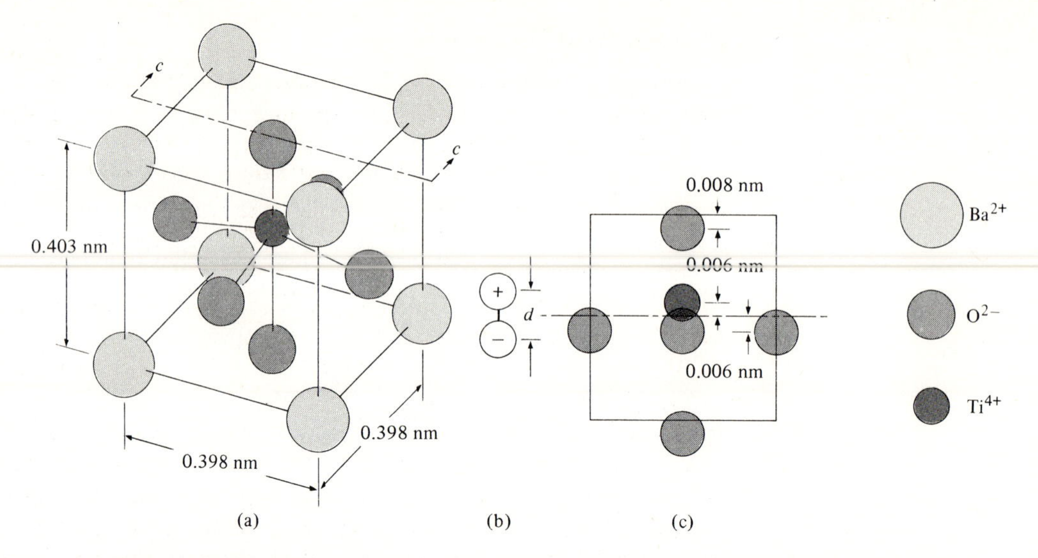

Figure 19-3.1 Tetragonal $BaTiO_3$. Above 120°C (250°F), $BaTiO_3$ is cubic (Fig. 11-3.1). Below that temperature, the ions shift with respect to the corner Ba^{2+} ions. Since the Ti^{4+} and the O^{2-} ions shift in opposite directions, the centers of positive and negative charges are not identical. The unit cell becomes noncubic.

19-3 POLARIZED CRYSTALS

We saw in Fig. 19-2.2 that some molecules are polar because their centers of positive and negative charges are not coincident. They possess a positive end and a negative end. Likewise, some crystals possess a polarity in the absence of an external electric field. This is illustrated in Fig. 19-3.1 by $BaTiO_3$ as it exists at 20°C. If we place the unit-cell corners at the Ba^{2+} ions, the O^{2-} ions are slightly below the center of each face, and the Ti^{4+} ion is 0.006 nm above the center of the cell. This locates the "center of gravity" for all of the negative charges below the center of the unit cell, and for all of the positive charges above the center of the cell.* Therefore, a dipole moment exists for the cell according to Eq. (19-2.1); and if we divide that dipole moment by the unit-cell volume, we have the polarization, $\mathcal{P}_p$, for the material;

$$\mathcal{P}_p = Qd/V = \Sigma Zqd/V. \qquad (19\text{-}3.1)$$

This is calculated in I.P. 19-3.1.

* The directions of up and down are arbitrary. We could have inverted Fig. 19-3.1, or we could have looked at it from the side and cited either a left or a right displacement of charges.

Piezoelectric materials

Polar crystals change their dimensions slightly when they are placed in an electric field. In addition, their polarization is changed if they are compressed (or stretched) to new dimensions. They are *piezoelectric,* literally pressure-electric. The above characteristics are useful because they permit the design of *transducers*—devices that transform electrical energy into mechanical energy and vice versa. In order to understand the principles, let us examine Fig. 19-3.2.

If a single crystal of $BaTiO_3$ is placed between two electrodes that are connected (Fig. 19-3.2a), electrons will move from one electrode to the other. They are repelled from the first electrode by the negative ends of the dipoles in the many unit cells (and attracted to the other electrode by the positive ends). Now, if a compressive stress, s, is applied (Fig. 19-3.2b) so that the dipole moment arm, d, is shortened, the polarization is decreased and some of the electrons return to the original electrode. Alternatively, if there is no electrical connection between the two electrodes when pressure is applied, the electrons cannot return; a voltage difference will be introduced between the two electrodes (Fig. 19-3.2c). Furthermore, if an added potential is

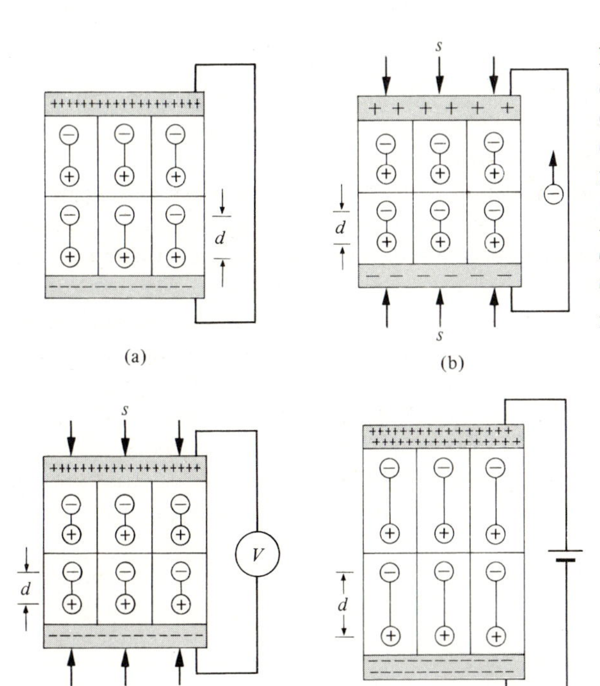

Figure 19-3.2 Piezoelectric material (schematic and exaggerated). (a) The electrical dipoles produce a charge on each electrode. (b) Pressure applied (with electrical contact). Compression decreases the dipole moment, and the charge density decreases. (c) Pressure applied (without electrical contact). A voltage difference is introduced. (d) Voltage applied. The distance between centers of charge is changed. The pressure-voltage changes of parts (c) and (d) are reversible.

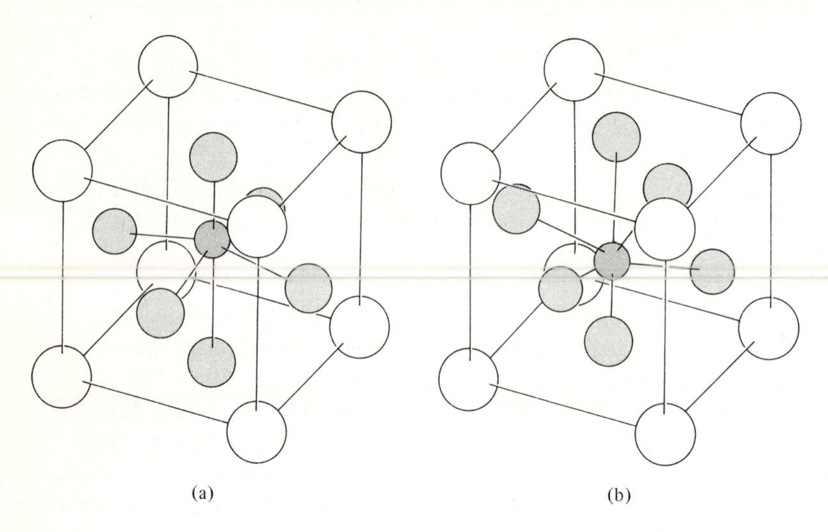

Figure 19-3.3 Ferroelectric crystal ($BaTiO_3$). (a) Same polarity as Fig. 19-3.1. (b) Reversed polarity. The Ti^{4+} is displaced downward and the O^{2-} ions displaced upward. Either polarity is stable until reversed by an external field.

applied (Fig. 19-3.2d), the dipoles of each unit cell will be stretched, because their positive and negative ends will be attracted to the electrodes.*

Crystals of piezoelectric materials can be used for pressure gages, for phonograph cartridges, and for high-frequency sound generators. A description of a ceramic phonograph cartridge will serve as an example. The stylus, or needle, follows the groove on the record. A small transducer is placed in contact with the stylus, which detects the vibrational pattern recorded in the groove. Both the frequency and the amplitude can be sensed as voltage changes (Fig. 19-3.2c). Although the voltage signal is small, it can be amplified through electronic circuitry until it is capable of driving a speaker.

The most widely used piezoelectric ceramics are $PbZrO_3$-$PbTiO_3$ solid solutions, called PZT's. They have the same structure as $BaTiO_3$ (Fig. 19-3.1). All of these materials lose their piezoelectricity at higher temperatures when they change from the structure of Fig. 19-3.1 to that in Fig. 11-3.1. Above that temperature (120°C for $BaTiO_3$, and 490°C for $PbTiO_3$), the unit cell is cubic and symmetrical, thus the centers of positive and negative charges are identical; the d of Eq. (19-3.1) is zero.

• Ferroelectric materials

Some dielectric materials have *reversible* polarity. These are called *ferroelectric.* This behavior is illustrated by Fig. 19-3.3. Part (a) shows the polarity of $BaTiO_3$ in Fig. 19-3.1. If a strong electric field is placed across this material with a negative electrode below, the central Ti^{4+} ion can "pop"

* A reversal of stress in Fig. 19-3.2(c) from compression to tension reverses the voltage; likewise, a reversal of voltage in Fig. 19-3.2(d) introduces a contraction.

past the four adjacent O^{2-} ions, which shift upward toward the positive electrodes. Thus, Fig. 19-3.3(b) is identical with Fig. 19-3.3(a) but with the reverse polarity. In fact, the polarity will cycle to match the frequency of an a.c. circuit.

The term ferroelectric is established in the technical literature; however, it is not a descriptive term, since iron is seldom, if ever, present in ferroelectric materials. The term was established because these materials possess an electrical hysteresis loop that resembles the hysteresis loop of ferromagnetic materials. We see this in Fig. 19-3.4. With a positive field, $\mathcal{E}$, so that all of the unit cells are polarized in the same direction, there is a maximum polarization that equals the value calculated in I.P. 19-3.2. If the field is removed ($\mathcal{E} = 0$), most of the polarity remains. We call this the *remanent polarization,* $\mathcal{P}_r$; it takes a reversed field of $-\mathcal{E}_c$, called *coercive field,* to drop the value of $\mathcal{P}$ to zero. A larger negative field will give saturation polarization in the opposite direction, $-\mathcal{P}_s$. The remaining part of the cycle duplicates the first half, except for signs. The above loop is repeated with each succeeding cycle.

The ferroelectric behavior just described has the capability of storing binary information; a positive field aligns the unit-cell dipoles in one direction; a negative field aligns the unit-cell dipoles in the opposite direction. In either case, the information is retained after the field is removed. With appropriate circuitry, the sign of the polarity can be "read" as required.

Piezoelectricity and ferroelectricity are related; in fact, we have used the same material, $BaTiO_3$, to describe both. *All ferroelectric materials are piezoelectric;* but not all piezoelectric materials are ferroelectric. This rela-

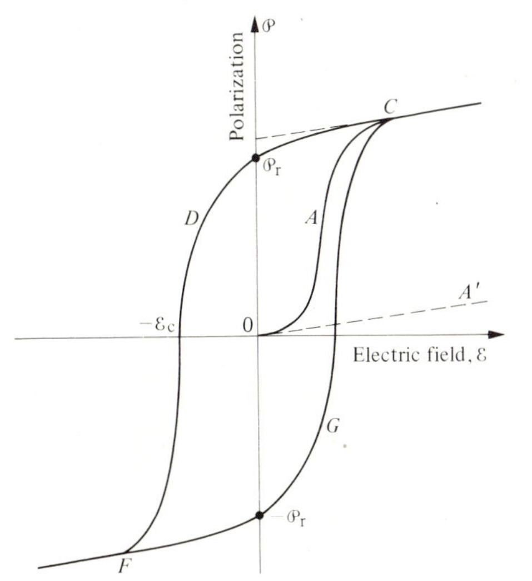

Figure 19-3.4 Ferroelectric hysteresis. Starting from 0, an electric field, $\mathcal{E}$, polarizes the material until saturation is reached at C. If the field is dropped to $\mathcal{E} = 0$, a remanent polarization, $\mathcal{P}_r$, remains. It takes a coercive field of $-\mathcal{E}_c$ to drop the net polarization to zero. The hysteresis loop, CDFGC, is followed for each succeeding cycle.

tionship applies because both types must be polar, and the piezoelectricity arises from the accompanying dipole moment. However, a polar material cannot be ferroelectric unless the dipole moment can be reversed. Quartz (SiO_2 of Fig. 11-6.3) has a polarity that cannot be reversed, so it is piezoelectric but not ferroelectric. The polarity of quartz cannot be reversed because bonds would have to be broken and new crystals formed in order to invert the polarity. The major amount of energy that is required cannot be introduced into the crystal electrically. In contrast, the polarity is reversed in $BaTiO_3$ by a simple displacement (Fig. 19-3.3). The energy requirement is the amount necessary to squeeze the Ti^{4+} ion through the space among the adjacent four 0^{2-} ions. The force of the electric field is capable of doing that.

Illustrative Problem 19-3.1 A piezoelectric material has a Young's modulus of 72,000 MPa (10,400,000 psi). What stress is required to change its polarization from 640 C·m/m³ to 645 C·m/m³?

SOLUTION $\Delta\mathcal{P}/\mathcal{P} = (645 - 640)/640 = +0.0078 = (\Delta Qd/V)/(Qd/V)$. However, Q and V do not change; therefore,

$$\Delta d/d = +0.0078 = e = s/E.$$

$$s = +0.0078\ (72{,}000\text{ MPa}) = +560\text{ MPa};$$

or $s = +81{,}000$ psi (tension).

ADDED INFORMATION The charge density also changes 5 C/m² with this pressure. Therefore, if the two electrodes of Fig. 19-3.2(b) are connected, there will be a transfer of

$$\text{electrons} = 5\ \text{C/m}^2/(0.16 \times 10^{-18}\ \text{C/el}) = 3 \times 10^{19}\ \text{el/m}^2 \qquad (\text{or } 3 \times 10^{13}\ \text{el/mm}^2). \blacktriangleleft$$

Illustrative problem 19-3.2 Calculate the polarization $\mathcal{P}$ of $BaTiO_3$, based on Fig. 19-3.1 and the information of this section.

SOLUTION With reference to the *center* of a cell cornered on the Ba^{2+} ions:

Ion		Q, coul	d, m	Qd, coul·m
Ba^{2+}		$+2(0.16 \times 10^{-18})$	0	0
Ti^{4+}		$+4(0.16 \times 10^{-18})$	$+0.006(10^{-9})$	3.84×10^{-30}
2 O^{2-}	(side of cell)	$-4(0.16 \times 10^{-18})$	$-0.006(10^{-9})$	3.84×10^{-30}
O^{2-}	(top and bottom)	$-2(0.16 \times 10^{-18})$	$-0.008(10^{-9})$	2.56×10^{-30}
				$\Sigma = 10.24 \times 10^{-30}$

From Eq. (19–3.1)

$$\begin{aligned}\mathcal{P} &= \Sigma Qd/V \\ &= (10.24 \times 10^{-30}\ \text{C}\cdot\text{m})/(0.403 \times 0.398^2 \times 10^{-27}\ \text{m}^3) \\ &= 0.16\ \text{coul/m}^2.\end{aligned}$$

ADDED INFORMATION This means that polarized $BaTiO_3$ can possess a charge density of 0.16 coul/m^2, equivalent to 10^{12} electrons/mm^2. ◀

• 19–4 MAGNETIC MATERIALS

It is easy to point out applications of magnets, ranging from transformer sheets to compasses (Fig. 19–4.1). Our purpose in this section is to understand some of the characteristics of magnets that lead to this rather unusual property.

Although every electron possesses a magnetic moment, relatively few elements of the periodic table have sufficient net magnetism from their complement of electrons to become useful to the engineer. These few elements include iron, cobalt, and nickel, plus some of the rare-earth elements.* Among these, iron is predominant and most familiar. This has led to terms such as ferromagnetic, ferrimagnetic, and antiferromagnetic. These terms will be most easily described when we consider ceramic magnets later in this section.

Two important categories of magnets include those we call permanent, or *hard*, magnets, and those called temporary, or *soft*, magnets. The former can retain their bulk magnetism indefinitely. The latter are not obviously magnetic except when they are placed in a magnetic field. We shall observe shortly that soft magnets contain many small magnetic domains with apparently random orientations. External magnetic effects are cancelled.

Magnetization†

Each electron possesses a *magnetic moment* of 9.27×10^{-24} A·m^2. We sometimes visualize this electron magnetic moment as a spinning charge and speak of an electron spin. From chemistry, we have learned that each orbital of an atom can possess two electrons, of opposite spins. As such, the *net*

---

* Elements with atomic numbers of 58 to 72 are called the *rare earths*. In reality, some of these, such as cerium and neodymium, are more prevalent in the earth's crust than are cobalt, lead, and zinc. The "rare earths" are difficult to extract and purify individually. Some, such as samarium, produce exceptionally strong magnets when alloyed with cobalt. These magnets are currently receiving considerable attention by engineers who are designing miniature, but powerful, electric motors.

† This subsection assumes a somewhat greater familiarity with chemistry than other sections in the text. If this section is among those assigned for a materials course, the student may want to have a chemistry book available for reference.

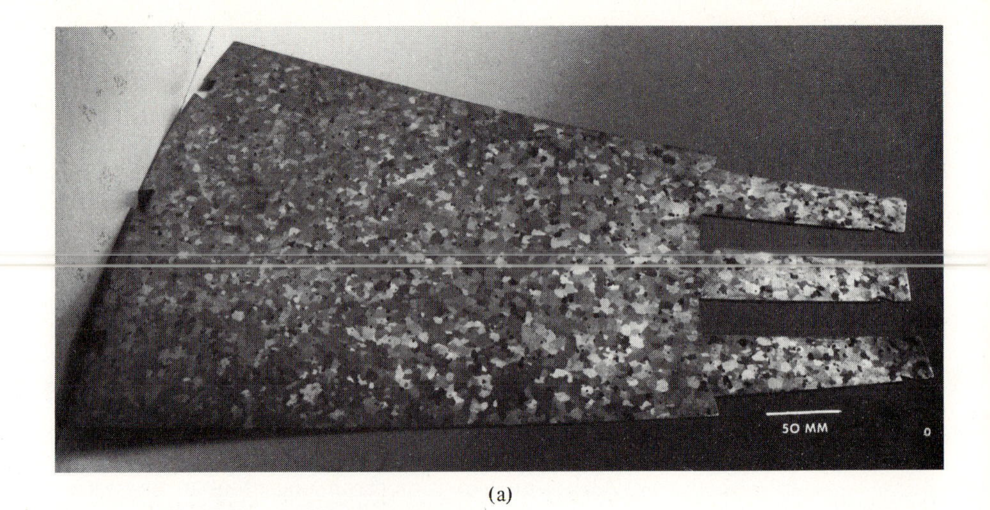

(a)

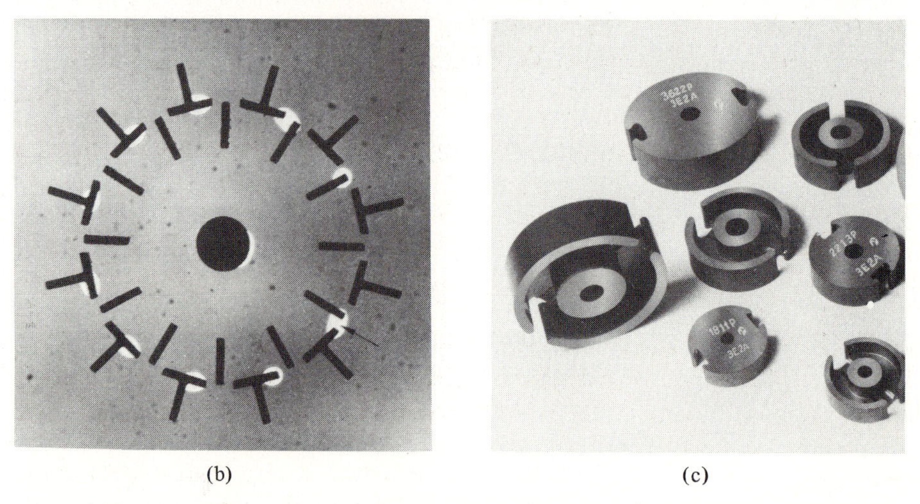

(b) (c)

Figure 19-4.1 Magnetic materials. (a) Grain-oriented silicon steel (stator core of a large electrical generator). As in Fig. 20-3.3, the individual crystals are aligned to give easy magnetization. (Courtesy of General Electric Co.) (b) Magnetic bubbles (ceramic garnets). A thin layer of magnetic garnet is deposited epitaxially on a nonmagnetic garnet single crystal. Opposite polarity (arrow) stores binary information. (Courtesy of A. H. Bobeck, Bell Telephone Laboratories.) (c) Soft ferrites for microwave applications. (Courtesy J. J. Svec, *CERAMIC INDUSTRY*, Cahner's Publishing Co.)

magnetic moment of each filled oribital is zero. This is the reason we do not observe magnetism in the vast majority of the elements. For example, oxygen, as an O^{2-} ion, has a *pair* of electrons in its $1s$, its $2s$, and its three $2p$ orbitals. As a molecule, O_2, the two covalent bonds have a *pair* of electrons

		K	Ca	Sc	Ti	V	Cr	Mn	Fe	Co	Ni	Cu	Zn
4s		↑	↑↓	↑↓	↑↓	↑↓	↑	↑↓	↑↓	↑↓	↑↓	↑	↑↓
3d	↑	–	–	1	2	3	5	5	5	5	5	5	5
	↓	–	–	–	–	–	–	–	1	2	3	5	5

Figure 19-4.2 Unbalanced magnetic spins (isolated atoms). The 4*s* orbital fills before the 3*d* orbitals when each additional proton is introduced to the nucleus.

each, as do the other orbitals. In each case the pair of electrons possess no net magnetic moment, because the two opposing spins cancel any externally observed effects. The electrons of the majority of all elements are similarly paired when their atoms are combined with other atoms, or as ions. The exceptions that lead to magnetism are associated with the unfilled subvalence orbitals. These are the 3*d* orbitals at transition-metal elements (and the 4*f* orbitals of the rare-earth elements).

Again, recall from your chemistry that the atoms in the first long row of the periodic table (Fig. 19-4.2) generally fill their 4*s* orbital before filling their 3*d* orbitals. When the five 3*d* orbitals of manganese are entered, the electrons align their magnetic spins as shown in Fig. 19-4.2. Thus, an individual manganese atom has five unbalanced electron spins; an iron atom has four; cobalt, three; and nickel, two. As a result, individual atoms of iron have a magnetic moment of $4(9.27 \times 10^{-24}\ \text{A}\cdot\text{m}^2)$. From this picture, we can easily move to the basis for ceramic magnets. Metallic magnets will be somewhat more complex.

Ceramic magnets

Ceramic magnets are ionic compounds, for example, the ferrites. Thus, any iron is present as Fe^{2+} or Fe^{3+}. A ferrous ion has lost two electrons. These are the two 4*s* electrons; the *six* 3*d* electrons remain to give four unpaired electrons (Fig. 19-4.3). A ferric ion has lost the two 4*s* electrons and *one* 3*d* electron; thus *five* unpaired electrons remain.

Nature's original magnet is the mineral magnetite (Fe_3O_4), formerly called lodestone. Magnetite is a natural ceramic phase with an fcc lattice of O^{2-} ions. The iron ions are in both 4-fold and 6-fold interstitial sites (Fig. 11-1.4). More specifically, the Fe^{2+} ions are in 6-fold sites; and the Fe^{3+} ions

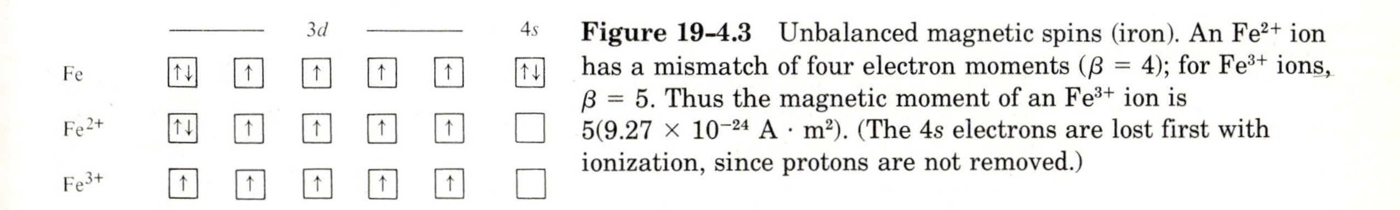

Figure 19-4.3 Unbalanced magnetic spins (iron). An Fe^{2+} ion has a mismatch of four electron moments ($\beta = 4$); for Fe^{3+} ions, $\beta = 5$. Thus the magnetic moment of an Fe^{3+} ion is $5(9.27 \times 10^{-24}\ \text{A} \cdot \text{m}^2)$. (The 4*s* electrons are lost first with ionization, since protons are not removed.)

are equally divided between 6-f and 4-f sites. The resulting structure was previously described as the $NiFe_2O_4$-type structure in Section 11–3.

The unit cell of this structure is magnetic because the magnetic moments of the ions in the 6-fold sites are all aligned in the same direction and those in the 4-fold sites are all aligned in the opposite direction. We can make an accounting in the compound Fe_3O_4, or $[Fe^{2+}Fe_2^{3+}O_4]_8$. For every 32 oxygen atoms, which are not magnetic, there are 8 Fe^{2+} ions and 8 Fe^{3+} ions with alignment in the ↑ direction. There are 8 Fe^{3+} ions with alignment in the ↓ direction. From this, plus the information in the previous two paragraphs:

Interstitial site	Spin alignment	Fe^{2+}	Fe^{3+}	β	Magnetic moment
6-f	↑	8		+32	$+8(4)(9.27 \times 10^{-24}\ A{\cdot}m^2)$
6-f	↑		8	+40	$+8(5)(9.27 \times 10^{-24}\ A{\cdot}m^2)$
4-f	↓		8	−40	$-8(5)(9.27 \times 10^{-24}\ A{\cdot}m^2)$
		Net:		+32	$+32(9.27 \times 10^{-24}\ A{\cdot}m^2)$

Magnetization, M, which may be described as the internal magnetic field of a material, is the total magnetic moment, p_m, per unit volume, V;

$$M = \Sigma p_m/V. \tag{19–4.1}$$

A unit cell of $[Fe^{2+}Fe_2^{3+}O_4]_8$ is cubic and has a lattice constant, a, of 0.837 nm. Therefore, the *saturation* (maximum) *magnetization*, M_s, for magnetite is

$$32(9.27 \times 10^{-24}\ A{\cdot}m^2)/(0.837 \times 10^{-9}\ m)^3 = 0.5 \times 10^6\ A/m.$$

This compares with 0.53×10^6A/m determined in laboratory experiments.

Nickel ferrite $[Ni^{2+}Fe_2^{3+}O_4]_8$ has the same structure as magnetite; however, Ni^{2+} ions have only two unpaired electrons. (Compare Fig. 19–4.2 with Fig. 19–4.3.) Therefore, it has a magnetic moment of only $(+8)(2)(9.27 \times 10^{-24}\ A{\cdot}m^2)$ per unit cell. Its unit-cell dimensions are close to that of magnetite; thus, its saturation magnetization is about half that of Fe_3O_4.*

The ceramic magnets just described are *ferrimagnetic*. That is, they possess an opposing, but unbalanced, alignment of the magnetic atoms; thus, they have a net magnetization. Some ceramic compounds, for example, manganous oxide (MnO) and nickel oxide (NiO) are *antiferromagnetic*. In NiO, which has the NaCl structure (Fig. 19–4.4), alternate planes of Ni^{2+} ions have antiparallel (opposite) orientations. Thus, while each Ni^{2+} ion has two unpaired electrons, the alternate planes lead to zero net magnetization.

* These two magnetic compounds, $[Fe_3O_4]_8$ and $[NiFe_2O_4]_8$, are called *ferrites* on the basis of their chemistry. They possess 32 O^{2-}, 16 Fe^{3+}, and 8 Fe^{2+} (or 8 Ni^{2+}) per unit cell, because the lattice constant must be twice that shown in Fig. 11–1.4 in order to obtain a matching of occupied interstices with full unit-cell translations. [See Van Vlack, *Elements of Materials Science and Engineering,* 4th ed., Addison-Wesley (1980), Section 3–5, and Fig. 8–6.5.]

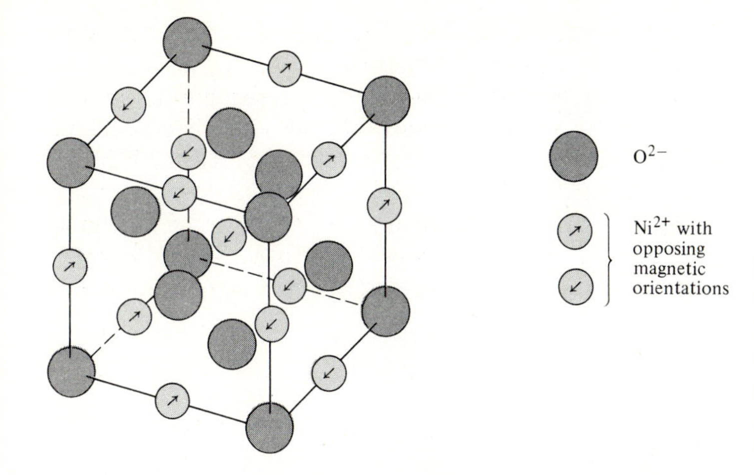

Figure 19-4.4 Antiferromagnetism (NiO). Magnetism exists; however, there is a balance of magnetic moments in the two opposing directions on alternate planes. Therefore, special laboratory procedures are required to detect its presence. (From Van Vlack, *Nickel Oxide.*)

Only by special procedures (making use of high-frequency detection or of neutron diffraction) is it possible to identify antiferromagnetism. In *ferromagnetic* materials, all of the magnetic atoms are said to have the same orientation. Such materials are limited to metals, and possess band structures.

Hard and soft magnets

The concept of a domain is required to understand the difference between "soft" and "hard" magnets. A magnetic *domain* is a part of a single crystal in which the unit cells possess the same magnetic orientation. This is illustrated schematically in Fig. 19-4.5. It may also be shown microscopically. Within a single domain, the magnetization matches that calculated earlier in this section from the number of atoms per unit cell and the magnetic moment of the electrons with each atom. However, most magnetic materials have numerous domains with various crystallographic orientations. Therefore, the net magnetization of the material will be much less than saturation, and even zero if there is a balance in the domain alignments. Such a material does not possess a magnetic polarity, since its internal magnetization is canceled out.

If the above material is placed in an external magnetic field, H, those domains that are aligned with the field will grow (and those domains with an unfavorable alignment will shrink) until the magnetic saturation occurs. The domain growth occurs by a sidewise movement of the *domain boundary.** If the external magnetic field is removed

a) from a *soft magnet,* the numerous domains form once again (by domain-boundary movements) and the net magnetization drops to zero;

* The atoms along the boundary invert their spin. In turn, the adjacent rows of atoms respond to their reoriented neighbors, etc.

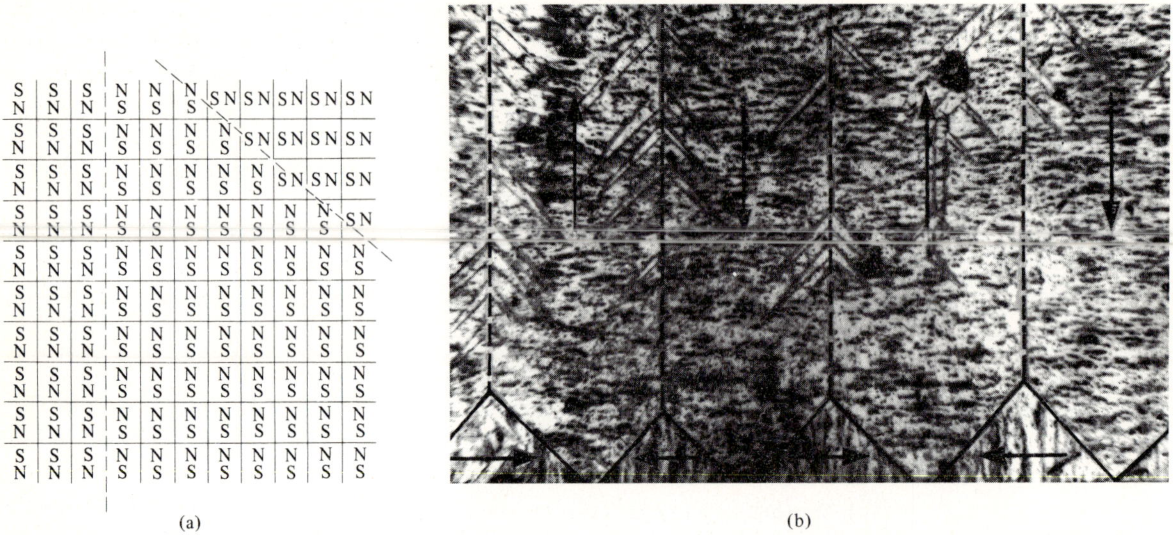

(a) (b)

Figure 19-4.5 Domains. Adjacent unit cells develop identical magnetic alignments to produce a domain. (a) Schematic. (b) Iron. Adjacent domains cancel the external effects, unless a magnetic field is encountered that moves the boundaries, enlarging favored domains and restricting others. Domains of a permanent magnet retain this magnetization after the external field is removed. (From *Ferromagnetism* by R. M. Bozorth. Reprinted by permission of Bell Telephone Laboratories.)

b) from a *hard magnet,* the total magnetization remains because the domain boundaries cannot move spontaneously (until a strong reverse magnetic field, called a *coercive field,* $-H_c$, is applied).

As seen from this contrast between soft and hard (permanent) magnets, the difference between the two arises from the ease with which the domain boundaries can spontaneously move to permit the randomization of the domains. Domain-boundary movements become anchored by small precipitated particles, by grain boundaries, and by dislocations. The latter explains the origin of hard and soft on a microstructural basis. The deformation of a material introduces a large number of dislocations. These not only increase the mechanical hardness but also restrict domain-boundary movements to "lock-in" the net magnetization that is developed. Hence, the term *hard* is used in two contexts.

There are two alternatives available to *demagnetize* a hard magnet. The first is to place the magnet in a strong 60-Hz magnetic field. This forces domain-boundary movements to produce magnetization, first in one direction, then the opposite direction. A random domain pattern develops as the magnet is removed from the center of the strong a.c. field into the weaker surroundings, leaving zero net magnetization.

The second method for demagnetization is to heat the magnet above its Curie temperature, T_c, and then cool it back to the ambient temperature. The *Curie temperature* is that temperature where the thermal agitation is sufficient to disrupt the magnetic coupling between adjacent atoms that permit domain formation. During cooling in the absence of an external magnetic field, domains nucleate at many sites with equal opportunity for orientation in each of the available crystal directions. Thus, no net magnetization results. Conversely, a permanent magnet may be *poled* by cooling it through the Curie temperature, while it is in a magnetic field. The domains oriented with the field grow more rapidly than the others, so that the whole magnet develops and retains an N–S polarity.

Metallic magnets

Bcc iron is the most common metallic magnetic material. It is a *soft* magnet with a Curie temperature of 770°C.* Many tons of magnetic iron products are made and used each year as sheets for transformer cores (Fig. 19–4.1b) and motor armatures. In these applications, the magnet must be soft in order to respond to the 60-Hz power sources. Typically, these steels contain 1 to 4 percent silicon in solid solution to increase their magnetic permeability; the steels are annealed; and they are permitted to develop large grain sizes. The annealing eliminates dislocations, and the grain-coarsening reduces the number of grain boundaries. Each of these changes facilitate domain-boundary movements.

Most *hard* metallic magnets are two-phase alloys. The oldest is high-carbon steel with bcc iron (ferrite) plus carbides. Alnico magnets are alloys of **al**uminum, **ni**ckel, and **co**balt that contain a very fine distribution of a second phase in an fcc matrix. The phase boundaries restrict domain-boundary movements and, therefore, assure the retention of the magnetization. In contrast, the typical hard ceramic magnet, for example, $BaFe_{12}O_{19}$, achieves its permanence through highly anisotropic crystals that give only limited options for domain-boundary movements.

Principal properties of magnets

A major property for soft magnets is their *permeability*, μ_r, which is the ratio of the flux density, that is, *magnetic induction, B*, to the total *magnetic field, H,* including magnetization, *M*, as shown in Fig. 19–4.6. Values of these ratios (maximum) for several common metallic and ceramic soft magnets

---

* At one time it was assumed that the iron formed a new polymorphic structure as it was heated past 770°C (1418°F), because it lost its magnetism at that temperature. It was called β-iron. We now know that bcc iron remains until 912°C (1673°F) where it changes to fcc iron. Thus, we only identify α, γ, and δ iron as crystal polymorphs (Section 6–2.)

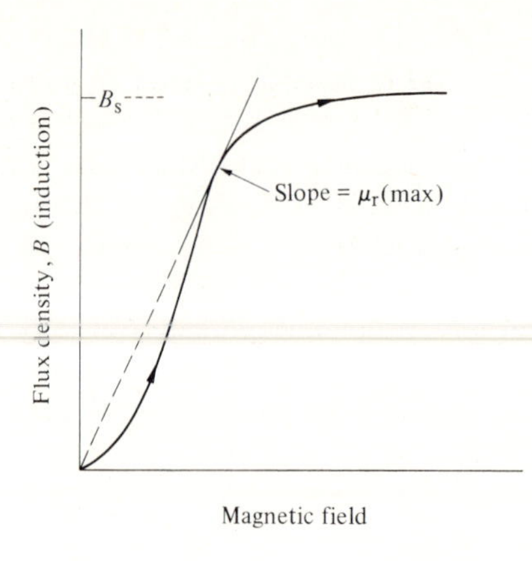

Figure 19-4.6 Magnetic permeability, μ_r. This is the ratio of flux density, or magnetic induction, B, to the magnetic field. Induction saturation, B_s, occurs when all of the domains are aligned in one direction. (See Table 19-4.1.)

Table 19-4.1
Properties of selected soft magnets (various sources)

Magnetic material	Saturation induction, B_s volt·sec/m²	Coercive field, $-H_c$ amp/m	Maximum relative permeability, μ_r(max)
Pure iron (bcc)	2.2	80	5,000
Silicon ferrite transformer sheet (oriented)	2.0	40	15,000
Permalloy, Ni–Fe	1.6	10	2,000
Superpermalloy, Ni-Fe-Mo	0.2	0.2	100,000
Ferroxcube A, (Mn, Zn)Fe_2O_4	0.4	30	1,200
Ferroxcube B, (Ni, Zn)Fe_2O_4	0.3	30	700

are shown in Table 19-4.1. Admittedly, permeability is not the only consideration, or else the two ceramic magnetic materials would not be used. At high frequencies, however, there must be a trade-off between permeability and electrical conductivity. The conductive metals become heated by induced currents at megahertz frequencies*; therefore, the electrical engineer cannot use metals but turns to ceramic magnets for deflection yokes in cathode-ray tubes and similar high-frequency electron applications.

There are two major properties for hard magnets. These are the *remanence induction*, B_r, and the *coercive field*, $-H_c$. Values of these for several hard magnets are shown in Table 19-4.2. The remanence induction is the induction remaining after the external field is removed (Fig. 19-4.7). The coercive field is the reverse field that must be applied to reduce the induction to zero. (Cf. polarization remanence, $\mathcal{P}_r$, and electric coercive field, $-\mathcal{E}_c$,

* Metallurgists use this principle to melt metals in induction furnaces.

Table 19–4.2
Properties of selected hard magnets (various sources)

Magnetic material	Remanence, B_r volt·sec/m^2	Coercive field, $-H_c$ amp/m	Maximum demagnetizing product, BH_{max}, joules/m^3
Carbon steel	1.0	0.4×10^4	0.1×10^4
Alnico V	1.2	5.5×10^4	3.4×10^4
Ferroxdur ($BaFe_{12}O_{19}$)	0.4	15.0×10^4	2.0×10^4

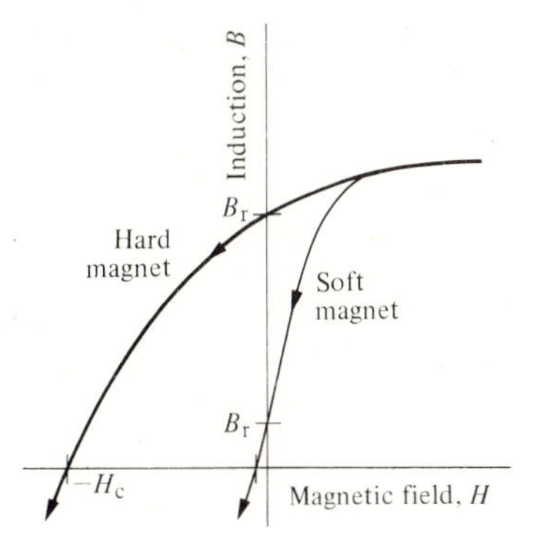

Figure 19–4.7 Soft versus hard magnets. When the field, H, is dropped to zero, the induction, B, of a soft magnet drops to nearly zero. With the external field removed from a hard magnet, the remanent induction, B_r, remains close to saturation; a reverse, coercive field, $-H_c$, is required to reduce the induction to zero.

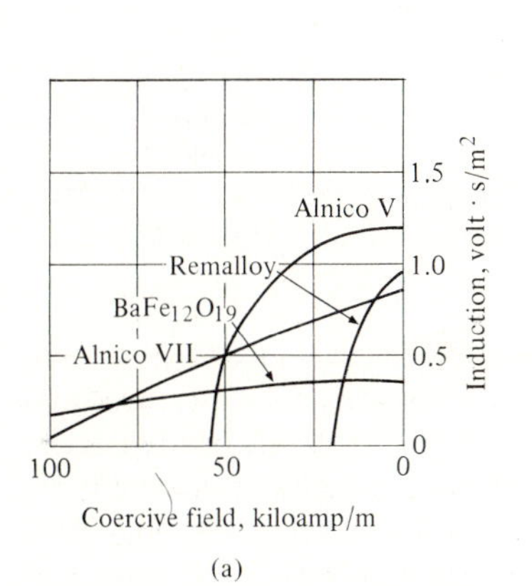

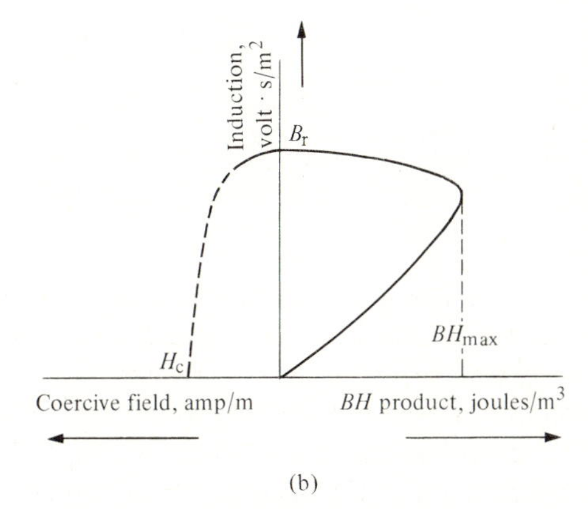

Figure 19–4.8 Demagnetization curves (second quadrant). (a) Hard magnets. (b) *BH* products. The maximum value is commonly used as an index of magnetic permanency.

in Section 19–3.) The *BH* product is an index of energy required to demagnetize and to reverse the polarity of a permanent magnet (Fig. 19–4.8).* Observe from Tables 19–4.1 and 19–4.2 that the coercive field for hard magnets are several orders of magnitude greater than for soft magnets.

* The value of BH_{max} does not correspond to $(B_r)(-H_c)$, but to the *BH* product at an intermediate point of the second quadrant of Fig. 19–4.8.

Illustrative problem 19-4.1 By substituting ($4\ Li^+ + 4\ Fe^{3+}$) for $8\ Fe^{2+}$, magnetite is altered from (a) $Fe_8^{2+}Fe_{16}^{3+}O_{32}$ to (b) $Li_4^+Fe_{20}^{3+}O_{32}$. Assume that there is negligible change in unit-cell size and that the Li^+ ions enter the 4-f sites and 6-f sites equally. What is the percent change in saturation magnetization?

SOLUTION The ferrite structure has 16 cations in 6-f sites and 8 cations in 4-f sites.

	6-f ↑	4-f ↓	fcc array
Mag:	$8\ Fe^{3+} + 8\ Fe^{2+}$	$8\ Fe^{3+}$	$32\ O^{2-}$
LiMag:	$14\ Fe^{3+} + 2\ Li^+$	$6\ Fe^{3+} + 2\ Li^+$	$32\ O^{2-}$

(O^{2-} and Li^+ are nonmagnetic; Fe^{2+} has 4β/ion; Fe^{3+} has 5β/ion.)

$$\beta_{Mag} = (8)(5\uparrow) + (8)(4\uparrow) + (8)(5\downarrow) = 32\ \beta/\text{unit cell};$$
$$\beta_{LiMag} = (14)(5\uparrow) + (6)(5\downarrow) = 40\ \beta/\text{unit cell}.$$

With the volume of the unit cell as V,

$$\Delta M_s = [(40 - 32)(9.27 \times 10^{-24})/V]/[32(9.27 \times 10^{-24})/V] = +25\% \blacktriangleleft$$

Review and Study

SUMMARY

Materials are used for a variety of electric circuit purposes, other than for the conductivity (or semiconductivity). More prominent among these are (a) insulation, (b) polarization, and (c) magnetization.

1. Insulators do not possess charge carriers, because there is a large energy gap below the conduction band. Stated in another way, the valence electrons are tightly attached to anions or in covalent bonds. At normal temperatures, technical insulators have resistivities above 10^{10} ohm·m, and dielectric strengths of more than 10,000 V/mm. Electrical breakdown most commonly occurs across contaminated surfaces or through internal cracks and pores.
2. Electrical polarization is the separation of positive and negative charges. Included are electric, ionic, and molecular displacements. In addition, the space charge of a conductive phase within a dielectric can produce the same effect. The electrical dipole moment is the product of charge and charge separation. Polarization is the dipole moment per unit volume.
3. Polarization within a dielectric material leads to a greater charge density on adjacent electrodes. The relative dielectric constant is the ratio of the charge density on an electrode, with and without an intervening dielectric. When the a.c. frequency is increased, the dielectric constant drops if the frequency exceeds

that which the polarization can follow. Molecular polarization ceases at lowest frequencies; ionic at $\sim 10^{13}$ Hz, and electronic at $\sim 10^{16}$ Hz. The dielectric constant is highest in amorphous materials just above their glass temperature where atom displacements become possible. Greater thermal agitation destroys the polarization.

4. Piezoelectric materials possess permanent polarization. Since pressure alters the polarization, these materials make useful transducers of mechanical and electrical energy. • The permanent polarization is reversible in ferroelectric materials. They possess a remanent polarization after the external electric field is removed and require a coercive field to eliminate the polarization. All ferroelectric materials are piezoelectric; not all piezoelectric materials are ferroelectric.

• 5. All electrons possess a magnetic moment; however, few atoms or ions have a net magnetic moment, since their electrons enter orbitals as pairs with opposing alignments. Exceptions involve transition metals with unfilled, subvalence orbitals. These elements, principally iron, cobalt, nickel and some of the "rare earths," provide our only magnetic elements. Atoms with a net magnetic moment reside in specific sites within a unit cell. Again, magnetism is not evident if the atomic magnetic orientations are balanced (antiferromagnetic). Useful magnets (ferrimagnetic) possess an unbalanced magnetic orientation with more magnetic moments aligned in one direction than another. The magnetization of the unit cells in these materials may be calculated.

• 6. Small parts of a crystal become a magnetic domain with each unit cell magnetized in the same direction. In a demagnetized material, there are an equal number of domains aligned in each of the crystallographic directions. An external magnetic field helps the favorably oriented domains grow and the others decrease in size. The domains of soft magnets regain their equal numbers when the external field is removed; the domains of hard (permanent) magnets retain their net magnetization. With the latter there is a remanent magnetization after the field is removed; and a coercive magnetic field is required to eliminate the net magnetization. The "permanency" of a magnet may be varied by microstructural factors such as grain size, second phases, etc.

TECHNICAL TERMS

• **Coercive field (electric, $\mathcal{E}_c$)** Electric field required to remove residual polarization.

• **Coercive field (magnetic, H_c)** Magnetic field required to remove residual magnetization.

• **Curie point (electric)** Transition temperature between symmetric crystal and polar crystal.

• **Curie temperature (magnetic)** Transition temperature for magnetic domain formation.

Dielectric An insulator. A material that can be placed between two electrodes without conduction.

Dielectric constant, relative (κ) Ratio of charge density arising from an electric field (1) with and (2) without the material present.

• **Dielectric strength** Electrical breakdown potential of an insulator per unit thickness.

Dipole moment (p) Product of electric charge and charge separation distance.

• **Domains** Microstructural areas of coordinated magnetic alignments (or of electrical dipole alignments).

• **Ferrimagnetism** Net magnetism arising from unbalanced alignment of magnetic ions within a crystal. (Antiferromagnetic materials have fully balanced alignment, and therefore no net magnetism.)

• **Ferrites** Compounds containing trivalent iron; commonly magnetic.

• **Ferroelectric** Materials with spontaneous dipole alignment.

• **Ferromagnetic** Materials with spontaneous magnetic alignment.

• **Induction (magnetic, B)** Flux density in a magnetic field.

• **Induction (remanent, B_r)** Flux density remaining after the external magnetic field has been removed.

• **Magnet, hard (permanent)** Magnet with a large ($-BH$) energy product, so that it maintains domain alignment.

• **Magnet, soft** Magnet that requires negligible energy for domain randomization.

• **Magnetization, (M)** Magnetic moment density, that is, magnetic moment per unit volume.

• **Permeability (magnetic, μ)** Ratio of induction to magnetic field.

Piezoelectric Dielectric materials with structures that are asymmetric, so that their centers of positive and negative charges are not coincident. As a result the polarity is sensitive to pressures that change the dipole distance, and the polarization.

Polar sites A local electric dipole within a polymeric molecule.

Polarization (electric, $\mathcal{P}$) The dipole moment, $p(=Qd)$, per unit volume.

• **Polarization (remanent, $\mathcal{P}_r$)** Polarization remaining after the external electric field has been removed.

Resistivity (ρ) Reciprocal of conductivity (usually expressed in ohm·m).

Space charge Polarization from conductive particles in a dielectric.

Transducer A material or device that converts energy from one form to another, specifically electrical energy to or from mechanical energy.

CHECKS

19A Dielectric materials, which are used to separate two _***_, must have a large _***_ in order to serve as an insulator and not be a semiconductor.

• 19B Although the units for dielectric strength are _***_, the values for an insulator are generally not proportional to _***_, because _***_ . . . .

• 19C Dielectric strength can be reduced by ___***___, by ___***___, and by ___***___.

19D Several types of polarization may be described; these include ___***___, ___***___, and ___***___. Of these, ___***___ can occur at highest frequencies, and ___***___ is least rapid.

19E As a property, polarization is the summation of the ___***___ per ___***___; it also is the added ___***___ arising from the presence of displaced electrons, ions, and polarized molecules.

19F Relative dielectric constants are always greater than ___***___; in general, they have ___***___ values at higher frequencies.

19G With d.c., the highest dielectric constant for a polymer is obtained just above the ___***___ temperature. It decreases at higher temperatures because ___***___. . . . At lower temperatures, ___***___ polarization is the principal contributor to the dielectric constant.

19H The polarization of a ___***___ material can be changed by a stress that introduces an elastic ___***___.

19I The interrelationship between polarization and strain provides a basis for ___***___ devices that convert mechanical energy to electrical, and vice versa.

• 19J For a material to be either ___***___ or ___***___, it must lack a center of symmetry so that the centers of positive and negative charges are separated.

• 19K All ___***___ electric materials are ___***___ electric, but not all ___***___ electric materials are ___***___ electric because ___***___ . . . .

• 19L ___***___ polarization is that amount of polarization remaining after any external field is removed; the ___***___ is the reverse field required to drop the polarization to zero.

• 19M The property of ___***___ is the summation of the magnetic moments per ___***___. The magnetic moment per electron is ___***___ $A \cdot m^2$; for two electrons in the same orbital, the net magnetic moment is ___***___.

• 19N Ferrites are illustrative of ceramic magnets in which iron atoms are present as ___***___ and ___***___ ions; the former has ___***___ unpaired electrons; the latter, ___***___.

• 19O In magnetite, a ferrite found in nature, the iron ions are in ___***___ sites and ___***___ sites. The iron ions in the two sites have ___***___ magnetic orientation.

• 19P Since the net magnetic moment of a unit cell of $[MnFe_2O_4]_n$ is 370×10^{-24} $A \cdot m^2$, there are ~___***___ unpaired electrons per unit cell. Since the cubic unit cell has $a =$ ___***___ nm, the magnetization of this compound is 600,000 A/m.

• 19Q The term, ___***___ magnetic, refers to a material that possesses ions with magnetic moments, but one in which the magnetization is not obvious since there are equal numbers of ions with opposing magnetic orientations. In a ___***___ magnetic material, ions have opposing magnetic orientations, but the magnetic moments are not balanced.

• 19R A ___***___ magnet has significant remanent magnetic induction after the external magnetic field has been removed; a ___***___ is necessary to lower the induction to zero.

- 19S A magnetic material possesses _***_, which are small regions containing unit cells, all with the same magnetic orientations. The _***_ between these are readily moved in a _***_ magnet.
- 19T A hard magnet is evaluated on the basis of its _***_ product, which is a measure of _***_ to demagnetize the material.
- 19U A cold-worked metal is mechanically hard because of the introduction of _***_ that become tangled and restrict slip. A cold-worked metal is magnetically hard because their introduction also restricts _***_ movements.
- 19V Domain boundaries may become anchored by _***_, by _***_, and by _***_.
- 19W A material loses its magnetization at its _***_ temperature because _***_ . . . .
- 19X The permeability of a magnet is the ratio of _***_ and _***_. In comparing hard and soft magnets, this property receives more attention in _***_ magnets.

STUDY PROBLEMS

19–2.1. A capacitor of two parallel plates must carry a charge of 10^{-9} coulombs at 5 volts. With no spacer it is impractical to bring the plates closer together than 1 mm. What area is required for the capacitor? (Assume $\kappa_{dc} \cong \kappa_{60Hz}$.)

Answer: 230 cm^2

19–2.2. Refer to S.P. 19–2.1. An alternate capacitor may be made by metallizing (with aluminum) the two sides of an 0.5-mm nylon sheet. What area is required?

19–2.3. A radio-frequency (10^6 Hz) capacitor has been made with an 0.5-mm plasticized PVC spacer. It is replaced with an 0.33-mm polyethylene spacer. How much will the charge density be affected, assuming that other factors remain unchanged?

Answer: Within the accuracy of the data, they are comparable.

19–2.4. The relative dielectric constants for PVC (C_2H_3Cl) and PTFE (C_2F_4) show the following changes with frequency:

Frequency, Hz	d.c.	10^2	10^4	10^6	10^8	10^{10}
PVC	7.4	6.5	4.7	3.3	2.8	2.6
PTFE	2.1	2.1	2.1	2.1	2.1	2.1
Vacuum	1.0	1.0	1.0	1.0	1.0	1.0

Plot the capacitance (that is, C/V = farads) versus frequency for a capacitor with a 100 cm^2 effective area and an 0.5-mm plate separation: (a) with PVC; (b) with PTFE; (c) with a vacuum. (d) Explain the differences between the three curves. [*Hint:* By rearranging Eqs. (19–2.3) and (19–2.4), we get

$$\text{Capacitance} = \mathfrak{D}A/E = \kappa\epsilon_0 A/d.]$$

Answer: (a) PVC at 10^2 Hz, $C = 1.15 \times 10^{-9}$C/V = 1.15 nf; etc.

19–3.1. (a) What dimensional change (%) is realized in a piezoelectric quartz crystal (E = 300,000 MPa) when an electric field changes the polarization from 35.1 C/m^2 to 35.4 C/m^2? (b) What stress would be required to produce the same strain?

Answer: (a) 0.85% (b) 2600 MPa

19–3.2. A piezoelectric crystal has a Young's modulus of 130,000 MPa (19,000,000 psi). What stress must be applied to reduce its polarization from 560 to 557 C/m^2?

19–3.3. Refer to I.P. 19–3.2. A ten-millimeter cube of $BaTiO_3$ is compressed 1%. The two ends receiving pressure are connected electrically. How many electrons travel from the negative end to the positive end?

Answer: 10^{12} electrons

19–3.4. Refer to I.P. 19–3.2 What is the distance between the centers of positive and negative charges of each unit cell?

19–3.5. Potassium niobate ($KNbO_3$) has the same structure as $BaTiO_3$, but with K^+ replacing Ba^{2+}, and Nb^{5+} replacing Ti^{4+}. Assume that the dimensions of the unit cell of $KNbO_3$ are approximately the same as those given for $BaTiO_3$ in Fig. 19–3.1. (a) What is the distance between the center of positive charges and the center of negative charges? (b) What is the dipole moment, $p = Qd$, of the unit cell? (c) What polarization is possible?

Answer: (a) 0.0117 nm (b) 11.2×10^{-30} C/m^2 (c) 0.175 C/m^2

19–4.1. Nickel ferrite $[Ni^{2+}Fe_2^{3+}O_4]_8$ has the same structure and the same unit-cell size as $[Fe^{2+}Fe^{3+}O_4]_8$. What is the saturation magnetization of the nickel ferrite?

Answer: 250,000 A/m

19–4.2. Manganese ferrite $[Mn^{2+}Fe_2^{3+}O_4]_8$ has the same structure as magnetite, Fe_3O_4. Its unit-cell size, a, is 0.84 nm. What is the saturation magnetization of the manganese ferrite?

19–4.3. Refer to S.P. 19–4.1. Assume that 20% of the unit cells are in domains with 180° reversed magnetic alignment, that is, 80% ↑ and 20% ↓. What is the magnetization?

Answer: 150,000 A/m

19–4.4. A magnetic ferrite has a unit cell with 32 oxygen ions, 16 ferric ions, and 8 divalent ions. If the divalent ions are Zn^{2+} and Ni^{2+} in a 3:5 ratio, what weight fraction of ZnO, NiO, and Fe_2O_3 must be mixed together for the product?

19–4.5. Refer to S.P. 19–4.4. In this particular magnetic ferrite, the Zn^{2+} ions preferentially choose four-fold sites and force a corresponding number of Fe^{3+} ions into six-fold sites. How many unbalanced electron spins does this produce per unit cell of $(Zn_3,Ni_5)Fe_{16}O_{32}$?

Answer: Net 40 ↑

CHECK OUTS

19A conductors
energy gap

19B volts/meter
thickness
pores, impurities, structural imperfections

19C cracks, impurities, pores

19D electronic, ionic, molecular, space charge
electronic, molecular

19E dipole moment/unit volume
charge density

19F one
lower

19G glass transition
thermal agitation destroys the orientation
electronic

19H piezoelectric
strain

19I transducer

19J ferroelectric, piezoelectric

19K ferro-
piezo-
piezo-
ferro-
some piezoelectric dipoles are not reversible

19L remanent
coercive field

19M magnetization
unit volume
9.27×10^{-24}
zero

19N ferric, ferrous
five, four

19O 4-f, 6-f
opposite

19P forty
0.85

19Q antiferro-
ferri-

19R hard
coercive field

19S domains
boundaries
soft

19T *BH*
energy

19U dislocations
domain boundary

19V dislocations, pores, second phases, grain boundaries

19W Curie
thermal agitation destroys the cooperative orientation

19X induction, magnetic field
soft

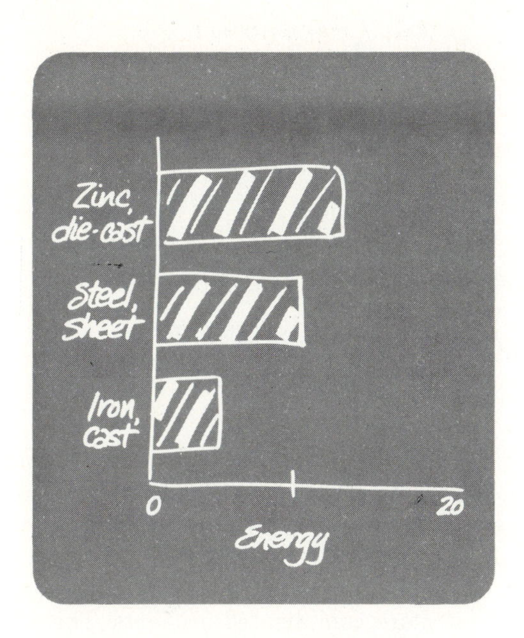

UNIT TWENTY

MATERIALS AND ENGINEERING: 2000-2020

In addition to being a vital contributor to technical products and systems, materials have a direct interaction with nontechnical considerations in modern society. For example, availability of raw materials, materials for energy conversion, and product liability each interface the technology of materials with economic, political, environmental, and consumer considerations. These interactions are certain to become more important in the years ahead when our resources diminish, and if life-quality goals are to be achieved. This unit introduces the broader picture of materials and engineering considerations that today's students will encounter in tomorrow's engineering.

Contents

Prerequisite Unit 1; and an interest in the future of technology.

From Unit 20, the engineer should

1) Look at materials in a broader perspective than simply a structure-properties-performance relationship.

2) Extrapolate the trends of materials as they affect her or his engineering discipline for the next 20 years.
3) Extrapolate socioeconomic trends as they affect the materials and the products of his or her engineering discipline for the next 20 years.
4) Prepare a term paper on at least one of the topics for further study at the end of the unit, requiring library research.

20-1 INTRODUCTION

You students who are studying this text will pass your mid-career point near the year 2010. What considerations must an engineer give to materials at that median date? You would like to know because half of your engineering activities will still be ahead of you.

Expectedly, a forward step of twenty or more years will find more sophisticated materials available for more complicated products. The student of today who best understands the principles that control the properties and behavior of materials will be in the best position to develop and apply such materials. That has been the prime purpose of this text. When looking ahead, however, we also need to observe some of the nontechnical aspects of materials—some of the factors that speak to their availability, that optimize their performance, that affect their dependability, and restrict their usage. In the decades ahead, these factors may become as important as the technical aspects of structure and properties. To do this, we shall make a quick review of *materials and civilization,* since the two are so closely associated; followed by a description of the *materials cycle.* This includes the sequence through which materials are obtained, processed, and discarded (and even reused). We will then look more specifically at *materials resources;* at two technical aspects of *materials and energy;* and finally at *designing with materials* for product optimization.

Materials and civilization

Materials have always been closely associated with human progress. This is because the human species makes things—clothing, tools, homes, weapons, vehicles, electronic devices, etc.; and, of course, all of these products must be made out of the appropriate materials, with the required properties, in the specified shapes, and often with a desired appearance.

With improvements in materials by the artisans and the technologists across history, it was possible to make more sophisticated products. Clothing became more protective against the elements or more attractive; tools were made more available to relieve toil; homes were designed for more comfort; weapons became more destructive and lethal in an effort to gain superiority over the opponents' advanced weapons; vehicles achieved longer ranges and quicker transport and gave more ready access to food, supplies, and enjoyment. This type of listing could go on and on. The obvious point is that materials and human achievements are very closely associated.

There are numerous records of early civilizations that show the products they made and the materials they used. We defined this progress and culture in terms of the Stone Age, the Bronze Age, and the Iron Age. Not only the tools and weapons but also the arts and crafts were influenced by the artisans' ability to work (process) these materials into objects that were desired.

In early civilizations, as today, materials affected a broad spectrum of other human activities. During the Paleolithic period, the sites of the better flint deposits became the sites of villages and cross roads of primitive barter. In ensuing centuries, the requirements for materials to meet societal needs brought forth exploration, rudimentary monetary systems, trade centers, advances in transportation, political downfalls, as well as wars. The transition from the Bronze Age to the Iron Age, which started some 3500 years ago, introduced additional nontechnical changes. Iron was a "democratic material" in these early civilizations, because its widespread availability produced an impact on the common person through the introduction of implements, tools, and utensils in every clan. No longer were the benefits of available metals limited to the affluent. Probably the most significant feature of this new democratic material was that the numbers of users exploded. More craftspeople were required in the materials trades. Advances now occurred in centuries and decades rather than in millenia. This acceleration continues to the present day.

Closer to modern times, we can cite the American and British railroad systems as an example of the symbiosis of materials technology and socioeconomic changes. As early as 1830, attempts were made to use steam power for land transportation. Rails were necessary for several obvious reasons; however, rails at that time were merely straps of soft wrought iron nailed to planks. An economical metal was not available with the required characteristics. Were it not for the developments of Kelley (U.S.) and of Bessemer (U.K.) in steel production, the railroad system could not have developed to open the West of the United States and to industrialize England. Conversely, were it not for the industrial and agricultural demand for transportation, the incentives, the capital, and the technology for steel development would have been missing.

A current example of the interplay of technological and nontechnical areas involves silicon. Economically, it has introduced a multibillion dollar industry. Communication has been facilitated at all distances from hearing aids to extraterrestial telemetry. Our day-to-day life is becoming altered via the availability of in-home entertainment, and by the introduction of computers at all levels of personal records. Changes are not solely technical.

The materials cycle

With the above injection of materials into the whole of civilization, we need to look more closely at the flow of materials through our economy. Figure 20-1.1 depicts the global sequence of materials within a modern society. The products of nature—ores, plant products, fuels, water, and air—are first extracted, then refined and processed into metals, chemicals, fibers, etc., to form alloys, ceramics, polymers, composites, and other materials to meet performance requirements. These modified materials are shaped and as-

Figure 20-1.1 The materials cycle. All materials of engineering originate in nature, are extracted, refined, processed, and assembled into products of engineering design. The life span of these materials eventually ends when the product is either worn out or superseded by new devices, machines, and structures. A high level of technology will be required in the decades ahead in order to recycle larger amounts of these discarded materials back into production and into useful functions. (*Materials and Man's Needs,* COSMAT. Reproduced with the permission of the National Academy of Sciences.)

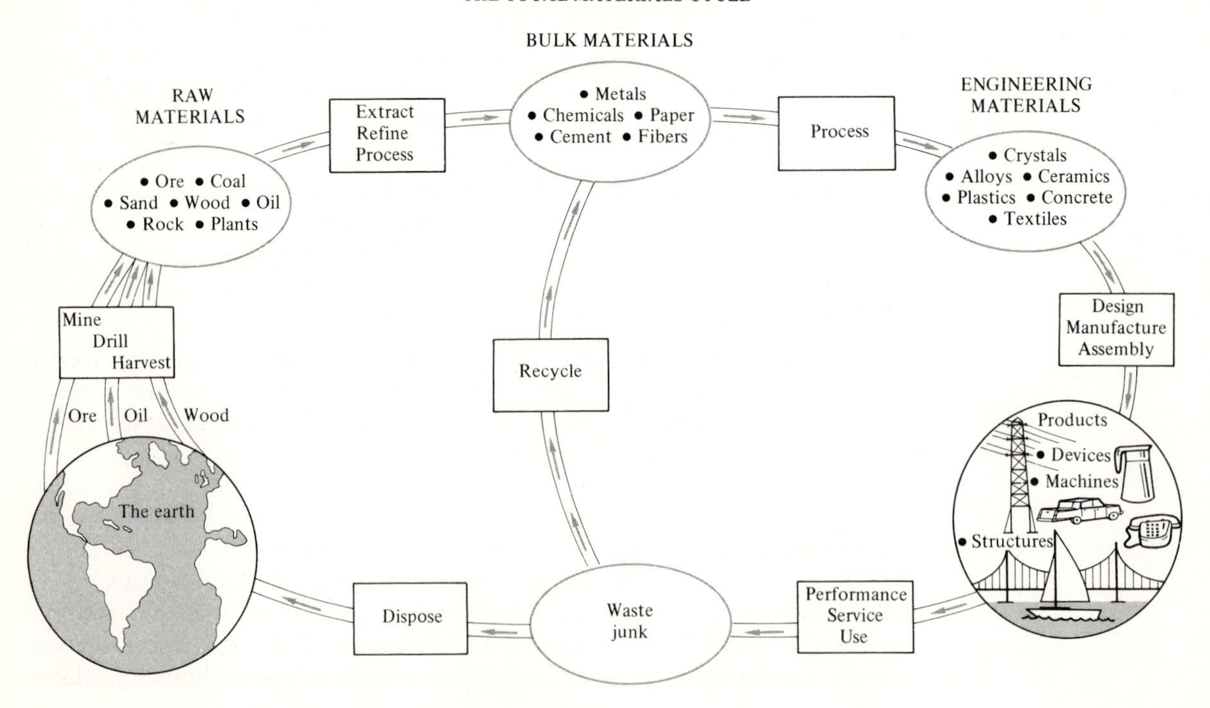

sembled into the products and systems that have been designed by the engineer—TVs, solar cells, ships, highways, etc. Sooner or later, but inevitably, the useful life of the product is ended, either because it fails to perform its duty or because it is superseded by a newer design with greater capabilities or economies. Those products (and materials) may then be returned to nature as a discard or recycled for a second circuit through the materials cycle. For a variety of reasons, which will certainly become more critical in the upcoming decades, the latter route is becoming mandated. Also, we shall see in a later section that recycling brings with it an increased complexity in the required technology.

20–2 MATERIALS RESOURCES

The ultimate source of all materials is nature. These resources are presumably available for use; however, there are limitations.

1. The raw materials must be won from nature. Although a few sand and gravel deposits are available for the scooping, or timber for the cutting, there is generally considerable work and technology required to obtain the raw material. This is particularly true with those raw materials such as copper and gold that are distributed in grams per kilogram or in grams per ton quantities.
2. Energy is required not only for extraction but also for refining. It takes as much energy to extract and refine a kilogram of aluminum as it does to light a home for nearly a week. Of course, energy supplies have been progressively more expensive with no expectation for a reversal. As a result, the recovery of used raw materials becomes important.
3. Raw materials are not uniformly distributed. Automatically this involves transportation, sometimes great distances and from country to country, as is the case for chromium and many other ores. In addition, these raw materials may be in a political arena where embargoes, wars, changes in governments, tariffs, and environmental considerations are encountered.
4. Finally, substitute materials provide a resource that the engineering designer must have available. This often requires significant technology since the service performance always differs among materials.

Natural raw materials

Table 20–2.1 lists the relative abundance of the more common elements in nature's warehouse. This list could be misleading since all of those in the list except oxygen must be mined as compounds, commonly single and multiple oxides. None can be obtained directly as metals. The list can also be discouraging. We associate very few of these elements with technology. Absent

Table 20-2.1
Abundance of elements in the earth's crust

Oxygen	47%	Potassium	3%	Fluorine	0.06% (600 ppm)
Silicon	27%	Magnesium	2%	Barium	400 ppm
Aluminum	8%	Titanium	0.4%	Strontium	400 ppm
Iron	5%	Hydrogen	0.1%	Sulfur	300 ppm
Calcium	4%	Phosphorus	0.1%	Others	<200 ppm
Sodium	3%	Manganese	0.1%		

from this list, for example, are copper, zinc, and tin, the components of brass and bronze. Also, the noble metals—silver, gold, etc.—fall substantially below the 200-ppm level in the table. Even carbon is a minor element; and chromium, nickel, cobalt, niobium, and tungsten—all critical components of steel—are below 100 ppm. Few materials are sufficiently abundant so that we can assume unlimited supplies.

Fortunately, the ores of many raw materials are localized. Thus, we find large taconite deposits of iron ore containing major percentages of iron. With 13 ppm of lead in the earth's crust, it is not necessary that 75,000 kg of rock must be mined to get 1 kg of product, since there are deposits with 13,000 ppm (>1% Pb). However, as our prime mines are depleted, we must extract materials from leaner ores.

Here we need to distinguish between *reserves* and *resources*. The former are known ore deposits that can be economically extracted. The latter includes all potential sources. The two are differentiated by the McKelvey diagram of Fig. 20-2.1. Increased exploration can increase the reserves, as can (1) improved technology for extraction, or (2) changed economic demands for the material. For example, high copper prices can bring previously unprofitable mines into operation. In reality, United States data show

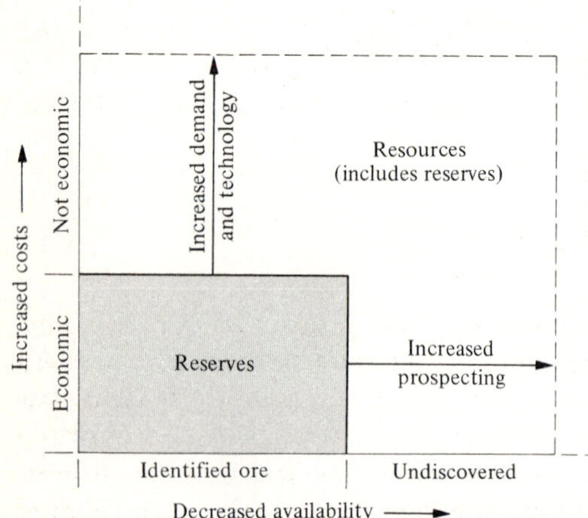

Figure 20-2.1 Resources and reserves. Estimated resources include many raw materials that are not economically viable currently. Future demand and improved technology can add some of the resources to the reserves; however, their costs will undoubtedly increase. (Modified from V. E. McKelvey, *Technology Review*.)

Table 20-2.2
World metal reserves (millions of metric tons)*

Date of estimate	Cu	Pb	Sn	Zn
1950	164	33	4.6	41
1964	118	26	4.6	58
1974	270	98	3.2	92

* These reserves have increased in part from added exploration, and in part from demand and technology that have made previously noneconomic resources viable. The new reserves will invariably be more expensive because of additional energy and processing requirements, and increased energy costs. (Data collected from U.S. sources by A. Hurlich, *Metals Progress.*)

that world resources of several familiar metals have increased over the past three decades (Table 20-2.2). Of course, there is some ultimate limit, and the costs of extraction will be high.

Copper is an example of a *depletable* raw material. A century ago, ores with five percent copper were being mined. Today, ores with less than 0.5 percent copper are being mined. Future mining must draw upon still lower percentages. The technology has been developed (1) to beneficiate the ore and discard the unwanted gangue material, and (2) to extract the metal from the concentrated ore. Both of these steps must be improved to keep abreast with the changing ore supplies. The technology must also take into account the increased energy requirements and costs, the handling of additional solid wastes, and the elimination of environment pollutants.

Assuming unrestricted trade, the concern of the engineer is not the complete depletion of a raw material; rather, it is whether the costs will force substitutions.

Wood is an example of a *renewable* raw material. As with copper, the past century has introduced a major change in the technology of timber resources. Efforts are being made to completely replace the former cut-and-burn practices with modern tree-farming procedures to make timber and wood a fully renewable resource. Concurrent developments also promise a better raw material (Fig. 10-1.4).

Recoverable raw materials (recycling)

The materials cycle of Fig. 20-1.1 shows that those products that have ended their initial life may be reintroduced into the materials stream. This is not a new phenomenon. Broken glass (cullet) has long been remelted with new glass. Not only does this save raw materials, it also modifies the melting operation. Reprocessed wool contains fibers from former discards. Admittedly, the fibers are shorter and there is a consequent change in product characteristic. For decades the diet of a steel furnace has included a major percentage of scrap iron and steel—the amount depending upon the economics of scrap availability and the collection and transportation costs. Through this market, 95 to 98 percent of all automobiles ever discarded have been returned to the steel furnace.*

Aluminum is an example of a *recyclable metal*. Let us examine this metal more closely since it is a major element—8 percent of the earth's crust. The incentive to recycle aluminum therefore does not come from depletion.† Rather, it arises from the energy balance. It takes approximately 60 kWh/kg (~25 kWh/lb) to extract and refine aluminum metal. It takes only about 10 percent of that energy to remelt the scrap. Aided by the above cost benefits, there is now an active scrap market for aluminum. However, the whole recycling picture is more complicated than simply that of cost. We do not see a comparable scrap market for titanium, which has energy requirements similar to aluminum. The volume of recyclable aluminum is now great enough to warrant collection centers, and to have sorting operations to remove objectionable scrap. A generation ago there was not enough discarded aluminum to support these required activities. That is the situation today for titanium and a number of other materials.

Automobiles are an example of a *recyclable product*. The primary incentive for recycling automobiles to date has been the iron and steel that they contain;‡ however, other factors are becoming increasingly important. Until the last decade most of the scrap sorting followed hand dismantling. Today the majority of discarded vehicles are processed by auto shredders, giant machines that can consume up to 1000 cars per day. After tires, fuel tanks, batteries, and radiators are removed, the entire car is fed into the shredder shown schematically in Fig. 20–2.2. The result is a collection of fragments, none larger than 250 g (<0.5 lb). The lightweight fragments (urethane foam, some rubber, upholstery, thin-gauge aluminum, etc.) are air separated. Then the magnetic fragments (mainly iron and steel) are segregated from the heavy, nonferrous fraction. The ferrous fraction supplements virgin metal in steel-melting furnaces. However, problems exist. Any tin or copper that is not removed completely from the iron and steel scrap cannot be removed later from the molten steel; it leaves the steel mill within the steel product. These *residuals* gradually build up in the system as recycling becomes the established procedure. At relative low amounts (0.1%–0.2%),

---

* Of course auto "grave yards" remain. However, sooner or later the former vehicle finds its way to the scrap-preparation yards. In the meantime, some of these serve as an inventory of spare parts with greater value than the simple scrap value.

† The availability of current ores is a factor, however. Three fifths of the current high-grade bauxite ores are located in three countries—Australia, Guinea, and Jamaica. The aluminum content of clays, syenites, etc., while high in comparison with the metal content in other ores, is very difficult to extract economically by present-day technology.

‡ The reader is referred to J. J. Harwood, "Recycling the Junk Car—A Case Study of the Automobile as a Renewable Resource," *Materials and Society* (1977).

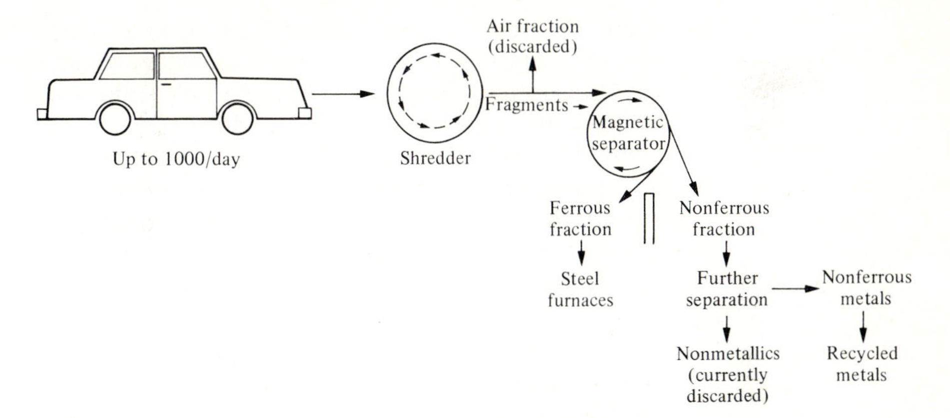

Figure 20-2.2 Auto shredder. In order to be economically viable, large operations are required with mechanical sorting. Further technological improvements are necessary to salvage the remaining discarded fractions.

they reach the level where they are detrimental to the steel product, especially in its capability of being processed. Thus, the steel mill operator will naturally resist the inclusion of major additions of scrap of this nature.*

The nonferrous fraction from auto shredders includes everything from rubber, fibers, and reinforced plastics, to copper wire, stainless steel,† and glass. This nonferrous fraction does not directly reenter the materials cycle as does the iron and steel scrap. Rather, it passes to other processors who use more sophisticated separation procedures. Usually the first step is a dense-medium separation where materials with densities of less than 2 g/cm^3 float and those that are heavier sink. This may be repeated for a 4 g/cm^3 separation. Thus, it is possible to segregate organic materials from aluminum (2.7 g/cm^3), from copper, bronze, and stainless steel, each of which exceed 7 g/cm^3. Final separation of the metal values involves chemical treatments and high-temperature refining. More than 1000 tons of aluminum and zinc are now returned to the materials cycle per week through this scrap industry. Future increases are possible as efficiencies of collection and separation are improved. This will involve both technological developments and economic incentives.

The nonmetallic fraction, both air-separated and liquid-floated, is not efficiently recovered at the present, and the majority goes to landfills. Extensive efforts are being made for the use of the discarded organic materials in fuels; other attempts are being made to process the organic materials so they may be used as feedstocks in polymer manufacture. Neither may be

---

* Similar considerations are encountered in any recycling process, for example, iron in aluminum scrap, impurities in glass cullet, clay fillers in recycled paper.

† Austenitic stainless steels are nonmagnetic.

called successful as yet, except where selected municipalities subsidize the operation to reduce other disposal costs. Here we see the conflicting forces of our socioeconomic system. The drive for lighter-weight, fuel-saving cars leads to the use of more plastics, which in turn presents a problem in disposal and resource recovery. Likewise, the efforts to improve safety has led to a twentyfold increase in the use of polyurethane foam for passenger protection. These foams are among the most difficult materials to dispose from the junk automobile. They are bulky for landfill discard; they are not biodegradable; and they are difficult to use as secondary fuel stock because of environmental considerations.

Distribution of raw materials

Raw materials are very nonuniformly distributed in the crust of the earth. Historically, many industrial cities have grown up in the vicinity of specific natural resources. Ore carriers ply the lakes and oceans from the mines to the industrial centers. Some major countries, such as Japan and the smaller European countries, are without ore deposits; others like Chile have an

Figure 20-2.3 Reliance on imported raw materials (United States). Similar data can be collected for all other industrial countries. Each country must develop technical alternatives in the event these sources are interrupted. (U.S. Bureau of Mines data adapted by E. Verink.)

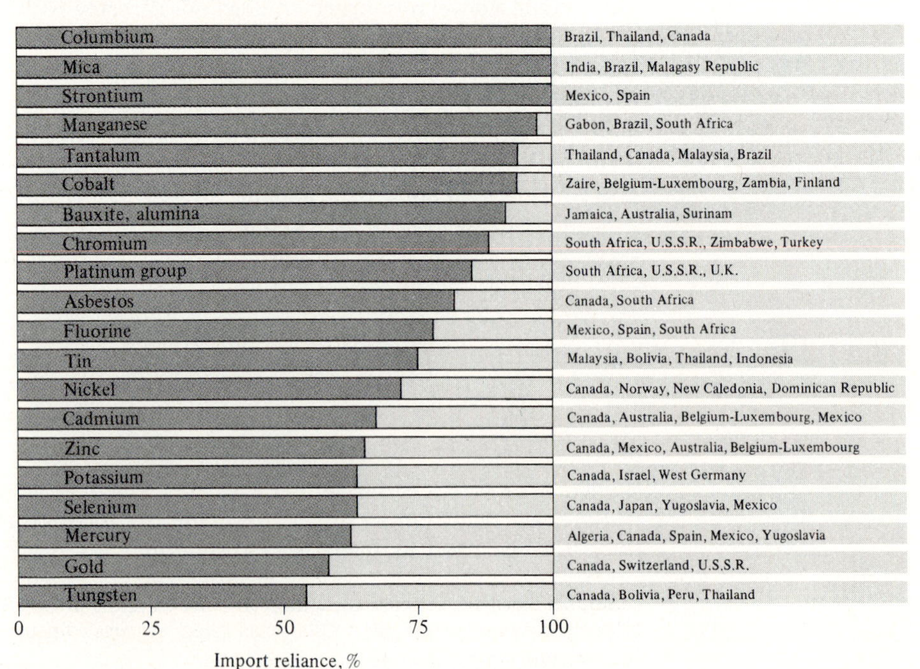

economy predominantly associated with a specific raw material—copper. No country is fully self-sufficient in all of the desired raw materials. These situations have led to colonialization, to international tensions, and even to wars. They also lead to cooperation and economic co-dependence. All countries wish for a greater self-sufficiency and the accompanying effects on balance of payments.

The United States depends extensively on imports for its raw materials (Fig. 20-2.3). Similar histograms could be drawn for other industrial nations. Conversely, the major supplies of some raw materials are located in only a few countries. A few examples are listed in Table 20-2.3. An examination of Fig. 20-2.3 and Table 20-2.3 reveals how economic and political considerations can enter the materials supply picture, whether the country is the United States or another industrial nation. How does a country adjust its usage, which means changing its industrial products if economic conditions dictate a correction in the balance of trade payments? How does an engineer replace chromium, for example, in a heat-resistant steel for an exhaust converter if a political action in the form of an embargo is taken against a principal-supplier country? Should a country build a plant to refine its aluminum ore to aluminum metal to increase the value of exports when that requires capital that either (1) would have been used for housing developments, or (2) must be obtained abroad, and gives partial ownership of the refinery to a foreign company? These and dozens of other questions are related to the supply and processing of materials and the distribution and uses of technical products made of industrial materials.

Materials substitution

The history of technology is replete with examples of materials substitution. Several quickly cited examples include fiber-glass composites in place of wood for hulls of small boats, nylon for silk, high-strength, low-alloy (HSLA) steels for mild steel in bridge beams, plastics for hard rubber in automobile battery cases, and glass curtain walls for brick; but there are many more. A major goal of materials engineering has been to provide new and better materials than those presently available in order to meet two objectives—performance and cost. A substitute material is readily accepted in making consumer products if it is cheaper and will perform as well as the previous material, or if it performs better than the previous material at no extra cost. If it performs better, and also costs more, the acceptance must involve the consumer's reactions.

The engineer is concerned not only with the above type of materials substitution where performance and cost are factors but also with the increasing numbers of cases of "forced substitution" that come about for other reasons such as raw material unavailability, safety requirements, environmental impacts, or energy embargoes. For example, the use of chromium was restricted a decade or so ago because of an embargo against a supplier

Table 20-2.3
Sources of important raw materials

Material	Total yearly products*	Production by principal suppliers
Asbestos	5.8×10^6 s.t.	USSR: 46%; Canada: 30%; So. Africa: 7%; others: 17%
Bauxite (Aluminum ore)	81×10^6 m.t.	Australia: 32%; Guinea: 14%; Jamaica: 14% Surinan: 6%; USSR: 6%; others: 28%
Chromite (Chromium ore)	10.8×10^6 s.t.	So. Africa: 33%; USSR: 22%; Zimbabwe: 7%; Finland: 7%; Turkey: 7%; Philippines: 6%; others: 18%
Cobalt	32.6×10^3 s.t.	Ziare: 36%; New Caledonia: 14%; Australia: 12%; Zambia: 8%; USSR: 6%; others: 24%
Niobium	22.7×10^6 lbs.	Brazil: 78%; Canada: 17%; Nigeria: 3%; others: 2%
Fluorspar	5.15×10^6 s.t.	Mexico: 20%; USSR: 11%; Spain: 9%; France: 8%; So. Africa: 8%; China: 8%; others: 36%
Lithium	85,000 s.t.	USSR: 65%; China: 13%; Zimbabwe: 12%; S.W. Africa: 3%; others: 7%
Mercury	199,000 flasks	USSR: 29%; Spain: 19%; USA: 14%; Algeria: 13%; China: 10%; Mexico: 8%; others: 7%
Niobium (columbium)	22,700 lbs.	Brazil: 78%; Canada: 17%; others: 5%
Platinum-group metals	6.4×10^6 oz.	So. Africa: 46%; USSR: 45%; Canada: 7%; Columbia: 0.5%; others: ~1%
Tungsten	93×10^6 lbs.	China: 22%; USSR: 19%; Bolivia: 7%; USA: 6%; So. Korea: 6%; others: 40%

* 1977, U.S. Bureau of Mines (s.t.: short tons; m.t.: metric tons; oz.: troy)

country; cobalt-based alloys for jet aircraft have been prohibitively expensive because of the imbalance of supply and demand; and asbestos has been effectively outlawed by the restrictions it receives from health and safety codes. These substitutions generally carry a time factor that has not been encountered in the past. Furthermore, the substitution must often be made with a penalty of cost and/or performance. The edicts for more energy-efficient cars provide an example. To meet timetables, aluminum can be substituted directly for steel in automobile bumpers in order to reduce weight; but it is at an added expense and at some loss of function. Of course, it may be pointed out that equal performance can be attained by redesign (with increased production costs), and that the eventual costs to a country may be more without such conservation measures. In any event, considerable materials technology is required to optimize the substitution (Illustrative Example 20-2.1).

There are four steps in the sequence of events if materials substitution is to be successful. These are outlined in Table 20-2.4.

Table 20-2.4
Engineering steps for successful substitutes

Step	Comments
Modification of the substituent material	It would be uncommon for a shelf item to be applicable directly without some tailoring, either by the primary manufacturer or by the using industry. The filled plastic composites for automotive grills are an example. The manufacturing and service requirements are sufficiently unique in the automobile so as to require extensive modification and redevelopment of preexisting plastics.
Redesign of the product	This is the situation required for many of the products that used asbestos, which is now deemed a health hazard. Its contributions to the braking operations could not be matched by alternative materials. As a result, specific changes have been necessary in the design of the braking system of a car.
Manufacturing processes and/or tooling	The two previous examples also demanded manufacturing changes; however, sometimes the effects are more subtle. Die modifications are required in deep-drawing operations with the substitution of alternate sheet materials for steel in an automobile, although the same final geometry is desired. Aluminum, and for that matter HSLA steels, require different thicknesses and have different drawing and spring-back characteristics than the former Al-killed sheet stock of steel. A wholly automated plant becomes particularly vulnerable to mandated changes that require material substitution.
Thorough testing	This final requirement is easily appreciated, particularly for complex products and with present-day interpretation of product liability. Testing that does not include real-life, service operations places the industry in a very vulnerable position. Special problems arise with this, because the service life may extend over many years. Accelerated tests always must be suspect, particularly if there are a variety of service conditions by different users. This complicates the substitution picture and places a premium on advanced planning for long-range materials policies by individual companies, by industries, and by governments.

Adapted from W. J. Arrol, *Materials and Society*

Illustrative example 20-2.1 The optical waveguide (Section 13-4) is an example of a currently developing materials substitution. By substituting light-transmitting fibers for metallic conductors in ground-based communication systems, it is possible (1) to reduce our dependence on copper, (2) to increase severalfold the number of simultaneously transmitted messages, and therefore, (3) to reduce the demand for public utility's rights-of-way in heavily congested metropolitan regions. The following documentation describes the technical coordination that has been involved.*

CASE RECORD "Communications systems engineers, with backgrounds in electrical engineering and physics, first set basic performance objectives for optical fibers. They pursued analytical studies of optical-waveguide configurations, invented various possible implementations and made preliminary explorations of a number of them. Other scientists with different experience backgrounds, and in different parts of the organization, were drawn into the program. . . . Absorption losses had to be reduced; practical ways to build in predetermined radial profiles to the refractive index to minimize dispersion losses had to be devised; means for drawing preforms into fibers had to be carefully engineered so as to avoid contamination, and ways had to be found to ensure that tight dimensional and concentricity tolerances were met; nondamaging methods for coating fibers on-line in the drawing process with protective polymeric materials had to be developed, including exploration of the materials themselves, otherwise the strength of the fibers rapidly deteriorated; ways of packaging individual fibers into multifiber ribbons or bundles had to be invented and developed so as to make cables practicable; the cables required working-out of suitable fiber-splicing and connecting techniques; cables and their fibers had to be able to sustain the mechanical stresses imposed when they are drawn through conduits under city streets, and they also had to be able to withstand the hostile environmental conditions generally found down such conduits and manholes; and all the while the systems engineers and network specialists had to keep reassuring themselves that optical communications would be economic and competitive with other transmission media, such as copper wires and coaxial cables.

The optical cable, perhaps deceptively simple in its appearance, is in fact an exceptionally sophisticated piece of materials science and engineering." ◄

20-3 MATERIALS AND ENERGY

Energy is required for the processing of materials. Conversely, all energy conversion systems require materials for their construction and operation. An example of the first statement is the energy requirement cited in the last section for the extraction and refining of aluminum. Examples of

* R. S. Claassen and A. G. Chynoweth, "Materials Science and Engineering: Its Evolution, Practice, and Prospects, Part II," *Materials Science and Engineering,* Vol. 37, (1979), p. 67.

Table 20–3.1
Energy requirements for materials processing

Aluminum (ingots)	11%*
Cement, portland	22%
Copper (ingots)	14%
Steel (ingots)	21%
Steel (finished products)	11%

* Percent of product value. Ingots are the initial bulk solid form of the metal. The ingots subsequently receive secondary processing (Section 4–2). (Adapted from *Materials and Man's Needs,* COSMAT, National Academy of Sciences.)

the converse statement are the roles of materials in solar cells, in steam-power plants, in nuclear reactors, in gas turbines, etc. We shall examine these two aspects of materials and energy more thoroughly.

Energy requirements for materials production

The amount of energy required to produce a material varies significantly. Table 20–3.1 lists the energy bill as the fraction of the total cost for selected materials. Figure 20–3.1 shows the amount of energy required to produce 1 kg of several common engineering materials from their original raw materials. The contrasts are striking. It forces us to ask whether there is a true saving in substituting aluminum for steel in an automobile for weight reduction and, therefore, fuel conservation. It immediately becomes apparent that the analysis must go deeper. What if we compared production energy

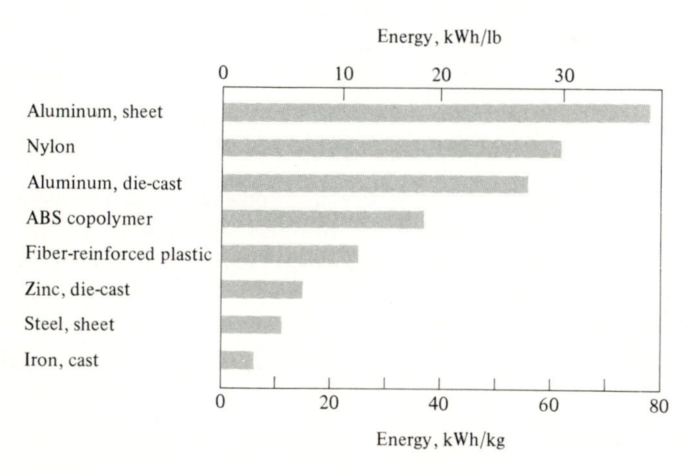

Figure 20–3.1 Energy requirements for processing selected materials of engineering (per unit mass). The relationships will be altered on a per-unit-volume basis. (Adapted from data by Harwood, "Recycling of Junk Cars," *Materials and Society.*)

Table 20-3.2
Energy requirements to manufacture a 3600 lb (1640 kg) automobile (1974 model)

Metallic materials (by supplier)	24,000 kWh
Other materials (by supplier)	2,000 kWh
Transportation of materials	600 kWh
Fabrication and assembling	9,000 kWh
Transportation of finished car	200 kWh
Production Total	35,800 kWh
Fuel used (50,000 miles)	~160,000 kWh

Adapted from Harwood, "Recycling the Junk Car," *Materials and Society*.

requirements on the basis of volume instead of weight? In some applications where rigidity is important the E/ρ ratio is the comparative parameter (Section 14–5). How will the energy impact vary?

Table 20–3.2 shows the energy consumption for a 1974 automobile. Two trends will modify this for cars of the present decade. The first is shown in Fig. 20–3.2 for the weight of cars where there is about a 25% decrease anticipated between 1980 and 1990. This brings about a 15%–20% reduction in energy requirements for processing.*

The second energy reduction in processing of all types of materials has come about by more efficient fuel management. With the increase in energy costs, industry has made changes to minimize heat losses, to salvage waste heat, and to improve processing efficiencies. In 1973, it took 60,000 Btus (18 kWh) of energy to produce $1 of national product in the United States. In 1979, that figure had dropped 10% to 54,000 Btu (~16 kWh) on a constant, inflation-adjusted dollar basis.† These are average GNP figures. The savings were greater than average in the industrial (materials processing and production) sector. Even with this improved energy efficiency, the engineer must expect to improve technology for a lower energy input, both for cost reasons and for conservation purposes.

Materials for energy conversion

It is impossible to produce energy; however, it is both possible and necessary to *convert* energy from one form to another for human use. Thus, we change the chemical energy in coal to thermal energy in steam and, hence, to

* The energy saving is not proportional to the weight reduction because most of the latter is achieved through a decrease in the amount of iron and/or steel, which requires less processing energy accordingly (Fig. 20–3.1). However, as aluminum is eventually recycled, and as ways are developed to obtain energy credits for plastic scrap after the automobile is junked, the two reductions will become more comparable.

† "Five Critical Issues in Engineering," National Academy of Engineering (1980).

electrical energy. In turn, this energy may be converted to mechanical energy by a motor, to light energy, or even back to thermal energy for heating purposes. Each of the above conversions must be performed within specifically designed equipment or devices. The engineering sophistication of these conversion devices will of necessity become rather complex, because of the attention that must be given in the future to the efficiency of the conversion. As an example, the distribution of electrical energy requires the conversion of wattage in a transformer to high voltage for cross-country trans-

Figure 20-3.2 Automobile weight reductions, 1980-1990. The principal weight reductions arise from down-sizing. Additional reductions arise from the substitution of lightweight materials for iron and steel.

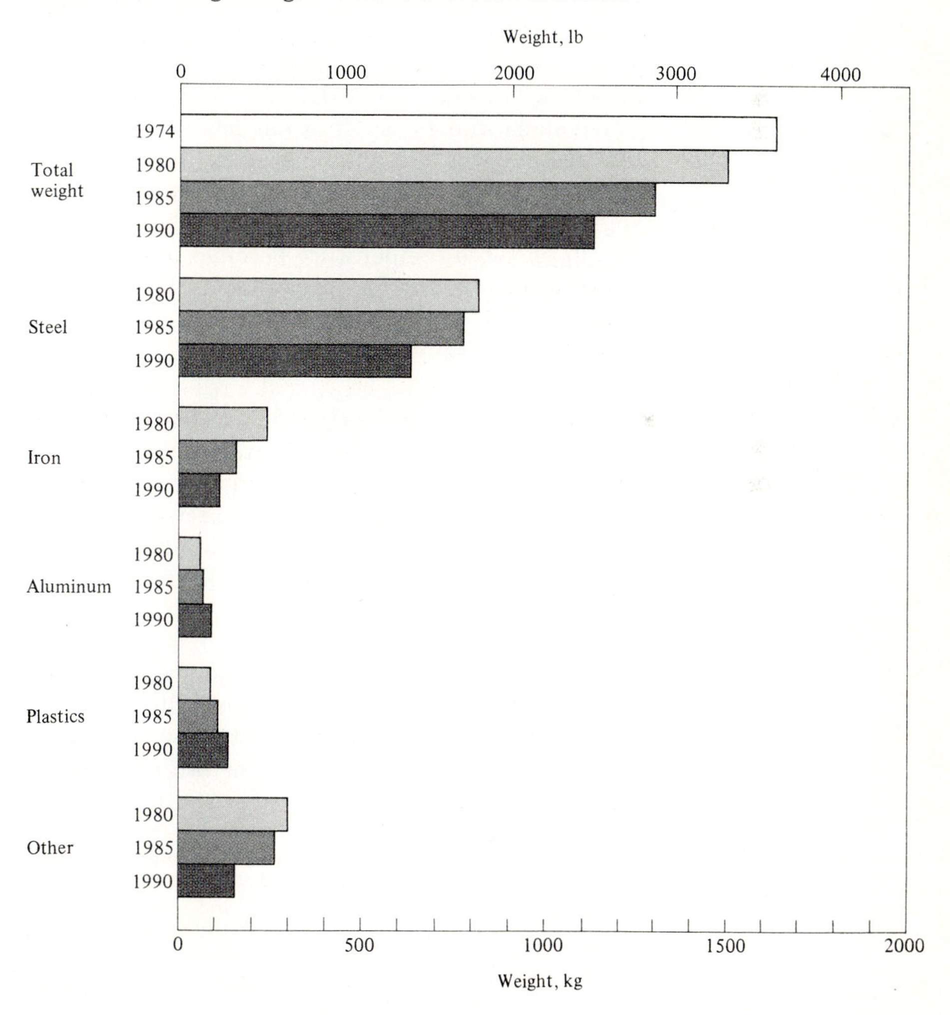

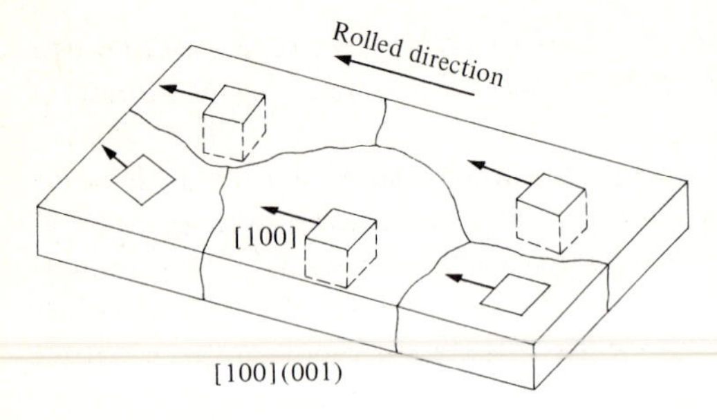

Figure 20-3.3 Grain-oriented steel (transformer sheet). (Cf. Fig. 19-4.1.) Since ferrite (bcc iron alloy) is magnetized more readily in the <100> directions, the materials engineers have learned how to recrystallize silicon-containing iron to align most of the grains close to the direction of rolling. The transformer manufacturer orients the sheet so that the rolled direction lies parallel to the magnetization direction within the core. This procedure has produced major savings of energy.

mission, then step-down transformers in two or three steps to the 120V–220V level in the home and business. None of these several transformers is 100% efficient;* however, their conversion efficiency was increased 1% (from ~97 to 98%) when grain-oriented transfer sheet became available to the transformer designer and producer (Fig. 20-3.3).† Since each of the billions of watts of power that we use must go through several such voltage conversions, the 1% savings per pass becomes significant. The balance of this section will present several examples of how advances in materials technology are required to improve energy conversion technology as the engineer looks ahead to the year 2000 and beyond. We will see that operation at elevated temperature becomes a critical factor in many of the processes when striving for efficient conversion. The basis for this arises from thermodynamic relationships that will be studied by most engineering students in other courses.

High-temperature gas turbines. High-temperature gas turbines are already with us, as energy converters for high-speed aircraft. They are also used by large electric utility companies for "peaking power," that is, for power production during high-use periods of the day when the capacity of the principal power generating equipment is exceeded. In general, they have not been used for continuous power production, nor for ground transportation purposes, which represents a quarter of our energy consumption. The principle of gas-turbine operation is the same as the farm windmill of a half-century ago. In place of wind, however, the rotor is turned by the hot gases of combustion. As fuel is burned, its volume must increase many fold. This expansion is directed toward a revolving turbine wheel (Fig. 16-6.1) that drives a generator, pump, or other energy conversion device. With

* This is the reason every power transformer is designed to provide cooling. The lost energy must be removed.

† Iron magnetizes more readily in the directions paralleling the edges of the bcc unit cell (Fig. 2-2.1). Therefore, the magnetization of the transformer core occurs more efficiently if the sheets of steel that go into the transformer core are processed so that (1) the grains are large (Fig. 2-7.5), and (2) each crystal is oriented so that its unit cell edges approximate the direction of magnetic flux.

combustion gases at the present 870°C (1600°F) level, the thermal efficiency is 25%–30%; with 1100°C–1300°C (2000°F–2400°F), a thermal efficiency of 40%–45% can be expected!

The problem of engineering design for a gas turbine is not how to achieve the higher gas temperature, but how to have turbine blades (Fig. 20–3.4), heat exchangers, gas seals, etc., withstand the hostile environments. Requirements include oxidation resistance, thermal-shock durability, and toughness, as well as high-temperature strengths and ease of fabrication. Metals cannot meet the requirements at these higher temperatures. Candidates receiving current attention are carbides, nitrides, and oxides (Fig. 20–3.5). In fact, the material that is currently in the forefront is a solid solution of **si**licon and **al**uminum **o**xide and **n**itride—sialon. However, toughness (impact resistance) and fabricability are still unsatisfactory. Also (and equally) important is the need to develop new design alternatives that can accommodate the unforgiving characteristics of brittle fracture (Fig. 11–8.3 and Section 14–2). Designs must be modified to reduce tensile loading and to build in opportunity for arresting the progress of nonductile fractures.

Figure 20–3.4 Gas turbine rotors. (Cf. Fig. 16–6.1.) (a) High-temperature alloy buckets mounted in a gas turbine wheel. (Courtesy of General Electric Co.) (b) Injection-molded turbine rotor (silicon nitride ceramic). This material is in the developmental stage; if it is successful, it will raise the operating temperature for significantly greater energy efficiency. (Courtesy of T. Whalen, Ford Motor Co.)

(a)

(b)

Figure 20-3.5 High-temperature ceramics (silicon nitride). An engineer checks the dimensions of a pump plunger. The dark-colored parts have not been fired (sintered); the light-colored parts have been and are therefore 18%–20% more dense. (Courtesy *CERAMIC INDUSTRY*, Cahner's Publishing Co.)

When the success of the 1200°C gas turbine is realized, the mandate will be given to reach for still higher, more efficient operating temperatures. The materials engineer will always have a job working with the design engineer.

Solar cells. Photovoltaic cells (Figs. 18–4.3 and 20–3.6) have come a long way in the recent past. It was not too long ago that more energy was re-

Figure 20-3.6 Solar energy conversion (5 kW photovoltaic array). This demonstration unit consists of 72 panels, 65 cm × 130 cm each, and feeds 150 batteries, which in turn power fifteen gasoline dispensors and lights for a service station. (Courtesy of Solarex Corp.)

quired to produce a cell than would ever be realized in its lifetime. That has changed. Unfortunately, the date is still beyond the horizon when such cells can be used for day-in/day-out power generation. Solar energy not only has a nocturnal interruption, but also its variability in many parts of the industrial world handicaps its usage. However, intensive research and development activities are bringing the cost down, increasing the efficiencies, and providing backup systems for weather interruptions. There is hope that modest usage of this energy source will be available for the engineer in the 2000–2020 period. Materials developments are required.

Fission reactors. Present-day nuclear reactors (which obtain energy from fissioning—splitting—heavy atoms) place very critical demands upon materials of engineering. As shown in Fig. 20-3.7, the design is complex and sophisticated materials are used—stainless steel, zirconium cladding, uranium dioxide, boron carbide, etc. Media coverage of potential disasters emphasizes the technical importance of good design and materials engineer-

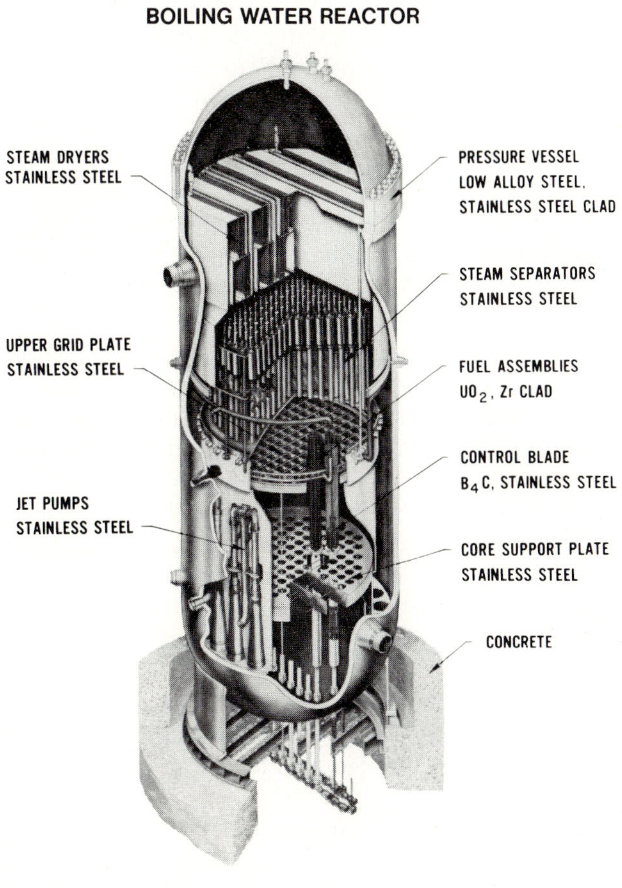

BOILING WATER REACTOR
STEAM DRYERS STAINLESS STEEL
UPPER GRID PLATE STAINLESS STEEL
JET PUMPS STAINLESS STEEL
PRESSURE VESSEL LOW ALLOY STEEL, STAINLESS STEEL CLAD
STEAM SEPARATORS STAINLESS STEEL
FUEL ASSEMBLIES UO_2, Zr CLAD
CONTROL BLADE B_4C, STAINLESS STEEL
CORE SUPPORT PLATE STAINLESS STEEL
CONCRETE

Figure 20-3.7 Fission reactor (boiling-water). An integrated design such as this reactor must be thoroughly tested for performance, safety, environmental impact, and cost effectiveness in addition to the normally required physical properties. (Courtesy of General Electric Co.)

ing. The majority of these problems have been surmounted. These include fuel-element stability as it develops fission products, both solid and gases; heat-transfer consistency as the fuel element is spent; and the corrosion resistance of the metal cladding. The current problems that are still being encountered include equipment malfunctions, safety procedures in emergency, and similar engineering factors that are presumably subject to control.

In order to achieve greater use of fission energy by the year 2000+, it will be necessary (1) to dispose of spent nuclear fuels, and (2) to develop a breeder reactor. A number of procedures have been promoted for containing the residual, long-life isotopes that are in spent fuels. A very promising one is to incorporate these elements as oxides into permanent glasses. While technically feasible, many people are not convinced that they want these residues stored near the place where their great grandchildren will live and work. There must be assurance that the short-time tests of the inertness of these glasses are truly indicative of the long-term behavior.

The "breeder" reactor, which generates fissionable atoms, is like the current "burner" reactor in that it will contain uranium-plutonium oxides encased in thin-walled tubing of metallic alloys, for example, zirconium alloys, or 18Cr–8Ni–2Mo stainless steels. As pointed out in reviews,* service conditions include "(a) recoiling ions and atoms with energies up to 120 MeV; (b) accumulation of up to 20% of about 40 different fission-product atoms; and (c) temperature gradients approaching 10,000 °C/cm". Engineering considerations such as elastic and plastic deformation, heat production, heat flow, temperature and thermal gradients come into play. There is confidence these problems are solvable; the question is whether a 99.9+% confidence in the answer is sufficient.

Fusion reactors. The fusion reactor, a cousin of fission energy, releases energy by the combination of light atoms, specifically hydrogen, into heavier elements. This source of energy is attractive because hydrogen is plentiful (from water) and it is "clean," that is, there are no contaminants to enter the environment. Predictions in 1982 suggest that commercial reactors may be available by the turn of the century, or in the decade thereafter. Although this prediction lacks certainty, many of today's engineering students will work with this reactor, or with energy it obtains.

The materials requirements encountered in the fusion reactor arise from the extremely hot ($>10^6$ K) plasmas of charged ions that are developed. No container could hold them except for the fact that they can be focused by magnetic fields and avoid direct contact with the container walls. Even so, interactions occur between the hot plasma and the materials of construction. In addition, magnetic materials, electrodes, and numerous other com-

* *Materials and Man's Needs,* COSMAT, National Academy of Sciences.

ponents must be selected by the design engineer to exacting materials specifications.

Superconducting transmission. A significant portion (~10%) of the cost of electrical energy is that of distribution. Part of the cost is the ohmic heat loss along the transmission line. A greater cost occurs in regions where space is limited, for example, in the underground matrix of metropolitan areas, because expensive provisions are required for heat dissipation. Superconductive materials (Section 17–2) become attractive because they have negligible resistance. A cryogenic (ultra-low temperature) transmission cable (Fig. 16–7.1) could well be less expensive than the elaborate heat-removal equipment that is anticipated as the required power levels rise. The threshold of technical success in this area will be reached when superconductors are available that will operate on a practical basis at liquid-H_2 temperatures (<20 K) rather than liquid-He temperatures (<4.5 K). A second requirement for engineering success involves joining superconductive cables into the distribution system. Welding presents a problem since melting is required; the melting alters the microstructure that makes superconducting cables feasible. The alternative that calls for long continuous cables without joints also encounters difficulties, since superconductive materials are inherently inflexible and present installation problems.

Synthetic hydrocarbon fuels. One goal of an energy program is to make fluid fuels out of solid, or semisolid, fuels. These include coal, tar sands, oil shales, and even biomass. Let us mention only one, the use of coal. There are three routes that the engineer can take to change coal to a fluid fuel: (a) by *pyrolysis* to form liquids, tars, and coke (which in turn can be gasified); (b) *by gasification* that uses O_2 and H_2O to form CO, CH_4, and H_2; and (c) by *direct liquefaction* through a coal-plus-H_2 reaction (Fig. 20–3.8). Two types of materials involvements are required. The first requires the adaptation of presently available refractories (Section 16–6) and metals to massive conversion units. Minimal costs and long-life stability will be required to contain capital costs and operating expenses. The second materials challenge is with catalysts that can be developed, which will increase the gasification reaction rates and efficiencies. The field of catalytic materials is complex (beyond the scope of this book) but one that continually demands greater technical attention.

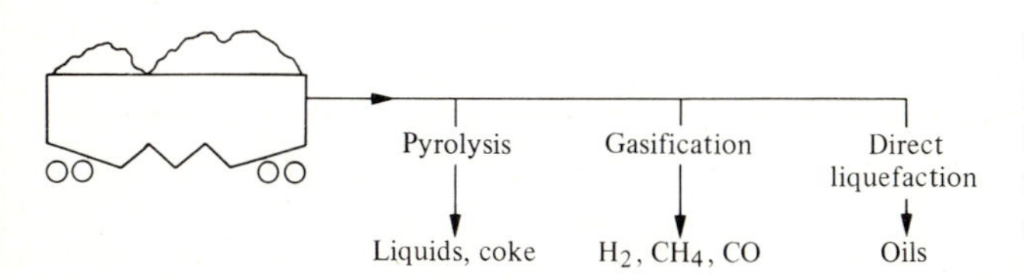

Figure 20–3.8 Fluidization of coal. There are three principal processes for converting coal to fluid fuels. Each process subjects reactor materials to severe service conditions. Also, catalytic materials play an important role.

20-4 "THE BOTTOM LINE"

In final analysis, the engineer has major responsibilities for the successful use of the materials in a product or system. Both over-design and premature failure are to be avoided. The former adds unneeded costs and wastes resources, while the latter introduces liabilities ranging from interrupted service, to replacement costs, and even to concerns for personnel safety. Because of these design and materials-selection requirements, the engineer must depend on product testing to assure the desired service. The engineer must also be able to analyze failures of products, components, and materials. Since premature failures can develop from improper usage, incorrect design, inadequate instruction, or defective materials, it is necessary to interpret the causes that are responsible for fracture and similar phenomenon. Only in that manner can future corrections be made and can necessary liabilities be assigned.

Design optimization

The essence of optimum design was described in the last century by the poem in which "The one-hoss shay was built in such a logical way, that it ran a hundred years to the day,"

. . .

He would build one shay to beat the taown
'N' the Keounty 'n' all the kentry raown';
It should be so built that it couldn' break daown:
–"Fur," said the Deacon, 't's mighty plain
Thut the weakes' place mus' stan' the strain;
'N' the way t' fix it, uz I maintain,
Is only jest
T' make that place uz strong uz the rest."

. . .

(One Hundred Years Later)

. . .

There were traces of age in the one-hoss shay,
The general flavor of mild decay
But nothing local, as one may say
There couldn't be,—for the Deacons art
Had made it so like every part
That there wasn't a chance for a crack to start.

. . .

(Earthquake)

. . .

. . . it went to pieces all at once,—
All at once, and nothing first,—

. . .

OLIVER WENDELL HOLMES

Now let us replace the one-hoss shay with an automobile, a calculator, or a comparable technical product. Can engineers of the year 2000 design a car so that there isn't "*a chance for a crack to start*" prior to the time that an

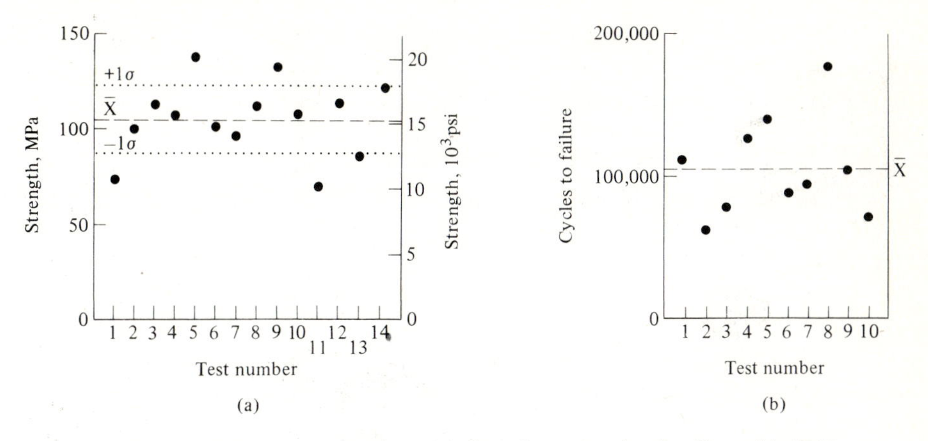

Figure 20-4.1 Variance in testing: (a) flat glass, 4-point loading; (b) 4340 normalized steel, 400 MPa (57,000 psi). All samples of each material had the same composition and were prepared and tested by identical procedures. The variations must be incorporated into the design calculations.

unpredictable event destroys it? Such an achievement is, of course, idealistic, since the real world contains many variables. The service conditions in which any particular component "*must stand the strain*" (sic) are not identical in any two cars. Furthermore, there are variances in the strengths of duplicate test samples of the same material, particularly those materials that are nonductile in nature and those that are loaded cyclically (Fig. 20-4.1). We seldom find identical matches in service lives of two or more components of a given product.

The engineer has a design alternative in certain products. That alternative is to make the individual components replaceable within the product. Thus, the automobile windshield is designed to be replaced as necessary, since we cannot predict its service life. However, engineering judgement concludes that a storage battery of a car should not have replaceable grids when they deteriorate. Rather, the practice is to replace the total battery. This is partly a matter of cost; however, other factors such as acid disposal and the quality of the repaired battery favor the established practice of total replacement. This anti-repair practice becomes even more pronounced in complex products, to the point that a pocket radio is never repaired if a transistor malfunctions or a soldered connection cracks, but is discarded and replaced by a new radio.*

The one-hoss shay may not be an optimum model for design in another respect. Its service life was from 1755 to 1855—"*one hundred years to the day.*" A comparable product designed in 1900 would find no use for transpor-

* Economically, it is cheaper to replace the whole radio than to repair it. However, should the radio be recycled? If so, how should its materials be recovered, if indeed they can be when we use milligrams of GaAs, micrograms of gold-plated connectors, etc.?

Figure 20-4.2 Stainless steel car body. This automobile, built in 1936, is not subject to normal corrosion. However, in addition to corrosion, the design engineer must consider cost, availability of raw materials (chromium), and the unused life when the car must be replaced for other factors. (Courtesy of Allegheny–Ludlum Steel Corp.)

tation at the present as we approach 2000. It would have been impossible for an engineer of 1900, when horse-drawn transportation was still prevalent, to have been foresighted enough to have predicted the complete changes that occurred in the succeeding two or three decades. A more technical aspect of the problem of obsolesence and optimum design is represented by Fig. 20-4.2. This stainless steel car was built in 1936 and represents the ultimate in materials achievements that would have eliminated a major problem for the car owner—corrosion. This car is still running!! However, would our present society and government be tolerant of 10 to 20 million cars of that vintage on today's highways with their excessive exhaust emissions, their relatively low fuel per passenger-mile efficiency, and numerous unsafe (for today) characteristics?

Product dependability

Since products of engineering are conceived and made for society's needs, users expect performances that are dependable and not subject to interruptions or failures. *Failure* need not be fracture, leakage, or excessive wear but simply a point at which *the product is not able to fulfill its intended purpose.* Thus, failures range from the fracture of a welded bridge beam to the clouding of a rear-view mirror in an automobile; from potholes in streets to the oxidized contact of a microswitch in a home thermostat; and from a punctured tire to a faded color print. Some failures are nearly instantaneous; the intended purpose of others is lost gradually, extending over months and years. Some failures are catastrophic with potentialities for human injury and/or excessive monetary losses; others may be said to be "fail-safe" with minimal concerns except for inconveniences or for replacement costs. There are two extremes of cost responsibility: (1) With normal usage for the intended purposes, a gradual deterioration is realized, and the user expects to include replacement costs as part of normal economic planning. (2) A product guarantee provides the user with assurances against faulty workmanship or materials during a specific period after purchase. Between these extremes, there are many degrees of cost responsibility in case of product failure.

Product liability. Since the *costs* of product failure must be borne by someone, we find that legal cases abound for the purposes of assigning the responsibilities for failure. The legal interpretations of the responsibility have changed over time; furthermore, it is to be expected that they will change more between now and 2000–2020. For example, early liability cases required a breach of contract for the courts to settle for the plaintiff. Thus, an 1842 case ruled against an injured driver of a defective mail coach, because there had been no direct contract written between the user of the product and the manufacturer. Today, such a contract is no longer required and the seller of the product can be broadly interpreted as the original manufacturer plus essentially everyone connected with the sale of the defective produce—the retailer, the wholesaler, the advertiser, the distributor, etc. Today the plaintiff need not be the owner; in fact, the product may have been used illegally and without the owner's permission. Furthermore, the product may have been used for purposes other than those *intended by the manufacturer*. Rather, the producer must now consider the *user's intentions*, even if they represent misuse by the manufacturer's standards. Finally, the expiration of a guarantee period does not release the manufacturer, distributor, retailer, etc., from liability if negligence is proven.* These changes in liability interpretations place major responsibilities on the future engineer in product development and manufacture. It is necessary to draw upon all of the technical expertise one can obtain.

Product testing. Testing always accompanies the development and manufacture of any successful technical product. Nondestructive evaluations (Fig. 20–4.3) are performed where possible because they do not destroy the product. However, tests to failure (destructive tests) must be used when it is necessary to know the limits of material performance or product quality. Such tests are necessary (1) on prototype units during the development stage, and (2) on samples of materials used in production in order to maintain quality control.

Because all tests are statistical in nature, both in sampling (Fig. 20–4.1) and in the details of failure, it is impossible to ensure absolute perfection in the final product. In extreme cases, such as in space vehicles and aeronautical hardware, tests are extensive and thorough, and design redundancies are included. These result in both a smaller than usual probability of failure

* In fact, in 42 states, prior negligence need not be proven if a product is judged unsafe by today's standards.

(a)

(b)

Figure 20-4.3 Nondestructive evaluation (NDE). Polarized light shows the locations of residual stresses and strains without destroying the product. (a) Glass bottle (reheated to $T > T_g$, and cooled rapidly). (b) Styrene glass. (The thermoplastic vinyl, $T > T_g$, was injected into a cold mold.)

and a greatly increased cost. For most manufactured products, common usage has established a working balance between economic cost and the probability of failure. The equilibrium point of this balance was formerly established between manufacturers and the market place. More and more, society is using the courts, through the medium of product-liability lawsuits, to argue for a decrease in the probability of failure, especially when the probability of personal injury is high if failure occurs, though this invar-

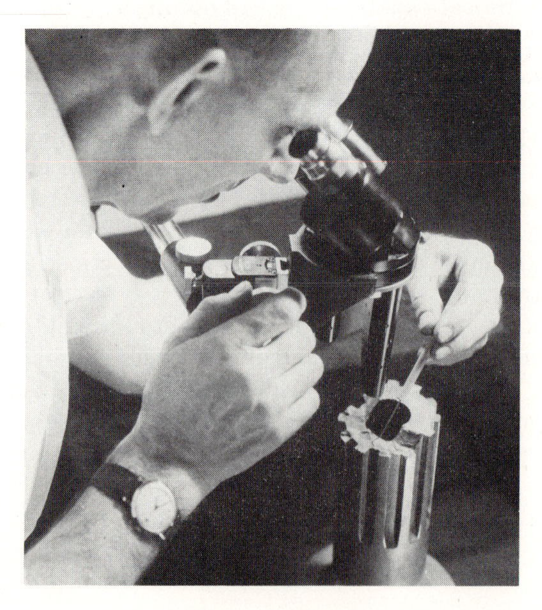

Figure 20-4.4 Failure analysis. Microscopic examination must be accompanied by any additional procedures that are available to interpret mechanisms of failure in prototype testing, in quality control, or in service losses. (Courtesy of D. Krashes, *Metals Progress,* November 1980.)

iably entails a larger economic cost. As in environmental legislation, engineers will be intimately involved with these problems in years ahead.

Failure analysis For materials development, for product design, or for establishing quality control bases, this involves a post-mortem examination of the failure itself, utilizing every means at the engineer's disposal (Fig. 20–4.4). In such situations, the engineer must be systematic in exploring and evaluating all possibilities. Necessary steps and reiterations are sequenced in Table 20–4.1. For example, it is apparent that the mode of failure for the broken shaft in Fig. 20–4.5 was a fatigue crack that started at the set-screw hole. Cracking progressed by the cyclic loading of the rotating shaft until final failure occurred catastrophically by fracture. The important question is yet to be answered. Specifically, what *caused* the crack to start? Was the shaft overloaded by the user? Was there faulty design by permit-

Table 20–4.1
Engineering steps in failure analyses

Step	Questions
A) What is the purpose of the failure analysis?	Redesign? Litigation? Documentation? Quality control?
B) What circumstances produced the failure?	What part(s) failed? What were the measures of failure? Service conditions that led to failure?
C) What scope should the analysis include?	Pertinent factors—service factors, human factors, related mechanisms? Limits—time, cost, safety, codes, policies? Trade-offs and modifications?
D) Choosing a theory or model of failure	Cause–effect relationships? Related parameters? Reliability of collected data?
E) Possible alternatives	Use of all available information? Use of all available explanations?
F) Gathering additional information	Additional witnesses? Special tests? Validity of information?
G) Evaluating alternatives	Does *all* evidence fit the hypothesis? What explanations are certain, probable, possible, unlikely, impossible? Rerun necessary tests based on most probable mechanism. Rank alternatives.
H) Implementation	Recommend conclusions or actions.
I) Evaluation	Does the redesigned product or system no longer fail? Is the quality control improved? Is the legal stance more secure?

Adapted from W. L. Larsen, *J. Engr. Educ.*

Figure 20-4.5 Product failure (14-cm steel shaft). The failure mechanism was by fatigue in which a crack slowly progressed through 90% of the rotating shaft. The final 10% fractured abruptly. The engineer also needs to know whether it was design, materials, manufacturing, or service that permitted the crack to start. (Courtesy of H. Mindlin, Battelle Memorial Institute.)

ting a set-screw hole to be located in a stress-sensitive position? Was the wrong steel used? Or, was there faulty heat treatment of the steel during manufacture? The engineer must systematically consider each alternative, plus any other plausible cause.

Failure analysis commonly requires the combined detective work of various experts. Illustrative Case 20-4.1 describes a litigation in which structural panels failed. The panels were plywood with an aluminum veneer laminated to the surface. The product, which is not a recent design, has several attractive features. The plywood provides a lightweight, rigid panel for construction, while the aluminum sheets that are bonded to the two surfaces are water proof, are easily cleaned, and do not require painting. In this case, however, the aluminum and plywood became delaminated, and the aluminum corroded excessively, leading to complete failure and significant monetary losses. Therefore the case was brought into court.

The question in this case was not the mode of failure. That was easily deduced—delamination and corrosion. The questions were what caused the delamination and the corrosion, and how were they related, so that the responsibility could be assigned. Only by knowing the causes was the court able to arrive at a judgement as to who would bear the accumulated expenses—new materials, replacement costs, delayed availability of the structures, etc.

Testing included microscopic and fluorescent samplings, and x-ray diffraction, plus consultation with experts on wood, corrosion, and adhesives. The conclusion reached by the court was that there was an absence of quality control that permitted excess moisture in the plywood prior to lamination, that this moisture transported chloride ions from the adhesive in the plywood to the aluminum where it caused delamination and corrosion. The case is being appealed; in any event the stakes are high, and technical people must play critical roles for both the plaintiff and the defendant.

Illustrative case 20-4.1 Obtain the article "Forensic Engineering: a Case Study," from the August 1980 issue of *Metals Progress*.

a) Outline the required "Engineering Steps in the Failure Analysis" (Table 20-4.1).
b) Cite the testimony that was given without the correct grasp of the underlying principles. (Unfortunately too many engineers fall in this trap.)
c) Detail the chain of technical observations that led to the interpretation of the mechanism of failure.

ADDED INFORMATION This failure analysis involved litigation. Such situations will involve increased numbers of engineers in the future; however, many more engineers must consider failure analysis as a tool in product design, in process development, with materials substitution, and for production control. ◄

20-5 CONCLUSION

Technical solutions are required to solve the materials problems that come with resource depletion, increased recycling, materials substitution, more efficient energy conversion, and greater product dependability. However, the engineer cannot be solely technically oriented. The raw materials for those products must be obtained within the constraints of political, economic, and social realities. The products must be dependable and responsive to the uses assigned to them; they must not contribute to personal, economic, or environmental liabilities.

Review and Study

TECHNICAL TERMS

Depletable resources Nonreplaceable resources.

Energy conversion The change of energy from one form to another; for example, chemical energy (in fuels) to mechanical energy (in motion), or to thermal energy (in heat).

Failure analysis Systematic examination of (1) the nature of service termination, (2) the synthesis of the cause(s), followed by (3) positive recommendations for future improvements.

Materials cycle The sequence of extraction, refining, manufacturing, use, and discard, or eventual recycle of materials.

Nondestructive evaluation (NDE) Inspection by methods that do not destroy the part to determine its suitability for use.

Product liability Responsibility assignment for product failure.

Raw materials The precursor of processed materials. Raw materials are usually from natural sources but may include recoverable resources.

Recycle The return of discarded products to the materials cycle.

Reserves Raw materials identified and capable of economic extraction.

Resources Raw materials, including those not yet identified nor currently economical for extraction.

FOR FURTHER STUDY

The following topics are suggestions for more detailed studies. Your final product is to be a report. It should relate materials to the concurrent society. For the most part, the reports will involve library searching. The required length for reports will not be the same for any two topics. However, the report should draw upon at least six, and preferably a dozen, separate sources, if the report is to accurately reflect the multiple aspects of the selection and use of materials.

20–1.1 Toolmaking during the Stone-Age.
20–1.2 The Artisan and Early Glassmaking.
20–1.3 Metals for Coinage in Alexander's Time.
20–1.4 The Origin and Early Uses of Sterling Silver.
20–1.5 Clay, Fire, and Earthenware Containers.
20–1.6 The Bessemer Process and the Industrial Revolution.
20–1.7 Expansion of the American Rail System in the 19th Century.
20–1.8 Your Choice.

20–2.1 The Sources of Iron Ore in Colonial America.
20–2.2 The Resources for Stainless Steels.
20–2.3 Wood Availability for Eliminating Foreign Fuel Dependence.
20–2.4 (a) Copper, or (b) Iron Resources and the Environment.
20–2.5 Manganese Resources in the 1990s.
20–2.6 Technology for the Recycling of (a) Paper; (b) Glass; (c) Aluminum.
20–2.7 The Junk-automobile Market.
20–2.8 The Use of Petroleum for Plastics: Pro and Con.
20–2.9 Substitution of Aluminum for Steel in Cars.
20–2.10 Substitution of Fiber-reinforced Plastics for Aluminum in Boats.
20–2.11 Your Choice.

20–3.1 Energy Requirements for Manufacturing Down-sized Cars.
20–3.2 The Choice of Glass, Metal, or Plastics for Beverage Containers.
20–3.3 Temperature and the Efficiency of Energy Conversion.
20–3.4 Materials for High-temperature Conversion of Fuels (a) to Electrical Energy; (b) to Mechanical Energy.

20-3.5 Materials Requirements (a) for Solar Cells; (b) for the Solar Home.
20-3.6 The Containment of Radioactivity in Glass.
20-3.7 Materials for Superconducting Transmission Cables.
20-3.8 Your Choice.

20-4.1 (a) The Automobile Muffler: 1950 vs. 1980. (b) The Automobile Tire. (c) The Automobile Bumper.
20-4.2 Materials Failure or Design Failure? (a) The Liberty Ships of World War II. (b) The British Comets. (c) The DC-10.
20-4.3 The Origin and Development of ASTM Specifications for ________.
20-4.4 The Legal Responsibilities of (a) the Product Manufacturer; (b) the Product User.
20-4.5 The Pattern of Product Liability Judgments (a) before 1900; (b) before 1950; (c) in the 1980s.
20-4.6 Accelerated Testing for (a) High-temperature Service; (b) Ultra-violet Exposure; (c) Polymer Stability.
20-4.7 An Analysis of the Failure of (a) a Broken Garage Door Spring; (b) a Broken Bicycle Axle; (c) a Part or Component That You Have Had to Replace.

APPENDIXES

Appendix A
Properties of selected engineering materials (20°C)*

Material	Density Mg/m^3 ($=g/cm^3$)	Thermal conductivity, $\left(\frac{watts}{mm^2}\right)\Big/\left(\frac{°C}{mm}\right)$†	Linear expansion, $°C^{-1}$‡	Electrical resistivity, ρ ohm·m§	Average modulus of elasticity, $\bar{E}$	
					MPa	psi
Metals						
Aluminum (99.9+)	2.7	0.22	22.5×10^{-6}	29×10^{-9}	70,000	10×10^{6}
Aluminum alloys	2.7(+)	0.16	22×10^{-6}	$\sim 45 \times 10^{-9}$	70,000	10×10^{6}
Brass (70 Cu-30 Zn)	8.5	0.12	20×10^{-6}	62×10^{-9}	110,000	16×10^{6}
Bronze (95 Cu-5 Sn)	8.8	0.08	18×10^{-6}	$\sim 100 \times 10^{-9}$	110,000	16×10^{6}
Cast iron (gray)	7.15	—	10×10^{-6}	—	140,000(±)	$20 \times 10^{6}\pm$
Cast iron (white)	7.7	—	9×10^{-6}	660×10^{-9}	205,000	30×10^{6}
Copper (99.9+)	8.9	0.40	17×10^{-6}	17×10^{-9}	110,000	16×10^{6}
Iron (99.9+)	7.88	0.072	11.7×10^{-6}	98×10^{-9}	205,000	30×10^{6}
Lead (99+)	11.34	0.033	29×10^{-6}	206×10^{-9}	14,000	2×10^{6}
Magnesium (99+)	1.74	0.16	25×10^{-6}	45×10^{-9}	45,000	6.5×10^{6}
Monel (70 Ni-30 Cu)	8.8	0.025	15×10^{-6}	482×10^{-9}	180,000	26×10^{6}
Silver (sterling)	10.4	0.41	18×10^{-6}	18×10^{-9}	75,000	11×10^{6}
Steel (1020)	7.86	0.050	11.7×10^{-6}	169×10^{-9}	205,000	30×10^{6}
Steel (1040)	7.85	0.048	11.3×10^{-6}	171×10^{-9}	205,000	30×10^{6}
Steel (1080)	7.84	0.046	10.8×10^{-6}	180×10^{-9}	205,000	30×10^{6}
Steel (18 Cr-8 Ni stainless)	7.93	0.015	16×10^{-6}	700×10^{-9}	205,000	30×10^{6}
Ceramics						
Al_2O_3	3.8	0.029	9×10^{-6}	$>10^{12}$	350,000	50×10^{6}
Brick						
Building	2.3(±)	0.0006	9×10^{-6}	—	—	—
Fireclay	2.1	0.0008	4.5×10^{-6}	1.4×10^{6}	—	—
Graphite	1.5	—	5×10^{-6}	—	—	—
Paving	2.5	—	4×10^{-6}	—	—	—
Silica	1.75	0.0008	—	1.2×10^{6}	—	—
Concrete	2.4(±)	0.0010	13×10^{-6}	—	14,000	2×10^{6}
Glass						
Flat	2.5	0.00075	9×10^{-6}	10^{12}	70,000	10×10^{6}
Borosilicate (fibers)	2.4	0.0010	2.7×10^{-6}	$>10^{15}$	70,000	10×10^{6}
Silica	2.2	0.0012	0.5×10^{-6}	10^{18}	70,000	10×10^{6}
Vycor	2.2	0.0012	0.6×10^{-6}	—	—	—
Wool	0.05	0.00025	—	—	—	—
Graphite (bulk)	1.9	—	5×10^{-6}	10^{-5}	7,000	1×10^{6}
MgO	3.6	—	9×10^{-6}	10^{3}(1100°C)	205,000	30×10^{6}
Quartz (SiO_2)	2.65	0.012	—	10^{12}	310,000	45×10^{6}
SiC	3.17	0.012	4.5×10^{-6}	0.025 (1100°C)	—	—
TiC	4.5	0.030	7×10^{-6}	50×10^{-8}	350,000	50×10^{6}
Polymers						
Melamine-formaldehyde	1.5	0.00030	27×10^{-6}	10^{11}	9,000	1.3×10^{6}
Phenol-formaldehyde	1.3	0.00016	72×10^{-6}	10^{10}	3,500	0.5×10^{6}
Urea-formaldehyde	1.5	0.00030	27×10^{-6}	10^{10}	10,300	1.5×10^{6}
Rubbers (synthetic)	1.5	0.00012	—	—	4–75	600–11,000
Rubber (vulcanized)	1.2	0.00012	81×10^{-6}	10^{12}	3,500	0.5×10^{6}
Polyethylene (L.D.)	0.92	0.00034	180×10^{-6}	$10^{13}-10^{16}$	100–350	14,000–50,000
Polyethylene (H.D.)	0.96	0.00052	120×10^{-6}	$10^{12}-10^{16}$	350–1,250	50,000–180,000
Polystyrene	1.05	0.00008	63×10^{-6}	10^{16}	2,800	0.4×10^{6}
Polyvinylidene chloride	1.7	0.00012	190×10^{-6}	10^{11}	350	0.05×10^{6}
Polytetrafluoroethylene	2.2	0.00020	100×10^{-6}	10^{14}	350–700	50,000–100,000
Polymethyl methacrylate	1.2	0.00020	90×10^{-6}	10^{14}	3,500	0.5×10^{6}
Nylon	1.15	0.00025	100×10^{-6}	10^{12}	2,800	0.4×10^{6}

* Data in this table were taken from numerous sources.

† Alternatively, $(W/mm^2)/(K/mm)$. Multiply by 1.92 to get $Btu/(ft^2 \cdot s)/(°F/in.)$.

‡ Or, K^{-1}; divide by 1.8 to get $°F^{-1}$.

§ Multiply ohm·m by 39 to get ohm·in.

Appendix B
Table of selected elements

Element	Symbol	Atomic number	Atomic mass, amu	Melting point, °C	Density (solid), Mg/m^3 (= g/cm^3)	Crystal structure, 20°C	Approx. atomic radius, nm†	Valence (most common)	Approx. ionic radius, nm‡
Hydrogen	H	1	1.0078	−259.14	—	—	0.046	1+	Very small
Helium	He	2	4.003	−272.2	—	—	0.176	Inert	—
Lithium	Li	3	6.94	180	0.534	bcc	0.1519	1+	0.068
Beryllium	Be	4	9.01	1289	1.85	hcp	0.114	2+	0.035
Boron	B	5	10.81	2103	2.34	—	0.046	3+	~0.025
Carbon	C	6	12.011	>3500	2.25	hex	0.077	—	—
Nitrogen	N	7	14.007	−210	—	—	0.071	3−	—
Oxygen	O	8	15.999	−218.4	—	—	0.060	2−	0.140
Fluorine	F	9	19.00	−220	—	—	0.06	1−	0.133
Neon	Ne	10	20.18	−248.7	—	fcc	0.160	Inert	—
Sodium	Na	11	22.99	97.8	0.97	bcc	0.1857	1+	0.097
Magnesium	Mg	12	24.31	649	1.74	hcp	0.161	2+	0.066
Aluminum	Al	13	26.98	660.4	2.70	fcc	0.14315	3+	0.051
Silicon	Si	14	28.09	1414	2.33	*	0.1176	4+	0.042
Phosphorus	P	15	30.97	44	1.8	—	0.11	5+	~0.035
Sulfur	S	16	32.06	112.8	2.07	—	0.106	2−	0.184
Chlorine	Cl	17	35.45	−101	—	—	0.101	1−	0.181
Argon	Ar	18	39.95	−189.2	—	fcc	0.192	Inert	—
Potassium	K	19	39.10	63	0.86	bcc	0.2312	1+	0.133
Calcium	Ca	20	40.08	840	1.54	fcc	0.1969	2+	0.099
Titanium	Ti	22	47.90	1668	4.51	hcp	0.146	4+	0.068
Chromium	Cr	24	52.00	1875	7.20	bcc	0.1249	3+	0.063
Manganese	Mn	25	54.94	1246	7.2	—	0.112	2+	0.080
Iron	Fe	26	55.85	1538	7.88	bcc	0.1241	2+	0.074
						fcc	0.1269	3+	0.064
Cobalt	Co	27	58.93	1494	8.9	hcp	0.125	2+	0.072
Nickel	Ni	28	58.71	1455	8.90	fcc	0.1246	2+	0.069
Copper	Cu	29	63.54	1084.5	8.92	fcc	0.1278	1+	0.096
Zinc	Zn	30	65.37	419.6	7.14	hcp	0.139	2+	0.074
Germanium	Ge	32	72.59	937	5.35	*	0.1224	4+	—
Arsenic	As	33	74.92	~809	5.73	—	0.125	3+	—
Krypton	Kr	36	83.80	−157	—	fcc	0.201	Inert	—
Silver	Ag	47	107.87	961.9	10.5	fcc	0.1444	1+	0.126
Tin	Sn	50	118.69	232	7.3	bct	0.1509	4+	0.071
Antimony	Sb	51	121.75	630.7	6.7	—	0.1452	5+	—
Iodine	I	53	126.9	114	4.93	ortho	0.135	1−	0.220
Xenon	Xe	54	131.3	−112	2.7	fcc	0.221	Inert	—
Cesium	Cs	55	132.9	28.4	1.9	bcc	0.267	1+	0.167
Tungsten	W	74	183.9	3410	19.4	bcc	0.1367	4+	0.070
Gold	Au	79	197.0	1064.4	19.32	fcc	0.1441	1+	0.137
Mercury	Hg	80	200.6	−38.86	—	—	0.155	2+	0.110
Lead	Pb	82	207.2	327.5	11.34	fcc	0.1750	2+	0.120
Uranium	U	92	238.0	1133	19	—	0.138	4+	0.097

* Diamond cubic

† One half of closest approach of two atoms in the elemental solid. For noncubic structures, the average interatomic distance is given; for example, in hcp, the atom is slightly ellipsoidal.

‡ Radii for CN = 6; otherwise, $0.97\ R_{CN=8} \approx R_{CN=6} \approx 1.1\ R_{CN=4}$. Patterned after Ahrens.

Appendix C
Constants and conversions

Constants*

Acceleration of gravity, g	9.80 . . . m/s^2
Atomic mass unit, amu	1.66 . . . $\times 10^{-24}$ g
Avogadro's number, N	0.6022 . . . $\times 10^{24}$ $mole^{-1}$
Boltzmann's constant, k	13.8 . . . $\times 10^{-24}$ J/K
Capacitivity (vacuum), ϵ	8.85 . . . $\times 10^{-12}$ C/V·m
Electron charge, q	0.1602 . . . $\times 10^{-18}$ C
Electron magnetic moment, β	9.27 . . . $\times 10^{-24}$ $A \cdot m^2$
Electron volt, eV	0.160 . . . $\times 10^{-18}$ J
Fe-Fe_3C eutectoid composition	0.77 w/o carbon
Fe-Fe_3C eutectoid temperature	727°C (1340°F)

Conversions*

1 ampere	= 1 C/s
1 angstrom	= 10^{-10} m
	= 10^{-8} cm
	= 0.1 nm
	= 3.937×10^{-9} in.
1 amu	= 1.66 . . . $\times 10^{-24}$ g
1 Btu	= 1.055 . . . $\times 10^{3}$ J
1 Btu/°F	= 1.899 . . . $\times 10^{3}$ J/°C
1 [Btu/(ft^2·s)]/[°F/in.]	= 0.519 . . . $\times 10^{3}$ [J/(m^2·s)]/[°C/m]
	= 0.519 . . . $\times 10^{3}$ (W/m^2)/(°C/m)
1 Btu·ft^2	= 11.3 . . . $\times 10^{3}$ J/m^2
1 calorie, gram	= 4.18 . . . J
1 centimeter	= 10^{-2} m
	= 0.3937 in.
1 coulomb	= 1 A·s
1 cubic centimeter	= 0.0610 . . . in^3.
1 cubic inch	= 16.3 . . . $\times 10^{-6}$ m^3
1°C difference	= 1.8°F
1 electron volt	= 0.160 . . . $\times 10^{-18}$ J
1°F difference	= 0.555 . . . °C
1 foot	= 0.3048 m
1 foot·$pound_f$	= 1.355 . . . J
1 gallon (U.S. liq.)	= 3.78 . . . $\times 10^{-3}$ m^3
1 gram	= 0.602 . . . $\times 10^{24}$ amu
	= 2.20 . . . $\times 10^{-3}$ lb_m

* All irrational values are rounded downward.

Conversions *(continued)*

1 gram/cm^3	= 62.4 . . . lb$_m$/ft^3
	= 1000 kg/m^3
	= 1 Mg/m^3
1 inch	= 0.0254 m
1 joule	= 0.947 . . . × 10^{-3} Btu
	= 0.239 . . . cal, gram
	= 6.24 . . . × 10^{18} eV
	= 0.737 . . . ft·lb$_f$
	= 1 watt·sec
1 joule/meter^2	= 8.80 . . . × 10^{-5} Btu/ft^2
1 [joule/(m^2·s)]/[°C/m]	= 1.92 . . . × 10^{-3} [Btu/(ft^2·s)]/[°F/in.]
1 kilogram	= 2.20 . . . lb$_m$
1 megagram/meter^3	= 1 g/cm^3
	= 10^6 g/m^3
	= 1000 kg/m^3
1 meter	= 10^{10} Å
	= 10^9 nm
	= 3.28 . . . ft
	= 39.37 in.
1 micrometer	= 10^{-6} m
1 nanometer	= 10^{-9} m
1 newton	= 0.224 . . . lb$_f$
1 ohm·inch	= 0.0254 . . . Ω·m
1 ohm·meter	= 39.37 Ω·in.
1 pascal	= 0.145 . . . × 10^{-3} lb$_f$/in.^{2}
1 poise	= 0.1 Pa·s
1 pound (force)	= 4.44 . . . newtons
1 pound (mass)	= 0.453 . . . kg
1 pound/foot^3	= 16.0 . . . kg/m^3
1 pound/inch^2	= 6.89 . . . × 10^{-3} MPa
1 watt	= 1 J/s
1 (watt/m^2)/(°C/m)	= 1.92 . . . × 10^{-3} [Btu/(ft^2·s)]/[°F/in.]

SI prefixes

giga	G	10^9
mega	M	10^6
kilo	k	10^3
milli	m	10^{-3}
micro	μ	10^{-6}
nano	n	10^{-9}
pico	p	10^{-12}

INDEX